Video Microscopy

Video Microscopy

Shinya Inoué

Marine Biological Laboratory
Woods Hole, Massachusetts
and University of Pennsylvania
Philadelphia, Pennsylvania

With contributions by

Robert J. Walter, Jr.
Michael W. Berns
Gordon W. Ellis

and

Eric Hansen

Plenum Press • New York and London

Library of Congress Cataloging in Publication Data

Inoue, Shinya.
Video microscopy.

Bibliography: p.
Includes index.
1. Video microscopy.
QH222.I56 1986 578'.4 85-28252
ISBN 0-306-42120-8

A Division of Plenum Publishing Corporation
233 Spring Street, New York, N.Y. 10013

Printed in the United States of America

Preface

Ever since television became practical in the early 1950s, closed-circuit television (CCTV) in conjunction with the light microscope has provided large screen display, raised image contrast, and made the images formed by ultraviolet and infrared rays visible. With the introduction of large-scale integrated circuits in the last decade, TV equipment has improved by leaps and bounds, as has its application in microscopy.

With modern CCTV, sometimes with the help of digital computers, we can distill the image from a scene that appears to be nothing but noise; capture fluorescence too dim to be seen; visualize structures far below the limit of resolution; crispen images hidden in fog; measure, count, and sort objects; and record in time-lapsed and high-speed sequences through the light microscope without great difficulty. In fact, video is becoming indispensable for harnessing the fullest capacity of the light microscope, a capacity that itself is much greater than could have been envisioned just a few years ago.

The time seemed ripe then to review the basics of video, and of microscopy, and to examine how the two could best be combined to accomplish these tasks. The Marine Biological Laboratory short courses on Analytical and Quantitative Light Microscopy in Biology, Medicine, and the Materials Sciences, and the many inquiries I received on video microscopy, supported such an effort, and Kirk Jensen of Plenum Press persuaded me of its worth.

This book addresses the reader already somewhat experienced in light microscopy, but perhaps with less background in video or electronics. Practical matters as well as some theory needed to select and use the equipment rationally are included, but circuit diagrams, for example, are deliberately omitted.* On the other hand, I have tried to explain the basics in sufficient detail so that the reader may be encouraged to explore applications of video to microscopy even in ways not attempted before.

Chapter 1 briefly reviews the invention of TV, its early applications to microscopy, and the present state of video microscopy.

Chapter 2 provides a glossary of terms commonly used in video and microscopy, and directs the reader to sections of the book dealing with each particular subject.

Chapter 3 guides the beginner through the preliminaries of getting started, especially in very practical terms. Common problems are listed, as well as pitfalls that can be avoided by knowing their warning signs.

TV is engineered to satisfy the human observer. In Chapter 4 we review the major features of human vision that relate to the design and use of TV as well as the light microscope.

In Chapter 5 we review the essentials of microscope image formation and examine the basic adjustments and special care of the light microscope needed to obtain high-quality images using video.

*Texts such as Ennes (1979), Hansen (1969), Showalter (1969), and van Wezel (1981) provide many examples of the actual circuits involved. See the service manual provided by the manufacturer for wiring diagrams of specific pieces of equipment.

In Chapter 6 we study the fundamental nature of the video signals and the essentials of video signal transmission from the camera to the monitor.

In Chapter 7 we examine the hardware at the front and back ends of video (the camera, including the video pickup device, and the video monitor, including the display device,) and compare their specific performances for various applications in light microscopy.

In Chapter 8 we examine the operation, choice, and care of video recorders, including video tape recorders and optical disk recorders, some with capabilities for time-lapse, freeze-frame, and high-speed analyses.

In Chapter 9 we survey the equipment and the methods available for video processing and analyses using analog video devices.

In Chapter 10 Walter and Berns examine the principles and applications of digital image processing and analysis, an approach that combines the power of computers with the advantages of video. Many examples of applications in microscopy are included.

In Chapter 11 I have sampled relatively recent applications of video, classified according to modes of contrast generation in microscopy (excluding those applications already discussed in the previous two chapters). Special advantages and requirements for applying video to the various contrast modes are discussed, together with suggestions for potential new applications.

In Chapter 12 we discuss the preparation of video for publication, or for tape or film presentation, including the means for stereoscopic displays.

In Appendix I Gordon Ellis discusses the removal of video scan lines by spatial filtration, and shows its striking effect on the appearance of picture details.

In Appendix II Eric Hansen discusses the fundamental relationship between resolution and contrast in the microscope and video. The relation between the optical diffraction pattern and the modulation transfer function of the system is examined.

In Appendix III I have revised an article that introduces the basics of polarized-light microscopy and its application to biology, a subject not adequately discussed in print elsewhere. This appendix contains an extensive, annotated bibliography on polarized-light microscopy and its applications to cell biology.

In Appendix IV I have attempted to provide an update of digital image processors and programs that have become available recently, and to make some recommendations on how one might get started doing digital image processing.

The organization of these chapters may appear somewhat unusual; I tried to develop the subject in a spiral fashion, interweaving the practical and the basics, hoping to help those who actually make use of video microscopy. For those using the book more as a reference source, I have included relevant section numbers for each entry in the Glossary. The glossary terms are boldfaced when they first appear in the text.

I have used footnotes liberally throughout the book. It should be emphasized that material discussed in the footnotes is by no means of secondary importance; rather, the arrangement was intended to minimize interruption in the flow of the discussion.

All references, except those for Appendix III, appear in a separate section at the end of the book, arranged alphabetically by author. Each entry is followed by bracketed notations indicating its location in the volume. Included is a separate list of pamphlets not readily available from libraries, along with addresses of their sources (p. 528).

Names and addresses of commercial firms mentioned in the text also have been compiled into a single list (pp. xxi–xxiii), except for those that appear in Tables 5-1, 7-6, and Table 1 of the Postscript.

These chapters were prepared mainly from the vantage point of a biologist, but it is hoped that they will prove useful to those in other fields as well.

As the outline shows, the discussions are heavily weighted toward video, much less being covered on modes of contrast generation and other matters of microscopy. Fortunately, several texts are in preparation at this time, and the compact paperback *Fundamentals of Light Microscopy* by Michael Spencer of University College, London, provides a concise overview of these subjects. I highly recommend Spencer's text as a companion to this volume.

The technology and the available equipment for video are advancing at a dizzying pace today. I have stressed the fundamentals wherever practical, so that these chapters will not, I hope, become outdated too quickly. The advances are, however, so rapid that I have made a few suggestions in the Postscript on how to keep current on the many exciting developments. Also, while the references to the literature and the tables showing types and sources of equipment are fairly extensive, they are clearly representative and are not meant to be all-inclusive. I would, however, appreciate learning of any major gaps in the presentations. Again, sources of additional information appear in the Postscript.

Shinya Inoué

Woods Hole, Massachusetts

Acknowledgments

Bob Walter and Mike Berns not only contributed an excellent Chapter 10, but waited patiently a whole year while I completed the other chapters. Gordon Ellis and E. Leitz, Inc., allowed us to use the material in Appendix I, and Eric Hansen prepared the concise discourse of Appendix II, which neatly ties together the optics and electronics scattered throughout the volume. Dave Cohen (Venus Scientific, now at Gull Airborne Instruments), Gordon Ellis, Eric Hansen, Kirk Jensen, Ernst Keller (Carl Zeiss, Inc.), Brian Laws (Crimson Camera), Doug Lutz (Marine Biological Laboratory), Ted MacNichol (Marine Biological Laboratory), Ted and Nancy Salmon (University of North Carolina), and Paul Thomas (Dage–MTI) reviewed various parts of the draft manuscript and provided much useful input.

In assembling the material for this book, many other individuals contributed original photographs, microscope slides, and other valuable information and material. In addition to those already mentioned, they include: David Agard (University of California, San Francisco), Bob Allen (Dartmouth College), Nina Allen (Wake Forest University), John Almen (Ampex), Anthony Alter (University of California, Irvine), Helène Anderson (Crimson Camera), Peter Anderson (University of Florida, St. Augustine), Peter Bartels (University of Arizona and University of Chicago), R. R. Beckman (General Electric), Gene Bell (Massachusetts Institute of Technology), Ken Braid (Sony), Sid Braginsky (Olympus), Mel Brenner (Nikon), Joe Bryan (Baylor College of Medicine), Fred Caspari (Dage–MTI), Ken Castleman (Perceptive Systems), Dick Chaison (Olympus), Jasmine Chow (University of California, Irvine), Cary Clayton (Instrumentation Marketing Corporation), Kenneth Cooper (University of California, Riverside), Michelle Crawforth (Eastman Kodak), Glenn Davis (Xybion), Jim Dvorak (National Institutes of Health), Barbara Ela (Sony), Kepie Engel (National Science Foundation), Ralph Eno (Hamamatsu), Fennell Evans (University of Minnesota), Fred Fay (University of Massachusetts, Worchester), Dennis Flanagan (Scientific American), Les Flory (RCA), Harold Fosack (Texas Instruments), Marty Frange (Panasonic), Maurice Françon (Laboratoire d'Optique), Henry Fuchs (University of North Carolina), Keigi Fujiwara (Harvard Medical School), John Grace (Crimson Camera), John Harshbarger (VII), Wyndham Hannaway (Colorado Video, now at G. W. Hannaway and Assoc.), John Hayden (Dartmouth College), David Hillman (Laser Concepts for Panasonic OMDR), Jan Hinsch (E. Leitz, Inc.), Ken Hori (MP Video), Andrew Huxley (Cambridge University), Hitoshi Iida (Hamamatsu), Marcos Intaglietta (University of California, San Diego), Colin Izzard (State University of New York, Albany), Ken Jacobson (University of North Carolina), Tedd Jacoby (For-A), Ron Jafcott (Texas Instruments), Lionel Jaffe (Marine Biological Laboratory), Carl Johnson (Harvard University), Bechara Kachar (National Institutes of Health), Harold Kidd (General Electric), Jerry Kleifgen (Dage–MTI), Chuck Koester (Columbia University College of Physicians and Surgeons), Michael Koonce (University of California, Irvine), Edwin Land (Rowland Institute for Science), Peter Lemkin (National Institutes of Health), Lawrence Levine (Panasonic), Cy Levinthal (Columbia University), Lew Lipkin (National Institutes of Health), Lenny Lipton (Stereographics Corporation), Doug Lutz (Marine Biological Laboratory), Arte Machia (Panasonic), Julie Madox (RCA), Ed Martins (Amperex), Y. Matsushita (Sony), Bryan Mayall (Lawrence Livermore), Curtis Mc Dowell (General Electric), Paul Mengers (Quantax), Dennis Neary (Spin Physics), Ken Orndorf (Dartmouth College), Barry Palewitz (University of North Carolina), Alan Penchansky (Geltzer & Co.), William Pratt (Vicom), Phil Presley (Carl Zeiss), Calvin Quate (Stanford University), George Reynolds (Princeton University), Ichiji Rikukawa (Nikon), George Robinson (RCA), Fred Sachs (General Electric), Dean Sadamune (University of California, Irvine), Masao Sakai (Hitachi), Kurt Scheier (Nikon), Lee Schuett (Nikon), Martin Scott (Eastman Kodak), John Sedat (University of California, San Francisco), Walter Seidl (Olympus), Hy Shaffer (Crimson Camera), Lawrence Sher (BBN Laboratories), S. Shigemasa (Toshiba), Art Shoemaker (Reichert Scientific Instruments–A. O. Reichert), David Shotton (University of Oxford), Ann Siemens (University of California, Irvine), Tony Silvestri (Sony), Jesse Sisken (University of Kentucky), Glen Southworth (Colorado Video), Ken Spring (National Institutes of Health), Bob Squires (WHOI), Ray Stephens (Marine Biological Laboratory), Art Sterling (Dage–MTI), Dick Taylor (Colorado Video), Lans Taylor (Carnegie–Mellon University), Ron Tomczyk (Panasonic), Ron Vale (National Institutes of Health and Marine Biological Laboratory), Fran Valenti (F. M. Valenti, Assoc.), Barbara Walter (University of California, Irvine), John Wampler (University of North Carolina), Bob Wang (Imaging Technology), Watt Webb (Cornell University), John White (Medical Research Council, Cambridge University), Bob Worbecki (Precision Echo), and Don Yansen, (Interactive Video).

Illustrations and tables without specific acknowledgment of source are originals, produced or compiled in the authors' laboratories.

Thanks to the many publishers and authors of cited sources for their permission to reproduce figures, and to those who provided original photographs and drawings for reproduction.

The rapid development of video microscopy would not have been possible without the willing and helpful participation by many commercial organizations. Although trademarks are not designated in the text, it should be pointed out that the following are registered trademarks or copyrights of the indicated companies. Chalnicon, Toshiba; Image-I, Universal Imaging Corporation; Kimwipe, Kimberly-Clark; Kleenex, Kimberly-Clark; Newvicon, Matsushita; Plumbicon, Phillips–Amperex; Saticon, Hitachi; Scotch Cover-Up Tape, 3M Company; SIT, RCA; ST Vidicon, RCA; Ultricon, RCA; Ultricon II, RCA; Videotherm, ISI Group.

I am grateful to these companies, and others, for permission to reproduce illustrations from their publications and product specification sheets. In this regard, extensive use has been made of the *RCA Electro-Optics Handbook* and the Conrac *Raster Graphics Handbook*.

Last but not the least of the many who helped put this volume together: NIH Grant GM-31617, NSF Grant PCM-8216301,* and Olympus Corporation of American provided generous support for the preparation of the manuscript. Nikon Instrument Company, Carl Zeiss, Colorado Video, and Dage–MTI provided especially helpful cooperation. The MBL staff and administration supported my task in many ways as usual, but especially Judy Ashmore of the Library, who helped by tracking down numerous references and difficult-to-locate books. Ed Horn, of our laboratory, produced excellent photographs at the last minute in addition to designing and fabricating the many superior parts for our video, high-extinction polarizing microscope. Ted Inoué developed the Image-I digital image-processing program and opened my eyes to the world of digital computers. Linda Pellechia (North Babylon, New York) and Bob Golder of our Photo Lab prepared the many fine line drawings, and Linda Golder and Chris Inoué, the photoprints. Mary Safford at Plenum lightened the pain of editorial corrections with her patience and empathy with the author; Kenneth McCandless and Sylvia strained their eyes proofing; Bertha Woodward, Doug Lutz, and the students kept the laboratory running productively. Jane Leighton (STARS) tirelessly typed the manuscript through its many revisions. Mort Maser (Woods Hole Educational Associates) performed a heroic task in the last month of helping to bring the manuscript into final shape. And Sylvia kept my spirits up and the cozy home fire burning. To all my deepest thanks.

S.I.

*Any opinions, findings, and conclusions or recommendations expressed in this publication are those of the authors and do not reflect the views of the National Science Foundation or the National Institutes of Health.

Contents

Chapter 4
Physiological Characteristics of the Eye 71

Chapter 5
Microscope Image Formation 93

Chapter 6
The Video Image and Signal 149

List of Tables

List of Firms Cited in the Text*

Alan R. Liss 150 Fifth Avenue, New York, New York 10011. (Publisher of video disk supplements to *Cell Motility*)

American Optical (*see* Reichert)

Amperex Electronics Corporation Providence Pike, Slatersville, Rhode Island 02876. (Plumbicon and other vidicon tubes)

Ampex Corporation 401 Broadway, Redwood City, California 94063. (1-inch VTRs)

Arlunya Division, The Dindima Group Pty., Ltd. P.O. Box 106, Victoria 3133, Australia. (Digital image processor)

Bausch & Lomb 820 Linden Avenue, Rochester, New York 14620. (Microscopes, Omnicon digital image analyzer)

BBN Laboratories 10 Moulton Street, Cambridge, Massachusetts 02238. (Consultants on three-dimensional imaging, etc.)

Carl Zeiss, Inc. One Zeiss Drive, Thornwood, New York 10594. (Microscopes, IBAS image processor)

Colorado Video, Inc. Box 928, Boulder, Colorado 80306. (Analog processors, digital frame store, slow-scan digitizers)

Comprehensive Video Supply Corporation 148 Veterans Drive, Northvale, New Jersey 07647. (Porta-Pattern test charts, video supplies, test equipment)

Conrac Division, Conrac Corporation 600 North Rimsdale Avenue, Covina, California 91722. (Monitors)

Crimson Camera Technical Sales 325 Vassar Street, Cambridge, Massachusetts 02139. (Distributor of industrial video equipment, including Videotek monitors)

Dage–MTI 208 Wabash Street, Michigan City, Indiana 46360. (Video cameras, high-resolution monitor, image processor)

DeAnza (*see* Gould–DeAnza)

Drexon, Drexler Technology Corporation 2557 Charleston Road, Mountain View, California 94043. (Optical memory disks)

Eastman Kodak Company Rochester, New York 14650. (Lenses, film, filters, photographic emulsions, chemicals, videotape)

Edmund Scientific Company 7082 Edscorp Building, Barrington, New Jersey 08007. (Ronchi gratings)

E. Leitz, Inc. Rockleigh, New Jersey 07647. (Microscopes, 35-mm cameras, TAS image processor, DADS photometer)

EMI 80 Express Street, Plain View, New York 11803. (High-gain image intensifier tube, photomultipliers)

EOIS (Electro Optical Information Systems) 710 Wilshire Boulevard, Suite 501, Santa Monica, California 90401. (Digital image processors)

*(See Also Tables 5-1, 7-6, and Table 1 of the Postscript)

EOS Electronics AV, Ltd. Weston Square, Barry, South Glamorgan, CF6 7YF, United Kingdom. (Modification of Sony 5850 for animation)

Fish–Shurman Corporation 70 Portman Road, New Rochelle, New York 10802. (Fluorescent uranium-glass blocks, filters)

For-A Corporation of America 49 Lexington Street, West Newton, Massachusetts 02165. (Analog image processors, time–date generators, time-base correctors)

General Electric Company, Medical Systems Milwaukee, Wisconsin 53201. (High-resolution 1-inch VTR)

General Electric Company, Projection Display Products Operation Electronics Park, Syracuse, New York 13201. (Light valve video projector)

Gould–DeAnza, Inc. 118 Charcot Avenue, San Jose, California 95131. (Digital image processors)

Grinnell Systems Corporation 6410 Via del Oro Drive, San Jose, California 95119. (Digital image processor)

GYYR 1335 South Claudina Street, Anaheim, California 92805. (½-inch time-lapse VCR)

Hitachi Denshi America, Ltd., Broadcast and Professional Division 175 Crossways Park West, Woodbury, New York 11797. (1-inch VTRs, video cameras, recorders, monitors)

Ikegami 37 Brook Avenue, Maywood, New Jersey 07607. (Video equipment, high-resolution monitors)

Image Magnification Systems New Smyrna Beach, Florida 32069. (Catadioptric video projector)

Imaging Technology, Inc. 600 West Cummings Park, Woburn, Massachusetts 01801. (Image-processing boards)

International Imaging Systems 1500 Buckeye Drive, Milpitas, California 95035. (Digital image processors)

Interpretation Systems, Inc. 6322 College Boulevard, Overland Park, Kansas 66211. (Digital image analysis systems)

ISI Group, Inc. 211 Conchas, S.E., Albuquerque, New Mexico 87123. (Digital image processor)

Joyce-Loebl Vickers Marquisway, Team Valley, Gateshead, Tyne and Wear, NE11 0Q3, United Kingdom. (Scanning recording microdensitometer)

JVC, Professional Video Division 58–75 Queens Midtown Expressway, Maspeth, New York 11378. (Video cameras, VCRs, monitors, etc.)

Karl Lambrecht 4204 North Lincoln Avenue, Chicago, Illinois 60618. (High-extinction coated calcite polarizing prisms)

Kontron Bildanase Distributed in the U.S. by Carl Zeiss, Inc. (IBAS image-analyzing system)

Machlett Laboratories 1065 Hope Street, Stanford, Connecticut 06907. (Sniperscope)

MCI/LINK, Link Systems (USA), Inc. P.O. Box 50810, Palo Alto, California 94303. (Intellect 100/200 image analyzers)

Merlin Engineering Works 1880 Embarcadero Road, Palo Alto, California 94303. (Refurbishing and custom modification of 1-inch VTRs)

Motorola, Inc.,Communications Systems Division 9733 Coors Road, N.W., Albuquerque, New Mexico 87114. (PLZT materials)

NAC, Instrument Marketing Corporation 820 South Mariposa Street, Burbank, California 91506. (High-speed video cameras and recorders)

NEC America, Inc., Consumer Products Division 130 Martin Lane, Elk Grove Village, Illinois 60007. (Video cameras, VCRs, monitors, etc.)

Nikon, Inc., Instrument Division 623 Stewart Street, Garden City, New York 11530. (Microscopes; CF, rectified lenses; 35-mm cameras)

Olympus Corporation, Scientific Instruments Division 4 Nevada Drive, Lake Success, New York 11042. (Microscopes, 35-mm cameras)

Panasonic Electronic Components One Panasonic Way, Secaucus, New Jersey 07094. (Video cameras, VCRs, monitors, etc.)

Panasonic Industrial Company, Optical Disk Department 333 Meadowland Parkway, Secaucus, New Jersey 07094. (Optical memory disk recorders)

Perceptive Systems, Inc. 5231 Whittier Oaks, Friendswood, Texas 77546. (Course on image processing, digital image analyzers)

Photonics Microscopy, Inc. 2625 Butterfield Road, Suite 204-S, Oak Brook, Illinois 60521. (Hamamatsu Photonic image processors, low-light-level cameras)

Polaroid Corporation 575 Tech Square, Cambridge, Massachusetts 02139. (Sheet Polaroid, Polaroid cameras, film)

Precision Echo 3105 Patrick Henry Drive, Santa Clara, California 95054. (Video magnetic disk recorder/reproducer systems)

Quantex Corporation 252 North Wolfe Road, Sunnyvale, California 94086. (Digital image processors)

RCA 600 North Sherman Drive, Indianapolis, Indiana 46201. (VCRs)

RCA, New Products Division New Holland Avenue, Lancaster, Pennsylvania 17604. (Video pickup devices, intensifiers, cameras, monitors, etc.)

Recognition Concepts, Inc. P.O. Box 8510, Incline Village, Nevada 89450. (Digital image processors)

Reichert Scientific Instruments P.O. Box 123, Buffalo, New York 14240. (Microscopes)

Rolyn Optics Company 738 Arrowgrand Circle, Covina, California 91006. (Ronchi gratings, lenses, prisms, filters)

Sanyo Electric, Inc. Electronics Division, Communication Products, 1200 West Artesia Boulevard, P.O. Box 5177, Compton, California 90220. (Video cameras, monitors)

Seiler Instrument & Mfg. Co. 170 East Kirkham Avenue, St. Louis, Missouri 63110. (U.S. distributor for aus Jena microscopes)

Sierra Scientific 2189 Leghorn Street, Mountain View, California 94043. (Video cameras, monitors)

Sony Corporation of America 700 West Artesia Boulevard, Compton, California 90220. (Sony video courses for instruction in video principles, service, and repair)

Sony Corporation of America, Video Communications 5 Essex Road, Paramus, New Jersey 07652. (Industrial and broadcast video equipment and supplies)

Spatial Data Systems 420 South Fairview Avenue, Goleta, California 93116. (Log E digital image processors)

Spin Physics 3099 Science Park Road, San Diego, California 92121. (Very-high-speed video cameras and recorders)

Stereographics Corporation P.O. Box 2309, San Rafael, California 94912. (Three-dimensional video, PLZT goggles)

3D Technology Corporation 4382 Lankereshim Boulevard, North Hollywood, California 91602. (Stereo-Color TV Projection System)

Tektronix, Inc. Tektronix Industrial Park, P.O. Box 500, Beaverton, Oregon 97077. (2048 × 2048 pixel CCD imager)

Toshiba Corporation 1101A Lake Cook Road Dearfield, Illinois 60015. (Chalnicon, consumer video products)

TriTronics, Inc. 2921 West Alameda Avenue, Burbank, California 91505. (High-speed video cameras)

Universal Data Systems, Division of Motorola Information Systems Group 5000 Bradford Drive, Huntsville, Alabama 35805. (Modem)

Vickers Instrument Company P.O. Box 99, Malden, Massachusetts 02148. (Joyce-Loebl Magiscan image analyzer)

Vicom Systems 2520 Junction Road, San Jose, California 95134. (Digital image processors)

Videotek 125 North York Street, Pottstown, Pennsylvania 19464. (High-resolution color monitors)

Visual Information Institute, Inc. P.O. Box 33, Xenia, Ohio 45385. (Video test equipment, test charts)

Xybion Electronic Systems Corporation 7750-A Convoy Court, San Diego, California 92111. (High-speed video and Gen-II CCD cameras)

Color Plates

FIGURE 1-8. Very-low-light-level fluorescence microscope image of 1.25-μm-diameter fluorescent microspheres, captured with a **SIT** camera. By digital image processing we can: (E) display intensities in **pseudocolor;** and (F) display intensity contours in three dimensions. 40/0.95 Plan Apo. Digital image processing with Image-I system developed for the author's laboratory. See Fig. 1-8A (p. 10) for unprocessed, single frame image of the same field of view. (From Ellis *et al.*, 1985.)

FIGURE 4-19 (insert). CIE chromaticity diagram. Only the smaller figure from Fig. 4-19 (p. 89) is reproduced here. This is the 1931 CIE Chromaticity Diagram shown schematically in Figs. 4-18 (p. 88) and 7-28 (p. 239). The 1976 version shown on p. 89 better reflects the response of the average human eye to light of different hues. (See Miller, 1985.) (Courtesy of Photo Research, Burbank, California.)

FIGURE 10-55. Use of pseudocolor transformations to enhance contrast in fluorescent cell images. Monochrome images similar to that of Fig. 10-54a were subjected to pseudocolor enhancement to accentuate changes in cellular fluorescence with time. Panels (a)–(d) show changes in fluorescent patterns occurring over a period of 45 sec. (Photographs courtesy of Ann Siemens, Department of Developmental and Cell Biology, University of California, Irvine.)

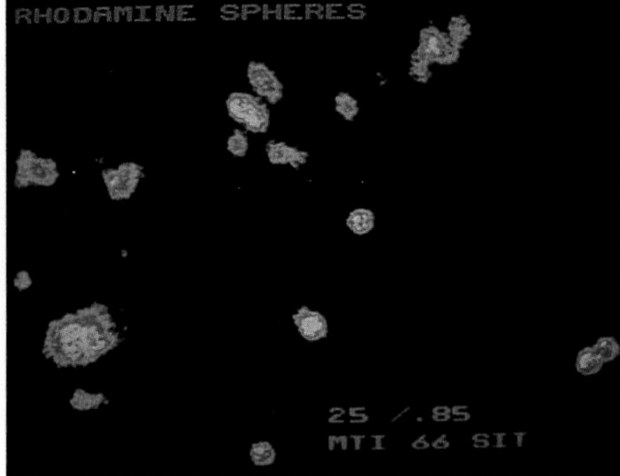

FIGURE 1-8E

FIGURE 1-8F

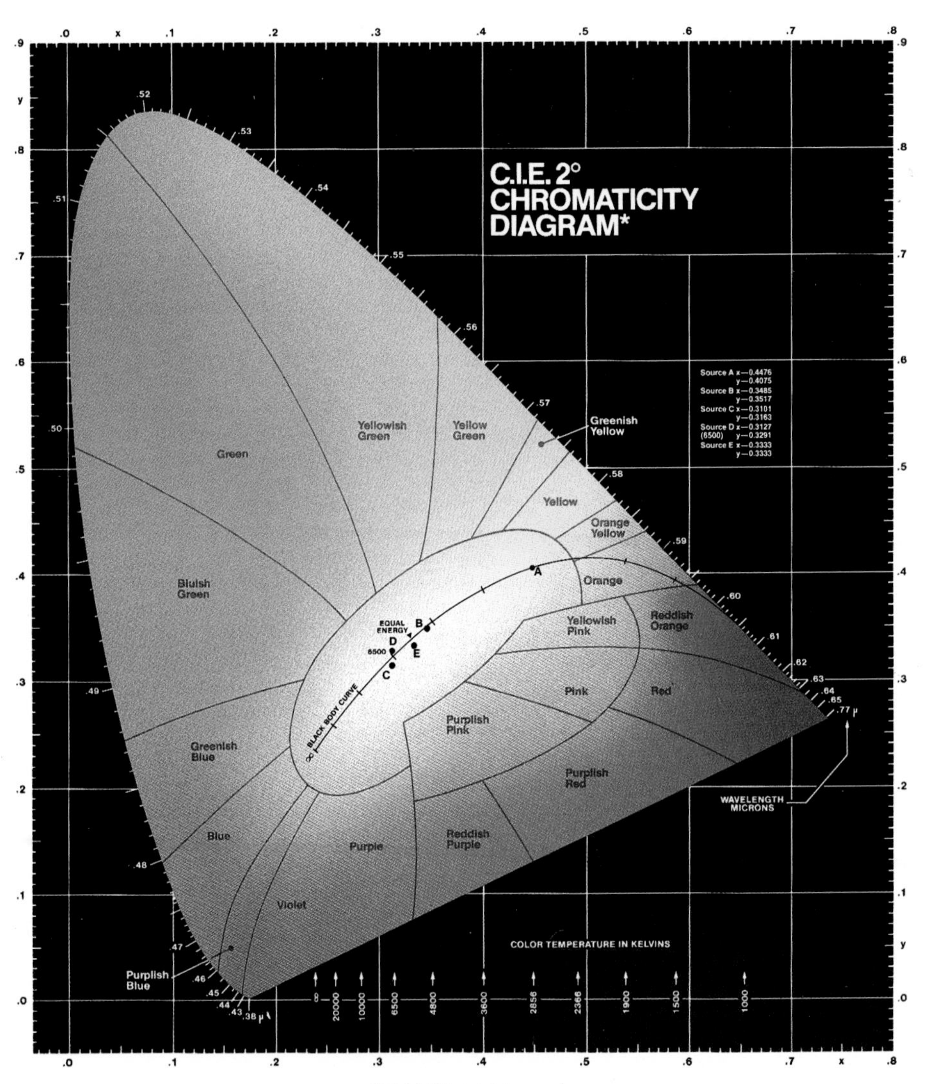

FIGURE 4-19 (insert)

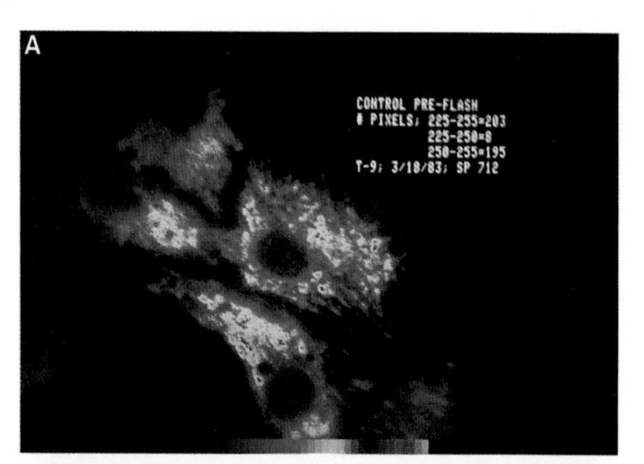
A
CONTROL PRE-FLASH
PIXELS; 225-255=203
225-250=8
250-255=195
T-9; 3/18/83; SP 712

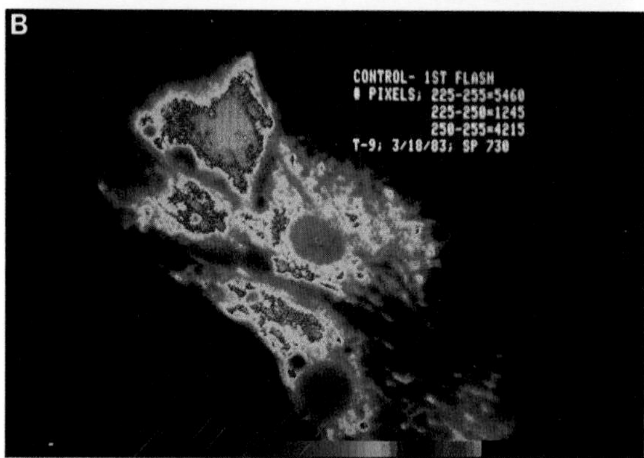
B
CONTROL- 1ST FLASH
PIXELS; 225-255=5460
225-250=1245
250-255=4215
T-9; 3/18/83; SP 730

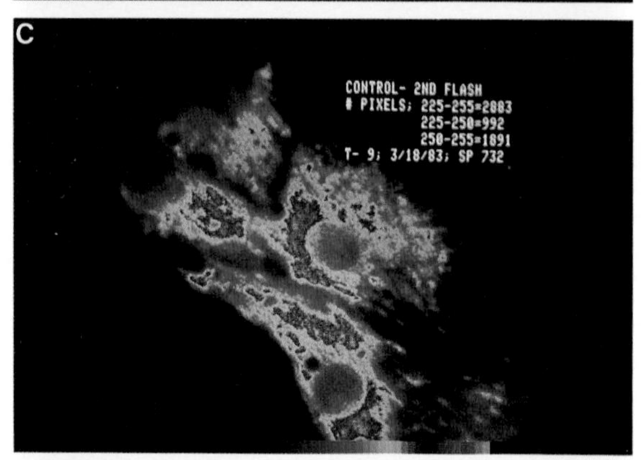
C
CONTROL- 2ND FLASH
PIXELS; 225-255=2883
225-250=992
250-255=1891
T- 9; 3/18/83; SP 732

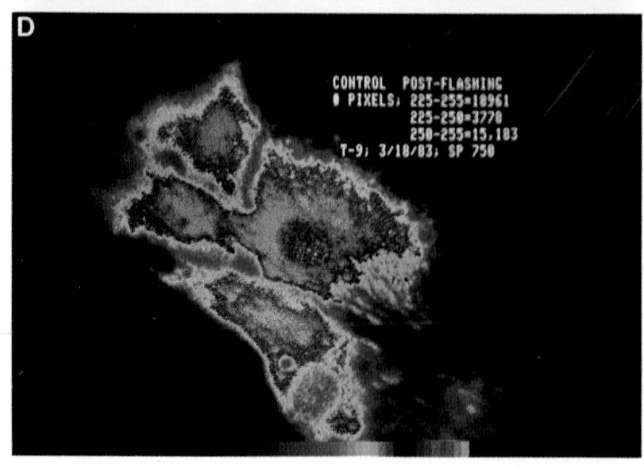
D
CONTROL POST-FLASHING
PIXELS; 225-255=18961
225-250=3778
250-255=15,183
T-9; 3/18/83; SP 750

FIGURE 10-55

Video Microscopy

1

Why Video?

This chapter examines the early evolution of television, and surveys how television was applied to the light microscope in the three decades up to the mid 1970s, a period of exploration and early development for **video microscopy.** Following this is a brief glance at the striking improvements in light microscopy that became possible with the introduction of modern video and related technology. The introductory survey concludes with a few general remarks on the chapters that follow, whose overall plan of organization was outlined in the Preface.

1.1. PRE-1940: THE INVENTION OF TV

In a landmark paper published in 1934, Zworykin (of RCA Victor in Camden, N.J.) announced the successful development of a photoelectric vacuum device that directly converted an optical **image** into an electrical signal. His paper, entitled ''The Iconoscope—A Modern Version of the Electric Eye,''* began as follows:

> Summary—This paper gives a preliminary outline of work with a device which is truly an electric eye, the iconoscope, as a means of viewing a scene for television transmission and similar applications. It required ten years to bring the original idea to its present state of perfection.
>
> The iconoscope is a vacuum device with a photo-sensitive surface of a unique type. This photo-sensitive surface is scanned by a cathode ray beam which serves as a type of inertiales commutator. A new principle of operation permits very high output from the device.
>
> The sensitivity of the iconoscope, at present, is approximately equal to that of photographic film operating at a speed of a motion picture camera. The resolution of the iconoscope is high, fully adequate for television.
>
> The paper describes the theory of the device, its characteristics and mode of operation.
>
> In its application to television the iconoscope replaces mechanical scanning equipment and several stages of amplification. The whole system is entirely electrical without a single mechanically moving part.
>
> The reception of the image is accomplished by a kinescope or cathode ray receiving tube described in an earlier paper.
>
> The tube opens wide possibilities for applications in many fields as an electric eye, which is sensitive not only to the visible spectrum but also to the infra-red and ultra-violet region.
>
> The idea of being able to observe far-away events is a fascinating one. A device which will enable a person to do so has been for centuries the dream of inventors and for decades the goal of earnest scientific workers.
>
> The goal of television is to make this dream a reality. The problem, however, is a difficult

*Iconoscope: icon, Greek for image; scope, to observe. The ''electric eye'' in the title indicated the photoelectric cell. A photoelectric cell responds to light (passing electrical current in proportion to light intensity), but does not respond to images, whereas the iconoscope does. (See Zworykin, 1931.)

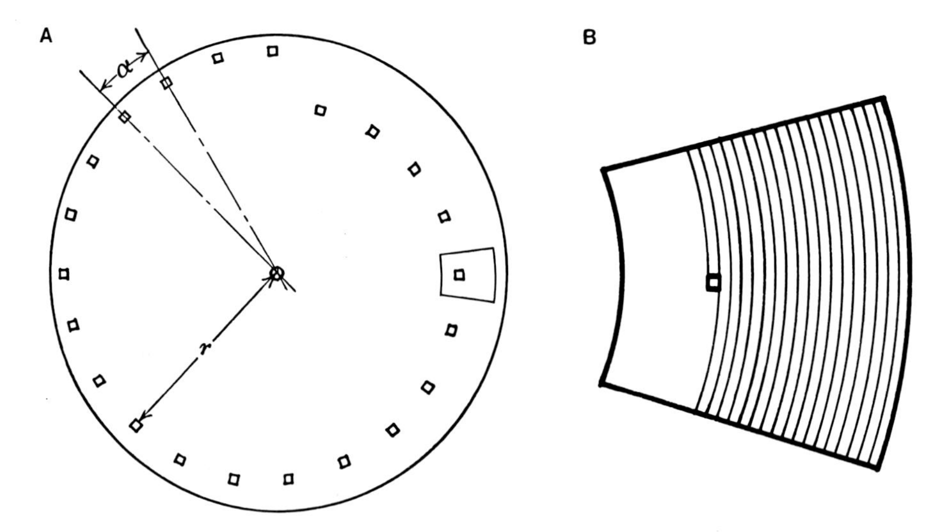

FIGURE 1-1. (A) A Nipkow disk. Identical rectangular holes are arranged circumferentially at a constant angle (α) and radially at constantly decreasing distances from the center (*r*). When the disk is spun, each hole sweeps across an image through the window at a different radial location, producing a raster scan, as shown in B. (After Zworykin and Morton, 1954.)

> one and requires for its solution a great many component elements, most of them unknown up to quite recent years.
>
> The meaning of seeing over a great distance can be interpreted as sending instantaneously a picture through this distance. This requires means of communication extremely rapid and free from inertia. The discovery of electricity and the development of electrical communication, therefore, laid the foundations for the future realization of television.

Zworykin concluded the paper with the following:

> With the advent of an instrument of these capabilities, new prospects are opened for high grade television transmission. In addition, wide possibilities appear in the application of such tubes in many fields as a substitute for the human eye, or for the observation of phenomena at present completely hidden from the eye, as in the case of the ultra-violet microscope.

Thus, even as Zworykin announced the development of the first practical TV **image pickup tube**, he was aware of the far-reaching potentials of TV to extend the limits of human vision, including into the microscopic world.

Retracing the history of TV farther back, we find several proposals and attempts to take advantage of the photoconductive property of selenium that was discovered in 1873.* In some of the proposals, mosaics of selenium cells were arranged to pick up the brightnesses of discrete parts of an image focused onto the mosaic, and some simple patterns could in fact be transmitted. But the early devices required many parallel wires and were much too elaborate and slow to be of practical use for TV.

Paul Nipkow, a natural sciences student in Berlin, discovered in 1883 how to rapidly convert an optical image into an electrical signal that could be transmitted through a single cable. Rather than trying to send all of the elements of the picture at once, he could transmit the picture point-by-point in rapid succession by scanning the image with a spinning disk that was appropriately perforated (Nipkow, 1884).

The small (rectangular) holes in the **Nipkow disk** are arranged as shown in Fig. 1-1A. The holes are located circumferentially a constant angle (α) apart, and with their radii (*r*) varying as an Archimedes spiral (i.e., *r* varies in arithmetic progression). Thus, when a motor spins the Nipkow

*The electrical conductivity of (the gray form of) the element selenium increases up to a thousandfold on exposure to light; but the response is actually too sluggish for use in TV.

disk, the successive holes in the disk scan the image in a **raster** pattern (Fig. 1-1B). A photoelectric cell behind the disk senses the brightnesses of the image scanned through the successive holes and gives rise to a series of corresponding electrical signals. A neon discharge tube, connected to the electrical signal and seen through a second Nipkow disk (with the same pattern of holes and spinning at the same speed as the first disk, or seen through another part of the first disk), produces a **picture** of the image that was scanned by the first Nipkow disk.

While Nipkow's conception laid an important foundation for the development of TV, the amount of light that could be transmitted through the tiny holes was so small that his device did little more than demonstrate the principle of his ideas. In 1910, Ekstrom partly overcame this difficulty by reversing the arrangement. Instead of using the Nipkow disk to scan the image of an object formed by a lens, he scanned the object with a focused image of the Nipkow disk that was illuminated from behind with a carbon arc lamp. The light scattered by the object scanned in a raster by a bright spot of light was collected by photoelectric cells, which gave rise to the electrical signal.

Ekstrom's scheme overcame some of the difficulties inherent in Nipkow's, but several additional developments were needed before ''distant electric vision'' could be made a reality. In 1908, A. A. Campbell Swinton suggested the use of ''kathode rays'' at the transmitting and receiving stations in place of mechanical switching devices to achieve synchronized switching of some 160,000 image elements/sec. In 1911 and 1924 he elaborated on a **pickup tube** embodying **cathode**-ray scanning (see Zworykin and Morton, 1954, pp. 245–250). And in 1934, Zworykin announced the development of the iconoscope described in the paper extracted earlier in this chapter (see also Zworykin, 1931, 1933; Zworykin and Morton, 1954).

In the iconoscope, the optical image is **focused** onto the **target** in an evacuated glass flask (Fig. 1-2). The target is **scanned** in a raster by a sharply focused beam of electrons emitted from an **electron gun.** The beam is steered electromagnetically by two sets of **deflection coils.** The target is made up of a very thin sheet of mica that is coated on the side away from the electron gun with an electrically conductive layer (the signal plate). The side of the target that faces the lens and the electron gun, is coated with ''a myriad of tiny photo-sensitized silver elements'' that are electrically insulated from each other (the mosaic). When illuminated, these tiny silver elements release photoelectrons [which are picked up by the **anode** (A)]. Where electrons were released from the silver

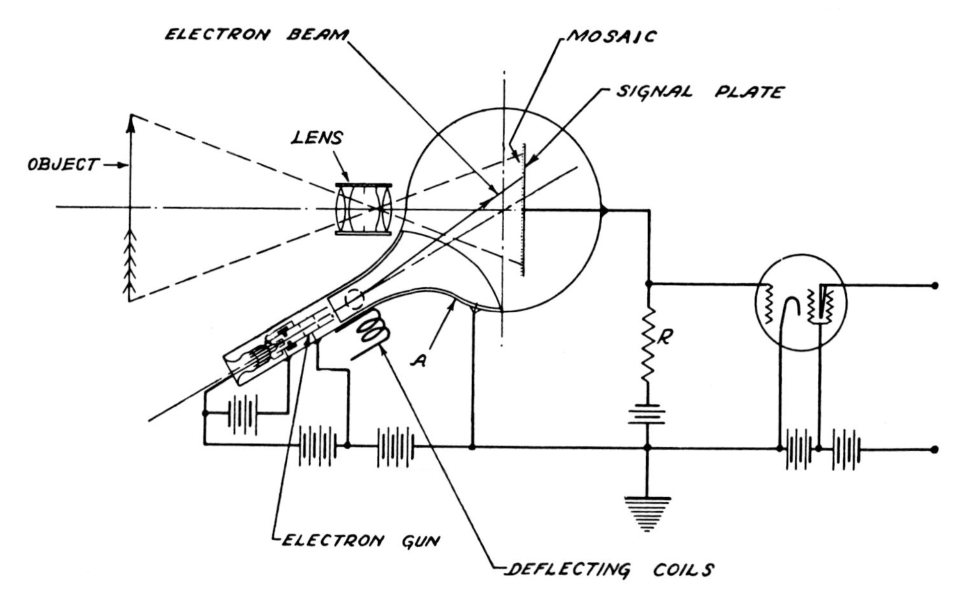

FIGURE 1-2. Zworykin's 1934 Iconoscope, an image-storage type of pickup device, which, along with the kinescope, enabled practical video transmission to be accomplished. See Fig. 1-3 and text. [From Zworykin, 1934; copyright 1934, IRE (now **IEEE**).]

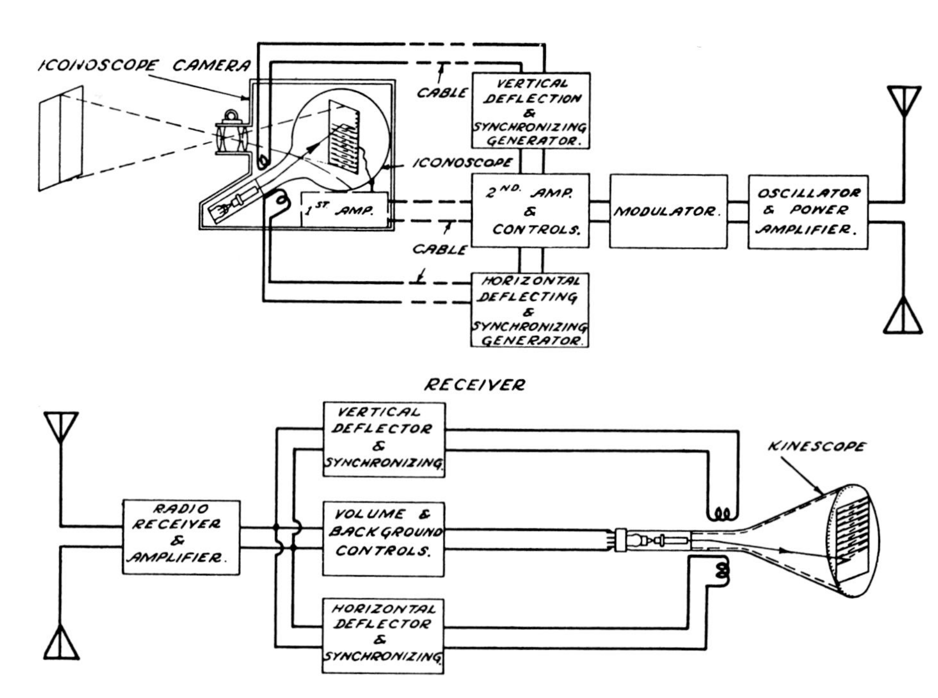

FIGURE 1-3. Component diagram of a complete practical TV system of the mid 1930s. [From Zworykin, 1934; copyright 1934, IRE (now IEEE).]

elements, an electrical charge accumulates across the thin mica plate. In the bright regions of the image, the charging continues until the electron beam reaches the region and neutralizes the charge. Then the cycle begins anew as the electron beam repeatedly scans the target. Thus, the target in the iconoscope acts as a highly sensitive *image storage device* with vastly improved sensitivities compared to the earlier, nonstorage type of image pickup device.

With these developments, coupled with the use of thermionic vacuum tubes to obtain high-speed electronic amplification, practical TV became a reality by the mid 1930s (Fig. 1-3).* Then came World War II.

1.2. EARLIER APPLICATIONS OF VIDEO TO MICROSCOPY

After the war, TV broadcasting and entertainment surged forward. The **vidicon** camera tube, which used a photoconductive target rather than a photoemissive one (as in the iconoscope), was introduced by Weiner *et al.* in 1950.† **Closed-circuit TV** (CCTV) was applied to commerce, education, industry, medicine, science, and the military (Zworykin *et al.*, 1958). The application of CCTV to microscopy was a natural part of this development. **Video** was coupled to the microscope for classroom instruction, enhancement of **contrast,** and the extension of the spectral range from the visible to the ultraviolet (UV) and infrared (IR) regions of the spectrum (Flory, 1951; Land *et al.*,

*Triode vacuum tubes were invented by Lee DeForest in 1907 (e.g., see Wheeler, 1939). See Chapters VII and VIII in Zworykin and Morton (1954) and Chapter I in Zworykin *et al.* (1958) for further accounts on earlier image pickup and display devices that had been tried or proposed. See also Kazan and Knoll (1968) for an interesting account of electronic devices that permit refreshable image storage.

†Chapter 7 describes the characteristics of the **vidicons** and other modern video image pickup tubes.

1948; Parpart, 1951; Zworykin and Flory, 1951). I can recall the exciting displays of a giant amoeba, twice my height, crawling up on the auditorium screen at Princeton as I participated in a demonstration of the RCA projection video system to which we had coupled a **phase-contrast** microscope. Impressive as that sight was, CCTV in those days did not appear quite ready to help us improve the actual quality of the image seen through the microscope.

Much has happened to video microscopy in the three decades that followed.* In 1951, Young and Roberts showed that a *flying-spot UV microscope* (capturing the output signal with a photoelectric tube) would provide a sensitivity two orders of magnitude greater than with photography, and also that a flying-spot microscope would reduce **flare** and improve the microscope image and, above all, introduce the possibility of quantitative analysis. Montgomery *et al.* (1956), Flory (1959), Freed and Engle (1962), and others did successfully develop and apply flying-spot UV microscopes, but the damage to the living cells by UV, while considerably less than with the exposure needed for photography, was nevertheless significant.

In 1968, Petráň *et al.* introduced a tandem-scanning reflected-light microscope that used a Nipkow disk to scan both the illuminating and image-forming beams. By illuminating and transmitting only a single focused spot at a time, they were able to visualize the structure of three-dimensional objects, such as excised dorsal root ganglia, without losing the image in unwanted light scattered from out-of-focus regions. While there is much of potential interest in the **confocal optical system** developed by Petráň *et al.* (Section 11.5), these early attempts were limited by the lack of bright light sources; e.g., they were forced to use reflected sunlight as the source.

In the meantime, **image tubes** were developed to directly convert invisible images into visible ones. Depending on the type of image tube, UV, IR, or very-low-light-level visible-wavelength images could be converted to pictures, on a phosphor screen, that could be seen with the eye or photographed (Section 7.1). George Reynolds in particular pioneered the application of the latter type of image tube, or the **image intensifier,** to *low-light-level* imaging in microscopy (Reynolds, 1964). He and co-workers captured many low-light-level events that would otherwise have been impossible to see or photograph (Eckert and Reynolds, 1967; Reynolds, 1972; Section 11.2).

Caspersson and co-workers, who for many years had been exploring the nucleic acid metabolism of cells with the UV microscope, used *fluorescence microscopy* to identify human chromosomes stained with quinacrine mustard, a DNA-binding fluorescent dye. They displayed the intensity scans (of a TV line running through the length of the chromosome) on an oscilloscope or graphics plotter to measure the banding pattern and ratio of chromosome arm lengths (Caspersson *et al.*, 1970; Caspersson, 1973). Usually, they scanned photographic negatives of chromosomes taken through a fluorescence microscope, but also pointed out that a TV camera could be attached directly to the microscope as well. They adjusted the TV camera **black level** to coincide with the **pedestal** of the TV signal to boost the contrast of the fluorescent images.

As the power and speed of digital computers grew in the 1960s to early 1970s, a major effort was launched to develop *automated microscopy*. Taking advantage of the "pattern recognition" capabilities of **digital** computers, video images of blood cells, Papanicolaou smears, chromosomes, etc. were automatically scanned, sorted, and counted under a microscope that was also controlled by the computer (e.g., Bacus and Gose, 1972; Bartels and Wied, 1973; Butler *et al.*, 1966; Carman *et al.*, 1974; Castleman and Wall, 1973; Flory and Pike, 1953; Lemkin *et al.*, 1974; Lipkin and Rosenfeld, 1970; Mendelsohn *et al.*, 1968; Stein *et al.*, 1969). These efforts complemented the development of flow cytometry.

Together with computers, video equipment such as cameras, processors, recorders, **monitors,** etc. were also improved by the introduction of solid-state devices. Many new camera tubes were introduced, **solid-state pickup devices** were explored, and cameras and monitors generally became smaller and lighter as their performances improved. After the transverse- and helical-scan **video tape recorders** (VTRs) were introduced in the mid to late 1950s, half-inch reel-to-reel VTRs costing 1/100th of the studio recorders were seen in the 1960s, followed by **videocassette recorders** and **azimuth** recording by 1973. In the 1960s, compact **digitizers** and analyzers were developed for *quantitating* light intensity, distances, flow, etc. in **real time** with CCTV (e.g., Blifford and

*Developments in video microscopy since around 1975 will be described later.

Gillette, 1973; Pike, 1962). Some were used, for example, to measure dimensions of microvasculature and blood flow under the microscope (e.g., see literature reviewed in Intaglietta *et al.*, 1975; Sections 9.5, 9.6).

1.3. WHERE WE ARE TODAY

The pace of progress has quickened since the late 1970s. The remarkable developments in video and in computer technology have brought us into a new era. Our notions of what we can see, and what we can measure with the light microscope, are indeed undergoing dramatic changes. I will illustrate some of the points with a few photographs (Figs. 1-4 to 1-8).

As these photographs illustrate, video can now reveal structures that previously were too fine to be seen, movements too fast to be recorded, and scenes too vague, too dim, or too **noisy** to be clear. They can all be displayed brilliantly on the monitor today. Naturally, the brightness of the monitor

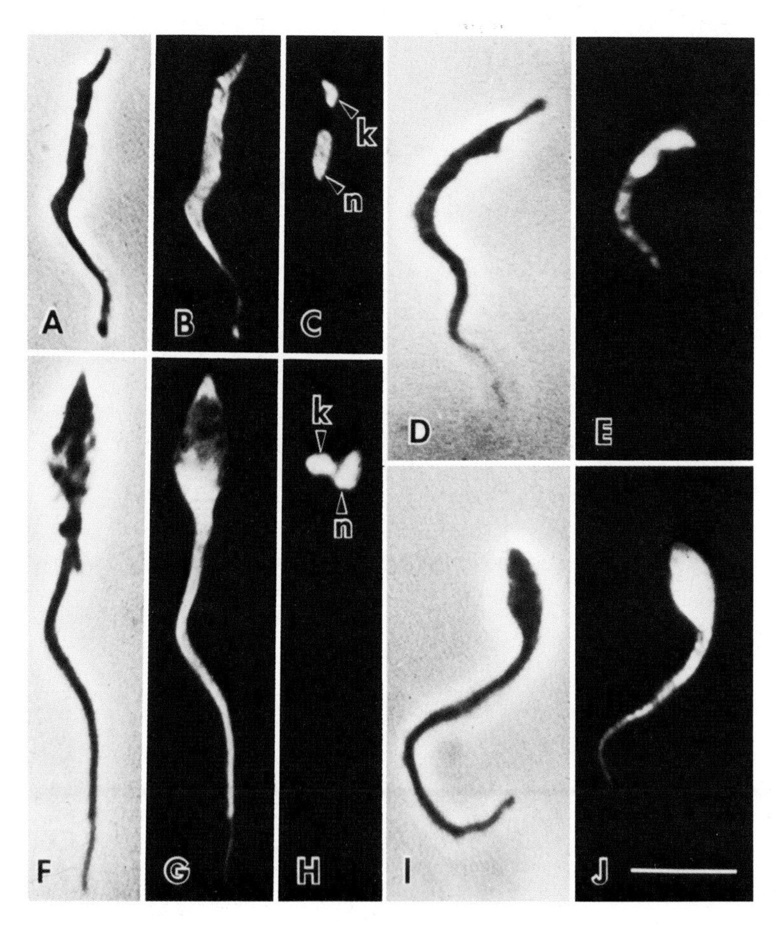

FIGURE 1-4. Phase-contrast (A, D, F, I) and fluorescent micrographs of the carriers of Chaga's disease (*Trypanosoma cruzi*) observed with a video camera equipped with an **image isocon** tube. The Image Isocon is even more sensitive to low levels of light, and provides better resolution, than the **image orthicon,** which was the tube used by Dvorak *et al.* to capture a time-lapse sequence of a malaria parasite in the process of invading a human blood cell (Dvorak *et al.*, 1975; Chapter 11). A–E show the metacyclic trypomastigote form of the organism, and F–J the epimastigote. Different fluorescent stains (FITC for total protein in B and G; propidium iodide for DNA in C and H) localize different chemical components of the cell, but at light levels too low for capture on common photographic emulsions. Scale = 5 μm. (Courtesy of J. A. Dvorak.)

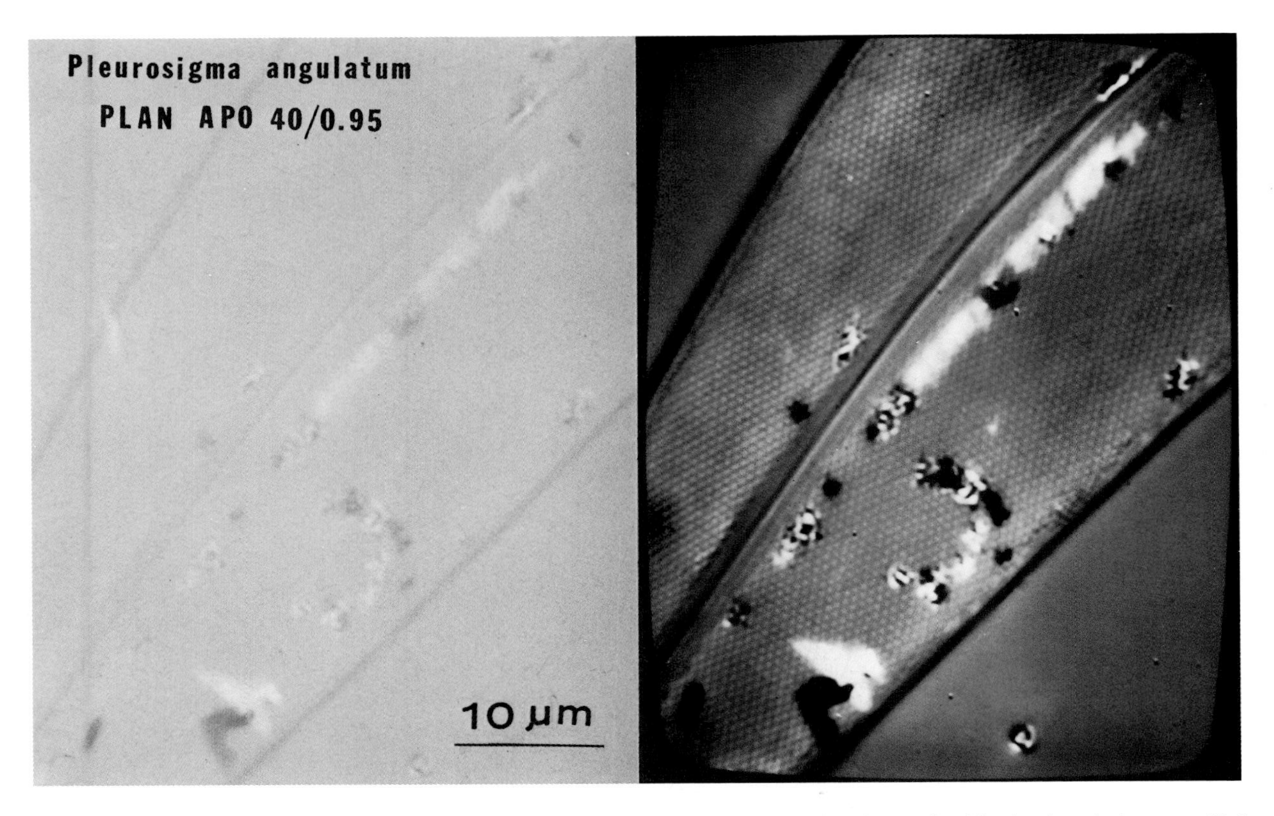

FIGURE 1-5. A diatom frustule in polarized-light microscopy, seen through a **Plan Apochromatic objective lens** (using a rectified condenser with **NA** matched to the objective lens), with and without **analog** video enhancement. The contrast and image detail resulting from the video enhancement (right) are striking indeed (Inoué, 1981b; Chapters 7, 9, 11). Without the **background subtraction** and **contrast enhancement** provided by video, we could not have dreamed of taking advantage of the excellently corrected Plan Apo lenses (which contain **birefringent** crystalline elements that ordinarily mask those of the specimen) for polarized-light microscopy to observe such weakly birefringent objects.

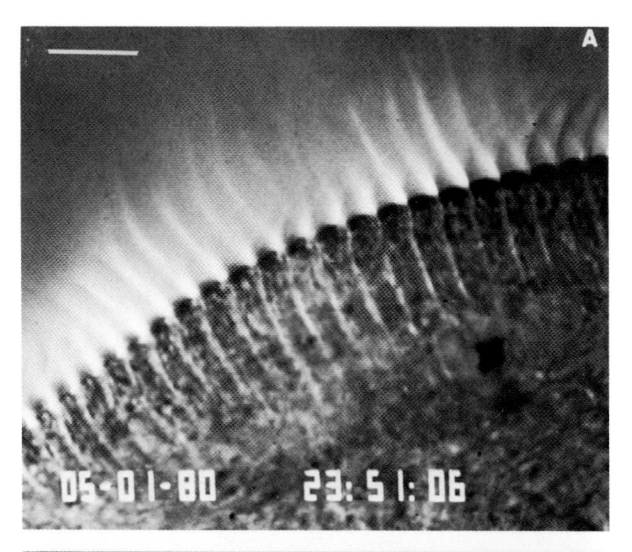

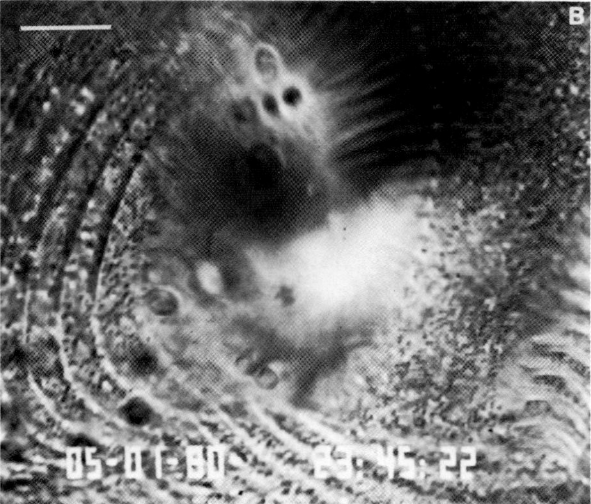

FIGURE 1-6. Beating rows of cilia in a swimming *Stentor* observed in polarized light under optical conditions similar to Fig. 1-5. The weak birefringence of the beating cilia, as well as of their rootlets, is clearly captured in this (1/60 sec) frozen-field image. (A) Side view. (B) Top view of oral region. Such low-light-level, dynamic images were virtually impossible to capture photographically in the past. They can now be recorded readily on a VTR and analyzed field-by-field (Chapters 7, 8). Rectified-polarization optics; 40/0.95 Plan Apo; 546-nm illumination. Scale = 10 μm. (From Inoué, 1981b.)

can be adjusted independently of the microscope **illumination,** and the scene can be viewed by several individuals participating in an observation.

Video recording is now simple. Recorders are easy to operate, videotape can be overwritten many times if desired, and very slow changes, as well as quite rapid motions, can be recorded and displayed repeatedly in **freeze frame** or at playback speeds appropriate for analysis (Chapter 8). Recording of exact time and date, and simultaneous or later addition (**dubbing**) of sound, are likewise straightforward.

The capability and sophistication of computers that can process video images are advancing at a very rapid pace. NASA has treated us to spectacular, computer-enhanced images of our sister planets and their satellites, assembled and processed after telecasting over vast distances in space. Many of these techniques can now also be applied to microscopy as discussed in Chapter 10.

Not only can computers improve the image better and faster, but they can extract specific image information as well as control the microscope. These developments are taking place at an accelerating pace today (Chapter 10; Baxes, 1984; Castleman, 1979).

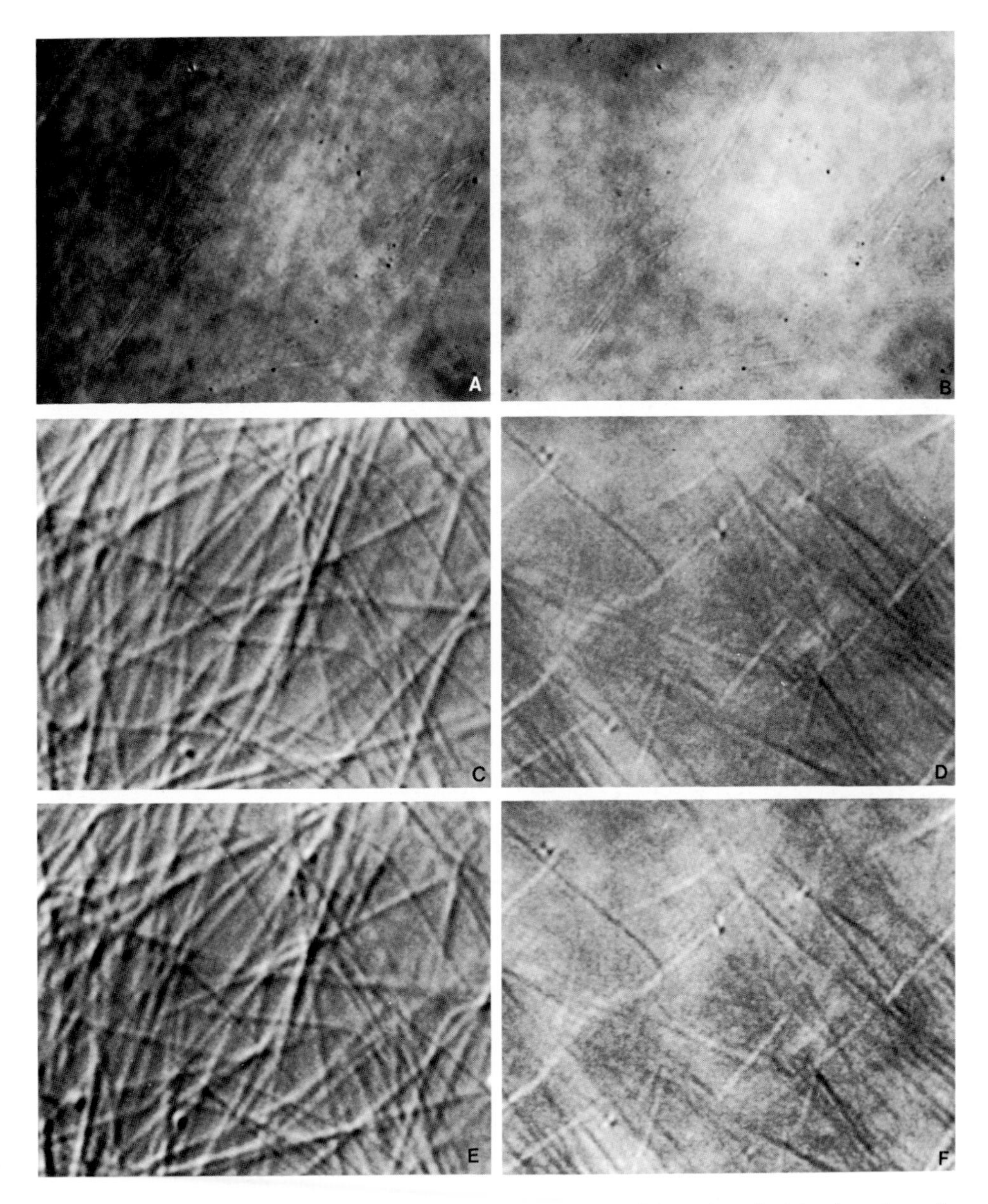

FIGURE 1-7. Contrast-enhanced micrograph of a slurry of microtubules polymerized *in vitro,* shown in **differential interference contrast** (A, C, E) and polarized-light microscopy (B, D, F). C and D have had **hot-spot,** shading, and mottle digitally subtracted from out-of-focus images of A and B. E and F have been additionally enhanced, electronically (field height = 12 μm) with the Photonic "AVEC" system. Field height is 12 μm. (Courtesy of R. D. Allen; from Allen, 1985.)

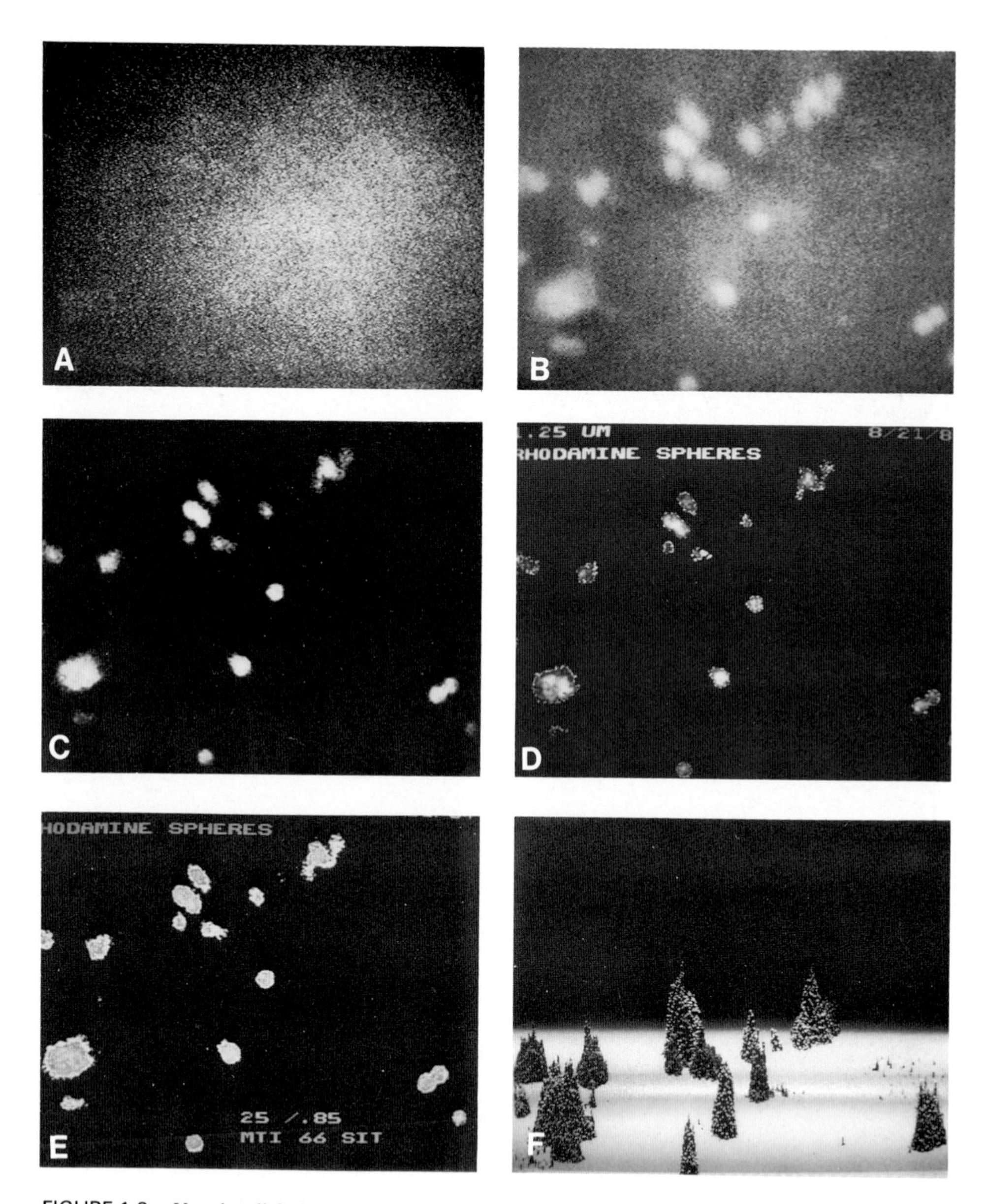

FIGURE 1-8. Very-low-light-level fluorescence micrographs of 1.25-μm-diameter fluorescent microspheres, captured with a SIT camera. In a (1/30 sec) frozen frame, the image of the microspheres is buried in noise (A). By digital image processing we can: (B) dramatically reduce noise by **frame** summation combined with background subtraction; (C) suppress background by thresholding; (D) add outline emphasis; (E) display intensities in **pseudocolor;** and (F) display intensity contours in three dimensions. 40/0.95 Plan Apo objective. Digital image processing with ''Image-I'' system developed for the author's laboratory. For color reproductions of Fig. 1-8E,F see color plate following page xxvi. (From Ellis *et al.*, 1988.)

1.4. A FEW GENERAL REMARKS

Video shows exciting potential and promises for microscopy, and allows us to extract information hidden in the dark, low-contrast, changing scene. But what can be extracted is only that that was actually hidden and unseen. What was not signaled by the original distribution of photons, or image-forming light waves, cannot, and *should* not, appear in the video-enhanced image.

Thus, on the one hand, the light microscope can provide images that are much better and contain more information than we have been accustomed to. On the other hand, it is not difficult to understand that the quality of the image can be disappointing and the information can even be erroneous if video is applied improperly.

To take good advantage of video microscopy, the rule of thumb should be: start with a microscope capable of providing the highest-quality image, adjust it well, *then* enhance it with video. Make sure the video is processing the image reliably and optimally, not introducing undecipherable artifacts of its own. As explained in Chapters 3 and 6 through 10, *every piece of video equipment—camera, mixer, recorder, processor, analyzer, monitor—contains circuits that must process the video signal.* The microscope itself is also a sophisticated **image-processing** device (Chapter 5, Appendixes II, III).

It is therefore as important to understand how the microscope works, how to align and use it well (Chapter 5), as it is to know how to select, hook up, and adjust the video equipment (Chapters 3, 6–12). Likewise, it is helpful to understand some basic characteristics of human vision, the ultimate target for which video was developed (Chapter 4).

2

Glossary

The world of video is fraught with confusing jargon and acronyms. One can become lost just reading articles about video or looking through catalogs of video equipment. The glossary that follows should take some of the mystery out of the jargon and at the same time serve as an index to sections of the text that deal with the terms in more detail.

While some of the terms used in video are clearly defined according to IEEE, EIA, or other national and international standards, definitions of other terms vary with the user. (Many acronyms used in electrical engineering stand for several different terms as can be seen, for example, in the IEEE dictionary of standard terms.) I have tried to choose reasonably common terms and usage, but some authors and manufacturers do use their own special jargon. Experienced technical or sales personnel may have to be consulted for clarification of the terms used to describe the function of specific equipment or software.

In addition to terms used in video, including digital image processing, the glossary includes selected terms used in microscopy. Some terms in the glossary are not used elsewhere in the text; they are included because they often appear in product descriptions of video and image-processing equipment.

The spiral organization of this book is designed to guide the reader through a web of practice and theory, in much the way I believe many of us learn in a new situation. I recommend that you scan the glossary the first time through, more to get a feel for the terms rather than to study them in detail. Then return to the list as though it were a dictionary as the need arises.

Aberration Failure of an optical or electron-optical lens to produce exact geometrical (and chromatic) correspondence between an object and its image. In a video camera tube or cathode-ray tube, aberrations arise when the (electrostatic or electromagnetic) lens does not bring the electron beam to sharply focused points uniformly on the target or screen, or to correct geometrical positions, as the beam is deflected. *See also* Distortion. [Sections 5.9, 7.2]

Abbe limit Ernst Abbe's specification for the limit of resolution of a diffraction-limited microscope. According to Abbe, a detail with a particular spacing in the specimen is resolved when the NA of the objective lens is large enough to capture the first-order diffraction pattern produced by the detail at the wavelength employed. *See also* Rayleigh criterion, Sparrow limit. [Section 5.9]

Accommodation (of the eye) The act of adjusting the eye, to bring objects that are closer to the eye into focus. Contraction of the ciliary muscles relaxes the tension on the crystalline lens, which rounds up by virtue of its elasticity while also moving forward slightly. The net effect is to reduce the overall focal length of the eye. [Section 4.2; Fig. 4-2]

Achromatic aplanatic condenser A well-corrected microscope condenser lens; corrected for chromatic and spherical aberrations and satisfying the sine condition. [Section 5.10]

Achromatic lens *Also* achromat. A lens cluster whose foci and power are made the same for two wavelengths (commonly for the red hydrogen C line, $\lambda = 6563$ Å, and blue F line, $\lambda = 4861$ Å). The simplest achromat is a doublet that combines two single lenses with different dispersions and curvatures to achromatize the combination. Even the simplest "achromatic' microscope objectives contain two such doublets. *See also* Apochromatic objective lens. [Section 5.10]

Active picture area The area of the video scan that is actually visible on an underscanned monitor. Starts with the end of the V blank for the current frame and ends at the start of the V blank for the next frame. *See also* Underscan, V blank. [Section 6.2; Fig. 6-17]

Active scan lines The H-scan lines that cover the active picture area. [Section 6.2; Fig. 6-17]

Adaptation *See* Dark adaptation.

ADC *See* A/D converter.

A/D converter (analog-to-digital converter; ADC) A component, or piece of equipment, used to convert video and other analog signals into discrete digital signals (usually binary). A necessary component preceding digital image processing by a computer. [Sections 9.6, 10.2]

Add editing *See* Assemble editing.

AGC *See* Automatic gain control.

AHFC *See* Automatic horizontal frequency control.

Airy disk The bright disk of light (surrounded by alternating dark and bright diffraction rings) that is formed by a perfect diffraction-limited lens, focusing an image of an infinitely small source of light. For a minute absorbing spot, the diffraction pattern is a dark Airy disk surrounded by brighter and darker diffraction rings. Since the Airy disk is the smallest unit that makes up the image of a luminous or absorbing object (formed by a properly corrected microscope lens in focus), the radius of the disk determines the limit of resolution of the microscope. *See* Abbe limit, Rayleigh criterion. [Section 5.5, Appendix II; Fig. 5-19]

Aliasing A pattern of image sampling error in digital systems. Aliasing forces spatial frequency components higher than a critical value (the Nyquist frequency) to be displayed at progressively *lower* frequencies. Aliasing introduces an undesirable moiré pattern when the spatial frequency of the signal exceeds the sampling rate in a digitizer. [Section 10.4, Appendix II]

Alignment tape A special videotape containing picture and sound reference signals; used for adjusting a VTR to the manufacturer's specifications. [Section 8.7]

ALU *See* Arithmetic–logic unit.

AM (amplitude modulation) A mode of signal transmission in which the amplitude of a carrier wave is varied in accordance with the instantaneous value of the modulating signal. *See also* FM. [Section 8.2]

Amplitude modulation *See* AM.

Amplitude response (of video image pickup device) The value of the amplitude response curve of a video image pickup device for spatial frequency at 400 TV lines per picture height. 60% is a high number indicating excellent picture resolution; 25% is a low number indicating poor resolving power. *See* Amplitude response curve. [Section 7.2; Figs. 7-10, 7-39]

Amplitude response curve A curve that depicts the ability of an electronic device to relay the frequencies in the input signal to its output. For video image pickup devices, it shows the ability to retain image contrast at various spatial frequencies. The curve is plotted as the percentage modulation of the signal amplitude versus spatial frequency. [Section 7.2]

Anaglyph (red–green) A pair of superimposed pictures produced in red and green (blue), which, when viewed through complementary colored glasses (placed appropriately in front of the left and right eyes), gives rise to a stereoscopic image. Its effectiveness is not affected by color blindness, although a certain fraction of the population is nonetheless unable to perceive stereoscopic images. [Section 12.7]

Analog An information representation scheme with continuous amplitudes. Contrasts with digital, where information is quantized into discrete steps. [Sections 9.1, 10.2]

Analog-to-digital convertor *See* A/D convertor.

Anisotropy The difference in physical properties such as velocity of wave propagation, elasticity, refractive index, absorbance, etc., for the same material measured along different directions. *See also* Birefringence, Dichroism. [Section 11.3; Appendix III]

Anode The electrode to which a major flow of electrons takes place internally (as in a cathode-ray tube) or to which an external positive voltage supply is connected. The converse of cathode. [Sections 1.1, 7.1, 7.5]

Aperture In a video camera tube or monitor, the aperture refers to the size and shape of the electron beam that lands on the target or phosphor. In an optical instrument, the opening of a lens or aperture stop. [Sections 5.4, 7.1]

Aperture correction An electronic process used in some high-resolution video cameras and monitors that compensates for the loss in sharpness of detail due to the finite dimensions (aperture) of the scanning beam. [Sections 7.3, 7.5]

Aperture function In a diffraction-limited optical system, the function that determines the relationship between the image and each point in the object. Modifying the aperture function changes the image according to the modified Fourier-filtering (or optical filtration) property of the aperture. *See also* Fourier transform. [Section 10.12, Appendix III; Figs. 10-49, III-23]

Aperture mask *See* Shadow mask.

Aperture plane In a microscope adjusted for Koehler illumination, the conjugate planes that include the light source, the condenser iris diaphragm, the objective lens back aperture, and the eyepoint. Spaces in the aperture planes are the reciprocal of those in the field planes. [Section 5.3; Figs. 5-14, 5-15]

Aphakia The absence of a crystalline lens in the eye. While accommodation is lost, the UV transmission of the ocular media increases. [Section 4.3]

APL *See* Average picture level.

Apochromatic objective lens A microscope objective lens designed to provide the same focal length for three wavelengths and freedom from spherical aberration for two wavelengths of light. The magnification can still vary with wavelength in which case a compensating eyepiece is used to cancel the colored fringes. *See also* Fluorite objectives, Plan Apo. [Section 5.9]

Arithmetic–logic unit (ALU) In a digital image-processing computer, the component that performs a variety of arithmetic and logic operations (sum, difference, exclusive or, multiply by constant, etc.) on the image signal at extremely high speeds. [Section 10.14]

Array processor A specialized digital signal-processing unit capable of handling large-scale arrays (matrices) of computation (such as those encountered in fast Fourier transforms, matrix inversions, etc.) used in some image-processing tasks. [Section 10.2]

ASCII (American Standard Code for Information Interchange) An 8-bit code (7 bits plus 1 parity bit) commonly used to designate alphanumeric and other characters and symbols for computers.

Aspect ratio The ratio of width to height for the frame of the video picture. The U.S. broadcast standard (NTSC) is 4 : 3. [Section 6.2; Fig. 7-32]

Assemble editing *Also* Add editing. One of two methods for editing a video- or audiotape. New sequences are consecutively added to the tail end of a previously edited scene. *See also* Insert editing. [Section 12.3; Fig. 12-11].

Astigmatism A lens aberration in which the focal length varies with the ray's incident plane into the lens; can also result from the birefringence of calcite prisms. The same point of light, or electron source, becomes imaged as a bar elongated in one of two orthogonal directions depending on the level of focus. [Sections 5.9, 7.5; Fig. III-21]

Attenuation A reduction in signal amplitude.

Audio dub In video, the recording of sound onto a videotape without disturbing the previously recorded picture. [Section 12.3]

Audio head A magnetic recording head that records or plays back sound. [Fig. 8-4]

AUDIO IN/OUT Audio input/output jack. Connectors for the sound signals on VTRs and other equipment. May also be labeled MIKE IN OR LINE IN/OUT to designate connections to a microphone or to an audio amplifier. [Section 8.7]

Audio track The portion of the videotape that stores the sound signal. [Section 8.2]

Auto black A video camera circuit that automatically adjusts the signal black level to the darkest region of the picture regardless of its absolute brightness. [Section 7.3]

Auto iris A motor-driven iris in the lens of a video camera that automatically regulates (to a preset level) the amount of light reaching the target of the camera tube. [Section 7.4]

Automatic bandwidth compression A video camera circuit that automatically cuts down the amplifier bandwidth (and hence horizontal resolution) at lower light levels. The circuit reduces the output noise as the S/N ratio of the camera tube decreases. [Section 7.3]

Automatic gain control (AGC) An electronic circuit that automatically controls the loudness of sound or contrast of a video signal to an appropriate, preset level. [Sections 7.2, 7.3, 8.2]

Automatic horizontal frequency control (AHFC) A circuit in a video monitor that maintains the synchrony of the horizontal sweep in the same manner that a flywheel keeps the speed of a motor constant. Some monitors allow switching between slow and fast AHFC. A slow AHFC makes the H sweep more immune to electrical noise, while a fast AHFC allows the picture to synchronize better to the signal from a VTR. [Section 7.5]

Automatic pedestal control A circuit that automatically adjusts the level of the pedestal (the height of the blanking level) in the video signal. *See also* Auto black. [Sections 7.3, 9.3]

Automatic target control A circuit that automatically controls that target voltage in sulfide vidicons and some other image pickup tubes so that the tube sensitivity is adjusted optimally for the scene brightness. For certain types of vidicon tubes (Newvicon, Chalnicon), the target voltage is fixed and cannot be varied. [Section 7.3]

AUX IN (AUXILIARY IN) On VTRs, the input terminal for the audio line. It is used to connect the audio signal from an amplifier, not from a microphone. On a special effects generator, AUX IN is the video input terminal that is connected to a VTR or other device that acts as the sync source. [Section 12.14; Fig. 12-13]

Average picture level (APL) The averaged voltage value of a video picture signal. The APL determines whether the DC level of the picture signal during an H sweep would or would not change without DC restoration [Section 7.5]

Azimuth In a VTR, the angle that the gap in the read/write head makes relative to the direction at right angles to the recording track. [Section 8.3; Fig. 8-4]

Back aperture The exit pupil of a microscope objective lens. The objective lens back aperture, which can be examined with a phase telescope or by inserting a Bertrand lens, displays the conoscopic interference figure and diffraction patterns. [Section 5.3]

Background subtraction A procedure used in video image processing to subtract away persistent noise patterns generated in the video camera or the optical system. Also used to display motion as changes between the current and old "background" video images. [Sections 9.3, 10.8]

Back porch The (zero volt) portion of a composite video picture signal that lies between the trailing edge of a horizontal-sync pulse and the trailing edge of the corresponding blanking pulse. The color burst, if used, is placed on the back porch. [Sections 6.3, 7.4; Figs. 6-8, 7-29]

B & W (black and white) Monochrome as contrasted to color video. [Section 7.4]

Bandwidth The difference between the upper and lower limit of a frequency band expressed in number of cycles per second (hertz). In video, the signal bandwidth directly affects the horizontal resolution of the picture; 1 MHz of bandwidth is required for every 80 TV lines of resolution. *See* Resolution, Limiting resolution. [Section 6.4; Table 6.4]

Barrel distortion A distortion, or aberration, that makes the image appear to bulge outward on all sides like a barrel. *See also* Pincushion distortion. [Section 5.9]

Barrier filter (fluorescence) An optical filter (usually a short-wavelength cut filter) that prevents the wavelengths used to excite the fluorescence from reaching the eye, or photographic or photoelectric sensor. The barrier filter extinguishes the undesired background light and also prevents injury to the sensor.

Baud rate The number of bits per second at which a digital signal is transmitted from one digital computer to another, e.g., over a telephone line. [Section 9.1]

Bertrand lens A small telescope objective lens that flips into the body tube of a microscope in front of the ocular. The Bertrand lens and ocular together form a telescope that is focused onto the objective lens back aperture for conoscopic observations. [Section 5.4]

Beta One of the two common formats for recording 1/2-inch videocassette tape; the other format being VHS. Beta I, II, and III refer to Beta-format tape of decreasing thicknesses that runs at tape speeds providing 1, 2, and 5 hr of recording per full cassette. [Section 8.3; Figs. 8-4, 8-11; Table 8-3]

Bias light A small amount of light that illuminates the target in Plumbicon- and Chalnicon-type camera tubes. The bias light reduces the lag that appears at low light levels in these tubes. [Section 7.2; Figs. 7-21, 7-22]

Birefringence The presence of different refractive indices for light waves vibrating in different planes in a medium. All crystals except those with a cubic lattice exhibit intrinsic birefringence. Regular arrays of submicroscopic rodlets and platelets exhibit form birefringence whose value varies with the refractive index of the immersion medium. Isotropic substances can be made birefringent by deformation, flow, or by the application of electric fields, magnetic fields, or mechanical forces. [Appendix III]

Bit A contraction of *b*inary dig*it*, the smallest unit of information in a notation using the binary system (1 or 0). A byte is commonly made up of 8 bits. [Section 10.2]

Black and white *See* B & W.

Blacker than black In video, the condition where the signal corresponding to the darkest part of the image is made to dip below the optical (or reference) black level down to the blanking level.

Black level The small video signal voltage (usually around 0.05 to 0.1 V) that corresponds to a specified limit for black peaks. In a black-negative, composite video signal, the blanking, sync, and other control signals that are not part of the picture signal have voltages below the black level. [Section 6.3; Fig. 6-8]

Black negative A standard signal format used in CCTV in which the voltages of the picture black level and sync signals are lower than the white levels of the picture signal. [Section 6.3; Figs. 6-7, 6-8]

Blanking The time interval in the video signal in which the scanning beam is turned off in the camera tube and video monitor. During the blanking period, the beam is scanned back (flyback) to its starting point without leaving a trace. At the end of blanking, the beam starts to display a new line (after H blank) or field (after V blank). [Section 6.2]

Blanking level The voltage level (taken as 0 V, or IEEE scale 0) of a composite video signal that separates the picture information from the synchronizing information. [Section 6.2]

Blooming A spilling of the white region of the video picture into adjacent areas. Blooming may arise in the camera tube with excessive illumination or in a monitor whose CONTRAST and BRIGHTNESS controls are adjusted too high. [Sections 7.2, 7.5; Figs. 7-23 to 7-25]

BNC connector One of the several types of connectors used to couple coaxial cables to video and other high-frequency electronic equipment. [Section 3.1; Fig. 3-3]

BR Designation indicating broadcast-quality videotape. [Section 8.5]

Build-up lag The converse of decay lag. The slow buildup of the signal current that is seen after some video camera tubes have remained in the dark. [Section 7.2; Fig. 7-22]

Burn A transient or permanent afterimage on the camera tube or a picture tube. Caused by prolonged exposure to a fixed scene or exposure to an excessively bright source of light or electron beam. Transient burn in a camera tube may be dissipated by exposure to a moderately bright, uniformly white scene. [Section 7.2]

Bus A parallel set of wires in a computer over which control and data are exchanged between the various devices (terminal, CPU, memory, ALU, etc.) connected to the bus. The bus can be standardized following IEEE recommended standards or may be a special bus (such as a video bus) that allows rapid interchange of specific information. A system-level bus will generally contain four component buses: the data bus, address bus, control or status bus, and power bus. The bus provides great flexibility to a computer, permitting different (internal and peripheral) devices to be plugged in, actuated, and signals to be exchanged.

Camera tube *Also* Image pickup tube. The vidicon or other type of vacuum tube in a video camera that converts the input optical image on its target to an output, electronic, picture signal. [Section 7.1]

Candela (cd) The luminous intensity emitted by a standard source of light. One candela emits one lumen per steradian. (See *RCA Electro-Optics Handbook,* Section 2.1.) [Section 7.2]

Capacitance electronic disk (CED) RCA's video disk system. Contrasts with the laser video disk or optical memory disk.

Capstan A roller and motor-driven rotating shaft that determines the speed at which the tape is transported through a VTR or other tape recorder. [Section 8.2]

Carrier wave *Also* subcarrier. A standard, high-frequency sine wave whose amplitude, frequency, or phase is modulated according to the amplitude of the signal that is to be carried by the wave. *See* AM, FM, Color burst. [Section 8.2]

Catadioptric An optical system in which both reflecting and refracting curved surfaces are used to form an image. Some "reflecting" objective lenses, as well as video projection systems, are catadioptric; the latter uses a Schmidt plate to correct the spherical aberration introduced by the spherical reflecting mirror. [Section 12.5]

Cathode The terminal on an electrical device from which a major flow of electrons takes place inwards (as in a cathode-ray tube) or from which electric current flows outward. The converse of anode. [Sections 1.1, 7.1, 7.5]

Cathode-ray oscilloscope (CRO) A display device that uses an electron beam to trace out electrical waveforms on a calibrated, phosphor faceplate. [Sections 6.3, 7.5]

Cathode-ray tube (CRT) A vacuum tube in which the electrons emitted by a heated cathode are focused into a beam and directed toward various points on a phosphor-coated surface. The phosphor becomes luminous at the point where the electron beam strikes. [Section 7.5]

CCD *See* Charge-coupled device.

CCIR Comité Consultatif International des Radio Communications (International Radio Consultative Committee). Also used to designate the TV formats (A to M inclusive) used in Europe and elsewhere outside of the United States and Japan. [Section 6.3]

CCTV *See* Closed-circuit TV.

cd *See* Candela.

CED *See* Capacitance electronic disk.

Central processing unit (CPU) The part of a computer that includes the circuitry for interpreting and executing instructions. [Section 10.2]

CFF *See* Critical flicker frequency.

C-format One of the formats for recording a video signal in helical scan on a 1-inch videotape. [Section 8.3]

Chalnicon Toshiba's trade name for a type of vidicon image pickup tube. Equipped with a cadmium selenide photoconductive target, the tube has moderately high sensitivity, low blooming of highlight details, low dark current, and good resolution. [Section 7.2; Table 7-1]

Charge-coupled device (CCD) A light-sensitive silicon chip used in place of a pickup tube in miniature video cameras. Also used in an electronic-only mode in comb filter circuitry, and as a volatile memory device in a time-base corrector. *See also* CID, CPD, MOS. [Section 7.1]

Charge injection device (CID) One of several types of solid-state chips similar to a charge-coupled device that are used in place of a video image pickup tube or as a transient memory device. [Section 7.1]

Charged priming device (CPD) One of several types of solid-state chips similar to a charge-coupled device that are used in place of a video image pickup tube. [Section 7.1]

Chip Short for integrated circuit chip. *See* Integrated circuit.

Chroma *See* Chrominance.

Chroma detector A circuit used in some VTRs, time-base correctors, etc. at the color encoder input that detects the absence of chrominance information. The chroma detector automatically deletes the color subcarrier from the color encoder output when chrominance is absent. [Section 7.4]

Chromatic aberration An optical defect of a lens causing different colors of light, or wavelengths (voltages) of electrons, to be focused at different distances from the lens. Chromatic aberration gives rise to colored fringes or halos around edges and points in the image. [Section 5.9]

Chrominance *Also* Chroma. The saturation, or purity, of color as differentiated from its hue, and spectral quality, and luminance, or brightness. *See also* Hue. [Section 7.4]

Chrominance signal The portion of the encoded color TV signal that contains the color information. *See also* Luminance signal, Color burst. [Section 7.4; Fig. 7-29]

CID *See* Charge injection device.

CIE (Commission Internationale de l'Éclairage) An international committee that has established a method for specifying and quantifying color based on psychophysical measurements. [Section 4.7]

Clamping The process that establishes a fixed black level for the video signal at the beginning of each scan line. [Section 6.5]

Closed-circuit TV (CCTV) A video system that does not broadcast TV signals over the air but rather transmits them over a coaxial cable. [Sections 6.1, 6.3]

C mount Standard screw-in lens mount found on most video cameras and 16-mm cine cameras. The male thread on the lens is 1 inch in diameter with 32 threads/inch. [Section 3.3; Fig. 3-6]

Coaxial cable *Also* Coax cable. An electrical cable with a central conductor surrounded by a low-loss insulating sleeve and insulated ground shield. A coaxial cable is capable of passing very-high-frequency electronic signals with low signal loss and noise pickup. *See also* Seventy-five-ohm cable. [Section 3.1; Fig. 3-3]

Color bars A series of eight vertical, colored bars (white, yellow, cyan, green, magenta, red, blue, and black) or a corresponding series of calibrated video signal. The bars and signal are used as references for adjusting picture brightness, contrast, color intensity, and correct color balance. The signal can be generated optically or electronically with a color bar generator. [Section 7.4]

Color burst In NTSC color video, a burst of approximately nine cycles of 3.58 MHz (the subcarrier) that is placed on the back porch of the composite video signal. The color burst serves as a synchronizing signal to establish the frequency and phase reference for the chrominance signal. *See also* Color subcarrier, Hue. [Section 7.4; Fig. 7-29]

Color camera A video camera capable of responding to color and producing a corresponding RGB or NTSC color video signal. [Section 7.4]

Color encoder A device that transforms an RGB color signal into an encoded signal such as NTSC. The encoded signal can be transmitted over a single line instead of the three lines required for RGB using a chrominance signal with a limited bandwidth. *See also* Decoder. [Sections 7.4, 7.5; Fig. 7-26]

Color killer An electronic circuit used in color VTRs to suppress the 3.58-MHz color carrier frequency when a B & W tape is played back. The same circuit in some B & W VTRs suppresses the color carrier frequency when a color recording is played back. Without a color killer, the color carrier is displayed as a fine periodic stripe on a B & W picture. [Section 8.4]

Color monitor A video monitor capable of displaying pictures in color; may require an RGB or NTSC input depending on the type of color monitor. [Section 7.4]

Color subcarrier In NTSC color video, the 3.579545 (≈3.58)-MHz carrier whose modulation sidebands (the portion of a modulated carrier wave that contains the desired information) are added to the luminance signal to convey color information. [Section 7.4; Fig. 7-29]

Color temperature A measure of the bluishness or reddishness of "white" light expressed as the absolute temperature (°K) of a black body that would radiate light whose wavelength compositions are essentially identical to the light in question. At high color temperatures, a greater fraction of the radiated energy lies in the blue end of the spectrum relative to the red and infrared, which dominate at low color temperatures. The sensitivity of video pickup devices is specified for light with a color temperature of 2856°K. [Section 7.2]

Coma A lens aberration in which the off-axis beams do not form single focused spots but rather comet-shaped patterns. [Section 5.9]

Comb filter An electronic circuit that separates (combs out) the chrominance video signal in some VTRs and color monitors. This sharply-tuned filter removes the chrominance signal selectively without appreciably affecting the high-frequency luminance signal. [Section 7.5]

Commission Internationale de l'Éclairage *See* CIE.

Compatibility The ability of one piece of equipment to interface and function with another. [Sections 6.3, 6.6, 12.3]

Composite video signal A video signal that contains a picture signal, vertical and horizontal blanking, and synchronizing pulses. [Sections 6.3, 7.3; Fig. 6-7]

Compression Reducing the amplitude of sound or picture signals that are too high. The opposite of expansion. [Sections 9.3, 10.6]

Condenser iris diaphragm The "substage" iris diaphragm located at the front focal plane of the condenser lens of a microscope. With Koehler illumination the iris lies in a plane conjugate with the rear focal plane of the objective lens. Rays emerging from each point at the condenser iris diaphragm exit the condenser into the specimen space as parallel rays, or plane waves. The iris controls the working NA of the condenser and should not be used to adjust the brightness of the microscope field. [Sections 5.3, 5.4, 5.8, 11.3, 11.10]

Confocal optics A (microscope) optical system in which the condenser and objective lenses both focus onto one single point in the specimen. Generally, the image of a pinhole source is focused onto a point in the specimen, and that point is focused by the objective lens onto a point detector or through a mask with a pinhole aperture. With confocal optics, the Abbe limit of resolution can be exceeded since only a limited region of the specimen is viewed at any one time. [Sections 1.2, 11.5]

Conjugate planes/points Planes (or points) that are in focus relative to each other. In a microscope adjusted for Koehler illumination, there are two sets of conjugate planes: the aperture planes and the field planes. *See* Aperture planes, Field planes, Koehler illumination. [Sections 5.2, 5.3; Figs. 5-14, 5-15]

Conoscopic image The interference pattern and diffraction image seen at the back aperture of the objective lens. The conoscopic image provides a two-dimensional projection of the rays traveling in three dimensions in the specimen space. [Section 5.4; Fig. 5-17]

Contrast A measure of the gradation in luminance that provides gray scale (or color) information. Contrast is expressed as the ratio (difference in luminance)/(average luminance) in adjoining areas of the scene. Under optimum conditions, the human eye can just detect the presence of 2% contrast. [Sections 3.5, 3.7–3.9, 4.5, 7.3, 7.5, 8.4, 9.3, 10.6, 11.3]

CONTRAST A manual control on a monitor that affects both the luminance and contrast of the picture. [Sections 3.2, 7.5, 12.1]

Contrast enhancement/stretching In digital image processing, the enhancement of contrast by using an image histogram and look-up table. Can also be achieved with analog devices. [Sections 7.3, 9.3, 10.6, 10.7; Figs. 10-22, 10-23]

Contrast range The range of gray between the lightest and darkest parts of a scene; expressed as a ratio of light to dark. More appropriately, it should be called brightness ratio. *See also* Dynamic range. [Section 10.5]

Contrast transfer function (CTF) *Also* Square wave response curve. *See* Modulation transfer function.

Control track A track on one edge of the videotape containing the pulses (one per frame) that control the playback speed of the tape. [Section 8.2; Figs. 8-3, 8-4]

Convergence In color video cameras and monitors, the precise alignment of the images or pictures in the three primary colors (red, green, and blue). A lack of convergence gives rise to asymmetrically colored fringes. [Section 7.5]

Convolution *See* Image convolution.

Convolution kernel, mask The 3 × 3, 5 × 5, etc. matrix of numbers that are used to perform an image (or spatial) convolution in digital image processing. The kernel determines whether the convolution results in image sharpening, smoothing, edge detection, etc. [Section 10.11]

Correction collar An adjustment collar provided on some high-NA, microscope objective lenses. Rotation of the collar adjusts the height of certain lens elements in the objective lens to compensate for variations in coverslip thickness or immersion media. At high NAs, even a small deviation of the coverslip thickness (by as little as a few micrometers in some cases), or refractive index of the immersion medium from the designated standard, can introduce significant aberrations. [Section 5.9]

CPD *See* Charge priming device.

CPU *See* Central processing unit.

Critical flicker frequency (CFF) The frequency of light fluctuation that just barely evokes a sensation of flicker for a given luminance. Depending on luminance level, wavelength of light, and size and location of the fluctuating field, the CFF can vary from a few hertz to over 100 Hz. [Section 4.6; Figs. 4-15, 4-16]

CRO *See* Cathode-ray oscilloscope.

Cross talk Interference between adjacent video tracks or audio channels. [Section 8.2]

CRT *See* Cathode-ray tube.

CTF (contrast transfer function) *Also* Square wave response curve. *See* Modulation transfer function.

Cue A warning or signal for an event that is about to take place. [Section 12.3; Fig. 12-11]

DAC *See* D/A converter.

D/A converter (digital-to-analog converter; DAC) A circuit used at the output of a digital computer or processor to provide analog signals or power. DAC is used to drive nondigital electrical devices including standard video monitors. [Section 10.2]

Dark adaptation The phenomenon of becoming able to see after one enters a darker environment, eventually by switching from photopic to full scotopic vision. Adaptation proceeds in several steps, the longest of which may take up to an hour. *See* Photopic vision, Scotopic vision. [Section 4.3; Fig. 4-4]

Dark current In video, the signal current that flows in a photodetector such as a vidicon when it is placed in total darkness. [Section 7.2]

dB (decibel) A unit for comparing the strengths of two signals. In video, decibels are the ratio of the two signal voltages expressed as $20 \times \log_{10}$ (ratio of the amplitude of the test signal as compared to a second, usually a standard, signal). In some applications, power ratios are used; dB (power) is defined as $10 \times \log_{10}$ (ratio of powers). [Section 6.5; Table 6-5]

dBm A unit of measure indicating the number of decibels that the power of a signal is above or below 1 mW.

DC restoration The reestablishment of the DC and low-frequency components of a video signal after being lost by AC transmission. [Section 3.5]

Decibel *See* dB.

Decoder In a color monitor or TV receiver, the circuit that transforms the encoded NTSC signal into an RGB signal, which operates the color picture tube. [Section 7.4; Fig. 7-26]

Definition The degree of detail or sharpness in a video picture. [Section 8.4]

Deflection coils The electromagnetic coils that steer the electron beam in a cathode-ray tube or image pickup tube, e.g., in a raster scan. [Sections 1.1, 7.1, 7.5]

Degaussing (of monitor) An operation that is activated by pressing the DEGAUSS button on a high-resolution color monitor. Residual magnetism and charge are removed by this operation so that the electron beam can land on the proper phosphor and render correct colors. [Section 7.5]

Depth of field The distance between the closest and farthest objects in focus within a scene as viewed by a lens at a particular focus and with given settings. The depth of field varies with the focal length of the lens and its f-stop setting or NA; and the wavelength of light. Depth of fields only a small fraction of a micrometer can be achieved at 546 nm with microscope lenses of N.A. > 0.9. [Sections 5.5, 11.3, 11.5; Fig. 11-9, 11-10]

Depth of focus The range of distances between a lens and image plane (target in the video pickup device) for which the image formed by the lens at a given setting is clearly focused. With a high-NA microscope objective, the depth of field is very shallow, but the depth of focus can be quite deep and reach several millimeters. [Section 3.9; Fig. 3-15]

DIC *See* Differential interference contrast.

Dichroism An optical phenomenon in which a crystalline structure exhibits two different colors depending on the direction of observation or the plane of vibration of polarized light waves traveling through the crystal. It is a form of pleochroism (exhibiting many colors), which reflects the difference of absorption curves for chromophores oriented in different directions within the crystalline structure. [Section 11.3, Appendix III]

Differential interference contrast (DIC) A mode of contrast generation in microscopy that yields an image with a shadow relief. The relief reflects the gradient of optical path difference. DIC, which is a form of interference microscopy that uses polarizing beam splitters, can be of the Smith or Nomarski type. [Sections 5.6, 11.3]

Digital (signal) A signal whose units are represented by either one of only two states: on or off, yes or no, 1 or 0. Since no gradations in between are permitted, digital signals are precise, unambiguous, and quite immune to noise. *See also* Analog. [Sections 9.1, 10.1–10.4]

Digital multimeter (DMM) An electronic device for measuring voltage, current, resistance, signal power, etc. The readings appear as numerals on an LCD, LED, or glow-discharge-tube display panel rather than with a pointer and meter. [Section 12.4]

Digital-to-analog converter *See* D/A converter.

Digital voltmeter (DVM) *See* Digital multimeter.

Digitizer *See* A/D convertor. *See also* Digitizing tablet, Optical digitizer. [Section 10.3]

Digitizing tablet A tablet with embedded sensors for rapidly converting the x–y coordinates of a pointer (placed on a drawing, photograph, projected image, etc.) into a digital signal that can be processed by a digital computer. Resolutions as high as 0.1 mm can be achieved on a 30×30-cm tablet. [Section 10.14]

Dilation In digital image processing, the operation that expands the outline of objects. The opposite of erosion. Both are used in preparation for image segmentation. [Section 10.13]

DIN (Deutsche Industrie Norm) German standard for electronic and other industrial products. DIN plugs can be three- to six-pin connectors with the same outer diameter and appearance. Photographic film with DIN ratings double their speed with each increment of 3.

Direct memory access (DMA) An arrangement used in digital computers, including digital image processors, where some device can take control of placing in or taking out the contents of some peripheral device, frame memory, etc. without intervention of the central processing unit while the particular operation is taking place.

Disk, disc A magnetic or optical, disk-shaped recording medium used for storing a large amount of digitized information. [Sections 8.9, 10.2]

Dissolve In video or motion picture production, the gradual emergence of one picture as the other fades out. [Section 12.4]

Distortion In signal processing, the lack of correspondence of the waveforms in the output relative to the incoming signal caused by undesired nonlinear processing. The amplitudes or phases of certain frequency components have changed relative to other frequency components. In imaging, geometrical distortion refers to the lack of correspondence of the shape (proportional distances) of the image relative to the image produced by an idealized lens or device. Geometrical distortions are expressed as a percentage of distances departing from the distances in a nondistorted image. *See also* Barrel distortion, Pincushion distortion. [Sections 5.9, 6.5, 7.2, 7.5; Figs. 6-22, 7-36, 8-20, 10-31]

DMA *See* Direct memory access.

DMM *See* Digital multimeter.

Dropout compensator A video circuit that senses picture signals that are missing locally, and substitutes the preceding line for the line with the dropout. Single horizontal scan lines are continuously placed in memory, and are available to replace the defective following lines. Many VTRs are equipped with dropout compensators. [Section 8.2]

Dubbing Adding new material to an existing recording; often adding a sound track, or a substitute sound track, to a videotape. Sometimes refers to the addition of a scene to a recorded videotape. [Section 12.3]

DVM (Digital volt meter) *See* Digital multimeter.

Dynamic focusing An automatic focusing adjustment of the electron beam in high-quality cathode-ray or video image pickup tubes. The beam is made to land with the properly shaped minimum-sized spot regardless of its position in the raster scan. [Sections 7.2, 7.5]

Dynamic range In video and other electronic equipment and in photographic emulsions, the ratio of the maximum to minimum signal levels that introduce no more than acceptable levels of signal amplitude distortions. *See* Intrascene dynamic range, Usable light range. [Sections 7.2–7.4]

Dynamic tracking An automatic adjustment of the read/write heads in a helical-scan VTR, such that the heads follow the recorded tracks precisely even at nonstandard speeds of tape transport. The dynamic tracking heads are mounted on a piezoelectric actuator that dynamically adjusts the head position. [Section 8.4; Tables 8-1, 8-3]

Dynode The electrodes in a photomultiplier tube, microchannel plate, or multiplier section of a return-beam-type video image pickup tube that amplifies the flow of electrons by secondary emission. [Section 7.1]

Echo *Also* Ghost. The formation of double or multiple images by reflection of the RF video signal. [Section 3.6]

Edge detection A mode of digital image processing that displays the boundaries of regions with different luminances, i.e., the high-spatial-frequency components of the image, selectively. *See* Image convolution. Laplacian filter. [Sections 10.11, 10.13; Fig. 10-42]

Edge enhancement *Also* edge sharpening. Treatment of a video signal by analog or digital means to sharpen the appearance of edges. Achieved by subtracting the second derivative of a video signal from itself, or by the use of a high-frequency pass filter. [Sections 9.3, 10.11, 10.13; Figs. 9-7, 10-45]

Editing deck A specially constructed VTR that has, in addition to play and record circuitry, circuitry and controls enabling assemble and insert editing. For editing, another VTR that plays back the mastertape (synchronized to an external sync generator) is used in conjunction with the editing deck. [Sections 8.4, 12.3]

Editor *See* Editing deck.

EIA (Electronic Industries Association) An American committee that sets engineering standards for TV, computers, etc.; for example, EIA RS (recommended standard)-170, -330 for video sync formats, and EIA RS-232C for telephone-line-compatible communications between computers and related equipment. *See also* IEEE. [Section 6.3]

EIAJ Electronics Industries Association of Japan.

Eidophor Trade name of an oil film type of video projection system. *See* Light-valve projector. [Section 12.5]

EJECT A push-button switch or lever used for unloading or loading a VCR. Generally, the power must be on for the EJECT to work properly. [Section 8.2]

Electric displacement vector In electromagnetic waves, including light waves, the vector that designates the direction and amplitude of the electric field generated in a material medium. [Appendix III]

Electron gun The assembly inside a video camera tube and monitor picture tube that contains a heated cathode. Electrons emitted by the gun are focused to produce the scanning beam. [Sections 7.1, 7.5; Figs. 7-1, 7-30]

Electronic Industries Association *See* EIA.

Electronics-to-electronics *See* E-to-E.

Encoder *See* Color encoder.

ENG; EFP Electronic news gathering; electronic field production.

EPROM (erasable PROM) A PROM that can be erased by exposure to strong UV or by applying a high electrical current (EEPROM). *See also* PROM.

Equalizing pulses Synchronizing pulses that are present at twice the horizontal line scan frequency in the RS-170 video signal. The pulses start 3-H intervals before, and end 3-H intervals after, the vertical-sync pulse. Equalizing pulses improve the interlacing between the even and odd fields (each with 262.5 H-scan lines) in a TV monitor. [Section 6.3; Figs. 6-10, 6-11]

Erase head A magnetic head that erases the signal recorded on tape. The erase heads are usually stationary, but in editing decks and some other VTRs, are mounted directly in front of the read/write head on the scanner to prevent the generation of noisy frames. [Sections 8.2, 8.4, 12.3; Figs. 8-6, 8-11; Table 8-1]

Erosion In digital image processing, an operation that shrinks the boundary of an object. *See* Dilation, Image segmentation. [Section 10.13]

E-to-E (electronics-to-electronics) *Also* loop through. A method of hooking up monitors and other video equipment so that the signal is looped through a Hi-Z circuit rather than becoming terminated. *See also* Hi-Z. [Sections 3.6, 12.4; Figs. 3-10, 12-6, 12-13]

ETV Educational Television.

Extinction factor (EF) In a polarizing or DIC microscope, the ratio of the amount of light that is transmitted with the axes of the polars parallel over the amount transmitted with their axes crossed. Unless the lenses are rectified, the EF drops exponentially as the NA is increased. [Section 11.3, III.9]

Faceplate In a video camera tube, the glass plate that supports the target and converts the optical image into a pattern of electrical charges. In a monitor, the glass plate coated with the phosphor that converts the impinging electrons into an observable picture. [Sections 7.1, 7.2, 7.5; Figs. 7-1, 7-30]

Fade In video or audio editing, a gradual change in the picture or sound intensity, such as in fade-in and fade-out. [Section 12.4]

False color Representation in colors differing from the original scene. In the false-color display of a processed video image, colors are assigned to spectrally separated images before recombination to allow multispectral analysis. *See also* Pseudocolor display.

FCC (Federal Communications Commission) The U.S. federal agency responsible for making policy and exercising control over the use of wired and wireless remote communication and related electromagnetic interference.

F connector One of the several types of connectors used for fastening the end of a coaxial or antenna cable to a piece of video equipment. [Section 3.1; Fig. 3-3]

Feature extraction A set of operations in digital image processing whereby certain features (such as boundaries of objects, areas, shape and size classes, fluorescence, color, etc.) are selectively enhanced in preparation for image analysis, quantitation, or pattern recognition. [Section 10.13]

Federal Communications Commission *See* FCC.

FET *See* Field-effect transistor.

Fiber optics coupling Used, for example, as a fiber optics plate (a thin slice of a fiber optics bundle) contacted with immersion oil to optically couple the output of an image intensifier to the faceplate of a vidicon tube. Considerably less light is lost than by coupling the two with a high-aperture lens. [Section 7.1; Fig. 7-5]

Field In video, one vertical sweep of a raster scan. In 1:1, 2:1, and 4:1 interlaced video, one, two, and four fields respectively make up a video frame. *See also* Frame. [Section 6.2; Figs. 6-5, 6-8]

Field blanking *See* V blank.

Field diaphragm The iris diaphragm that is located in front of the collecting lens of the light source. With Koehler illumination, the condenser focuses the image of the field diaphragm onto the image plane. [Sections 5.3, 5.4, 5.8; Figs. 3-12, 5-15]

Field-effect transistor (FET) An extremely sensitive high-frequency transistor used for amplifying weak electrical signals such as the signal from a video camera tube. Can be damaged by static discharge simply by touching its input circuit (gate) with a finger even when the FET is not powered up. [Section 3.4]

Field planes The set of planes in a microscope adjusted for Koehler illumination that are conjugate with the focused specimen. They include the plane of the specimen, the field diaphragm, the intermediate image plane, and the image on the retina, photographic emulsion, or the faceplate of the video pickup device. [Sections 5.3, 5.4, 5.8; Fig. 5-15]

Field sequential playback A mode of playback used in some special VTRs that displays successive fields rather than interlaced frames. While sacrificing vertical resolution, field-sequential playback doubles the time resolution, e.g., of a 525/60 video recording from 33 to 16.5 msec. [Sections 8.4–8.6]

Flagging A waving, S-shaped distortion that can appear at the top of a picture during playback of videotape. Commonly caused by incorrect tape tension, flagging is especially prominent in playback of time-lapse VTRs. The degree of flagging can be reduced by using a monitor with a fast AHFC, and by fine-tuning the H-HOLD control. [Sections 6.3, 8.5; Fig. 8-16]

Flare Unwanted light in an optical instrument that arises by reflection at lens surfaces (including the observer's eyeglasses) and lens barrel, etc., and sometimes from lens aberration. Flare reduces image contrast and may form undesirable focused images and hot spots. Flare in a microscope is reduced by immersion of the condenser and objective lenses, antireflection coating of the optical surfaces, proper design of lens curvature, and appropriate use of aperture stops. *See also* Hot spot. [Sections 5.3, 5.4, 5.8, 5.9]

Flicker *See* Critical flicker frequency.

Fluorite objectives Microscope objective lenses considerably better corrected than achromats but not quite as well corrected as the apochromats. By using fluorite crystals (which have lower dispersion) in place of some of the glass elements, a fluorite objective corrects for spherical aberrations in three wavelengths at considerably lower cost than the apochromats. [Section 5.10; Fig. 6-3]

Flyback In a video camera or monitor, the rapid movement of a scanning spot back from the end of a line or field to the beginning of the next line or field. Horizontal and vertical blanking pulses shut off the electron beam from just before to a short time after flyback so that a retrace line does not appear and degrade the image. [Sections 6.2, 7.5; Fig. 6-3]

FM (frequency modulation) A signal impressed on a carrier wave in such a way that the carrier frequency changes in proportion to the signal amplitude. FM signals are less susceptible to noise and waveform degradation than AM signals. *See* AM [Section 8.2]

f number *See* f stop.

Focus The point or plane in which light rays or an electron beam form a minimum-sized spot that has the proper intensity distribution. Also the act of bringing light or electron beams to a fine spot. [Sections 5.2, 7.1, 7.5; Fig. 5-21]

Fold-over frequency *See* Aliasing.

Footcandle (fc) A measurement of illuminance, or illumination, expressed in lumens per square foot. It is the amount of illumination from 1 international candle (the candela) falling on a 1 ft^2 surface at a distance of 1 ft. In SI units, 1 fc = 10.764 lux (lx). (See *RCA Electro-Optics Handbook,* Section 2.1.)

Footlambert (fL) A unit of luminance or brightness. 1 fL = $(1/\pi)$ cd/ft^2. In SI units (preferred), 1 fL = 3.426 nit. (See *RCA Electro-Optics Handbook.)*

Format In video, the standard conventions used to specify the shape of the picture, and the amplitudes and arrangements of the picture, sync, and blanking signals. For VTRs, format refers to the pattern of recording and width of tape used. [Sections 6.2, 6.3, 8.3]

Fourier transform The separation of an image or a signal into its spatial or temporal sinusoidal frequency components. Once separated, the components can be separately filtered, so that the inverse transform gives rise to an image or signal with selected frequencies accentuated or missing. *See also* Frequency. [Sections 5.6, 10.12, Appendix II; Figs. 10-46 to 10-52]

Fluorescence recovery after photobleaching (FRAP) A new technique in light microscopy using a pulse from a focused laser microbeam to deplete the fluorescence in a local region in a living cell. The subsequent recovery of fluorescence in the irradiated region is measured to establish the mobility of the molecules that carry the fluorescent tag. [Section 11.4]

Fovea The part of the primate eye with the highest concentration of cone cell outer segments. This part of the retina gives the highest image resolution given adequate scene luminance. The central fovea is devoid of rods. The fovea is located in a slight depression on the retina (the macula lutea) and is free from overlying layers of cell nuclei and blood vessels, which would otherwise scatter the image-forming rays. [Section 4.4; Figs. 4-2, 4-7]

FPS Frames per second.

Frame In the standard, 2 : 1 interlaced video format, the frame is a complete video picture composed of two fields. In the NTSC format, a frame made up of 525 scan lines (of which at most 480 are visible) appears 30 times a second. [Section 6.2]

Frame buffer In a digital image processor, the hardware in which the frame memory resides. The frame memory is a RAM that stores a full frame of the video picture signal. [Section 10.2]

Frame grabbing The acquisition and storing of a single video frame into a frame buffer of a digital image-processing computer.

FRAP *See* Fluorescence recovery after photobleaching.

Freeze frame *Also* Still frame. A feature of some VCRs and video disk players that allows freezing the playback upon command. A single image is maintained on the screen until normal playback is resumed. In many cases, a VCR displays a single frozen field rather than a single frame. *See also* Frame grabbing. [Sections 8.5, 8.6, 12.1, 12.4; Figs. 8-19, 8-20]

Frequency The number of times per second a repetitive signal undergoes a full cycle of vibration. Frequency units are hertz (Hz). For spatial frequency, the number of cycles of image brightness variation along a scan direction, generally expressed per millimeter. [Sections 5.7, 6.4, 6.5, 7.4, 8.2, 10.11, 10.12, Appendix II]

Frequency modulation *See* FM.

Frequency response The range, or band, of frequencies to which a piece of electronic equipment or circuit responds without appreciable loss of amplitude or shift in phase. [Sections 6.4, 6.5, 7.5, 8.2]

Front porch The portion of a composite video signal that lies between the leading edge of the horizontal (or vertical) blanking pulse and the leading edge of the horizontal (or vertical) sync pulse. [Section 6.3; Fig. 6-8]

f stop *Also* f number. The f number is inversely related to the speed of the lens, or ability of the lens to gather light. The f stop is calculated by dividing the focal length of the lens by the diameter of the limiting aperture (commonly the iris). [Section 12.1]

Gain Amount of signal amplification. In a video camera, turning up the GAIN control boosts the contrast and effectively also its brightness. [Sections 7.3]

Gamma In video, the exponent of the function that relates the output signal to the input signal. Sulfide vidicons respond with a gamma of approximately 0.65, while many other camera tubes have gammas of about 1. Monitors generally have a gamma between 2 and 3. A gamma of 1 indicates that the device has a linear transfer characteristic. *See also* Light transfer characteristics. [Sections 7.2, 7.5, 9.3]

Gamma compensation A circuit that adjusts the gamma in a piece of video equipment. In a color monitor, the gamma is adjusted to about 2.2 to compensate for the lower gamma of the Vidicon tube. Without compensation, the RGB signals would become disproportionate, and the hue would shift, as the scene changes brightness. [Section 7.2, 7.5; Fig. 7-34]

Gen I intensifier *See* Image intensifier.

Gen II intensifier An image intensifier using microchannel plates in place of, or in addition to, the gain mechanism in the intensifiers of the Gen I type. [Section 7.1; Fig. 7-5]

Genlock A circuit used to synchronize an internal sync generator to an external composite video signal. The signal from the sync source (a sync generator, VTR, or a camera) is fed to the GENLOCK input of the camera or special effects generator so that two or more video signals can be synchronized and mixed. [Sections 6.6, 7.3, 12.4; Figs. 3-4, 6-26, 12-6, 12-13, 12-23]

Geometrical distortion A lens aberration in which the image is distorted relative to the object. *See* Aberration, Barrel distortion, Distortion, Pincushion distortion.

Ghost *See* Echo.

Glitch A distracting disruption or noise-associated defect in the video picture. Common in the playback of poorly edited tapes or tape played back on an incompatible VTR or at different speeds. Also can be introduced by electromagnetic interference, poor cable connection, or defective components. [Sections 6.6, 7.3, 8.5, 12.3; Figs. 3-13, 8-7, 8-15 to 8-17]

Gray level *Also* Gray value. The brightness of pixels in a digitized video image; commonly expressed in integers ranging from 0 (black) to 255 (white) for an 8-bit digital signal. [Sections 10.3–10.7]

Gray level histogram In digital image processing, a histogram that depicts the number of pixels at each gray value. The histogram can be used to measure the areas that have given ranges of gray values or to adjust image contrast by histogram stretching or equalization. *See also* Image histogram. [Sections 10.5, 10.6]

Gray scale The various shades of gray or luminance values in a video picture. As industrial test standards, gray wedges are used with discrete steps incrementing in brightness by factors of $\sqrt{2}$. [Sections 7.3, 7.4, 9.3]

Gray value *See* Gray level.

Gray wedge An elongated rectangular pattern whose brightness changes from black through shades of gray to white along its length. In calibration wedges, the brightness may vary linearly or logarithmically in discrete steps. *See also* Gray scale.

Guard-band noise *Also* Noise bar. A horizontal band of hash occupying about a tenth of the video picture height that is sometimes seen on playback of videotape. The guard-band noise is caused by the magnetic head scanning the tape at the wrong angle (mistracking) and thereby reading over the unrecorded guard band between the video tracks. [Sections 8.2, 8.5; Fig. 8-7]

Guard bands The narrow separations between video or audio tracks on a videotape that guard against interference from adjoining tracks. [Sections 8.2, 8.3; Fig. 8-3]

H (horizontal) Short for 1-H period, the time interval between (the leading edges of) the successive H-sync pulses. For 525/60 2 : 1 interlace video, 1 H = 10^6 μsec ÷ (525 × 60/2) = 63.5 μsec. *See also* Scan rate. [Sections 6.2, 6.3]

Hardware Major items of electronic equipment or their components. Contrasts with software such as computer programs. [Section 10.1]

Head An electromagnetic device with a narrow gap and high-frequency response that records or reads information from magnetic tape. [Sections 8.2, 8.3, 8.5; Figs. 8-1 to 8-4]

Helical scan recording A pattern of video tape recording used in Beta, VHS, U-matic, C, and other common formats. The video tracks, each a field long, are laid down diagonally at a precisely defined small angle (unique to the format) to the length of the tape. *See also* Segmented scan recording. [Sections 8.2–8.4; Fig. 8-2]

H HOLD (HORIZONTAL HOLD) The control on a video monitor that varies the free-running frequency of the horizontal deflection oscillator. Allows adjustment of the horizontal scan rate to achieve proper H sync. [Sections 3.5, 7.5; Fig. 3-1]

High-extinction microscopy Polarized-light, interference, (fluorescence), and other modes of microscopy using polarization rectifiers and other devices to achieve a high degree of background extinction in order to bring out the signal originating from a very small degree of birefringence, optical path difference, (fluorescence), etc. [Section 11.3, Appendix III]

High impedance *See* Hi-Z.

Highlight The region of a scene or video picture with the maximum brightness. [Section 7.5]

Histogram *See* Image histogram.

Hi-Z (high impedance) In a video input circuit, an impedance of 800 to 10,000 ohms. Used when the input signal is looped through E-to-E to another piece of equipment. *See also* Seventy-five-ohm termination. [Section 3.6; Figs. 3-5, 3-10]

Holographic microscopy A mode of light microscopy in which a highly coherent, laser beam is split into a reference and main beam, with the reference beam (usually traveling outside of the microscope) being made to interfere with the main beam that has passed through the specimen. The interference of the two mutually coherent beams forms a hologram. The depth of field gained by viewing the hologram is essentially infinitely great, and the contrast mode or observation can be switched to darkfield, phase contrast, interference contrast, etc., after the hologram has been formed by the microscope in brightfield. [Section 11.5]

Horizontal *See* H.

Horizontal blanking Reduction of the video signal voltage to blank out the horizontal retrace. *See also* blanking level. [Sections 6.2, 6.3; Figs. 6-4, 6-8 to 6-12, 6-14]

Horizontal hold *See* H HOLD.

Horizontal resolution The number of black and white vertical lines that can just be detected at the center of the video picture, measured for a distance equal to the height of the picture. The horizontal resolution is limited by the bandwidth of the system or signal, and is commonly considered to be 80 TV lines/MHz when the video amplifier is the limiting factor. [Sections 6.4, 7.3, 7.5; Fig. 6-17]

Horizontal scan (H scan) A single horizontal line of a raster-scanned video signal. Occurs 525 times per frame in the NTSC format. The odd and even H-scan lines alternate in successive fields and are interlaced in the two successive vertical scans that make up the frame. [Section 6.2; Fig. 6-5]

Horizontal-sync pulse (H-sync pulse) The sync pulses that control the start of horizontal scanning in video. Occurs at a 15.75 kHz rate in EIA RS-170 and -330 format monochrome video and at a 15.734 kHz rate in NTSC color format. [Section 6.3; Figs. 6-6 to 6-14]

Hot spot An undesirable bright patch that intrudes at the center of the video microscope picture. Commonly arises from reflections in the lens barrel or coupler that is placed in front of the camera tube. [Section 3.8; Fig. 3-13]

H scan *See* Horizontal scan.

H-sync pulse *See* Horizontal-sync pulse.

Hue The dominant (or complementary) wavelengths of light that give rise to the sensation of color such as red, blue, yellow, green. Black, gray, and white are colors but not hues. [Sections 4.7, 7.4]

Hum Electrical disturbance that appears at the power-line frequency or its harmonics. In video, such interference is minimized by synchronizing the field scan rate to the power-line frequency [Section 6.5]

Hz (hertz) Cycles per second. MHz (mega hertz) is a million cycles per second.

I and Q vectors In encoded color video, where the R, G, B, and luminance signals are combined into a single signal such as the NTSC, I and Q are the two orthogonal vectors that represent the orange–cyan and magenta–green axes, respectively. *See also* Color encoder, Decoder. [Section 7.4]

IC *See* Integrated circuit.

IEEE (Institute of Electrical and Electronic Engineers) An organization that (together with EIA and IRE) has helped establish industrial standards for electrical and electronic equipment and signals to promote interchangeability. The IEEE publishes many journals and conference proceedings related to video engineering and image processing. [Section 6.3]

IEEE scale A scale that expresses the levels of voltages in a video signal according to IEEE standards and the recommendations of the TV Broadcasters and Manufacturers for Coordination of Video Levels. 140 IEEE units corresponds to the peak-to-peak amplitude of the composite video signal, including 100-unit-high picture and 40-unit-high sync signals. For a 1-

V peak-to-peak video signal, the picture signal occupies 0.714 ≃ 0.7 V and the sync signal 0.286 ≃ 0.3 V [Section 6.3; Fig. 6-8]

I-isocon *See* Intensifier isocon.

Illuminance *Also* Illumination. The density of luminous flux incident on a uniformly illuminated area, measured in footcandles (lumens per square foot) or lux (lumens per square meter). [Sections 4.3, 7.2]

Illumination *See* Illuminance.

Image An optically reproduced scene. In video, the image formed by a lens on the camera tube is transformed into an electronic signal that gives rise to a corresponding picture on the monitor. [Sections 5.2, 6.2]

Image analysis The use of digital computers to derive numerical information regarding selected image features, such as contour lengths, areas, shape, size distribution, etc. *See also* Feature extraction. [Section 10.13]

Image averaging A way of reducing snow and other random picture noise by averaging the pixel brightnesses in several successive video frames. Achieved with a digital image processor or by photographic integration. [Sections 9.4, 10.4, 12.1; Figs. 1-7, 1-8, 6-23, 6-24, 9-8, 10-5, 10-40]

Image convolution *Also* Convolution. In manipulating an image with a digital image processor, the substitution of the gray value of each pixel with another gray value that takes into account the gray values of the neighboring pixels. The convolution mask, or kernel, used to calculate the influence of the neighbors, determines the degree to which the picture is sharpened or smoothed by the convolution process. Contrasts with point operation, where the gray value of each pixel is transformed without considering the neighbors. *See* Convolution kernel, Intensity transformation function. [Section 10.11; Figs. 10-38 to 10-45]

Image dissector An electro-optical device that consists of a linear array of sensors (and memory) such as CCDs, which yield good linearity and high dynamic range. As the dissector scans the image (or vice versa), very-high-resolution video pictures can be produced, generally at scan rates slower than in conventional video.

Image enhancement A procedure for manipulating the video signal to sharpen or otherwise improve the video picture. Some form of image-enhancing circuit is included in most video equipment including cameras, VTRs, monitors, etc. Striking enhancements are possible with analog and especially digital image processing. *See also* Contrast enhancement. [Sections 6.5, 7.3, 7.5, 9.2–9.4, 10.1, 10.3, 10.5–10.9, 10.11, 10.12, Chapter 11]

Image histogram *Also* Histogram, Gray level (value) histogram. A graph that depicts the number of pixels displaying each (range of) gray value. Histograms can be used for manipulating look-up tables that control the image contrast, and for measuring the number of pixels (i.e., the areas) that have selected gray values. *See also* Look-up table. [Sections 10.5–10.7; Figs. 10-7 to 10-15, 10-18 to 10-24, 10-26, 10-36]

Image intensifier *Also* Image tube. A device coupled (by fiber optics or lenses) to a video camera tube to increase sensitivity. The intensifier is a vacuum tube with a photocathode on the front end that emits electrons according to the image focused upon it, an electron lens that focuses the electrons onto a phosphor at the back end, and a high-voltage accelerator that increases the energy of the electrons. The excited phosphor gives a picture 50–100 times brighter than the original image. Can be single or multiple stage. [Sections 7.1–7.3, 11.2; Figs. 7-3, 7-4]

Image isocon A video image pickup tube of the return beam type with high sensitivity. Characterized by low blooming, high resolution, low lag, and large intrascene dynamic range (although the gamma curve shows a characteristic knee). Can produce excellent-quality images in low-light applications, but is very expensive and fragile. [Section 7.1; Fig. 7-7]

Image orthicon An older style of TV image pickup tube of the return beam type. Was used in color cameras for low-light-level broadcasting but now mostly replaced by modern vidicons. [Section 7.1]

Image pickup tube *Also* Camera tube. *See* Camera tube. [Section 7.1]

Image processing Generally refers to digital or analog enhancement and geometric manipulation of the video signal. Contrasts with image analysis which emphasizes the measurement of image parameters. *See also* Image analysis, Image enhancement. [Chapters 9, 10, Appendix IV]

Image segmentation In digital image processing, the partitioning of the image into nonoverlapping regions according to gray level, texture, etc. [Section 10.13; Fig. 10-54]

Image thresholding A procedure that eliminates shades of gray from regions of a video picture above or below particular gray levels, replacing them instead with solid white or black. Used in digital image processing; also in analog circuits to ''key in'' a second scene with a special effects generator. [Sections 9.3, 10.13, 12.4]

Image tube *See* Image intensifier.

Immersion medium In microscopy the medium used to immerse the specimen, the space between the objective lens and coverslip, or the condenser lens front element and the slide. For the latter purposes, cedar or synthetic oils with refractive indices and dispersions approximating the front elements of the lens are used for homogeneous immersion. Homogeneous immersion provides high NA, less light loss and depolarization, and generally improved correction of aberrations. Glycerol or water may be used for immersion of particular lenses, but for high-NA lenses designed to be used with a variety of immersion media, proper adjustment of the correction collars is necessary; the refractive index, thickness, and dispersion of the immersion media and coverslips all enter into the corrections for aberrations of high-NA objective and condenser lenses. [Section 5.8; see Section 5.10 for cleaning immersion media off lenses]

Impedance The resistance to flow of an electrical current. The impedance of a circuit can vary with frequency depending on the values and arrangements of its inductive and capacitative components. For a circuit to receive or transmit a signal with minimum loss, especially important in the RF range, the impedances of its input and output circuits must match the impedances of the devices, including cables, to which they are connected. *See also* Hi-Z, Seventy-five-ohm termination. [Sections 3.1, 3.6]

I^2N, I^2V, I^3V Abbreviations for two-stage intensifier Newvicon, two- and three-stage intensifier Vidicon. [Section 11.2; Fig. 7-12B; Table 11-1]

Inches per second *See* IPS.

Index pin A short pin at the base of a vidicon tube that indicates the direction of the H sweep. It is used to properly orient the tube within the yoke. When the tube is not properly oriented, diagnonal moiré-like patterns or vertical, ringing patterns appear at the left edge of the picture. *See also* Ringing. [Section 7.1; Fig. 7-1]

Inlay insert *See* Keying.

Insert editing In the editing of videotape, the insertion of a segment into an already recorded series of segments. Requires an editing deck with this capability. Contrasts with assemble editing, which can be carried out by using the PAUSE button on some recorders. [Section 12.3]

Institute of Electrical and Electronic Engineers *See* IEEE.

Instrumentation camera In video, the type of camera used in scientific and industrial imaging as distinct from cameras that are designed for surveillance applications, as consumer products, or for studio and electronic news gathering. [Sections 3.1, 7.3]

Integrated circuit (IC) A very compact electronic component containing photoetched, miniature circuits.

Intensifier isocon (I-isocon) An Image isocon with an image intensifier coupled in front to provide increased sensitivity. [Section 11.2; Table 7-2, 7-4]

Intensifier vidicon (IV) A standard vidicon-type image pickup tube coupled (with fiber optics) to an intensifier to provide increased sensitivity. [Sections 7.1, 11.2; Table 11-1]

Intensifier Silicon-Intensifier Target tube *See* ISIT.

Intensity transformation function (ITF) In digital image processing, a function that specifies how a look-up table transforms the gray value of each pixel in one image into another gray value in a second image. The process is a form of point operation. *See also* Image convolution, Image histogram. [Sections 10.6, 10.7; Figs. 10-9 to 10-18, 10-20 to 10-23, 10-26]

Intensity transformation table *See* Look-up table.

Interactive Computer hardware or software that requires or allows the user to participate in the particular activity.

Interlace The arrangement of the raster lines in standard video where the alternate (odd- and even-) scan lines generated by the successive vertical sweeps fall in between each other. With 2 : 1 interlace, the odd and even scan lines can be made to fall exactly halfway between each other. With random interlace, the relationship between the two sets of scan lines varies randomly with time. *See also* Noninterlace. [Sections 6.2, 6.3, 7.5; Fig. 6-5, 6-16]

Intermediate image plane In a light microscope, the plane into which the objective lens directly focuses the image of the specimen. The plane is usually located a set distance (commonly 10 mm) below the shoulder for the ocular and another fixed distance (generally 160 mm) behind the rear focal plane of the objective lens. The ocular forms a virtual image of the intermediate image for visual observation, or projects a real image for photography and video microscopy. Note: The objective lens, combined with the coverslip of proper thickness, is corrected for projecting the primary image to the specified intermediate image plane only. [Sections 5.2, 5.3, 5.9; Figs. 5-9, 5-10, 5-15]

Intrascene dynamic range The greatest ratio of highlight to shadow brightness within a single scene that a video camera (tube) can handle usefully. Contrasts with the "usable light range" of a camera, which is considerably greater than the intrascene dynamic range. [Sections 7.2, 7.3]

IPS (inches per second) Speed of tape transport or writing speed in a VTR. In helical or segmented scan recording, the tape speed is only a small fraction of the writing speed, which is the relative velocity between the tape and the spinning read/write head. [Section 8.3]

IRE scale Now called the IEEE scale, since the IRE (Institute of Radio Engineers) was absorbed by the IEEE. *See* IEEE scale.

ISIT (Intensifier Silicon-Intensifier Target tube) RCA's trade name for a video image pickup tube designed for extremely low-light applications. Essentially a SIT tube with an additional intensifier coupled, by fiber optics as a first stage, to provide the increased sensitivity. [Sections 7.1, 11.2]

ITF *See Intensity transformation function.*

ITV Instructional Television.

IV *See* Intensifier Vidicon.

Jaggies In digitally processed images or raster graphics display, the discrete steps that appear on a diagonal line. Particularly pronounced in the display or printout of computer graphics or video freeze-field display, that use limited numbers of H-scan lines. [Fig. 9-11b,c, 10-54c–g]

Jitter An instability of the video picture where the lines jump up and down for short distances (or appear to do so) at several hertz. Jitter may be seen on individual lines or on the entire picture. The jitter may be inherent in the video signal (e.g., in freeze frame when there is movement between the fields), may arise out of misadjustment of the H and V HOLD controls, or may be caused by raising the monitor BRIGHTNESS and CONTRAST controls too high. [Sections 3.5, 7.5]

Kell factor A number, commonly between 0.7 and 0.8, by which the number of scan lines or pixels, per distance equal to the picture height, is multiplied to calculate the practically attainable vertical resolution in a video system. [Section 6.4]

Keyed insert *See* Keying.

Keying *Also* Keyed insert, Inlay insert. A method for blanking out, or montaging, another picture having the shape of the "key" into a limited region of the main picture. In a video mixer, the shape of the key is generated electronically or derived from the thresholded highlights of a scene viewed by the keying camera. [Section 12.4]

Keystone distortion *See* Keystoning.

Keystoning *Also* Keystone distortion. A camera or projector distortion in which a rectangle appears as a trapezoid with a wider top than bottom, or vice versa.

Kinescope The process of, or equipment for, recording a video picture by photographing a monitor with a motion picture camera synchronized to the video. In the early days of TV, kinescope referred to cathode-ray tubes with a suppressor grid to control beam brightness, now called picture tube. [Sections 1.1, 12.6]

Koehler illumination Mode of microscope illumination in which the light source is imaged onto the condenser iris diaphragm and the field diaphragm (in front of the lamp collector lens) is imaged by the condenser onto the plane of focus of the specimen. With Koehler illumination, the aperture and field can be regulated independently to provide maximum resolution and optimum contrast. Also, a field with uniform illumination is obtained, circumscribed by the image of the field diaphragm. [Sections 3.7, 5.3, 5.4; Figs. 5-14, 5-15]

Lag *Also* decay lag. The persistence of the image highlights for two or more frames after excitation is removed from a video pickup tube or monitor. Commonly expressed as percent picture retention in the third field. *See also* Buildup lag. [Section 7.2; Figs. 7-17 to 7-22; Tables 7-1, 7-2]

Laplacian filter In digital image processing, one of the convolution masks that yields high-spatial-frequency pass filtration which provides an omnidirectional edge enhancement. The coefficients of the 3 × 3 matrix in the common Laplacian mask are a central 8 surrounded by eight −1s. *See* Image convolution. [Section 10.11; Figs. 10-42, 10-43]

Large-scale integrated circuit (LSI) A solid-state chip that incorporates a large number of ICs. *See* Integrated circuit.

Lateral magnification *See* Magnification.

LCD (liquid crystal display device) A thin, flat-panel display device with liquid crystal material sandwiched between two sets of transparent electrodes (and commonly, sheet Polaroids). The voltage, imposed selectively on the electrodes, orients the liquid crystals and causes a change in the material's optical property. [Section 7.5]

LED (light-emitting diode) A small solid-state device that efficiently converts low-voltage DC into cool light of various colors specific to the device. Used as signal lights; also potentially in flat-panel video display devices.

Light-adapted vision *See* Photopic vision.

Light-emitting diode *See* LED.

Light transfer characteristics In an image pickup device, the curve that depicts the relationship log (signal current) versus log (illumination of faceplate). The slope of the straight line portion of this curve is the gamma of the device. The transfer characteristics are measured under electrical conditions specified for the tube type with illumination from a standard incandescent source that has a color temperature of 2854 or 2856°K. [Section 7.2; Fig. 7-14]

Light valve projector A video projector that uses an external high-luminosity (xenon) light source, a light valve instead of the phosphor of a cathode-ray tube, and catadioptric optics, to project a video picture. The surface of the continuously refreshed oil film in the light valve, or Eidophor, is contoured by the charge that is deposited on it by the scanning electron beam. The picture is produced by the powerful xenon arc lamp projecting a schlieren image of the embossed oil film. [Section 12.5]

Limiting resolution A measure of horizontal resolutions in video; usually expressed as the maximum number of black *plus* white vertical lines, per distance equal to the picture height, that can be seen on a test chart. On the amplitude response curve of a video camera, the limiting resolution is taken to be the number of black plus white lines that yield a 3% contrast. [Sections 6.4, 7.2, Appendix II; Figs. 7-9, 7-11; Tables 7-1, 7-2]

LINE IN An audio input for VTRs and audio mixers. This input is used instead of the MIKE, or microphone, input when the audio signal is already amplified. Not to be confused with AC power input. [Section 12.3; Fig. 12-7]

LINE OUT Audio output on a tape recorder. The audio signal at the output has been amplified to provide a moderate power for driving other amplifiers but not a loudspeaker directly. The amplitude is too high and will distort the signal if the audio from the LINE OUT is connected to the MIKE terminal of a second recorder. [Section 12.3; Fig. 12-7]

Line scan The display, on a monitor or cathode-ray oscilloscope, of the picture signal of a single H scan of video, or of signal amplitudes sampled along a vertical or some other line. [Section 9.6; Figs. 6-23, 9-8, 9.14]

Liquid crystal display device *See* LCD.

LLLTV *Also* L^3TV. Low-light-level TV. [Sections 7.3, 11.2]

Longitudinal magnification *Also* axial magnification. *See* Magnification.

Look-up table (LUT) *Also* Intensity transformation table. In digital image processing, a function stored in memory that directs the conversion of the gray value of each pixel into another gray value or color that is specified by the table. The LUTs can be programmed to manipulate image contrast, threshold a picture, apply pseudocolor, etc. *See also* Image histogram, Intensity transformation function. [Sections 10.6, 10.14; Fig. 10-15]

Loop through *See* E-to-E.

LSI *See* Large-scale integrated circuit.

Luma *See* Luminance signal.

Lumen A unit of luminous flux, equal to the flux through a unit solid angle (steradian) from a uniform point source of 1 cd. (See *RCA Electro-Optics Handbook,* Section 2.1.) [Section 7.2]

Luminance Photometric brightness, or the brightness of light calibrated for a sensor whose spectral response is similar to the light-adapted human eye; measured in nits (cd/m^2) or footlamberts. *See also* Photometric units, Radiometric units. [Sections 4.3, 7.4]

Luminance signal *Also* Luma. The portion of the NTSC video (color) signal that contains the luminance or brightness information. *See also* Chrominance signal. [Section 7.4; Fig. 7-29]

Luminous density Luminous output per unit area of a source of light. Expressed in cd/cm^2, cd/$inch^2$, etc. [Section 5.8; Table 5-1]

LUT *See* Look-up table.

Lux The amount of visual light measured in metric units at the surface that the light is illuminating. One lux equals one lumen per square meter. One footcandle equals 10.764 lux. (See *RCA Electro-Optics Handbook,* Section 2.1.) [Section 7.2]

Magnetic focusing The focusing of an electron beam by a magnetic field. [Sections 7.1, 7.5]

Magnification In microscopy, the ratio (distance between two points in the image)/(distance between the two corresponding points in the specimen). The apparent size of the specimen or primary image at 25 cm (≃ 10 inches) from the eye is considered to be at 1×. Commonly, magnification refers to the ratio of the distance between two points in the image, lying in a plane perpendicular to the optical axis of the microscope, to the corresponding two points in the specimen. This is the lateral magnification. The longitudinal or axial magnification is measured along the optical axis and (for small distances from the focal plane) is given by the square of the lateral magnification. [Section 5.2]

Maximum white *See* White peak.

MCP *See* Microchannel plate.

Metal oxide semiconductor (MOS) A type of transistor. MOSs are also used as the basis of some integrated circuits and solid-state image pickup devices. *See also* CCD. [Section 7.1]

M-format A 1/2-inch VCR system used in electronic news gathering that provides exceptionally high-performance video and audio recording. Using T-120 VHS cassettes at high speed (8 inches/sec), the M-format provides only 20 min of recording, but with greater than 550 lines of H resolution and S/N ratio of close to 60 dB. [Section 8.3]

Microchannel plate (MCP) *Also* Gen II intensifier. A thin plate made up of a large bundle of capillary channels, each of which acts as an electron multiplier. An MCP is a compact, efficient intensifier of two-dimensional electron images. [Section 7.1; Fig. 7-5]

Mixer *See* Special effects generator.

Modem (*mo*dulator/*dem*odulator) An electronic device that is used to link digital computers or slow-scan video equipment, allowing them to communicate with each other through a low-frequency circuit such as a telephone line. [Section 9.1]

Modulation The process of modulating the amplitude, frequency, or phase of a standard, high-frequency carrier wave by an audio or video signal. *See* AM, FM. [Sections 5.7, 6.4]

Modulation transfer function (MTF) A mathematical function that expresses the ability of an optical or electronic device to transfer signals faithfully as a function of the spatial or temporal frequency of the signal. The MTF is the ratio of percentage modulation of a sinusoidal signal leaving to that entering the device over the range of frequencies of interest. The MTF is usually presented as a graph of MTF versus log (frequency). For a square wave signal, the function is known as the CTF. [Sections 4.5, 5.7, 6.4, 7.2, Appendix II; Figs. 6-20, 7-39, II-8]

Monitor A device that converts the video signal from a camera, VTR, computer, etc. into a picture that is displayed on its cathode-ray tube. Compared to a TV receiver that must first decode an RF-modulated video signal, the monitor has no tuner and accepts the signal from the source without demodulation, thus permitting the use of a higher bandwidth and providing greater resolution. *See also* Receiver-monitor. [Sections 3.1, 3.2, 3.5, 3.6, 7.5, 12.1]

Monochrome signal A black-to-white (black and white) video signal containing only luminance information. The monochrome signal can be displayed as a black-to-white picture on a color or monochrome monitor. [Sections 7.4, 7.5]

MOS *See* Metal oxide semiconductor.

MTBF (Mean time between failures) An expression of reliability of equipment or systems, usually expressed in hours.

MTF *See* Modulation transfer function.

Multiplex To interleave, or simultaneously transmit, two or more messages on a single channel.

NA (numerical aperture) For a microscope, (the half angle of the cone of light accepted by the objective lens) × (refractive index of the medium between the specimen and the lens). Also similarly for the cone of light emerging from the condenser lens. *See* Abbe limit. [Section 5.4, 5.5; Fig. 5-16, 5-17]

National Television Systems Committee *See* NTSC.

Neutral-density filter A light-absorbing filter whose absorption spectrum is moderately flat. Depending on the type, the absorption curve is flat primarily in the visible spectral range, or may extend to varying degrees beyond the visible range. For video microscopy, this is an important point since the absorbance may or may not extend into the near-infrared region where the sensitivity of many video image pickup devices is very high. [Sections 3.2, 5.8]

Newvicon Panasonic's trade name for a vidicon-type image pickup tube designed for relatively low-light applications. Shows some lag, but otherwise has desirable characteristics such as good resolution, high sensitivity, nonblooming of high-brightness details, and relative freedom from burn-in. [Sections 7.2, 7.3, 11.3; Table 7-1]

Nipkow disk An opaque circular disk perforated with small holes arranged at equal angular separations and in an Archimedes spiral. The holes trace a raster scanning pattern when the disk is spun around its center. The Nipkow disk was used in early experiments on television and more recently in Petráň's confocal microscope. [Sections 1.1, 1.2, 11.5; Fig. 1-1]

Nit (nt) An SI (Système International) unit of luminance; the candela per square meter. (See *RCA Electro-Optics Handbook,* Section 2.1.)

Noise Unwanted interference in any signal or display. Noise in the video signal and picture (snow, hash, glitches, etc.) may be picked up from outside of the equipment through the AC line or by RF radiation, or may originate in the equipment (e.g., the camera or VTR). It may also be caused by defective connecting cable and connectors or poor grounding. The limiting noise from vidicon-type cameras at reduced light levels arises in the preamplifier that follows the vidicon tube. With intensified cameras, the limiting noise may be due to photon statistics (for the most sensitive tubes) or may arise in the tube itself. *See also* Optical noise, Signal-to-noise ratio. [Sections 6.5, 7.2, 8.2, 9.1, 10.4, 10.11, 11.2; Figs. 1-7, 1-8, 3-13, 8-7, 8-15, 8-17, 9-8, 10-40, 11-3, 11-7]

Noise bar *Also* Video guard-band noise. An unsightly, broad horizontal stripe that appears on the monitor picture when a videotape is played back at the wrong speed. The noise bar is generated when the read/write head crosses a video guard band. [Sections 8.2, 8.5; Fig. 8-7]

Noncomposite video signal A video signal containing picture and blanking information but no sync signals. Sync signals are added separately. *See also* Composite video signal. [Section 6.3]

Noninterlace *Also* One-to-one interlace. A scanning format in closed-circuit TV in which a single V scan, rather than the conventional alternating fields, makes up a full frame. Used in some high-resolution video formats (often at slow scan rates) to avoid the limitations arising from interlacing. Also used for the output of many personal computers, where the frame rather than the field is repeated at the power-line frequency rate and with approximately one-half the number of scan lines. *See also* Interlace, Random Interlace. [Section 6.3; Tables 6-1, 6-2]

Nonsegmented recording A system of video tape recording that uses the single swipe of a head to scan one uninterrupted field of video signal on the tape, generally in the helical scan mode. *See also* Segmented scan recording. [Section 8.2]

Notch filter A special type of electronic filter designed to reject a narrow band of frequencies without adversely affecting the other frequencies. A comb filter is even more selective. *See also* Comb filter. [Section 7.5]

nt *See* Nit.

NTSC (National Television Systems Committee) A U.S. broadcast engineering advisory group. NTSC also refers to the 525-line 60 field/sec, color video standard adopted by the Committee in 1953 and used in standard broadcast color TV in the United States, Japan, and other countries. *See also* PAL, SECAM. [Section 6.3; Fig. 6-10; Table 6-3]

Numerical aperture *See* NA.

Nyquist frequency *Also* Fold-over frequency. *See* Aliasing.

OEM *See* Original equipment manufacturer.

OMDR *See* Optical memory disk recorder.

One-inch (25 mm) vidicon A vidicon tube with a nominal outside diameter of 1 inch for the main part of the glass envelope. For 1-inch tubes, the effective target area is generally 16 mm in diameter. [Sections 3.1, 7.3; Table 7-5]

One-to-one interlace *See* Noninterlace.

Optical copying In video, a method used for copying one tape to another tape of an incompatible format. The picture from the first tape recorder is displayed on a monitor and is then copied by the second recorder optically, through a camera pointed at and synchronized with the first monitor. Also the act of copying motion picture film, charts, titles, etc. with a video camera. [Section 12.2]

Optical digitizer An electro-optical device for converting the brightness and spatial coordinates of a picture or image into digital form. Also a device that provides digital electrical readout of positions, distances, or angles measured on a precision optical scale. [Sections 10.2, 10.3]

Optical memory disk recorder (OMDR) A video disk recorder using a laser beam for writing and reading video frames. *See also* Video disk recorder. [Section 8.9; Fig. 8-18, Table 8-7]

Optical sectioning The use of high NA objective and condenser lenses on a microscope to achieve a shallow depth of field. With a very shallow depth of field, objects above and below focus contribute little to the in-focus image, so that a clean optical section is obtained. *See also* Depth of field. [Sections 5.5, 11.3, 11.5; Figs. 11.9, 11.10]

Optical scrambler An optical device for scrambling the image of a non-uniform light source so that it now fills the condenser aperture uniformly without appreciable loss of total luminous flux through the microscope. The scrambler can be a simple loop of a single otical fiber with its ends appropriately polished. [Section 5.8; Fig. III-21]

Optical noise Image defects that become especially conspicuous when the image is enhanced. In video microscopy, includes hot spots, mottle, uneven illumination, etc. [Sections 3.8–3.10, 10.8; Figs. 1-7, 1-8, 3-12, 3-13, 10-28]

Optical transfer function (OTF) The relationship between the image produced by an optical instrument and the amplitude and phase of a periodic specimen, measured at various spatial frequencies. The OTF curve, which shows how well contrast is maintained for finer specimen details, is a complex function, of which the real term gives the ratio of amplitudes, and the imaginary term the phase relationships. When the brightness of the periodic specimen varies as a sine wave, the modulus or absolute value of the OTF becomes the MTF (modulation transfer function); when the brightness varies as a square wave, the MTF is known as a CTF (contrast transfer function). *See* Modulation transfer function [Section 5.7, Appendix II]

Optional An item not included in the standard package but available for an additional price.

Original equipment manufacturer (OEM) A company that manufactures, or assembles from components, substantial pieces of equipment for sale. [Appendix IV]

Orthoscopic image The image normally observed in a light microscope (at the intermediate image plane and its conjugate planes). Contrasts with the conoscopic image, which appears at the back focal plane (near the back aperture) of the objective lens. *See* Conoscopic image. [Section 5.4]

OTF *See* Optical transfer function.

Overscan A method used on most TV sets and monitors to hide the defects at the edges of the active picture area. Some 5–15% of the picture width and height that may be useful in closed-circuit TV are lost by overscan. Some monitors allow one to view the full picture area by toggling an UNDERSCAN switch. [Sections 6.3, 7.5]

Pairing The overlapping of scan lines from the alternate video fields that results in a twofold loss of vertical resolution in a 2 : 1 interlace system. Commonly arises from faulty adjustment of the V HOLD control on the monitor. With a random interlace signal, pairing occurs intermittently by chance. [Sections 6.4, 12.1; Fig. 6-16]

PAL (phase-alternating line system) A color TV system in which the subcarrier derived from the color burst is inverted in phase from one line to the next in order to minimize errors in hue that may occur in color transmission. *See also* NTSC, SECAM. [Section 6.3]

Pan Movement of a video camera or the image displayed on the monitor horizontally from left to right or vice versa. *See also* Scroll. [Section 10.14]

Pause Momentarily stopping the tape while recording or playing back with a VTR. In playback, used for closely examining single frozen fields. In recording, the PAUSE control can be used for glitch-free assemble editing on some VCRs. Most VCRs automatically exit the pause mode after several minutes, in order to prevent damage to the tape. [Sections 8.4, 8.7, 12.1, 12.3]

PC Personal computer.

Peak-to-peak (P-to-P, P–P) The difference between the most positive and the most negative excursions (peaks) of an electrical signal. For RS-170 and -330 video signals with proper termination, EIA specifies that the P–P voltage be 1 V from the tip of the positive-going reference white to the tip of the negative-going sync signal. For separate sync pulses, a P–P voltage of 4V is commonly specified. [Sections 6.3, 6.5, 7.5; Figs. 6-8, 6-25]

Pedestal In a composite video signal, the voltage level that separates the picture signal from the sync pulses. The pedestal is at the blanking level, just below the optical black level, at the base of the sync pulses. [Section 6.3; Fig. 6-8]

PEL *See* Picture element.

Perpendicular magnetic recording *See* Vertical field magnetic recording.

Per picture height *See* PPH.

Persistence In a cathode-ray tube, the period of time a phosphor continues to glow after excitation is removed. *See also* Lag. [Sections 7.2, 7.5]

Phase-alternating line system *See* PAL.

Phase contrast An optical method devised by F. Zernike for converting the focused image of a phase object (one with differences in refractive index or optical path but not in absorbance), which ordinarily is not visible in focus, into an image with good contrast. [Section 5.6]

Phase-locked loop (PLL) An oscillator circuit that locks in phase and regenerates an input signal frequency, but without any associated jitter that may have been present. Used in practically every piece of video equipment from camera, VTR, special effects generator, to monitor as well as in image-processing computers. [Sections 6.3, 6.6, 8.2]

Phase plate The plate used near the back focal plane of a microscope objective lens (in conjunction with an annulus at the front focal plane of the condenser lens) to achieve phase contrast. The phase plate selectively shifts the phase of the waves diffracted by the specimen by a quarter wave and reduces the amplitude of the undeviated, direct beam. [Section 5.6]

Photocathode The cathode in a photocell or video image pickup tube that converts the incoming light to a corresponding change in electrical potential or current. [Sections 7.1, 7.2]

Photodiode array A linear or two-dimensional array of diodes that become conductive upon illumination with light of appropriate wavelengths. The amount of light falling on each and all of the photodiodes can be measured through many parallel circuits with very short time delay, or sequentially by a bucket-brigade mechanism to provide a standard or slow-scan video signal. [Sections 7.1, 11.4]

Photometric units The units (lm, lx, cd/m^2, etc.) that are used to measure the amount of light (illuminance) or the brightness of an object or image (luminance), taking into account the sensitometric characteristics (wavelength-dependent response) of the human eye. Contrasts with radiometric units (e.g., W/m^2), which measure light in terms of the physical energy independent of the sensitometric characteristics of the eye. [Sections 4.3, 5.8, 7.2]

Photon limited operation *Also* Photon counting mode, Quantum limited mode. The use of chilled photomultipliers, charge-coupled devices, or intensifier camera tubes at levels of light and noise so low that individual photons can give rise to clearly detectable bursts of electrons. Commonly, about 10% of the absorbed photons trigger a signal, and the level of energy of the electrons is statistically distributed. [Sections 7.1, 11.4; Table 7-4]

Photopic vision *Also* Light-adapted vision. Vision that occurs at moderate and high levels of luminance and permits distinction of colors; attributed to the retinal cones in the eye. Contrasts with scotopic (twilight) vision. [Section 4.3]

Pickup tube The image pickup tube in a video camera. [Chapter 7]

Picture In video, the final picture that is displayed on the monitor or other display device. *See also* Image. [Sections 3.5, 6.2, 7.5]

Picture element (PEL) Any segment of a video scan line whose dimension along the line is equal to the line spacing. *See* Pixel.

Picture signal The part of the video signal that contains the picture and blanking information. In a sync-negative, composite video signal, the portion of the signal above the pedestal level. [Sections 6.2, 9.3]

Picture tube The cathode-ray tube in a video monitor or TV receiver. The picture is produced on the phosphor on the faceplate of the tube, by variations of beam intensity as the electron beam is scanned in a raster. [Section 7.5]

Pincushion distortion A geometrical distortion in a video picture, or a form of optical aberration, that makes the middle of all sides of a square appear to bow inward. The distortion can amount to several percent of the picture height in some intensified camera tubes. *See also* Aberration, Distortion, Barrel distortion. [Sections 5.9, 7.2, 7.5, 10.9]

Pipeline processor A special kind of arrayed digital processor that can handle several parallel or repetitive sequences of processing tasks at very high speed without requiring specific instructions from the central processing computer at each step. [Section 10.14]

Pixel Picture element (*pix* for picture, *el* for element). A single, finite-sized element of a digitized video picture. A pixel is defined by its *X* and *Y* coordinates and its gray level (luminance), commonly expressed by binary numbers. [Sections 9.1, 10.2–10.4]

Plan Apo (plan apochromatic objective lens) A modern, high-NA microscope objective lens designed with high degrees of corrections for various aberrations. It is corrected for spherical aberration in four wavelengths (dark blue, blue, green, and red), for chromatic aberration in more than these four wavelengths, and for flatness of field. A single Plan Apo objective may contain as many as 11 lens elements. [Section 5.9]

Plan apochromatic objective lens *See* Plan Apo.

Playback head An audio or video magnetic head used to obtain a signal from the videotape. Some heads are capable of playback and record, others of playback only. The video heads of most helical-scan VTRs serve as both record and playback heads. [Section 8.2]

PLL *See* Phase-locked loop.

Plumbicon Phillips's trade name for a special vidicon-type tube that uses a lead oxide target. Shows very low lag and distortion and a sensitivity that is greater than a sulfide vidicon but less than a Chalnicon or Newvicon. [Section 7.2; Table 7-1]

PLZT A transparent, piezoelectric ceramic (lanthanum-modified lead zirconate titanate) whose birefringence changes with the application of an electric potential. Sandwiched between a pair of crossed polarizers and light-transmitting electrodes, PLZT is used in shuttered goggles, which very rapidly transmit or shut off the light reaching the eye according to an electronic signal. [Section 12.7; Figs. 12-27, 12-28]

Pointing device A mechanical device, such as a joystick, mouse, or digitizing tablet, that is used with a digital computer to move a marker and point to a desired location on the display screen. [Section 10.14]

Point operation *See* Image convolution, Intensity transformation function.

Point-spread function The mathematical representation of the image of a point source. For a diffraction-limited optical system operating in the absence of aberrations, the point-spread function is the Airy disk. *See also* Three-dimensional diffraction pattern. [Sections 5.5, 10.4, Appendix II; Figs. 5-19, 5-21, II-3]

Pot *See* Potentiometer.

p-n Junction diode A solid-state photosensitive device with a junction between a p-type and an n-type semiconducting material. In the target of a SIT tube, a tightly spaced, fine matrix of p-n junction diodes are formed on a very thin wafer of silicon. They amplify the photoelectrons generated in the intensifier section. [Section 7.1]

Potentiometer (Pot) A center-tapped variable resistor, generally used for adjusting the signal or bias voltage levels.

P–P *See* Peak-to-peak.

PPH (per picture height) The distance over which horizontal resolution is measured in video. The number of vertical black and white lines (twice the number of line pairs) that can be resolved is measured in the center of the picture for a horizontal distance equal to the height of the active picture area. [Sections 6.4, 7.2, 7.3, 7.5; Fig. 6-17]

Preamplifier An amplifier, usually with very high input impedance, that is used to increase the output of a low-level signal for further processing of the signal. [Section 6.5]

Principal plane In geometrical construction of ray diagrams through a perfect lens, the single plane that can be considered (disregarding the refraction of the rays at each optical interface) to convert paraxial rays into rays converging onto the rear focal point. For thin symmetrical lenses, there is one principal plane that bisects the lens. For thick or complex lenses, there is a principal plane for the paraxial ray to the left, and another for the paraxial ray to the right, of the lens. [Figs. 5-7 to 5-9]

Proc amp *See* Processing amplifier.

Processing amplifier (proc amp) A nonlinear amplifier used for many purposes in virtually every piece of video equipment. Processing amplifiers are used to clamp or clip off parts of the picture signal, separate off the sync pulses, anticipate or correct for distortion, etc. [Sections 6.6, 7.5]

PROM (programmable read-only memory) A ROM into which a nonvolatile program can be burned in by the user. EPROM is an erasable PROM; EEPROM is an electrically erasable PROM; UVEPROM is a UV-erasable PROM. *See also* ROM.

Pseudocolor display In video image processing, the assigning of selected colors to levels of gray values in order to designate areas with similar gray values or enhance subtle differences of gray values in the picture. *See also* False color. [Sections 9.3, 10.6, Figs. 1-8, 10-55]

P-to-P *See* Peak-to-peak.

Pulse-cross display The display of the edges of the video picture shifted in such a way that the blanking and sync pulse regions (which are ordinarily hidden) appear as a darker cross through the middle of the monitor. The pulse cross displays the degree of synchrony of the various timing signals as deciphered by that monitor. [Section 6.3; Figs. 6-13, 6-14, 6-27]

Quantum limited mode *See* Photon limited operation.

Rack mounting A method of mounting a series of (electronic) equipment stacked vertically in a 19-inch-wide standard metal rack.

Radio frequency (RF) High-frequency electrical signal that can be carried through coaxial conductors that have properly matched impedance, but which otherwise can easily be radiated into, or picked up from, space. Since much of video and computer signals occur at RF, proper impedance matching and electromagnetic shielding are crucial. [Sections 3.4, 6.5, 8.2]

Radiometric units Units (W/m^2, J, etc.) that are used to measure the quantity of electromagnetic radiation. Contrasts with photometric units (used for the measurement of visible light), which take into account the sensitometric characteristics of the human eye. *See also* Photometric units, Color temperature. [Sections 4.3, 7.2]

RAM (random-access read/write memory) A solid-state storage device in computers into which, or from which, the user can routinely store or retrieve information. *See also* ROM.

RAM disk An array of RAM chips mounted on a single board that the computer treats similarly to a single magnetic disk. Provides a volatile memory with very-large-scale storage capacity available for rapid, random access.

Ramsden disk The small circular patch of light that appears at the eyepoint above the ocular lens. This is the exit pupil of an optical instrument that, in a microscope adjusted for Koehler illumination, lies in a plane conjugate with the objective rear focal plane, condenser iris, and light source. Alteration of the Ramsden disk (e.g., by the observer's iris) modifies the aperture function, diffraction pattern, and direction of view of the specimen. [Sections 3.9, 5.3, 12.7; Fig. 3-15]

Random access Ability to immediately find, and gain direct access to, any particular video frame or recording, or to specified portions of a computer memory. [Section 8.9]

Random interlace A scanning format that is often used for low-cost closed-circuit TV, e.g., in surveillance applications. With random interlace, there is no fixed relationship between the beginning positions of the H-scan lines in successive fields. Thus, the interlace is imprecise, leading to some loss of resolution as well as potential trouble for syncing to high-precision video equipment. *See also* Two-to-one interlace, Noninterlace. [Sections 3.1, 6.3]

Raster The sequential scanning pattern used in video that dissects the image in the pickup device and lays down the picture in the display device. [Sections 6.2, 6.3]

Raster line removal *See* Scan line removal.

Rayleigh criterion A criterion chosen by Lord Rayleigh to define the limit of resolution of a diffraction-limited optical instrument. It is the condition that arises when the center of one diffraction pattern is superimposed with the first minimum of another diffraction pattern produced by a point (or line) source equally bright as the first. For a microscope under this condition, a 26.5% dip in brightness appears between the two maxima, giving rise to the sensation (or probability) of twoness. *See also* Abbe limit, Sparrow limit (of resolution). [Section 5.5, Appendix II; Fig. 5-19]

Real time Video display or image processing that is accomplished within, or nearly within, the framing rate of video. Today, increasing numbers of image-processing functions can be accomplished in real time, but many others still involve very-large-scale computations or signal storage that take much longer. [Sections 7.5, 9.1, 10.1]

Receiver–monitor A combined TV receiver and video monitor. For closed-circuit TV, the TV channel can be used for displaying an RF-converted NTSC signal. Alternatively, monochrome, or RGB or NTSC color, signals can be coupled directly to the video input without conversion to RF, thus avoiding the attendant loss of picture quality and resolution. [Section 7.5]

Recommended standards *See* RS-170, RS-232C, RS-330.

RECORD A push-button switch on a VTR that activates the video and audio recording when pressed together with the PLAY or FORWARD switch. When the RECORD switch alone is pressed, the live video and audio appear on a monitor that is connected to the video and audio output terminals without initiating the VTR recording functions. [Sections 8.5, 12.3]

Rectified optics Microscope lens system correcting the rotation of polarized light that takes place at high-incidence-angle interfaces between the polarizer and analyzer. Rectification provides high extinction for polarized-light and DIC microscopy at high NAs, thus permitting the detection of weak birefringence or phase retardation combined with high image resolution. In the absence of rectification, the diffraction pattern of a weakly birefringent or low-phase-retarding object viewed at high NA between crossed polars becomes anomalous. [Section 11.3, Appendix III; Figs. 8-19, 11-5, III-21, III-23 to III-25]

Red, green, and blue *See* RGB.

Registration The proper overlapping of the red, green, and blue electron beams in a color pickup tube or monitor to form a correctly colored picture. *See also* Convergence. [Sections 7.4, 7.5]

Resolution A measure of how fine a detail can be detected, in terms of distance in space or passage of time. Note that the convention used to measure spatial resolution in video differs from that used in optics. In optics one measures the distance between (or frequency of) line *pairs*, whereas in video *every* black and white line is counted. *See also* Limiting resolution. [Sections 5.5, 6.4, 7.2–7.5, 10.4, Appendix II]

Responsivity Similar to sensitivity; a rating of the output current from a camera tube divided by the incident flux of light, but expressed in different units. [Section 7.2; Tables 7-1, 7-2]

Retrace *See* Flyback.

RF *See* Radio frequency.

RF converter A device that converts video and audio signals from a VTR or computer into RF so that the signals can be played back on a standard TV receiver. This step degrades the signal quality but may be necessary if a proper video monitor is not available. [Section 7.5]

RGB (red, green, and blue) A form of color video signal, or type of color camera or monitor. Does not use an encoded signal such as NTSC, which can be transmitted over a single cable or channel, but instead uses three separate cables for carrying the R, G, and B signals. Without the loss due to encoding and decoding, the picture quality and resolution of an RGB system is superior to encoded color video. [Sections 7.4, 7.5; Figs. 7-26A, 7-27]

RGB-sync, RGBY Forms of RGB color video signals that require a fourth connection to carry a separate sync signal (RGB-sync) or luma signal (YRGB). [Sections 7.4, 7.5]

Ringing An oscillatory transient that occurs as a result of a sudden change in input. In video monitors, ringing gives rise to narrow, reversed-contrast stripes to the right of sharp vertical lines or edges. The camera tube can also contribute to ringing, especially if (the index pin of) the tube is not properly oriented relative to the H-scanning direction. *See also* Index pin. [Sections 6.4, 7.3, 7.5; Figs. 6-19, 7-1]

Rise time The transient response of an electronic device or circuit to a rising square wave pulse. The rise time is measured as the time interval that elapses between a 10% and 90% rise of the steady-state value. [Section 6.4; Fig. 6-19; Table 6-4]

rms (root-mean-squared) A method of expressing the amplitude of a signal in relation to its energy. For sine wave signals, its rms value is given by (peak amplitude)/$\sqrt{2}$. [Section 6.5]

Roll A loss of vertical synchronization, causing the video picture to spontaneously slip or flip, up or down, on a receiver or monitor. [Sections 3.2, 3.5, 6.3]

ROM (read-only memory) A solid-state, computer storage device from which preprogrammed information can only be read. One cannot write information into a ROM as one can into a RAM. *See also* PROM.

Root-mean-squared *See* rms.

RS Recommended standards.

RS-170 The 525/60, 2 : 1 interlace standard format recommended by EIA for broadcast monochrome video in the United States. For NTSC color, the temporary standard RS-170A is used. For closed-circuit TV, RS-170 is common, but other formats including RS-330 are also used. [Section 6.3; Fig. 6-10]

RS-232C A standard recommended by EIA specifying basic guidelines for a class of relatively slow (up to 20,000 baud), serial, communications links between computers and peripherals. In RS-232C, up to 25 data lines are available but commonly only 3 to 5 are used.

RS-330 A standard recommended by EIA for signals generated by closed-circuit TV cameras scanned at 525/60 and interlaced 2 : 1. The standard is more or less similar to RS-170, but H-sync pulses are absent during V sync. Equalizing pulses are not required and may be added optionally during the V-blanking interval. [Section 6.3; Fig. 6-9]

RS-343A EIA standards for high resolution monochrome CCTV. [Section 6.3; Table 6-2]

Sampling interval The short time interval or distance that is used as a unit for converting an analog signal into a digital signal. If the sampling interval becomes greater than half of the (sinusoidal) period in the original signal (the Nyquist criterion), the representation may not be accurate, and false signals such as aliasing can arise. *See also* Aliasing. [Section 10.4, Appendix II; Fig. II-5]

Saticon Hitachi's trade name for a vidicon-type image pickup tube. Exhibits low lag, high resistance to blooming, low dark current, near-unity gamma, and well-balanced spectral sensitivity. Popular for use in color cameras. [Table 7-1]

Saturation In color, the degree to which a color is undiluted with white or is pure. The vividness of a color, described by such terms as bright, deep, pastel, pale, etc. Saturation in the color video signal increases proportionately with the amplitude of the chrominance signal. [Section 7.4; Fig. 7-29]

Scan In video, the movement of the electron beam across the target in a camera tube or across the phosphor screen in a monitor. Also the movement of the read/write head relative to the tape in a VTR. [Sections 6.2, 7.1, 7.5, 8.2]

Scan line removal Eliminating the raster lines from, or making the lines less intrusive in, (photographs of) the video monitor in order to make the subtle picture details more visible. Can be accomplished by optical spatial filtration, by electronic wobbling of the scanning spot, by doubling or quadrupling the number of scan lines and interlacing with proportioned signals, or by the use of a Ronchi grating. [Section 12.1, Appendix I; Figs. 11-5, 12-3, I-1, I-2]

Scan rate The number of horizontal-scan lines per frame and vertical scans per second that are repeated in video; e.g., 525/60, 625/50. In 525/60, 2 : 1 interlaced video, the V scan is repeated at the field rate (which is half of the frame rate for 2 : 1 interlaced video) so that 525 H scans take place 30 times a second. The H-scan rate is therfore 525 × 30 = 15.75 kHz. With 525/60, 1 : 1 interlace, the H-scan rate would be twice this value. [Sections 6.2, 6.3]

SC IN (Subcarrier In) A connector found on some VTRs that accepts a standardized color subcarrier wave. The color bursts generated by an external TBC eliminate the shift of hue that could result from time-base errors in the chrominance signals.

Scope Short for cathode-ray oscilloscope. *See* Cathode-ray oscilloscope.

Scotopic vision Vision that occurs in dim light as a result of dark adaptation; attributed to the function of the retinal rods in the eye. Contrasts with photopic (daylight) vision. [Sections 4.3, 11.2]

Scroll Movement of the video picture up or down the video screen at a moderate even pace to expose the lower or upper portions of a scene that previously were not visible. *See also* Pan.

SEC Westinghouse's trade name for a secondary electron conduction camera tube of a non-return beam type; characterized by high sensitivity and low lag. [Figs. 7-16, 7-18; Table 7-2]

SECAM Abbreviation for Systéme Electronique Couleur Avec Mémorie. Also sequential color with memory. A 625/50-format color TV system developed in France. Differs in several respects from both the NTSC and PAL color systems. [Section 6.3; Table 6-3]

SEG *See* Special effects generator.

Segmented scan recording A system of video tape recording that divides each field into several segments and lays them in parallel tracks nearly at right angles to the tape length. *See also* Nonsegmented recording, Helical scan recording. [Sections 8.2, 8.3; Fig. 8-1]

Sensitivity Similar to responsivity of an image pickup tube but numerically different. The signal current developed in an image pickup tube per unit incident radiation density (watt per unit area) or illuminance (lux on the faceplate). Unless otherwise specified, the radiation is understood to be that of an unfiltered incandescent source at 2854 or 2856°K. *See also* Responsivity, Light transfer characteristics. [Sections 7.2, 7.3; Fig. 7-14; Tables 7-1 to 7-4]

Serrations The horizontal timing pulses that appear as positive-going notches in the negative-going vertical sync pulse in some video timing signals. While the polarity of the serration pulses are reversed from the equalizing pulses, it is still the negative-going voltage change that signals the start of an H interval. [Section 6.3; Figs. 6-9 to 6-11, 6-14]

Setup In a video picture signal, the difference between the black and blanking levels expressed in IEEE units. Also the ratio (expressed in percentages) of this value to the difference between the reference white and blanking levels. [Section 6.3; Fig. 6-8]

Seventy-five-ohm cable A coaxial cable commonly used to carry video signals. The characteristic impedance of the cable for a signal frequency greater than 4 kHz is 75 ohms. [Sections 3.6, 6.5, 6.6]

Seventy-five-ohm termination The impedance with which the input and output circuits of standard video equipment (connected to a 75-ohm cable) should be terminated. Proper termination is needed for the video signal to be at the correct voltage level, to prevent reflections that would otherwise occur at unterminated ends, and to prevent distortion of waveforms. *See also* Hi-Z, impedance. [Sections 3.6, 6.5, 6.6, 7.5; Figs. 3-5, 3-9, 3-10]

Shading A large-area brightness gradient in the video picture, not present in the original scene.

Shading can arise from camera lens falloff, from nonuniform sensitivity of the camera tube target or monitor phosphor, or from poor low-frequency responses in one or more pieces of video equipment. [Sections 6.5, 7.2–7.5; Figs. 1-7, 7-36D, 10-4, 11-7C]

Shading compensation Dynamic sensitivity control of the picture signal found in some precision video cameras and monitors. The compensation reduces the amount of change in the video level from center to edge, or edge or edge, of the picture. Shading can also be compensated during analog or digital image processing by adding suitable waveforms, or by subtraction of, or division by, the background image. [Sections 7.2, 7.5, 9.3, 10.3, 10.8, 10.10; Figs. 9-6, 10-4]

Shadow mask *Also* Aperture mask. In color video monitors, a finely perforated or striped metal plate that is located between the three electron guns and the phosphor screen. The mask ensures that the electron beams carrying the R, G, and B signals land on the corresponding phosphors at every point in the picture. [Section 7.5; Fig. 7-41]

Sharpening filter An analog circuit or digital signal-processing operation that sharpens transitions, or the edges of objects, in an image. *See* Edge enhancement, Laplacian filter. [Sections 9.3, 10.11, 10.13; Fig. 9-7]

Signal-to-noise ratio (S/N ratio) In video, the ratio of the peak amplitude of the signal to the rms value of the noise. Usually expressed in decibels. The higher the S/N ratio, the cleaner the video picture or reproduced sound. [Sections 6.5, 8.4, 8.9, 9.4, 10.4; Tables 6-5, 8-1, to 8-5, 8-7; Fig. 7-16]

Silicon *Also* Silicon vidicon, S-T Vidicon. Short for silicon target vidicon tube. *Not* the same as SIT. The Silicon is a vidicon-type camera tube with a silicon target having a very high sensitivity in the near infrared. Has quite good resolution but is subject to blooming. *See also* Ultricon. [Section 7.2; Figs. 7-13, 7-24; Tables 7-1, 7-2]

Silicon Intensifier Target tube *See* SIT.

Silicon target vidicon *See* Silicon.

Silicon vidicon *See* Silicon.

SIT (Silicon-Intensifier Target tube) RCA's trade name for a video camera tube of the direct readout type designed for low-light applications. Has a silicon target and an integral intensifier section, with direct electron input to the silicon target. Not the same as silicon target vidicon tube or ISIT. Has high sensitivity, and under adequate faceplate illuminance, moderately low lag and good resolution. *See also* p-n junction diode. [Section 7.1; Figs. 7-6, 7-11 to 7-18, 7-24; Table 7-2]

Single-sideband edge enhancement microscopy (SSEE) A mode of contrast generation in microscopy that provides an especially high modulation transfer at high spatial frequencies. SSEE takes advantage of the fact that the two sidebands (orders diffracted to the left and the right) are not phase shifted, upon diffraction at refractive index gradients, by exactly a quarter of a wavelength of the illuminating light. [Section 11.3]

Skew In a VTR, the tape tension between the supply reel and first rotary idler in the tape path around the head assembly. The picture becomes unstable or skewed (the top of the picture bends to the left or right) when the skew changes, e.g., as a result of the tape slackening on the reel, stretching by repeated playback of the same tape region, by playing tape on a different VCR, or by excessive humidity or temperature. *See also* Tracking. [Sections 8.2, 8.4, 8.7]

SKEW *Also* SKEW ADJUST, SKEW CONTROL. A knob or lever available on some VTRs to adjust skew (tension on the tape) during playback. It should automatically return to the neutral position during recording. *See* Skew. [Section 8.7; Fig. 8-13]

Skip-field A recording or playback process used in some time-lapse VTRs. Instead of reproducing both the even and odd fields and interlacing them, only one field is reproduced. [Section 8.5]

Slave A piece ot video equipment whose synchronizing signals are derived from another piece of equipment that serves as the sync source. [Sections 6.6, 12.4]

Slow scan A system of video scanning in which the time used to read each line has been increased in comparison to standard video. The bandwidth needed to faithfully transmit or record the signal is reduced in inverse ratio to the scanning time. Slow scan allows the video signal to be transmitted over a telephone line, or line scans to be registered on a chart recorder, without loss of spatial resolution.[Sections 9.1, 9.4, 9.6]

Smoothing filter Generally a high-frequency cutoff filter that reduces noise and smooths out sharp contours. Can be part of an analog circuit or a convolution mask used in digital image processing. [Sections 9.4, 10.11; Figs. 9-8, 10-39, 10-40]

SMPTE (The Society of Motion Picture and Television Engineers) A U.S. standards-setting committee. Also SMPTE (pronounced sem-tee) time code. *See* Time code generator. [Section 8.3]

S/N (signal-to-noise ratio) Also sometimes used as an abbreviation for serial number; can be somewhat confusing in the case of electronic equipment. *See* Signal-to-noise ratio.

Snow Heavy random noise seen on the video screen. May arise from the camera at low light levels, from defective cables or connectors, or from a VTR with a dirty head or which is not operating properly. [Sections 7.2, 8.4, 8.7; Figs. 1-8, 8-17; Table 7-4]

Software Digital computer programs that carry out various operations. Contrasts with hardware. [Appendix IV]

Solid-state pickup device *Also* Solid-state sensors. In video, image pickup devices that use solid-state sensors such as CCD, CID, MOS, etc., rather than vacuum-tube-type sensors such as the vidicons or intensifier camera tubes. [Section 7.1]

Solid-state sensors *See* Solid-state pickup device.

SP, LP, SLP Standard play, long play, and super long play. Recording and playback speeds of 2, 4, and 6 hr used on VCRs of VHS format. [Section 8.3]

Sparrow limit (of resolution) The spatial frequency at which the modulation transfer function just becomes 0. *See also* Abbe limit, Rayleigh criterion. [Appendix II]

Spatial convolution *See* Image convolution, Convolution kernel.

Spatial filtration In optics, the removal of selected spatial frequency components from a picture by masking out their diffraction patterns at the aperture plane. *See also* Image convolution. [Sections 10.11, 10.12, Appendix I; Fig. I-2]

Spatial frequency The frequency in space in which a recurring feature appears, e.g., 20 line pairs/mm. In a video signal, spatial frequency along the horizontal axis is proportional to 1/(time interval between the recurring signal), or to frequency (Hz, MHz, etc.) of the electrical signal. *See also* Spatial filtration. [Sections 5.5–5.7, 6.4, 10.11, 10.12, Appendix II, Fig. 5-21B, II-8]

Special effects generator (SEG) A video device that allows switching and mixing between several cameras and other inputs. The inputs, synchronized through the SEG, can be mixed using a variety of special effects such as dissolve, wipe, keyed insert, etc. [Section 12.4; Figs. 12-12 to 12-14]

Spectral response curve In a photodetector, the curve that relates the electrical current, or voltage, generated per unit power input for various wavelengths of light. [Section 7.2; Fig. 7-13]

Spherical aberration The aberration caused by (near-paraxial) monochromatic light rays or electron beams passing through different radii of a lens not coming to the same focus. [Section 5.9]

Spike In electronics, a sharp brief voltage pulse. Powerful RF spikes that can damage video equipment and computers are generated when mercury or xenon arc lamps are started. To a limited extent, spikes entering through the power line (but not those radiated as RF) can be suppressed with spike arrestors (transient suppressors). [Sections 3.4, 6.5]

Square wave response curve *Also* CTF. *See* Modulation transfer function.

Standard signal A signal whose peak-to-peak voltage is appropriate for the system. For EIA RS-170 and -330 composite video signals, it is 1.0 Vpp; for noncomposite picture signals, 0.7 Vpp; and for sync-only signals, 4V. [Section 6.3; Figs. 6-7, 6-8]

Stand by lamp Used in some VCRs to warn the operator that execution of other functions is forbidden while this lamp is lit. [Section 8.7]

Stepper motor *See* Stepping motor.

Stepping motor *Also* Stepper motor. An electric motor that rotates in discrete increments in the specified direction every time an appropriate current pulse is applied. Allows precise control of angular rotation and other motions by output from a digital computer. [Section 10.13]

Stereoscopic acuity The minimum angle by which the image seen by the left and right eye must differ in order to perceive a difference in distance along the line of sight. Stereoscopic acuity turns out to be several times better than would have been predicted from our ability to resolve spacings laterally. [Section 4.8; Figs. 4-20, 4-21]

Still frame *Also* Freeze frame. Recording on, or playing back from, an individual ''frame'' of videotape. Some VCRs yield an (interlaced) single video field rather than a single full frame. Other VTRs and the OMDR play back full video frames. [Section 8.4, 8.9; Figs. 8-19, 8-20]

Streaking *Also* Reverse polarity smear. In video, a picture condition in which very bright or dark objects appear to be extended horizontally beyond their normal boundaries, often in reverse contrast. May arise in the camera, in overdriven amplifiers, or by improper adjustments of the monitor CONTRAST and BRIGHTNESS controls. [Section 7.5; Fig. 7-37]

S-T vidicon *See* Silicon.

SUBCARRIER IN *See* SC IN.

Sulfide vidicon *See* Vidicon.

Surveillance camera In video, a camera that is used in commerce, industry, and elsewhere to monitor transactions, observe intrusion, etc. Generally, these are inexpensive cameras with random rather than 2 : 1 interlace, but intensifier cameras are also used for low-light surveillance applications. [Section 3.1]

Sync (synchronization) The H- and V-timing pulses that drive the video scanning system. Also the act of synchronizing. *See* Sync pulse. [Sections 6.3, 6.6, 7.3, 7.5, 12.4]

Sync generator An electronic circuit or device that produces the sync pulses for the H and V drives in video. [Sections 6.3, 6.6]

Synchronization *See* Sync.

Sync level The level of the tips of the video synchronizing signals. *See also* Sync tip.

Sync pulse *Also* Sync signal. The sharp square-wave pulses whose leading edges are used to synchronize the video scanning, blanking, etc. [Section 6.3; Figs. 6-7 to 6-14, 6-25]

Sync signal *See* Sync pulse.

Sync source A piece of video equipment that provides the sync signals that drive other (slave) video equipment in synchrony. [Sections 6.6, 12.4]

Sync stripper A video circuit or device that separates the sync signal from the picture signal in a composite video signal. After the sync pulses are reshaped, they are made available for further use. [Sections 6.6, 9.2; Fig. 6-25]

Sync tip The level of the tips of the H- and V-sync pulses. In EIA RS-170 and -330 video signals, the level of the sync tip with proper termination is specified to be −40 IEEE units, or −0.3 V. [Fig. 6-8]

Target In video image pickup tubes, the light-sensitive structure deposited on the faceplate of the tube. Scanned by the electron beam, the target generates a signal output current corresponding to the charge-density pattern formed there by the optical image. The target usually includes: the photoconductive storage surface, which is scanned by the electron beam; the transparent, conductive backplate (target electrode); and the intervening dielectric. [Sections 1.1, 7.1; Figs. 7-1, 7-2, 7-6, 7-7]

Target integration A means of increasing the sensitivity and reducing the noise of a low-light-level scene in a video camera when viewing a static scene. The beam is shut off for a predetermined number of frames and read out in the first frame after the beam is turned back on. In vidicon tubes, the target automatically integrates the signal for 16.7 msec at standard scanning rates. [Sections 7.1, 9.4]

TBC *See* Time-base corrector.

Td *See* Troland.

Tearing A condition of the video picture where (the top of) the picture is distorted horizontally. Usually caused by horizontal sync problems, especially in VTR playback. Improved by toggling the monitor to the fast AHFC setting, adjusting the H-HOLD, or by switching to another monitor with a faster AHFC. *See also* AHFC, Flagging. [Sections 6.3, 8.5; Figs. 6-27, 8-16]

Termination The insertion, or presence, of a proper load at the end of a signal cable. For closed-circuit TV, an impedance of 75 ohms provides proper termination and impedance matching for commonly used video equipment and cables. An improperly terminated end of a cable can reflect the signal back as an echo and also alter the signal voltage. Fifty ohms is commonly used to terminate RF- and cable-TV signals; 600 ohms is commonly used to terminate audio lines. *See also* Hi-Z, Impedance. [Sections 3.6, 6.4, 6.6, 7.5; Figs. 3-9, 3-10]

Three-dimensional diffraction pattern The diffraction pattern (of a point source) that appears in the three-dimensional space in and near the focal plane. For an aberration-free, diffraction-limited system, the slice of the diffraction pattern in the focal plane is the Airy disk and its surrounding diffraction rings. Above and below focus, the pattern changes periodically along the axis of the light beam so that bright and dark Airy-disk-like patterns appear alternately. The axial period of repeat is spaced twice as far apart as the radial period of repeat in the Airy disk and its diffraction rings. [Section 5.5; Fig. 5-21A]

Three-tube color camera A color video camera that uses beam splitters and three color filters and image pickup tubes to register the R, G, and B components of the image spectrum. The output can be an RGB signal or an encoded NTSC signal. [Section 7.4; Fig. 7-26]

Threshold The signal that can just barely be recognized or that gives a certain percentage of affirmative response in psychological testing. A high threshold signifies a lower sensitivity. [Sections 4.4–4.6; Figs. 4-10, 4-16]

Time-base corrector (TBC) A device that electronically corrects the improper timing relationships of the sync signals created by mechanical and electronic errors in a VTR or other video source. [Sections 6.6, 8.5, 12.3; Fig. 6-28]

Time code generator An electronic device that creates a special address or time code to be recorded on video or data recording tape, for example using the SMPTE standard code, which provides hour/minute/second/frame number. On a video editing deck, the desired video frame can be located (e.g., for accurate editing), by electrically searching through the time code recorded on the designated audio track. [Section 8.4]

Time-date generator An electronic device that is used to display the time of day and date on the video picture. Generally it is either a separate component that can be plugged in (e.g., into a VTR) or a separate piece of equipment that can be inserted E-to-E anywhere in the video train. [Section 8.5; Fig. 3-11]

Time-lapse recorder A VTR that allows the recording (on special-quality tape) of each video field or frame at intervals of lapsed time. Upon playback at standard speed on the same recorder, the original scene is speeded up inversely proportionally to the recording interval. Usually the tape cannot be played back on a non-time-lapse VTR. [Section 8.5]

Tracking In a VTR, an electronic control of the servoloop that allows precise following of the narrow, recorded tracks by the read/write head. Imprecise tracking, which leads to weaving or breakup of the picture, can be corrected on some VTRs by adjusting the TRACKING control. [Sections 8.4, 8.5; Figs. 8-13, 8-15]

Transmittance In optics, the ratio of the transmitted to the incident flux of light, usually expressed as percentage. The transmittance can vary with definition (radiant, luminous) and the method of measurement (specular, diffuse, etc.). *See* Radiometric units. [Section 5.8]

Trinicon tube A Sony image pickup tube that incorporates three sets of vertically striped color filters in front of the target. A single Trinicon tube can replace the function of the three separate tubes commonly used in color cameras. *See* Three-tube color camera. [Section 7.4]

Troland (Td) A unit of visual stimulation. To standardize on the amount of light that stimulates the retina, one takes an extended source of luminance equal to 1 cd/m^2. *Seen through a pupil area of 1 mm^2* this source produces a retinal illumination of 1 troland. [Section 4.3]

TTL (transistor, transistor logic) A computer interface component that responds with a logic 0 or logic 1 depending on the voltage of the input. In TTL, 0 to 0.8 V signals a logic 0, and 2.4 to 5 V signals a logic 1.

Tube length In light microscopy, the length of the body tube that defines the plane of focus for which the aberrations of the objective lens are minimized. The *optical tube length,* measured from the back focal plane of the objective lens to the intermediate image plane, is commonly around 160 mm for biological microscopes but can vary depending on the objective lens. The *mechanical tube length,* measured between the lower shoulder of the nose piece and the upper shoulder of the body tube, is gradually becoming standardized to 160 mm for biological microscopes. [Section 5.9; Figs. 5-13, 5-26]

Two-to-one interlace *See* Interlace. *See also* Noninterlace, Random interlace.

U format The tape recording format adopted for VCRs using 3/4-inch tape. *See also* U-matic. [Section 8.3, 8.4; Figs. 8-2, 8-3, 8-6]

UHF connector One of several types of RF connectors used on video equipment. [Section 3.1; Fig. 3-3]

Ultor The second anode in a cathode-ray tube. A high DC voltage is applied to the ultor to accelerate the electrons in the beam prior to its deflection. [Section 7.5]

Ultricon RCA's trade name for an improved type of Silicon vidicon that has very high sensitivity in the near infrared and good resolution. It is less subject to blooming of the image highlights than standard Silicons but not quite immune to blooming. [Section 7.2; Fig. 7-24; Table 7-1]

U-matic Sony's trademark for its 3/4-inch VCRs. *See also* U format. [Section 8.3]

Underscan The display of the entire video picture by shrinking the scan area on a monitor. *See also* Overscan. [Sections 6.3, 7.5; Fig. 6-17]

Usable light range The ratio of the maximum to the minimum levels of illuminance over which a video camera or camera tube can provide a usable signal. Being aided by automatic irises, gray-wedge wheels, etc., in addition to varying electrode voltages where permissible, the usable light range can be several orders of magnitude greater than the intrascene dynamic range. *See also* Intrascene dynamic range. [Sections 7.2, 7.3]

V blank *Also* Vertical blanking, Field blanking. The voltage that suppresses the beam current during vertical retrace in the camera tube and monitor; also the time interval of the V blank. V- and H-sync pulses and other signals that are not displayed on the monitor are added during this interval. *See also* Flyback. [Sections 6.2, 6.3; Figs. 6-4, 6-8, 6-14]

VCR *See* Videocassette recorder.

VDU (video display unit) A circuit that provides CRT control and graphics capabilities.

Vertical blanking *See* V blank.

Vertical frequency The number of video fields scanned every second: 60 Hz for monochrome in the 525/60, 2 : 1 interlace format; 59.4 Hz for color video in the NTSC format. [Sections 6.2, 7.4]

Vertical field magnetic recording *Also* perpendicular magnetic recording. A format of magnetic disk or tape recording that achieves a very high density of signal recording. In vertical field magnetic recording, the magnetic field being oriented perpendicular to the substrate surface (rather than parallel to the surface as in conventional magnetic recording), adjacent, oppositely polarized fields stabilize each other so that a higher density of stable magnetic fields can be recorded. [Section 8.2]

Vertical resolution The number of black-and-white horizontal lines per picture height that can be resolved vertically on the video monitor or in a video signal. Commonly, vertical resolution is given by the number of active H-scan lines × the Kell factor, but is somewhat less for random interlace video. *See also* Horizontal resolution. [Section 6.4; Fig. 6-15]

Vertical retrace At the end of each field scan in a video camera or picture tube, the returning of the scanning beam from the bottom of the picture to the beginning position of a new vertical scan. Also the time interval for the V retrace, which takes place over 12–15 H-scan intervals. Thereafter, the first several lines of the new V scan are still hidden at the top of the picture until the V blank ends. *See also* Flyback, V blank. [Section 6.2; Fig. 6-3]

Vertical scan (V scan) In raster-scanned video, the slow, downward (as seen on the monitor) deflection of the electron beam that makes each active H-scan line appear in succession below the previous line. In NTSC video, the V scans alternate 60 times a second with the starting phases 0.5 H apart, thus giving rise to the odd and even fields. *See also* Scan, H scan, Scan rate [Section 6.2; Figs. 6-3 to 6-5]

Vertical-sync pulse (V-sync pulse) The negative-going, V-timing pulse that occurs during the V-blank interval in a composite video signal. The pulse triggers the vertical-sweep generator in the monitor. In the NTSC color signal, the V-sync pulse which is 3 H long, is interrupted by six serrations. [Section 6.3; Figs. 6-6, 6-8 to 6-14]

Very-large-scale integrated circuit (VLSI) A compact, solid-state chip containing a very large number of circuit elements and circuits.

V HOLD (VERTICAL HOLD) The control on a video monitor or TV set that is used to adjust vertical sync and proper interlace of the scan lines. *See also* Interlace. [Sections 3.2, 3.5, 6.3, 7.5, 12.1; Fig. 3-1]

VHS (Video Home System) One of two incompatible VCR recording formats using 1/2-inch tape; the other being the Beta format. Used at different tape speeds including SP, LP, SLP. [Section 8.3, 8.6; Tables 8-3, 8-5, 8-6]

Video The picture portion of TV as contrasted to audio. [Sections 1.2 to 1.4]

Videocassette recorder (VCR) A VTR with the tape, supply, and take-up reels housed in a self-contained, removable cassette. When the cassette is inserted into the VCR (or for some VCRs when the FORWARD or PLAY button is pressed), the tape is automatically drawn out of the cassette and wound around over half the circumference of the recording drum. The reverse action takes place when the STOP, (FAST FORWARD, REVERSE), or EJECT button is pressed. *See also* Video tape recorder. [Sections 8.2 to 8.8; Figs. 8-6, 8-10, 8-11, 8-13, 8-14, 8-17]

Video disk recorder A device for recording and playing back video frames onto a magnetic or optical memory disk. The disk allows fast, random access and repetitive playback of the frames without wear. *See also* Optical memory disk recorder, Video tape recorder. [Section 8.9; Fig. 8-18]

Video display unit *See* VDU.

Video gain The control on a video camera, monitor, processor, SEG, TBC, etc. that allows the user to alter the amplitude of the video picture signal. Results in a change in picture contrast and brightness. [Section 7.3]

Video guard-band noise *See* Noise bar.

Video head The video read/write head used in a VTR. [Sections 8.2; Figs. 8-1 to 8-4, 8-8, 8-10, 8-11, 8-17]

VIDEO IN The input connector for the video signal. Contrasts with VIDEO OUT, except in a video monitor where the two are connected E-to-E and can be used interchangeably. [Sections 3.2, 12.3; Figs. 3-2, 3-5, 3-9, 3-10, 9-10, 12-5 to 12-8, 12-13]

Video microscopy Microscopy that takes advantage of video as an imaging, image processing, analyzing, or controlling device. [Sections 1.2–1.4, 3.1, 10.1, Chapter 11]

Video mixer A device for switching between several video input and output devices as well as for mixing the signals in desired proportion or according to static or dynamic key features. *See also* Special effects generator. [Section 12.4]

VIDEO OUT The connector on a piece of video equipment that provides an output of a composite video signal. When properly terminated, the full, EIA RS-170, RS-330, or NTSC signal should measure 1 Vpp. [Sections 3.2, 12.3; Figs. 3-2, 3-4, 3-9, 3-10, 7-38, 9-10, 12-5 to 12-8, 12-13]

Video tape recorder (VTR) A device that records and stores video and audio signals sequentially onto magnetic tape. Most VTRs also allow playback of the tape. The VTR may be a VCR that uses a video cassette, or may be of the reel-to-reel type. *See also* Video disk recorder. [Sections 8.1–8.7, 12.3; Figs. 8-6, 8-10 to 8-14; Tables 8-1 to 8-6]

Video waveform The display of the (composite) video signal onto a cathode-ray oscilloscope triggered by the V- or H-sync pulse. The amplitudes and waveforms, as well as precision of timing, of the signal components are made visible. [Sections 6.3, 12.3; Figs. 6-12, 6-23, 7-29, 7-39]

Vidicon Refers to the standard, or sulfide, vidicon tube rather than to a family of vidicon-type camera tubes. Using antimony trisulfide as the photoconductive element in the target, the Vidicon has a relatively low sensitivity and high dark current, but high resolution, AGC compatibility, and low gamma. The low gamma signifies nonlinear response to light but also results in a high dynamic range. [Sections 7.1, 7.2; Tables 7-1, 7-2]

Vidicons A family of video camera tubes in which an electrical charge-density pattern is formed by photoconduction in the target and stored there. The signal current is generated as the electron beam scans the target in a raster. Includes the standard (sulfide) Vidicon, as well as the Newvicon, Chalnicon, Plumbicon, etc. [Sections 7.1, 7.2; Figs. 7-1, 7-2; Tables 7-1, 7-2]

Vignetting An unintentional, shaded loss of the edges of an image or picture by an optical component clipping the peripheral beams can lead to loss of contrast in video. [Section 3.7]

Visual acuity The ability to detect fine details or small distances with the eye. Visual acuity can vary substantially depending on the definition used and method of measurement chosen. Under favorable conditions, the resolution of the human eye, or ability to distinguish the twoness of adjoining lines, is about 1 minute of arc while the threshold detectability, or the detection of misaligned steps in a line (which is also a form of visual acuity), can be as low as a fraction of a second of arc. [Section 4.4]

VLSI *See* Very-large-scale integrated circuit.

Volatile memory A computer component that retains the contents of its memory only if equipment power is kept on. In contrast, a nonvolatile memory device, such as bubble memory, disk, or tape, retains its memory contents even with the power turned off.

Vpp Volts peak-to-peak. *See* Peak-to-peak.

V scan *See* Vertical scan.

V-sync pulse *See* Vertical-sync pulse.

VTR *See* Video tape recorder.

VU meter A meter that measures audio levels in volume units.

Waveform monitor A cathode-ray oscilloscope that can conveniently display the video signal. *See also* Video waveform. [Sections 6.3, 12.3]

White balance A color video camera circuit that allows simple adjustment of the RGB signals to produce a correctly balanced white. It is activated by pointing the camera to a reference white surface and pressing the WHITE BALANCE button. [Section 7.4]

White compression/clipping Amplitude compression/clipping of the signals in the brighter regions of the picture. Results in differential gain that expands the dark and middle regions of the gray scale. [Section 7.3]

White level *Also* reference white level. The IEEE scale, or voltage level, of the video picture signal that corresponds to a specified maximum limit for white peaks. [Section 6.3; Fig. 6-8]

White peak *Also* Maximum white. The highest amplitude or peak in white reached by the video picture signal; measured in IEEE scale or voltage. [Section 6.3; Fig. 6-8]

White punch Reversing the polarity of video signals at very high levels of white. Very bright areas are displayed increasingly darker inside a white rim. [Section 7.3]

Wipe The use of a special effects generator to replace the present scene by smoothly expanding the area covered by a new scene. [Section 12.4]

Write protect Devices for preventing a video or audio cassette tape or computer disk from being written over accidentally. [Sections 8.7, 12.3; Fig. 8-17]

Xtal Abbreviation for crystal, as in xtal oscillator. [Section 6.6]

Y Symbol for the luminance portion of the picture signal in NTSC video. [Section 7.4]

Yoke The assembly of electromagnetic coils that are placed over the camera tube or the neck of a picture tube or a cathode-ray tube of the magnetic focusing type. The focusing coil in the yoke controls the convergence of the electron beam. The H- and V-deflection coils deflect the electron beam so that it sweeps the target or phosphor in a standard video raster. [Sections 7.1, 7.5; Figs. 7-1, 7-31]

3

Getting Started

Some Practical Tips

This chapter gives some practical tips on getting started with video microscopy, and surveys the hookup and mechanics needed to operate the video equipment. Some problems that are commonly encountered in video microscopy will also be examined.

A minimal knowledge and experience with the operation of video equipment will be assumed at this point, and detailed explanations of the reasons for the recommended procedures are deferred to later chapters.

I will assume that both a compound microscope with capabilities for **Koehler illumination** (Chapter 5) and a video camera with its lens, monitor, and cable are available.

3.1. THE MINIMUM ESSENTIAL COMPONENTS

Stripped to its bare essentials, a video microscope consists of a video camera mounted to a microscope through a lighttight coupler, a monitor for displaying the image, and a shielded cable connecting the two (Fig. 3-1).*

As shown schematically in Fig. 3-2, a real image of the specimen is projected by the microscope onto the camera tube in the video camera. The camera converts the two-dimensional optical image into a linear train of high-frequency electrical impulses, the *video signal* (Chapter 6). The video signal travels through a shielded **coaxial cable** (Fig. 3-3) to the monitor. Finally, the monitor converts the linear signal back to a two-dimensional optical image, the *picture*.

The choice of *video camera* depends on the type of microscope to which the video is to be applied. Some cameras are highly sensitive but produce noisy images, while others are less sensitive but can provide clean images with a good **gray scale** (Chapter 7).

Except in low-light-level applications, the video camera of choice for (monochrome) video microscopy would probably be an **instrumentation camera** with a **1-inch Newvicon** or **Chalnicon** tube (Allen *et al.*, 1981a,b; Inoué, 1981b). The characteristics of various camera and image tubes and video cameras are examined in Chapter 7. Their suitability for different modes of microscopy is discussed in Chapter 11.

If the basic video equipment is not available, the following exercises can be carried out with: a simple **surveillance camera** with a lens of about 18- to 25-mm focal length (e.g., Panasonic Model WV 1100A, or equivalent); an inexpensive 9-inch **black and white** monitor (e.g., Sanyo Model VM 4209, or equivalent); and 6 to 10 feet of **75-ohm coaxial cable** equipped with **BNC** or BNC and

*In order to reduce vibration, the camera should be supported by its own stand (Figs. 5-28, 5-34A) rather than by the microscope. If the microscope must be used to support the camera, as shown here, it should be as small and lightweight as possible.

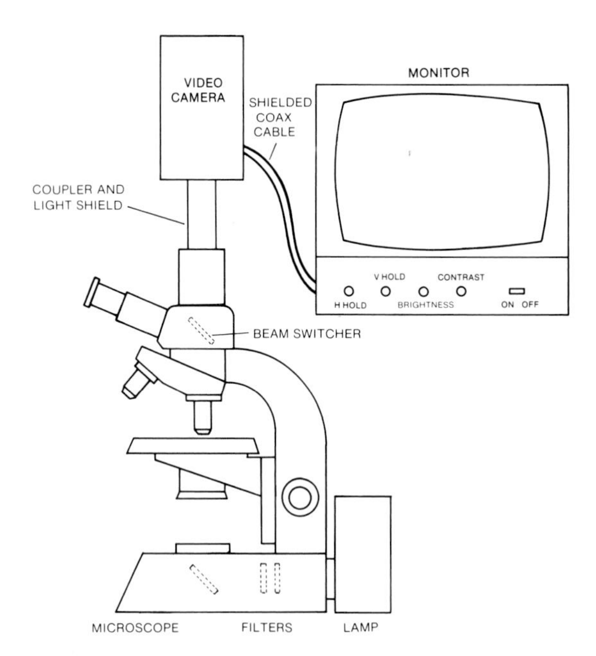

FIGURE 3-1. Schematics of the minimum setup needed for video microscopy. A video camera is linked by a coupling tube to the microscope and connected by a shielded coaxial cable to the monitor. More stable, recommended arrangements are shown in Figs. 5-28 and 5-34A.

UHF connectors. It is possible to purchase this minimum set of equipment for under $1000, 1985 U.S. price. The price of instrumentation cameras ranges from $1500 upwards.

While I do not recommend such a surveillance camera* for use as the primary camera in video microscopy, they have been used effectively for some applications in microscopy (e.g., see Schatten, 1981). A simple camera with its lens is also useful for inserting data, with the aid of a **special effects generator,** or **video mixer,** onto a video scene taken through the microscope with the main camera (Chapter 12).

For microscopy, the camera lens is removed from the video camera. The lens is, however, used for adjusting the internal focus of the camera off the microscope, as well as the controls on the monitor as described later.

The monitor of the type mentioned above is used routinely in video microscopy. In fact, such inexpensive monitors tend to respond better to time-lapsed and other video microscope signals than the more expensive ones made for instrumentation and studio applications (Chapters 6–8).

The 75-ohm-**impedance** coaxial cable is needed to conduct the high-frequency video signal without loss or **distortion.** It also prevents radiation of the **radio frequency** video signal, and minimizes pickup of unwanted electromagnetic interference.

The connectors on the camera, monitor, etc. may be of the BNC, UHF, **F**, or one of the multiple-pin types depending on the equipment (Fig. 3-3). Adapters are available for coupling between BNC, UHF, and F male and female connectors.

3.2. INITIAL ADJUSTMENT AND HOOKUP OF THE VIDEO

Before attaching the video system to the microscope, a few calibrations are in order.

For these calibrations, attach the camera lens to the video camera. With the power to the camera turned off and the lens capped, hook up the VIDEO OUT terminal of the camera (Fig. 3-4) to the VIDEO IN terminal of the monitor (Fig. 3-5). The hookup is made with a 75-ohm coaxial cable.

*Surveillance cameras tend to have **random,** rather than **2 : 1, interlace,** resulting in somewhat lower **resolution** and imprecise **syncing** capability than with instrumentation cameras (Section 6.4).

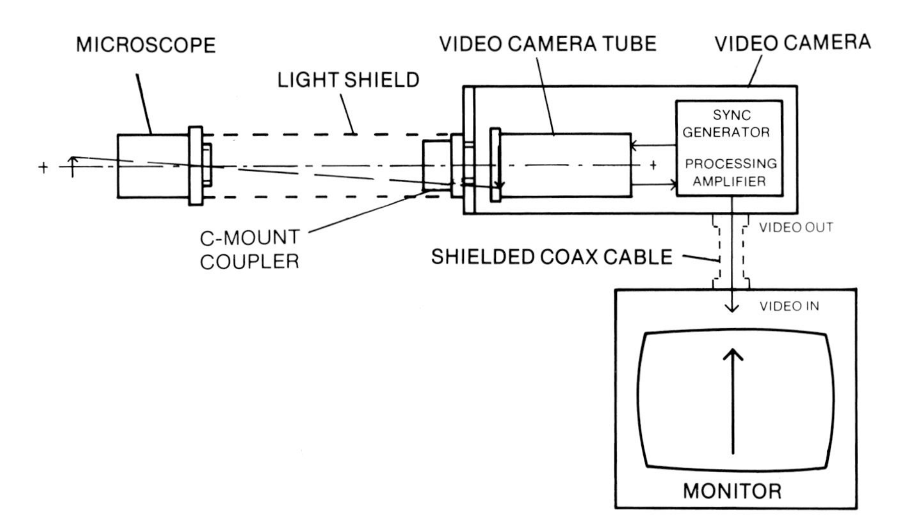

FIGURE 3-2. Image and signal schematics. The microscope image is projected onto the video camera tube where it is converted to an electrical signal. The signal is reconverted into a picture, which is displayed on the monitor.

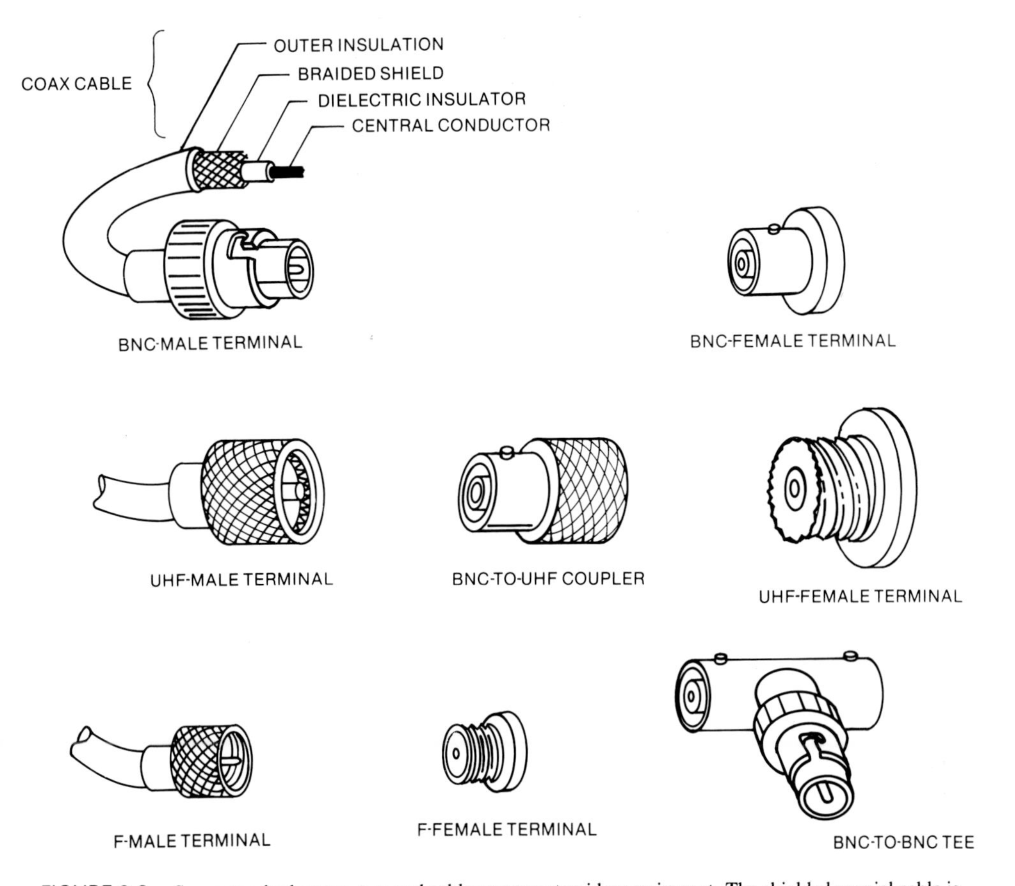

FIGURE 3-3. Some standard connectors and cable common to video equipment. The shielded coaxial cable is required to prevent distortion or loss of the signal and to minimize interference from external sources of electromagnetic radiation. Several types of connectors are used in video, with the BNC and UHF types being common for professional equipment.

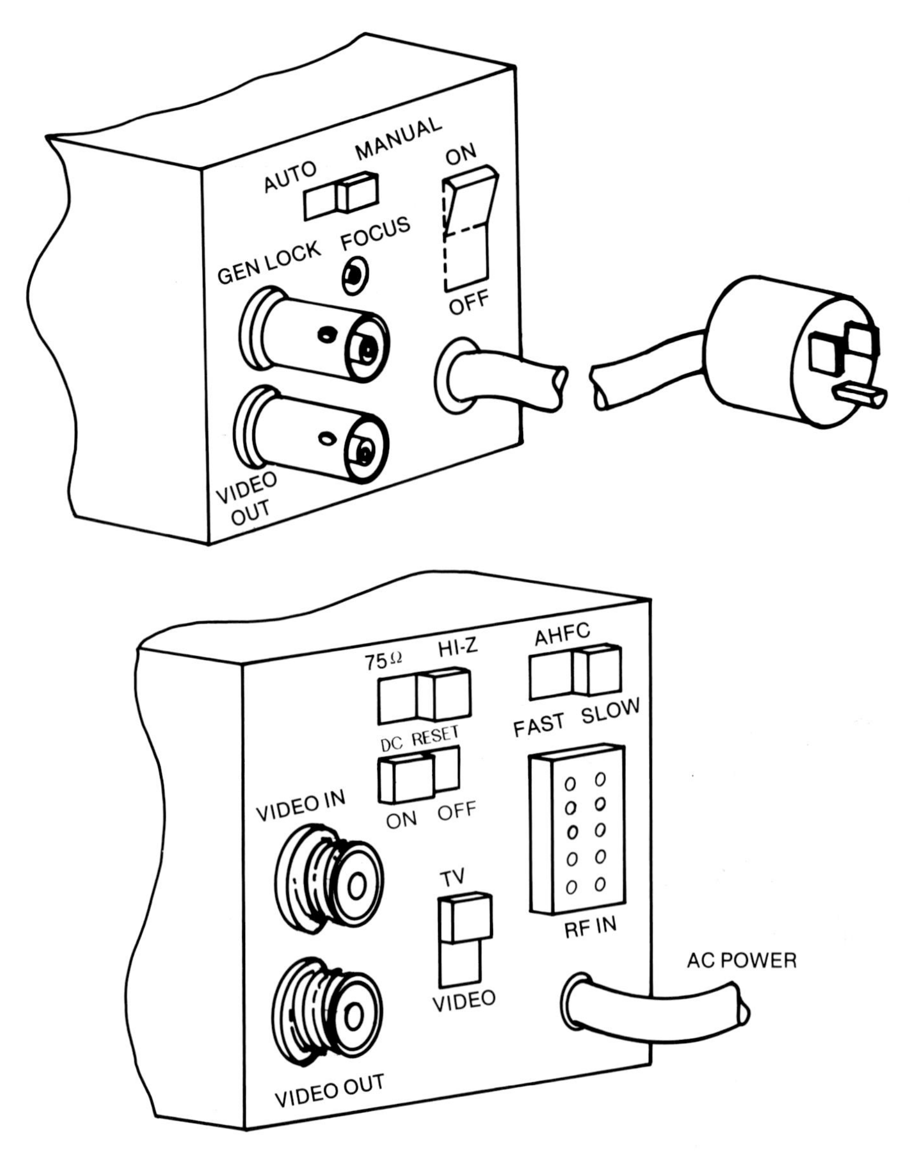

FIGURES 3-4 (top) and 3-5 (bottom). Common rear-view appearance of video cameras (top) and monitors (bottom) showing switches and connections. See text for recommended settings.

On the back panel of the monitor, set the **DC RESTORATION** or **BLACK RESET** switch to OFF and the **75 OHM terminator** to the ON, or 75 OHM, position. If the camera is equipped with AUTO–MANUAL switches, set them to the AUTO position.

With the lens cap still in place, and the lens iris closed all the way down (to the largest **f stop,** turn on the monitor and camera power. Turn up the **CONTRAST** control on the front panel of the monitor clockwise.

After the camera and monitor have both warmed up (in about 1 min), adjust the BRIGHTNESS control on the monitor so that the monitor brightness becomes a uniform, dark gray. This provides the *home setting for the BRIGHTNESS control* (Fig. 3-1). If the picture **rolls** or a dark band **scrolls** over the picture tube, adjust the **V HOLD** and **H HOLD** controls so that the monitor shows a stable, uniformly dark gray field.

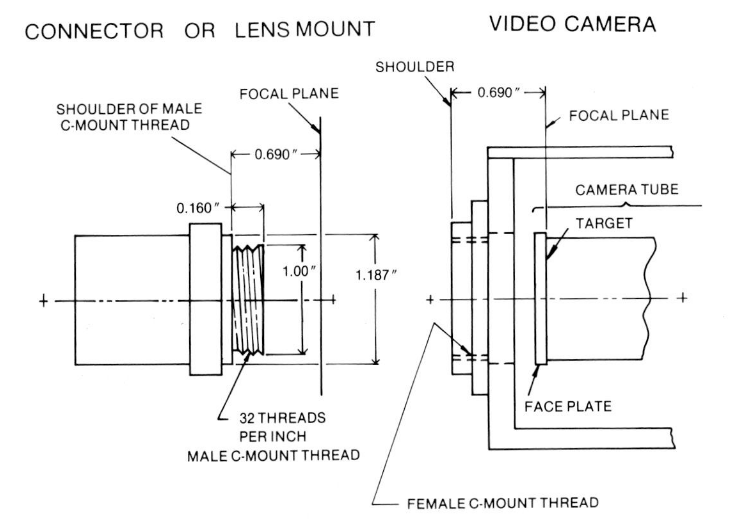

FIGURE 3-6. Detail of C-mount connections to the video camera. Note that the image is projected or focused on the target of the camera tube (comparable to the film plane in photography), which is 0.690 inch from the shoulder of the C-mount.

Remove the lens cap and point the video camera to a well-lit room scene. Open the iris on the camera (toward a lower f stop) until an image appears on the monitor. Focus the camera lens and adjust the camera iris until a sharp bright image is seen on the monitor. If an iris is not present on the camera lens, adjust the room lighting or place an appropriate **neutral-density filter** in front of the lens to control the light entering the camera.

With the monitor BRIGHTNESS control still in its home position, adjust the CONTRAST control on the monitor counterclockwise until the details show clearly in the picture, including the highlight, or bright regions of the image. This gives the *home setting for the CONTRAST CONTROL.**

Mark the home positions of the BRIGHTNESS and CONTRAST control knobs, for example with a short vertical line or a dot on the upper half of the knob faces or with adhesive labels placed on the face of the knobs and similarly marked.

Next, adjust the internal focus of the video camera. The CONTRAST and BRIGHTNESS controls on the monitor need not be in their home positions and the DC RESTORATION or BLACK RESET may be set to ON or OFF for these adjustments.

The camera iris should be fully opened to minimize the **depth of field.** If necessary, dim the room light or place a neutral-density filter in front of the lens to prevent the image from **blooming.** The image is said to bloom if the **highlights** expand into the darker surrounding areas.

Place a high-contrast object with fine detail at a known distance from the camera, i.e., from the camera focal plane, which is located about two-thirds of an inch (0.690 inch, to be exact) behind the face of the C-lens mount (Fig. 3-6). Set the focusing scale on the camera lens to the distance to the object. If the image seen on the monitor is not in focus, adjust the internal focus of the camera with a small screwdriver through the FOCUS or LENS opening on the back panel of the camera (Fig. 3-4).

Thus far, the home positions for the monitor CONTRAST and BRIGHTNESS controls have been determined, and the location of the camera tube has been adjusted. Turn off the power to the video camera and put a dust cap in place on the lens. In the next step, the camera is to be coupled to the microscope.

*Strictly speaking, these adjustments should be made by using electronically generated signals rather than signals from the camera. Alternatively, the voltage of the video signal should be monitored with a high-frequency oscilloscope (Chapter 6). However, the adjustments described here using the camera signal for a standard scene provide a good starting point.

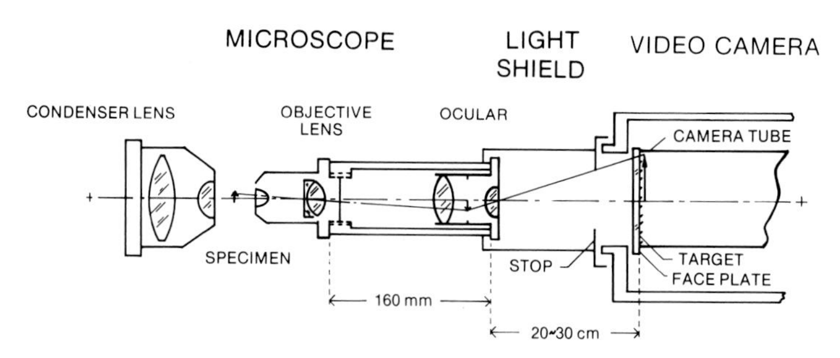

FIGURE 3-7. Video microscopy without a camera lens. In this scheme, the microscope image is focused on the video camera target by the (projection) ocular.

3.3. COUPLING THE VIDEO CAMERA AND THE MICROSCOPE

In coupling the video camera and the microscope, the microscope image can be projected by the ocular directly onto the *target* in the video camera tube (Fig. 3-7). Alternatively, the image can be focused by an infinity-focused lens as in conventional photomicrographic equipment (Fig. 3-8; see also Chapter 5).*

Most video cameras are equipped with a standard female **C-mount** thread for attaching a 16-mm cine lens (Fig. 3-6). The video camera without its lens could thus be mounted on a photomicrographic coupler equipped with a male C-mount thread. On the other hand, the beam splitter and focusing telescope on the photomicrographic coupler are generally not needed in video since the monitor itself displays the image. Commercially available coupling tubes may use either the scheme shown in Fig. 3-7 or that in Fig. 3-8.

For trial purposes, simply mount the video camera *without* its lens on a tripod or other suitable support. Placed the camera 20–30 cm from the ocular on axis with the microscope. Shield the space between the camera and the microscope with a light baffle made of rolled-up black paper, or a tube of metal, cardboard, or opaque plastic. Alternatively, place a right-angle prism on the ocular, and place the camera horizontally. An arrangement similar to the one in Fig. 5-34A would be even better.

Selection of the microscope optics including the ocular, the **magnification** for the video image, and the matching of the video and microscope image resolutions are complex but important matters that will be discussed in detail in Chapters 5, 11, and 12. For initial setup, practically any research microscope equipped with a medium-power (say ×20 to ×40) objective, a trinocular tube with beam switcher (Fig. 3-1), and a ×5 to ×10 ocular could be used. The specimen could be a stained section or a stage micrometer.

3.4. WAIT! READ BEFORE TURNING THE MICROSCOPE LAMP ON

Before turning the microscope lamp on, make sure the following adjustments have been made whenever the video camera is coupled to the microscope:

A. The beam-switching prism or mirror is positioned to send the light beam to the viewing ocular and not the camera.
B. In polarized-light, interference, or differential interference contrast (DIC) microscopy, the analyzer is properly crossed with the polarizer.
C. In DIC, phase contrast, or darkfield, or on a microscope equipped with a DIC condenser, the objective lens and the condenser (turret setting) match each other.

*Video camera lenses intended for general application usually are not suitable for microscopy because they produce **vignetting** as a result of their pupil location. In recent years, several microscope manufacturers have offered lenses specially designed for video microscopy.

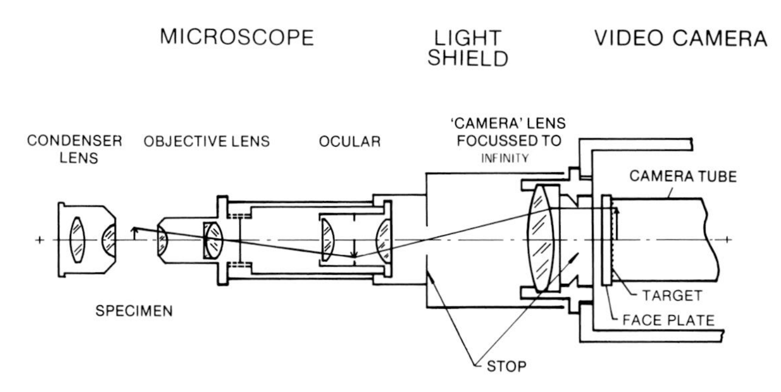

FIGURE 3-8. Video microscopy using a camera lens. The microscope is focused to infinity, so that the infinity-focused camera lens focuses the image on the target. When a camera lens is used, it should be one specially designed or selected for video microscopy, in order to avoid vignetting.

D. In fluorescence microscopy, the correct **excitation** and **barrier filters** are inserted in place.
E. For incandescent illuminators, the intensity regulator is turned to the lowest setting.
F. On the video camera, the AUTO–MANUAL control switches are all placed in the AUTO position.
G. All electronic equipment located in the vicinity of arc lamps, lasers, igniters and their cables, are turned off. Check all video cameras, electronic volt- and ohmmeters, oscilloscopes, **time–date** generators, digital clocks, digital counters, video mixers, processors, recorders, monitors, and computers.

Precautions A through E help *prevent* the video camera from being exposed to excessive light and possibly damaged.

Precaution F helps to *protect* the video camera and image tubes, especially in high-sensitivity cameras such as the SIT or intensified camera, from being damaged by excessive illumination. The automatic electronic circuitry in the camera can respond very much faster than we can, should the camera accidentally be blasted by light. The circuits automatically cut down the camera **gain**, as well as the high-voltage power supply, within a small fraction of a second. In contrast, manual fumbling with a shutter, beam switcher, filter holder, etc. might take several seconds after realization that the camera has been overexposed.

Precaution G protects the sensitive electronic components, especially **field-effect transistors, integrated circuit** chips, counter logic, etc., from damage by voltage surges induced by *high-intensity electromagnetic fields.* When a mercury or xenon arc lamp or a high-powered laser is ignited, strong radio frequency fields are momentarily generated by the power supply, the lamp or laser tube, and the cable connecting the igniter and lamp. The fields induce voltage **spikes** in the electronic circuits, even those that are not connected to the lamp circuit. The surge from the igniter can also propagate through the AC power line.

These high-frequency surges can enter the sensitive electronic devices of the type listed in G, and cause costly, and sometimes difficult to localize, internal damage to the equipment *in an instant.*

Keep these power supplies, lamps, and cables at least several feet away from the video and other electronic devices, properly grounded and well shielded whenever possible, and supplied through separate AC power lines.

To summarize:

- Protect the video camera and image tubes from intense illumination.
- Check the optics to ensure that excessive light will not enter the camera.
- Switch the camera controls to AUTO.
- Take a look at the image brightness before diverting the beam to the camera.
- Protect the sensitive electronic and video equipment from damage by electromagnetic surges.

- Turn off all electronic and video equipment in the vicinity, before turning on mercury or xenon arc lamps.

3.5. ADJUSTMENTS OF THE MONITOR

The video camera is now coupled to the microscope, and the microscope adjusted through the eyepieces to provide a focused and adequately illuminated image. A picture should appear on the monitor once the video equipment is switched on, the camera and monitor have warmed up (in about 1 min), and the beam switcher is positioned to send the beam to the camera.

If the picture is distorted or unstable and flips or rolls, adjust the H HOLD and V HOLD controls on the front panel of the monitor to stabilize the image (Fig. 3-1).

If the picture cannot be stabilized, make certain that: the cable from the VIDEO OUT terminal of the camera is securely and correctly connected to the VIDEO IN terminal of the monitor; the INPUT SELECT switch on the monitor is on VIDEO and not on TV or RF (if the monitor can accept either *TV or video input;* Fig. 3-5); the SYNC SELECT switch is on INTERNAL and not EXTERNAL (if the monitor is equipped with *external sync* capability); both the central conductor and ground connection of the coaxial cable are *conducting properly* and *not shorted.*

Check the coaxial cable both visually and with an ohmmeter. The resistance of the cable through the (insulated) braided shield (Fig. 3-3) and outer metal body of the connectors, which together form the ground connection, should be less than 0.1 ohm. The resistance of the central conductor measured between the two pins at the ends of the cable must also be less than 0.1 ohm. The DC resistance between the central conductor and ground connection should be infinite, and none of these three measurements should change when the cable and connectors are wiggled.

If the microscope had been set up properly for Koehler illumination (Section 3.7) and focused by eye before the light beam was switched to the video camera, the picture on the monitor should have come into focus with just a slight turn of the microscope fine adjustment. The internal focus of the video camera should *not* be adjusted at this point, since the microscope image has a large **depth of focus** (amounting to several millimeters). Adjust the internal focus of the camera *before* coupling the camera to the microscope as described earlier.

Once the image is focused, adjust the illumination of the microscope so that the image on the monitor acquires a reasonable contrast. See Chapters 5 and 12 for details on this important adjustment.

Actually, it is possible to adjust the contrast and brightness of the picture over a wide range with the *monitor controls,* especially in a darkened room. These adjustments can bring out details of the image that would otherwise be difficult to see. However, we should remember that when the monitor controls are offset from their home positions in this fashion, the video signal that provides the image details may not be optimal for recording, or for playback on another monitor (see Chapters 8 and 12).

If the video signal is to be recorded, adjust the microscope and camera rather than the monitor so that a good image appears on the monitor with the monitor CONTRAST and BRIGHTNESS controls *set to their home positions.*

The DC RESTORATION or BLACK RESET switch (Fig. 3-5) restores the DC level of the AC video signal arriving at the monitor. It is normally left on for the average scene. It can be turned off when one desires to have substantial control over the brightness of the background seen on the monitor. However, when that option is exercised, the same precaution must be taken as with the nonstandard adjustments of the CONTRAST and BRIGHTNESS controls.

When a discrete patch of the scene is very bright, the picture of that patch on the monitor may be flanked by dark horizontal bars. Also, the whole scene may **jitter** up and down, and the right margin of the scene may be distorted into a wave. These are symptoms of an *overdriven monitor* (see Chapters 6, 7). Reducing the contrast and brightness of the monitor can often cure these problems. Otherwise the image on the camera may have to be dimmed.

Sometimes the scene on a monitor displaying a standard camera signal can be too bright or too dark when the monitor controls are set to their standard positions. In such cases, the DC RESET may not be on or the cable may not be properly terminated at the monitor.

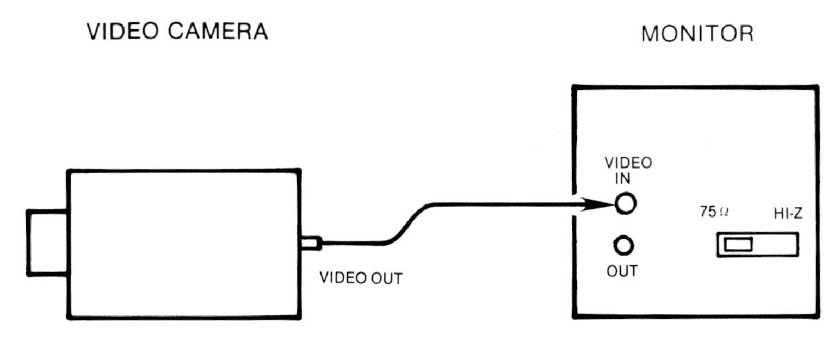

FIGURE 3-9. Termination scheme A. In order for a proper video image to be displayed, the last component receiving the video signal must be properly terminated, i.e., a 75-ohm load must be provided. In the simplest case, shown here, the monitor switch is set to the 75-OHM position.

3.6. TERMINATION

Termination, or proper loading of the electrical circuit, is especially important in video. Without proper termination, the signal can acquire inappropriate amplitudes, become distorted, and the open end of a coaxial cable can *reflect* the signal and introduce **echoes** or **ghosts** into the picture. *The last piece of equipment to receive the video signal must always be terminated to provide a 75-ohm impedance load.*

Thus, if a single monitor is connected to a camera or tape player, provide the 75-ohm termination to the monitor by setting the 75 OHM–**HI-Z** control to the 75 OHM position (Figs. 3-5, 3-9). If there are two or more monitors to be fed from a camera, *only the last monitor should be terminated.* The other monitors should be *unterminated* and connected E-to-E as shown schematically in Fig. 3-10. Do not connect terminated circuits in parallel with a T- or Y-adapter.

In video, it is quite simple to hook up a number of components such as extra monitors, video recorders, time–date generators, image processors, etc. (Sections 6.6, 9.5, 12.3, 12.4). Most of this equipment is reasonably compact, and can be connected together with standard coaxial cables. Video equipment and coaxial cables are generally equipped with the type of terminals shown in Fig. 3-3, commonly of the BNC or UHF types.

In video tape recorders and video processors, the signal is essentially generated anew. In such

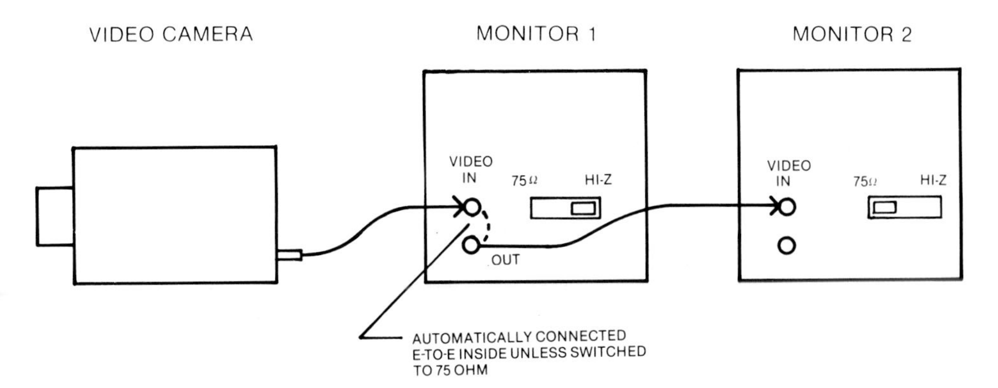

FIGURE 3-10. Termination scheme B. If a second monitor (or certain other components) is added to the circuit after the first monitor, only the last device should be terminated. Care must be taken that the load is *not* terminated before the last device. Rather, the selector switch of the first monitor must be in the HI-Z position so that the signal is passed along "E-to-E" through that monitor. Compare with Fig. 3-11.

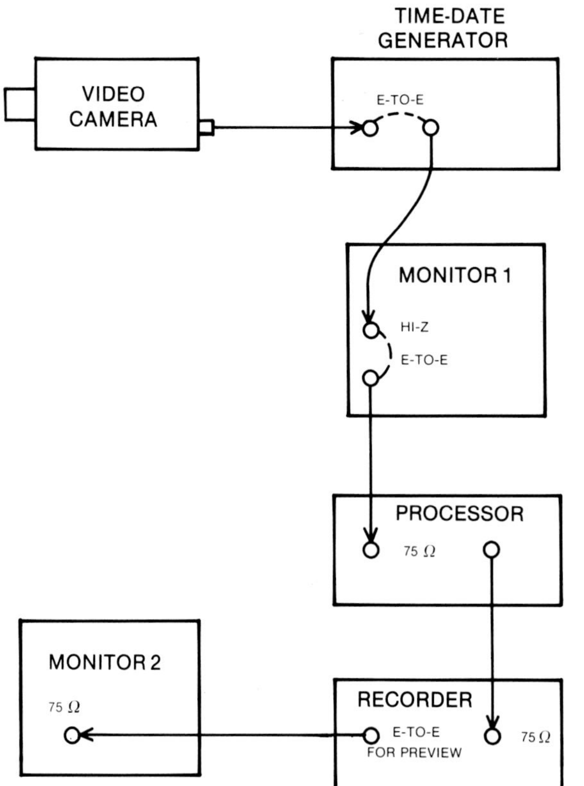

FIGURE 3-11. Termination scheme C. Exceptions to the "termination rule" are video tape recorders and processors. Since these components generate new video signals, the incoming circuit must be terminated in them even when they are not the last device in the chain.

pieces of equipment the inputs must be terminated with a 75-ohm impedance.* Their outputs are again treated as though they had come from a camera or similar video signal-generating device (Fig. 3-11).

On the other hand, monitors and time–date generators do not act as receivers of the video signal in the same way. In the unterminated HI-Z mode (Fig. 3-11), these pieces of equipment can simply tap in, or add, to the signal without changing the loading of the circuit. In fact, the VIDEO IN and VIDEO OUT terminals in pieces of equipment are wired in parallel inside the case, so that one can use either terminal for the input and the output. Of course, the same is not true for video recorders and processors.

3.7. IMPORTANCE OF ALIGNING THE VIDEO CAMERA WITH THE MICROSCOPE

For video imaging, the microscope should be carefully adjusted for Koehler illumination. This form of illumination not only provides the brightest, uniformly illuminated field achievable, but also allows one to separately regulate the field and **aperture** of the microscope (Shillaber, 1944). Independent regulation of the field and aperture is essential for optimizing contrast and resolution (Chapter 5).

Even with Koehler illumination, the video monitor may not be fully covered by the microscope image. Part of the image may be cut off or *masked,* or the image may be *vignetted,* i.e., shaded by some out-of-focus obstruction.

Usually, masking and vignetting are caused by one or more misadjustments of the microscope. Such defects include:

*Most such pieces of equipment come with properly terminated input connections.

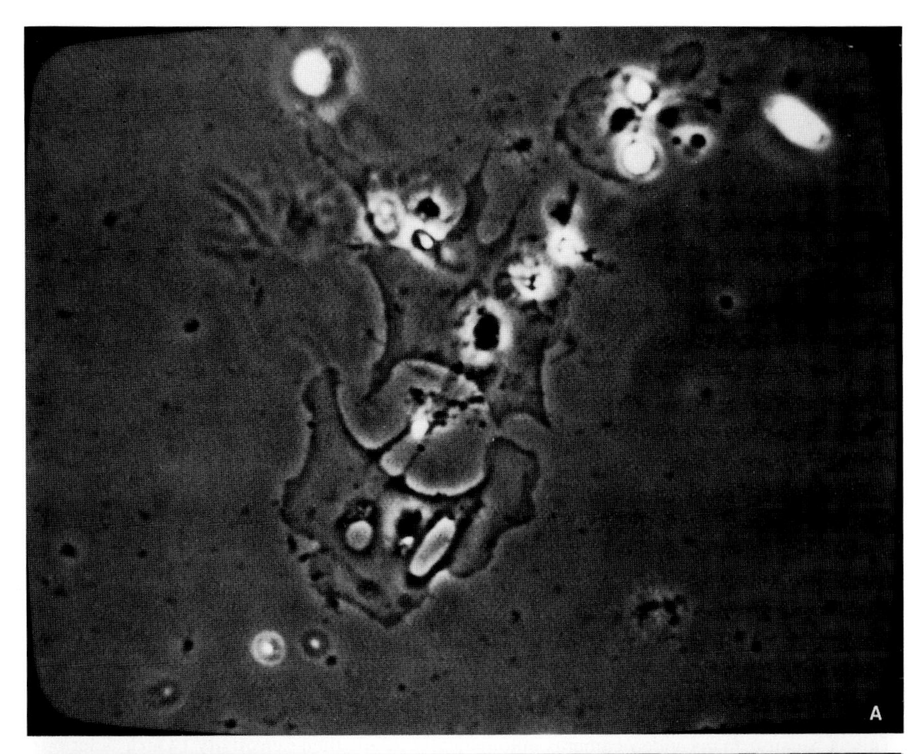

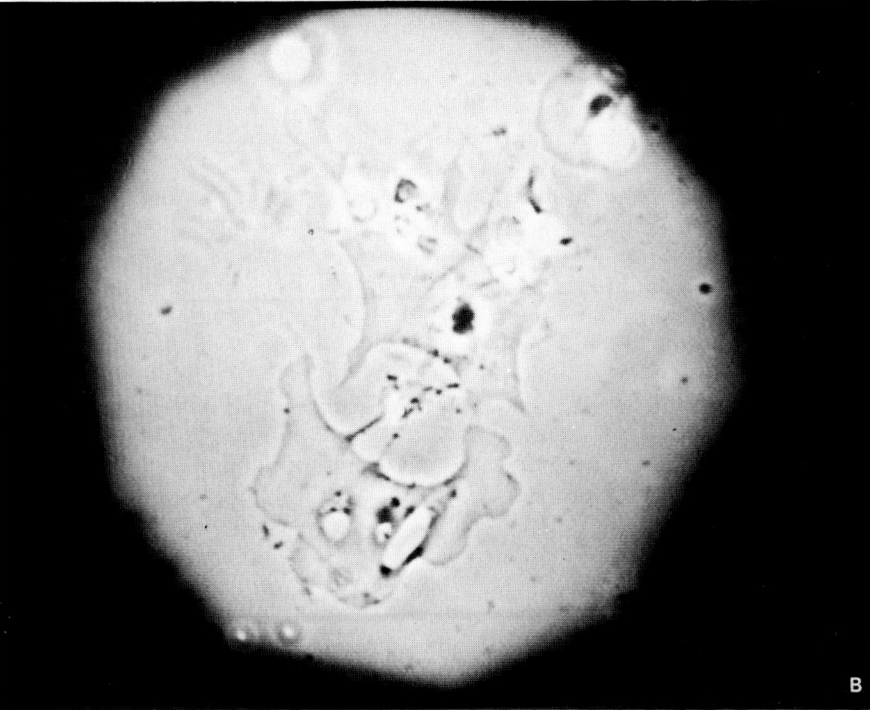

FIGURE 3-12. Effect of automatic camera controls on contrast. (A) A properly adjusted video scene of a sea urchin hemocyte shows a high-contrast image. (B) The contrast drops when the field diaphragm is closed too far. The automatic-black-level control of the camera took the image of the field diaphragm to be the darkest part of the picture and adjusted the contrast accordingly.

a. The microscope image being too small to cover the camera tube's active area, perhaps because the **field diaphragm** (or condenser iris if not adjusted for Koehler illumination) is closed too far (see Fig. 3-12B).
b. The illuminator and/or condenser being off-center.
c. Improper focus of the lamp or condenser.
d. Poor quality of lamp collector and/or condenser lenses.

These conditions can be avoided by using high-quality optics and by adhering strictly to the principles of Koehler illumination, discussed in detail in Chapter 5.

Vignetting can also be caused by the microscope if:

e. Beam switchers or filter holders are too small or not set in their "home" positions.

The video camera or its connection to the microscope can also contribute to vignetting if:

f. The video camera is not on axis with the microscope.
g. The video camera lens is limiting the field. Unless the lens is specially designed for microscopy, it should be removed when the camera and microscope are coupled.
h. The electron beam is misaligned within the video camera.
i. Some material of the light baffle is protruding into the light path.

An image that is partially masked or vignetted not only is esthetically unpleasant and hides information in the shaded area, but can also cause trouble as explained below.

For visual observation and photomicrography, contrast is improved by closing down the field diaphragm and restricting the illuminated area to the central region of the field. In video, the dark shadow produced by the field diaphragm, or other darkness introduced by vignetting, or by dirt particles, can reduce the contrast (Fig. 3-12).

This is because many video cameras automatically adjust their sensitivity (which determines the contrast) to span the brightest and darkest parts of the image (Chapter 7). The darkest part of the image is commonly not the darkest part of the scene of interest, but rather the masked or vignetted areas. The video camera cannot discriminate between a naturally dark part of the scene and a masked or vignetted region, and adjusts its black level to the darkest part wherever it may be. Thus, the automatic black adjusting feature, which is meant to improve the contrast, can work against us if the scene is masked or vignetted by a shadow darker than the scene.

Vignetting can affect the quality of the video microscope image, as well as its contrast. This is especially true when the vignetting is asymmetric. Asymmetric vignetting can arise from (b), (e), (f), and (i) in the list above. The axis of the image-forming beam becomes tilted, which is equivalent to illuminating the specimen obliquely (Chapter 5). Oblique illumination produces asymmetric diffraction and shadowing of image detail. Introduced inadvertently, these asymmetries can lead to erroneous interpretation of the microscope image.

Problem (h) is not encountered too often. If suspected, the camera can be checked out by screwing on a standard camera lens and imaging a test target that is placed along the axis of the video camera (Section 3.2). Many of the problems in the list can be eliminated by careful alignment of the microscope and judicious adherence to the principles of Koehler illumination.

3.8. OPTICAL NOISE, HOT SPOTS

A major advantage of video is its ability to enhance image contrast. However, when the image contrast is enhanced, certain image defects, also become more intrusive. I shall refer to such defects collectively as **optical noise.** Optical noise can take the form of hot spots, discrete dark specks, mottles, regular patterns not present in the specimen, or unevenness of field brightness.

Hot spots generally appear in the middle of the video image, sometimes surrounded by radiating bright rays. The hot spot becomes especially noticeable as the image contrast is raised by video (Fig. 3-13).

A hot spot is not simply an irritating blemish. Like masked areas, the bright, hot spot can also lower image contrast. The effect is especially damaging to dark, low-contrast scenes, since a video camera with automatic contrast control would interpret the hot spot as the highlight of the image.

Sometimes the hot spots arise from multiple reflections at the lens and other glass surfaces, especially those that are not treated with antireflection coating (Dempster, 1944). The internal lens elements of some *inexpensive* zoom oculars often are uncoated, which precludes their use in video microscopy.

More commonly, a hot spot is introduced into the middle of the field by reflections from the

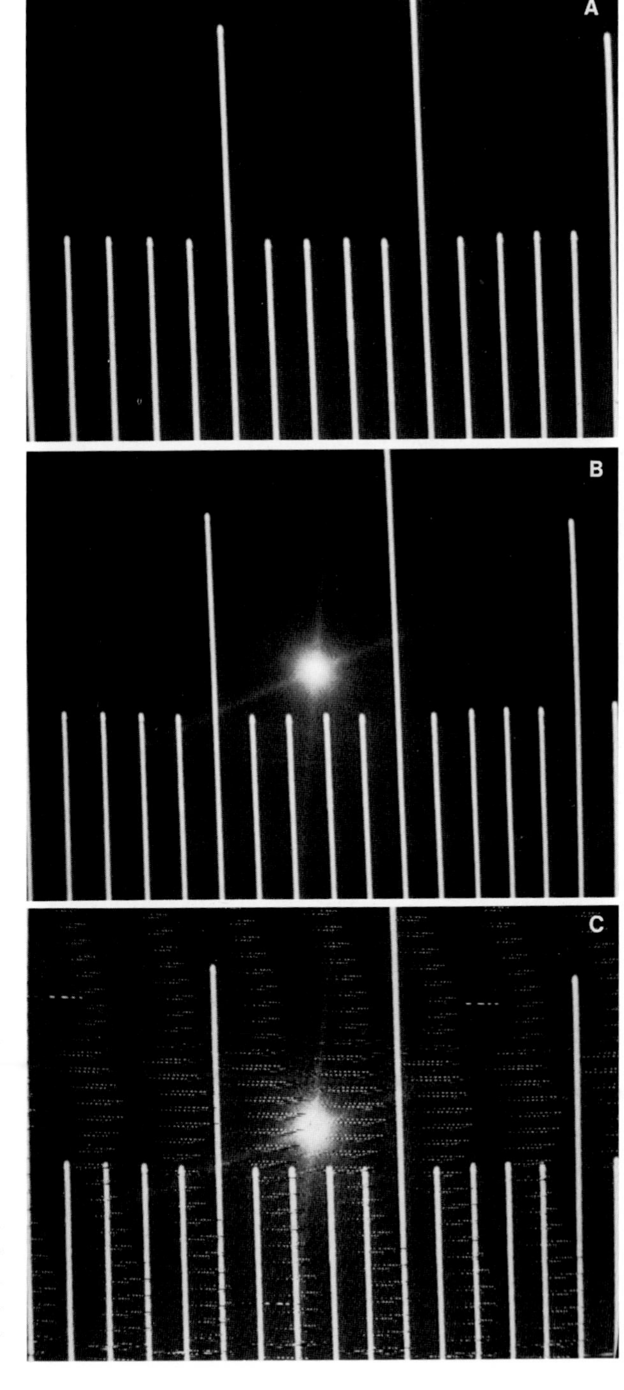

FIGURE 3-13. Hot spot. (B) A hot spot is introduced by inadequate baffling of the tube between the camera and the microscope. (A) Same field with proper baffling, or with the size of the field diaphragm critically adjusted. (C) Same field showing electronic noise picked up by the camera. The light baffle or lens mount screwed into the camera C-mount, acts as an antenna when it is not grounded to the camera. This happens often when the C-mount thread is anodized (since the anodized oxide layer is an insulator).

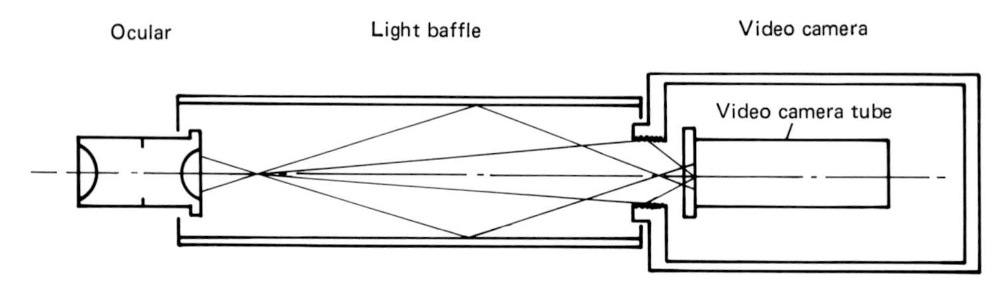

FIGURE 3-14. Source of hot spot. Hot spots may be produced on the video camera tube by reflections from the light baffle between the camera and microscope, or from the camera lens mount, as well as from within the microscope. This problem can be minimized by the use of high-quality and anti-reflection coated microscope optics and by the judicious placement of stops as shown in Figs. 3-7 and 3-8.

cylindrical surfaces of the lens barrel and microscope tubes (Dempster, 1944). Reflection from the tube that couples the microscope and the video camera can be especially pronounced (Fig. 3-13). The interior of the threaded lens mount on the coupler can also be responsible (Fig. 3-14).

With Koehler illumination, the light rays striking and reflecting from these surfaces can be reduced by critical adjustment of the field diaphragm (on the illuminator). However, since the rays strike these surfaces at near-grazing angles, the hot spots can be difficult to eliminate. Even flat-black paint on the reflecting surfaces is not very effective at grazing angles. These reflections and hot spots can be eliminated by placing properly sized stops or iris diaphragms at strategic locations in the coupling tube (Figs. 3-7, 3-8; see Chapter 5).

3.9. OTHER SOURCES OF OPTICAL NOISE

In video microscopy or photomicrography, thin pencils of light form the projected image. As shown in Fig. 3-15, the image-forming beam converges to a narrow waist, the **Ramsden disk,** at the eyepoint as it exits the ocular. Compared to this small spot of light, the projection distance is large. Each image point is thus formed by nearly parallel rays rather than by highly convergent rays as is the case with video or photographic cameras using regular camera lenses.

The depth of focus in the image is accordingly large, and can reach several millimeters. The large depth of focus relaxes the tolerance for locating the target of the video camera tube (Fig. 3-6, Chapter 5). At the same time, any structure in the vicinity of the focal plane will also appear to be in focus.

Thus, in video microscopy, dirt (e.g., dust particles), lint, or smears of oil or other substances, on the surface of the camera tube can produce intrusive, sharp shadows. In addition, slight structural defects, interference patterns, and inhomogeneities in the **faceplate** of the camera tube and adjoining electrodes (Fig. 3-7), which would not be serious in conventional uses of video, adversely affect the image in video microscopy (Sections 3.10, 7.2).

The offending sources of optical noise can usually be spotted by moving or turning the suspected component around the axis of the optical train.

For example, when the video camera is turned around its optical axis, the optical noise arising from the camera does not move on the monitor. As the camera is turned relative to the projected microscope image, the microscope scene turns. In contrast, the images of dirt particles, mottles, grid pattern, etc. arising on, or in, the camera tube faceplate remain *stationary on the monitor* (since they turn with the camera). The same is true for noise arising from any component turned together with the camera.

A common source of optical noise is dirt or oil/grease smear on the slide and coverslip. These can be detected by turning or moving the microscope stage or the slide itself. The noise image

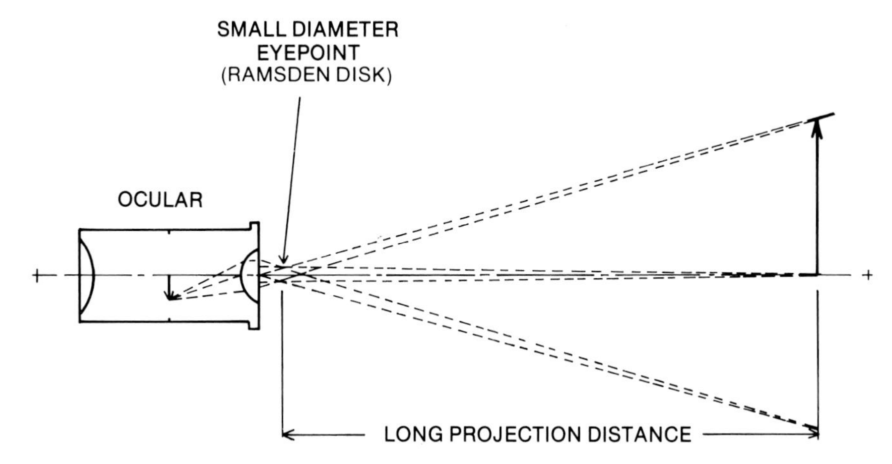

FIGURE 3-15. Ray paths from Ramsden disk. The long projection distance and very narrow cones of light between the Ramsden disk above the microscope ocular and the camera target result in a large depth of focus. Although this allows a large tolerance in the positioning of the target, it also results in the imaging of dust and defects that would otherwise be out of focus.

arising from these sources is displaced, although not always at the same rate as the image of the specimen.

In video microscopy, as in other forms of **high-extinction microscopy,** it is good practice to start off with carefully cleaned slides and coverslips, and to acquire the habit of not laying the slide or coverslip directly on the bench top. Always rest one end of the slide or coverslip on an applicator stick or some other thin support. This simple precaution keeps the glass surfaces from touching the bench top and picking up dirt. Keep these surfaces covered, so that airborne dust, lint, or other particles do not land on them. Remember that clean surfaces can also *attract* dust and lint particles electrostatically.

Other common sources of optical noise includes dirt and smears of immersion oil and test solutions on the objective and condenser lenses or oculars, and air bubbles and other inclusions in the immersion oil. Optical noise from these sources, and noise arising in the illuminating and beam-relay trains, can produce annoying, conspicuous shadows.

In modern microscopy, one is dealing with a large number of optical surfaces, so that the source of optical noise is not always easy to detect. Carefully study the optical train and conjugate foci described in Chapter 5, and turn or move the suspected components in turn. Examination of the optical train with a phase telescope or with the aid of a **Bertrand lens** can often be helpful.

Once isolated to a given optical component, dirt can be spotted by examining the isolated component with a magnifier under good illumination (Section 5.10). A quartz–halogen, fiber-optic ring light, of the type used for illuminating a dissecting microscope, is quite effective for this purpose. Naturally, the video camera must be turned off before inspecting its image tube with such a bright illuminator.

While optical noise can be a serious impediment to video microscopy, one should not, beyond blowing off the dust with an air bulb, attempt to clean the optical faces without proper instructions. Recommended cleaning methods and precautions in cleaning the delicate faces of the microscope lenses, mirrors, and video camera tube faceplates are listed in Section 5.10.

3.10. ELECTRONIC REDUCTION OF NOISE

While it is obviously preferable to reduce optical noise and **shading** at their source, this can be a frustrating experience in microscopy when the contrast is greatly enhanced by video.

Persistent optical noise in video microscopy can, however, be subtracted from the image electronically. While expensive, the equipment is available commercially and the result can be very striking. Allen and Allen (1983) have made effective use of this mode of image processing coupled with digital **image enhancement** (Figs. 1-7, 1-8; see also Chapter 10 and Ellis *et al.*, 1988).

In brief, the image arising from the noise alone is first obtained by slightly defocusing the specimen. The video signal of the noise image is converted into digital form and stored on a **frame buffer** or ''digital frame store'' in a computer. The contrast of the optical noise image is inverted and the inverted-polarity noise signal is added back onto the digitized in-focus image of the specimen. The sum of the two provides a clean image of the specimen on a uniform background free of optical noise. Background subtraction also reduces shading of field brightness (Figs. 1-7, 1-8, Chapter 10).

In addition to persistent optical noise, *random* optical noise is encountered in low-light-level video microscopy (Chapters 7, 11). Such noise, arising from photon statistics, is superimposed with statistical electronic noise generated in the video camera.

To a certain extent, random noise arising from photon statistics is integrated (the image is averaged), normally in the target of the video camera tube (Section 7.1). Random optical and electronic noise can be reduced further by signal or **image averaging**. For example, the signal can be averaged by digital and analog processing (Figs. 1-7, 1-8, Chapters 9–11, Appendix IV).

The physiological characteristics of the human eye also play important roles in image integration and noise averaging (chapter 4). Photography can also be used to integrate the image over time and reduce random noise (Chapter 12).

4

Physiological Characteristics of the Eye

4.1. INTRODUCTION

Much of the equipment and technology used in closed-circuit TV, including video microscopy, is borrowed from broadcast TV. It should be noted that broadcast TV has never sought to produce a perfect optical image from a remote location, but rather to provide an instant facsimile of lifelike scenes economically to an audience scattered over wide geographic areas.

Many practical considerations entered into the design of video systems, including the characteristics of the human visual system. A knowledge of the physiology and psychophysiology of human vision thus helps in understanding the rationale for the design, as well as the limitations, of video. These aspects of the performance of the human eye, and some fundamentals of visual physiology that relate to microscopy, are reviewed in this chapter. For further detail, and a considerable amount of quantitative information relevant to these discussions, as well as to optical systems design in general, see Levi (1980, Chapter 15), Rose (1947, 1973), Cornsweet (1970), Barlow and Mollon (1982), and *RCA Electro-Optics Handbook.*

I shall start with a brief review and some remarks on the structure and function of the human eye.

4.2. STRUCTURE, REFRACTION, AND ACCOMMODATION OF THE EYE

In our eye, the primary *refraction* that produces an inverted image of the external world takes place at the convex, front surface of the cornea. Therefore, in order to yield a good image, the cornea must curve with proper uniform radii in all directions.* In addition, the corneal surface, including the layer of tears, must be kept smooth to the level of optical tolerances.

As shown in Fig. 4-1, the rays refracted at the air–corneal interface successively traverse the cornea, aqueous humor in the anterior chamber, the crystalline lens, the gelatinous vitreous body that abuts against the retina, and the vascular and neuronal layers of the retina before they reach the photosensitive outer segments of the cone and rod cells (Fig. 4-2). These thin outer segments, radiating outward at the rear surface of the retina, are in intimate contact with protuberances of pigment cells in the choroid.†

*The eye is kept turgid by an intraocular pressure of 20–25 mm Hg. The turgor is maintained by the ocular fluids, which are present in a steady state of continuous production and removal.

†The pigment cells provide nutrients, exchange metabolites, and phagocytize the terminal disks spent by the cones and rods. New photosensitive disks are continuously assembled at the base of the cones and rods at a rate of 80–90 disks/day (e.g., see Bok and Young, 1979).

The rays refracted at the corneal surface are further converged by the crystalline lens. **Accommodation** of the lens adjusts the focal length of the eye to bring the image exactly into focus onto the photosensitive layer of the retina. Accommodation relaxes the tension, applied through the zonule fibers to the crystalline lens, and allows the anterior surface of the lens to increase its curvature. The increased refraction, coupled with a slight forward shift in the position of the lens, brings objects that are closer to the eye into focus.

When the illumination changes, the diameter of the pupil, positioned just in front of the

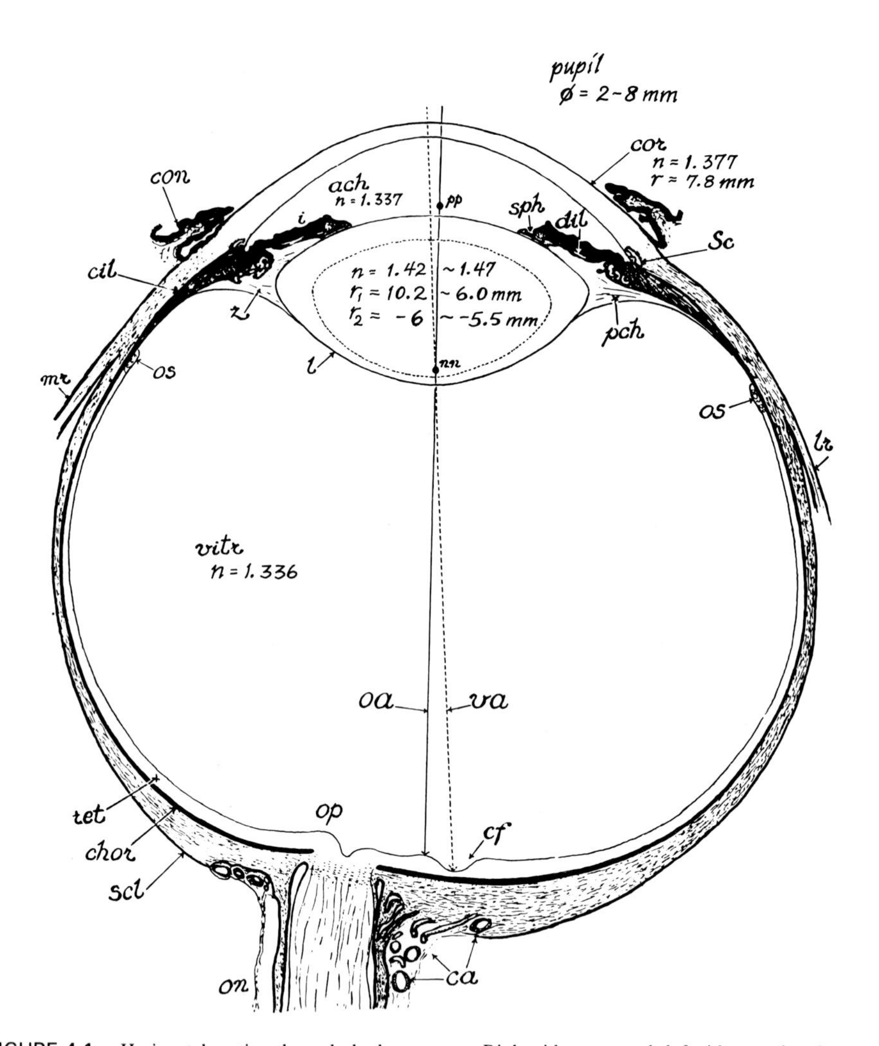

FIGURE 4-1. Horizontal section through the human eye. Right side, temporal; left side, nasal. ach, anterior chamber; ca, ciliary arteries; cf, central pit or fovea; chor, choroid membrane; cil, ciliary body; con, conjunctiva; cor, cornea; dil, dilator muscle of pupil; i, iris membrane; l, crystalline lens; lr, lateral rectus (tendon); mr, medial rectus (tendon); nn, double nodal point of eye; oa, optical axis; on, optic nerve; op, optic papilla; os, ora serrata; pch, posterior chamber; pp, double principal point; ret, retina; Sc, Schlemm's canal; scl, sclera; sph, sphincter muscle of pupil; va, visual axis; vitr, vitreal body and chamber; z, suspensory ligament (zonule fibers) of lens.

Approximate length of the visual axis is 21.5 mm. Refractive indices and radii of curvature of several structures are given by n and r, respectively. Note the variation in thickness of the layers: they are thickest posteriorly. The sclera is thinnest along the nasal equator; the choroid, along the entire equator; the cornea, along its vertex. The scleral radius of curvature varies from point to point, and is different on the nasal and temporal sides, making the shape of the eyeball asymmetric. (After Polyak, 1957, p. 208.)

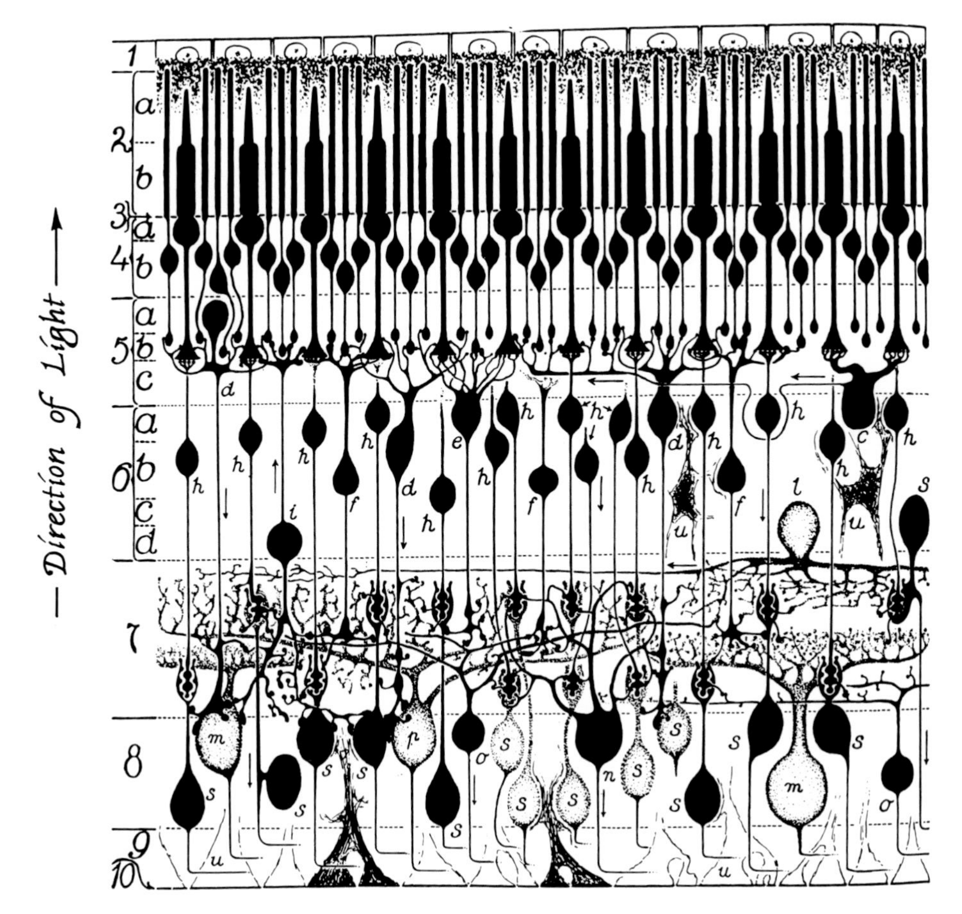

FIGURE 4-2. Schematic representation of the primate retina, in a region near the fovea, as seen after Golgi staining. The layers and zones are: (1) pigment layer; (2a) outer zone and (2b) inner zone of the rod and cone layer; (3) outer limiting membrane; (4a) outer zone and (4b) inner zone of the outer nuclear layer; (5) outer plexiform layer with its three zones; (6) inner nuclear layer with its four zones; (7) inner plexiform layer; (8) ganglion cell layer; (9) optic nerve fiber layer; (10) inner limiting membrane.

In the upper part of the diagram are photoreceptors, both rods (a) and cones (b), the bacillary parts sticking into the pigment layer, the inner fibers of the rod cells terminating with round spherules, and those of the cone cells, with conical pedicels, which form zone 5b. Here the synaptic contacts are made with upper or dendritic expansions of horizontal cells (c), along which the impulses spread in adjoining parts of the retina, and with the bipolars of centripetal varieties (d, e, f, h), along which they pass to the inner plexiform layer where synaptic contacts are established with the ganglion cells (m, n, o, p, s). From the ganglions arise nerve fibers, which, as the optic nerve, enter the brain. Other types of neurons, called "centrifugal bipolars" (i), may condition photoreceptors under the influence of the ganglion cells, hence in a direction opposite to that of the centripetal bipolars (d, e, f, h). There may be still other neuron varieties whose connections are lateral, spreading in the inner plexiform layer (7). This indicates that the primate retina is a complex organ whose function is not only to react to physical light as an external stimulus, but also to sort the generated impulses in many ways before they are further transmitted to the brain. (After Polyak, 1957, p. 254.)

crystalline lens, reflexively varies between about 8 mm and 2 mm, adjusting the amount of light that reaches the retina.* At the same time, as the pupil narrows with brighter illumination, the peripheral parts of the refractile elements are excluded from the optical path. Thus, fewer *aberrations* are encountered by the image-forming rays and the image on the retina becomes sharper.

On the other hand, too narrow a pupil would increase *diffraction*. As the pupil closes down to 2 mm, the retinal image of a point source, spread by ocular aberration, becomes roughly matched with its spreading due to diffraction by the narrowed pupil.

4.3. VISUAL SENSITIVITY AND ADAPTATION

In broad daylight, we can see in the glaring light of sunlit snow scenes (ca. 3×10^4 **cd**/m^2), while at night we can detect large objects by starlight when the moon is dark. At **threshold** sensitivity, the human eye can detect the presence of some 100–150 photons of 507-nm light entering the pupil. At that level, the luminance of the object is below 1×10^{-6} cd/m^2 (e.g., see Pirenne, 1962, Chapter 6). This represents an incredible brightness range of over ten decades (Fig. 4-3). By **adaptation,** we can see under such extremes of brightnesses; but during the brief time before adaptation occurs, we can sense a range of brightness covering only about three decades (the intrascene dynamic range).

Several mechanisms enter into our ability to adapt to a high range of brightnesses. As commonly experienced, depending on the level of brightness change, adaptation can take place in seconds (by initial pupillary reaction) or may take many minutes (for **dark adaptation**). For the upper seven decades of brightness, or in **photopic vision,** the retinal cones are primarily responsible for photoreception, while in the lower four decades, or **scotopic vision,** the rods are exclusively responsible. Whereas we reach full cone sensitivity in about 5 min, it takes from one-half to three-quarters of an hour to adapt from moderate photopic sensitivity to reach the full scotopic sensitivity given by the rod cells (Fig. 4-4).

The light-adapted human eye responds to wavelengths of light ranging from around 400 to 700 nm, with the peak sensitivity at 555 nm. The dark-adapted eye responds to ranges between 380 and 650 nm, with the peak at 507 nm (Fig. 4-5). For both photopic and scotopic vision, these cutoff wavelengths are not absolute, but vary with the intensity of light (Levi, 1980, pp. 1032–1035). The long-wavelength cutoff follows the absorption curves of the cone and rod photopigments, while the short-wavelength cutoff is also affected by the *transmission* characteristics of the optical media in the eye.† The transmission is progressively lower at shorter wavelengths; at 500 nm, 50% of the

*When the pupil opens up, the amount of light that effectively stimulates the retina does not increase proportionately with the area of the pupil as would be expected, but at a lesser rate. This "Stiles–Crawford effect" reflects the diminished sensitivity of the cones to rays entering at larger angles to their axes (Pirenne, 1962, Chapter 3).

†Dr. E. F. MacNichol, Jr. (personal communication) points out that if the illumination is sufficiently intense, the eye will respond as far into the red as 750 nm, although the sensitivity there is only about 10^{-4} of that at peak photopic sensitivity (Fig. 4-5). Given enough photons, their cumulative energy will eventually produce a response at even longer wavelengths, even though more and more photons are required as the energy of each becomes progressively lower. In the absence of light, rhodopsin (a visual pigment) is excited thermally at a very low rate. This is thought to be responsible, at least in part, for the "self-light" or "Eigengrau" experienced in the absence of any light. The high-wavelength limbs of the curves in Fig. 4-5 continue as approximately straight lines as the relative luminous efficiency (sensitivity) becomes vanishingly small. Unlike the case at short wavelengths, there is no problem with intraocular absorption out to λ of about 2 μm; thus, at wavelengths longer than that of maximum sensitivity, the eye's scotopic response coincides with the absorbance of rhodopsin.

On the blue (short wavelength) side of peak scotopic sensitivity, its rapid drop is due to lens absorption. Below 400 nm, lens fluorescence becomes greater than direct excitation of photoreceptors; images are not formed on the retina, but rather the whole visual field is filled with whitish light. On the other hand, **aphakics** (those with lensless eyes) are able to see images in the UV. (George Wald once reported that an aphakic subject could accurately focus his microscope in the UV.) Fluorescence probably accounts for the UV sensitivity shown in Fig. 4-5.

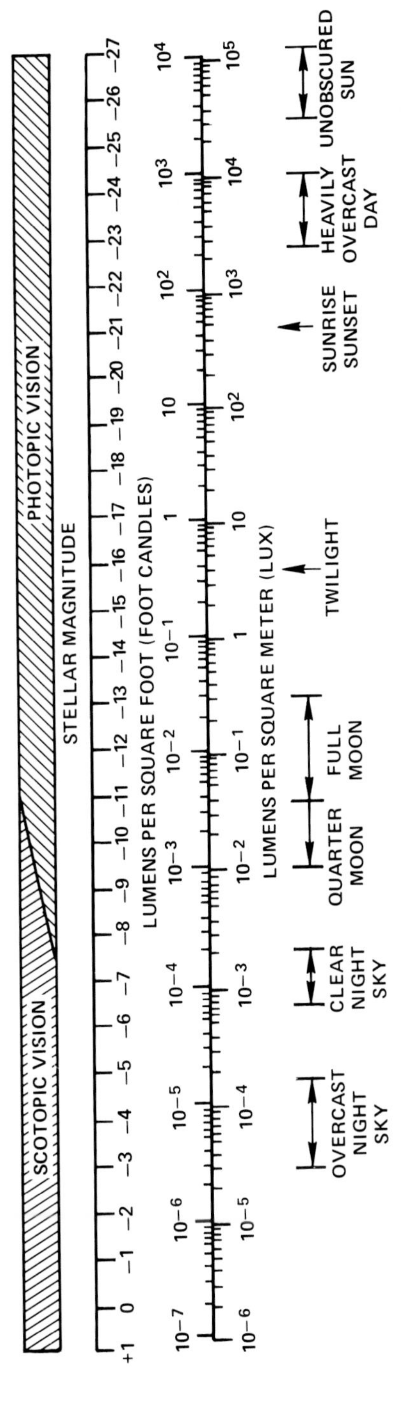

FIGURE 4-3. Range of natural illuminance levels. Vision of scenes darker than a moonless night is produced entirely by rod cells (scotopic vision). (Courtesy of RCA, from ''Imaging Devices,'' Publication IMD 101.)

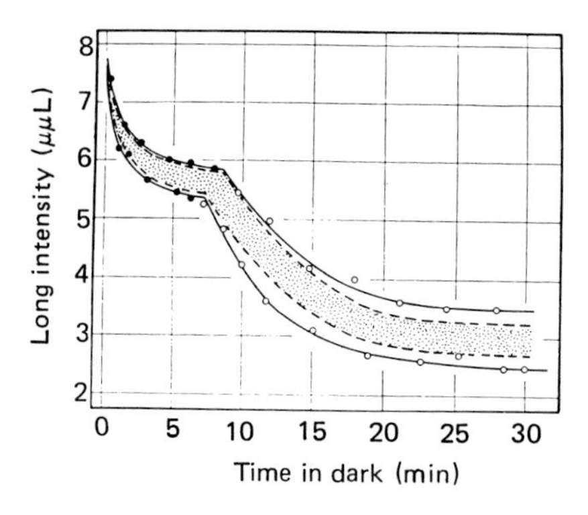

FIGURE 4-4. The course of dark adaptation (the ability to detect a 3° test field) in 110 normal subjgcts. Natural pupil. The circles show measurements from the two subjects having the highest and lowest values of the final threshold. The stippled area contains the measurements for 80% of the subjects. The luminance of light adaptation is 1500 millilamberts. (From Hecht and Mandelbaum, 1939.)

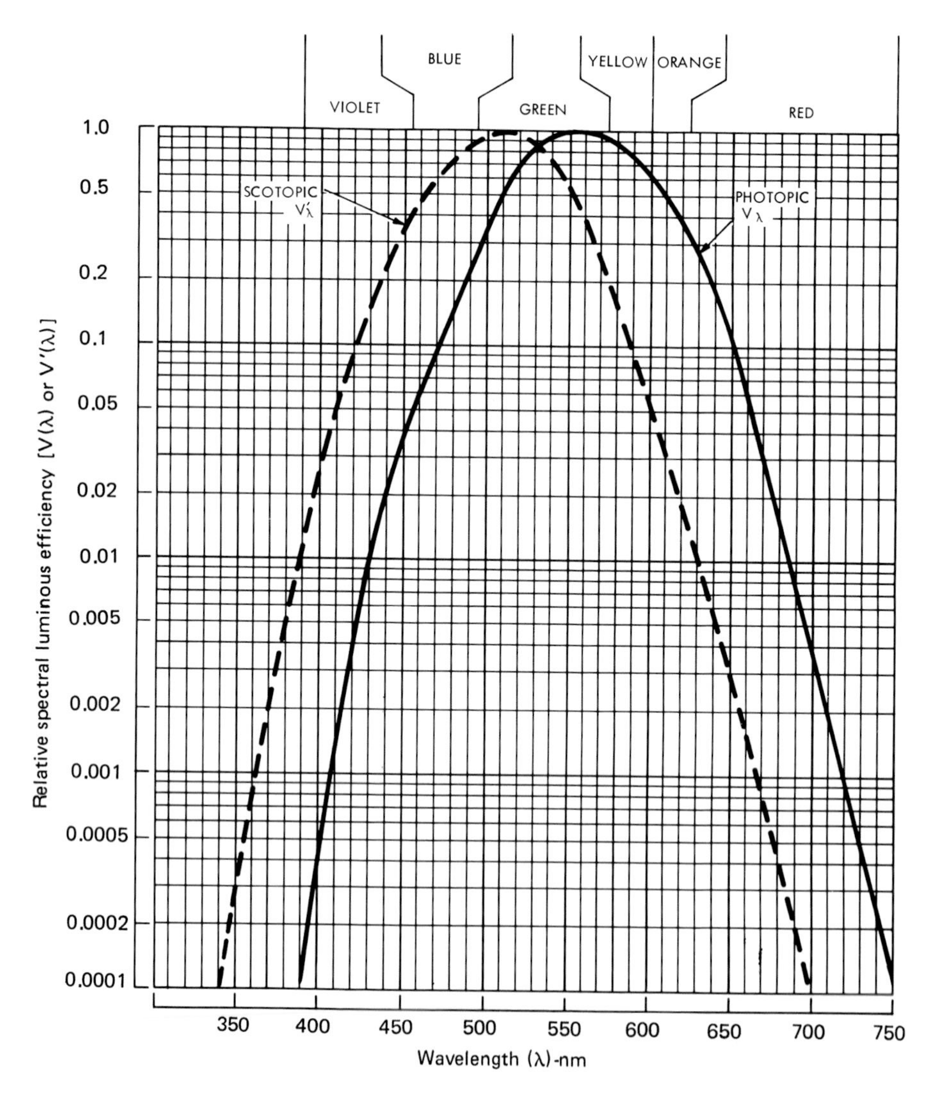

FIGURE 4-5. The normalized relative responses of the human eye to radiation of different wavelengths, for light- and dark-adapted eyes. (Courtesy of RCA. From *RCA Electro-Optics Handbook,* 1974, p. 56.)

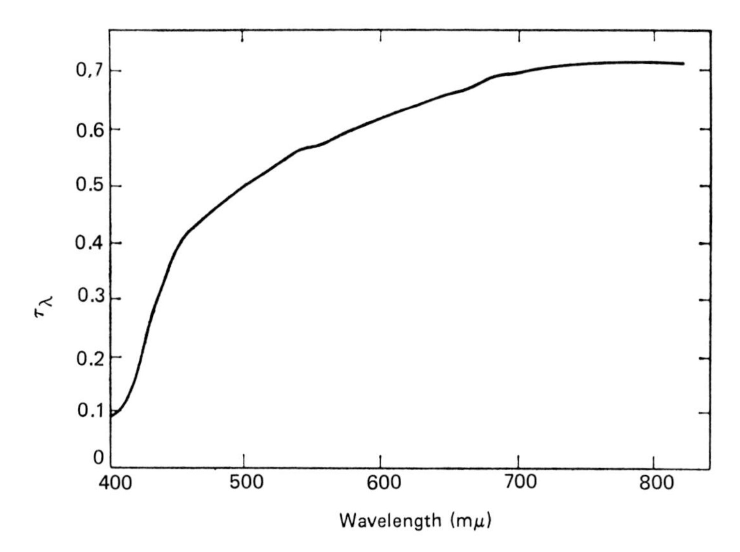

FIGURE 4-6. The fraction of light transmitted (τ_λ) as a function of wavelength, for young human eyes. Note the low transmission at short wavelengths. A table of numerical values is given in the original paper. (From Ludvigh and McCarthy, 1938.)

light entering the eye reaches the image point on the retina, but at 400 nm that value is only 10%, even in a young eye (Fig. 4-6). Light scattering and absorption by the crystalline lens both contribute to the diminished transmission. With aging, the yellowing of the crystalline lens contributes to a further loss of sensitivity in the far blue (as well as better protection from UV?)

As noted, the human eye responds to a limited range of the electromagnetic spectrum. Within the visible range, the sensitivity depends on the wavelength and adaptation level. Therefore, rather than using units that measure the absolute energy content of the light wave, it is convenient to establish a scale for measuring the brightness, or more properly the *luminosity,* of light that can be sensed by the human eye. Thus, we use **photometric units** such as candelas per square meter for measuring luminosity, in contrast to **radiometric units** such as watts per square meter for measuring radiant energy. Luminosity and radiometric units are related by a sensitometric curve defined as being that of an average human eye.*

The amount of light reaching the retina varies with the area of the pupil. Therefore, visual stimulation, or **illuminance** of the retina, is sometimes defined in terms of a special unit called the **troland,** which designates retinal illumination with light that has passed through a standardized pupil area of 1 mm^2.

The spectral distribution of energy from an illuminator also varies according to various conditions, including its temperature. Therefore, standards for **luminances** are defined in terms of specific types of light sources with given **color temperatures** (Section 7.2).

4.4. RESOLUTION, VISUAL ACUITY

For photopic vision, photoreception takes place primarily in the outer segments of the cone cells, which are densely packed in the **foveal** region of the retina. In the human *central fovea* (measuring from 500 to 550 μm across), the center-to-center distance of the cones is reduced to

*Unless otherwise designated, luminosity generally refers to *photopic luminosity*. The calibration curve for luminosity is the one for the light-adapted eye in Fig. 4-5. Since the wavelength dependence of sensitivity for photopic vision differs from scotopic vision, the latter has its own separate scale for luminosity, the *scotopic luminosity,* and its corresponding calibration curve.

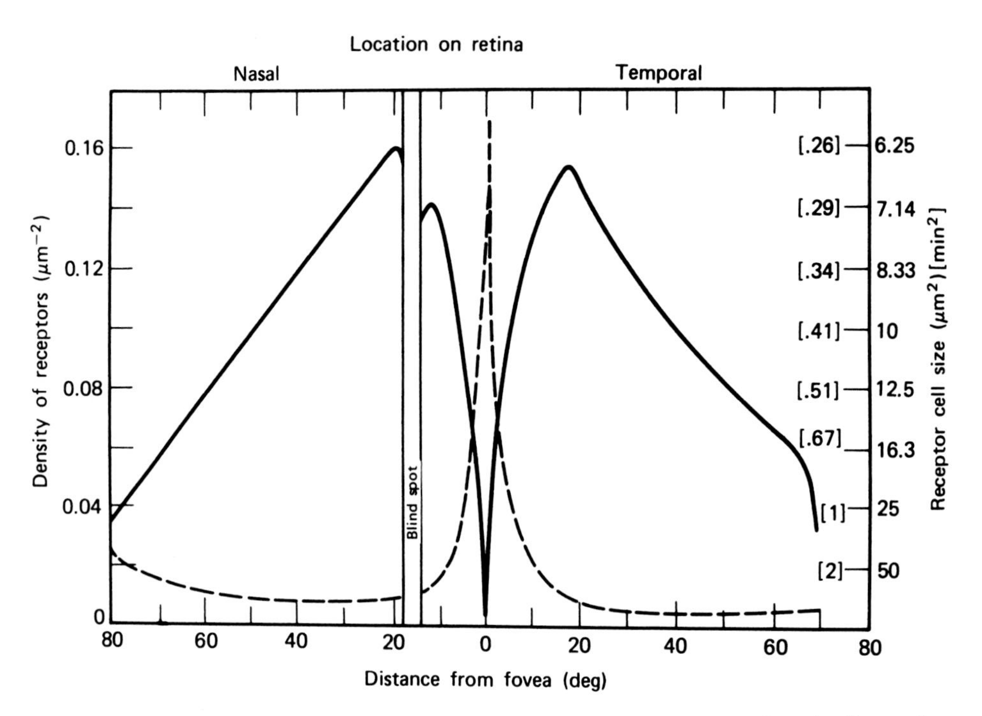

FIGURE 4-7. Distribution of rods (solid line) and cones (dashed line) as a function of angular distance from the fovea. The size of the receptor cells follows similar distributions, as indicated by the ordinate scale on the right. (From Levi, 1980, p. 348.)

between 1.5 and 2 μm. This region of the retina, which gives the highest **visual acuity** in the light-adapted eye, is devoid of blood vessels and most of the nerve cell layers. In other parts of the retina, the vessels and cell layers scatter the light approaching the photosensitive outer elements.

The central fovea is devoid of rod cells. Outside of the central fovea, cones and rods are intermixed, with the density of the cones rapidly decreasing away from the fovea (Fig. 4-7). As shown in Fig. 4-7, the density of the rods reaches a maximum at about 20° from the fovea. For a dark-adapted eye, the sensitivity is greatest at about 10°, i.e., where we see with "averted vision." As also shown in Fig. 4-7, rods and cones are missing from the optical papilla, or the blind spot centered about 17° to the nasal side, where blood vessels and axons enter and exit the eye (Fig. 4-1).

These arrangements of the retinal outer segments partly define the limit of *resolution* in different regions of our eye. It is generally held that in order to resolve an image, one needs a row of less-stimulated photoreceptors between two rows of photoreceptors that are stimulated (Fig. 4-8). Otherwise, one could not distinguish whether the stimulus originated from two closely spaced images or from a single image that spanned the two rows of receptors. With a center-to-center spacing of 1.5–2 μm for the cones in the central fovea, a separation of roughly 3–4 μm should give a resolvable set of stimuli. Based on an average figure of 16.7 mm for the distance between the retina and the nodal point of the eye, the angle subtended by two rays 3–4 μm apart on the retina would subtend an angle of 37–49 sec of arc in the ocular medium. The minimum angle resolvable by the human eye is usually taken to be about 1 min of arc, although under ideal conditions a resolution of 28 sec of arc has been reported (Shlaer *et al.*, 1942). In comparison, the radius of the first minimum of the diffraction pattern that would be formed on the retina (by a point source with a wavelength of 555 nm) is 4.6 μm when the diameter of the pupil is 2 mm.

These arguments show how the arrangements of the sensory elements in the retina could determine the limiting resolution of the eye. However, visual acuity, the ability of the eye to detect small objects or resolve their separation, varies with many factors including the definition of acuity

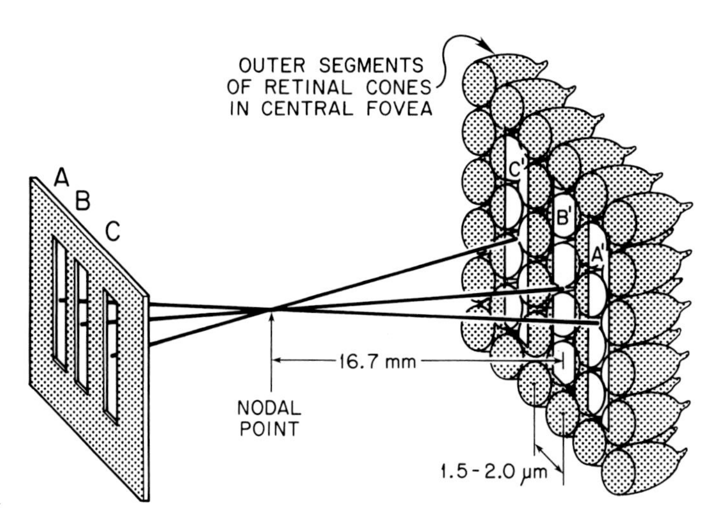

FIGURE 4-8. Effect of photoreceptor spacing on image resolution. If the object ABC is focused on the retina as A′B′C′, the lines C′ and B′ will be distinguished (resolved) because they are separated by less-stimulated photoreceptors. However, lines B′ and A′ will not be resolved because they are so close together that they stimulate adjacent photoreceptors nearly uniformly. In ''high-resolution'' (foveal) regions of the retina, the center-to-center distance of cone cells is 1.5–2.0 μm.

and the method by which acuity is measured. Over the retina, acuity is generally greatest in the central fovea, which spans a visual field of about 1.4°, and especially at its central 0.2° field where, under favorable conditions, one can *detect* dark lines only 0.5 sec of arc wide against a brighter background. Also, misalignment of parallel sharp edges of dark bars by only 0.2 sec of arc can be detected. These values for *threshold detectability* are far smaller than the threshold for resolution described above. Figure 4-9 shows the dependence of visual acuity on field luminance.

As these examples show, visual acuity cannot be assessed simply by considering the eye to be a stationary camera with a large number of isolated receptors fixed in the focal plane. For one thing, the eye is continuously undergoing *involuntary rapid movements*. The optical image on the retina is not stationary but is constantly being scanned. In fact, when the image is prevented from moving relative to the retina by an optical fixation device, the image is no longer sensed after just a few seconds (e.g., see Fiorentini, 1964). Naturally, too rapid a motion would blur the image, but a moderate rate of scanning of the retinal image is essential for vision.

In addition, the neurons that signal the visual information from the retina to the brain are not simply connected one to one with the sensory cells. Each cone and rod cell in the fovea sends signals

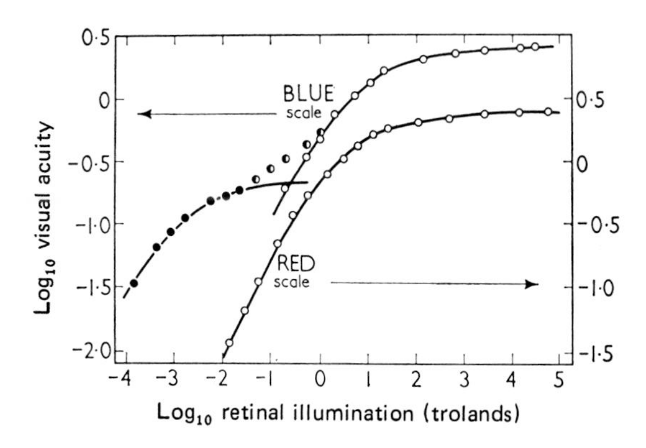

FIGURE 4-9. Variation of visual acuity with field luminance in blue (left ordinate scale) and red (right ordinate scale) light. Filled circles represent values attributable to rod function alone; half-filled circles, to combined rod and cone function; and open circles, to cone function alone. The limiting value for acuity is about the same for both red and blue light. (From Shlaer *et al.*, 1942.)

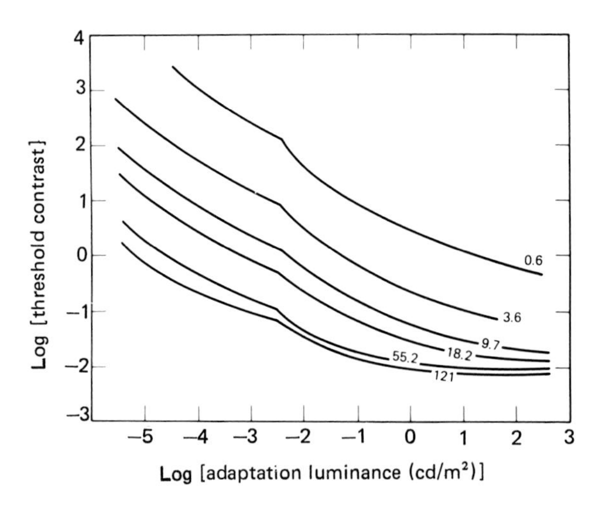

FIGURE 4-10. Threshold modulation contrast as a function of adaptation (background) luminance for bright circular disks of different diameters (expressed in minutes of arc). The threshold is higher for lower luminance and for smaller test objects. The inflections indicating change from photopic to scotopic vision are clearly shown. (From Levi, 1980, p. 370.)

to some three bipolar cells, while in the more peripheral regions of the retina, the signals from very large numbers of rods converge to a single ganglion cell.* Other neuronal cells interconnect the sensory and bipolar cells or ganglion cells, providing a complex network of inhibitory and excitatory pathways within the retina (Fig. 4-2; Michael, 1969). Thus, the signals from the 4–7 million cones and 125 million rods in the human retina are *processed* and carried by only about 1 million myelinated optical nerve fibers, via the lateral geniculate body, to the visual cortex. Ganglion cells in the lateral geniculate body send impulses to the ocular muscles.

During embryonic development, the retina arises as an outpocketing of the brain. In it, the three layers of neuronal cells are linked up in a complex network of excitatory and inhibitory pathways. These circuits and feedback loops result in excitatory and inhibitory effects that cause edge sharpening, contrast enhancement, spatial summation, noise averaging, and other forms of signal processing, and perhaps others that have not yet been discovered. In human vision, much additional image processing does take place in the brain, but the retina itself also is involved in a wide range of image-processing tasks.

In addition to location on the retina, visual acuity varies with the contrast of the object and the level of illumination. Visual acuity and resolution are thus intimately coupled with contrast discrimination.

4.5. CONTRAST DISCRIMINATION AND THE MODULATION TRANSFER FUNCTION

Our ability to discriminate luminance differences in adjoining fields, or *contrast discrimination,* depends on the average luminance of the fields, the size of the test object, and wavelength. Given a test object of fairly large diameter and with no separating lines at its boundary, the minimum contrast $\Delta I/I$ that can be detected, or the *threshold contrast,* reaches about 2% when the luminance is reasonably high (Fig. 4-10).†

As the luminance is decreased, the threshold contrast rises until about 4×10^{-3} cd/m², the luminance level encountered at dusk. There, an *inflection* appears on the curve and the rate of rise of the threshold suddenly becomes less than at slightly higher luminances. As the luminance is reduced further, the threshold for contrast discrimination rises again at a more rapid rate so that eventually at

*By having the signals from a large number of rod cells converging onto a single ganglion cell, spatial resolution is lost from the rods. On the other hand, by many sensory cells participating in the capturing of weak signals, threshold sensitivity (or relative freedom from noise or mistakes) is much improved.

†The constancy of $\Delta I/I$ (at higher luminance) fits the Weber law stating that we respond linearly to a constant incremental ratio of stimuli. That is the basis for the approximation that we respond logarithmically, e.g., to light intensity.

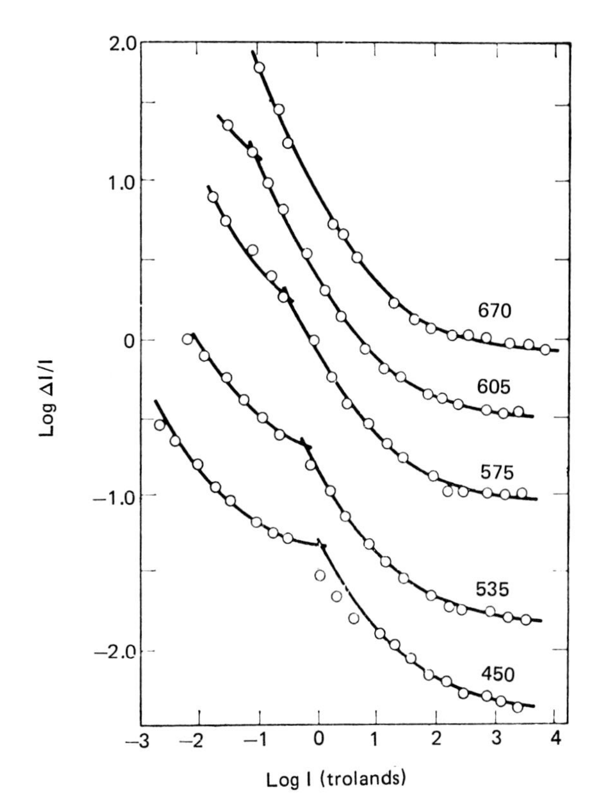

FIGURE 4-11. Intensity discrimination for the red, orange, yellow, green, and blue parts of the spectrum. The ordinates apply to the results for yellow light, in the middle of the diagram. The results for orange and red have been raised 0.5 and 1.0 log unit, respectively; those for green and blue have been lowered 1.0 and 0.5 log unit, respectively. Data are for a 12° test field, exclusive of the fovea. (From Pirenne, 1962, p. 161.)

very low luminances, a contrast of 100% or more is needed before we can detect a difference in the two adjoining fields.

When the same test is carried out with light of a different wavelength, the inflection point shifts to lower luminance levels as the wavelength is increased (Fig. 4-11). At 670 nm, an inflection point is no longer observed. For each wavelength, *the inflection point signifies the light intensity at which scotopic vision takes over from photopic vision.*

In addition to the field luminance, contrast threshold is affected by the *size* of the object. As the size of the test object decreases, the contrast threshold rises so that greater contrast is required before the object can be detected (Fig. 4-10). In fact, Rose (1947), who has contributed much to defining the image parameter needed for video systems, has shown that the behavior of the human eye can be modeled to a close approximation by the relationship

$$B \cdot C^2 \cdot \alpha^2 = \text{const.}$$

Here, B is the luminance of the scene, C is the threshold contrast, and α is the angular size of the object.*

Turning to a different type of test object, one can graphically visualize the contrast needed to just detect a series of stripes in a modulated sinusoidal grating (Fig. 4-12). The **spatial frequency** of the sinusoidal grating increases linearly to the right while the contrast between the brighter and darker regions of the stripes is made to decrease logarithmically from top to bottom. Interestingly, the boundary of the stripes that can just be distinguished runs pretty much as shown by the solid line

*According to this model, the human eye shows an apparent storage time of 0.2 sec, and threshold **S/N ratio** of 5. For levels of illumination between 10^{-6} and 10^2 lamberts, contrast threshold of 2–100%, and α of 2–100 min of arc, Rose further shows that the performance of the human eye is matched with an ideal sensing device with 5% quantum efficiency at low light and 0.5% quantum efficiency at high light levels.

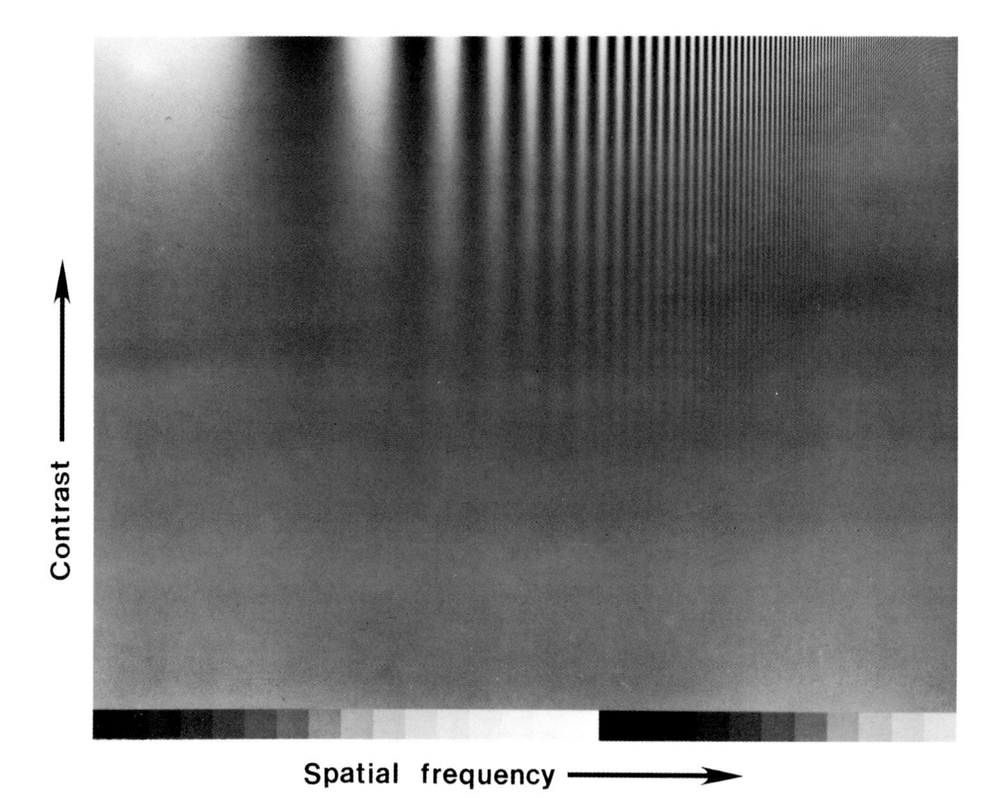

FIGURE 4-12. Modulated sine wave grating, with the spatial frequencies of the vertical lines increasing toward the right and the log of the contrast increasing toward the top. (Courtesy of Dr. W. K. Pratt, from Pratt, 1978, p. 40.)

in Fig. 4-13. For the achromatic case, where the spatial frequency is very low (broad spacing), high contrast is required to detect the sinusoidally varying intensity. As the spatial frequency rises, we can detect sinusoidal periods with less contrast, reaching a peak at a period of 8 cycles per degree in the visual field. Beyond that, again higher contrast is required to detect the finer sinusoidal stripes until the stripes become undetectable altogether.†

Curves such as those in Fig. 4-13 represent the **modulation transfer function** (MTF) of the human visual system.‡ Figures 4-12 and 4-13 show that the contrast needed to detect the luminance variation in a sinusoidal grating increases at both higher and lower spatial frequencies. In other words, our visual system behaves quite differently from a simple imaging device. The MTF curve of a simple, focused, imaging device, for example a camera or a brightfield microscope, shows a maximum modulation at zero spatial frequency, with the modulation dropping off more or less monotonically to zero at the system's cutoff frequency.

Figure 4-14A shows an interesting visual phenomenon related to the present discussion. The rectangle on the right appears brighter than the one on the left until a pencil (or even a much narrower object) is placed over the junction of the two, masking the sharp discontinuity. Then the

†When colored gratings and background are used, the relative response is shifted toward lower spatial frequencies, as in the chromatic response of Fig. 4-13. However, the curves vary according to the particular colors used.

‡The MTF curves were in fact not derived in the manner described above, but by other means. Figure 4-12, which appears in Pratt (1978, p. 40), is meant to be illustrative and not to be used as a quantitative test target. See Sections 5.7, 6.4, 7.2, and Appendix II for further discussions about MTF curves and their utility.

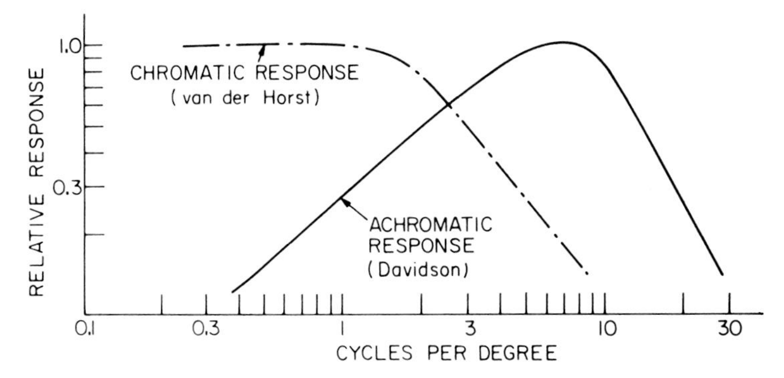

FIGURE 4-13. Modulation transfer functions (MTF) of the human visual system. The achromatic response curve corresponds approximately to the boundary of contrast detectability in Fig. 4-12. See footnote ‡, p. 82. The chromatic response is shifted toward lower spatial frequencies and is complicated by dependence of the response on the particular wavelengths of both the grating bars and the background. (From Pratt, 1978, p. 46; after Davidson, 1968, and van der Horst *et al.*, 1967.)

two rectangles, which in fact have equal *average* luminance, although both are darker toward the right (as shown in Fig. 4-14B), do appear equally bright.*

This striking demonstration, as well as the Mach bands,† reflect some aspects of the processing capabilities of the human visual system. The suppression of contrast discrimination at low spatial frequencies allows us to disregard some shading over wide areas, while accentuation of contrast discrimination at edges and higher frequencies enhances our ability to detect fine details as well as boundaries of objects.

4.6. FLICKER

When the luminance of a scene fluctuates periodically several times a second, we perceive an irritating sensation, as though the sequential scenes were disjointed. As the **frequency** of fluctuation increases, the irritation rises and reaches a maximum at around 10 **Hz,** especially when bright flashes of illumination alternate with darkness. At higher frequencies, the scene no longer appears disjointed, and figures displaced discontinuously from one scene to the next are perceived to be moving smoothly. However, the annoying sensation of fluttering of light, or **flicker,** can persist up to several tens of hertz.‡ Beyond a certain frequency, known as the **critical flicker frequency** (CFF), the flicker is no longer perceived.

The CFF varies with luminance, rising linearly with the log of luminance over a wide range (Fig. 4-15A). At very high luminances the CFF can exceed 100 Hz, but commonly does not rise beyond 60 Hz.

As the luminance level is reduced, the CFF versus illumination curve shows an abrupt inflection. Below the inflection point, flicker continues to be observed at lower luminance levels. The shorter the wavelength of light involved, the more the flicker is observed at the lower luminance levels. As in the curve relating log contrast threshold to log luminance (Fig. 4-10), the inflection for the CFF versus luminance curve occurs at the transition of photopic to scotopic vision. Again, the major inflection in the curve disappears for light of 670-nm wavelength, as would be expected if the low-intensity branch of the curve is ascribed to rod function.

*This demonstration was devised by Land and McCann (1971) to illustrate an important attribute of visual perception relating to color invariance. See Section 4.6.

†The Mach bands are the bright band that appears on the lighter side and the dark band that appears on the darker side of the junction of two homogeneous fields that have a moderate difference of luminance.

‡Subjectively, the fluttering sensation occurs at a lower frequency than the frequency at which the luminance is actually changing.

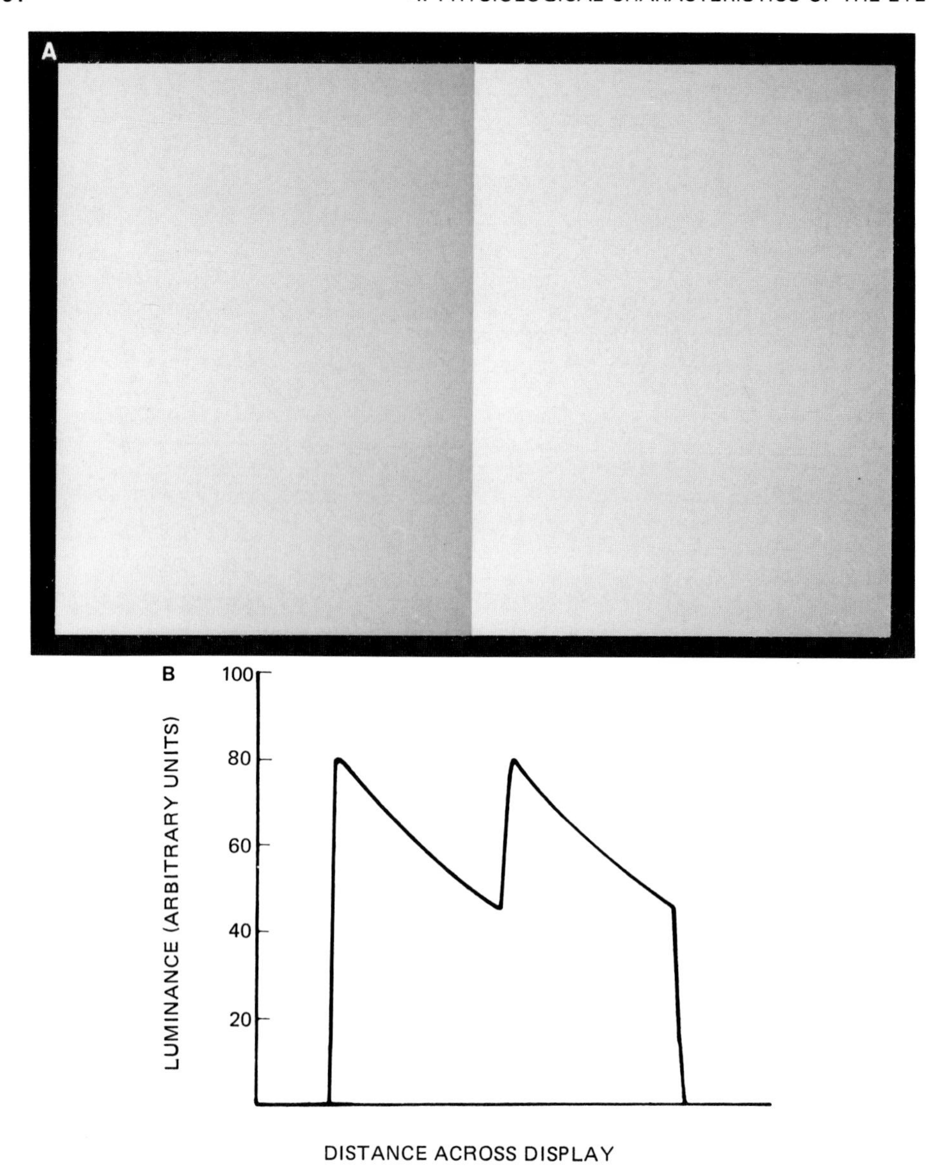

FIGURE 4-14. An interesting visual phenomenon. The two patches in A are identical, as shown by the horizontal luminance trace in B. Yet the patch on the right appears strikingly brighter. The effect is destroyed if a pencil, or similar object, is placed along the boundary between the patches. (A: courtesy of Dr. E. H. Land; B: from Land and McCann, 1971.)

The CFF also varies with the size of the stimulus diameter and the location of the flickering stimulus on the retina. As shown in Fig. 4-15B, the CFF increases with increasing diameter of the stimulus. For small stimuli, the CFF progressively decreases with distance from the fovea (Fig. 4-15C).

The threshold modulation of luminance that gives rise to a sensation of flicker that can just be detected occurs at 6–7% at low frequencies of light modulation (Fig. 4-16). As the frequency rises, the threshold modulation decreases to a minimum at between 5 and 20 Hz depending on the luminance level. At higher luminances, modulation of even less than 1% can produce flicker.

Since video is comprised of a series of repeating scenes as in motion pictures (the flicks), but

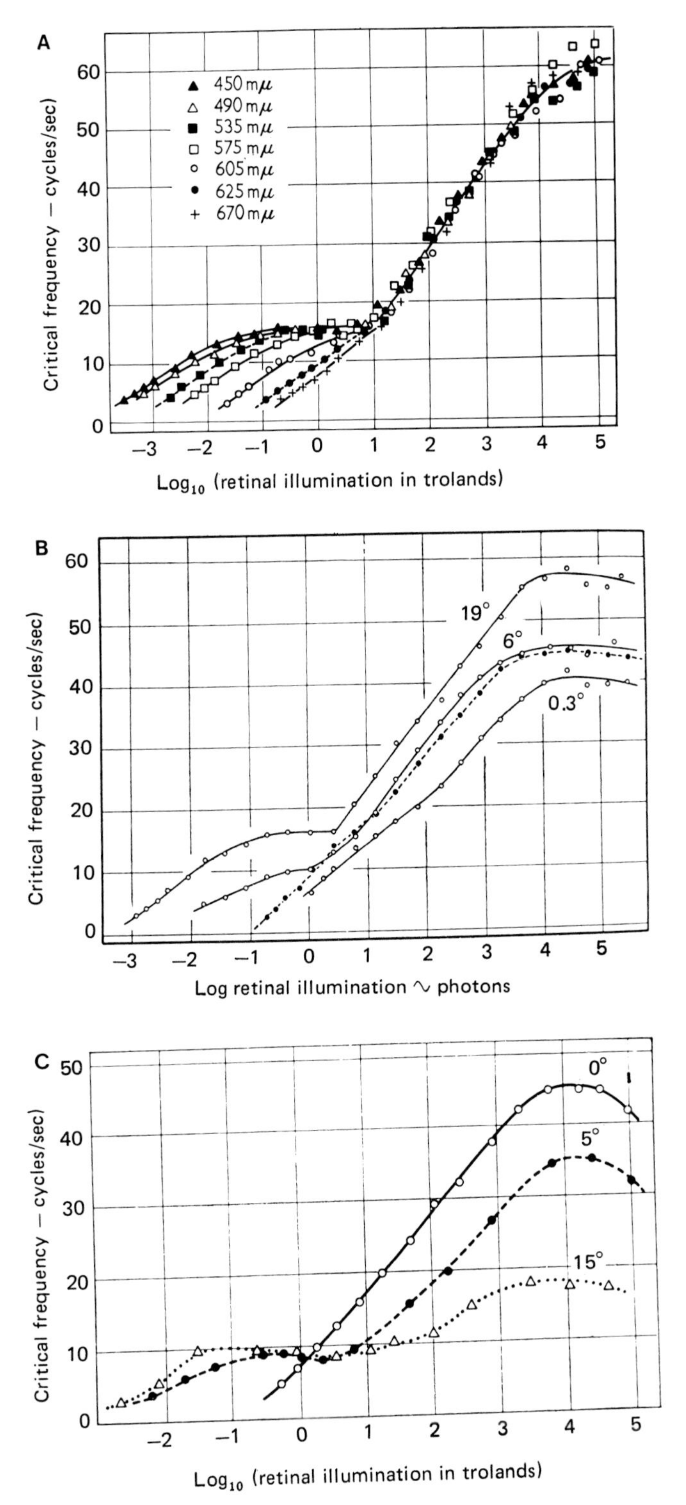

FIGURE 4-15. (A) Relationship between critical (flicker) frequency (CFF) and luminance at different wavelengths. In general, flicker is less noticeable at lower luminance levels. (B) The CFF also increases with the diameter of the stimulus. Here each curve represents a test object of different diameter (expressed in degrees of arc). (C) Different regions of the retina exhibit different CFFs. In general, the CFF decreases with distance from the fovea, expressed here as angular distance. (A: from Hecht and Shlaer, 1936; B: from Hecht and Smith, 1936; C: from Hecht and Verrijp, 1933.)

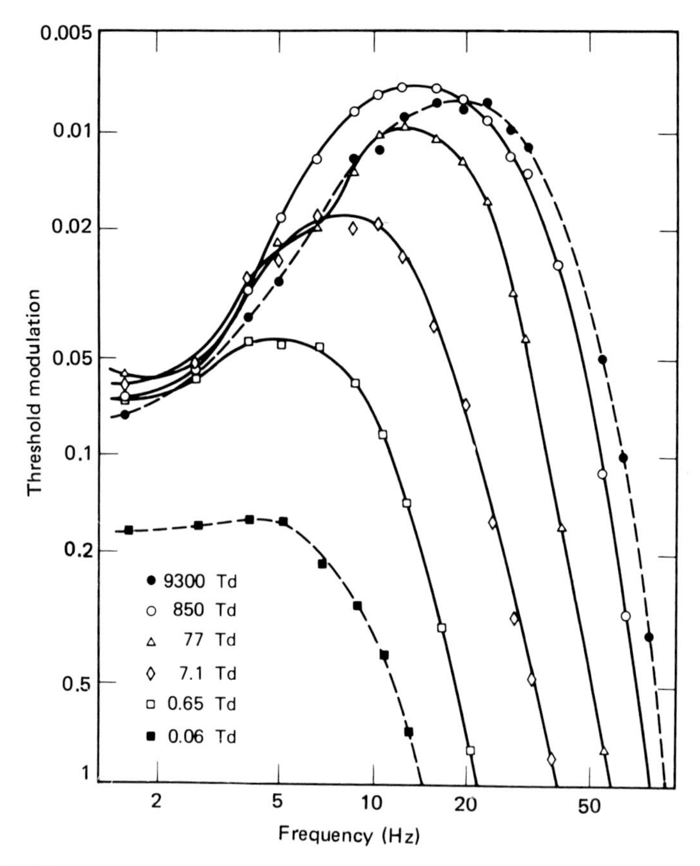

FIGURE 4-16. Flicker modulation threshold as a function of frequency for several levels of luminance. At lower luminances, the low-frequency threshold is higher than at high luminances, and the sensitivity rise with frequency is far less pronounced. (From Kelly, 1961.)

with scenes that are often brighter than movies, flicker poses an important constraint in the design of video (Section 6.2).

4.7. COLOR VISION

Color, or the spectral quality of light, yields important visual cues above and beyond the information that can be gained from luminance alone. In daily life, color (which reflects the chemical and physical makeup and sometimes the temperature of a material) often provides attractive or warning signals.

When light is dispersed by a spectroscope, we perceive a unique **hue** for each major region of the visible spectrum. Each hue, such as blue, green, yellow, orange, and red, is associated with a particular wavelength range in the electromagnetic spectrum. On the other hand, the perception of a given hue does not necessarily signal the presence of light with a particular wavelength. Light of two different wavelengths could be mixed in appropriate ratios to match the hue in question. For example, a mixture of blue and green light (but not pigments) can give rise to cyan; a mixture of red and green light can give rise to orange or yellow; and so on. In fact, an appropriate mixture of light with the three primary hues of red, green, and blue can yield essentially any hue of the spectrum as well as paler colors such as pink, pale blue, etc., or shades of gray from black to white.

These observations prompted Young (1802) and Helmholtz (1866) to theorize that color vision was based on the presence of three types of photoreceptors with differing spectral sensitivities. In

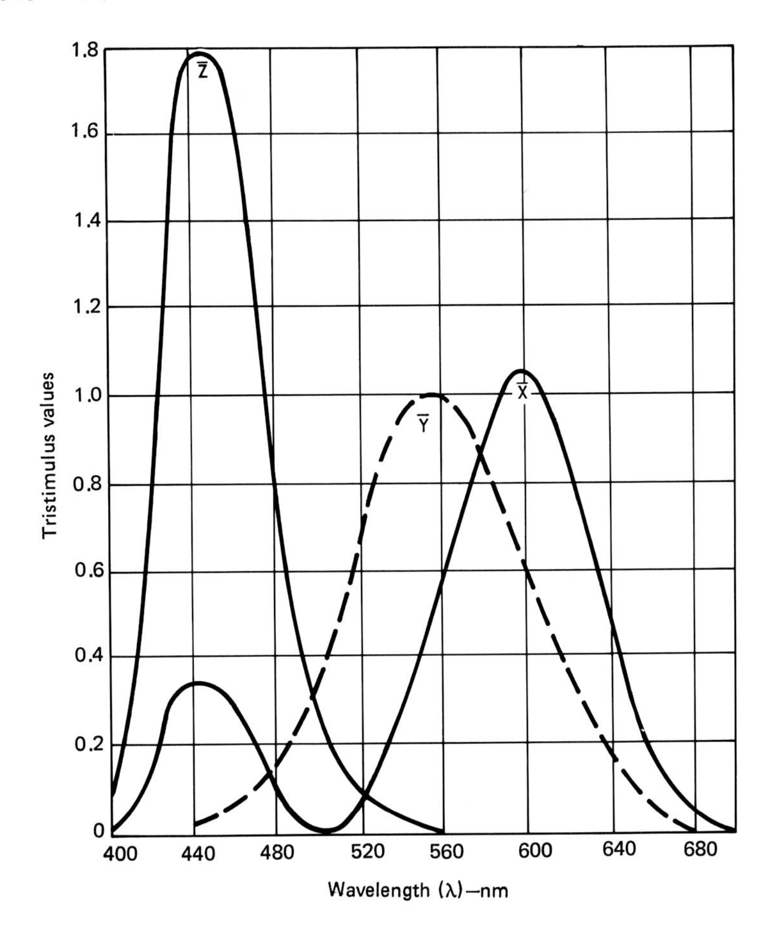

FIGURE 4-17. CIE standard color-mixtures curves. The curves represent the amounts of idealized primary colors required to match any of the pure spectral colors represented by wavelength. The *Y* curve represents the relative spectral luminous efficiency curve for photopic vision in Fig. 4-5, and the *X* and *Z* curves are drawn to provide appropriate matching chrominance. (Courtesy of RCA, from *RCA Electro-Optics Handbook,* 1974, p. 51.)

the 1960s, a century and a half after Young's original proposal, three types of cones with absorption peaks of about 430, 540, and 575 nm were indeed found in the human retina. Very interestingly, electrophysiological recordings showed that information about hue was not transmitted from the retina to the brain as the red, green, and blue signals but, instead, as a pair of signals: the yellow–blue and the red–green.

Long before the Young–Helmholtz theory was verified by these discoveries, an international commission met in France to establish the **CIE** chromaticity standard.* The standard allows determination of color by photometry, which depends on judgments of equality, rather than quantity, and which can be made with great precision. Since light of three hues with the spectral characteristics X, Y, and Z could be mixed at a ratio of x, y, and z to produce any color, a color C can be expressed by the formula

$$C = x \cdot X + y \cdot Y + z \cdot Z$$

*Commission Internationale de l'Éclairage, 1931 (e.g., see Wright, 1940.) The CIE system depends only upon data obtained from psychophysical measurements and is complete within itself. Knowledge of the details is irrelevant; in fact, this knowledge is still so incomplete that it is not yet possible to derive the color-matching equations from physiological data.

By defining X, Y, and Z as in Fig. 4-17, one could then mix these in the ratio of x, y, and z to obtain the color C. Conversely, any color could be determined by photometry, using calibrated sources and sensors.

Since x, y, and z are ratios, they can be set so that

$$x + y + z = 1$$

In this way, only two of the three fractions need to be specified to yield all three values, since $z = 1 - (x + y)$. The *CIE chromaticity diagram* can thus be plotted as a two-dimensional graph with y as the ordinate and x as the abscissa (Figs. 4-18, 4-19; see also Section 7.4 and Fig. 7-28).

Each point on the CIE chromaticity diagram specifies a unique color. Starting with the hue, blue, at 380-nm wavelength at the lowest point ($x = 0.17$, $y = 0$, $z = 0.83$) and moving clockwise along the spectrum locus, the wavelength increases through cyan (480–500 nm) (a mixture of blue and green; see Fig. 7-28) to green (520 nm) at the top of the curve ($x = 0.07$, $y = 0.83$, $z = 0.09$).

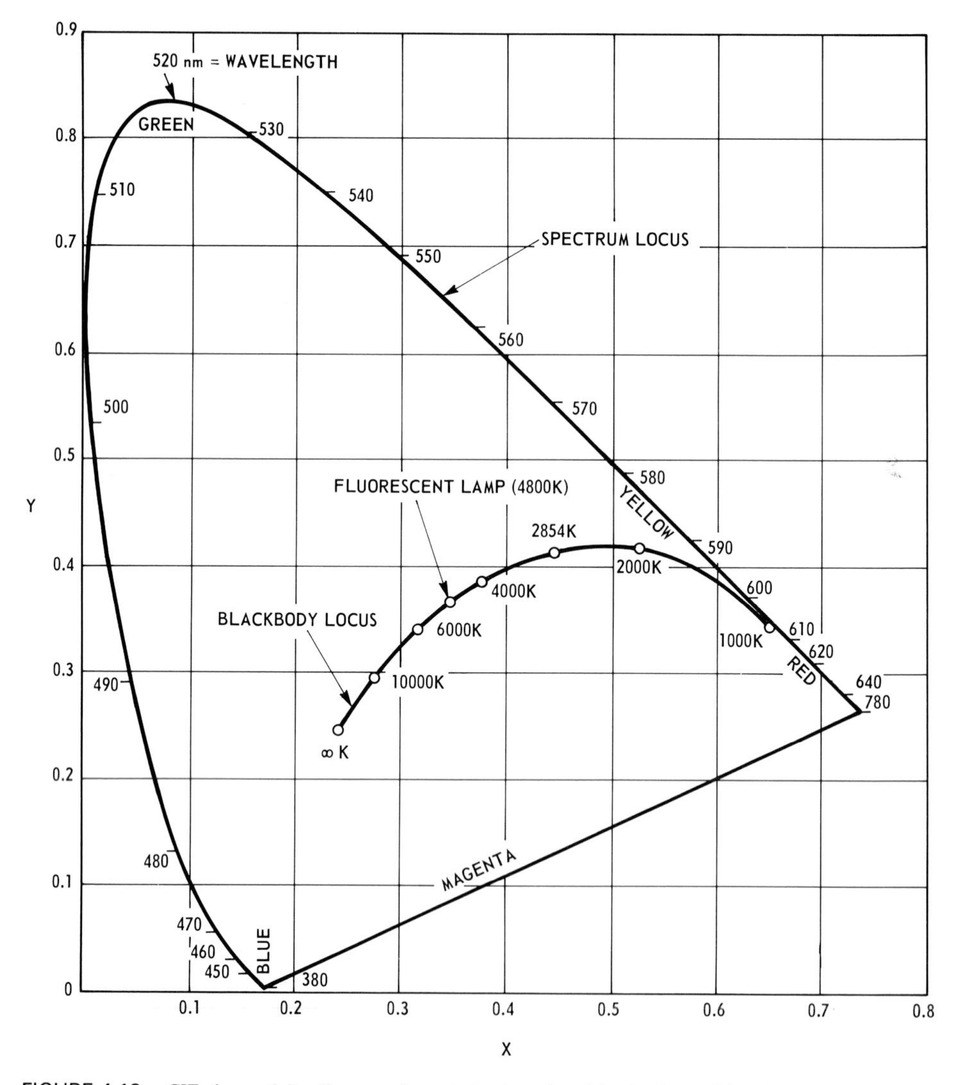

FIGURE 4-18. CIE chromaticity diagram of spectral colors closed (at the bottom) by "nonspectral" colors. See text for description. (Courtesy of RCA, from *RCA Electro-Optics Handbook*, 1974, p. 52.)

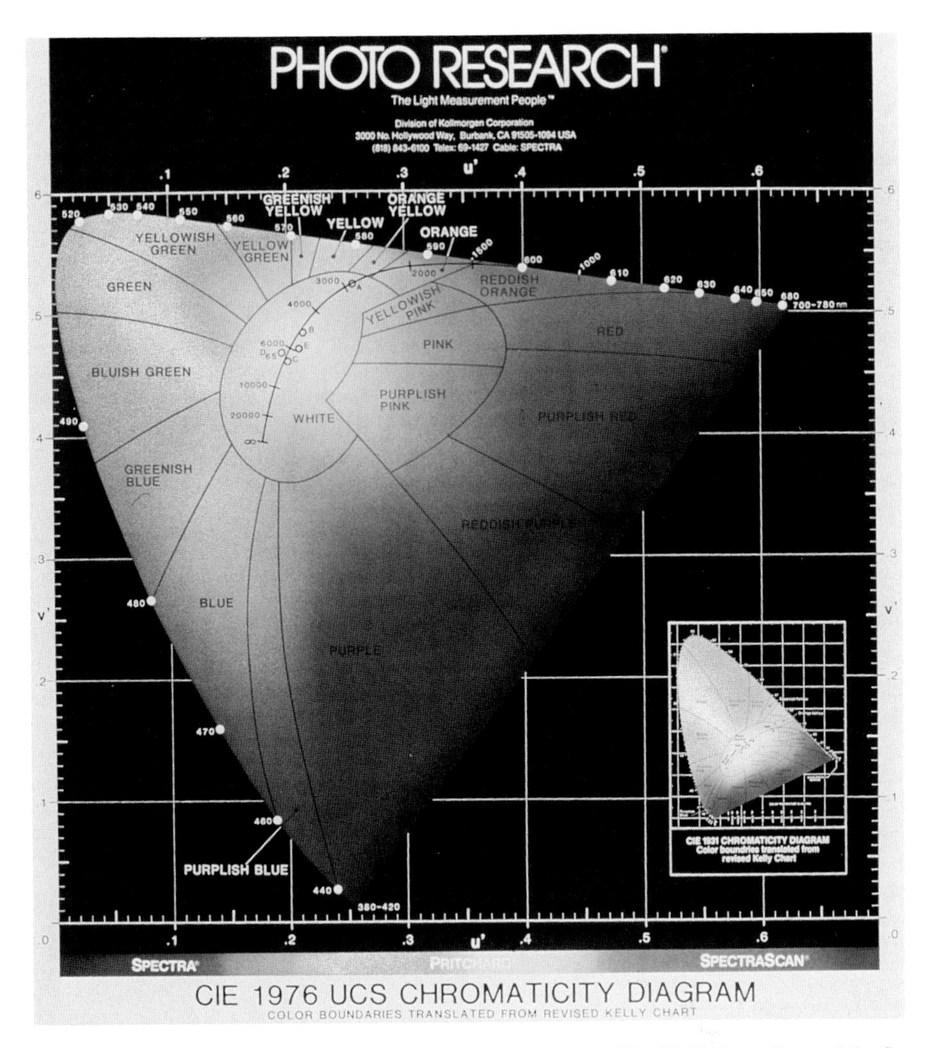

FIGURE 4-19. CIE chromaticity diagrams. The large figure is the 1976 CIE Uniform Chromaticity Space Diagram. The smaller figure is the 1931 CIE Chromaticity Diagram shown schematically in Fig. 4-18. The 1976 version better reflects the responses of the average human visual system to light of different hues. For color reproduction of the 1931 CIE Chromaticity Diagram see color plate following page xxvi. (See Miller, 1985.) (Courtesy of Photo Research, Burbank, Calif.)

Then descending along the spectrum locus, the wavelength continues to increase through yellow (570–590 nm), orange (590–610 nm), to red (610–700 nm). Abruptly, the curve takes a sharp turn, and the hues shift along the magenta axis (white minus green) back to blue again.*

The fully saturated colors, or hues, lie along the line that we just traced. The point of least saturation is CIE white ($x = 0.33$, $y = 0.33$, $z = 0.33$). Colors described by points closer to the

*The blackbody loci between 2500 and 12,000°K shown in Fig. 4-18 are recognized as "white" depending on the level of adaptation of the observer.

saturation line are more saturated (less whitish) than those represented by points closer to CIE white. The degree of saturation, or the lack of whiteness, is known as the **chrominance** of the color. Thus, color is characterized by its hue (spectral wavelength of the saturated color) and chrominance (the degree of saturation).

Interestingly, so long as we view color under adequate luminance, a colored patch (including shades of gray and white) does not change its color when the luminance of the scene is changed. More remarkably, even a gradient of luminance across the scene does not alter the perceived color or grayness (whiteness) of a patch. These observations of *color invariance* led E. H. Land to propose his retinex theory of color vision (Land, 1977, 1983; Land and McCann, 1971).

Once the luminance drops to the level of scotopic vision, the sensation of color vanishes; apparently, we have only one type of rod cell. However, even at the photopic luminance level a significant portion of individuals (about 8% of men and 0.4% of women) are unable to discriminate between certain colors. Because the different types of color blindness reflect the types of cone cells that are missing, visual responses of color-blind individuals have been extensively measured. Such studies have helped to uncover the roles of the different sensory cells in various visual mechanisms [see e.g., Marriott (1962) for further discussions on color vision].

4.8. STEREOSCOPY

Many cues give rise to our visual sense of depth, or of the third dimension. They include: scenes hidden by opaque objects lying closer to the observer, foreshortening of more distant objects by perspectives, shadows cast by oblique illumination, shading of the surface luminance, accommodation and convergence of the eye, rotation of the object, and binocular parallax.

While these cues are used routinely in daily life, we enter an unfamiliar world when we look through a microscope. With transparent objects that are illuminated from behind, many of these cues disappear. Furthermore, shadings that resembled the three-dimensional cue, but which do not necessarily signify the same spatio-optical relationships, can appear with oblique illumination or when asymmetric **aperture functions** are present (e.g., see Section 11.3).

In the absence of some means for generating other types of cues, binocular parallax (and the

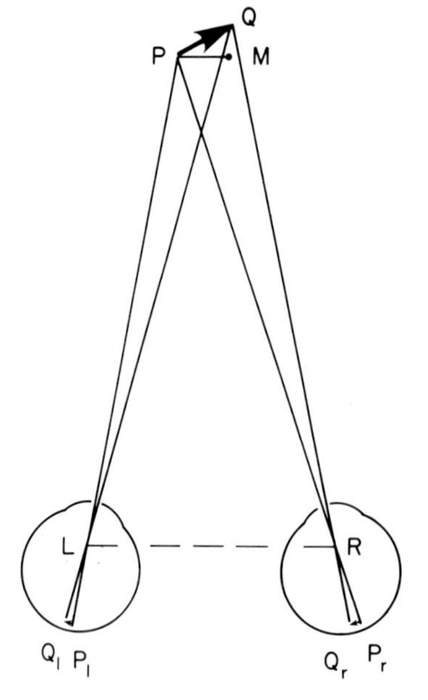

FIGURE 4-20. Stereoscopic acuity (how well one can detect small distances along the line of sight) may be specified by the minimum detectable difference between the angles PLQ and PRQ, which is indicative of the distance between points Q and M. Many individuals can detect differences as small as 10 sec of arc (Fig. 4-21).

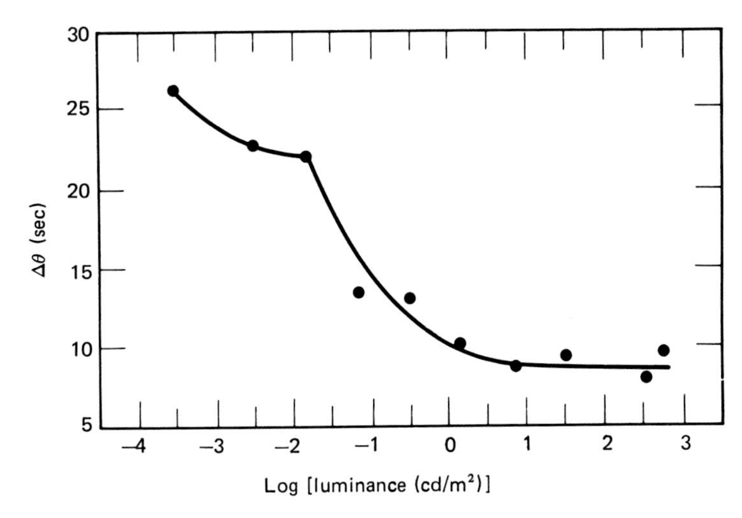

FIGURE 4-21. Stereoscopic acuity (ΔΘ) as a function of luminance. The familiar inflection point marks the transition between photopic and scotopic vision. (From Mueller and Lloyd, 1948.)

related phenomenon of specimen rotation) become the main cues that can potentially provide stereoscopy through a microscope.* Several means for generating binocular parallax and stereoscopic images with microscopes using high-NA objectives are described in Section 12.7.

Here we shall briefly examine **stereoscopic acuity,** namely the sensitivity of our visual sense for resolving small distances along the line of sight.

Consider two points P and Q, or an object PQ, that lies in front of our eyes as shown in Fig. 4-20. The question is: how far does PQ have to lie away from PM, which lies in a plane parallel to LR, the interocular axis, before we can tell that point Q is farther away from us than point M?

Since the binocular cue arises from the apparent differences of the distance P_1Q_1 produced on the left retina and P_rQ_r on the right retina, the question reduces to: how sensitively can we detect the size differences between these two images formed *in the two eyes*? In order to make the calculations independent of the absolute values of PQ and the distance to the object, we express our measurements in angles ∠PLQ and ∠PRQ. The acuity is given by the minimum difference of (∠PLQ - ∠PRQ) that tells us that Q is behind M.

As discussed in Section 4.4, our visual acuity for resolving two points or lines lying next to each other is about 1 min of arc (or in very favorable cases about 30 sec of arc). This limiting resolution matched the expectation from the geometrical distribution of the cones in the central fovea.

When stereoscopic acuity is actually measured, it turns out to be better than 10 sec of arc as shown in Fig. 4-21! (Stereoscopic acuity improves as the luminance of the object is raised; again, there is the familiar inflection point between photopic and scotopic acuity .)Neither the differences in visual acuity between the two eyes nor of contrast of images presented to the two eyes are found to have much effect on stereoscopic acuity. And we can retain this stereoscopic acuity even when wearing glasses that present the two eyes with different size images.

All of these observations suggest the intricate processing that must be taking place in the retina and the brain. The two eyes in concert are providing the brain with incredibly precise information that would even tax the acuity of a single eye alone. In addition, the brain has to instantly establish which points in the left and right eye are congruent with each other.†

*Clearly, many of the other cues can also exist when we use a dissecting microscope with epi-illumination.
†For a very interesting analysis and challenging accounts on human vision, including stereoscopic processing, see Marr (1982), Gregory (1970), and Julesz (1971).

4.9. ADDITIONAL REMARKS

As noted in Section 4.1, the goal of TV is to provide an instant facsimile of lifelike scenes to a widely scattered audience. Thus, the early designers of TV took many attributes of the human visual (and auditory) system into account in specifying TV signal and equipment parameters. The very fact that one can use a raster scan with a limited number of lines, presented only 50 or 60 times/sec, with rather limited frequency ranges, to transmit the video signal (which limits the **horizontal resolution**; see Chapter 6), reflects the finite resolving power, critical flicker frequency, and integrating capacity of the human visual system. The sensitometric characteristics of the camera and monitor, the mode of color presentation, and the choice of color-signal format (Chapter 8), all reflect the trichrome response and the wavelength-dependent variation in acuity of our visual system (e.g., Bedford, 1950). The degree of tolerable noise and the degree of fidelity needed to present acceptable or pleasing pictures (Chapters 7–12) were not only matters of major concern to the video engineer, but also led to many studies on visual acuity, contrast discrimination, signal detection, and processing capabilities of the eye and visual system (see especially Rose, 1973).

These characteristics of the eye and visual system not only provide important foundations and constraints for the TV engineer, but are useful to keep in mind when we apply video to microscopy. By being aware of these characteristics and of the limitations and capabilities of the video equipment (and, or course, of the microscope), we have a better chance of getting more out of video microscopy. To take a few simple examples: we can dramatically improve the visibility of details in some low-contrast scenes by dimming the room light, by reducing the monitor BRIGHTNESS control, and by turning up its CONTRAST control. Or, in another scene, intrusive flicker might be tamed by turning down the monitor BRIGHTNESS control. In a rapidly changing scene, the jitter appearing in a frozen-frame display could disappear if the latter were presented as a frozen **field** (Chapter 6).

It is useful to remember that TV may not provide a reasonable signal to a viewer who (or receiver that) does not have average human visual and auditory characteristics. An animal whose characteristics differ from our own might perceive the signal quite differently. To a photographic emulsion, photoelectric cell, or video camera, with sensitometric characteristics and spatial and temporal integrating powers unlike those of our visual system, the signal would indeed appear very differently. This kind of awareness not only can help us avoid pitfalls, but can provide additional opportunities for conveying or extracting information that otherwise might not be apparent.

5

Microscope Image Formation

5.1. INTRODUCTION: ESSENTIAL OPTICAL TRAIN OF THE LIGHT MICROSCOPE

The compound light microscope provides a two-dimensional magnified image of the specimen that allows us to resolve and measure fine details of the specimen structure. The specimen can be positioned (oriented) and focused precisely, and the contrast and brightness of the image can be adjusted to bring out desired features of the specimen structure.*

Mounted on a stable mechanical base, the optical components of the microscope can be rapidly exchanged and precisely centered, oriented, and focused. The base supports and couples the optical components and the specimen, on axis, into an *optical train.*

The optical train of a microscope usually consists of: (1) the illuminator, including the light source and collector; (2) the condenser; (3) the specimen, including the slide and coverslip; (4) the objective lens; (5) the ocular, or eyepiece; and (6) the camera, or observer's eye.

In addition, a conditioning device is often inserted between the illuminator and the condenser and a complementary filtering device is placed between the objective lens and the ocular. The conditioning and filtering devices together modify the image contrast as functions of spatial frequency, phase, polarization, wavelength, etc.

Whether or not specific devices are inserted to condition the entrant waves and to filter the image-forming waves, some conditioning and filtering naturally takes place in any microscope.

Therefore, we will consider the sequence of components in the optical train of the light microscope to be: (1) the illuminator; (2) the conditioner; (3) the condenser; (4) the specimen; (5) the objective lens; (6) the image filter; (7) the ocular; and (8) the receiver (Fig. 5-1).

Some of these components act as imaging elements, while most also have filtering or transforming functions. These points will be explained in this chapter.

5.2. IMAGING COMPONENTS IN THE LIGHT MICROSCOPE: MAGNIFICATION

The components that form images in the microscope optical train are: the collector in the illuminator, the condenser lens, the objective lens, the ocular, and the refractive elements of the eye or the lens on the camera.

As their names imply, some of these components are not commonly thought of as imaging components. Nevertheless, their properties are important in determining the final quality of the microscope image.

Before examing the individual imaging components, we shall first review the imaging characteristics of a *perfect lens,* which is an ideally corrected lens, free of **aberrations** (Section 5.9).

*Among the many excellent general works on light microscopy, the following are especially useful: Beck (1938), Belling (1930), Françon (1961), Gage (1941), Loveland (1981), Martin (1966), Piller (1977), Shillaber (1944), Spencer (1982).

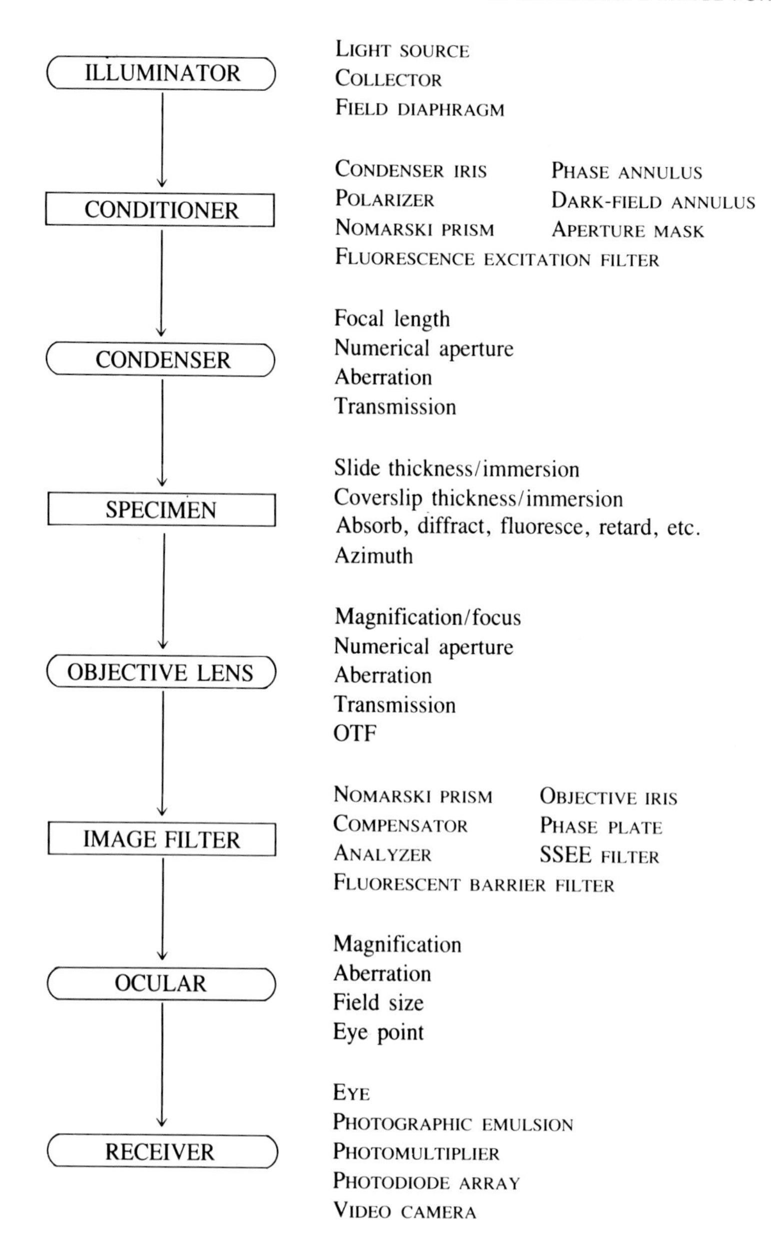

FIGURE 5-1. Components of the optical train in a light microscope. The components shown in rounded outlines perform imaging functions. The primary effect of the other components is to modify the light by absorption, aperturing, diffraction, fluorescence, retardation, polarization, etc. Elements that may make up the component (upper case), or selected properties of the component (lower case), are shown to the right.

A perfect lens converts a paraxial, parallel beam of light into a concentrated, fine spot of light at its focus (F in Fig. 5-2). Conversely, a point source of light located at F emerges as a paraxial, parallel beam of light from the lens (with the direction of the arrows reversed in Fig. 5-2).

The same phenomena, considering light as trains of waves, or as wave optical phenomena, are seen as follows.

In a parallel beam, the light waves (for each monochromatic wavelength) are vibrating in phase. Therefore, the wave front lies in a plane that is normal to the direction of propagation of the

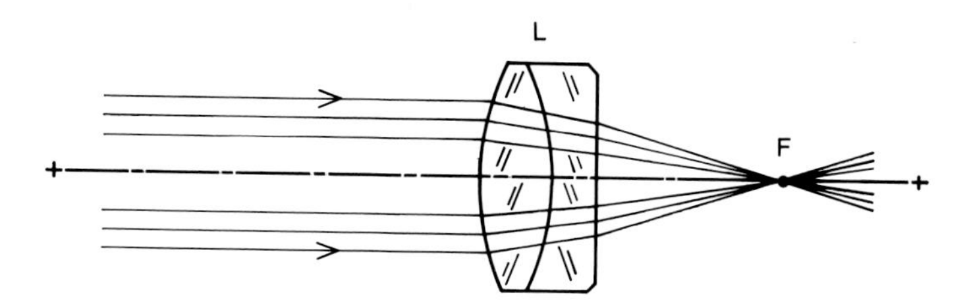

FIGURE 5-2. Ray diagram through a perfect lens. L = lens; F = focus. See text.

parallel beam (Fig. 5-3). The plane wave is converted to a spherical wave by a perfect lens. The front of the spherical wave is centered at the focus of the lens (left to right in Fig. 5-3). The waves all arrive in phase and constructively interfere with each other at the focus. Conversely, a spherical wave front, emanating from the focus of a perfect lens, is converted by the lens into a plane wave (right to left in Fig. 5-3).

The plane wave front need not be normal to the axis of the lens, but can be tilted by an angle α (Fig. 5-4). The center (S) of the spherical wave then lies at a distance δ from the axis of the lens, where

$$\sigma = f \sin \alpha \tag{5-1}$$

f being the focal length of the lens (Fig. 5-4).*

In general, a perfect lens converts one spherical wave into another spherical wave. When the radius of one spherical wave is infinite, as in the examples just given, the radius of the other becomes equal to the focal length of the lens (Figs. 5-3, 5-4).

In considering a point source S_1 that does not lie in the focal plane of the lens, the lens L can be thought of as being made up of two lenses, L_a and L_b (Fig. 5-5). The lenses are chosen so that S_1 lies at the focal distance f_a away from L_a, and S_2 at the focal length f_b away from L_b.

Then, a spherical wave emanating from S_1, located at a distance δ from the axis of the lens, is converted by L_a into a plane wave that is tilted with respect to the lens axis normal by α (as in Fig. 5.4). δ and α are again related by equation (5-1), with f being replaced by f_a.

Returning to Fig. 5-5, the second lens (L_b) converts the plane wave emerging from L_a into a spherical wave. The center of this spherical wave is located at S_2.

In other words, the perfect lens L (= $L_a + L_b$) focuses S_1 onto S_2, and also focuses S_2 onto S_1. S_1 and S_2 are known as **conjugate points.**

If S_1 is considered as a set of points, then a perfect lens would focus each point of the set to a conjugate point in set S_2. If the set S_1 lies in a plane normal to the lens axis, then the set S_2 would lie in a corresponding plane, also normal to the lens axis, or in some specific corresponding surface.

Likewise, the perfect lens would focus every point S_2 lying in the second plane (or surface) onto corresponding points S_1 in the original plane. Such corresponding planes (or surfaces) are known as **conjugate planes** (or surfaces).

Light can be viewed as a train of waves, or alternatively, the propagating waves can be represented by rays that are oriented normal to the wave fronts.

Thus, we can simplify the situation described in Fig. 5-5 to that shown in Fig. 5-6. Using the geometrical construction shown in Fig. 5-6, we can readily determine the size and location of the images formed by the lens; we take two representative light rays, one paraxial, and the other traveling through the center of the lens L. Let the specimen, or source of light, S_1, be located at a distance a (greater than the front focal length f') to the left of the lens. An inverted image, S_2, then forms at a distance b to the right of the lens.

*Figure 5-4 shows $\delta = f \tan \alpha$, rather than $\delta = f \sin \alpha$. In fact, the latter is correct since, strictly speaking, f refers to the radius of the arc centered on S and going through the middle of the lens; that is so because microscope objectives are designed to satisfy the sine condition, which minimizes the aberrations.

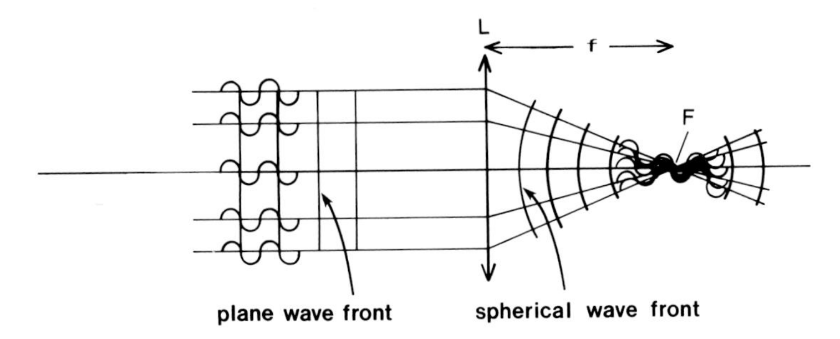

FIGURE 5-3. Wave diagram through a perfect lens. L = lens (the double-headed arrow representing the lens replaces, for example, the **achromatic** doublet shown in Fig. 5.2.); F = focus; f = focal length. See text.

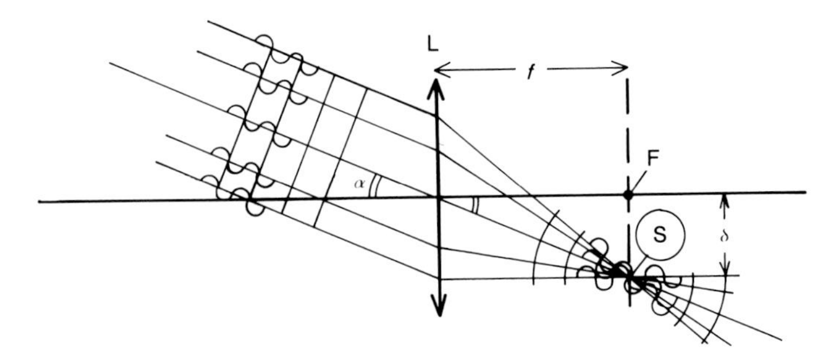

FIGURE 5-4. Wave diagram for an oblique beam through a perfect lens. L = lens; F = focus; f = focal length; α = angle of tilt of light beam with respect to lens axis; S = center of focused spherical wave; δ = distance of S from lens axis. See text.

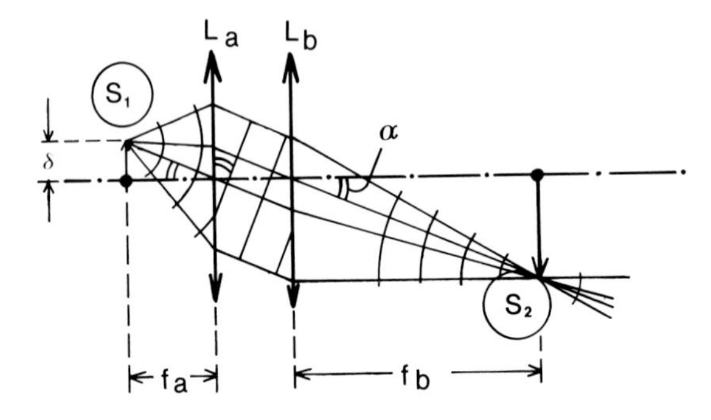

FIGURE 5-5. Wave diagram for an oblique beam through a perfect lens. L_a, L_b = the two lenses of which the perfect lens can be considered to be composed; f_a, f_b = focal lengths; α = angle of tilt of a central ray with respect to the lens axis; S_1, S_2 = conjugate points: either can be an image of the other. See text.

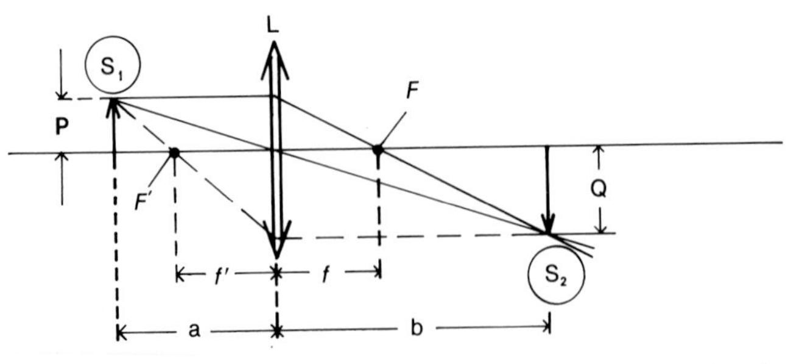

FIGURE 5-6. Ray diagram for an oblique beam. L = lens; F, F′ = foci; f, f' = focal lengths; S_1, S_2 = source, or object, and its (inverted) image; P, Q = heights of S_1 and S_2; a, b = distances of S_1 and S_2 from the lens. See text.

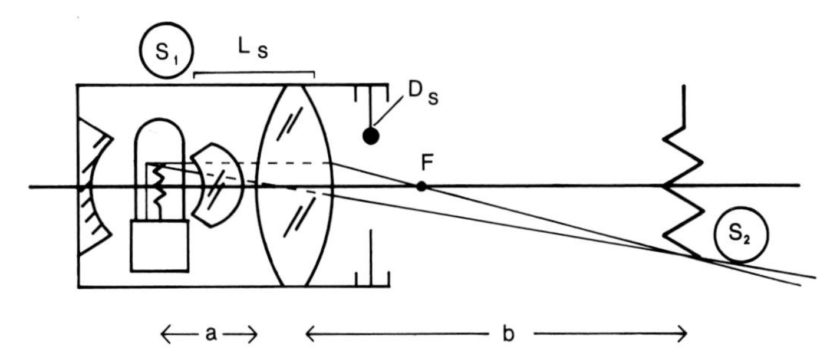

FIGURE 5-7. Image planes of the collector lens. L_s = collector lens; F = focus; S_1, S_2 = the light source and its image, which, in Koehler illumination, is placed in the plane of the condenser iris diaphragm; a, b = distances of S_1 and S_2 from the **principal planes** of the lens; D_s = field diaphragm.

The distances a, b, and the rear focal length f of the lens are related to each other by

$$1/a + 1/b = 1/f \quad (5\text{-}2)$$

Q, the height of the image S_2, is magnified M times with respect to P, the height of S_1, where the **lateral magnification** M (approximating to a thin lens) is given by

$$M = Q/P = b/a \quad (5\text{-}3)$$

Since S_1 and S_2 lie in conjugate planes, S_2 would likewise be focused at S_1 by the same lens. The focal length f would then be changed to f' (equation 5-2), and the magnification M to $1/M$.

The **longitudinal** (or axial) **magnification,** which is the ratio of distances between two image points along the lens axis and between their two conjugate points in the source, is the square of the lateral magnification (for small distances from the image plane).

These general relationships hold for all of the imaging components in the microscope: the collector in the light source (Fig. 5-7); the condenser (Fig. 5-8); the objective lens (Fig. 5-9); many oculars in the projection mode (Fig. 5-10); the camera (Fig. 5-11); and the refractile elements of the eye (right half of Fig. 5-12).

In looking directly through the microscope, we view the image through the ocular (Fig. 5-12) rather than using the ocular to project an image. When we view the image through the ocular, an *intermediate image* (I_3) is first projected by the objective lens (Fig. 5-9) in such a way that I_3 is

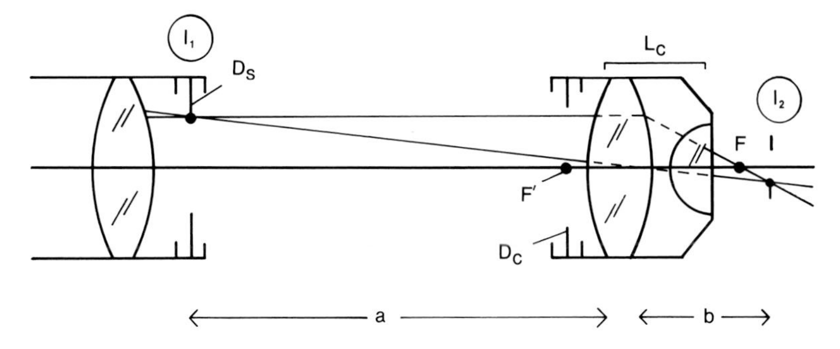

FIGURE 5-8. Image planes of the condenser lens. L_c = condenser lens; F, F′ = foci; I_1, I_2 = conjugate images (in Koehler illumination, the field diaphragm, D_s, and specimen, respectively); a, b = distances of I_1 and I_2 from the principal planes of the lens; D_c = **condenser iris diaphragm.**

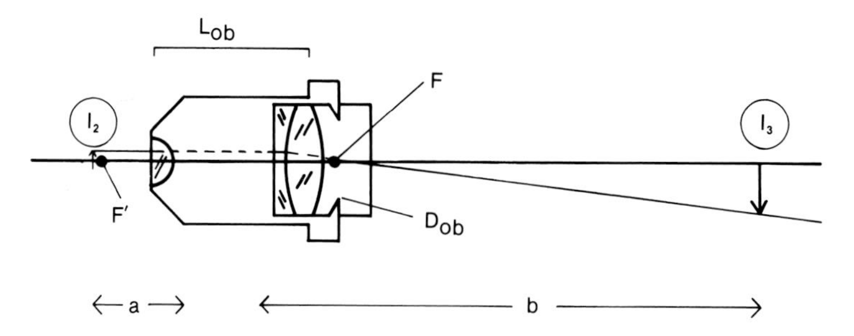

FIGURE 5-9. Image planes of the objective lens. L_{ob} = objective lens; F, F′ = foci; I_2, I_3 = conjugate images (the specimen and **intermediate image plane** located at the ocular field stop, respectively); *a*, *b* = distances of I_2 and I_3 from the principal planes of the lens; D_{ob} = back aperture of the objective.

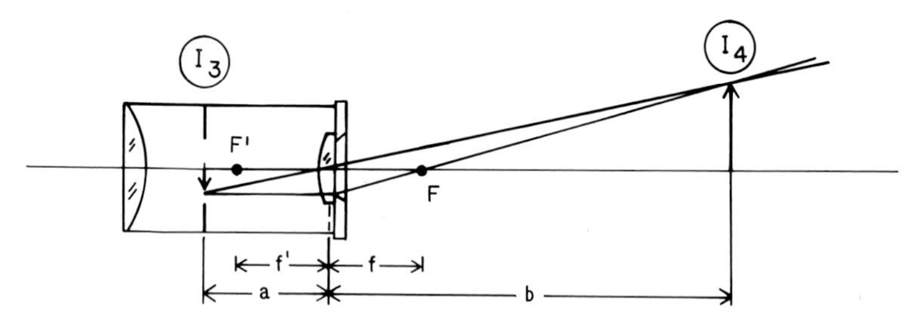

FIGURE 5-10. Image planes of the ocular (in projection mode). F, F′ = foci; *f*, *f′* = focal lengths; I_3, I_4 = the intermediate image and its conjugate; *a*, *b* = distances of I_3 and I_4 from the lens of the ocular. Since $a > f'$, I_4 is a real image. See text.

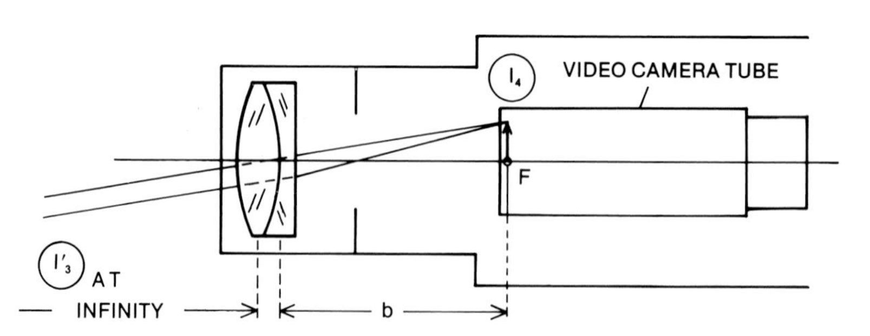

FIGURE 5-11. Use of a special lens for imaging onto the video camera tube. If a lens is used on the video camera after the ocular, it converges the virtual image, I'_3, which the ocular has focused to infinity, onto the faceplate of the camera tube, at I_4. *b* = focal length of lens. A standard video camera lens can usually not be used because of vignetting. See text.

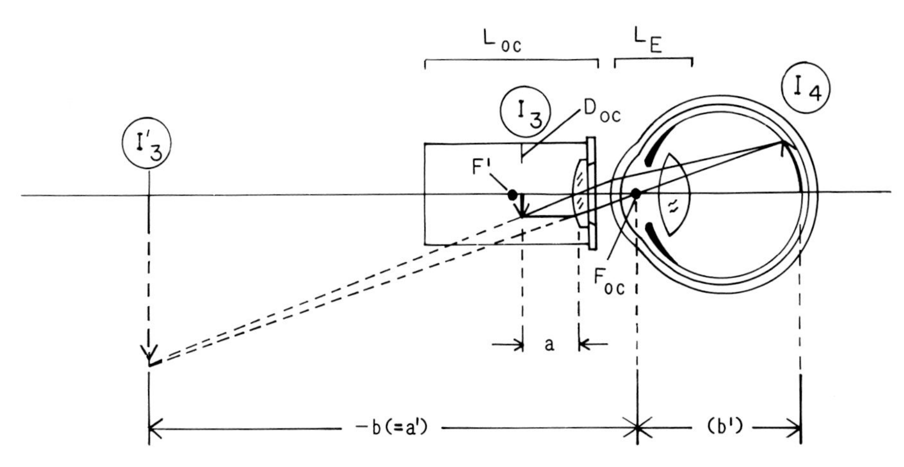

FIGURE 5-12. Image formation on the retina. When the specimen is viewed through the ocular, an intermediate image, I_3, is formed (by the objective lens) at a distance, *a*, closer to the ocular than its front focal length, precluding the formation of a real image after the lens as in Fig. 5-10. Rather, the ocular and eye together form an image, I_4, on the retina as though the eye were seeing the virtual image, I'_3. L_{oc} = ocular lens; L_E = refractile elements of the eye (corneal surface and crystalline lens); F_{oc}, F' = foci of the ocular; D_{oc} = ocular field stop.

formed at a distance (a) slightly shorter than the front focal length of the ocular (Fig. 5-12). If equation (5-2) is applied to the ocular used in this manner, we see that when $a < f$, then $b < 0$.

Therefore, a real image cannot be formed to the right of the ocular (in the absence of the eye or a camera). Instead, a virtual image (I'_3) is formed at a distance $-b$ to the right, or b to the left, of the lens (using the definitions of a and b in Fig. 5-6). In looking through the ocular, the image-forming beam (diverging out of the ocular) would appear to come from a virtual source I'_3 (Fig. 5-12).

Geometrically, I_2, I_3, I'_3, and I_4 are related to each other as shown in Fig. 5-13. In all of the imaging steps, except for the formation of I'_3, the image is real and inverted (Figs. 5-8 to 5-11, and the right half of Fig. 5-12). In the case of the ocular used for direct viewing, the image I'_3 is not real but virtual, and is not inverted relative to the intermediate image (Figs. 5-12, 5-13).

In our eye, the image (I_4) on the retina is inverted relative to I_3 and I'_3 (Fig. 5-12), but we do not perceive it to be inverted (Chapter 4).

From equation (5-3) and the geometrical construction in Fig. 5-13, we can deduce the total magnification of the microscope as follows.

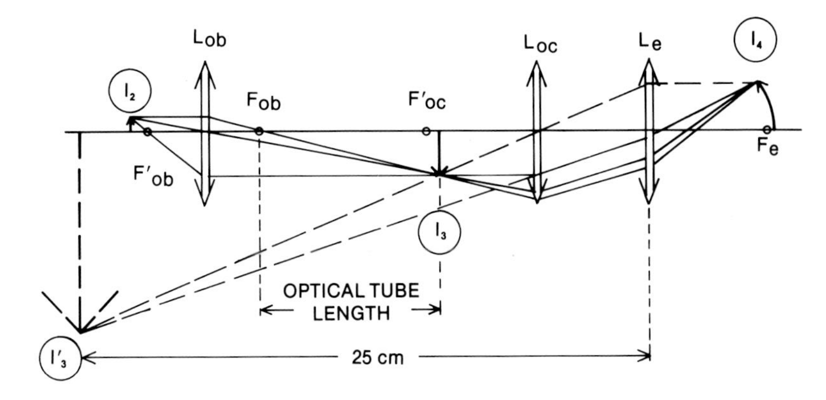

FIGURE 5-13. Relationships among the conjugate field planes shown separately in Figs. 5-9 through 5-12.

The magnification of the objective lens is specified for a particular projection distance for which the objective lens is corrected (Section 5.9). In many microscopes, the projection distance for the objective lens (approximately the optical tube lengths) is about 160 mm. Therefore, a 4-mm-focal-length objective lens, for example, would have a lateral magnification of about 40 (160/4). (The longitudinal magnification would be 1600 times.)

For visual observation, unit magnification is assumed when an object or image is placed 250 mm from the eye. Thus, a 25-mm-focal-length ocular would be designated as having a power of 10 times (250/25).

The total lateral magnification of the microscope for visual observation is given by the product of the objective lens and ocular magnifications. In the present example, the total lateral magnification would be about 400 ×.

For video microscopy or photography, a special positive lens (not the standard camera lens; see Sections 3.3, 3.7) is often placed after the ocular. The lens converges the image, which is focused to infinity coming out of the ocular, onto the plane of the **photocathode** (target) or the emulsion (Fig. 5-11). In this case, the lateral magnification (M_{proj}) of the projection system (without the objective lens) i.e., of the relay optics, is given by

$$M_{proj} = f_{camera}/f_{ocular}$$

where f_{camera} and f_{ocular} are respectively the focal lengths of the lens used after the ocular and the ocular itself.

With such a projection system, the total lateral magnification (M) on the faceplate of the video camera tube or photographic emulsion would be

$$\begin{aligned} M &= M_{obj} \cdot M_{proj} \\ &= M_{obj} \cdot M_{ocular} \cdot f_{camera}/250 \text{ mm} \end{aligned} \qquad (5\text{-}4)$$

When a converging lens is not used after the ocular, but rather the ocular itself is used to project the image onto the video image tube or photographic emulsion, the total lateral magnification becomes

$$M = M_{obj} \cdot D_p/f_{ocular} \qquad (5\text{-}5)$$

where D_p is the projection distance from the ocular to the image. Unless a special projection ocular is used, D_p should be at least 20–30 cm to avoid image distortion (Sections 5.9, 5.10).

The magnifications inscribed on the objective and oculars are nominal. The exact magnification of the final image is calibrated against a standard stage micrometer, which is imaged under the identical optical condition as the specimen.

Occasionally, the faceplate of the camera tube is placed directly in the intermediate image plane, and no ocular is used. This method is not recommended unless the performance of the video system is limited by the absolute amount of light available. With this method, image magnification on the faceplate is fixed at the designated magnification of the objective lens. Such a fixed magnification imposes a severe limitation since it prevents us from adjusting the magnification and optimizing the quality of the final video image (Section 5.10, Chapter 11). In addition, a proper ocular should be used since in most microscopes, the ocular is made to correct for the residual aberration in the objective lens.*

The next section examines how the various imaging components of the microscope are related to each other in Koehler illumination.

*The Nikon CF (chrome free) lenses and Zeiss Axiomat lenses are exceptions to this rule. The objectives and oculars are independently corrected in these systems.

5.3. PRINCIPLES OF KOEHLER ILLUMINATION

The condenser in the light microscope was initially developed (as the name implies) for collecting ample light to illuminate the specimen. While many individuals still view this as the main function of the condenser, in fact the ''illuminating system'' plays a far more important role in microscopy than is commonly recognized. The role of the illuminating system is indeed so important that one can quickly spot the degree of sophistication of the user of a microscope by the way the condenser and illuminator are adjusted.

One can achieve a superb image by properly aligning the axes, and adjusting the foci and diaphragm openings in the light source and the condenser. On the other hand, by poorly adjusting the optical train that illuminates the specimen, one can use a $2000 objective lens and produce an image that is worse than that produced by a $200 objective lens used skillfully. Of course, a more expensive, better-corrected objective lens can provide an image that is vastly better than a lower-priced lens *if* the illuminating system is adjusted well.*

In adjusting the illuminating system, or the light waves that illuminate the specimen in a microscope, Koehler illumination (Köhler, 1893; Dempster, 1944) provides several advantages:

1. The field is homogeneously bright.
2. The working NA of the condenser and the size of the illuminated field can be regulated independently.
3. The specimen can be illuminated by a converging set of plane wave fronts, each arising from separate points of the light source imaged in the condenser aperture.
4. This gives rise to the maximum lateral resolution and very fine optical sectioning, which yields maximum axial resolution.
5. The front focal plane of the condenser becomes conjugate with the rear focal plane of the objective lens, a condition needed for optimal contrast enhancement of the finer specimen details.
6. Flare, arising from the microscope optics and their barrels, is reduced without any vignetting.

For Koehler illumination, the various lenses of the microscope are arranged as shown in Fig. 5-14 and 5-15 and focused as follows:

1. The collector lens (L_s) focuses an image of the light source onto the condenser iris diaphragm (D_c).
2. The condenser lens (L_c) focuses an image of the field diaphragm (D_s) in the plane of the specimen.
3. The objective lens, the ocular, and the refractive elements of the eye (or camera) together focus an image of the specimen onto the retina (or camera image plane).

These adjustments give rise to the two sets of optical paths and image planes represented in Fig. 5-14 and 5-15. Together, the two sets of optical paths and image planes, characterize Koehler illumination.

Each set of optical paths and image planes will now be examined in turn, starting with Fig. 5-14.

As shown in Fig. 5-7 and 5-14, the collector lens of the illuminator produces an enlarged inverted image (S_2) of the light source (S_1) onto the condenser iris diaphragm.

Then the condenser and the objective lens together form an image (S_3) of S_2 onto the rear focal plane of the objective lens in the following manner.

The condenser of the microscope is designed so that its iris diaphragm is located at the front

*One can equip modern research microscopes with condensers that are well corrected for aberrations (e.g., aplanatic achromats). In addition to selecting a superior-quality condenser, it is important to use the condenser under the specified optical conditions: dry or oil-immersed; proper distance between the condenser and the field diaphragm; and fully adjusted for Koehler illumination; with the aperture uniformly and fully illuminated.

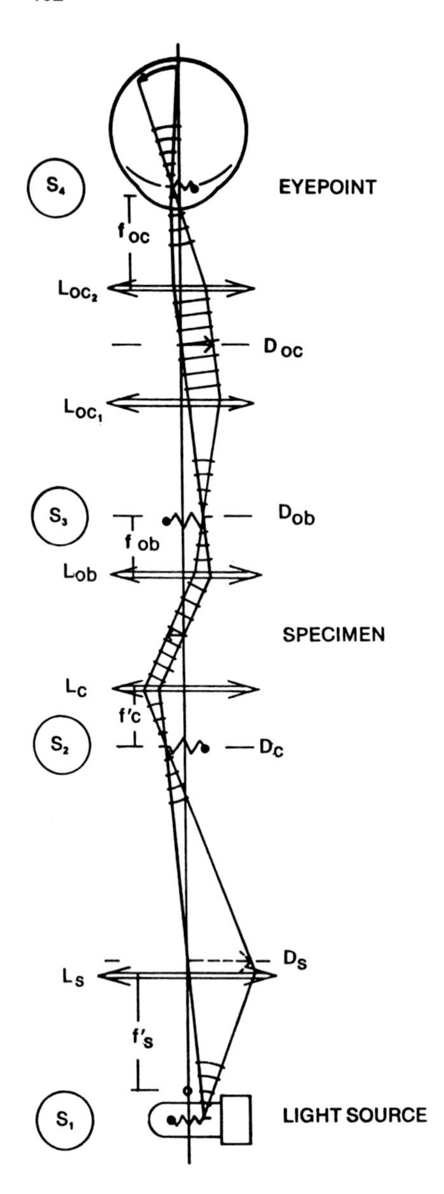

FIGURE 5-14. The optical train adjusted for Koehler illumination, illustrating the conjugate aperture planes, S_1–S_4. See text.

focal plane (F′) of the condenser lens (Figs. 5-8, 5-14). Therefore, light emanating from a point in the plane of the condenser iris emerges from the condenser as parallel rays, or as a plane wave (right to left in Fig. 5-4).

This plane wave traverses the specimen space (Fig. 5-14) and enters the objective lens. The objective lens converts the plane wave to a spherical wave (as does L_b in Fig. 5-5). The spherical wave converges to the rear focal plane of the objective lens (S_3, Fig. 5-14).

Thus, each point at the rear focal plane of the objective lens is conjugate with a corresponding point in the plane of the condenser iris diaphragm, and the condenser and objective lenses together form an inverted real image, S_3, of S_2 at the rear focal plane of the objective lens (Fig. 5-14).

The rear focal plane of the objective lens or, to a close approximation, its **back aperture** (D_{ob} in Fig. 5-9), is in turn focused by the ocular (Fig. 5-14), which forms a small inverted image (S_4) of D_{ob} ($\simeq S_3$) at the eyepoint.

The eyepoint is located just beyond the rear focus (F_{oc}) of the ocular; it is where the pupil of the eye is placed (Fig. 5-12).

In this manner, a series of planes containing S_1, S_2, and S_3 are successively focused and relayed to the eyepoint by the lenses L_s, L_c plus L_{ob}, and L_{oc}. At the eyepoint, these images are all superimposed in the same plane and appear as the Ramsden disk.

For each lens (or lens group), the image and its source can be interchanged as discussed in Section 5.2; each image and its source lie in conjugate planes. This means that a source of light, an image, or an object placed in any one of the four planes (S_1–S_4) would also be focused to the other three planes.

Therefore, in Koehler illumination, the following four structures lie in conjugate planes: the light source, the condenser iris diaphragm, the rear focal plane of the objective lens, and the eyepoint. These four lie in the **aperture planes** of the light microscope.

Simultaneously, another set of conditions represented in Fig. 5-15 is fulfilled in Koehler illumination.

As shown in Figs. 5-8 and 5-15, the condenser lens (L_c) produces a small inverted image of the

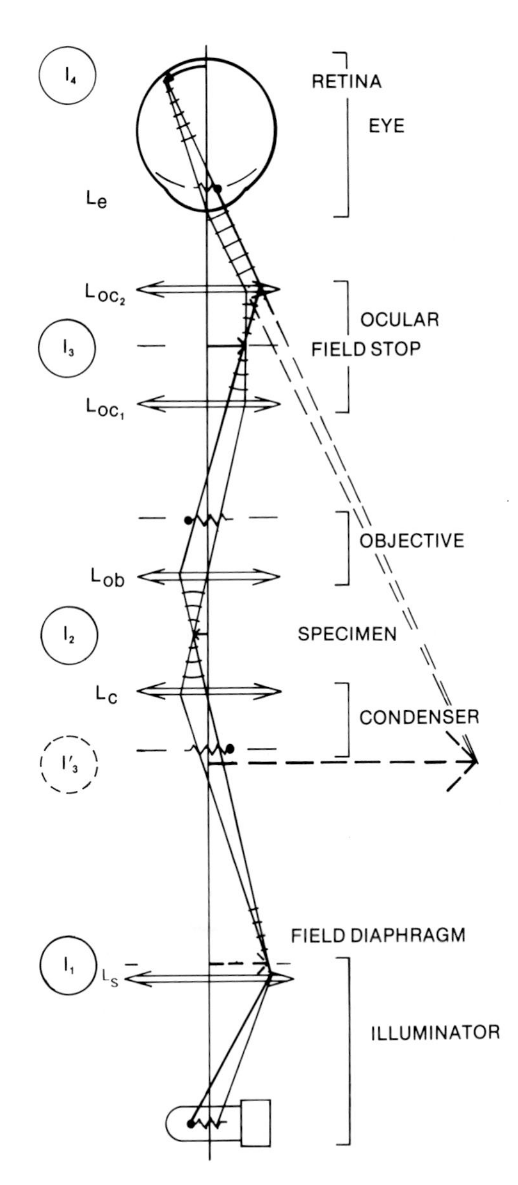

FIGURE 5-15. The optical train adjusted for Koehler illumination, illustrating the conjugate field planes, I_1–I_4. See text.

field diaphragm onto the plane of the specimen. The field diaphragm is located at I_1 and the specimen at I_2.

The specimen is then focused by the objective lens (L_{ob}: Figs. 5-9, 5-15), which produces a magnified inverted image (I_3) of the specimen. I_3 appears in the intermediate image plane containing the field stop (D_{oc}) in the ocular.

Next, as in Figs. 5-12 and 5-15, the lenses of the ocular (L_{oc}) and the refractive elements of the eye (L_e) together form an image (I_4) on the retina. I_4 is inverted relative to I_3.*

Therefore, in Koehler illumination, the following four structures also lie in conjugate planes: the field diaphragm, the specimen, the ocular field stop, and the retina. These four conjugate planes are the **field planes.**

In Koehler illumination, then, we have two sets of conjugate planes: the aperture planes S_1–S_4, and the field planes I_1–I_4.

The two sets of conjugate planes are related to each other in an important fashion as follows. In Fig. 5-14 the spherical wave fronts converge, and the rays focus, onto the aperture planes (S_1–S_4). In Fig. 5-15 the wave fronts converge and focus onto the field planes (I_1–I_4).

On the other hand, in Fig. 5-14 the wave fronts are planar, or nearly so, as they traverse the field planes. Likewise, in Fig. 5-15 the wave fronts are planar, or nearly so, as they traverse the aperture planes.

Therefore, spherical waves in one set of conjugate planes become (nearly) plane waves in the other set of conjugate planes, and vice versa. Rays that are focused in one set of conjugate planes are (nearly) parallel rays in the other.

The two sets of conjugate planes are thus reciprocally related to each other.

This reciprocal relationship explains how the various diaphragms and stops in a microscope affect the cone angle of illumination and the size, brightness, and uniformity of the microscope field (Section 5.4). Whatever changes can be made to the beam of light in the field planes (i.e., by manipulating apertures or angles of the light path or by inserting conditioners) also can be made in the aperture planes by changing the angle or opening for the beam. The relationship is reciprocal.

More fundamentally, the reciprocal relationship between the two sets of conjugate planes explains how the light waves illuminating the specimen and diffracted by the specimen relate to each other (Section 5.5). These relationships in turn explain the function and adjustments of the devices that condition the entrant wave and filter the specimen image (Sections 5.5–5.7).

All of these parameters together determine (or affect) the resolution, contrast, and fidelity of the microscope image.

5.4. ADJUSTING THE MICROSCOPE FOR KOEHLER ILLUMINATION

This section will consider, in turn, three practical matters: How do we align the microscope for Koehler illumination? How do we focus the lenses to achieve Koehler illumination? And how do we adjust the field and aperture diaphragms in Koehler illumination.

In translating the principles of Koehler illumination into practice, we need to note that the lenses in a microscope are not, in fact, corrected to be free of all aberrations. Rather, each lens group is corrected to fit a given task at a specific location under specified conditions of focus in the optical train.

Although these design features constrain how we select and position the lenses in the microscope (as will be discussed in Section 5.9), we can take advantage of the residual lens aberrations to critically align the lenses for Koehler illumination.

Our concern is to align all of the optical components on axis, to focus the lenses correctly for Koehler illumination, and to adjust the diaphragms appropriately. The illuminator will be considered first.†

*Subjectively, I_4 would appear to reside at I'_3 rather than at I_3 or I_4 (Chapter 4, Section 5.2).

†See Piller (1977) for additional useful information and illustrations regarding the functions and adjustments of the microscope parts.

Alignment and Focus of the Illuminator

Misalignment of the illuminator can give rise to an unevenly illuminated field of view or an unevenly lit aperture. An unevenly illuminated field can be particularly troublesome in video microscopy where image contrast is enhanced electronically; unevenness of the field is accentuated together with the contrast of the specimen.

Although an unevenly illuminated aperture will not necessarily affect field uniformity, it can critically affect the quality of the image (Sections 5.5–5.7). Therefore, the light source and the whole illuminator need to be properly aligned, as well as correctly focused, for Koehler illumination.

For various reasons, the collector lenses in an illuminator are rarely corrected very well. We will take advantage of this imperfection to center the light source with respect to the collector lenses.

The following description assumes that the microscope is illuminated, via a substage mirror, from an external illuminator. In many modern microscopes, that is not the arrangement; rather, the illuminator is built into, or attached to, the microscope base. For such microscopes, the procedures for achieving Koehler illumination are somewhat different (as discussed later) from those described below.

However, I recommend that the reader do the following exercises and become thoroughly familar with the associated procedures. These exercises will lead to a better understanding of the significance of the adjustments than simply adjusting a microscope equipped with a built-in illuminator. The reader will then be in a better position to spot and, where possible, to correct troubles in the adjustment and design of the microscope. The exercises follow.

Place a target, such as a plain stiff piece of paper, about 25 cm in front of the illuminator. Close down the field diaphragm on the illuminator. After removing any ground-glass diffuser and color filters, focus the collector lens until a sharp image of the light source is projected onto the target (Fig. 5-7).

Open and close the field diaphragm and carefully observe the projected image of the source. If a light area expands and contracts uniformly around the image, the source and the collector lenses are aligned.*

If the light area does not appear symmetrically around the central image, the source and the collector are not aligned with each other. In that case, the source is brought into alignment by adjusting the two centering screws that support it.

If a concave collector mirror is present (in the lamp house behind the source), it is also adjusted now. By turning the appropriate screws, the tilt and focus of the mirror are adjusted so that an inverted image of the source, formed by the mirror, lies next to, or meshed with, the direct image of the source. The two images together should form a relatively uniform, nearly square, image of the source.

Once the collector and source are aligned, the illuminator is placed 10–20 cm from the substage mirror of the microscope. With the field diaphragm closed down, the image of the source is focused onto the substage mirror. The whole illuminator (without recentering the lamp and mirror in the illuminator) is oriented until the image of the source is centered on the substage mirror.

The tilt of the substage mirror is then adjusted so that the image of the source is centered on the condenser iris diaphragm. (The image on the condenser iris can be seen by looking through the mirror.)

Focusing the Condenser; Alignment of the Illuminator Continued

A microscope slide containing a stained section or some other high-contrast, thin specimen is placed on the microscope stage. The specimen is brought into focus using a 10× or other low-power

*The light area around the image of the source arises from the **spherical aberration** of the collector. In the presence of such aberration, rays from the outer zones of the collector lens converge to different foci than rays from the central zones of the lens. The rays from the outer zones are therefore spread out in the plane where the rays from the central zones are focused.

objective lens. The condenser iris diaphragm is closed down and the height of the condenser is adjusted until a sharp image of the field diaphragm is superimposed on the specimen.

Make sure that you see the image of the field diaphragm and not that of the condenser iris diaphragm. To ensure this, place a pencil or other slender object in front of the field diaphragm on the illuminator: a dark *image* of the object should appear in the microscope field of view. Also, when the field diaphragm is opened and closed, the area of the specimen illuminated should expand and contract, but the brightness of the image should not change.

Conversely, when the condenser iris is opened and closed, the brightness of the field seen through the microscope should change, but, in principle, the area illuminated should remain constant. In practice, some (colored) light might be noticed spilling outward into the dark image of the field diaphragm; the spherical and **chromatic aberrations** of the condenser increase with greater dilation of the condenser iris.

If the image of the field diaphragm is not centered in the field, adjust the tilt of the substage mirror until the image is centered. This adjustment assumes that the condenser is properly centered relative to the objective lens. In fact, that may not be the case, as discussed later.

Once the image of the field diaphragm is centered, realign the axis of the illuminator by pivoting the illuminator around the center of the field diaphragm, which is kept as stationary as possible. Then recenter the image of the field diaphragm by trimming the tilt of the mirror. Open up the field diaphragm while looking through the microscope and note whether the field of view is uniformly illuminated. If the illumination is not uniform, but is symmetrical around the center of the field diaphragm, try refocusing the light source.

If the illumination is asymmetrically nonuniform, continue to adjust the alignment of the illuminator until the illumination becomes symmetrical with respect to the center of the image of the field diaphragm. With each adjustment of the illuminator, a slight readjustment of the substage mirror will be required, but the light source should not have to be recentered relative to its mirror or collector lens.

If repeated adjustments fail to yield a symmetrically illuminated field, the reason may be that the quality of the collector lenses is too poor or that the condenser is not aligned with the objective lens.

Centering the Condenser

The condenser can be centered with respect to the objective lens using either of the following approaches*:

A. With the field diaphragm closed down and the condenser iris wide open, inspect the color fringes around the image of the field diaphragm in the specimen plane. Are the fringes symmetrical in color, brightness, and width? Assess the fringes carefully by closing and opening the condenser iris diaphragm, and by slightly refocusing the condenser. Make the colored fringes symmetrical by adjusting the condenser centration with the two centering screws.

B. Observe the back aperture of the objective lens conoscopically, i.e., with a phase telescope (or by inserting a Bertrand lens if one is available).† With the field diaphragm closed, open and close the condenser iris diaphragm. Does the image of the condenser iris disappear from, or occlude, the objective back aperture symmetrically? Or is the occlusion asymmetric? If asymmetric, make sure that the image of the light source covers the condenser aperture symmetrically: trim the illuminator axis or bring the illuminator closer to the substage mirror as necessary. Then adjust the condenser centration until occlusion of the objective back aperture by a closed condenser iris is perfectly symmetric.

*For microscopes equipped with a revolving stage, adjust the centration of the objective lens to the stage, or vice versa, before adjusting the centration of the condenser.

†The ocular is replaced with a telescope used for phase-contrast microscopy adjustments or, alternatively, the Bertrand lens is inserted into the optical path so that it and the ocular together function as a telescope.

With both approaches, some trimming of the mirror and illuminator orientation is also necessary.*

With the source, and the axes of the collector, condenser, and objective lenses aligned, we are ready to optimize the size of the illuminated field and the cone angle of illumination.

Before proceeding to a discussion of these important adjustments, we will briefly examine the alternate method of aligning the illuminator and condenser in a microscope whose illuminator is built into the microscope base.

Alignment of Microscopes with a Built-in Illuminator

For a microscope with a built-in illuminator, we first center the condenser† so that the image of the field diaphragm appears in the middle of the field of view, superimposed on the focused specimen. The collector mirror, if present in the lamp house, is adjusted as described earlier. Next, the light source is focused so that its image is formed on the condenser iris diaphragm. The image of the source is then centered to the condenser iris diaphragm by adjusting the lamp centering screws. These simple steps should, in principle, bring all of the optical components into alignment. Unfortunately, that is often not the case; but, if so, the exercises described above for setting up microscopes equipped with separate illuminators should help pinpoint the trouble.

Adjustment of the Field Diaphragm

Once the light source, and the collector, condenser, and objective lenses are aligned and focused for Koehler illumination, the field and condenser iris diaphragms must be adjusted. *These adjustments must be made any time you switch an objective lens.*

The main function of the field diaphragm is to regulate the flare in a microscope. Flare is the haziness or brightness that is superimposed on the image; it reduces image contrast.

Flare arises from two main sources in a light microscope: (1) multiple reflection from the surfaces of lenses and other optical elements, and (2) reflection from the interior of the lens barrels, microscope body tube, and couplers (Dempster, 1944; Shillaber, 1944). Sometimes, poor correction of the lenses or dirty lenses can also contribute to flare.

In modern microscopes, several features have been incorporated to reduce flare. Improved antireflection coatings minimize the reflection of light at the surfaces of the optical elements; some of the coatings are effective over a wide range of wavelengths of light.‡

Furthermore, modern microscope lenses are designed to reduce the reflection that does remain after coating; the curvature of the optical interfaces is designed so that only a small fraction of the rays that are multiply reflected rejoin the image-forming rays. Computer-optimized design (which is used together with the new types of glass now available) not only provides lenses with better correction but with reduced intrusion of flare. Also, greater care is exercised by microscope designers to reduce light scattering from lens barrels and the body tube.

Despite these improvements, the user still must adjust the field diaphragm to eliminate flare. In general, the illuminated area of the specimen should be kept as small as practical. For visual observation and photography, the field diaphragm is set so that its image appears in the field, just surrounding the essential area of the specimen. At the very least, the image of the field diaphragm

*With some high-intensity illuminators (such as concentrated mercury or xenon arcs), one may not be able to fill the aperture completely or uniformly. For the purpose of the present exercises, you may have to insert the ground-glass diffuser that is commonly found in such light sources. However, much light is lost, defeating the whole purpose of using the high-intensity source. To remedy the problem, a light scrambler (consisting, e.g., of a loop of optical fiber) can be used (see footnote on p. 127).

†See footnotes on p. 106.

‡Without the coating, 4–5% of light is reflected at each optical interface, even at normal incidence. The amount reflected is given by $(n_2 - n_1)^2/(n_2 + n_1)^2$, where n_1 and n_2 are the refractive indices on both sides of the interface. At higher angles of incidence, the reflection becomes considerably greater.

should be visible at the edge of the field of view, which otherwise is limited by the field stop in the ocular.

For video microscopy, adjustment of the field diaphragm is even more critical. *The video camera should not see its image* at all, since the camera would then adjust its "black level" to the darkness of the field diaphragm rather than to the darkest part of the specimen (Section 3.7). Contrast within the specimen image would then be lost. *The field diaphragm should not be opened too far* either, since flare and hot spots arising from reflections in the various microscope tubes can ruin the image (Section 3.9).

Adjustment of the Condenser Iris

Proper adjustment of the condenser iris diaphragm is even more important than that of the field diaphragm. The condenser iris (or its conjugate, the light source) determines the cone angle of light that illuminates the specimen; as described later, this cone angle directly affects image resolution as well as contrast (Section 5.6).

In Koehler illumination, the condenser lenses focus an image of the field diaphragm in the plane of the specimen (Fig. 5-15). Before entering the condenser lenses, the beam of light passes the plane of the condenser iris diaphragm, where the rays that comprise it are nearly paraxial (Figs. 5-16, 5-17).

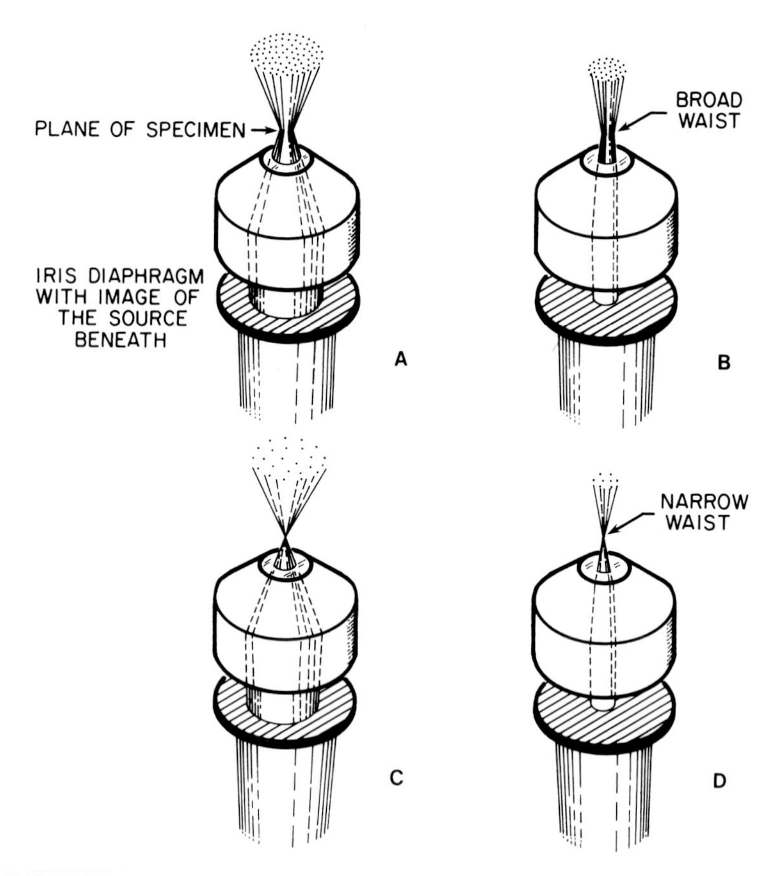

FIGURE 5-16. Effects of changing the field diaphragm and condenser iris diaphragm settings. A and C have identical condenser iris settings, as do B and D. Closing the condenser iris results in an illuminating cone of small angle (lower working N.A.; B and D); changing the field diaphragm results in a broad (A and B) or narrow (C and D) waist, without affecting cone angle. Effects such as these can be visualized by placing a fluorescent uranium-glass block on the condenser (or above the ocular).

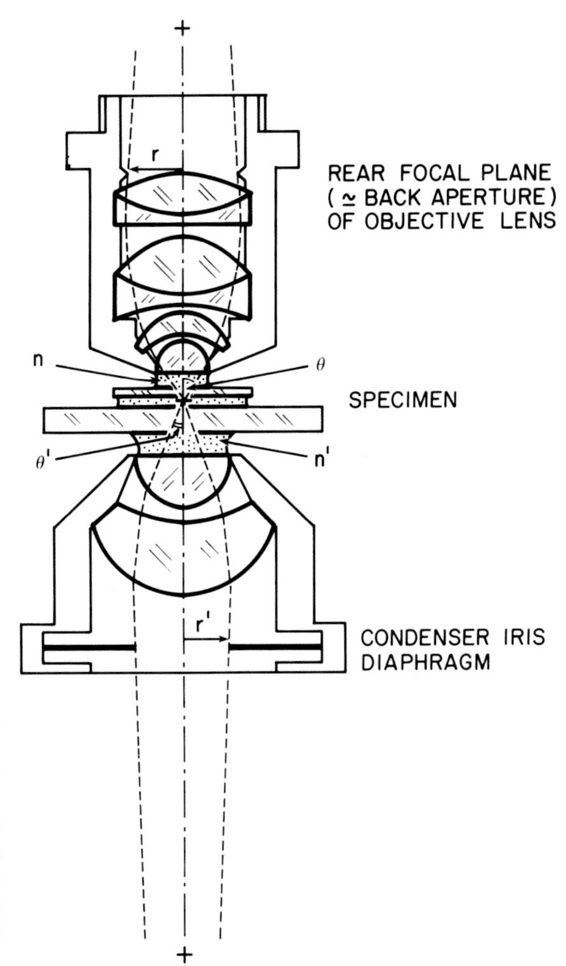

FIGURE 5-17. The working numerical aperture of the condenser, NA_{cond}, is $n' \sin \theta'$, where n' is the refractive index of the medium between condenser and specimen, and θ' is the angle shown. NA_{cond} is proportional to r', the radius of the condenser iris opening (equation 5-6). Thus, changing the condenser iris setting changes NA_{cond}. Analogous considerations for the objective lens also are illustrated.*

The fraction of the paraxial rays that enters the condenser lenses is regulated by the condenser iris diaphragm; the smaller the opening of the condenser iris, the narrower the beam that enters the lenses and the smaller is the cone angle of the beam that emerges from the condenser. The cone angle of the rays illuminating the specimen is thus regulated by the size of the opening in the condenser iris diaphragm (compare Fig. 5-16A to B and C to D).

Note, however, that the diameter of the narrow waist in the cone does not change as the condenser iris is opened and closed (Fig. 5-16); rather, it is the amount of light traversing the waist that increases and decreases. The narrow waist in the cone is where the light rays have been focused by the condenser. In fact, a focused image of the field diaphragm is seen when a target is placed at the narrowest section of the waist.†

When the condenser iris is kept at a constant setting, but the field diaphragm is opened and closed, the cone angle of the light rays illuminating the specimen does not change. Instead, the

*In order to clearly show the paths of light rays, the slide and coverslip have been depicted with discontinous surfaces.

†These events can be strikingly visualized in a darkened room by placing a fluorescent uranium-glass block above the condenser. The reciprocal behavior of the light beam at the aperture plane (as represented by the eyepoint) also can be observed by placing the fluorescent block above the ocular. Uranium-glass blocks for these demonstrations can be purchased from Fish–Shurman Corp.

whole cone becomes brighter or dimmer, as the diameter at the narrow waist of the cone expands and contracts. At the narrowest section of the waist (the focused image of the field diaphragm), the brightness per unit area remains constant, even though its diameter changes (compare Fig. 5-15A to C and B to D).

Quantitatively, the radius (r') of the opening in the condenser iris is related to the half-cone angle (θ') of light emerging from the condenser (Fig. 5-17) by the following equation:

$$r' = f'n' \sin \theta' \tag{5-6}$$

where n' is the refractive index of the medium between the condenser and the specimen. The quantity $n' \sin \theta'$ is called the *working numerical aperture of the condenser* (NA_{cond}).

With Koehler illumination, the condenser iris is in a plane conjugate with the rear focal plane of the objective lens (Section 5.3). Therefore, an inverted image of the condenser iris appears at the rear focal plane of the objective lens (Fig. 5-17).

The rear focal plane of the objective lens can be conveniently inspected conoscopically with a phase telescope or a Bertrand lens. With the microscope focused for Koehler illumination, and the telescope focused on the objective lens rear focal plane, the image of the condenser iris diaphragm should be clearly visible. The image of the light source should also appear in the same focus. As the condenser iris is opened and closed, its image also opens and closes.

Since the radius of the opening in the condenser iris is proportional to the NA_{cond} (equation 5-6), and since the same equation (without the primes in equation 5-6) also holds for the objective lens, *the ratio of the condenser NA to the objective NA (NA_{obj}) is directly given by the ratio of their diameters seen at the objective back aperture.*

In fact, each ray in the cone of light passing through the specimen, or emerging from the condenser, is represented by a unique point at the rear focal plane (or to a close approximation, at the back aperture) of the objective lens. For this reason, observation of the rays at the objective back aperture is said to provide a **conoscopic image.***

Conoscopically then, we can inspect the cone angle covered by the rays from the condenser, and we can measure the ratio of NA_{cond} to NA_{obj}.

As shown in the next section, image resolution increases proportionately with the sum of the objective and condenser NAs. From that point of view, we should always use the largest NA_{cond} possible. However, flare also increases (and, therefore, image contrast is reduced) as the NA_{cond} is increased by opening up the condenser iris diaphragm. (As the NA_{cond} is increased, image contrast is reduced for several other reasons as well. See Section 5.6 and Chapter 11).

Therefore, *the NA_{cond} is chosen to provide a reasonable compromise between image resolution and contrast.* In practice, the NA_{cond} is chosen to be some fraction of the NA_{obj}, depending on the nature of the specimen and the mode of contrast generation.

For *visual observation* of stained, or otherwise strongly light-absorbing specimens viewed in brightfield, the NA_{cond} is set to approximately nine-tenths of the NA_{obj}. In other words, the diameter of the condenser iris is set so that the image of its opening is nine-tenths the diameter of the objective back aperture.

For visual observation of low-contrast, nonabsorbing specimens, the optimal NA_{cond} is four- to five-tenths of the NA_{obj} for observations in DIC, polarization optics, etc.

However, when video with enhanced contrast is used, the NA_{cond} can be made nearly as large as the NA_{obj} under many modes of contrast generation. Thus, *video can significantly increase the practical resolution* that can be achieved with a particular objective lens.

The adjustments of the field and condenser iris diaphragms, just discussed, constitute the final important steps in setting up the microscope for Koehler illumination.

*In contrast to the conoscopic image seen at the back aperture objective lens, the regular image of the specimen, observed without a Bertrand lens or phase telescope, is the **orthoscopic image** (see also Fig. III-18).

5.5. IMAGE RESOLUTION AND WAVE OPTICS

The Diffraction Pattern

As discussed in Sections 5-2 and 5-3, the *magnification* and *general imaging properties* of the light microscope can be derived by following the paths of the light rays geometrically; the image produced by the objective lens is conjugate with the specimen. Each image point is geometrically related to a corresponding point in the specimen, and each point in the specimen is represented by a corresponding point in the image. Light emitted from a point in the specimen is focused to the corresponding point in the image plane, and any light emitted from a point in the image plane would be focused to the corresponding point in the specimen.

Unlike those properties, image *resolution* and *contrast* in the microscope can only be fully understood by considering light as a train of waves. From wave optics, light emitted from a point is not in fact focused to an infinitely small point in the conjugate plane. Rather, upon passing a perfect lens, the waves converge together and interfere with each other around the focus to produce a **three-dimensional diffraction pattern** (e.g., Born and Wolf, 1980; Cagnet *et al.*, 1962). The diffraction pattern arises in the following manner.

The image wave arising from a point source emerges from the back aperture of the objective lens (a perfect lens) as a spherical wave centered on the image point (Fig. 5-5). According to Huygen's principle of wave optics, this spherical wave front, shown as Σ_i in Fig. 5-18A and B, is itself made up of a series of point sources of light. Each point emits a spherical wave known as a Huygen wavelet. In Fig. 5-18A, an image of the point source A appears at A_0' because the wavelets arising from all points on the spherical wave front arrive in phase and constructively interfere with each other at A_0'. (The figure illustrates the situation by taking two points M_0 and M on the wave front Σ_i.) As we move away from point A_0' in the focal plane π, the phase of the wavelets emanating from M_0 and M gradually become out of phase, until at A_1', the two wavelets are out of phase by 180° (Fig. 5-18B). There the two wavelets destructively interfere with each other and are canceled. If we consider the contribution of wavelets from all points of the aperture, rather than from M_0 and M alone, the sum of the wavelets becomes zero and no light is present at A_1'.* Since this situation is symmetric around the axis AA_0', A_0' is thus surrounded by progressively less bright zones ending up in a dark ring. Once past the dark ring, the wavelets interfere constructively to some degree, and a somewhat brighter ring appears. Beyond that brighter zone, another dark ring appears where the wavelets destructively interfere and cancel each other.

We have just described the diffraction pattern formed by a perfect lens in the image plane of a point source of light. Figure 5-19A shows a photograph of this diffraction pattern and Fig. 5-19D its intensity distribution.

Returning to Fig. 5-18A and B, it is not only in the focal plane, π, that the wavelets emanating from the spherical wave front Σ_i periodically come in and out of phase. In fact, if we add the contribution of the wavelets from the aperture, their sums alternate along the axis of the light path AA_0', just as it did in the plane π. Thus, when we sum the phases of the wavelets in the three-dimensional space around A_0', we find that *the diffraction pattern is periodic along the axis of observation as well as in planes perpendicular to that axis.* This gives rise to the three-dimensional diffraction pattern that we are discussing. Furthermore, such a three-dimensional diffraction pattern is not only present for a point source of light, but the same relation, with the brightnesses reversed, also holds for an absorbing point in the object.

*For points (A_n') between A_0' and A_1', the Huygen wavelets from Σ_i add together to produce a distribution of intensity that falls off in a bell-shaped curve, i.e., as $(\sin \Theta/\Theta)^2$, where Θ is the angle $A_n'M_0A_0'$. Beyond A_1', the "Bessell function," $(\sin \Theta/\Theta)^2$, becomes somewhat greater than zero again, and oscillates with a progressively diminished amplitude (Fig. 5-19D).

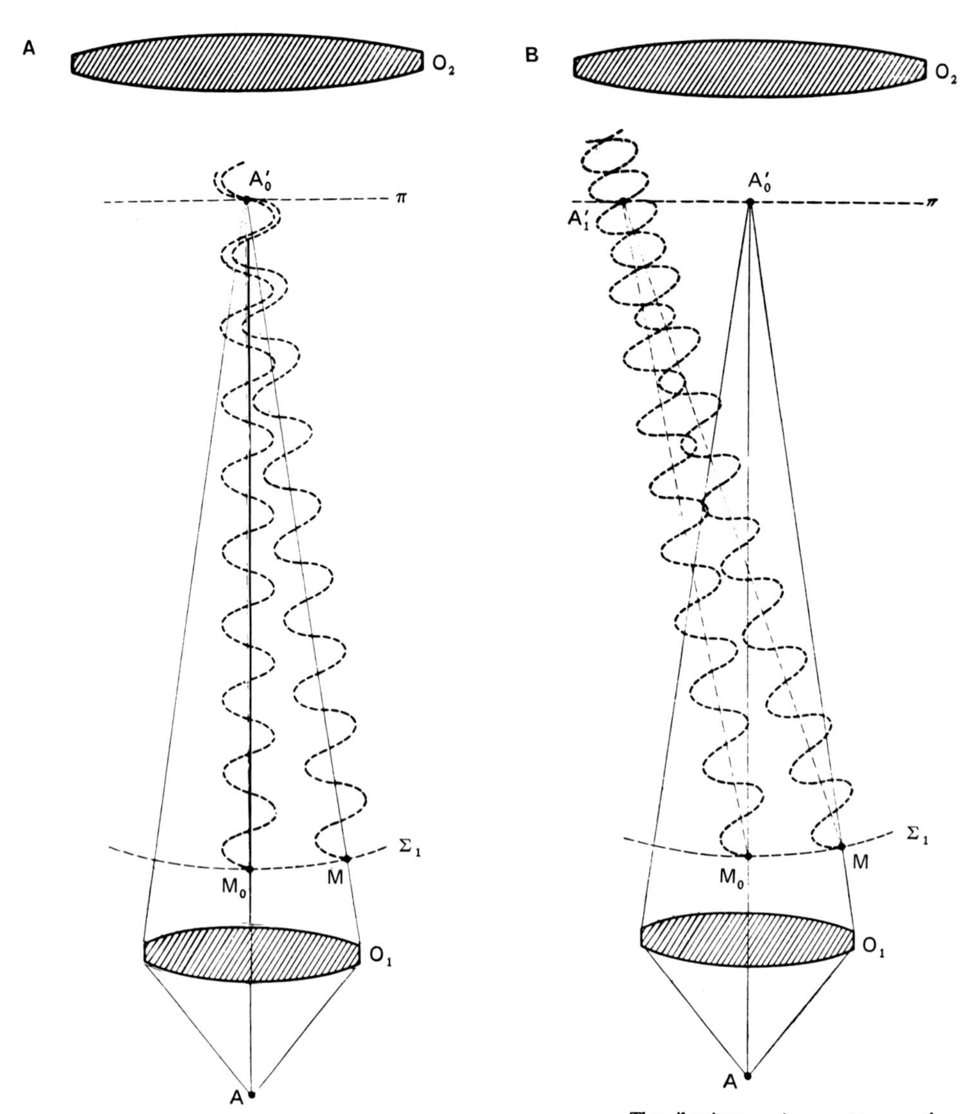

FIGURE 5-18. Formation of an Airy disk. According to geometrical optics, light coming from a point (A) on the specimen is focused (at A'_0) in the intermediate image plane (π) by the objective lens (O_1). However, according to wave optics, the light leaving O_1 is not only a spherical wave centered on A'_0; Huygen wavelets emanate from each point along that front (Σ_i). For the conditions illustrated in panel A, wavelets leaving M and M_0, and which converge at A'_0, are in phase: they constructively interfere at A'_0 and result in a bright image. Conversely (panel B), wavelets leaving M and M_0, and which converge at A'_1, are 180° out of phase: they destructively interfere at A'_1 and produce a dark image. Summation of the mutual interference of all waves from a particular specimen point that converge in the image plane (π), results in an alternating light-and-dark concentric pattern, observed as the Airy Disk, Fig. 5-19A. (From Françon, 1961.)

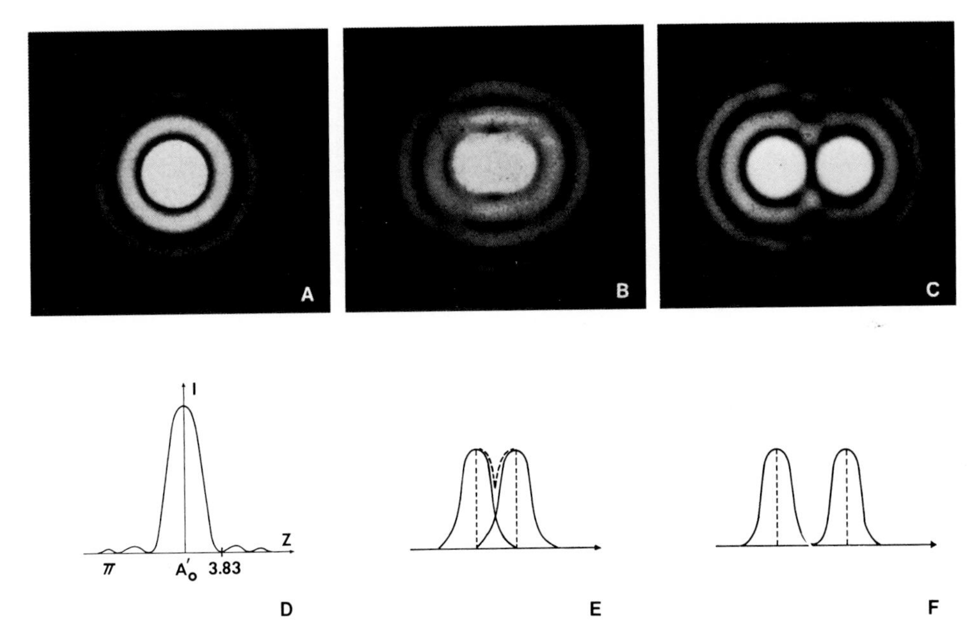

FIGURE 5-19. Airy disk and intensity distribution in the diffraction pattern. (A) Photograph of Airy Disk; (D) intensity distribution. (B, C, E, F). Images of neighboring Airy Disks (B, C), and their corresponding intensity distributions (E, F), separated by different distances. In C (and F), where the distance between the disk centers is larger than their radii, the disks (and intensity peaks) are clearly separate, or resolved. In B, the center-to-center distance equals the radius, the disks are just barely distinguishable, and the images are just resolved (the Rayleigh criterion).* If the disk centers are closer together than their radii, their intensity distributions merge into a single peak (D), and they are said not to be resolved. (After Françon, 1961.)

Image Resolution and the Diffraction Pattern

For a well-corrected microscope objective lens, the three-dimensional diffraction pattern near the focal point is periodic symmetrically along the axis of the microscope as well as radially around the axis. When the three-dimensional diffraction pattern is sectioned in the focal plane, it is observed as the two-dimensional diffraction pattern, the **Airy disk** (Fig. 5-19A).

In the image formed in the focused image plane, every point of the specimen is thus represented by an Airy disk diffraction pattern, rather than by an infinitely small conjugate point. Therefore, the resolving power of an objective lens can be determined by examining the size of the Airy disk formed by that lens. As will be discussed shortly, the radius (r) of the Airy disk s governed by the NA of the objective and condenser lenses.

For self-luminous bodies, or for specimens illuminated with a large-angle cone of light, the light waves that form adjacent Airy disks are *incoherent* and thus do not interfere with each other. We can therefore determine the minimum separation that can be resolved with a particular microscope objective lens by examining the total intensity distribution (i.e., the sum of intensities) of closely spaced, or overlapping, Airy disks in the image plane.†

Images, and their corresponding intensity distributions, of two Airy disks, separated by three different distances, are shown in Fig. 5-19. Let D equal the peak-to-peak distance of the intensity distribution curves (or the center-to-center distance of the Airy disk images) and let r equal the disk

*The curves in E and F are schematic only. Had they been drawn exactly as in D, the sum of the two curves in E (dashed line) would have dipped in the middle by only 26.5% from their peak.

†If the waves that make up the adjacent diffraction patterns were coherent, we would have to add amplitudes rather than intensities. The adjacent diffraction patterns would then interfere with, and modify, each other (e.g., see Martin, 1966).

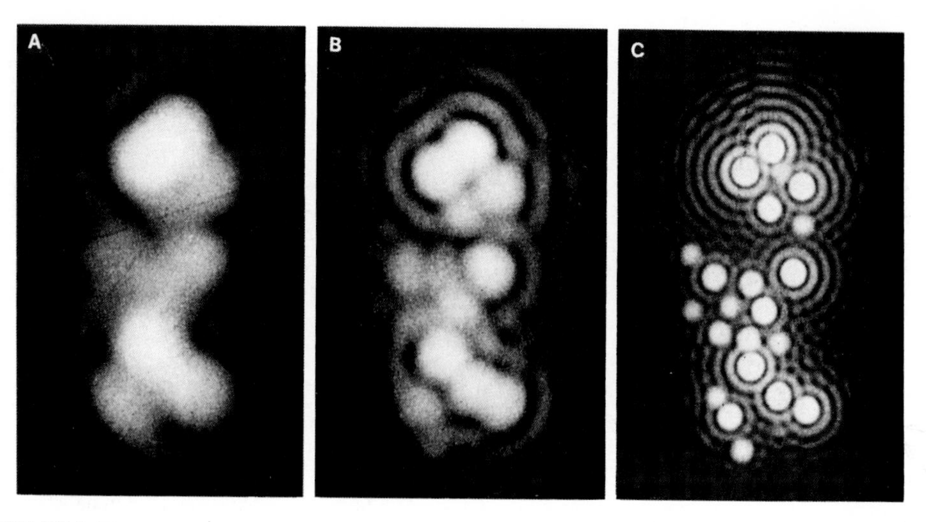

FIGURE 5-20. Image resolution versus numerical aperture. All specimens viewed through a microscope can be considered to be made up of a series of points, which become imaged as Airy disks. Therefore, the smaller the radius of the Airy disk, the greater is the resolution. The reduction of Airy disk radius (= greater resolution) as NA increases is demonstrated in these three images of the same test object. A, B, and C are low, medium, and high NA, respectively. The photos have been intentionally overexposed to accentuate the diffraction rings associated with the Airy disks. (From Françon, 1961.)

radius. The sum of the intensities of the pair of Airy disks clearly shows two peaks when $D > r$, and barely two peaks when $D = r$ (the condition known as the **Rayleigh criterion**). When $D < r$, the sum of the peaks merges into a single peak, and the images of the two points are said not to be resolved.

By observing the summed pattern (which is all that the observer can in fact detect), we can just barely see that two Airy disks had made up the observed pattern when $D = r$, i.e., when the Rayleigh criterion was met.

These discussions assume that the lenses are completely free of aberrations and that the unit diffraction pattern is in fact an Airy disk. The pattern is modified from an Airy disk to some other diffraction pattern by lens aberrations or by nonstandard aperture functions, which can be introduced intentionally or accidentally.*

Therefore, for well-corrected objective lenses with a uniform circular aperture, two adjacent points are just resolved when the centers of their Airy disks are separated by r.

For a self-luminous body, r is given by

$$r = 1.22\ \lambda_0/(2\mathrm{NA}_{\mathrm{obj}}) \qquad (5\text{-}7)$$

where λ_0 is the wavelength of the light in air and $\mathrm{NA}_{\mathrm{obj}}$ is the numerical aperture of the objective lens:

$$\mathrm{NA}_{\mathrm{obj}} = n \sin \theta \qquad (5\text{-}8)$$

where n is the refractive index of the medium between the specimen and the objective lens, and θ is the half-cone angle of light captured by the objective lens (Fig. 5-17). The influence of NA on r can be seen in Fig. 5-20, which illustrates how image resolution is affected by NA.

Often the specimen is neither self-luminous nor illuminated by a condenser whose NA is

*For photographs of various diffraction patterns, see Cagnet *et al.* (1962). See also Inoué and Kubota (1958) and Kubota and Inoué (1959) summarized in Appendix III, and Galbraith and Sanderson (1980), and Galbraith (1982).

identical to that of the objective lens. In that case, the minimum distance (d) in the specimen that can be resolved is given by

$$d = 1.22\ \lambda_0/(\mathrm{NA}_{obj} + \mathrm{NA}_{cond}) \tag{5-9}$$

where the working NA of the condenser (NA_{cond}) is given by equation (5-6) (Fig. 5-17).

Coherent illumination corresponds to the case where $\mathrm{NA}_{cond} = 0$. The resolution drops by a factor of 2 and the wave fronts from various focal levels interfere with each other to form a confusing series of diffraction patterns.

The Diffraction Pattern and Out-of-Focus Image

As mentioned previously, light from an infinitely small point in the specimen is converted by a perfect lens into a three-dimensional diffraction pattern (essentially, a three-dimensional **point-spread function;** see Chapter 10, Appendix II). With a perfect lens, this diffraction pattern is rotationally symmetrical about the optical axis, so that all meridional sections through it are identical. Figure 5-21A shows an intensity profile (contour lines of intensity) of a meridional section through such a diffraction pattern, in the vicinity of the focus.*

When the specimen is in focus, the image plane is at the level labeled ''Focal plane'' in the figure; the cross section of the intensity profile at this level gives rise to the Airy disk.†

As we move away from focus, the image plane transects the intensity profile at levels increasingly farther from the focal plane. The resulting two-dimensional diffraction pattern (at the successive ''out-of-focus'' image planes) periodically alternates between having dark and light centers. At the same time, more energy is concentrated away from the centers of the diffraction patterns, into the outer rings.

If, rather than observing the diffraction pattern of a single point, we observe a periodic specimen, then the changes in the diffraction pattern just described are manifested as contrast changes in the image. As we move away from focus, the contrast decreases and alternately *reverses,* first at fine spacings (high spatial frequencies), and subsequently at coarser spacings. This is shown graphically by the response [**contrast transfer function** (CTF); see Section 5.7] curves in Fig. 5-21B. Curve 1 is in focus; curve 5 is most out of focus. As discussed later, these CTF curves relate contrast to spatial frequency. The change of sign of the transfer function in curves 4 and 5 indicates reversal of contrast.

These phenomena are clearly illustrated in the video monitor pictures of Siemens Test Stars shown in Fig. 5-21C–E. Siemens Test Stars are made up of equally spaced, radial, black and white lines; their circumferential spacing increases proportionately with the radius. Spatial frequency is the number of periods per unit distance (in this case along any circumference of constant radius). Figure 5-21C shows an in-focus image, i.e., under conditions corresponding to curve 1 in Fig. 5-21B. Moving away from exact focus, Fig. 5-21D shows contrast reversal in the region of higher spatial frequency (approximating curve 4 in Fig. 5-21B). Still farther away from focus, in Fig. 5-21E, the contrast is reversed in regions with lower spatial frequency (farther from the center); it is reversed again in the region of higher spatial frequency (approximately the conditions shown in Fig. 5-21B, curve 5).

Note that these discussions apply to the case of perfectly corrected lenses with uniform circular

*The contours shown are for a perfect lens with a uniform circular aperture used under aberration-free conditions. If, as in phase contrast, DIC, etc., the objective lens does not have a uniform aperture, or if aberrations exist or have been introduced (e.g., by improper coverslip thickness or by not using rectified optics in high-extinction polarization microscopy; see Appendix III), the contours would be modified accordingly. They may become asymmetric above and below the focal plane, or their rotational symmetry may change. In addition, we assume here for non-self-luminous bodies, that the condenser aperture is uniformly illuminated.

†When phase objects are not viewed through conditioners that provide amplitude-modulated images, no contrast appears when the lens is exactly focused, and what seems to be the image plane in fact represents a pattern that is out of focus.

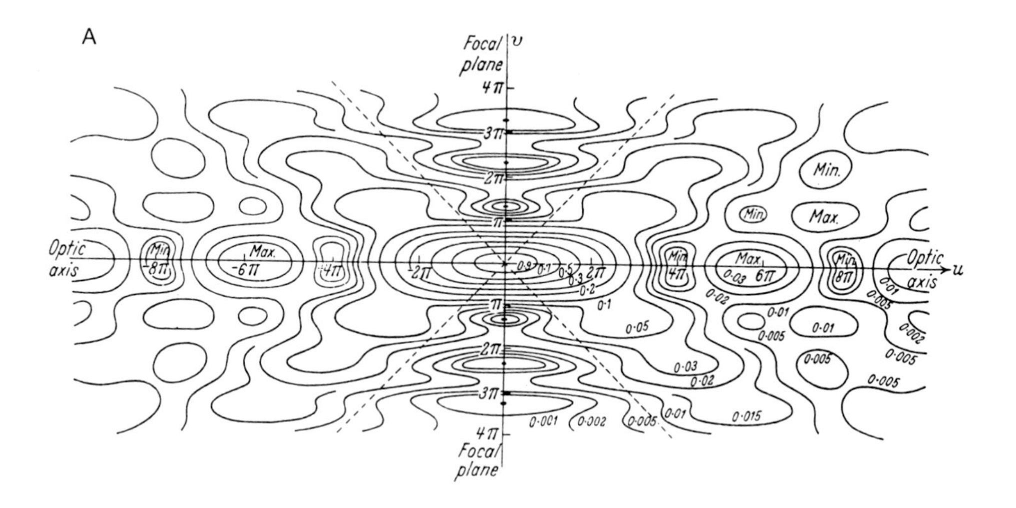

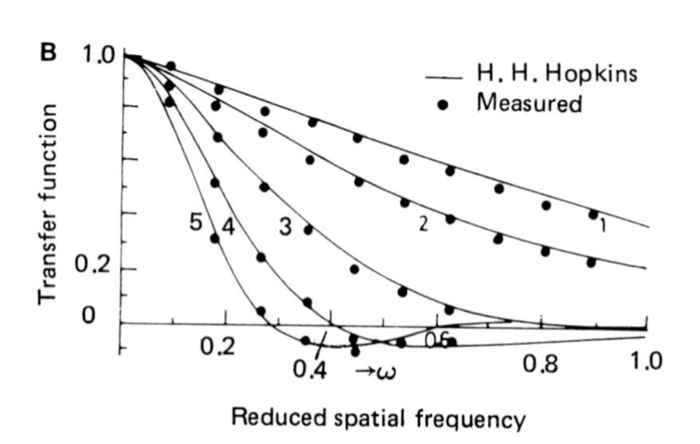

FIGURE 5-21. Three-dimensional diffraction pattern and the effect of focus on image detail.

(A) Three-dimensional diffraction pattern of a point source of light focused by a perfect lens with a uniform circular aperture. Illustrated is the intensity contour that appears in a meridional section through the diffraction pattern. A cross section through the pattern at the focal plane would give rise to the intensity contour (Fig. 5-19D) of an Airy disk (Fig. 5-19A).

(B) Degree of contrast retention by a series of black-and-white stripes, at varying degrees of defocus. These "contrast transfer function" (CTF) curves are plotted as transfer function (ratio of contrast of the image to that of the specimen) versus spatial frequency (number of line pairs per unit distance) of the object. Curve 1 is for the object in focus; curve 5 is most out of focus. Contrast is reversed where curves 4 and 5 dip below the abscissa and the sign of the CTF becomes negative.

(C–E) Monitor pictures of a test chart with Siemens Test Stars. (C) In focus. (D) Out of focus (approximately as curve 4 in B), showing contrast reversal in the finer spacings. The finest spacings are no longer resolved. (E) Out of focus (approximately as curve 5 in B), showing two contrast reversals and further decrease in contrast.

(A: from Born and Wolf, 1980; B: from Ghatak and Thyagarajan, 1978; C: original photographs.)

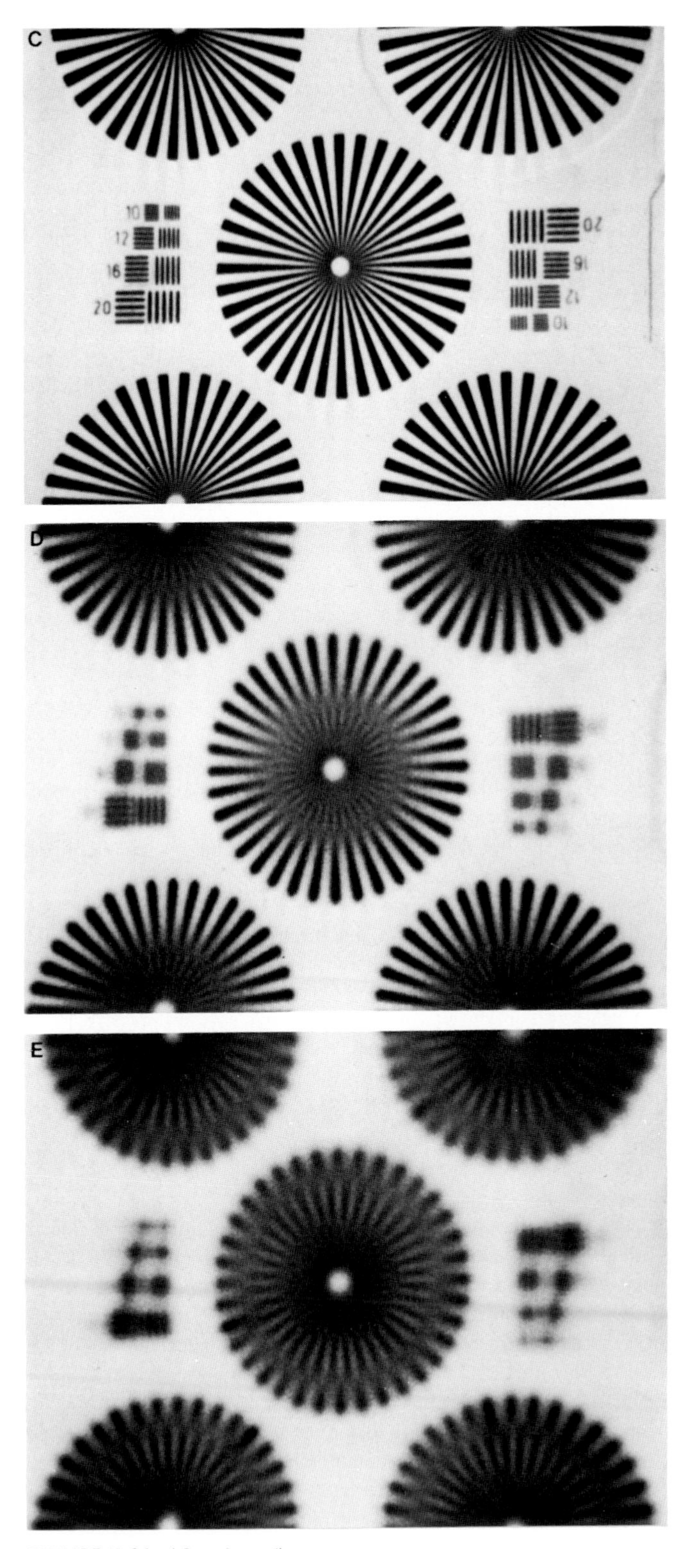

FIGURE 5-21 (*Continued*)

apertures that (for non-self-luminous objects) are illuminated with a condenser with its aperture uniformly and fully illuminated. Lens aberration, modified aperture function, and nonuniform illumination of the condenser aperture all modify the diffraction pattern and image (see footnotes * on pp. 114 and 115).

Depth of Field

Just as we can define lateral resolution for distances within the plane of focus, we can describe a **depth of field** along the axis of observation. This is the depth over which the FINE-FOCUSING control of the microscope can be adjusted without diminishing the sharpness of the image (of an infinitely thin specimen). At high NAs of the microscope, this depth is determined primarily by wave optics, while at lower NAs the geometrical optical "circle of confusion" dominates. Using different criteria for when the image becomes unsharp, several authors have proposed different formulaes to describe the depth of field (e.g., Françon, 1961; Martin, 1966; Michel, 1981; Piller, 1977; Shillaber, 1944). Françon defines the "setting accuracy" (2ξ) along the axis of the microscope as:

$$2\xi = \frac{\lambda}{4n \times \sin^2(u/2)} \qquad (5\text{-}10)$$

where λ is the wave length, n the refractive index of the immersion medium, and u the half angle of the cone of light that is captured by the objective lens (θ in Fig. 5-17). From Equation 5-10, we obtain setting accuracies of 0.29, 0.41, 0.40, 1.3 and 5.9 μm for 1.4 and 1.25 NA oil immersion, and 0.95, 0.65, and 0.30 NA dry lens systems, respectively at λ = 546 nm.

These setting accuracies correspond to the depth of field when a thin detector is used with no depth of accommodation (photographic emulsion, video camera). On the other hand, the human eye can normally accomodate from infinity to about 25 cm, so that the observed depth of field is greater than the values given above when one looks directly through the microscope. In addition, Françon points out that various other physical and psychophysiological factors affect the setting accuracy provided by high NA objectives. He says "It is but rarely that full use can be made of the setting accuracy provided by the high NA objectives" (Françon, 1961, p. 165).

In video microscopy, the shallow focal plane in the target of the camera tube, the high contrast achievable at high objective *and* condenser NAs, and the high magnification of the picture displayed on the monitor, all contribute to reducing the depth of field. Thus with video, we can obtain very sharp **optical sections** or high axial resolution, and we can critically define the focal level of a thin specimen (see Section 11.3, especially Figs. 11-9 and 11-10).

5.6. GENERATION OF IMAGE CONTRAST

With video, the contrast of the image can be substantially enhanced, and the image can be enlarged and made substantially brighter than the direct microscope image.

One may then ask: "Why should we worry about forming an image with good contrast in the microscope? Why not let video do all the work?" The answers to these questions consist of several parts, some very important ones of which have just been discussed and others to be given below.

Even with video, image contrast cannot be generated where none exists. A high degree of video enhancement or intensification is invariably accompanied by introduction of additional noise. The lower the contrast in the initial image, the greater the disturbance caused by noise, whether the noise is optical or electronic (Sections 3.8, 7.2).

Furthermore, image contrast tends to be lower for structures that are represented by higher spatial frequency. In other words, the finer the detail in the specimen, the lower the image contrast, even when the specimen otherwise would generate the same contrast. Expressed differently, as spatial frequency increases, the modulation transfer function (MTF) decreases, and contrast cannot be conveyed to the image as well (Section 5.7). The MTF drops off for higher frequencies both in the microscope and in video. Therefore, even with the opportunity to enhance contrast by video, we will want to start off with a microscope that provides adequate contrast particularly at high spatial frequencies.

Contrast in a microscope image is not an inherent property of the specimen. Rather, it is a product of (1) the interaction of the illuminating light waves and specimen structure and (2) the MTF and contrast-generating mode of the microscope. Point (1) depends both on specimen structure and on the condition of the illuminating light wave; point (2) depends both on the condition of the illumination and on how the waves leaving the specimen are treated.

Zernike's Phase Contrast

Perhaps the situation can be clarified by relating the elegant exposition by Fritz Zernike on the wave theory of microscope image formation (Appendix K in Strong, 1958), wherein Zernike explains the principles of phase-contrast microscopy. (He was awarded the 1953 Nobel Prize in physics for discovering the phase contrast principle; see Zernike, 1955.) The argument goes as follows.

First, a slide containing some minute, *light-absorbing* particles (say, some particles of graphite) is placed on the microscope stage. When one of the particles is carefully focused in brightfield under Koehler illumination, we see a small dark speck. The dark speck turns out to be a dark Airy disk (the negative of the bright Airy disk shown in Fig. 5-19A).

Next the condenser iris diaphragm is closed down to a pinhole, and a mask, complementary to the pinhole, is placed at the rear focal plane of the objective lens (Fig. 5-22A) where it covers the image of the pinhole and completely blocks out the illuminating beam. Therefore, when we look through the microscope, the background of the field is completely dark.

However, we find that the image of the specimen is not altogether absent. Rather, it appears as a bright speck on a dark background. The image of the minute absorbing particle is now a bright Airy disk!

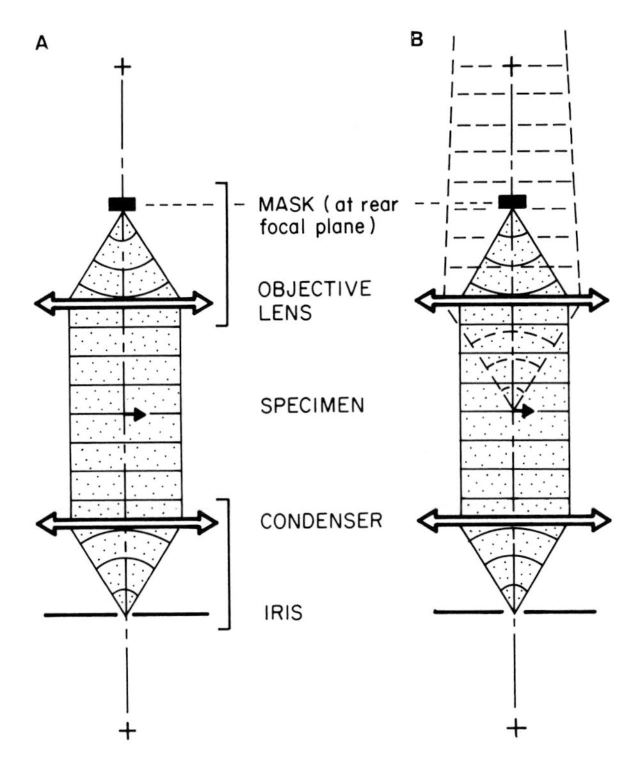

FIGURE 5-22. Scheme of Zernike's experiments. (A) In Zernike's second and fifth experiments (see text), a small mask is placed at the rear focal plane of the objective lens. The mask is complementary to the pinhole to which the condenser iris has been closed. (B) In the specimen space, light diffracted by the specimen is a spherical wave, centered on it, whereas the background illumination is a plane wave originating at the pinhole in the condenser iris. The mask effectively blocks out the direct beam, or background illumination, but permits passage of the diffracted wave.

What we see, with the background illumination masked in this way, is the image of the light scattered, or diffracted, by the specimen (Fig. 5-22B). The illuminating background wave is a plane wave originating from the pinhole at the front focal plane of the condenser. The light diffracted by the specimen is a spherical wave centered on the specimen.

And, Zernike continues, *the light diffracted from an absorbing object is half-a-wave out of phase with the illuminating background wave.* He comes to this conclusion because, without the mask at the rear focal plane of the objective lens, the specimen appears as a dark speck. On the other hand, we just learned that the specimen itself does diffract light. In brightfield microscopy, that diffracted light must, therefore, be destructively interfering with the background wave to produce the image—the dark Airy disk. Ergo, the two waves are half-a-wave out of phase with each other (Fig. 5-23).

In the third experiment, the small mask at the rear focal plane of the objective lens is replaced with a large mask having a small opening. The opening in the mask is matched to the image of the condenser iris closed down to a pinhole. Now, all of the illuminating beam, or the background wave, passes through the mask onto the image plane. How will the image of our small absorbing particle appear now? Viewed against the bright background, the image of the particle has all but disappeared. Since the objective back aperture was masked down, the working NA of the objective lens has decreased, with a concomitant increase in the radius of the Airy disk. Thus, the small dark Airy disk has expanded into a larger, pale Airy disk pattern.

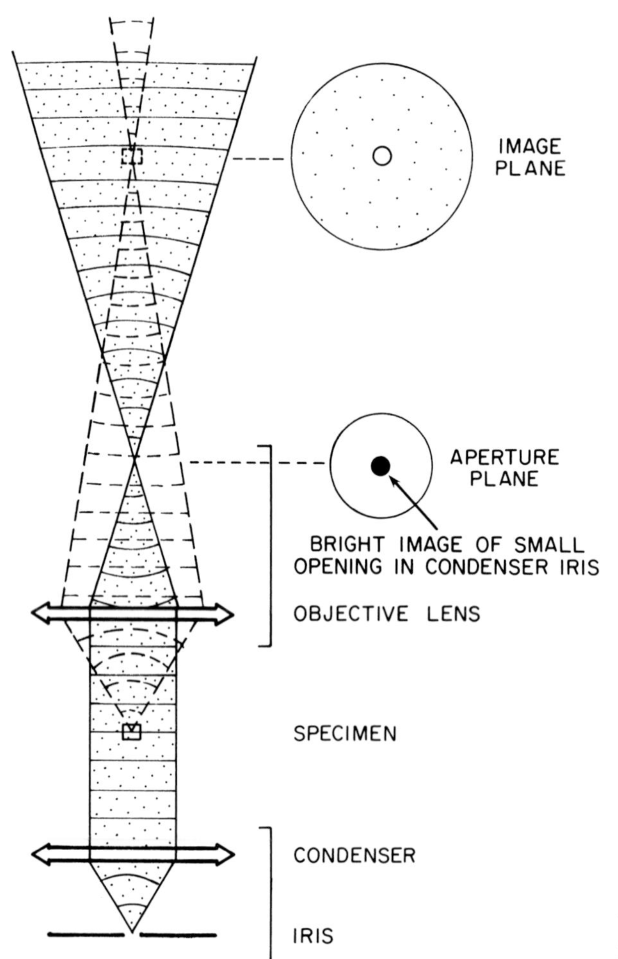

FIGURE 5-23. Paths of light diffracted by the specimen and of background illumination. The image is produced by the interference of these two waves. See text.

Zernike then considers a minute transparent specimen.

In the fourth experiment, a colorless, transparent specimen is viewed in normal brightfield with Koehler illumination. The illuminating wave is not modified, nor is the image-forming wave absorbed. Then, *when the microscope is focused precisely, the image of the thin, transparent specimen disappears.* It is only when the specimen is somewhat out of focus that we see a refracted shadow of the transparent specimen.

In the fifth experiment, the arrangement of the second experiment is repeated, i.e., the small mask, complementary to the aperture in the condenser, is placed at the objective back aperture (Fig. 5-22). But this time the specimen is transparent, not light absorbing. Now the transparent specimen is visible *in focus,* as a bright Airy disk on a dark background.* Again, this experiment shows that the specimen does in fact diffract and scatter light, even though its image was not visible when viewed in brightfield in focus.

Zernike reasoned that the light wave passing through the thin, transparent specimen is decomposed into two parts: the background unmodified wave; and a spherical wave that is *diffracted by the specimen and, in the process, has suffered a quarter-wave phase retardation.*† As we saw in the fourth experiment, the sum of the diffracted and background waves is indistinguishable from the background wave alone, and no visible contrast appears in the image. (We assume that, by traversing the specimen, the wave is retarded only a small fraction of a wavelength.)

We will now examine the paths taken by the waves diffracted by the transparent specimen and those of the background illumination (Fig. 5-23). The diffracted spherical wave passes the full aperture of the objective lens, while the background wave passes only a smaller region, the conjugate of the condenser opening. The two waves do not appreciably overlap, and they occupy distinct parts of the rear focal plane of the objective lens. This separation of the two waves in space allows them to be manipulated separately at the objective rear focal planes.

In the sixth experiment, Zernike introduces a **phase plate** at the rear focal plane of the objective lens. The phase plate converts the invisible image of the transparent specimen into one that is clearly visible, with its contrast reflecting the gradient of refractive indices in the specimen.

The phase plate is a flat glass plate made of two concentric parts. In the center is a partially absorbing mask, slightly larger than the conjugate of the pinhole at the condenser iris. The mask removes the major portion of the background light, allowing only a small fraction of the background waves to pass through. The mask is surrounded by a clear region, whose thickness is made slightly greater than that of the masked region, such that a quarter-wave retardation is added to the wave that passes through the clear region. In this way, the phase plate selectively retards the wave diffracted by the specimen further, by a quarter of a wave.

The effect of the phase plate is: (1) to selectively add a quarter-wave phase difference to the light waves diffracted by the specimen and (2) to reduce the amplitude of the background wave to the level of the amplitude of the diffracted wave. Since the diffracted wave suffers a quarter-wave retardation when it is scattered by the thin, transparent specimen,‡ the additional quarter-wave retardation introduced by the phase plate puts the diffracted wave a half-wave out of phase with the background wave. The diffracted light, retarded by a half-wave, thus interferes destructively with the background light, whose amplitude has been made to roughly equal the amplitude of the light diffracted by the specimen.

In this way, the ordinarily invisible phase-retarding object produces a dark, clear image in focus.

In the arrangement used in these experiments, the pinhole at the condenser aperture gives rise

*This is in fact a form of darkfield illumination. With a somewhat larger specimen, its edges would diffract light and shine brightly.

†The refractive index of the specimen is assumed to be higher than that of the medium surrounding the specimen. If the medium has a greater refractive index than the specimen, the diffracted wave would be *advanced* by a quarter-wave.

‡The experiments described here in fact prove that the light waves diffracted by a thin, transparent specimen are retarded (phase shifted) by a quarter of a wavelength. For mathematical derivations of the phase shift, see, e.g., Bennet *et al.* (1951) and Osterberg (1955).

to an NA_{cond} that approaches zero. To circumvent the resulting loss of resolution (equation 5-9), Zernike substitutes an annulus for the pinhole in the condenser; he uses a complementary ring-shaped absorber and phase plate. With this arrangement, the transparent small specimen is clearly visible in focus, with high contrast, and with good resolution.

The phase plate alters the imaging waves at the rear focal plane of the objective lens. The entrant wave is conditioned by the annulus at the front focal plane of the condenser lens. Together they give rise to Zernike's *phase contrast.*

One of the main lessons to arise out of Zernike's analysis is that *images are formed by the joint action of the illuminating background wave and the light wave modified by, and radiating from, the specimen. The two sets of waves converge at the image plane and interfere with each other to produce the image* (Fig. 5-23).*

Other Modes of Contrast Generation

As discussed in Chapter 10 and Appendix II, the diffraction pattern formed at the back focal plane of the objective is a **Fourier transform** of all of the spatial frequencies in the specimen. The intermediate and final images are inverse Fourier transforms of the diffraction patterns formed at the objective back aperture and eyepoint, respectively. Zernike, Nomarski, Ellis, and others have taken good advantage of these transform properties and enhanced the microscope image contrast by modifying the aperture function to introduce specific types of **spatial filtration.**

In addition to phase contrast, many other modes of contrast can be generated in the light microscope. Each mode uses an appropriate combination of a conditioner before the light strikes the specimen, and a complementary filter after the light has interacted with the specimen.

The alternate contrast modes include: phase contrast (Zernike, 1955, 1958; reviews by Bennet *et al.,* 1951; Osterberg, 1955); interference contrast (Dyson, 1950; Françon, 1961; review by Koester, 1961); differential interference contrast (Nomarski, 1955; Smith, 1956; reviews by Allen *et al.,* 1969; Galbraith and David, 1976; Padawer, 1967); modulation contrast (Hoffman and Gross, 1975); **single-sideband edge enhancement contrast** (Ellis, 1978); darkfield (reviews by Loveland, 1981); absorption (reviews by Shillaber, 1944; Chamot and Mason, 1958; Françon, 1961; Piller, 1977); reflectivity (review by Françon, 1961); the many modes of contrast generation in polarization optics (Schmidt, 1924, 1937; review by Bennett, 1950; principles in Hartshorne and Stuart, 1960; Ramachandran and Ramaseshan, 1961; also Appendix III); refractive index or optical path difference measurements (Françon, 1961); frustrated total internal reflection (reviews by Françon, 1961; Izzard and Lochner, 1976); **holography** (Ellis, 1966; review by Loveland, 1981); fluorescence (review in Thaer and Sernetz, 1973); luminescence (e.g., Eckert and Reynolds, 1967).

Any interaction of light and matter, or any process involving light emission, is, in principle, a candidate for the generation of image contrast in the light microscope.

Image contrast not only exposes specimen detail, but reveals the distribution of the underlying structures that gave rise to the contrast. Furthermore, the contrast is a measure of the quantity of structure, or substance, that altered the light wave and gave rise to the particular contrast.

Properly applied, *the light microscope thus ceases to be a simple imaging device. Instead, it becomes a tool for determining the physical characteristics of minute regions of the specimen.* The analysis can often be carried out *nondestructively* and reveal details of *structures or events even in the submicroscopic realm* (e.g., Schmidt, 1937; Inoué and Sato, 1966; Inoué and Ritter, 1975; Inoué and Tilney, 1982; see also Chapter 11, Appendix III).

5.7. MODULATION TRANSFER FUNCTION

As we saw in Section 5.5., the contrast between two image points decreases as they approach each other. When the separation between the two points becomes smaller than the radius of the Airy disk, little contrast is produced between the image points and they are said to be no longer resolved.

*Gabor (1948, 1949, 1951) (see also Stroke, 1969) has extended this concept further and laid the foundation for holography.

The relation between image contrast and the separation of the image points can be expressed quantitatively in a single graph.

For this purpose, we use test patterns (specimens) with several repeating periods, the frequency of which is known as the *spatial frequency*. Spatial frequency is the number of periods per unit distance, usually expressed as periods per millimeter or per micrometer.

For example, for a series of black and white pairs of lines that measure 2 μm/pair, or repeat 500 times/mm, the spatial frequency would be 500/mm.

If the test pattern is made of alternately absorbing and transparent line pairs, and we view its image through a microscope equipped with an objective and condenser, both with NAs of 1.4, a 1-μm period would be clearly resolved (equation 5-9). The specimen with the 1-μm period would produce an image with moderate contrast between dark and bright bars.

If the spatial frequency of the test pattern is 2000, its period is 0.5 μm, and its contrast would be less than that of the specimen with the 1-μm period. Similarly, the contrast from the 1-μm-period (1000 line pairs/mm) spatial frequency pattern would be less than that of a 5-μm-period (200 line pair/mm) pattern.

At a 0.2-μm period, or spatial frequency of 5000 line pairs/mm, we would reach the limit of resolution in the example of the lenses with NAs of 1.4 for a wavelength of 500 nm (equation 509). At this limiting spatial frequency, the contrast of the image would be just barely above zero. Depending on the size, brightness, and color of the image, our eye would cease to detect the periodic image at certain contrast levels, and we might or might not in fact reach the 0.2-μm limit of resolution (Chapter 4).

The ratio of image contrast to specimen contrast, and the phase shift in positions occupied by the actual and an idealized sinusoidal image, plotted as functions of spatial frequency, together are known as the **optical transfer function (OTF).** When the distribution of light from the specimen is sinusoidal, the modulus of the OTF becomes the *modulation transfer function (MTF)*.* If we substitute a series of periodic black and white stripes in place of the sinusoidal pattern, the MTF becomes the *contrast transfer function (CTF)*. Figure 5-24A shows MTF curves of a microscope at four different values of $NA_{obj} = NA_{cond}$, for a periodic absorbing specimen illuminated with light of $\lambda_0 = 546$ nm, the intense green line from the mercury arc lamp. Figure 5-24B shows MTF curves for the same conditions, except that the objective is kept at its maximum NA of 1.32, but with the NA_{cond} at progressively smaller values.

Each objective lens shows a specific OTF or MTF, depending on its design, mode of contrast generation, the wavelength used, and the NA_{cond}.

For example, phase-contrast microscopes, with their narrow annular illumination, show an MTF (Fig. 5-25, curve B) that is different from that of a system illuminated with a full circular aperture (Fig. 5-25, curve A). With DIC, we have a series of MTFs (e.g., Fig. 5-25, curve C) that vary with the angle between the period in the specimen and the shear direction of the Wollaston or Nomarski prisms. The MTF shown in curve D reveals the superior contrast obtained by Ellis's single-sideband edge enhancement microscope at high spatial frequency (Section 11.3).

Not only does a microscope respond with given MTFs, but the video systems, the photographic emulsion, and even our eye, each respond with their own MTFs (Chapter 4, Appendix II; see also Castleman, 1979; Hecht and Zajac, 1974; *RCA Electro-Optics Handbook,* 1974).

At each image-forming step, we tend to lose image contrast in certain frequency regions, generally at the higher end of the spatial frequency range. Conversely, each detector and processor can also function, or be used, to cut off or boost the MTF at certain frequencies. Just how much can be usefully boosted, and where it is best to cut off, depends on the type of image information desired. The choice of MTF also depends on the frequency dependence of noise levels in the image, as the noise introduced at various stages of image transfer and processing are also frequency dependent.

These considerations become especially important for video microscopy, where many MTFs interact to yield the final image, and where the devices can both limit and boost the MTF over quite a wide range.

*The OTF is a complex function. The MTF is the modulus, or absolute value of the real term, of the OTF (See Glossary, OTF; Appendix II, page 468; Hecht and Zajac, 1974).

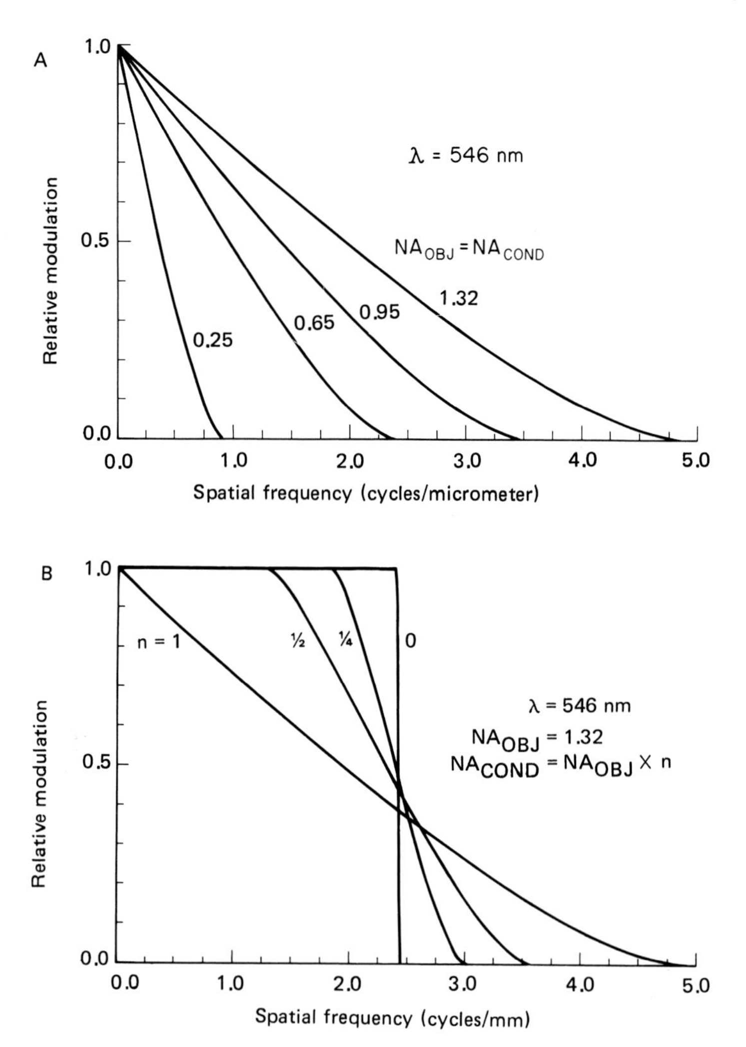

FIGURE 5-24. Modulation transfer function (MTF) curves for microscope lens, calculated for a periodic specimen in focus. (A) Each curve represents a different numerical aperture (NA), which is the same for the objective and condenser lens in these curves. Resolving power improves as the NA increases. (B) These MTF curves all represent an objective lens NA of 1.32, but with different condenser NAs; the conditions are otherwise the same as in A. (Courtesy of Dr. G. W. Ellis.)

The MTF is a mathematical concept, developed from information theory, that deals with the S/N ratio of information and information-handling devices. For a given type of message, the S/N ratio can be maximized by proper choice of the frequency characteristics, or MTFs, of the information-handling device.

In electronic devices that transmit the signal as a linear series of messages, one deals with frequencies of message units per unit of time. In video, the spatial information (distribution of intensity and color) is converted to a linear, time-dependent series of signals, so that spatial frequency in the image is converted to frequency of electrical signals per unit of time. Even though the spatial and temporal frequencies are based on signals with different dimensions, the MTFs describe the ability of the devices to transfer periodic information in both cases.

The MTF not only specifies the performance of a single imaging, or information-transferring, device, but allows one to calculate the overall MTF of a cascading series of such devices.

For example, in video microscopy, the following devices are often cascaded in series: the

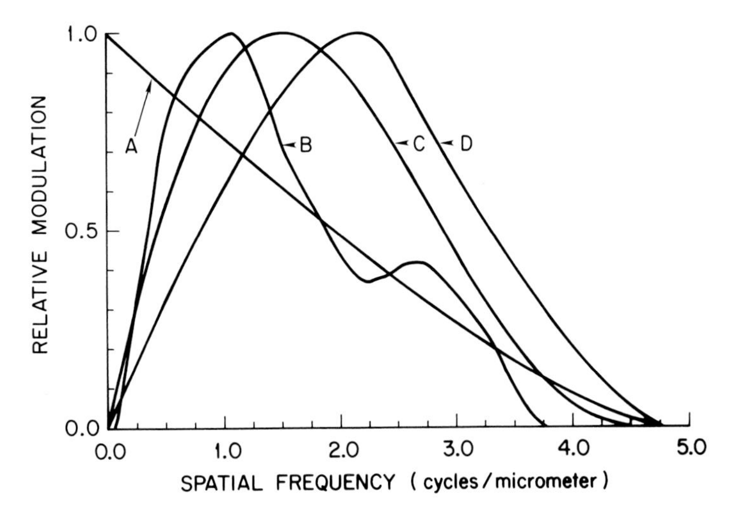

FIGURE 5-25. Modulation transfer function curves calculated for different modes of microscope contrast generation. A = brightfield (full circular aperture). B = phase contrast. C = differential interference contrast. D = single-sideband edge enhancement. The curves are plotted with their peak modulation normalized to 1.0. (Courtesy of Dr. G. W. Ellis.)

microscope, the video camera, video processors, the video tape recorder, the recorder as a playback device, the monitor, sometimes the photographic camera and the emulsion, and the eye.

*The overall system MTF, or frequency response, of such a cascading series of devices is given by the product of the MTFs of the component devices.** The MTFs of some of the devices can be calculated theoretically or measured empirically. The *RCA Electro-Optics Handbook* (1974) lists a number of useful tables for component devices. ''Kodak Professional Black-and-White Films'' gives MTF curves for a number of Kodak films.

MTFs have not yet been established for several contrast modes of the light microscope. Theories of image formation, especially in polarized light, would have to be further perfected, or appropriate test patterns or specimens would have to be found or developed, in order for those MTFs to be known.

5.8. INTENSITY OF LIGHT SOURCE AND OTHER FACTORS AFFECTING IMAGE BRIGHTNESS

In video microscopy, the brightness (luminance) of the image must be regulated far more critically than in other forms of microscopy or for other applications of video.

To expand on this point, when we observe the image directly through the microscope, our eye can adapt to a broad range of image luminances (Chapter 4). Also, in photomicrography, the time of exposure can be adjusted over a wide range. For standard, brightly lit scenes not taken through the microscope, a video camera (with its built-in filters and **automatic iris** and gain-controlling electronic circuits) can respond to variations in average scene luminances of 3–4 log units.

While such practical adjustments to luminance variations in between scenes are usually possible, the sensors themselves, such as the video camera tube or a photographic emulsion, can respond only to a limited range of absolute luminance levels. Below a critical luminance, the sensor ceases to respond, or the noise increases to an intolerable level. Above a limiting intensity, the sensor saturates and ceases to respond to intensity variations.

*The CTF does not have the same mathematical properties as the MTF: usually, the system CTF cannot be obtained by multiplying the component CTFs (Appendix II).

The useful operating range between the two extremes within a given scene, or the **intrascene dynamic range** of the video sensor, is often limited to less than 100 : 1 (Chapter 7). This **dynamic range** barely covers the range of intensities commonly encountered within a single common scene. (For example, see "Kodak Professional Black-and-White Films." The illuminance of the image impinging on the video camera tube must therefore be critically regulated, just as it is necessary to give exact exposure to a photographic emulsion with low latitude.

While in photography the total amount of light falling on the emulsion can be controlled by adjusting the duration of exposure, the exposure duration in video is essentially fixed. In the standard video **format** in the United States and Japan, the image, or frame, is refreshed 30 times/sec. The exposure must therefore be pretty much completed in 1/30 sec, or less, depending on the characteristics of the video camera tube.

Also in video microscopy, we cannot regulate the amount of light reaching the camera by adjusting iris diaphragms, as is done in camera lenses used for shooting standard video scenes. The field diaphragm in Koehler illumination does not affect image brightness; only the area of illumination changes (Sections 5.3, 5.4). Also, the opening of the condenser iris (or its conjugate apertures) is selected to optimize image contrast and resolution; the condenser iris must not be used to adjust the illuminance of the field. Narrowed stops in planes other than the conjugates of field or aperture planes simply vignette the image.*

Therefore, *neither the duration of exposure nor opening of the irises can be used in video microscopy to control the amount of light reaching the camera tube.*†

Yet, given a sensor that responds only to a limited range of illuminance, and a fixed duration of exposure for each scene, the brightness of the image falling on the video camera tube must somehow be fine-tuned in video microscopy.

If the image is too bright, the illumination can be reduced by regulating the current through the light source, or by interposing light-diffusing or -absorbing filters. If a diffuser is used, it is placed between the light source and the field diaphragm so that the image of the field diaphragm is not obscured. Better yet, a neutral-density filter can be used to reduce the intensity of illumination without shifting the color temperature or the paths of the rays.

If the image is too dark for the video camera, it has to be made brighter by: (1) raising the brightness of illumination, (2) increasing the signal from the specimen, (3) increasing the light-gathering power of the microscope, (4) reducing the projection magnification, or (5) increasing the light transmission through the microscope.

Alternatively, a video camera with a higher sensitivity would have to be used.

The sensitivity and related characteristics of video camera tubes and other sensors are described in Chapter 7. Practical means of optimizing the combined performance of the video equipment and microscopes, in the various contrast modes, are described in Chapter 11.

The basic factors that determine the luminance of the microscope image will be described next.

Brightness of Illumination

In a microscope equipped with well-corrected illuminator and condenser lenses, the degree of illumination (illuminance) of the field under Koehler illumination is governed by: (1) the intrinsic brightness (mean **luminous density**) of the light source; (2) the focal length of the collector lens of the light source; (3) the NA of the condenser lens; (4) the setting of the condenser iris diaphragm; and (5) the overall **transmittance** of the illuminating system.

Under Koehler illumination, light emanating from each point on the source uniformly illuminates the field stop, and hence also the microscope field (Fig. 5-14). Therefore, the size of the opening in the field diaphragm affects only the diameter of the field illuminated and not its brightness.

*However, an iris can be used to regulate the field illuminance of a microscope if the iris is one that is placed before the illuminating beam enters a "scrambler" (see footnote on p. 127).

†That is not to say that we cannot use flash exposures, e.g., to obtain stroboscopic records. This is possible because the video camera tubes do store the signal for some finite amount of time.

Contrary to expectation, the collecting power, or f value (diameter/focal length), of the collecting lens also does not, of itself, affect the illuminance of the field. Field illuminance would only be affected if the focal length of the collecting lens were not short enough to project an image of the source that covered the entire opening of the condenser iris diaphragm (Fig. 5-7, equation 5-3).

So long as the condenser iris opening is filled with the image of the light source, the illuminance of the field is determined by the mean luminous density of the light source and the square of the NA of the condenser. The size of the light source and the collecting power of the collecting lens affect the field illuminance only if the image of the source does not cover the entire condenser aperture.*

Stating it another way, *it is the mean luminous density, namely the light output per unit area, of the source that determines image brightness, and not its total light output, or the luminous flux. The luminous flux from the source is the product of the luminous density and the area of the source and, as mentioned above, the area plays only a secondary role.*

The luminous density of the source limits the field illuminance. In fact, theory shows that *field illuminance or luminance of the image can at best match, but never exceed (except perhaps by lazing), the luminous density of the source.* So long as the aperture is properly filled, the field can never be brighter than the source no matter what clever arrangements and combinations of mirrors, lenses, etc. are used.

Table 5-1 lists the luminous flux, arc size, mean luminous density, wavelength characteristics, etc. of some light sources that have high luminous density suitable for video microscopy.

At this point, a few comments are in order regarding the condenser. Some microscope condensers can be used with and without immersion. More often, better-quality condensers are designed to be used with a particular medium, having a specific refractive index and dispersion. The medium fills the space between the condenser and the specimen slide. Depending on the design, the medium can be oil, glycerol, water, or air.

Most darkfield, and some phase-contrast, DIC, and polarized-light, optics *require the condenser to be immersed* in order to achieve a high condenser NA. Also, the aberration of the condenser is corrected, and some functions of the condenser are planned, taking into account the refractive index and dispersion of the intended **immersion medium.**

Furthermore, immersion dispenses with the loss of light past the critical angle, eliminates additional refraction, reduces reflection losses of light, and cuts down the number of optical interfaces that could scatter or reflect light at high angles of incidence. Such scattering and reflection become a source of flare, and also alter the state of polarized light, thus reducing the extinction in high-NA polarization optical systems.

*Many concentrated arc lamps (Table 5-1) provide very high luminous density, whose distribution over the arc is highly nonuniform. Commonly, the arc is brightest in a minute spot right next to one of the electrodes, even within arcs whose overall dimensions are as small as 0.3×0.3 mm. Nevertheless, these arc lamps are essential for certain applications in microscopy because of their high *mean* luminous density.

When the image of such an arc is projected onto the condenser aperture, the aperture is no longer illuminated uniformly. Thus, the diffraction pattern produced by each point in the specimen departs from the ideal Airy disk. [Zernike pointed out that the aberration of the condenser lens has no influence on the resolving power of a microscope (e.g., see Born and Wolf, 1980, p. 522). However, he was emphasizing the degree of *coherence* of the light waves illuminating the specimen; a modified aperture function for the condenser could indeed affect the directional distribution of the illuminating wave and the diffraction pattern produced by a point object in and out of focus.]

The illumination of the aperture can now be made uniform, without appreciable loss of mean illuminance, by using an optical scrambler. The scrambler is a single optical fiber bent into an arc or loop (Figs. III-21, III-22). It forms an approximately 1:1 image of the light source. The light beam passing the curved fiber becomes scrambled so that the light exiting the scrambler is a homogeneous bright patch devoid of the heterogeneous distribution of intensities present in the original arc (Ellis, G. W., 1985).

This effective light source, at the exit of the scrambler, is focused by a well-corrected collector (we use a super-8-mm movie zoom lens) to fill the NA of the condenser selected. (By using the zoom projector, one can select between covering a large field at a lower NA_{cond} or a smaller field at higher NA_{cond}.) A device that performs the scrambling function (as well as controlling color temperature and image illuminance) is available from E. Leitz Inc.

TABLE 5-1 Selected Light Sources with High Luminous Density[a]

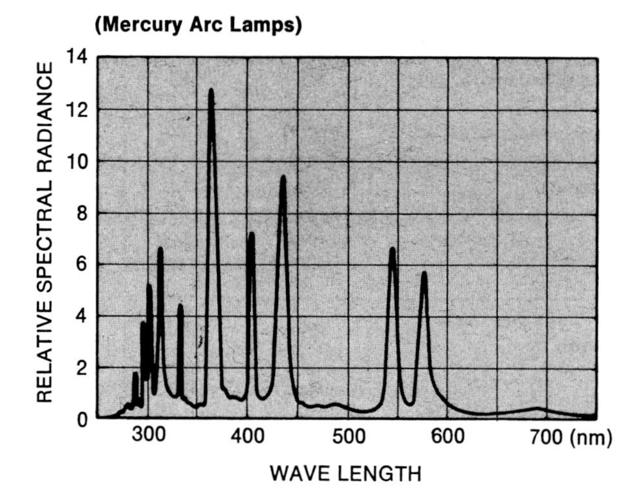

		Electric power (W)	Operating voltage (V)[b]
High-pressure mercury arc lamp	HBO 100 W/2[c]; III 110[d]	100	20 DC
High-pressure xenon arc lamp	XBO 75 W/2[c]	75	14 DC
	III X-74[d]	75	14 ± 2 DC
	959C1980[e]	500	14–20 DC
Tungsten halogen lamp	12 V 100 W	100	12 AC/DC

[a]The large variety of lasers now available are not included in this table. See *Lasers and Applications*, Volume 2, No. 13 (Dec. 1983), Designers' Handbook and Product Directory. As a microscope light source, the laser ouput must be made incoherent in time to reduce coherence length and to eliminate unwanted speckles (e.g., see Ellis, 1979).
[b]Arc lamps require high voltage and current (or high-frequency pulses) for starting.
[c]Osram. Distributed by Oswald Kerber, 7920 Fourth Ave., Brooklyn N.Y.
[d]Illumination Industries, Inc., 825 E. Evelyn Ave., Sunnyvale, Calif. 94086.
[e]Canrad-Hanovia, Inc., 100 Chestnut St., Newark, N.J. 07105.

Thus, immersion of the condenser affects the field illuminance as well as quality of the image. Condensers designed to be immersed should always be immersed with the correct immersion medium.

Although the illuminance of the field rises with the square of the NA_{cond}, opening the condenser iris diaphragm too far beyond matching the NA_{obj} produces flare, and we may also lose control over the image contrast.

In epifluorescence, where the objective lens serves also as the condenser, *the brightness of the image rises with the fourth power of the* NA_{obj}. That is so because the light-gathering power of the objective lens also rises with the square of its NA.

Increasing the Signal from the Specimen

In fluorescence microscopy, the brightness of the image is determined by the intensity of illumination, the quantum efficiency of the fluorochrome, and the collecting power of the micro-

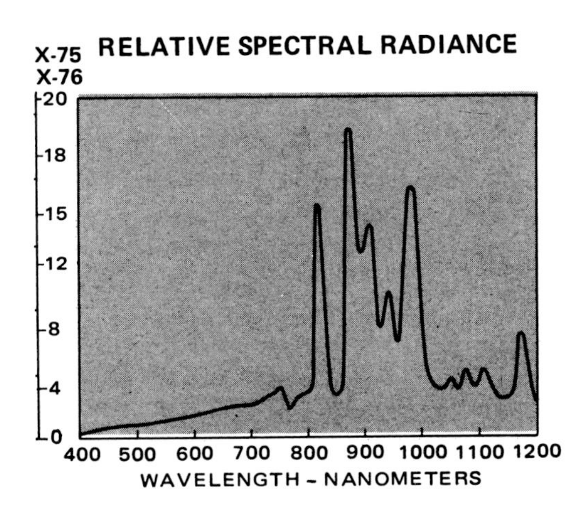

Lamp current (A)[b]	Luminous flux (lm)	Mean luminous density (cd/mm²)	Arc size (mm H × mm W)	Color temperature (°K)	Approx. life (hr)
5	2200	1700[f]	0.25 × 0.25	—[f]	200
5.4	850	400[f]	0.25 × 0.50	"6000"[g]	400
4.6–6.2	1125	800	0.50 × 0.38	"	300
30	9000	3500	0.30 × 0.30	"	200
8	2800	45	4.2 × 2.3	3100 @ 7.4 A	50

[f]Requires special high f number, heat-resistant collector lens, and optical scrambler (not ground glass; see footnote on p. 127) in order to fill condenser aperture without appreciable loss of luminous density. Elliptical and parabolic collectors, or arc lamp mounted in reflective collectors, are also available, e.g., from ILC Technology, 399 Java Dr., Sunnyvale, Calif. 94086; Optical Radiation Corp., 1300 Optical Dr., Szusa, Calif. 91702; Photo Research Assoc., Inc., 100 Tulsa Dr., Oak Ridge, Tenn. 37830.
[g]Relative spectral radiance shown above (varies with operating pressure of lamp; ca. 40 atm for Hg lamp). NB! Protect against *skin burn* and *eye damage*. High UV and IR output.

scope. The greater the intensity of illumination and the higher the quantum efficiency, the greater is the fluorescent signal and the brighter the image.

Likewise in polarized light, image brightness is governed by the illumination and birefringence retardation of the specimen. Up to a quarter-wavelength retardation, the larger the retardation, the greater is the birefringence signal.

In both of these cases, image brightness is governed by the signal from the specimen, the signal being a product of the intensity of illumination and the change in light introduced by the specimen.

For luminescence, or specimens that themselves emit light, the brightness of the image is clearly governed by the magnitude of light, or the signal, emitted by the specimen.

In these examples, the nature of the specimen itself affects the strength of the signal generated, and therefore the brightness of the image.

In other modes of microscopy, such as phase contrast, DIC, darkfield, etc., the intensity of illumination and image contrast can be varied independently. Nevertheless, the signal in the image, namely the increment of image brightness per unit of variation in optical parameter (e.g., difference in optical path), would still be given by the intensity of illumination × the contrast produced per unit variation in the optical parameter.

To detect small variations in any optical parameter, we should maximize the signal arising from variations in that particular parameter (Chapter 11).

Light-Gathering Power of the Objective Lens

Regardless of the mode of microscopy, image brightness is governed by the *light-gathering power of the objective lens.*

Just as the brightness of illumination is governed by the square of the working NA of the condenser, the brightness of the image is determined by the square of the NA of the objective lens.

Unlike the illuminating system, however, image magnification also determines image brightness. In fact, the image brightness is inversely proportional to the square of the lateral magnification.

The ratio $(NA_{obj}/\text{magnification})^2$ expresses the light-gathering power of the objective lens. Examples of objective lenses with high light-gathering power are listed in Table 5-2.

Generally, objective lenses with high NA are also better corrected for aberrations. Thus, for the same magnification, higher-NA objectives collect more light and give a brighter image, they give a better-corrected image, and the image is better resolved (equation 5-9).

We should note here that the brightness of the image decreases rapidly as the magnification increases. When the light level is limiting, we will want to use an objective lens with the highest NA, yet keep the magnifications of the objective lens and the ocular to the lowest level compatible with resolution as discussed below.

Image Brightness and Magnification

The area covered by the image is proportional to the square of the magnification. The image brightness therefore decreases with the square of the magnification.

Microscopes are generally used to bring out fine details of the specimen. At the same time, the microscope is a potent light-gathering instrument. Just as a telescope or binocular improves our vision at night, we can take advantage of the light-gathering power of the microscope to capture the image of faintly lit objects.

Depending on the features of the specimen we wish to visualize, the image may become more meaningful if we concentrate on collecting more light than on raising the magnification. For a number of luminescent and fluorescent objects, the light level may be so low that a highly magnified image becomes invisible or undetectable, and hence totally meaningless. In such cases, the image can be made more intelligible by collecting more light, by (1) integrating the light over time if the specimen is static (Chapter 9–11) or (2) by reducing the magnification of the image.

For extremely low-light-level situations, one can maximize image brightness by using the highest NA objective and the lowest overall magnification. Sometimes it is even beneficial to use the ocular to reduce, rather than increase, the magnification of the intermediate image.

Naturally, one would also use an image tube with the highest sensitivity. However, the higher the sensitivity of the image tube, the greater is the noise level, and the lower the resolution of the video camera (Chapter 7).

With video microscopy in general, the resolution of the image tube or camera tube is often limited to from a few hundred to about 800 TV lines (Chapter 7). Video tape recorders can further limit the resolution of the stored image to from 260 to somewhat over 400 TV lines (Chapter 8).

In order to match the video resolution to the resolving power of the objective lens, there is a general tendency to boost the magnification of the image projected onto the video camera tube. But since the brightness of the image drops by the square of the magnification, and since the video camera tube can function only within a restricted range of intensities, boosting the magnification can result in the image brightness dropping below the sensitivity of the camera tube.

TABLE 5-2. Microscope Objective Lenses with High Light-Gathering Power[a]

Manufacturer	Type	Mag	NA	$10^4 \times (NA/Mag)^2$	WD (mm)	Remarks
Nikon	CF Fluor (DL)	10	0.5	25	0.88	BF Ph Fl
Leitz	NPL FL Oil (PHACO)	10	0.45	20	0.33	BF Ph Fl Pol.
Olympus	S Plan Apo	10	0.40	16	0.55	BF
Olympus	UV FL	10	0.40	16	1.16	BF Fl
Zeiss	Plan Neofluar	16	0.5	9.8	0.15	BF Ph Int. Fl
Nikon	CF Fluor	20	0.75	14	0.66	BF Ph Fl
Olympus	S Plan Apo	20	0.70	12	0.55	BF
Olympus	UV FL	20	0.65	11	1.03	BF Fl
Zeiss	Plan Apo	25	0.80	10	0.32	Refl. BF Pol.
Zeiss	Plan Neofluar Oil/W/Gly corr.	25	0.80	10	0.13	BF Ph Fl
Leitz	NPL FL Oil	25	0.75	9	0.14	BF Ph Fl Pol.
Leitz	NPL FL Oil	40	1.30	11	0.21	BF Ph Fl Pol.
Olympus	UV FL Oil F/iris	40	1.30	11	0.11	BF Ph Fl
Nikon	CF UVF Gly	40	1.30	11	0.10	BF Fl
Zeiss	Plan Apo Oil/iris	40	1.00	6	0.38	BF Ph Fl
Reichert	Plan Apo Oil/iris	40	1.00	6	0.23	BF Ph Fl Pol.
Nikon	CF PLan Apo Oil/corr.	40	1.00	6	0.12	BF
Nikon	CF Plan Apo/corr.	40	0.95	5.6	0.10	BF
Zeiss	Plan Apo	40	0.95	5.6	0.09	BF Ph Fl
Olympus	S Plan Apo	40	0.95	5.6	0.13	BF
Zeiss	Plan Neofluar Oil/W/Gly corr	40	0.90	5.1	0.13	BF Ph Fl
Zeiss	Plan Neofluar	40	0.90	5.1	—	BF Pol.
Nikon	CF Fluor/corr.	40	0.85	4.5	0.39	BF Fl
Nikon	CF Fluor DF/corr.	40	0.85	4.5	0.37	BF Ph Fl
Olympus	UV FL/corr.	40	0.85	4.5	0.25	BF Fl
Zeiss	EPI Plan Pol	40	0.85	4.5	0.23	Refl. Pol. short mount (25 mm)
Reichert	Adv. Achrom.	40	0.80	4.0	0.50	BF Fl
Leitz	NPL FL W	50	1.00	4.0	0.68	BF Ph F
Leitz	NPL FL Oil	50	1.00	4.0	0.18	BF Ph Fl
Reichert	Plan Fluor ∞/0	50	0.85	2.9	0.48	Refl. BF DF Pol TL = ∞; no coverslip
Nikon	Plan Apo Oil	60	1.40	5.4	0.16	BF Fl
Olympus	S Plan Apo Oil	60	1.40	5.	0.12	BF
Zeiss	Plan Apo Oil	63	1.40	4.9	0.09	BF Ph Int. Fl
Olympus	NCD Plan FL	60	0.95	2.5	0.14	BF

[a]NA = numerical aperture; $10^4 \times (NA/Mag)^2$ = light-gathering power; WD = working distance; W = water; Gly = glycerol (Oil, W, Gly indicate an immersion lens); corr. = correction collar; BF = brightfield; Ph = phase contrast; Fl = fluorescence; Pol. = polarizing; Int. = interference; DF = darkfield; Refl. = reflecting; TL = tube length.

Therefore, *in video microscopy, there is a constant tug-of-war, and compromise, between an attempt to increase the resolution and to keep down the noise level, which rises as the image becomes dimmer.**

When the brightness, or signal, from the image is limited, one needs to carefully fine-tune the microscope magnification to strike the best balance.

*Some video cameras appear to be less noisy at very low light levels. That is because the **bandwidth** is purposely limited to cut down the noise when the signal level drops below a certain level.

Light Transmission through the Microscope

For a given condenser and objective NA, magnification, and illuminator brightness, the brightness of the microscope image can still vary depending on the light transmission through the optical components. The transmission depends on several factors including reflection losses at the optical interfaces, and transmittances of the lamp jacket, diffusing screen, filters, and polarizers.

The *transmittance* (intensity transmitted as a percentage of incident intensity) of some components can be wavelength dependent even within the visible range. Also, at wavelengths outside of the (middle) visible range, the transmittance of the lenses, prisms, and the mounting medium for the specimen, can drop appreciably.

As discussed in Section 5.4, all surfaces reflect some light and result in loss of transmission even when there is no absorption of light and the beam is incident normal to the interface. As the angle of incidence increases, the reflection and transmission losses increase on the average.*

Each uncoated air–glass interface can reflect 4–5% of the beam incident normal to the surface (see footnote‡ on p. 107). The transmittance per passage through an untreated interface would thus be 96–95% at normal incidence. By applying an *antireflection coating* (a quarter-wave thick interference film with an appropriate refractive index), reflection at the glass surfaces can be reduced to 1% or less for a moderate range of wavelengths and incidence angles.†

Some modern objective lenses with a high degree of correction can contain as many as 12 lens elements and 16 air–glass interfaces. If the lenses were uncoated, the reflection losses of axial rays alone would drop the transmittance of such a lens to about 50% ($= 0.96^{16}$). With all the surfaces coated, the transmittance could be improved to 85% ($= 0.99^{16}$).

In addition to those in the objective lens, there may be some two to four dozen optical interfaces in many microscopes (4 to 6 in the collector; 2 to 8 or more in mirrors, prisms, relay lenses; 2 to 8 in filters; 4 to 8 in condensers; 0 to 3 in specimen slide and coverslip; 4 to 6 in the ocular; 2 to 6 in the camera lens, etc.). In extreme cases, there can be as many as five dozen optical interfaces, when the objective lens is included.

Less than 5% of the axial rays would be transmitted through a microscope with 60 uncoated interfaces, and a little over 50% with all of the surfaces coated.

High quality of lens coating is essential not only for improving the transmission, but also for decreasing the flare due to multiple reflection at the glass surfaces.

However, even the best coatings cannot work beyond a certain wavelength range. For some wavelengths, the coating can even increase the reflectance (*a half-wave interference film is a perfect reflector*). This is an important point to keep in mind in video microscopy where the sensitivity of the photodetector can peak at wavelengths quite far from the peak for the human eye (Chapters 4, 7).

In addition to reflection losses at the optical interfaces, the transmission of the *lamp jacket* may reduce the image brightness. As the lamp ages, the glass or quartz jacket may craze or darken as it devitrifies or becomes coated or permeated by the metal evaporating from the filament or electrodes. It is worth noting that the transmittance of the lamp jacket may drop sooner in the UV than for visible light, so that the visible brightness, or luminance measured with a photometer, may be a poor indicator of the UV output of a mercury arc lamp, for example.

A ground-glass *diffusing screen* transmits only 10–15% of the light traveling within $+3°$ to its normal. Less light is transmitted at oblique incidence. Thus, ground-glass diffusers are to be avoided whenever the level of illumination needs to be maximized.

Filters and dichroic mirrors have transmissions that drop off as the passband is narrowed.

*The reflection losses for the light waves vibrating in and perpendicular to the plane of incidence vary differently as the angle of incidence changes (e.g., see Jenkins and White, 1957, on Fresnel coefficients of reflection and transmission).

†Multilayer antireflection coatings have also been developed, which can reduce the reflection at an air–glass interface to 0.1%. Multilayer coatings have a slightly greenish tint, as opposed to the purplish tint of single-layer coatings.

Many interference filters transmit only 15–30% of the incident energy at the wavelength for peak transmission. However, some selected, multilayer interference filters with half-transmission bandwidths as narrow as 50 Å can provide peak transmittance of 75% or greater.

The *polarizing filters* used in polarizing and DIC optics can also substantially reduce the transmission of the microscope. Even when the polarizer and analyzer axes are set parallel, namely at peak transmission, the transmittance for a pair of polarizing filters, each with a natural light transmittance of 20%, is only 8% [20% × (2 × 20%)]. In contrast, high-quality *calcite prisms* with antireflection-coated cover plates can transmit the theoretical maximum, namely 50% of the incident unpolarized light per pair.

As enumerated above, the transmission through the many optical components in a modern light microscope can be limited by a variety of factors. Where one is striving for the highest resolution, especially in contrast modes in which the image is inherently dim, all of the factors listed here should be examined carefully; *the transmittance through the whole microscope is given by the product of the transmittances of all of the optical components.*

High transmittance is especially important in video microscopy where high resolution requires high magnification, with the attendant loss of image brightness. Fortunately, the contrast-enhancing ability of video also allows the use of condenser and objective lenses with very high NAs.

5.9. ABERRATIONS

Given a long focal length and narrow aperture, a lens can be designed, without great difficulty, to produce an aberration-free image near its axis.

On the other hand, with a high-power microscope lens, the focal length may be as short as 2 mm or less, the NA may be as high as 0.95 in air (equivalent to an f/0.16 photographic lens), and the intermediate image may be as large as 10 mm, or even over 20 mm, in diameter.

Despite these stringent conditions, **Abbe** had worked out, by the end of the last century, the basic designs needed to correct the major aberrations in the central field of high-NA microscope objective lenses. The image produced by a microscope objective, corrected to satisfy the sine conditions according to Abbe's design principles, is limited by diffraction rather than lens aberrations.*

Indeed, Abbe's **apochromatic objectives** were remarkably free from **coma,** and were corrected for spherical aberration at two wavelengths, and longitudinal chromatic aberrations at three. Combined with compensating eyepieces, the lateral chromatic aberration was also substantially reduced, although the field remained quite curved.

In the past two decades, new rare-element glasses and high-quality low-reflection coatings have become available, and lens design in general has been speeded up by the use of computers. Not only are microscope lenses better corrected over wider fields, but the flare is so reduced and the transmission so improved that the image (e.g., of modern plan apochromatic objectives) is remarkably sharp, crisp, and bright.†

Even with the best microscope lenses, however, the user must care for and use the lenses properly, i.e., within the constraints imposed by their design. The care of lenses is described in Section 5.10. The means for properly aligning the lenses, achieving Koehler illumination, and adjusting the diaphragms were discussed earlier in this chapter.

*See Glossary (Chapter 2) for definition of aberrations such as chromatic aberration; spherical aberration, coma, **astigmatism,** pincushion and **barrel distortion,** etc. See Schillaber (1984) and Françon (1961) for excellent discussions of microscope lens aberrations and their detection. For photographs of diffraction patterns produced by lenses with specific aberrations, see Cagnet *et al.* (1962) and references in footnote, p. 114.

†According to a recent Zeiss catalog, modern plan apochromats are corrected for four wavelengths in the dark blue, blue, green, and red for spherical aberration, and for more than four wavelengths for chromatic aberration, in addition to providing a flat field of view. Also modern **fluorite objectives** are corrected for blue, green, and red for spherical aberration, and for more than three wavelengths for chromatic aberration.

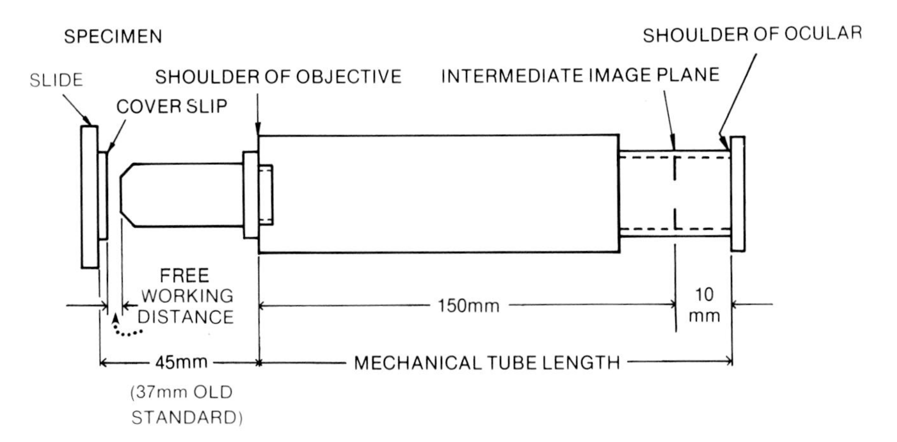

FIGURE 5-26. Standard microscope dimensions. See text.

Unlike photographic lenses, *microscope objective lenses are designed to function at a fixed distance to the intermediate image plane, and with standardized optical media of given thicknesses between the lens and the specimen.*

The distance between the shoulder of the nose piece, supporting the objective lens, and the intermediate image is gradually becoming standardized to 150 mm by the major microscope manufacturers. The intermediate image plane is also becoming commonly set to 10 mm below the shoulder of the body tube that supports the ocular (Fig. 5-26). The distance between the shoulders of the nose piece and the ocular tube is known as the mechanical **tube length**.*

For high-NA objective lenses, especially high dry objectives, variations in tube lengths of a few millimeters introduce detectable degradation of the diffraction pattern and loss of image detail and contrast.

Even more seriously, departure of the *coverslip thickness* (and depth of the specimen) from the designated value by as little as 10 μm or less can significantly degrade the image! A thorough, quantitative discussion of this topic may be found in Loveland (1981, Volume 1, Chapter 2).

Several high-NA objectives are provided with graduated **correction collars.** Turning the collar adjusts the position of certain lens elements in the objective to compensate for variations in the thickness of the coverslip and the refractive index of the immersion medium. The same adjustment also improves the diffraction pattern when the tube length departs from the standard.

For video microscopy with an NA 0.95 Plan Apo dry objective lens, I find that my adjustment of the correction collar falls reproducibly within +1 μm when I strive for the best image.

Standard coverslips are assumed to be 0.17 ± 0.01 mm thick (with a refractive index of 1.515). Number 1-1/2 coverslips are nominally selected for this standard thickness. However, for critical work, a micrometer or cover glass thickness gauge should be used to verify the thickness, and the flatness of the coverslip should be assessed by observing a reflected image of a diffuse source with straight edges or lines. Fluorescent lamps with good diffusers work well.

With increasing use of inverted microscopes for viewing tissue-cultured cells, objective lenses have become available that correct for thick slides or culture dishes (Table 5-3). Other objectives allow adjustment of the correction collar for immersion in water, glycerol, or oil (Table 5-3).

Generally, it is safer to use the ocular designated by the manufacturer for a given objective lens. The ocular may be designed to compensate the residual aberrations of the objective, and the tube lengths may not be strictly set to the common standard.

On the other hand, once the sources of aberration and the principles and practices of contrast

*For other design standards for microscopes and accessories, as well as a discussion of some problems involved in setting standards, see Keller and McCrone (1977).

TABLE 5-3. Microscope Objective Lenses with Long Working Distances

Ianufacturer	Type	Mag	NA	$10^4 \times (NA/Mag)^2$	WD (mm)	Remarks[a]
eiss	UD 16/0.17 Achr.	9.6	0.11	1.3	10.0	Mag, NA shown without hemisphere[b]
eichert	Plan Achr. Phase	10	0.25	6	9.1	
eiss	UD 20/0.57 Achr.	12.7	0.38	9	6.3	Mag, NA shown without hemisphere
eiss	EPI Plan	16	0.35	4.8	2.8	Refl. Pol.
eichert	LWD/corr.	20	0.50	6.3	1.2–1.7	BF Ph D=0.5–1.5
)lympus	ULWD-CD Plan	20	0.40	4.0	10.5	D=0 ~ 2
Iikon	ELWD/CFM Plan Achro	20	0.40	4.0	10.5	Refl. TL=210
Iikon	ELWD/CFBD Plan Achro	20	0.40	4.0	8.5	BF DF
Iikon	Phase Achro/corr.	20	0.40	4.0	0.45–2.16	BF Ph
eitz	EF L (PHACO)	20	0.32	2.6	6.83	BF Ph D=0.17–0.70
)lympus	LWD-CD Plan/corr.	20	0.40	4.0	3.00	BF D=0–2
eiss	UD 40/0.65 Achr.	25.8	0.41	2.5	6.8	Mag, NA shown without hemisphere
eitz	NPL FL L (PHACO)	25	0.35	2.0	13.5	BF Ph Fl D=0.6–1.6
eitz	NPL FL L (PHACO)	32	0.40	1.6	6.55	BF Ph Fl D=0.17–0.30
eiss	Achro Oil (Immersion)	40	0.85	4.5	0.35	BF Ph D=1.5
eiss	Achro W(ater Imersion)	40	0.75	3.5	1.6	BF Ph
)lympus	VLWD-CD Plan	40	0.60	2.3	8	D=0–2
eiss	Plan Achro LD Epi Plan	40	0.60	2.3	5.4	Refl. Pol.
)lympus	LWD-CD Plan	40	0.60	2.3	1.92	D=0–2
eitz	NPL FL L (PHACO)	40	0.60	2.3	1.65	BF Ph Fl D=0.6–1.6
Reichert	Adv Plan Achro	40	0.55	1.9	3.5	D=0
Iikon	DL LWD/corr.	40	0.55	1.9	0.17–2.37	BF Ph
Iikon	ELWD/CFM Plan Achro	40	0.50	1.6	10.1	Refl. TL=210
Reichert	LWD/corr.	45	0.66	2.2	0.6–1.1	
Iikon	CFM Plan Achro	60	0.70	1.4	4.9	TL=210
Iikon	CFBD Plan Achro	60	0.70	1.4	4.9	BF Ph
Nikon	CFBD Plan Achro	100	0.90	0.8	1.0	BF Ph
Nikon	CFM Plan Achro	100	0.80	0.6	2.0	TL=210
Nikon	CFBD Plan Achro	100	0.80	0.6	2.0	BF Ph

Many of these objectives can be used for DIC microscopy, as well. NA = numerical aperture; $10^4 \times (NA/Mag)^2$ = light-gathering power; WD = working distance; corr. = correction collar; BF = brightfield; Ph = phase contrast; Fl = fluorescence; Pol. = polarizing; DF = darkfield; Refl. = reflecting; TL = tube length (mm); D = coverglass thickness (mm).

See Fig. 12-20.

generation are well understood, and a good eye has been developed for judging the quality of the microscope image, one should not be overly inhibited about combining objectives, oculars, and microscope stands from different makers should the need arise.

Some oculars provide adjustments for *projection* distances ranging from several centimeters to infinity. (Naturally, the ocular should be set to infinity when the ocular is used in conjunction with a camera lens focused to infinity.) Without a camera lens, such oculars can directly project good images at the distance to which the lens is adjusted.

Other projection lenses, sometimes known as *amplifying lenses,* are negative lenses that shift the intermediate plane out of the microscope directly onto the image plane in the camera. Such lenses, which are designed to be used with specific objective lenses, can provide wide, well-corrected, flat fields.

5.10. CHOICE AND CARE OF LENSES FOR VIDEO MICROSCOPY

Objective Lenses

A superior-quality microscope image is achieved with modern, high-NA, plan apochromatic objective lenses. They not only yield a wide field in focus, but provide excellent correction, high resolution, and high light-gathering power.*

When the image is enhanced by video, these well-corrected high-NA objective lenses also perform outstandingly well. Even in polarized-light microscopy, where the need for high extinction previously precluded the use of lenses with textured crystalline elements, the best-corrected plan apochromats provide the highest-quality images in video.

On balance, perhaps one of the strongest points of video is the ability to enhance image contrast. Thus, if the objective lens has the inherent capability of providing the highest resolution and best-corrected diffraction pattern, video is able to compensate for some lack of contrast. With the contrast boosted, the overall MTF of the combined microscope–video system appears able to bring out the best in the better lenses.

While this observation still needs to be verified by measurements on standardized test samples, my experience has been that *ever since I have come to use video routinely on our polarizing microscope, my choice of objectives has dramatically shifted from the highest-extinction* **achromats** *to the higher-NA plan apochromatic objective lenses,* whenever such lenses with moderately high extinction were available. Likewise, video has permitted the use of higher NA_{cond}, with the attendant gain in resolution and image brightness.

The same principle seems to hold for contrast modes other than high-extinction polarization optics, such as in DIC, brightfield, fluorescence, etc.

Whenever the best image is needed in video microscopy, I would choose the best-corrected, highest-NA objective lens compatible with the generation of moderately high contrast in the selected contrast mode (Table 5-2). The same holds true with the condenser.

Oculars

As discussed earlier, to gain the best video image, it is most important that the resolution and brightness of the microscope image are harmonized with the resolution and sensitivity of the video sensor.

Therefore, it becomes necessary to fine-tune the magnification of the image projected onto the video camera tube. The image is projected either directly by the ocular, or by an infinity-focused ocular and a complementary lens on the camera (Section 5.2).

For some applications, the magnification can be adjusted closely enough with the magnification changer in the body tube, or by switching objectives, oculars, and the lenses on the camera tube.

In other cases, for any given objective lens, the final magnification of the projected image can be fine-tuned by sliding the video camera on an optical bench aligned parallel to the projection axis of the microscope. In such cases, the ocular alone is used to project the image on the video camera, and the camera is shielded from external light with a simple light baffle tube (Sections 3.3, 3.9).

For high-resolution images in video microscopy, the field of view is quite limited. It is therefore important to be able to readily move down in magnification in order to gain an overall view, or to localize the observed structure in the context of a larger part of the specimen. As the magnification is varied back and forth, it is desirable to keep the specimen in sight and maintain the axis of the objective lens with the specimen.

I have come to rely routinely on a *powered zoom ocular* to fine-tune the image magnification and brightness on our microscope. Not only does the zoom ocular yield the optimum balance of magnification and image brightness for each scene, but the zoom allows one to move to a lower-

*However, for epifluorescence microscopy a fluorite objective with greater transmission in the near UV may be preferred over a Plan Apo objective with somewhat greater NA and better correction.

resolution wider field to gain an overview of the field, or to a limited-field high-resolution image to inspect the details, without switching the objective lens or ocular. Aided by this capacity, cells can be micromanipulated and microinjected under high-power objectives, and the vibration introduced by switching the lenses is avoided.

We have adapted a 1.6 to 6.4 × powered zoom ocular (Fig. 5-27) designed to fit a Leitz

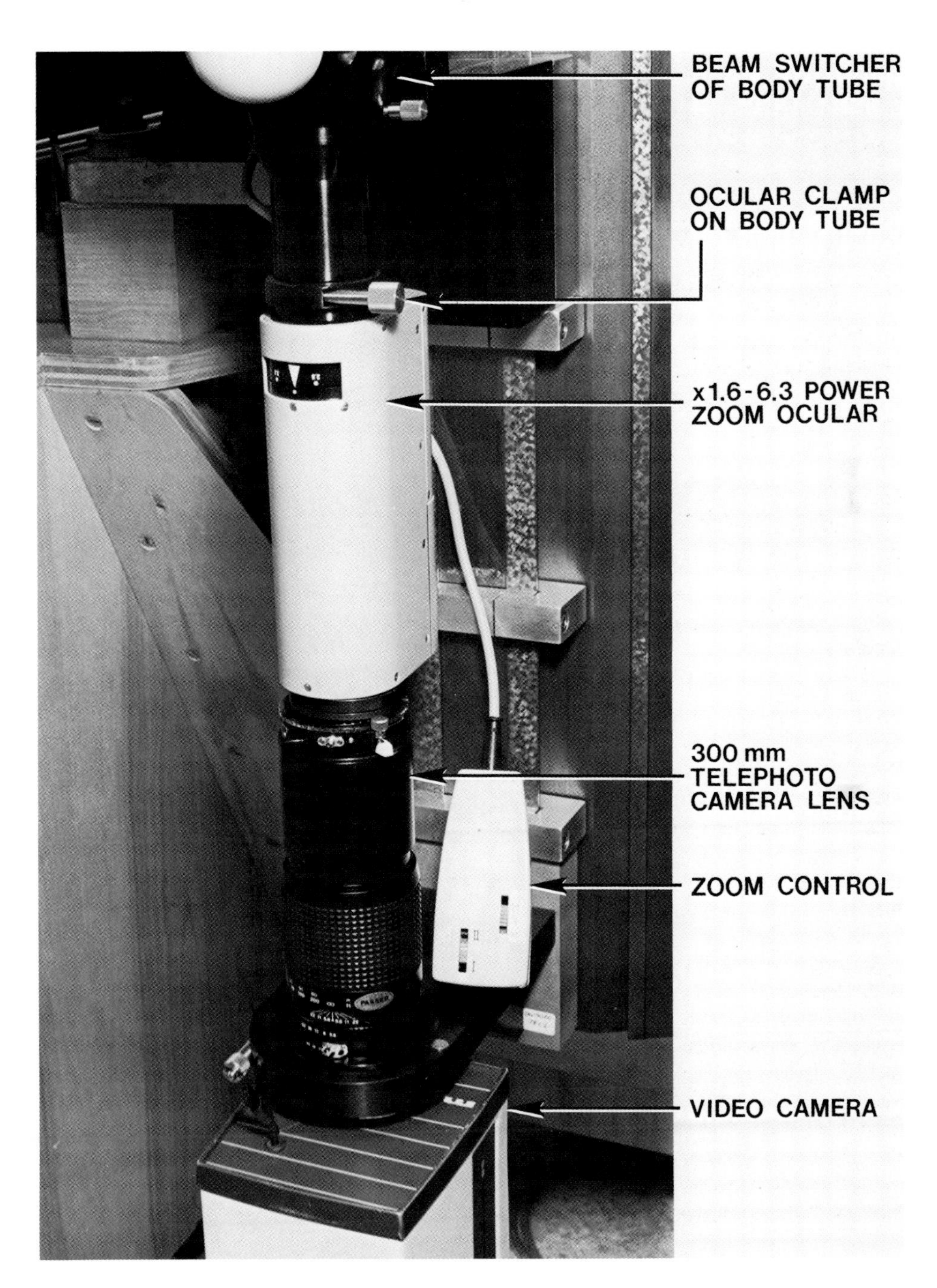

FIGURE 5-27. The Leitz motorized zoom eyepiece used as part of the author's inverted video microscope mounted on a 4-foot-long vertical optical bench (Figs. III-21, III-22).

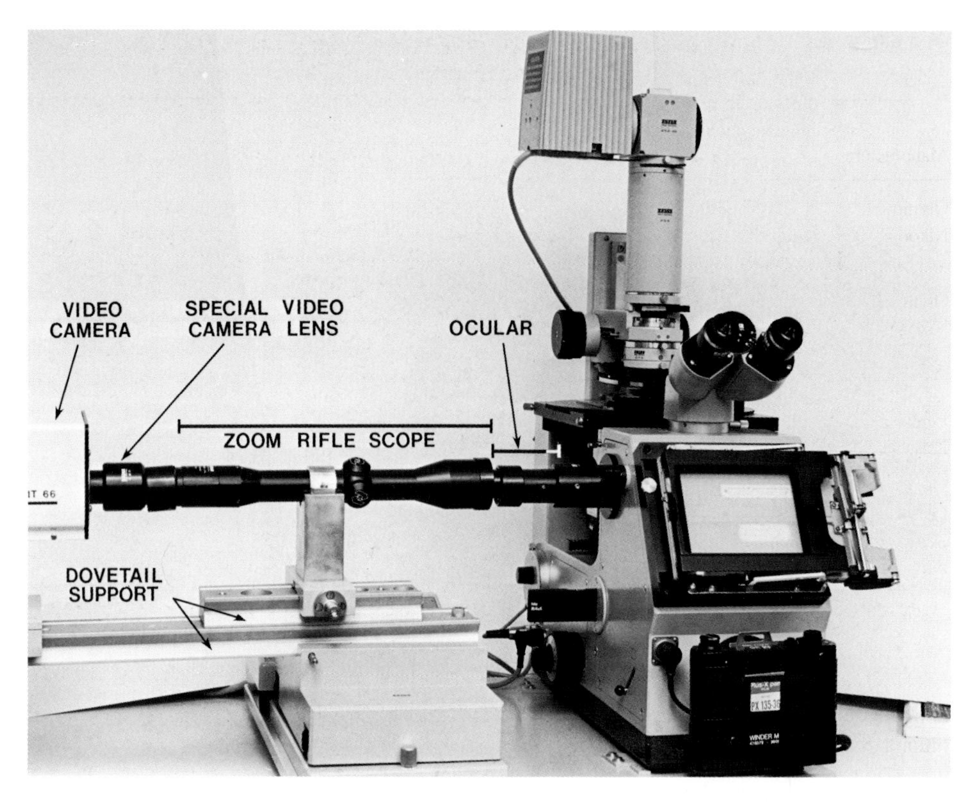

FIGURE 5-28. Rifle scope used as a zoom ocular. In this experimental configuration provided by C. Zeiss, the rifle scope mounted on a horizontal optical bench is coupled to an ICM inverted microscope. The rifle scope provides a zoom range of 1.5 to 6.0 power. Together with the ocular and camera lens, which are both focused to infinity, the total magnification of the projection system becomes: zoom ratio × focal length of ocular/focal length of camera lens.

Ortholux stand. A coated doublet or telephoto camera lens of about 300-mm focal length follows the zoom ocular. On our inverted, optical bench microscope, the powered zoom ocular introduces no vibration to the specimen or micromanipulation needle at the highest power.

Nikon also has developed a zoom ocular specially for video microscopy. It is shown in Fig. 5.34.

If space permits, an appropriate rifle scope with a zoom system can also be used in place of a zoom ocular. The rifle scope is placed in the infinity-focused beam between the ocular and the camera lens (Fig. 5-28).

Collector and Condenser Lenses

In video, inhomogeneity of field brightness is accentuated as the contrast of the specimen is enhanced. Thus, slight inhomogenieties of illumination, which might be overlooked in visual observations or photography, may become major distractions or actually limit the fidelity of the image in video microscopy.

It thus becomes especially important in video microscopy that the lamp collector lens be able to provide a homogeneously lit field. With critical use of the light microscope, one finds that the light source is often disproportionately poorly designed and constructed compared to the rest of the microscope.

In fact, many collectors are poorly corrected, even for spherical and chromatic aberrations, and may also contain focusable air bubbles and striae. Those collectors do not provide uniform illumina-

TABLE 5-4. Long-Working-Distance Condensers; Objectives Suitable as LWD Condensers[a]

Manufacturer	Type	Mag	NA	Focal length	WD (mm)	Remarks
Olympus	ULWD Phase		0.30		55	Ph. An.
Nikon	ELWD Phase		0.32		54	Ph. An. 4, 10, 20, 40
Olympus	LWD Phase, DIC		0.55		21	
Nikon	LWD Phase, DIC		0.52		20.5	Ph. An. 10, 20(4, 40); DIC 10, 20, 40
Leitz	L P	25	0.22		14.7	
Zeiss	Achro. Aplanat. Phase		0.63	15.3	11	
Leitz	L P	20	0.32		6.73	
Leitz	L P	32	0.40		6.55	
Zeiss	DF		0.7–0.85	8.5	6.5	
Zeiss	DF		0.8–0.95	8.2	6	

[a]Many of the objective lenses listed in Table 5-3 can be used for this purpose. NA = numerical aperture; WD = working distance; Ph. An. = phase annulus.

tion without the use of a diffuser. But, as discussed before, the diffuser reduces the intensity of illumination, unacceptably for some contrast modes.*

Compared to the collector, condensers with good correction have been available from a number of manufacturers for some time. For example, the **achromatic aplanatic condensers** can give very high NAs and are also well corrected. Naturally, these condensers must be used with the proper immersion medium and appropriate microscope slide as specified by the manufacturer.

Increasingly, the condenser, in addition to the objective, lens is being asked to play an important role in developing the contrast in wave optical microscope systems, such as in DIC. More condensers of high optical quality can be expected to become available in coming years.†

For certain applications involving extensive manipulation of the specimen, long-working-distance condensers are needed. A few condensers with long working distance are also well corrected (Table 5-4).

Inspecting and Cleaning the Optics and Video Camera Faceplate

As has been mentioned several times, blemishes such as dust, dirt, and smudges on the optical components, as well as scratches, pinholes, striae, etc. in the lenses, filters, prisms, mirrors, and faceplate of the camera tube, degrade the microscope image enhanced with video. Dirty, damaged, or defective *optical elements, at or near the conjugates of field plane* (sections 5.3, 5.4), *are especially likely to be disturbing.*

It is ironic that the better the optics, e.g., of the condenser and the collecting and relay lenses, and the higher the gain of the video system, the more these blemishes interfere and contribute to the optical noise.

Once the source of the optical noise is localized to a given optical component (by turning or moving the suspected components in turn), the dirt may be removed by gentle brushing with a *clean*

*The light-pipe scrambler described in the footnote on p. 127 would scramble out the visible defects of the lamp collector lens with little loss of light. Even then, much light would be wasted if spherical and chromatic aberrations prevented the collector from forming a concentrated image of the source onto the entrance of the light pipe.

†In certain cases, objective lenses can be substituted for the condenser.

camel's hair brush. Alternatively, an air blower or Freon duster could be used; but first make absolutely sure that no oil or similar spray is released from the device. Also, do not try to blow the dust off with your breath; most likely you will spray the surface with droplets of saliva.

Often, some oil or grease smear is present, or the dust particles cling to the glass surface (electrostatically and through wetting action) and cannot be blown or dusted off. If such problems are encountered on hard-coated surfaces (where the antireflection coatings are made hard to withstand a certain amount of wiping, such as the top lens of the ocular, the exposed faces of the collector and camera lenses, etc.), the lens can be cleaned in the following way.

First, blow or dust off the loose particles. Then breathe gently on the surface with your mouth wide open (not with the lips closed). Immediately wipe off the fogged surface with clean lens tissue or Kimwipes to remove the dirt and smeared material loosened by the moisture condensed from your breath.*

For cleaning the delicate lens surfaces, avoid using stiff lens paper. Always fold the clean lens tissue or Kimwipe into not less than four layers. The virgin, multilayered material keeps the oil and salts from our finger from penetrating the paper and contaminating the optical surface. Thoroughly washed pure cotton cloth can be better but the cleaning cloth has to be washed after each use.

To clean off immersion oil from the objective or condenser lenses, proceed as follows. Take a fresh, approximately 3 × 5-inch piece of lens tissue and hold it horizontally by one end. Using a medicine dropper equipped with a neoprene (not gum rubber) bulb, place *one* drop of reagent-grade benzene onto the lens tissue (Fig. 5-29A). The drop of benzene should hang below the lens tissue and not be completely absorbed.

Then (but not before), immediately bring the drop of benzene into contact with the tip of the immersion lens to be cleaned (Fig. 5-29B). Holding the lens tissue by its short edges, draw the tissue horizontally over the lens surface (Fig. 5-29C). Do not try to push the tissue toward the lens; instead, let the surface tension of the benzene and the oil hold the tissue and lens surface together as you draw the tissue.

As the tissue is drawn across the lens, the benzene and oil leave a streak on the tissue, and the benzene rapidly evaporates. Repeat three or four times using a clean drop of benzene and a fresh part of the lens tissue each time.

Benzene is highly flammable and its vapor is harmful to the liver. Use only a small dropper bottle and keep the cap tightly closed. Use adequate ventilation. Never use more than a small drop of benzene at a time, and keep your fingers away from the part of the lens tissue that contacts the benzene.

Benzene is a potent solvent for balsam and some other cements used on lenses and for mounting the coverslips. The trick is to finish the cleaning process (the streaking) so fast that the benzene remains on the lens for no more than a second or two. *Never soak the lenses with benzene,* or you may lose the lens altogether.

Xylene is a somewhat milder solvent but does not evaporate very rapidly. The residue could then penetrate and damage the lens. Some manufacturers prefer reagent-grade acetone, or a 1 : 1 mixture of ethyl ether and alcohol, but ether is even more flammable than benzene. Ether spilled on the bench top can even be ignited without an open flame, e.g., by the heat of a mercury or xenon arc lamp operating in its lamp house.

To clean water and glycerol immersion lenses, or salt crystals that have formed on objective or condenser lenses, wipe gently with lens tissue that has been folded to produce an angular corner and which is then dampened with distilled water (Fig. 5-30). After removing the water with a dry, folded corner, breathe on the lens and wipe off the freshly condensed moisture with another fresh, dry folded tissue. Start at the middle of the lens and work outwards in a spiral.

The front elements of the objective and condenser lenses can be inspected conveniently by using an inverted ocular as a magnifier as shown in Fig. 5-31.

*The suitability of a particular brand of tissue is not always obvious. For example, Kimwipes (Kimberly-Clark Co.) are recommended for lens cleaning, although they feel quite coarse to the touch. Conversely, the same company's Kleenex, manufactured to feel soft to the skin, contains hard substances that easily can damage lens surfaces. In general, *soft* lens tissues or Kimwipes should be used.

It is a good idea to inspect objective lenses this way every few days, since the working distance of high-power objectives is very short, and the changes of contamination are high. In general, *the most likely cause of a fuzzy microscope image is a dirty front element on the objective lens or a dirty coverslip.*

Recommended methods for cleaning microscope slides and coverslips are described in Fuseler

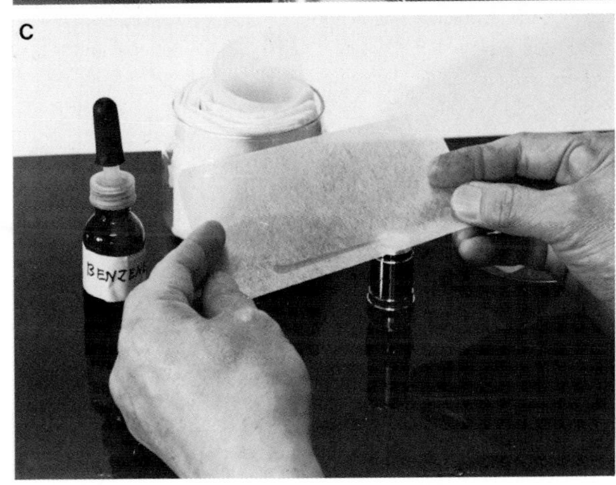

FIGURE 5-29. Method for cleaning immersion oil from an objective or condenser lens. See text.

FIGURE 5-30. Method for cleaning water-soluble material from lenses. See text.

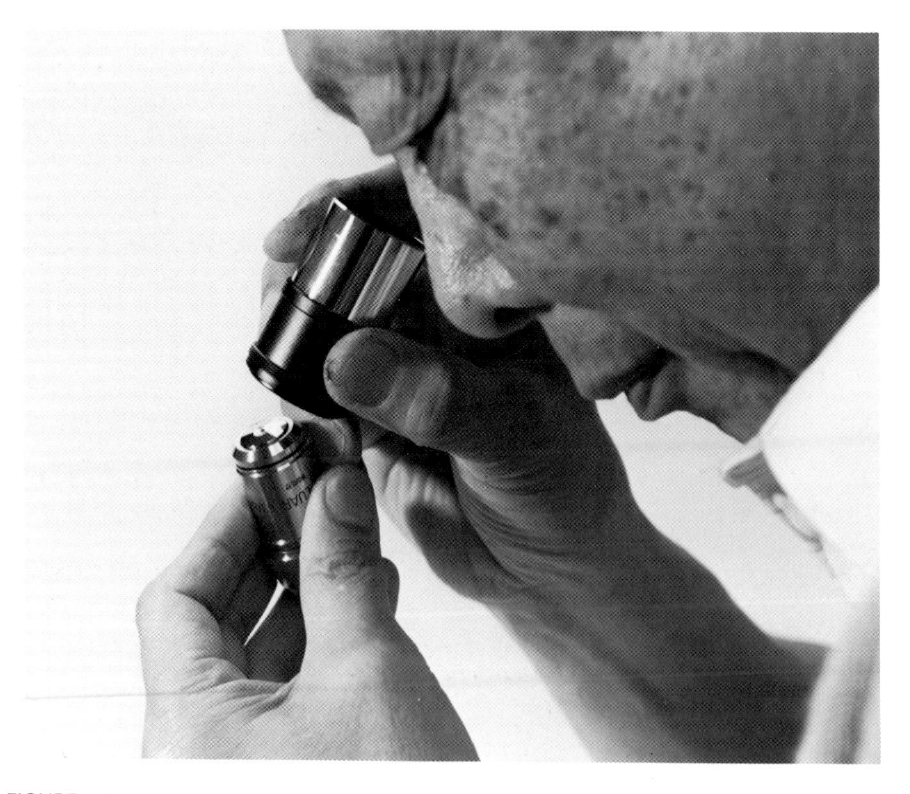

FIGURE 5-31. Use of an inverted ocular to inspect the front element of an objective or condenser lens.

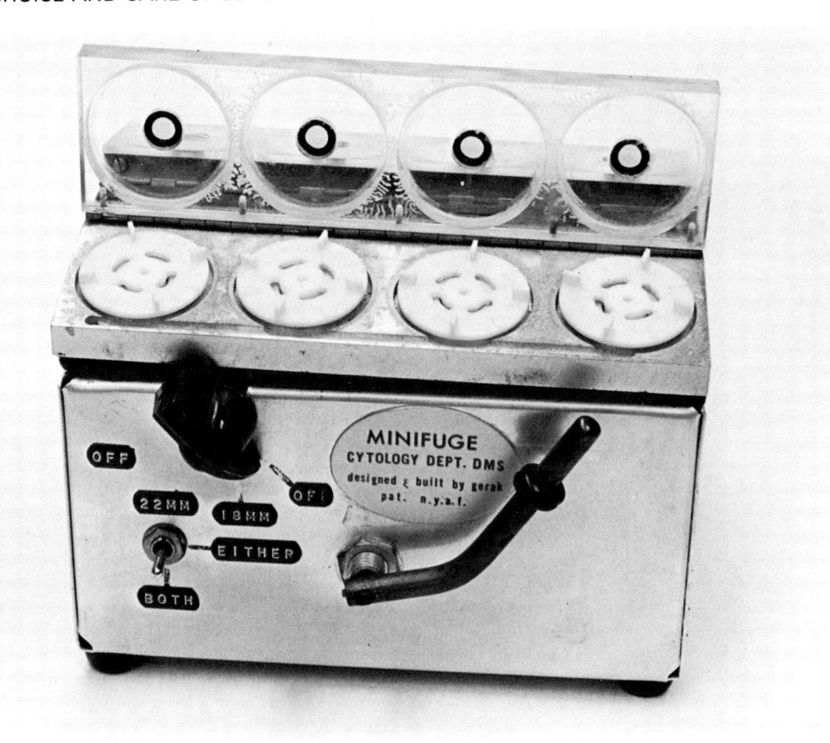

FIGURE 5-32. Miniature centrifuge for drying coverslips, without wiping. See text.

(1975). For critical work, coverslips can be dried off without wiping, on a miniature centrifuge (Fig. 5-32).*

Once the front elements of the objective and condenser lenses are cleaned, dust particles, and especially lint on the other lens surfaces may be found to be a source of optical noise. The problem can be especially severe in high-extinction polarization optical systems. Before attempting to cure such problems, however, one must remember to *never take apart an objective lens.*

The small lens elements in the objective are not only precisely spaced, but they are also centered critically. After the elements are assembled into the lens barrel, the manufacturer centers some of the inner elements with a special jig, through holes in the barrel, to provide the needed axial alignment between the lens elements.

If lint, flakes of paint, or other defects are located in any place besides the two outer faces of the objective lens, only the manufacturer can cure the problem.

To clean recessed surfaces such as the faceplate of the video camera tube, the back lens of the objective, the concave front surface of a high dry objective, or the inner surfaces of the oculars, the following cleaning method is often effective.

First inspect the surface under a bright light (of course the video camera must be turned off) and make sure that the cleaning is needed. A fiber-optic ring light made for dissecting microscopes works well for the inspection.

Make several fresh swab sticks by wrapping the tips of 2-mm-diameter wooden applicator

*This centrifuge, designed and built in the author's laboratory, will spin off ethanol from coverslips in less than a minute, thereby avoiding the residue that may remain after evaporation and the deposition of lint, smearing, and/or breakage that can come from wiping them dry. Cover slips are placed on any of the eight circular teflon tables (Fig. 5–32 shows a bank of four tables with their protective lid raised), which are ridged in order to suspend the coverslips above the flat surface. The tables are fashioned to accept various sizes of coverslips. Each table is spun at several thousand rpm by its own (battery-driven) miniature motor. The small circular objects on the protective lid are filters to remove debris from circulating air.

sticks with soft lens tissue (Fig. 5-33). A strip 2–3 cm wide, 10–12 cm long, torn (not cut) along the grain of long-fibered rice paper is preferred, but a similar torn strip of Kimwipe can also be used.

Lick and moisten a corner of the strip of lens tissue or Kimwipe and twirl it onto the applicator stick so that a clean portion of the tissue projects past the end of the applicator stick by 3–5 mm (A,B). Again, lick the upper free corner of the tissue so that it adheres to the roll (C). Then pinch off the tip of the rolled-up tissue to expose clean, virgin fibers (D). This part of the tissue acts as a disposable, soft ''brush'' that is *used only once*. The ''brush'' must be *absolutely fresh and clean*. Do not lay that part of the swab stick on the bench top.

First blow or dust off the loose dirt from the glass surface. Then moisten the tip of the tissue on the applicator stick with a drop of reagent-grade ethanol (E). Gently, without pushing the applicator stick itself to the glass surface, wipe the surface with the alcohol-dampened tissue. Start from the middle of the lens and expand outward in a spiral as you twirl the swab stick (F). Discard the applicator and immediately wipe off the residual alcohol from the glass surface with a clean, dry

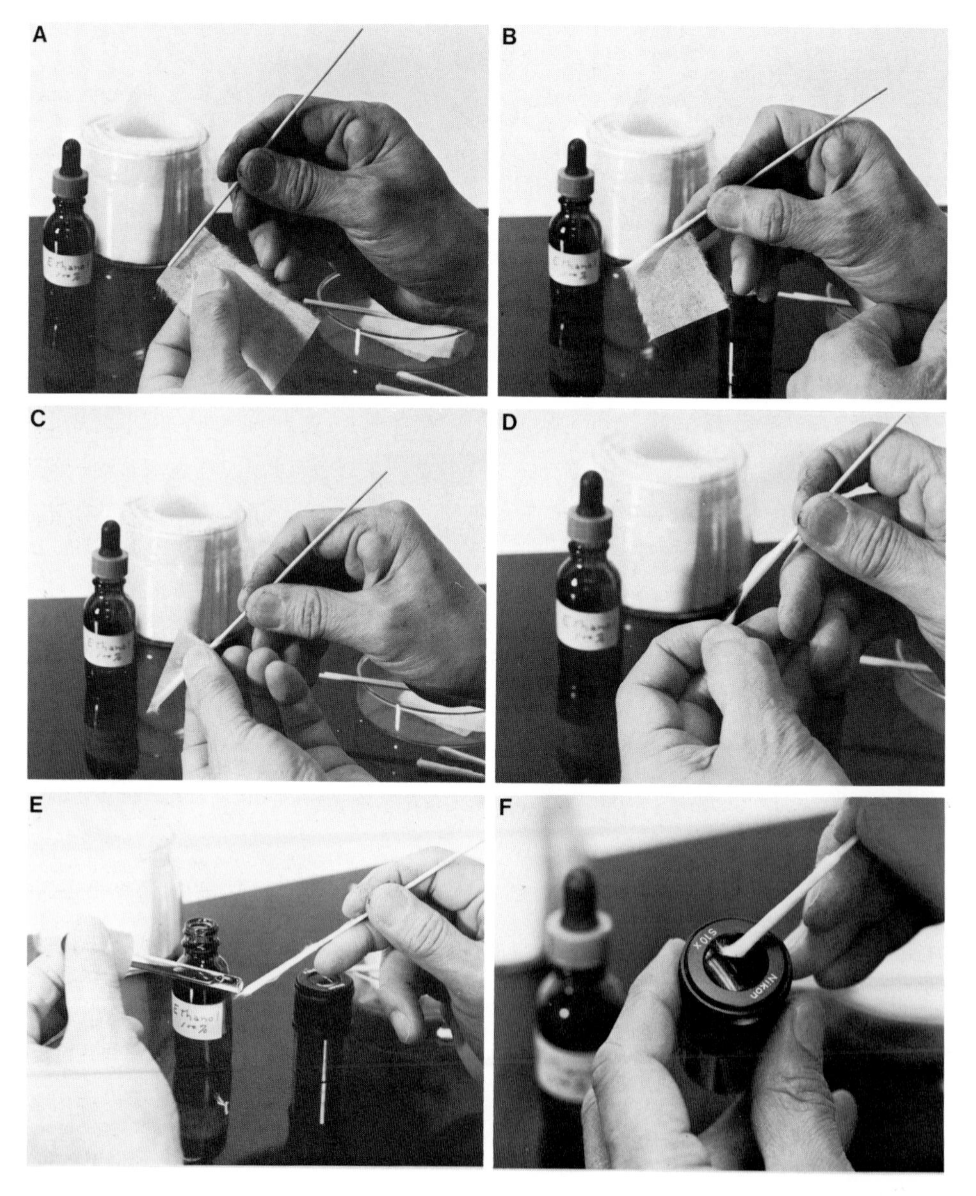

FIGURE 5-33. Construction of swab sticks used for cleaning recessed optical surfaces. See text.

swab stick. Again, twirl the swab stick so that fresh parts of the tissue wipe the surfaces as you work outwards from the center. *Do not rub.*

Inspect the surface with a bright light, and repeat the cleaning process if necessary. Note that an antireflection coating wetted with oil may look as though it had been peeled off from that region. A well-cleaned, coated surface will generally look uniformly deep purple. (See footnote * on p. 132.)

After cleaning with alcohol, blow off the residual lint with an oil-free duster, then immediately reinsert the cleaned component into the microscope, or place under a dust-free cover.

Some dust particles, such as ground-up sand, can scratch and damage glass surfaces. The lens coating is especially vulnerable to abrasion, although many modern coatings are durable enough to withstand *gentle* cleaning.

First-surface mirrors, i.e., mirrors with the reflective coating on the front surface, are especially vulnerable to abrasion. Never wipe, or even brush with a camel's hair brush, a first-surface mirror that is dry. Some first-surface mirrors can be cleaned with mild, neutral detergent followed by distilled water rinses. The final rinse of distilled water should be spun off on a centrifuge or very gently blotted off with clean lens tissue or cloth, but, again, *never wipe*. In general, first-surface mirrors should be protected from dust, and cleaning should be limited to occasional gentle puffs from an oil-free blower.

As a general principle, *keep the optical surfaces protected from dirt, dust, cigarette smoke, and other sources of contamination. Keep the room clean and the microscope under a dust cover whenever it is not in immediate use. Keep the optics clean, but clean only as needed.*

5.11. THE MICROSCOPE STAND

Microscope optics can function properly only if they are centered and positioned within design tolerances. Drifting focus and vibration can ruin an otherwise excellent image.

Several research microscope stands can be used effectively with video, but the weight of some video cameras and the high magnification involved favor some microscope designs over others.

Small video cameras can be mounted on top of an upright microscope (Fig. 3-1), but heavier cameras can cause problems.

Inverted microscopes are often more substantially built than upright microscopes and also bring the camera to a lower position. The lowered center of gravity increases the mechanical stability of the system and reduces vibration.

Examples of inverted microscopes with low video ports are shown in Figs. 5-28 and 5-34. The inverted optical bench microscope that I use is illustrated in Appendix II (Figs. III-21, III-22). The microscope has evolved over the years since around 1947, with input from many major optical companies.* The recent version of this microscope was designed by Gordon W. Ellis, Edward

*Special inputs from the manufacturers in developing the inverted optical bench microscope, more or less in historical order, include: specially ground and polished strain-free coverslips, by Tokyo University Department of Physics optics shop; selected strain-free optics, experimental coating of microscope objective lenses, special high-extinction polarizers, by Bausch & Lomb, Rochester, N.Y.; high-extinction coated calcite polarizing prisms by Karl Lambrecht, Chicago, Ill.; strain-free optics, rectified objective and condenser lenses, experimental thin film coating, high-precision goniometer for polarizer and compensator, high-efficiency collector for short-arc mercury lamp, etc., by American Optical Co., Southbridgem Mass., and Buffalo, N.Y.; Ortholux and Orthomat coarse and fine focusing blocks, centrable revolving nose piece, power zoom ocular, etc., by E. Leitz, Inc., Rockleigh, N.J.; selected strain-free optics by Olympus Corp. of America, New York, N.Y.; high-extinction rectified optics, rectified long-working-distance condensor, selected Plan Apochromatic objectives, rectified DIC plan apochromatic optics, etc., by Nikon Inc., Garden City, N.Y. and Tokyo, Japan; matched-image SIT and Newvicon cameras by Dage-MTI, Michigan City, selected strain-free optics, rifle scope zoom system, Venus intensifier camera with extended controls, etc., by C. Zeiss Inc., Thornwood, N.Y. and Venus Scientific Co., Farmingdale, N.Y.

Various parts or versions of the microscope were designed and constructed at the: Misaki Marine Biological Station; Biology Department, Princeton University; Department of Anatomy, University of Washington Medical School; Biology Department and the Institute of Optics, University of Rochester; Research Center, American Optical Co.; Department of Cytology, Dartmouth College Medical School; Department of Biology, University of Pennsylvania; Marine Biological Laboratory, Woods Hole.

A

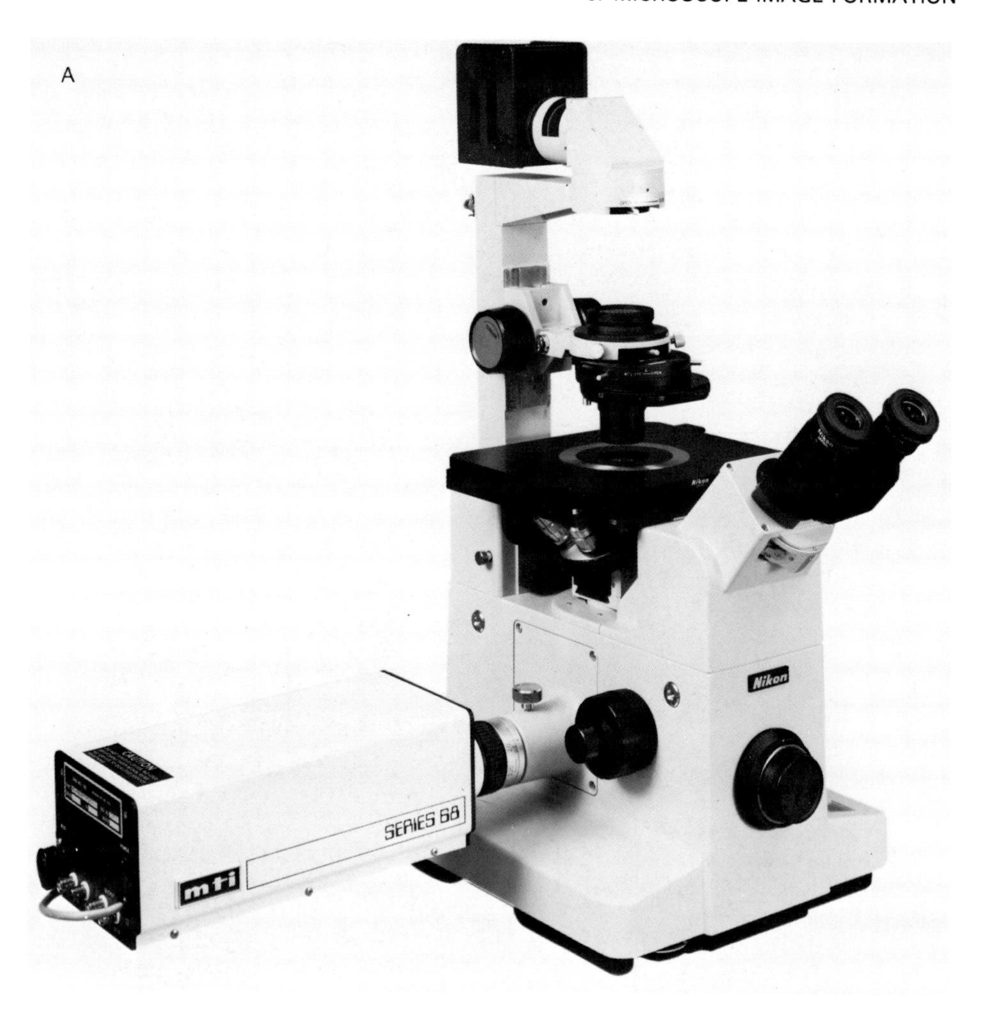

B

FIGURE 5-34. (A) Nikon inverted microscope with low video port. (The support beneath the video camera is not visible.) Configurations such as this are generally more stable for video microscopy than with the video camera mounted directly (without independent support) on top of an upright microscope. *See also* Figs. 5-28 and III-22. (B) Nikon zoom ocular for video microscopy. This special ocular, which incorporates a C-mount (at the left), can be seen linking the microscope and video camera in A. Other zoom configurations are shown in Figs. 5-27 and 5-28. (Courtesy of Nikon, Inc., Instrument Division, Garden City, N.Y.)

FIGURE 5-35. Beam splitter for dual video camera. In general, the best image can be obtained with a microscope by maintaining the optical path from the (imaged) light source to the camera on a straight line; some deterioration of the image is inevitable each time the path is bent. For certain applications, such as the use of dual cameras as illustrated here, the principle of a straight path may have to be sacrificed. In the illustration, the SIT camera on the right arm picks up low-level fluorescence, while the Newvicon camera on the left arm picks up the DIC or polarized-light image. The cameras are **genlocked,** and their outputs, balanced through a special effects generator, are displayed through the red and green channels of a high-resolution **RGB** monitor. Depending on the setting of the (white) selection knobs, the two images can be displayed independently or superimposed on each other, and with alternate filters selecting for images of the desired transmission or fluorescence. The dual camera system is especially useful for correlating the detailed distribution of fluorescence with very low levels of birefringence (and potentially for comparing the distribution of two fluorochromes) in motile cells and embryos (Inoué and Lutz, 1982).

Horn, and myself, incorporating commercial optical and mechanical components whenever suitable parts were available. Many parts, including some lenses, polarizing prisms, the optical bench, the U-shaped support desk, light source scrambler (footnote on p. 127), condenser mount, goniometer, body tube, precision revolving stage, dual camera optics (Fig. 5-35), etc., had to be newly designed or built in-house or on special order.

The inverted optical bench microscope has a rigid stable base, permits flexible arrangements of components on the high-precision Mehanite dovetail way (straight to 2.5 μm for the whole vertical 4 feet), and is well suited for excellent image production and convenient specimen manipulation in various contrast modes, including exceptionally high-extinction polarization microscopy.

The inverted optical bench microscope is especially well suited for video microscopy. An interesting design feature of this microscope is the absence of reflecting surfaces between the virtual light source and single video camera, thereby avoiding image degradation and noise often introduced by such surfaces. While its overall design has not been adopted commercially, several design principles and optical components developed on this microscope are becoming incorporated into commercially available instruments.

6

The Video Image and Signal

6.1. INTRODUCTION

The main task of video is to convert an optical image of a dynamic scene into an electrical signal that can be transmitted in real time to another location, and converted back without delay into an optical image that faithfully reproduces the original scene. Sound, synchronized to the scene, commonly accompanies the video image.

The main features to be considered are that: (1) the optical image is two-dimensional, (2) the scene is generally not static but dynamic, (3) the video signal is to be transmitted over some distance over a single channel or cable, (4) the signal may need to be processed or stored, (5) the monitor or receiver must immediately restore a two-dimensional, dynamic optical image from the signal received over the single channel, and (6) the image on the monitor must faithfully follow the original scene in real time.

These requirements, and the wide appeal of broadcast TV and other applications of video, have led to the development of an ingenious series of electro-optical and electronic devices. Video microscopy has profited much from these developments.

In video microscopy, as shown schematically in Fig. 6-1, the optical image (I_2) of the specimen (I_1), produced by the microscope, is converted to an electrical *video signal* (i_1), by a suitable video camera. The video signal may be enhanced, or otherwise processed or stored, but in general the signal is converted back to an optical image (the picture, I_3) displayed on a monitor in real time. The signal or picture may also be used for various types of analyses.

The signal in video microscopy is normally conducted for only a short distance through coaxial cables rather than through a wireless system, so that in video microscopy we are dealing with *closed-circuit TV* (CCTV). CCTV shares much in common with wireless or broadcast TV, but not all of the standards and conventions are identical. We will discuss the standards in Sections 6.2 and 6.3.

Depending on the purpose, the video system may be limited to monochrome, i.e., black and white, or may employ color. The color may be a near-natural representation of the colored image of the subject, a pseudocolor representation of certain features of the image, such as intensity, contour, movement, etc., or a **false-color** representation of the wavelength. To display the natural color of the specimen in video microscopy, one generally uses a color video camera and **NTSC** signal.* The U.S. broadcast-standard NTSC system is **compatible** with a black and white image and signal, but monochrome video equipment is not always compatible with an NTSC signal. In order to use NTSC signals, not only the camera, but the processor, recorder, and monitor must also be NTSC compatible. Compared to the monochrome system, the NTSC system is complex and the equipment tends to

*NTSC is a color video standard established by the National Television System Committee, sometimes expressed by video technicians as a euphemistic acronym for "never the same color." For details of the NTSC color fundamentals, see, e.g., Ennes (1979). This and other formats are discussed later.

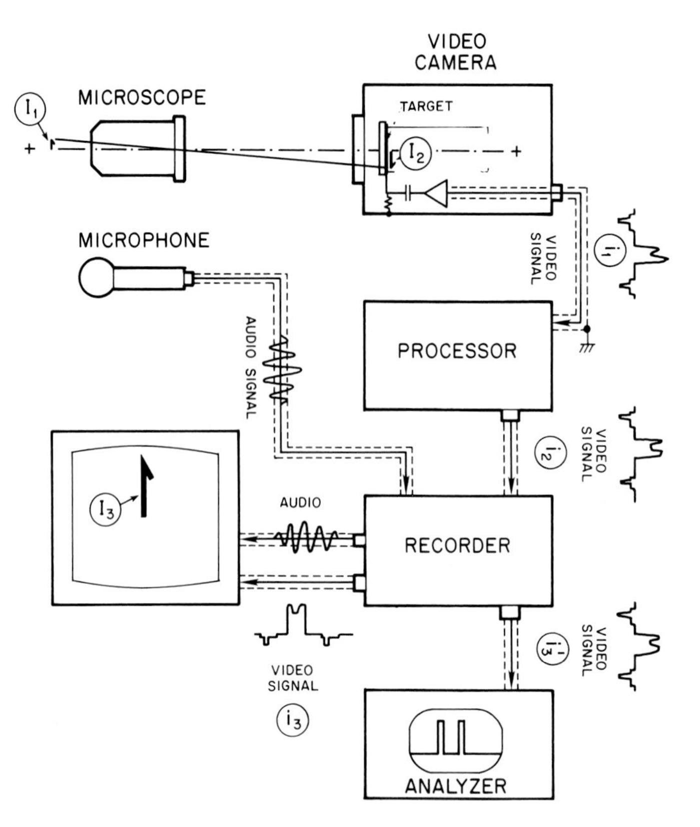

FIGURE 6-1. Schematic diagram of video for microscopy. Am image (I_2) of the specimen (I_1) is focused by the microscope onto the target of the video camera. The image is converted to a video signal (i_1) that may be processed or recorded, resulting in video signals i_2 and i_3. The video signal is converted back to a two-dimensional picture (I_3) on the monitor or displayed as a line scan on the analyzer.

be expensive. The color image is also limited in resolution, compared to the monochrome image obtainable on similarly priced equipment. We will defer a detailed description of color video to the next chapter, and deal primarily with the general principles of monochrome video here, touching on color only to the extent needed to recognize some of the basic features of color display and the NTSC signal.

We will first examine how a two-dimensional optical image is converted into a linear sequence of electrical signals, and then back again to a two-dimensional optical image (Section 6.2). Then we will discuss the waveform of the transmitted electrical signal, the nature and use of **synchronizing pulses,** and the generation of and standards for the **composite video signal** (Section 6.3).

Next we will discuss the general relation of image resolution and the electrical **frequency response** (Section 6.4), and the influences of electrical signal distortion and noise on the quality of the video image (Sections 6.5, 6.6). Finally we will examine an important practical problem: how to hook up the video equipment so as to synchronize the timing pulses among the several pieces of equipment in a video system. Without synchronized timing, we would be unable to obtain an intelligible picture from the video signal (Section 6.6).

In these sections, we will deal with the general principles of video imaging and the fundamentals of the video signal. Specifics of video cameras and monitors are covered in Chapter 7, and discussions on recording and processing of the video signal are deferred to Chapters 8 through 12.

6.2. HOW VIDEO WORKS

Generation of the Video Signal

As stated above, a video camera converts the two-dimensional optical image into a series of electrical impulses, the video signal. The video signal is converted by the monitor back into a two-dimensional picture that reproduces the dynamic scene of the original image in real time.

How does the conversion between the two-dimensional image and the recoverable train of electrical impulses, the video signal, take place?

Consider the image in Fig. 6-2A. To dissect the optical image so that it can be represented as a series of electrical signals, we scan narrow strips of the image. In principle, the image can be scanned by any electro-optical transducer that rapidly converts light intensities into electrical voltages or currents.

In conventional video cameras, the local brightness of the image is converted in the picture tube target to a proportionate electrical charge. The accumulated charges are converted to electrical currents by the scanning electron beam of the camera (Chapter 7).

In Fig. 6-2A, the scan CC′, for example, produces an electrical signal cc′ (Fig. 6-2B). The height (amplitude) of the electrical signal is proportional to the brightness of the part of the scene being scanned, so that the current or voltage signals at c_1, c_2, c_3, etc. correspond to the light intensities of the image at C_1, C_2, C_3, etc.

The narrow horizontal strip of the image, or the **horizontal scan line,** is scanned *at a constant speed* so that the distances CC_1, CC_2, CC_3, etc. in the scanned image are proportional to the time intervals cc_1, cc_2, cc_3, etc. in the electrical signal. The signal cc′ arising from a single horizontal scan line is thus a faithful, time-dependent, electrical representation of the narrow strip CC′ of the optical image.

In this way, *dimensions in space along the horizontal (or X) axis have been converted to time delays in the electrical signal, and the intensity of each small area of the image has been converted to the amplitude of the electrical signal* (e.g., in volts).

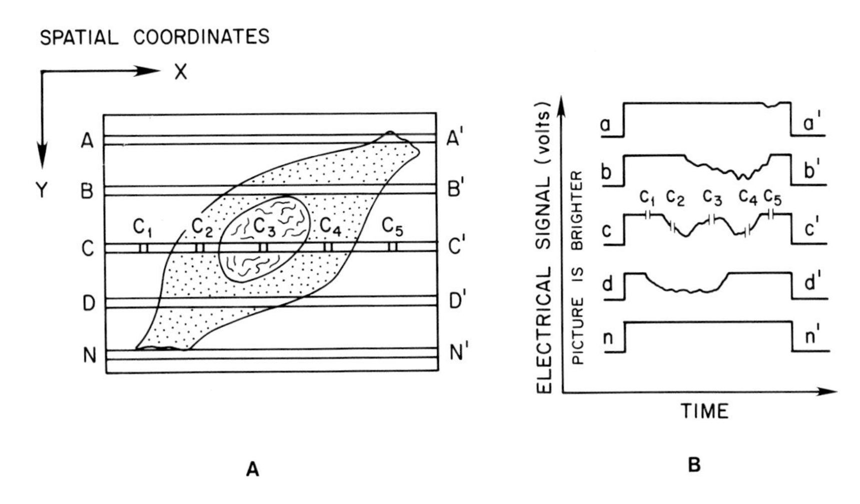

FIGURE 6-2. Relation between scan line and video signal. (A) The image on the camera target is scanned from left to right (e.g., A to A′, B to B′, etc.) slowly downwards in a series of horizontal strips. (B) The image brightness along the scan is converted to amplitude of an electrical signal (the video signal); the location of a given area on the image (e.g., C_1 to C_5) corresponds to its point in time (c_1 to c_5) in the video signal. Only five scan lines are shown here; in actuality, the entire image is scanned.

Sequential Scanning

In order to portray a complete two-dimensional image, the video scan lines must cover the whole picture area, and the electrical signal must contain the amplitude distribution of all points on all of the scan lines.

For CCTV, it should be possible, in principle, to use as many conductors or channels as there are scan lines and to conduct the video signals for all of the scan lines in parallel at the same time.* But as already mentioned, most of our video equipment is compatible with broadcast TV.

Instead of many parallel signals, the output from the scan lines is formatted *serially,* or in sequence, as a single linear signal. The scan lines AA′, BB′, CC′, etc. are combined sequentially into a single decipherable stream, or a single linear series of electrical pulses. The signal can then be conveyed along a single (coaxial) cable, or transmitted over the airwaves on a single channel.

H and V Scans

To convert the two-dimensional image into a linear series of electrical pulses, the convention is to scan the image from the upper left to the right, and then repeat on the next line down, as in reading English text (Fig. 6-3A).

Starting with the upper left corner (as translated to the orientation of the image on the monitor†), the image is scanned "horizontally" along AA′. Once the scanning spot reaches A′, it is made to **fly back** to the beginning of the next scan line at B. The flyback is much faster than the scan speed, and the signal is blanked during the flyback (or "retrace"). This **blanking** prevents the flyback trace A′B from contributing to the signal. The scan then continues, at the standard horizontal scanning velocity, along BB′ until it flies back again from B′ to C, and so on.

As the image is scanned from left to right, the scanning spot is also going through a **vertical (V) scan,** or moving downward at a slower, constant speed. Thus, the horizontal (H) scan is not actually horizontal, but slanted down slightly to the right [by a vertical distance equal to the separation of every *other* (i.e., every second) scan line, as will be explained shortly].

After the last horizontal scan NN′ (Fig. 6-3A), the scanning spot flies back from N′ at the bottom right corner of the field to the top left (again being blanked during flyback), and the process starts over again. This scanning pattern provides the video *raster.*

In this fashion, the spatial coordinates of the image on the video screen originate at the upper left-hand corner, where the values of the horizontal (or X axis) and of the vertical (or Y axis) are taken as 0. The X axis increases in value to the right, and the Y axis downwards (Figs. 6-2, 6-3). These are the same conventions used for computer video monitors, where the Y-axis direction is reversed from standard Cartesian coordinate. The coordinate for the starting point (A) of the scan is thus ($X = 0$, $Y = 0$), or (0, 0).

H and V Deflections

The movements of the scanning beams in the camera and monitor are generated by magnetic or electrostatic deflectors. The H and V deflectors provide: (1) the constant-velocity scan needed to drive the horizontal scan line and the vertical deflection, and (2) the rapid flybacks. These requirements result in the deflection waveforms being sawtooth shaped as a function of time (Fig. 6-3B,C).

The exact shape of the sawtooth wave is very important, since the shallower slope of the wave governs the velocity of scanning. If this part of the sawtooth were slightly curved and not exactly linear, or if the slope were not strictly constant, the video image would be distorted by the amount of that deviation.

*This method is in fact used for measuring rapid changes of brightness of many image points simultaneously through a microscope (see Grinvald *et al.*, 1981; Salzberg *et al.*, 1977).

†Notice that a camera lens ordinarily inverts the image formed on the target of the video camera tube, compared to the picture that should appear on the monitor. Therefore, the pattern of scanning in the camera tube must also be inverted relative to the scan in the monitor (see image orientations in Fig. 6-1).

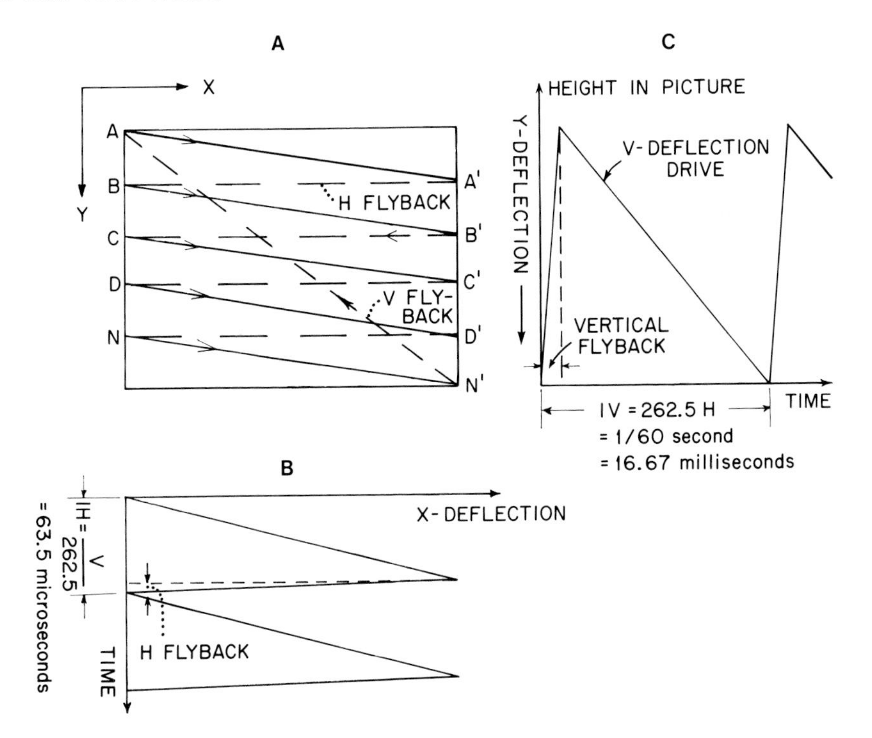

FIGURE 6-3. Schematic of horizontal and vertical scans. (A) As viewed on the monitor, the image is scanned beginning at the upper left (point A; coordinates 0, 0). The scan proceeds along a slightly sloping line to A′, where it rapidly "flies back" (nearly horizontally) to B, and the pattern is repeated. From point N′, the scanning beam flies back to A and begins a new scan.* (B, C). The sawtooth waveforms of the horizontal and vertical deflection drives, respectively. Their time axes illustrate that the horizontal scan is much faster than the vertical.

The frequencies of each H and V sawtooth wave determine the **scanning rates** in the H and V directions. Clearly, the H scan is much faster and repeats many times during a single **V scan.**

Blanking

During the H and V flybacks, the retraces are blanked by lowering the video signal to the **blanking level** so that the scanning electron beam is turned off. In this way, the retrace is prevented from adding extra lines that would degrade the picture.

The *blanking pulses* actually start a very short time before the H and V flybacks to ensure that the blanking is complete. The electrical signal that includes the blanking pulse (Fig. 6-4B) is presented schematically in Fig. 6-4C.

Standard Scanning Rates

In the North American and Japanese standard broadcast format, each scene is scanned by 525 horizontal lines 30 times each second. In the European and British format, the scene is scanned by 625 lines 25 times a second. The former is known as the 525/60 scan rate, and the latter as the 625–50 scan rate for reasons that will be given shortly. For our general discussions of video, we will assume a 525/60 scan rate.

When an image is presented to the eye intermittently at several frames per second, we perceive

*The **vertical retrace** between N' and A is not a straight line as shown here schematically but actually zigzags, since the retrace takes place over several H intervals.

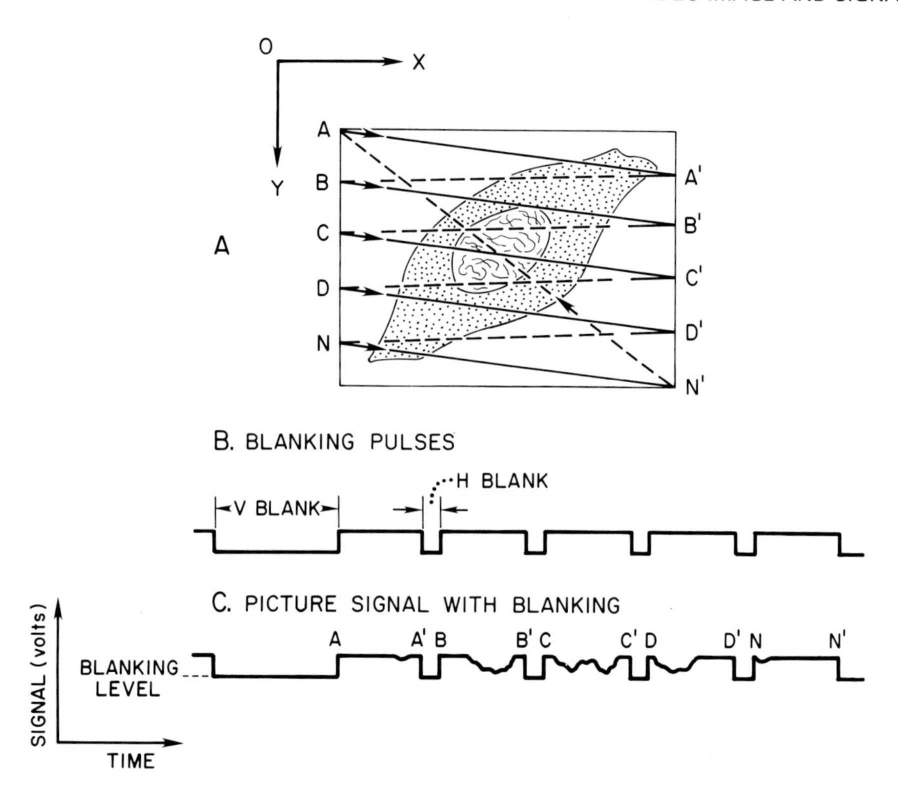

FIGURE 6-4. (A) The schematic scan pattern (raster) is shown superimposed on an image. (B) The blanking pulses, which turn off the scanning beam during flyback. (C) The picture signal, which includes a sequential signal of that shown in Fig. 6-2B and the blanking pulses.

the annoying sensation of flicker (Chapter 4). For a scene with a moderate luminance, the flicker becomes reasonably unobtrusive, and we reach the critical flicker frequency (CFF) at around 40 to 60 Hz (hertz, or cycles per second). Motion pictures are projected at 36 Hz for silent film (18 frames/sec, each frame interrupted once), and at 48 Hz (24 × 2) for sound film.

As the image becomes brighter, and for larger fields of view, the CFF rises substantially so that a bright image displayed at 30 Hz shows considerable flicker. In order to reduce the flicker, video images in the 525/60 scan rate are not presented at 30 Hz, but instead at twice that rate, or 60 times/sec.

Each *frame,* consisting of 525 horizontal scan lines, is dissected into two interlaced *fields,* the odd field and the even field. Each field has 262.5 scan lines and appears 60 times/sec. Thus, each frame, or a complete picture, appears 30 times/sec, but the flicker is essentially eliminated by using field rates of 60 Hz.

525/60 thus refers to a scanning system with 525 H scans/frame, and 60 fields/sec. The two fields, each with 262.5 H lines, and which are not identical, are **interlaced** along the vertical axis in the following manner.*

*One may ask: "If we need 525 lines in order to resolve the image vertically, and we have to repeat the scene 60 times a second to prevent flicker, why not repeat a frame with 525 scan lines 60 times each second?" The answer is that this could be done if there were enough bandwidth for the video signal. But in broadcast TV, it was necessary for various practical reasons to restrict the bandwidth allotted to each TV channel. We will discuss the relation between bandwidth of the signal frequency and horizontal and vertical resolution in Section 6.4.

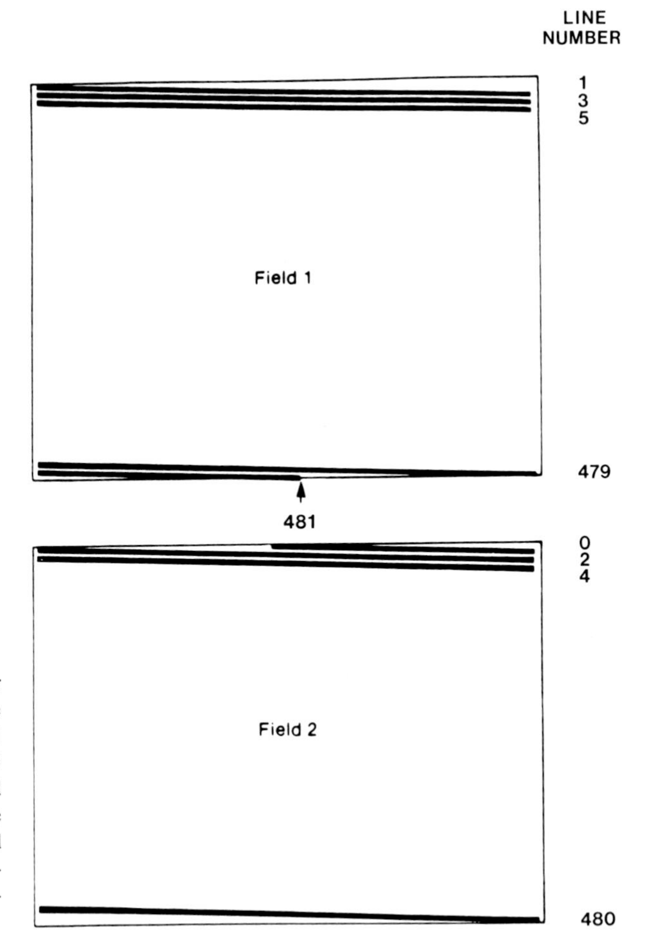

FIGURE 6-5. Interlaced raster scan. Pattern of scanning for the "odd" and "even" fields (fields 1 and 2, respectively, each with 262.5 scan lines), which are combined and interlaced to produce a complete frame of 525 lines (in the standard North American and Japanese systems). (After "Sony Basic Video Recording Course," Booklet 2, p. 3.)

Interlacing

As shown in Fig. 6-5, the scene is first scanned *every other line of the frame,* i.e., the 1st, 3rd, 5th, etc. lines. The scan continues for 262.5 times until the scan reaches the bottom of the field. Notice that the scan, which started at coordinate (0,0), ends at a point halfway across the H scan, or at (0.5 X, Ymax). This scanning pattern results in the *"first" of the two fields,* or the "odd field."

After the odd-field scan is completed, the scanning spot flies back to the top, and the *second,* or *even-field,* scan is started. The even field starts at the middle of the top of the frame at coordinates (0.5 X, 0). In other words, the even and odd fields both start their scans at $Y = 0$, the very top of the frame. The interlacing does not make the alternate fields start at different heights on the screen.

With the scan lines all slanting down to the right as discussed before, the even-field scan lines, whose starting point is shifted horizontally by H/2 relative to the start of the odd-field scan, fit neatly into the gap between the odd-field scan lines, and vice versa. In other words, because the field is not composed of an integral number of lines, but includes a half line, the even and odd fields become precisely interlaced. This format of interlacing is known as the *2 : 1 interlace.** Nonstandard interlace and other scanning rates will be described a little later.

*One can have an interlace of 4:1, 1:1, etc., depending on how many fields make up a frame.

In the 525/60 scanning system, a field is repeated 60 times/sec. Therefore, since there are 262.5 lines/field, the H-scan rate is (262.5 × 60) or 15,750 Hz. The H-scan interval (also called **H**) is thus 63.5 μsec (1/15,750). The V-scan, or **vertical frequency,** rate is 60Hz, and the V-scan interval (V) is 16.67 msec (262.5/15,750).

Active Picture Area

The blanking time for the V scan is approximately that of 21 H scans, so that in fact the number of **active scan lines** is [(262.5 − 21) × 2], or about 483 lines/frame. The active scan lines are those that contribute to the **active picture area,** which is the area not blanked in the final picture.

By convention, the **aspect ratio,** or the ratio of the width to height of the active video frame, for broadcast-compatible formats is 4 : 3. Adherence to a standard aspect ratio prevents gross distortion of the image, such as a circle appearing as an ellipse.*

The Picture Signal

We now have a linear series of electrical signals that includes the camera blanking pulses and which represents the brightness at each point in the original image (Fig. 6-4C). Such a **picture signal** could in principle be reconstructed into a two-dimensional display in the monitor by reversing the scanning process that produced the signal in the camera.

However, that would demand that every monitor receiving the signal from a camera be equipped with an extremely high-precision clock that somehow knows exactly when each H and V, and odd- and even-field, scan starts. In order to make the synchronizing task more practical, especially for broadcast TV, and to eliminate the need for super-high-precision clocks in the receiver, specific timing pulses are added to the image signal as described next.

6.3. TIMING PULSES AND THE COMPOSITE VIDEO SIGNAL

Sync Pulses and Composite Video

As already described, spatial information in the video image is carried along the time axis in the electrical signal. Therefore, we need to generate exactly timed H and V scans in the camera and reproduce them in the monitor in order to place the image elements of the two in register.

In order to reproduce the H and V scans in the monitor synchronously with those in the camera, video engineers have developed a timing system that does not require the presence of super-high-precision clocks in the monitor. Instead, the timing of each H and V scan is signaled by *sync pulses* that the camera sends out as part of the video signal.

As shown in Fig. 6-6, the H- and V-timing signals, or sync pulses, are brief, rectangular voltage pulses. They are repeated at exactly the periods of the H and V scans.

The sync pulses are generated by a precision master oscillator, the **sync generator,** which (in the 525/60 scan system) supplies the 15,750-Hz time base for the H scan. In the 2 : 1 interlace system, the same oscillator also provides the 60-Hz signal for the V scan, precisely synchronized to the H scan.† The oscillator is generally crystal controlled or **phase locked** to the 60-Hz power line.

*If the circle appears as an egg shape rather than an ellipse, the distortion is due to nonlinearity, or lack of constancy, of the scanning velocities. While an aspect ratio of 4H:3V is the standard recommended by **EIA** (Electronic Industries Association) for broadcast-compatible TV, aspect ratios of either 4:3 or 1:1 are used for CCTV (EIA RS-343).

†For exact 2:1 interlace, the output of a 31,500-Hz oscillator is divided by 2, an even number, to give the 15,750-Hz H scan for the first and second fields. Each field is triggered by the alternate, 60-Hz V scans, which are generated by dividing the 31,500-Hz oscillator output by 525, an odd number. Division of the master oscillator output by alternating even and odd numbers generates the needed phase relationship between the start of the H and V scans in the first and second fields.

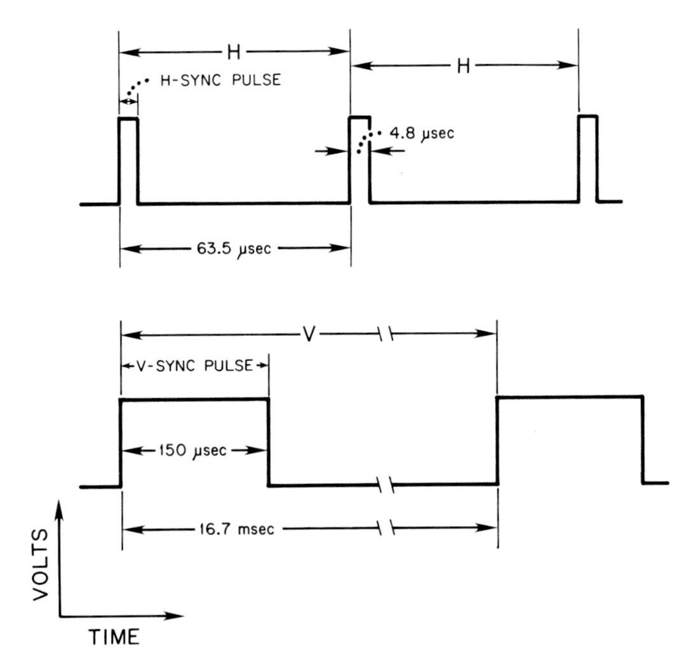

FIGURE 6-6. H- and V-sync pulses. These precisely timed signals (synchronizing pulses) are added to the picture signal to allow the camera and monitor scans to be in exact register.

For CCTV, the master oscillator is commonly built into the camera, but in more elaborate systems, the master oscillator may be present in a separate sync-signal generator that feeds the timing pulses to the camera(s), processor(s), recorder(s), monitor(s), etc.

In the camera, the **H-** and **V-sync pulses** shown in Fig. 6-6 are combined with the picture signal of the type shown in Fig. 6-4C. In combining the two, the electrical polarities of the sync pulses are inverted (Fig. 6-7B) relative to the image signal (Fig. 6-7A). By inverting the sync pulses, they can be combined with the picture signal, and yet be distinguished from it (Fig. 6-7C). A video signal that contains such a combination of picture signal and sync pulses is known as a *composite video signal.*

By choosing different voltage levels for the sync pulses and the picture signal in the composite video signal, the two can be separated, when necessary, and be processed independently. For example, in a video processor, recorder, or monitor, the picture signal "stripped" of the sync pulses and is reshaped, recorded, or displayed. The "stripped" and reshaped sync pulses are used to synchronize the timing circuits and tape drive in the recorder, and to trigger the H and V sweeps and flybacks in the monitor, etc.

In the monitor, the incoming sync pulses trigger and synchronize the otherwise "free-running" internal oscillators. Thus, the precision of the scan frequencies in the monitors is not limited by the quality of the internal drive oscillators. In fact, many monitors designated for a 525/60 scan rate can accept H-sync pulses that depart from the standard 15,750 Hz by as much as a few percent; say between 15 and 16 kHz.

The V scan in the camera and other sync-generating devices is phase locked to the 60-Hz power line. This not only provides a common reference for the various devices, but also minimizes interference from the pervasive 60-Hz electromagnetic stray fields emitted by transformers, motors, fluorescent lamps, dimmers, and many other kinds of common electrical equipment. By this means, the influence of power-line frequency noise is reduced (i.e., it does not appear randomly all over the picture), and also the picture can be prevented from weaving, or slowly oscillating on the monitor in a sea-sickness-inducing pattern.

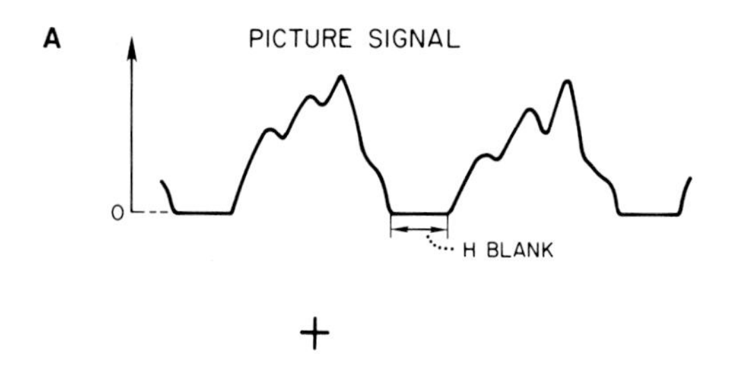

+

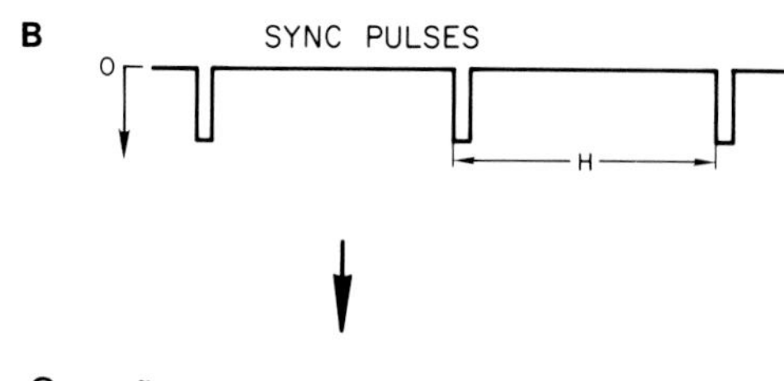

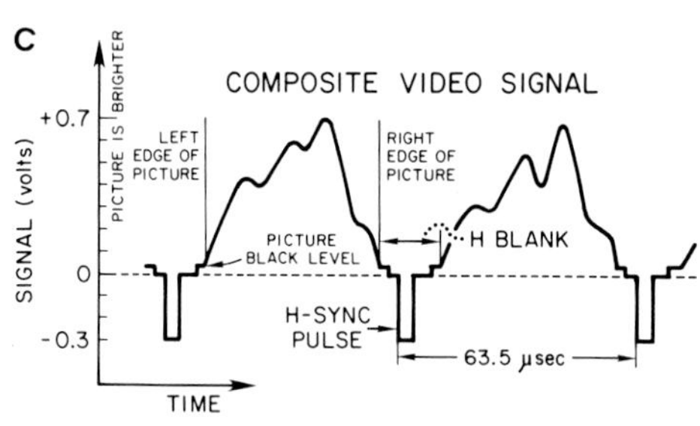

FIGURE 6-7. Generation of the composite video signal. The picture signal in A (Fig. 6-4C), when combined with sync pulse signals of reversed polarity in B (cf. Fig. 6-6), results in the composite video signal shown in C.

Conventions and Standards for CCTV

Recommended standards (RS-) have been set for video and related communication devices by the Electronic Industries Association. Some of the EIA standards specifying the timing frequencies, timing-pulse voltages and placements, etc. relevant to our discussion are listed in Table 6-1. These standards help to make various pieces of video equipment compatible with each other.

For video camera outputs, and for the inputs and outputs of most CCTV equipment, brighter

TABLE 6-1. TV Display-Signal Parameters Established by International Standards[a]

	Monochrome standards			
	U.S. broadcast	European broadcast	U.S. closed-circuit	U.S. high-resolution
Lines/frame	525	625	525	—[c]
Field rate (Hz)	60	50	60	60
V interval (μsec)	16,667	20,000	16,667	16,667
V blanking (μsec)[b]	833	1,200	1,250	1,250
V-sync pulse (μsec)	190.5	192	150	150
H interval (μsec)	63.5	64	63.5	—[c]
H blanking (μsec)[b]	11.4	12.8	10	—[c]
H-sync pulse (μsec)	5.1	5.8	4.8	2.8

[a]After *Raster Graphics Handbook*, 1980, pp. 8–13.
[b]Minimum values.
[c]See Table 6-2 (p. 165).

parts of the image are represented by higher voltages in the video signal. By convention, the sync pulses appear below the picture signal in the composite video signal (Fig. 6-7C). The polarity of such a composite video signal is said to be *sync negative* or **black negative.**

However, inside the circuits of the camera, recorder, monitor, etc., and in broadcast TV, the signal is often inverted from this convention. If the **sync signal** and the darker part of the image signal point up, and have a higher voltage, the signal is said to be *sync positive* or *black positive* (e.g., see Ennes, 1979).

The main definitions and standard values specified for the V- and H-scan portion of the **monochrome signal** in EIA **RS-170** (broadcast TV) and **-330** (CCTV) are summarized in Figs. 6-8 and 6-9. As shown, the **peak-to-peak** voltage of the composite video signal, measured with *the signal terminated by a 75-ohm impedance* (Section 3.6), *is standardized to 1.0 volt* (see **Standard signal**). Of the 1-volt peak-to-peak signal, *the sync signal occupies the lower 0.3 volt* (or to be more

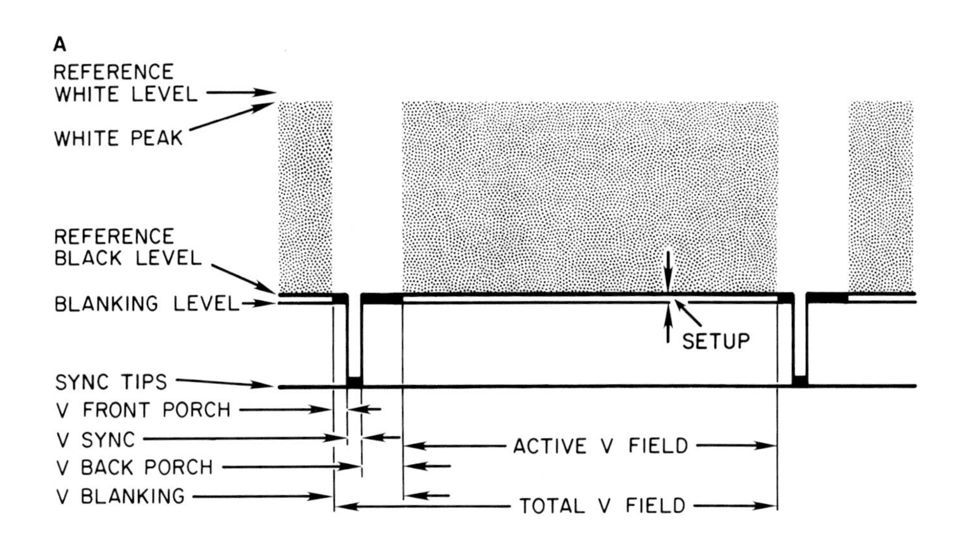

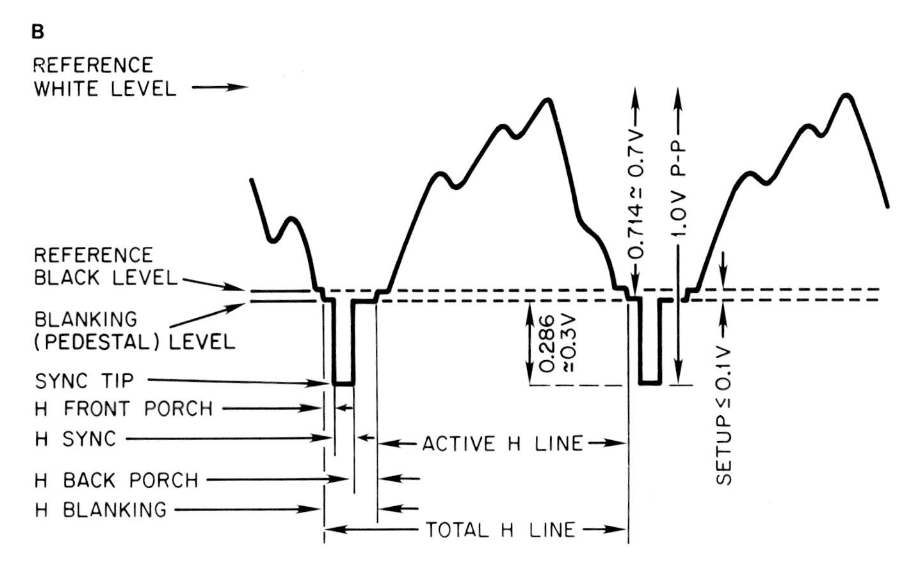

FIGURE 6-8. Main features of H- and V-scans specified in recommended monochrome standards RS-170 (broadcast) and RS-330 (CCTV) by the EIA. See text. (A) Vertical interval; (B) horizontal interval.

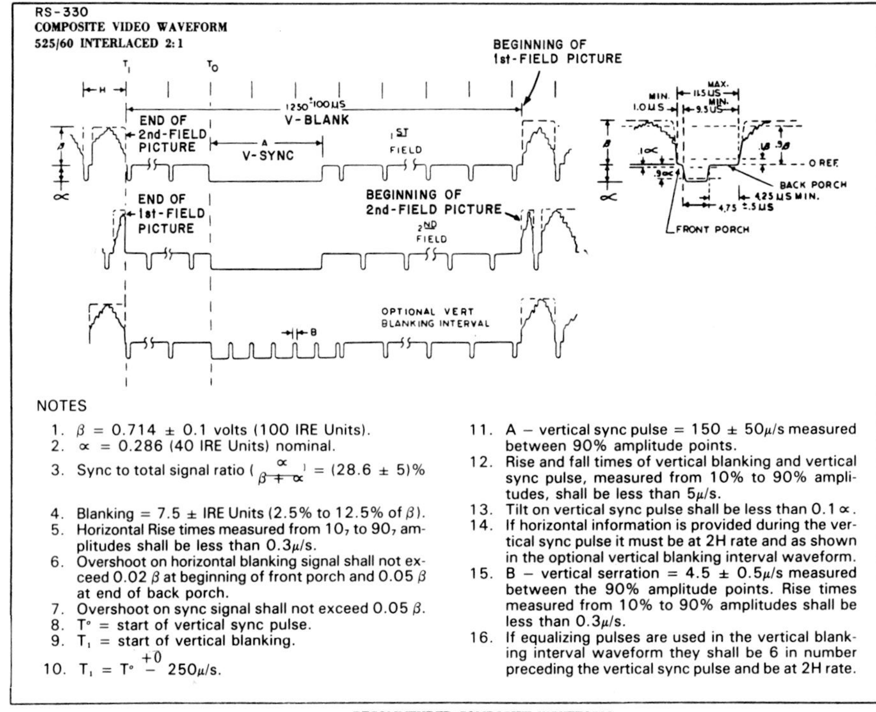

NOTES

1. β = 0.714 ± 0.1 volts (100 IRE Units).
2. ∝ = 0.286 (40 IRE Units) nominal.
3. Sync to total signal ratio (∝ / β + ∝) = (28.6 ± 5)%
4. Blanking = 7.5 ± IRE Units (2.5% to 12.5% of β).
5. Horizontal Rise times measured from 10₇ to 90₇ amplitudes shall be less than 0.3μ/s.
6. Overshoot on horizontal blanking signal shall not exceed 0.02 β at beginning of front porch and 0.05 β at end of back porch.
7. Overshoot on sync signal shall not exceed 0.05 β.
8. T° = start of vertical sync pulse.
9. T₁ = start of vertical blanking.
10. T₁ = T° +0 − 250μ/s.
11. A – vertical sync pulse = 150 ± 50μ/s measured between 90% amplitude points.
12. Rise and fall times of vertical blanking and vertical sync pulse, measured from 10% to 90% amplitudes, shall be less than 5μ/s.
13. Tilt on vertical sync pulse shall be less than 0.1 ∝.
14. If horizontal information is provided during the vertical sync pulse it must be at 2H rate and as shown in the optional vertical blanking interval waveform.
15. B – vertical serration = 4.5 ± 0.5μ/s measured between the 90% amplitude points. Rise times measured from 10% to 90% amplitudes shall be less than 0.3μ/s.
16. If equalizing pulses are used in the vertical blanking interval waveform they shall be 6 in number preceding the vertical sync pulse and be at 2H rate.

FIGURE 6-9. Standards for the RS-330 CCTV standard) H- and V-scans and waveforms adopted by the EIA. (From EIA Standard RS-330, copyright 1966.)

precise, 0.4/1.4 = 0.286 volt, between the base and tip of the sync pulse).* The base of the sync pulse occurs at the *blanking level,* or *pedestal.*

The picture signal should occupy the upper 0.7 volt (or 1.0/1.4 = 0.714 volt to be precise).* The peak of the signal corresponds to **maximum white** or **white peak** in the image. The *black level* of the picture, or the *optical black level,* is at a level slightly higher than the blanking level. The black level is raised above the blanking level by an amount known as the **setup.** The setup is less than 0.1 volt (usually around 0.05–0.075 volt).

For the same recommended standards, the actual duration of the H- and V-sync pulses are shown in Figs. 6-6 and 6-9. These values are also used in broadcast-format video.

As shown in Figs. 6-8 and 6-9, the sync pulses appear early during the vertical and **horizontal blanking.** Since the leading edge of the H-sync pulse triggers the flyback and sweep of the H-scan line, precision of the time intervals between the H-sync pulse and the **picture elements** is very important. Unless the intervals remain constant to about 0.0008 H (or 0.05 μsec) over many scan intervals, we would not obtain a stable, well-resolved picture.

The H-sync pulse must also maintain a stable phase relationship with the V-sync pulse. For good interlace, the period of the V sync must remain constant to within 0.05 H (or 3 μsec), or have a precision of 0.02% (Ennes, 1979, p. 280).

Sync Pulses during V Blanking

The H-sync pulses are generally present even during the **V blank** (Fig. 6-9). In the monitor, these timing pulses maintain the H-drive oscillator in synchronization during the V-blanking inter-

*For some broadcast applications, the picture signal used to be 1.0 volt, and the sync signal 40% of that value. For different video systems, the ratio of the two is held constant. The **IEEE scale** divides the peak-to-peak signal into 140 units (Ennes, 1979, Chapter 3).

val, so that the H sweep does not get out of register when the V blank ends and the picture in the next field starts.

The arrangement of the H-sync pulses during the V-blanking interval varies considerably with the application. For example, Fig. 6-9 shows the EIA RS-330 format, which is the monochrome standard for CCTV applications. Figure 6-10A shows the format of the sync pulses adopted by EIA for RS-170, the monochrome broadcasting standard. In RS-170, the vertical blanking starts 3-H intervals ahead of the V-sync pulse in order to guarantee complete blanking during V-flyback.* During this initial 3-H interval, or the V **front porch,** and during the V-sync pulse, which itself is 3 H long, as well as for a 3-H interval following it, **equalizing pulses** are added to the V blanking.

The equalizing pulses are forms of H-sync pulses that are spaced 0.5 H apart, or at twice the H frequency. By occurring at 0.5-H intervals, the equalizing pulses appear similar to both the even- and odd-field scans (the two being shifted in phase by 0.5 H).

During the V-sync pulse itself, the equalizing pulses appear as notches rather than spikes, and are referred to as **serrations** (Fig. 6-10, 6-11). Whether occurring as spikes or notches, it is the downward, negative-going voltage of the sync pulses, spaced apart by a 1-H interval†, that signals the H-timing cue.

After the 9-H intervals for the 18 equalizing pulses in RS-170, 9–12 synchronizing pulses occur at the standard H frequency. These H-sync pulses keep the H oscillator of the monitor "in sync" for the rest of the V-blanking period.

Figure 6-11 compares sync waveforms for broadcast, industrial, and military formats.

In the RS-330 for industrial monochrome CCTV cameras, H-sync pulses are present throughout the V-blanking period, except during the 3-H-long V sync where equalizing pulses may be added optionally (Fig. 6-9). Equalizing pulses are not always used, and serrations may or may not be present during the V-sync pulse.

In certain military formats, no serrations, equalizing pulses, or H front porch are used, and H-sync pulses are absent from the V front porch (Fig. 6-11).

In the NTSC color format (EIA RS-170A), **color bursts** are added to the **back porch** region (Fig. 6-10B), and two frames, made up of four fields altogether, make up a picture cycle. *The field rate for NTSC is 59.94 Hz rather than 60 Hz* (Section 7.4).

Other Interlace and Timing Conventions

Some low-cost CCTV cameras use *random interlace* rather than 2 : 1. In that case, the first and final rasters on the monitor may start or end at any point along the H-scan lines. In random interlace cameras, the H and V oscillators run independently and do not maintain a constant phase relationship with each other for any extended period. In other words, the V blank may end (at the top of the picture) or begin (at the bottom of the picture) at any point during the H-scan interval. Therefore, the scan lines from one field do not precisely interlace with the scan lines in the other field. Therefore, the ratio of the vertical spacing between the rasters of consecutive fields varies continuously, and the interlacing becomes random. Each H scan is, nevertheless, triggered by the leading edge of its H-sync pulse, which is held constant to 0.002 H averaged over 20–100 scan lines, and the V-sync pulse is held to within 3.5–3.9 H of the start of the V blanking (EIA RS-420). Therefore, even though the interlace is random, in the course of the camera's scanning several frames, all picture points will have been scanned so that the **vertical resolution** is still reasonably good, although not as good as with the 2 : 1 interlace system.

As seen from these examples, the timing and interlace conventions (which are specified for cameras, processors, recorders, monitors, etc.) can vary even among the 525/60 scan systems.‡

*V-flyback is triggered by the onset of the V-sync pulse.

†Even though the equalizing pulses occur at 0.5-H intervals, filters tuned to the H frequency select for the appropriate pulses spaced 1 H apart for each field.

‡While microcomputers can use the same monitors as CCTV and broadcast video, their interlace pattern and sync signals are often quite different from those used in regular CCTV equipment. Many microcomputers use a **1:1 interlace,** with each of the frames appearing at 60 Hz and with 264 (or some other number of) scan lines. For comments on combining microcomputer character outputs with video, see Section 6.6 and Chapter 12.

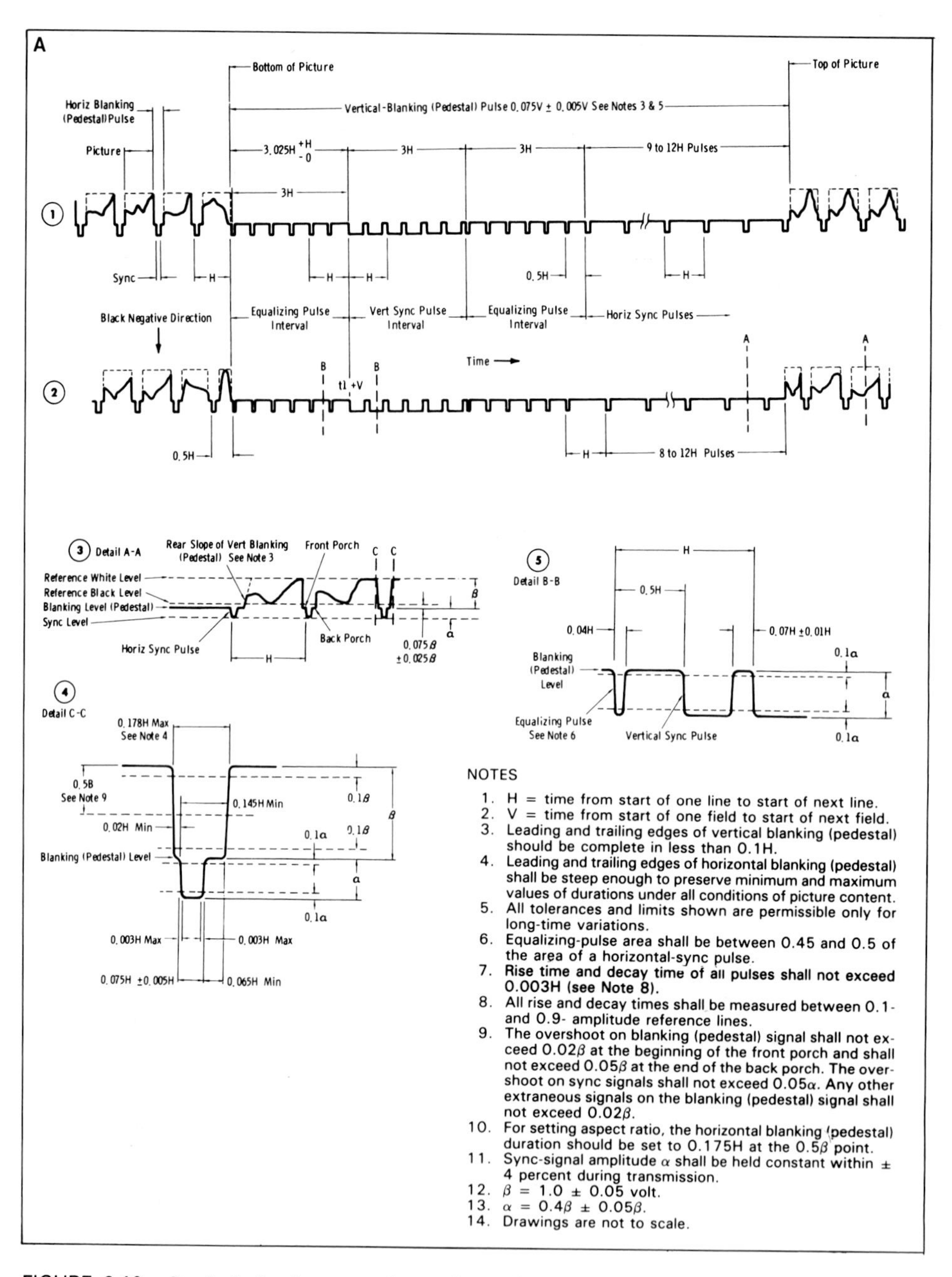

FIGURE 6-10. Standards for the sync pulses and waveforms (and color burst) adopted by the EIA for broadcast monochrome (A; RS-170) and NTSC color (B; RS-170A). (From EIA Standards RS-170 and RS-170A, copyright 1957 and 1977.)

B

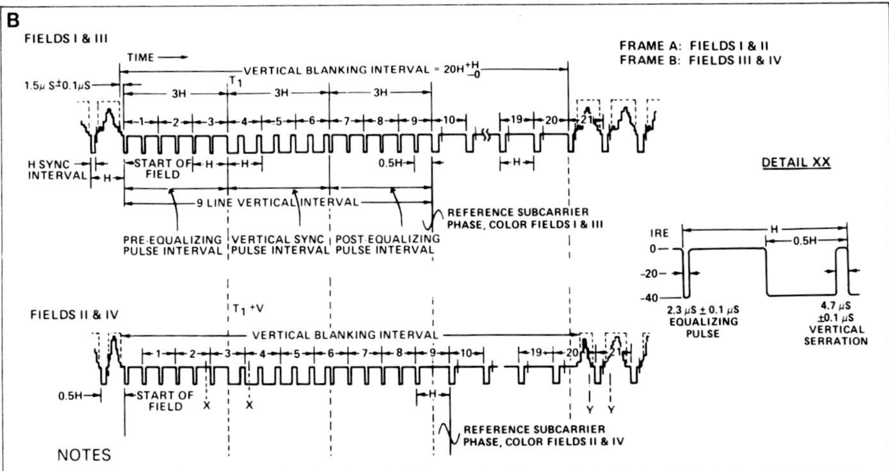

NOTES

1. Specifications apply to local system facilities. Common carrier and transmitter characteristics are not included.

2. All tolerances and limits shown in this drawing permissible only for long time variations.

3. The burst frequency shall be 3.579545 MH_Z ± 10 H_Z.

4. The horizontal scanning frequency shall be 2/455 times the burst frequency [one scan period (H) = 63.556 μS].

5. The vertical scanning frequency shall be 2/525 times the horizontal scanning frequency [one scan period (V) = 16.683 μS].

6. Start of color fields I and III is defined by a whole line between the first equalizing pulse and the preceding H sync pulse. Start of color fields II and IV is defined by a half line between the first equalizing pulse and the preceding H sync pulse. Color field I: That field with positive going zero-crossing sof reference subcarrier most nearly coincident with the 50_7 amplitude point of the leading edges of even numbered horizontal sync pulses. Reference subcarrier is a continuous signal with the same instantaneous phase as burst.

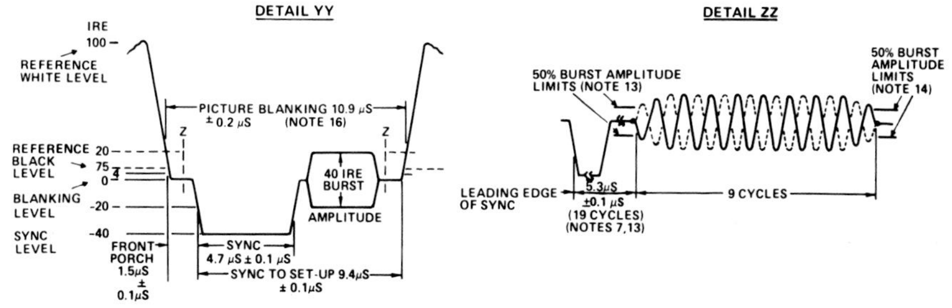

7. The zero-crossings of reference subcarrier shall be nominally coincident with the 50_7 point of the leading edges of all horizontal sync pulses. For those caes where the relationship between sync and subcarrier is critical for program integration, the tolerance on this coincidence is ± 40° of reference subcarrier.

8. All rise times and fall times unless otherwise specified are to be 0.14 μS ± 0.02 μS meaured from 10% to 90% amplitude points. All pulse widths are measured at 50% amplitude points, unless otherwise specified.

9. Tolerance on sync level, reference black level (set-up) and peak to peak burst amplitude shall be ± 2 IRE units.

10. The interval beginning with line 17 and extending through line 20 of each field may be used for test, cue and control signals.

11. Extraneous synchronous signals during blanking intervals, including residual subcarrier, shall not exceed 1 IRE unit. Extraneous non-synchronous signals during blanking intervals shall not exceed 0.5 IRE unit. All special purpose signals (VITS, VIR, etc.) when added to the ertical blanking interval are excepted. Overshoot on all pulses during sync and blanking, vertical and horizontal, shall not exceed 2 IRE units.

12. Burst envelope rise time is 0.3 μS + 0.2 μS − 0.1 measured between the 10% and 90% amplitude points. Burst is not present during the nine line vertical interval.

13. The start of burst is defined by the zero-crossing (positive or negative slope) that precedes the first half cycle of subcarrier that is 50% or greater of the burst amplitude. Its position is nominally 19 cycles of subcarrier from the 50% amplitude point of leading edge of sync. (see Detail ZZ)

14. The end of burst is defined by the zero-crossing (positive or negative slope) that follows the last half cycle of subcarrier that is 50% or greater of the burst amplitude.

15. Monochrome signals shall be in accordance with this drawing except that burst is omitted, and fields III and IV are identical to fields I and II respectively.

16. Occasionally, measurement of picture blanking at 20 IRE units is not possible beause of image content as verified on a monitor.

FIGURE 6-10 *(Continued)*

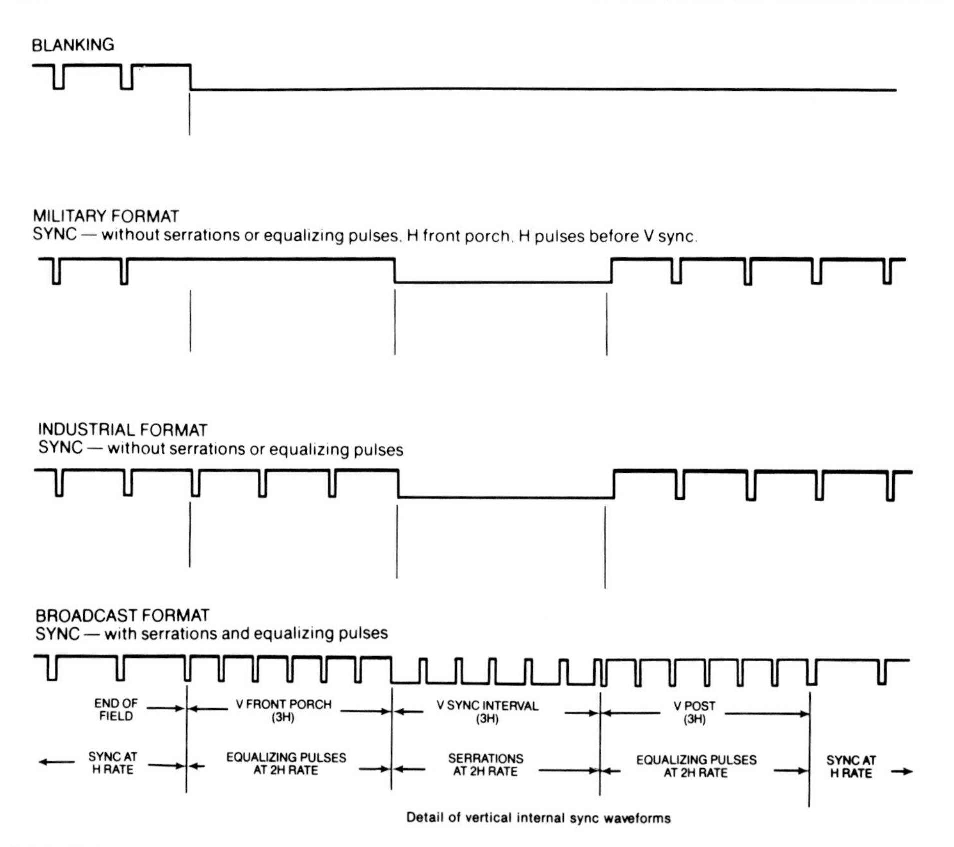

FIGURE 6-11. Comparison of standardized vertical-blanking intervals for different formats. (From "Applications Guide for VII Test Instruments," Visual Information Institute, Inc.)

In addition to the 525/60 scan rate, many other rates are used in CCTV, in order to raise image resolution or to simplify their interfacing with digital computers (Table 6-2).

While the 525/60 NTSC is the broadcast standard in North America, the northwestern coast of South America, Japan, and many of the Pacific islands, two other standards are commonly used around the world. Great Britain, central Europe, several African nations, Australia, China, and Argentina use the **PAL** 625/50. France, the Mideast, some African and many Soviet-bloc countries use **SECAM** 625/50. Brazil uses yet another standard, PAL 525/60. Each of these three latter systems uses color encoding schemes that differ from NTSC (Table 6-3).

In order to obtain a video image at all, timing conventions used in different video equipment must be *compatible*. Otherwise the image may **tear,** scroll, or just not take form. Compatibility does not necessarily imply that the scan rates must be identical, or that the sync signals must conform to the same standard. Nevertheless, it is important to remember that lack of compatibility can interfere with video operation in unexpected ways. *Compatibilities of video camera, processor, recorder, monitor, etc. should always be checked before acquiring such equipment, or before planning to use video equipment or recorded tape in another location.*

Visualizing the Waveform, Pulse-Cross Display

Should it become necessary to check the (**video**) **waveforms** of the video picture signal and timing pulses, they can be displayed on a 5-MHz or faster **cathode-ray oscilloscope** equipped with delayed trigger sweep. Examples of the oscilloscope displays are given in Fig. 6-12.

Alternatively, the timing pulses can be visualized on a video monitor that provides a **pulse-**

TABLE 6-2. TV Parameters for High-Resolution Systems[a]

						Fundamental generated frequency (MHz) (8)							
						R_hMHz(6)		$R_h = R_v$(9)		R_h = 800 lines		R_h = 1000 lines	
Lines/ frame	(1) Active lines	(2) Ver. res Rv	(3) f_h KHz	(4) t_h μsec	(5) t_{h_a} μsec	4:3(7)	1:1	4:3	1:1	4:3	1:1	4:3	1:1
675	624	425	20.25	49.38	42.38	63.6	84.8	6.69	5.01	12.6	9.44	15.7	11.8
729	674	475	21.87	45.72	38.72	58.1	77.4	8.18	6.13	13.8	10.3	17.2	12.9
875	809	575	26.25	38.09	31.09	46.6	62.2	12.3	9.25	17.2	12.9	21.4	16.1
945	874	600	28.35	35.27	28.27	42.4	56.5	14.1	10.6	18.9	14.1	23.6	17.7
1023	946	650	30.69	32.58	25.58	38.4	51.2	16.9	12.7	20.8	15.6	16.1	19.5

Vertical Blanking = 1250 μsecs nominal.
Horizontal Blanking = 7 μsecs nominal.

Notes:

(1) Active Lines = Lines/Frame less those occurring during vertical blanking.
(2) Vertical Resolution = Active Lines times Kell Factor (0.7). Vertical Resolution rounded to nearest 25 lines.
(3) f_h = Horizontal scanning fequency.
(4) t_h = Total horizontal line time.
(5) t_{h_a} = Total active horizontal line time (t_h − 7 μsecs.).
(6) R_h/MHz = Lines of horizontal resolution per MHz of bandwidth.
(7) Aspect Ratio.
(8) Fundamental generated frequency required to provide indicated resolution in lines per picture height.
(9) Fundamental generated frequency required to provide horizontal resolution equal to vertical resolution.
[a]From EIA **RS-343A**.

cross display. A push button or switch on such a monitor shifts the origin of the raster sweep from the upper left corner (1 in Fig. 6-13A) toward the middle of the screen.

As shown in Fig. 6-13B, the displaced scan now shows the four corners of the active picture area together with the part of the video screen that is ordinarily blanked. Now the blanked region appears as a broad dark cross on the screen. When we turn up the monitor BRIGHTNESS and turn down the CONTRAST, we see a dark cross displaying the various timing pulses within the dark-gray cross, and hence the term pulse-cross display (Fig. 6-13B).

The pattern in the cross shows the relation between the V- and H-blanking and sync pulses,

TABLE 6-3. Nations That Have Adopted the NTSC, PAL, or SECAM Color-Encoding Standards[a]

International television standards		
Lines/frame: 525 Field rate: 60 Hz Color coding: NTSC	Lines/frame: 625 Field rate: 50 Hz Color coding: PAL	Lines/frame: 625 Field rate: 50 Hz Color coding: SECAM
Antigua, West Indies	Algeria	Afars & Issas
Bahamas	Australia	Arab Republic of Egypt
Barbados	Austria	Bulgaria
British Virgin Islands	Bahrain	Czechoslovakia
Canada	Bangladesh	East Germany
Chile	Belgium	France
Costa Rica	Brazil (525/60)	Greece
Cuba	Brunei	Haiti
Dominican Republic	Denmark	Hungary
Ecuador	Federal Republic of Germany	Iran
El Salvador	Finland	Iraq
Guatemala	Hong Kong	Ivory Coast
Japan	Iceland	Lebanon
Mexico	Ireland	Luxembourg
Netherlands Antilles, West Indies	Italy	Mauritius
Nicaragua	Jordan	Monaco
Panama	Kuwait	Morocco
Peru	Malaysia	Poland
Philippines	Netherlands	Réunion
St. Kitts, West Indies	New Zealand	Saudi Arabia
Samoa (U.S.)	Nigeria	Tunisia
Surinam	Norway	USSR
Province of Taiwan	Oman	Zaire
Trinidad, West Indies	Pakistan	
Trust Territory of Pacific	Qatar	
United States of America	Singapore	
	South Africa	
	Spain	
	Sweden	
	Switzerland	
	Tanzania	
	Thailand	
	Turkey	
	United Arab Emirates	
	United Kingdom	
	Yugoslavia	
	Zambia	

[a]After *Raster Graphics Handbook*, 1980, p. 8–22.

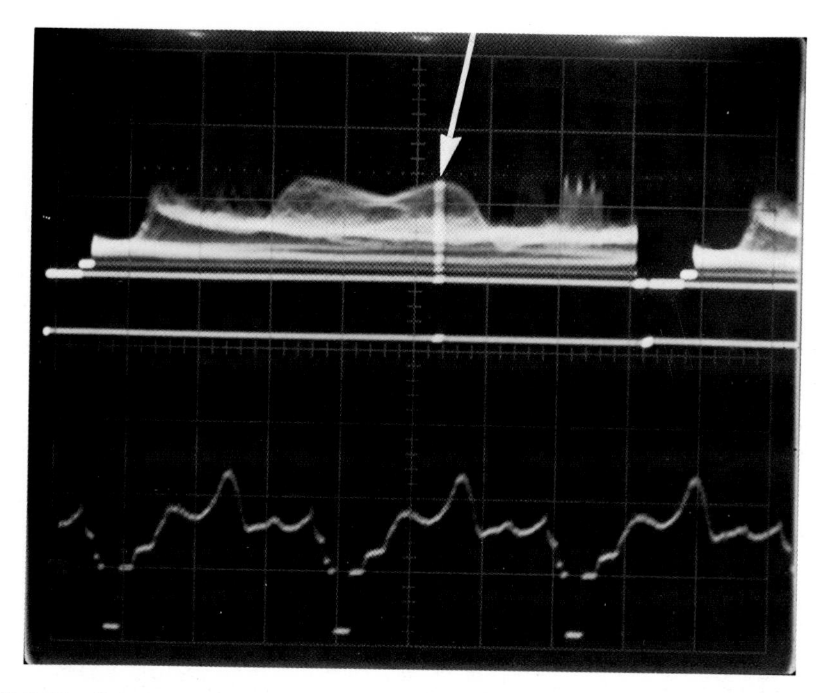

FIGURE 6-12. Cathode-ray oscilloscope display showing waveforms of video picture signals and timing pulses. (Top) Full-field display of picture shown in Fig. 9-14. (Bottom) Display, 3 H scans wide, of the narrow bright region indicated at the top. See Fig. 6-8 A,B (p. 159) for definitions.

both within and outside of the V-blanking interval (Fig. 6-14). The horizontal arms of the cross display the events during the V blanking, and the vertical part of the cross the events during the H-blanking interval.

The pulse cross also displays the stability of the H sync at the beginning of the picture scan. If the interlace is not stable, the start and end of the H sweeps wander about along the respective sweeps. If the H sweep is too far out of sync (as sometimes happens on playback from a video tape recorder), the first several H sweeps are unable to gain a properly synchronized trigger, and the top of the picture *tears,* or shows **flagging** (Fig. 8-16).

Most monitors do not show the whole active picture area but clip off some 5–15% of the picture area by **overscanning.** This hides minor flagging and misregistration at the edges of the picture. These imperfections can be seen with the pulse-cross monitor, or with a monitor with an UNDERSCAN button or switch.

If a monitor with pulse-cross display is not available, we can still inspect some of the events that take place during the V-blanking interval by making the following adjustments to the monitor. Turn the V HOLD control clockwise until the picture starts to roll down slowly. The broad, dark horizontal stripe (that appears between the bottom and top of the successive pictures) displays the V-blanking interval. Turn up the monitor BRIGHTNESS and turn down the CONTRAST. Then the V-sync pulse, and some of the serrations and equalizing pulses should appear within the rolling V-blank. With this display, much of the H-blanking interval is still not visible. More can be seen by switching to underscan, but we need the pulse-cross display in order to get a full view of the events during H blank.

Ennes (1979) gives a good description of the pulse cross and how its different parts relate to the various sync and blanking pulses. He also provides a thorough description of color video fundamentals including the nature of the timing pulses, color bursts, etc. These topics are also covered in the Sony Basic Fundamentals Library Programs (Sony Video Communications).

A very useful recent (January 1985) publication on performance evaluation of TV display

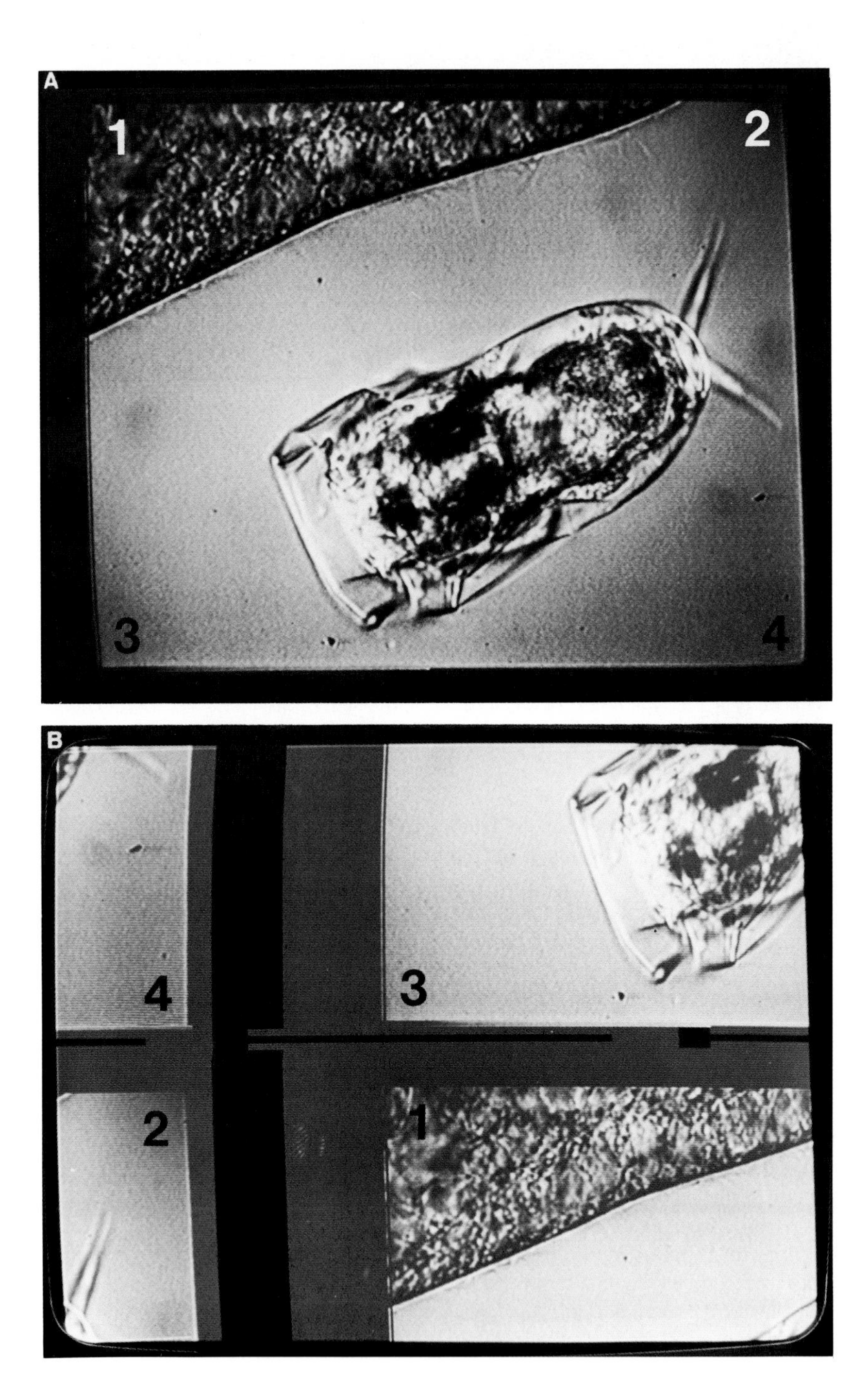

FIGURE 6-13. Pulse-cross display. On a monitor equipped for pulse-cross display, the origin of the raster sweep can be shifted from the upper left (1 in panel A) to the center of the screen (1 in panel B), thereby allowing the part of the picture that is usually blanked to be visualized. This blanked region contains displays of the various timing pulses, which are made visible by dropping the contrast and raising the brightness of the monitor (see Fig. 6-14). The corresponding points in the two monitor pictures are indicated by identical numbers.

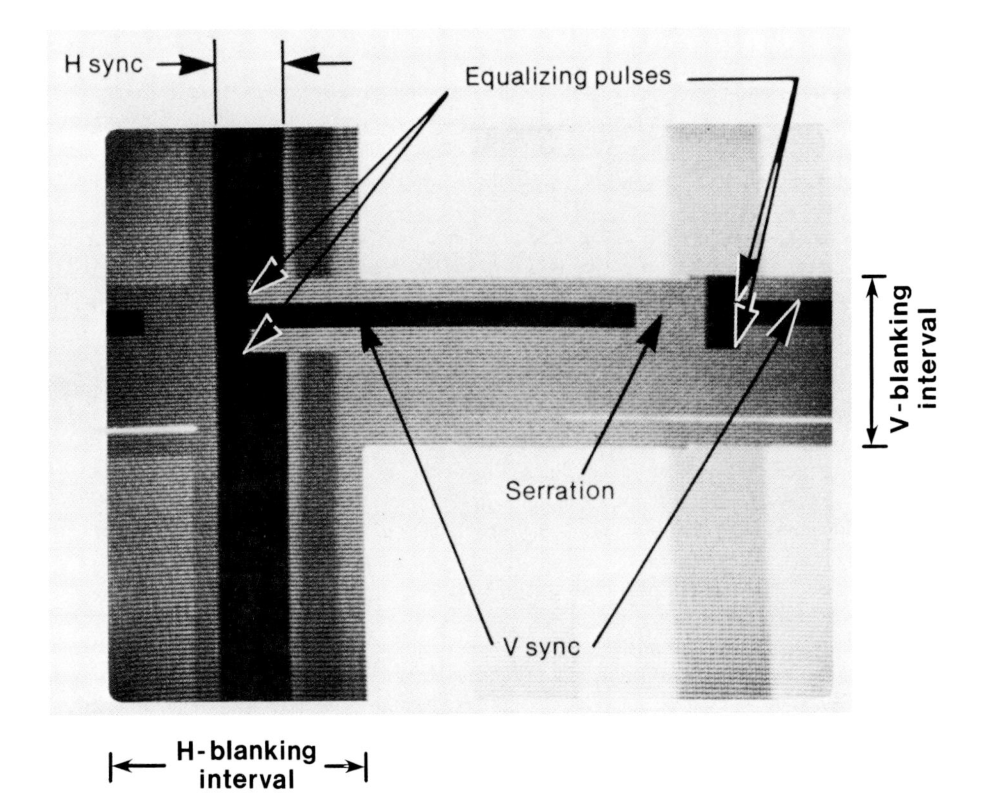

FIGURE 6-14. Schematic diagram of pulse-cross display. Since all parts of the pulse-cross display are at or below blanking levels, the display appears uniformly dark on a monitor adjusted to ordinary viewing conditions. However, by lowering the contrast and raising the brightness of the monitor, details within the pulse cross become visible, as shown here: the sync pulses and equalizing pulses, which have the lowest voltages, appear dark; the serrations appear lighter, at the blanking level. See text. Compare with Figs. 6-8 to 6-10, 6-13. (After "Sony Basic Video Recording Course," Booklet 2, p. 8.)

devices is "Performance Test for Medical Imaging Monochrome Display Devices" (Visual Information Institute, Inc.).

Synchronizing pulses used in CCTV are explained concisely in the pamphlet "Applications Guide for VII Test Instruments," also from Visual Information Institute, Inc.

Some basic standards recommended for CCTV by the EIA were listed in Table 6-1. Copies of the published standards are available from the EIA.

6.4. IMAGE RESOLUTION AND THE VIDEO SIGNAL

In the previous two sections, we saw how the video image is scanned in a raster, and how the video signal places each *picture element* in the final image in register with the corresponding picture element in the original image.*

We shall now examine the limits of image resolution in video, and how they relate to the frequency bandwidths of the video signal and signal-handling components.

*The term **pixel,** abbreviated from "picture element," is used to express the unit (rectangular) sample area in a digitized image (Section 10.2). In our present discussion, we shall use the expression *picture element* to mean the smallest segment of a raster line that can be resolved.

Definitions

In microscopy and photography, we measure resolving power of a lens, film, eye, etc. by the center-to-center spacing between *pairs* of lines or dots (or its inverse, the spatial frequency per unit distance) that can just be discerned. We ask: "How closely can two or more bright (or dark) lines, or dots, approach each other and still be perceived as being separate, rather than as having merged into one?"

For example, the minimum distance resolved by a particular microscope lens, for a high-contrast object in white light, might be 0.2 μm, or 5000 line pairs/mm (Sections 5.5–5.7). A photographic film may provide 20–50% image contrast at 100 line pairs/mm ("Kodak Professional Black-and-White Films," Eastman Kodak Co.).

Unlike in microscopy and photography, resolution in video is ordinarily not defined by the distance between line pairs or by spatial frequency of line pairs. Rather, *the convention is to express video resolution by the total number of black plus whitelines,* as explained below.

In video, the actual size of the frame varies with the active area of the camera tube and the size of the monitor while the number of scan lines remains the same. Therefore, a measure of absolute distance or spatial frequency is not meaningful in defining resolution. Instead, resolution is expressed relative to the size of the active picture, specifically in relation to its height. Thus, *in video, resolution is expressed as the total number of black and white lines that can be resolved over a distance equal to the height of the active picture area.*

For much common video equipment, the resolving power is not identical along the horizontal and vertical directions. Therefore, we define *horizontal* and vertical resolution separately. Furthermore, resolution tends to fall off quite sharply in the corners of the video picture. The horizontal resolution for a particular video camera or monitor may be 800 TV lines in the middle of the picture, but only 600 TV lines in the corners.

By convention, *horizontal resolution in video is measured for the midzone of the picture over a width equal to the height of the active picture area. Vertical resolution is measured through the middle of the picture for the active height of the picture.**

We shall now examine the limits of the vertical and horizontal video resolutions in turn.

Vertical Resolution

Vertical resolution is examined by checking how many black and white horizontal bars can be distinguished vertically through the middle of the picture.

As we have seen, a video image is made up of horizontal raster lines. The black and white bars in a horizontally oriented test pattern therefore run parallel to the raster lines. In this case, it seems intuitively obvious that the vertical resolution should rise in direct proportion to the number of raster lines—the larger the number of raster lines, the finer their spacing, and the higher would be the vertical resolution. This is in fact the case, as explained below.

For a test pattern, let us first consider a series of long black and white bars, the image of each bar on the video camera tube being as wide as the height of three scan lines (Fig. 6-15A). For checking vertical resolution, the bars are placed horizontally, so that three raster lines in the camera scan a dark region of the test target image and the next three lines scan a bright region of the test target image, and so on. In the video signal, the signal voltage for three successive H intervals would drop, reporting dark bars, followed by the next three rising and reporting light bars, and so on (Fig. 6-15a). This pattern in the signal drives the monitor where the image of the test target would be clearly resolved.

Next consider a test target whose black and white bars are each half as wide as the spacing of the raster lines (Fig. 6-15B). Now each raster line in the camera covers both a dark *and* a light region of the test target, so that the successive H intervals in the video signal would look alike (Fig. 6-15b). Clearly, the test target is not resolved.

*In order to measure resolution, we need to see the full active picture area. This requires a monitor with underscan capability.

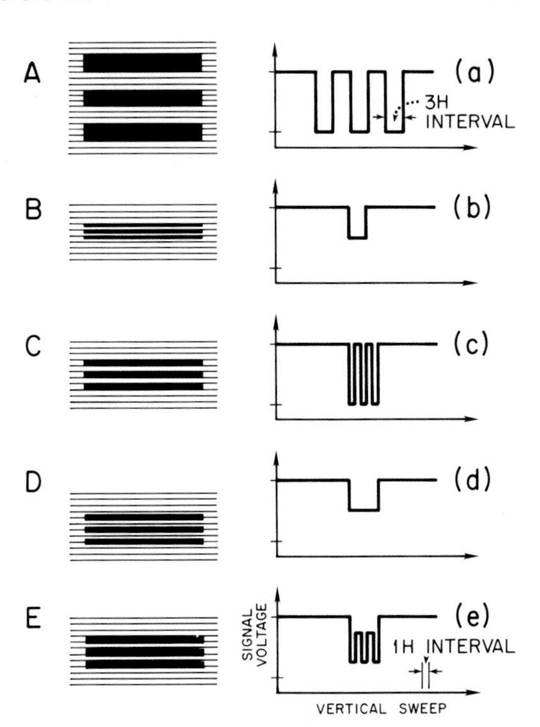

FIGURE 6-15. Vertical video resolution. In A, each dark bar of a test pattern is as wide as the height of three scan lines, resulting in the vertical sweep pattern shown in a: the bars are clearly resolved. In B, the test bar width is half the height of a scan line: the bars cannot be resolved (b). In C–E, the test bar width equals the height of a scan line. In C, the test bars happen to coincide with the scan lines and (in the ideal case) are resolved at full contrast (c). In D, the test bars happen to be exactly split on the scan lines, and are not resolved (d). In E, the usual case, the situation is at some intermediate position between C and D, and the scan lines are resolved at an intermediate contrast level (e). These figures are highly idealized; in fact, each raster scan line is not discrete as assumed here, but its brightness assumes a Gaussian distribution (Fig. 7-33).

Now consider another set of horizontal bars, with each black and white bar exactly as wide as the spacing of the raster lines in the camera. If, as shown in Fig. 6-15C, the black and white bars in the image of the test target line up exactly in register with the scan lines, alternate scan lines would be dark and light, so that the signal voltage for the alternate H intervals would drop and rise (ideally as shown in Fig. 6-15c). The monitor would then draw dark and bright lines on alternate scans. On the other hand, if the image of the same test target happens to fall on the camera raster as in Fig. 6-15D, the width of the bars would be exactly split by each raster scan. The successive H intervals in the video signal would then look alike and the test target would not be resolved (Fig. 6-15d).

These two cases, with the periodic image of the test pattern being either exactly in register with the raster scan in the camera, or being exactly out of phase by 180°, are chance events. Most of the time, the test pattern and raster lines would be neither exactly in register nor 180° out of phase (Fig. 6-15E). Thus, some intermediate degree of alternating brightness would be signaled by the alternate H intervals (Fig. 6-15e). Therefore, in the monitor, the contrast of the horizontal test target should fall somewhere between one and zero.

As shown here, when the spacing in the image of the horizontal bars equals the raster spacing, the bars may or may not be resolved. Peculiar moiré patterns could also occur at around this spatial frequency.*

This discussion suggests that the vertical resolution in video is somewhat less than the number of H-scan lines that appear in the active picture area. In practice, *the vertical resolution in video is found to be the number of active H-scan lines × 0.7. The factor of approximately 0.7, by which the number of scan lines is multiplied, is known as the* **Kell factor.**†

*In broadcast TV, the texture of the announcer's clothing can sometimes take on this chance relationship. A slight motion of the announcer or camera then gives rise to a striking flicker of color in that part of the scene. Actually, the situation is somewhat more complex, since even with a monochrome monitor a test target with periods *greater* than the scan line frequency can give rise to a moiré-like pattern through the phenomenon of **aliasing** (Appendix II). In color receivers, moiré patterns appear when the image texture approximates the spacing of the aperture mask in the monitor.

†The value of the Kell factor is related to the MTF function of the system. With electronically generated (digital) signals, the vertical resolution can equal the number of active scan lines (Kell factor = 1.0).

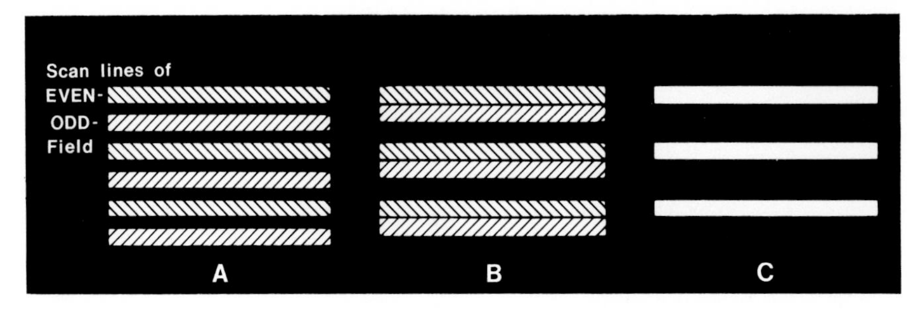

FIGURE 6-16. Schematic representation of interlace. (A) Proper 2:1 interlace of the even- and odd-field scan lines. (B) The even and odd scan lines have *paired* and the resolution is reduced. (C) The even and odd scan lines have exactly overlapped, and the V resolution has dropped to a half of A. H resolution is also reduced by the overlapping scan lines in the successive fields.

In the 525/60 2 : 1 interlace scan system, the number of scan lines in the active picture area is about 486 (total of 525 minus 7.5% for V blanking; Section 6.2). Multiplying this number by the Kell factor, *the effective vertical resolution with the 525/60 scan is 486 × 0.7 = 340.*

We should recall here that the alternate raster lines in the 525/60 scan arise from the interlaced even and odd fields (Section 6.2). Maximum contrast between the alternate raster lines can be maintained only with the interlace providing an exact 2 : 1 ratio of spacings (with about a 10% tolerance). The even-field raster lines should fall just halfway between the odd-field raster, and vice versa; otherwise, the even- and odd-field scans would **pair** or overlap (Fig. 6-16) and the contrast and resolution would drop (to as low as 50% of the maximum when the two field scans completely overlap).

For this reason, we would not be able to achieve quite as high a vertical resolution with random interlace as with a good 2 : 1 interlace scan.

As discussed in Chapter 7, the interlace on the monitor is altered by adjustments of the H and V HOLD controls. *These controls on the monitor should always be fine-tuned in order to achieve maximum vertical resolution.*

In addition to interlacing, the size and shape of the scanning spots that trace the raster in the camera and monitor affect resolution. The focus and astigmatism controls in both the camera and the monitor should be checked periodically to ensure that the scanning spots have the minimum size and proper shape.

Since vertical resolution in video is determined by the Kell factor times the number of scan lines, the number of scan lines would have to be increased in order to raise the vertical resolution. For CCTV, such nonstandard scans are possible, but the horizontal resolution must also be increased for the overall picture resolution to be improved (Table 6-2).

Horizontal Resolution

As defined earlier, horizontal resolution in video is measured in the middle of the picture over a width equal to the height of the active frame. We shall first examine what fraction of the complete H period is occupied by the active frame.

Since the aspect ratio, or the ratio of the width to the height of the frame, in standard video is 4 : 3 (Section 6.2), it follows that horizontal resolution is measured over the mid three-fourths of the active H-scan line (Fig. 6-17A).

The active part of the H scan occupies 82.5% of the total H-scan period, since H blanking masks 17.5% of each H scan (Fig. 6-17B, Section 6.2). Therefore, in order to determine horizontal resolution, we measure the number of black and white lines that can be discerned in 82.5% × ¾ × H = 0.62 H.

Horizontal resolution in the camera and monitor translates into frequency in the video signal in the manner discussed below.

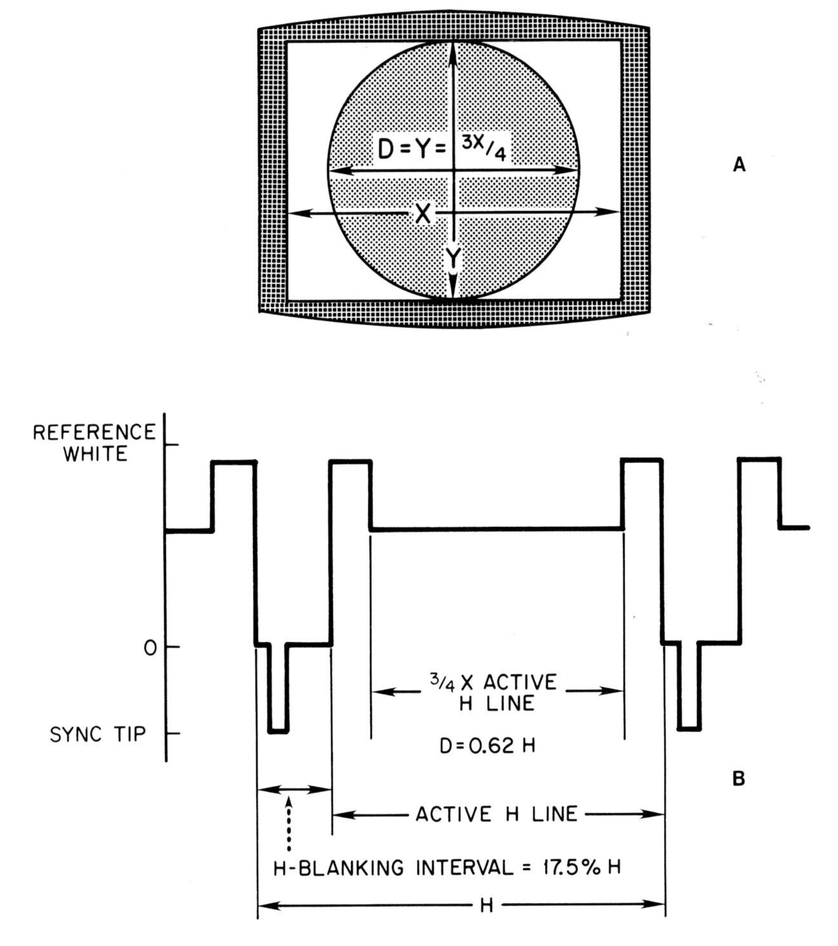

FIGURE 6-17. Active picture area and aspect ratio. (A) Picture of a gray disk on a white background seen on an underscanned monitor. (B) Horizontal line scan across the middle of A. The diameter of the circle (equal to the height of the active picture scan) defines the horizontal distances over which resolution is measured. The aspect ratio (X:Y) of the standard video picture is 4:3 (1:1 in some CCTV systems). See Fig. 7-32 for actual dimensions. Most monitors overscan by 10–15% so that the full active picture area is not visible without underscan.

Consider a series of black vertical bars on a white background (Fig. 6-18A). Let the dark bars and the white space between the bars each have the same width (d). We now scan horizontally across the bars, or along the X axis, with a small scanning spot whose diameter is P. If P is considerably smaller than d, the H-scan signal reflecting the vertical bars should appear as a rectangular wave (Fig. 6-18B).

For this relation to hold, not only most $P << d$, but the electronic system must be capable of responding faithfully to the rapid change in the signal. We shall detour briefly to consider this point.

Instead of alternating narrow vertical bars, let us for the moment consider a single bright vertical stripe, say with a width of 0.2 H, on a dark background. The ideal H-scan signal given by scanning across such a stripe would be as shown in Fig. 6-19A.

In fact, electronic circuits have a limited capability in responding to a rapid rise or fall in signal. Recall that a 1-H scan occurs 15,750 times/sec, or every 63.5 μsec. If the signal were to rise in 1/1000th of that time, the **rise time** would have to be 0.06 μsec, a very short time interval indeed.

Rather than responding as in Fig. 6-19A, the response is more characteristically as shown in

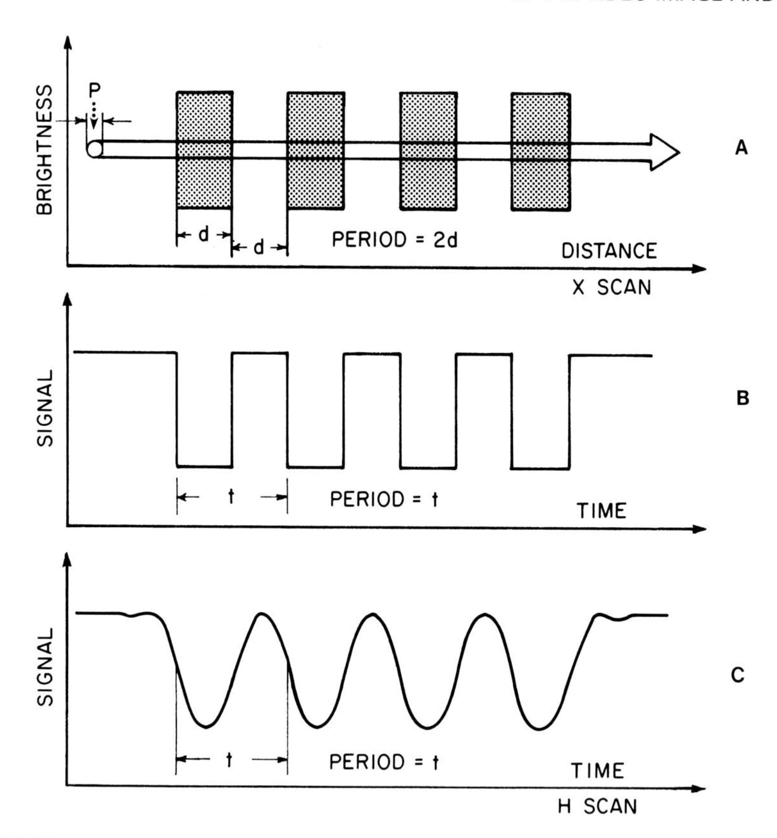

FIGURE 6-18. Horizontal test bars. (A) A test pattern is scanned along the *X* axis by a very small spot. (B) The appearance of an ideal brightness signal versus time. A less exact signal response, such as the fundamental sine wave of B (shown in C), is sufficient to resolve the spatial period (2*d*) of the test pattern.

Fig. 6-19B, C, or D. Depending on the resonant frequency of the circuit and its damping characteristics, the response may be an overshoot followed by oscillations with diminishing amplitude (Fig. 6-19B,D), or an asymptotic approach to the equilibrium value (Fig. 6-19C). The rise times vary accordingly.

Obviously, neither too vigorous an oscillation nor too strong a damping is desirable, since the former would result in **ringing,** and the latter in a sluggish response. In the video image, ringing gives rise to periodic vertical lines that should not be present, and a sluggish response results in longer rise time, which blurs the sharpness of vertical edges, reducing the contrast of image detail and lowering the horizontal resolution.

Whenever practical, the resonant frequency and damping characteristics are selected so that the response of the circuit to a step increase or decrease in signal is more or less as shown in Fig. 6-19D (with a rise time of less than 0.1 μsec). The rise time is reasonably short, and the waveform pretty closely follows the desired square wave.

Returning from the case of a single 0.2-H-wide broad stripe to that of a series of narrower stripes, the sample trace B in Fig. 6-18 was drawn to precisely follow the distribution of brightness across the bars. However, the task at hand is not really to trace the exact shape of each of the narrow bars. Rather, we are merely interested in whether or not we can *resolve* the bars. Just as in the discussion of image resolution in the microscope, we need only be able to tell that a structure with the spacing, or period, of the bars is present.

For examining the resolution of a video signal, then, we consider trace C, which shows the fundamental sine wave component of A and B (Fig. 6-18). Signal C, while no longer showing the detailed shape of the bars, nevertheless tells the receiver that a structure is present with the spacing (2*d*), or frequency, of that shown in A and B. Signal C can be transmitted through a circuit whose

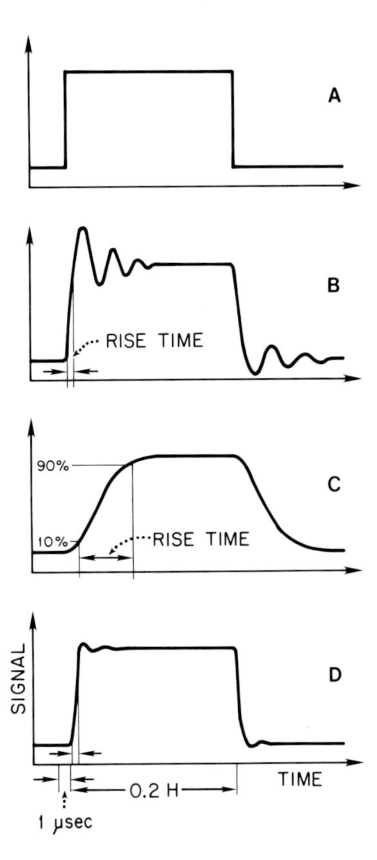

FIGURE 6-19. Horizontal scan across a broad white bar. An ideal H-scan signal across a single broad (0.2 H wide) bright vertical stripe on a dark background is shown in A. Departures from the ideal case are illustrated in B–D. B shows ringing, and C a blurring of the vertical edges. D is a practically satisfactory response, having a short rise time and a reasonably faithful reproduction of the brightness profile.

frequency response extends just far enough to include that frequency. In contrast, we would need a frequency response several times greater in order to transmit the signal shown in Fig. 6-18B without appreciable degradation.*

Our task, then, is to figure out what frequency video signal carries the signal for the spatial pattern in Fig. 6-18A, or the electrical signal in Fig. 6-18C.

As we saw earlier, three-fourths of the width of the active picture area is scanned by the raster in 0.62 H. Let us assume that we wish to resolve 800 *TV lines,* or a total of 800 black and white lines, in this part of the H scan. With a total of 800 black and white lines, we would have 400 cycles of the black and white paired patterns (compare Fig. 6-18C with A).

Since each H scan in the 525/60 scan is repeated 15,750 times/sec, the complete H scan takes 63.5 μsec, and 0.62 H takes $63.5 \times 0.62 = 39.4$ μsec. To fit 400 cycles into 39.4 μsec, the signal frequency would have to be $1/(39.4 \times 10^{-6}/400) = 10 \times 10^{6}$ Hz or 10 MHz.† Or, as often stated, *one needs a 1-MHz bandwidth for every 80 TV lines one wishes to resolve horizontally for a distance equal to the picture height.*

*A square wave is made up of the fundamental frequency sine wave superimposed with several higher harmonics of the fundamental frequency. To transmit the shape of the square wave, those higher-frequency harmonics would have to be included.

†What this says is that we need to generate and transmit a 10-MHz sine wave in order to signal 400 cycles of brightness change in the video image horizontally, or to resolve 800 TV lines horizontally over a distance equal to the picture height. But a signal made up only of a 10-MHz sine wave would only display a series of vertical stripes with spatial frequencies of 400 cycles on the monitor. The 10-MHz wave only signals the **limiting resolution** that we are after; we actually need frequencies up to 10 MHz and down to 30 Hz in order to represent the picture faithfully. That means that we need a *bandwidth* of 10 MHz to resolve 800 TV lines horizontally and to have a picture that faithfully reproduces the image captured by the camera pickup tube.

While this rule of thumb does, in fact, yield the horizontal resolution, detailed analysis shows that *phase,* in addition to amplitude of the resolved signal, is equally important. Unless a specific phase relation is maintained throughout the video circuit between signals of different frequencies, they become displaced relative to each other. The different frequency components of the signal would then appear at improper locations, displaced along the H scan in the monitor (Ennes, 1979, Chapter 3). Such a displacement would prevent proper resynthesis (e.g., of a broad square wave pattern, as in Fig. 6-19) so that patterns with even relatively low spatial frequency could not be properly reconstructed on the monitor.

Video engineers have carefully studied the response characteristics of circuit systems, and each piece of equipment is outfitted with combinations of band-pass *filters, peaking circuits,* etc., with specific response characteristics, to optimize the system performance. *Only if the equipment has adequate band-pass and phasing capabilities, and if the video cables are properly terminated (which the user must do), will the camera signal be transmitted in such a way that the image on the monitor appears with the proper spatial frequency, and with the spatial frequency components appearing in the proper location along the H scan.*

Each set of electronic circuits in the camera, processor, recorder, monitor, etc. must meet these stringent requirements. Given such circuits, and only then, do we have meaningful video resolution with a horizontal resolution of 80 TV lines for every 1-MHz bandwidth for the overall video circuit. Table 6-4 shows the relationships among bandwidth, rise time, and horizontal resolution.

Cascading a Number of Devices; System MTF

When several pieces of video equipment are cascaded, for instance as in going from a camera to a processor, recorder, and monitor, each stage contributes to the limitation of the overall system. The effect of a cascaded series of devices is to increase the rise time of the system. In other words, the system responds more sluggishly to a rapid change than do the individual pieces of equipment. According to Ennes (1979, p. 121), "The total rise time of a pulse through a series of cascaded stages is equal to the square root of the sum of the squares of the individual stage rise times." Ennes points out that since bandwidth is inversely proportional to rise time,

$$\frac{1}{\text{bandwidth of system}} = \left(\sum \frac{1}{(\text{bandwidth of component})^2}\right)$$

Two cascaded circuits with the same bandwidth, or rise time, are found to reduce the horizontal resolution of the system by 40% compared to each circuit individually.

TABLE 6-4. Rise Time and Horizontal Resolution at Various Video Bandwidths[a]

Bandwidth (MHz)	Rise time (μsec)	H resolution (TV lines)
1	0.35	80
2	0.18	160
3	0.117	240
4	0.087	320
5	0.07	400
6	0.058	480
7	0.05	560
8	0.044	640
9	0.039	720
10	0.035	800

[a]From Ennes (1979, p. 122).

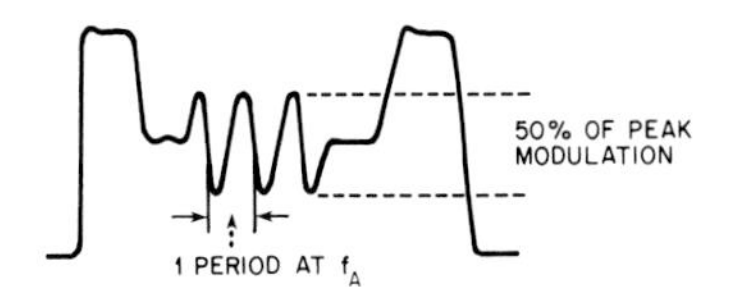

A. SIGNAL IN

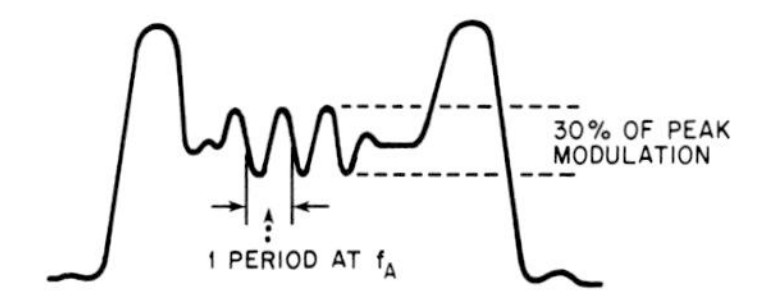

B. SIGNAL OUT

MODULATION TRANSFER = (30 ÷ 50)% = 60% AT FREQUENCY f_A

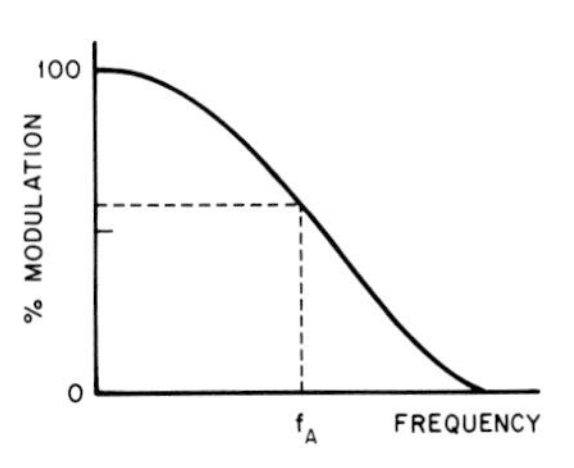

C. TYPICAL MTF CURVE

FIGURE 6-20. Derivation of MTF curve from percent modulation. A and B illustrate modulation transfer: If an incoming signal has 50% modulation of its peak range (for a sine wave with a frequency of f_A), and the modulation is reduced to 30% in the outgoing signal, then the modulation transfer is said to be 60%, that is, (30/50)%. C is an example of an MTF curve, which shows modulation transfer, as described above (% modulation), as a function of frequency. The 60% modulation transfer shown in A and B occurs at a frequency of f_A in the sample MTF curve.

An alternate way of looking at the bandwidth requirement is to examine the modulation transfer function (MTF). The MTF curve relates the percent **modulation** of the output signal compared to the incoming signal measured at various frequencies (Fig. 6-20C). For example, if a sinusoidal signal whose amplitude is modulated 50% (Fig. 6-20A) enters a piece of equipment and exits the device as a wave with a 30% modulation, the modulation transfer rate is (30/50)% = 60%. The percent of modulation transfer typically varies as a function of frequency as shown in Fig. 6-20C.*

From the definition of percent modulation transfer just given, it follows that when two or more video devices are cascaded, the system MTF becomes the product of the MTFs of the individual devices. For example, assume that device A provides a modulation transfer of a_1% and device B provides a modulation transfer of b_1% at frequency F_1. The rate of modulation transfer of the system at that frequency would be b_1% of a $_1$% or $a_1\% \times b_1\%$. At frequency F_2, the modulation transfer rates being a $_2$% and b_2%, the system modulation transfer rate would be $a_2\% \times b_2\%$, and so on. Therefore, for a cascaded series of (linear) devices, the system MTF is given by the product of the MTFs of all of the devices in the system (Appendix II).

In practice, if the rate of modulation transfer (or frequency response) of one of the device drops to zero at a particular frequency, the system MTF would also drop to zero at that frequency and the signal could not be transmitted or resolved. On the other hand, if the rate of modulation transfer for the one piece of equipment is, say, only 20% at a particular frequency, that does not mean that we should let the rate of modulation transfer of the other devices in that chain also drop to that low a value. For example, if the rate of modulation transfer of a particular video tape recorder (whose

*In a frequency response curve, the actual degree of signal **attenuation** is plotted against the frequency of the signal (Fig. 6-21). In the MTF curve, the rates of modulation transfer, expressed in percent, are plotted against frequency. (See also Sections 4.5, 5.7, and Appendix II.)

limiting resolution is, say, 340 TV lines) were 20% at 300 TV lines, it would still be worth using a camera and monitor with 800 TV lines of limiting resolution. They would prevent a further substantial drop of the percent modulation transfer (in essence the contrast) through the system for spacings that correspond to 300 TV lines **PPH** (per picture height).

6.5. DEGRADATION OF VIDEO SIGNAL, ELECTRICAL NOISE

Image Degradation and Frequency Response

In our discussion of horizontal resolution in video, we emphasized the need for adequate high-frequency responses. For example, in order to attain a horizontal resolution of 800 TV lines, we saw the need to work with signal frequencies as high as 10 MHz. We also touched upon the need to maintain proper phase relationships among the different frequency components; for example, for faithful reproduction of step changes in intensities at sharp edges in the image.

In addition to the higher-frequency responses, the electronic circuits in video must also respond faithfully to very-*low-frequency signals*. Without proper response to the lower-frequency signals, a uniformly lit large area in the image would not be represented by a constant voltage in the signal, but by a gradually changing voltage. That would introduce a gradation in brightness, or *shading*, in the reproduced image. Also, brightness at the edges of large areas would not be properly represented. Recalling that the video field repeats at 60 Hz, the low-frequency response must indeed extend to well below 30 Hz in order to keep vertical shading to a tolerable level.*

Thus, in video, the high-frequency and low-frequency responses both extend well beyond the responses that one encounters in audio systems (Fig. 6-21).† In other words, we need not only the very fast rise times to handle the high-frequency components of the signal, but also an enormous bandwidth that extends down practically to DC.

Several types of image degradation, including ringing, unsharp edges, trailing, shading, etc., can arise from inadequate response of the circuits. Poor responses to the high-, mid-, and low-frequency ranges can each give rise to one or another form of signal and image degradation. Examples of image degradation arising from limited responses of the electronic circuits are given in Chapter 7.

Distortion of Waveform

We have so far concerned ourselves with the ability of the electronic circuitry to transmit various frequency components of the video signal without appreciable loss of amplitude or alterations of phase relationships. *The waveform of the video signal can also be distorted by the signals being too strong.*

So long as we are dealing with video signals at the proper voltage levels (1.0 volt peak-to-peak of composite video signal into a 75-ohm load for CCTV; Section 6.3), the video equipment should respond linearly to the incoming signal.

However, if the brightness of the image on the video camera exceeds a reasonable level of illumination, the camera tube or circuit may no longer be able to respond linearly to image brightness. Or the CONTRAST or BRIGHTNESS controls of the monitor or processor may be set too high, so that either the amplifiers can no longer work within their specified ranges, or the **cathode-ray tube** itself behaves anomalously.

Video amplifiers generally respond linearly over a limited range of incoming voltages by

*Amplifiers in the camera and other video components incorporate a **clamping** circuit to restore the DC level of the signal to its proper value at the beginning of each horizontal sweep.

†In order to maintain the proper phase relationships for the different frequency components in the signal, the frequency response must be flat throughout the 30 Hz to 8 MHz video range (Fig. 6-21). The roll-offs at the lower and upper frequencies must also be gradual.

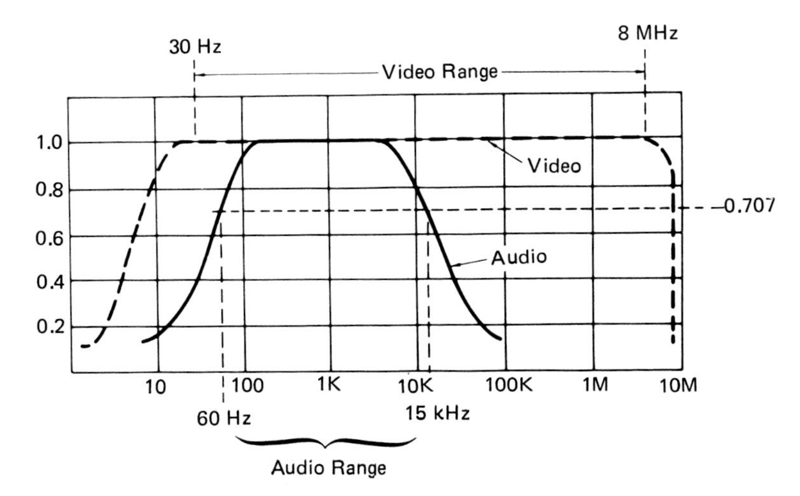

FIGURE 6-21. Frequency response curves for video and audio signals. Video requires a much greater frequency range than audio. In addition, the entire video range (30 Hz to 8 MHz in this example) must have a flat peak response. Audio is much less demanding, in that a roll-off of 3 dB (0.707 relative response) is perfectly acceptable. (From Ennes, 1979.)

providing an output current or voltage proportional to the input (Fig. 6-22b).* So long as the amplitude or voltage of the incoming wave lies within the rated range (as with the interval t_1 to t_2 for the wave in Fig. 6-22a), the output current (t_1 to t_2 in Fig. 6-22c) is proportional to the input. However, when the input voltage or amplitude exceeds the rated range (t_2 to t_7 in Fig. 6-22a), the output current is no longer proportional to the input and the waveform is distorted (t_2 to t_7 in Fig. 6-22c).

Assume that the wave in Fig. 6-22 has a relatively low frequency, reflecting a gradual change of scene brightness over a moderate fraction of the H scan. The rise in image brightness signaled between t_2 and t_3 would give rise to a proportionate output current. But after t_3, the voltage exceeds the linear range of the response curve, so that between t_3 and t_4 the output signal is "saturated," and the brighter part of the image is clipped. The darkest shadow regions (t_5 to t_6) would also be clipped so that shadow details would disappear.

If such alterations of waveform are introduced intentionally, the limited range of the amplifier is not an impediment. Indeed, *white* and *black clipping,* as well as other nonlinear responses, are used in the electronic circuits in the camera, processor, etc. to suppress or accentuate particular features of the image (Chapters 7, 9–12).

However, if the signal had been accidentally boosted to an unreasonably high level, and the amplifiers had been driven beyond their intended ranges, the waveforms would in fact be distorted as described above. Distortions of waveform would provide misleading signals or images, and the picture on the monitor would not properly represent the scene picked up by the camera.

Proper termination of the video cable is most important in this regard since an *unterminated cable can give a signal that is twice as high as that intended* for the input of the device receiving the signal.

Distortion of the video waveform would also introduce spurious harmonics and further degrade the signal. Examples of image degradation arising from nonideal responses of the video circuitry and other causes are given in Chapter 7.

*Other amplifiers, also found in video circuits, are intentionally designed to respond nonlinearly. However, we are not discussing amplifier characteristics as such here, but the distortions that can arise by exceeding rated input voltages.

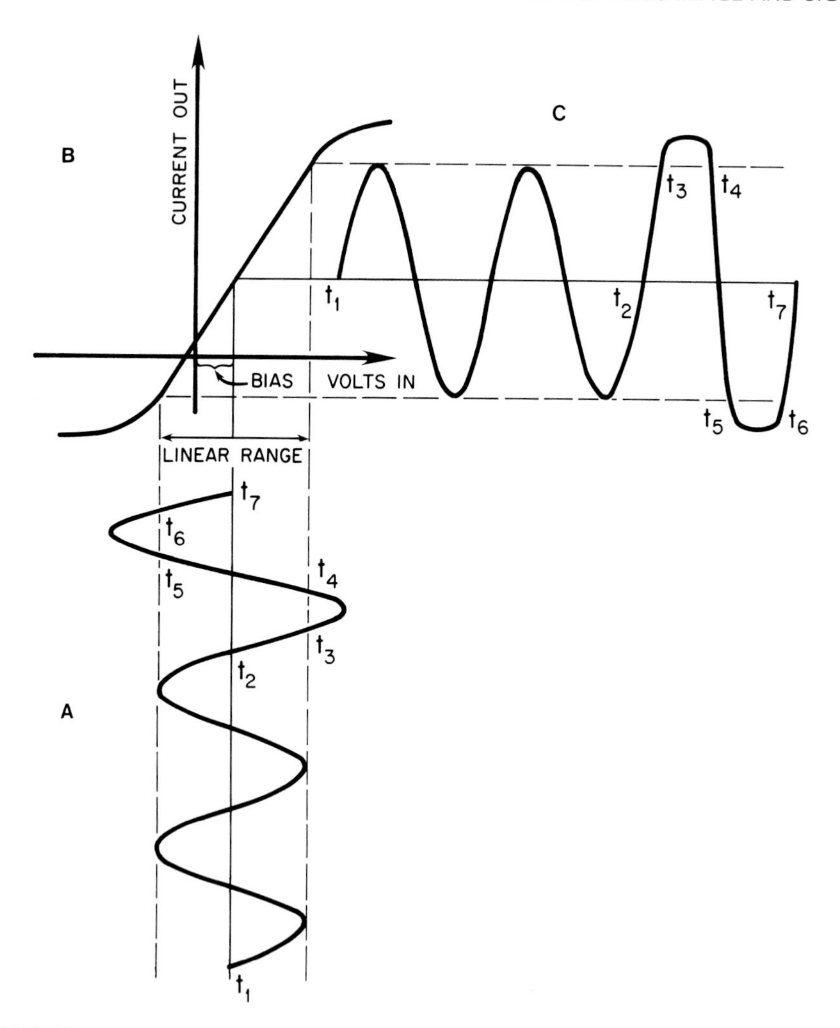

FIGURE 6-22. Amplifier characteristics; clipping. (a) Input signal. (b) Response curve of the amplifier. (c) Output signal. The output of this video amplifier is linear only over a limited, specified range. Incoming signals having amplitudes within the specified range of the amplifier are faithfully converted to output voltages (t_1 to t_2). However, when the signal is too strong (t_2 to t_7), the output is distorted, and the peaks of the signal are "clipped." The signal is sometimes clipped intentionally, for example, to emphasize or suppress parts of the signal, or as in stripping the sync pulses.

Electrical Noise

In any electronic system, an increase in bandwidth is also accompanied by an increase in *noise* level.

Electrical noise is generated in amplifiers, detectors, switches, contacts, and even across a simple resistor (not only by virtue of defective contact, but also by the very presence of the resistance itself). The noise is an unpredictable change in the electrical voltage or current, and in the limiting cases reflects random thermal agitation or statistical fluctuation of current flow in some electrical component.

Noise is commonly measured by the **root mean square** (rms) of the amplitude of the noise voltage (or current).* The rms voltage essentially represents the averaged energy of the noise.

The noise level generally rises proportionately with the input impedance and the square root of the bandwidth of a circuit. Thus, a basic video amplifier with a bandwidth 10,000 times greater than a typical audio amplifier would have an rms noise voltage that is 100 times greater, if comparable input stages were used. In video, special low-noise **preamplifiers** are used, the input impedance is kept low, and the signal voltage is kept at a moderately high level.

The limiting noise in CCTV employing a vidicon-type pickup tube commonly arises from the preamplifier in the camera. When the image brightness decreases below a threshold level, the rms noise voltage from the preamplifier approaches or exceeds the weak signal voltage from the camera or image tube (Chapter 7). Noise then appears as **snow** or granularity of the video image (Fig. 11-7C). In intensifier camera tubes, the noise can arise from thermal fluctuation of electron potentials in the photocathodes or electron multipliers, or may reflect photon statistics in the case of extremely high-sensitivity tubes (e.g., Fig. 11-1). Noise such as this is usually random (*white noise*),† which can be averaged out by temporal or spatial integration (Figs. 1-8, 6-24, 10-40). The camera or image tube itself may also have blemishes that are unresponsive to light or are over-responsive hot spots. These blemishes contribute to *nonrandom noise.*

Video tape recorders (VTRs) can contribute video noise, either as random noise or as discrete "dropouts." Video equipment commonly incorporates circuits to reduce white noise and, in the case of VTRs, to compensate for dropouts.

Fluorescent fixtures, transformers, motors, and especially arc lamps tend to generate interfering radio frequency (RF) spikes or noise pulses. The noise becomes conspicuous when the equipment is not properly grounded; for example, when different pieces of equipment are connected to separate AC supply lines. Such noise, which appears as prominent hash or unsteady vertical stripes, is introduced by "ground loop" problems. Camera lenses or C-mount coupling tubes, inadequately grounded to the camera can act as antennae and introduce extreme and intermittent hash (Fig. 3-13C).

In some cases, the electrical noise is synchronized and occurs at the AC power-line frequency. The degradation of the video picture arising from such sources of noise is minimized (as is done in video cameras) by synchronizing the video field scan rate to the line frequency. The disturbance that occurs in phase with the power line is then localized to a particular zone (height) of the picture, rather than spreading randomly all over the picture. In addition, the clamping circuit, which restores the DC voltage level at the beginning of each horizontal sweep (at 15,750 Hz), eliminates the picture signal or sync voltage fluctuation that could arise from spurious pickup of low frequency AC such as the 60 Hz **hum** from the line voltage (Ennes, 1979). Partial breaks in coaxial cables and connectors can also be the source of noise and unsteady pictures.

S/N Ratio

The level of noise is generally expressed as a ratio with the signal rather than as an absolute value. The peak-to-peak signal voltage is divided by the rms noise voltage and expressed as the *signal-to-noise* (S/N) *ratio.* Thus, if the rms noise level is 1% of the peak-to-peak signal voltage, the S/N ratio would be 1/1% = 100 : 1.

The S/N ratio is commonly expressed in **decibels** (dB) rather than as the ratio, R. The dB value

*For a practical discussion of how to measure the rms noise voltage with an oscilloscope, see Franklin and Hatley (1973). Mathematically, the rms value of a signal is defined as an integral function. Evaluation of the integral for a sine wave yields the rms value equal to $\sqrt{2} \times$ peak value. For "white" (or purely random) video noise, it is common to approximate the rms value of the noise as one-sixth of the generally highest repetitive peak-to-peak noise signal.

†White noise is an aperiodic, random noise that shows as a hash on an oscilloscope (Fig. 6-23). In audio, white noise can be heard as a rushing hiss when the amplifier gain is turned up.

for voltages is 20 times $\log_{10}$ R, so that a ratio of 100 : 1 would be 40 dB, and a ratio of 10 : 1 would be 20 dB, etc. (Table 6-5).

As a unit of measurement, the decibel is not just limited to noise but rather is used widely in communications engineering to express the ratio of AC signals. For example, in the frequency response characteristics shown in Fig. 6-21, the video response is down $1/\sqrt{2} = 0.707$, or 3 dB down, at around 7 Hz and 8 MHz. The response of the audio circuit in the same figure is down 3 dB at 60 Hz and 15 kHz.

Video signals with progressively better S/N ratios are illustrated in Fig. 6-23A–D and E,F. Monitor images produced by signals with more or less comparable S/N ratios are shown in Fig. 6-24A–D. When viewed live on the monitor, signals with an S/N ratio of 36 dB, or an rms noise level of 1.6% (Table 6-5), look quite grainy. Signals with an S/N ratio of 42 dB, or an rms noise level of 0.8%, give quite a "clean" video image.

Various electronic equipment, including video equipment, is designed to maximize S/N ratios within the constraints of needed performance, available technology, and price. Ultimately, the signal becomes limited by the noise in the original signal (e.g., photon statistics in the low-light-level optical image), by the bandwidth required, and by thermal fluctuation.

Given a particular noise level, there is a limit to the meaningful message content of the signal per unit of time. Nevertheless, we can often extract meaningful signals of selected types *if we are prepared to sacrifice information of another type*. For example, purely statistical, random noise can be integrated out by averaging many frames of the same video scene, either through digital image averaging or summing (Figs. 1-7, 1-8, Chapter 10), or by long photographic exposures (Chapter 12). In these cases, temporal resolution is sacrificed in order to gain spatial resolution.

For some quantitative analyses, the S/N ratio of the video signal may have to be greater than that (say 42 dB) that visually gives a clean image. For example, if the brightness of each pixel is to be measured to one part in 2^8 or 1/256, one would require an S/N ratio of 48 dB to reduce the rms noise to the 0.4% level. The statistical fluctuation of measuring 256 counts at a time would still be

TABLE 6-5. Decibels versus Voltage- and Power-Ratios[a]

dB	Voltage ratio	Voltage fraction = down to (%)	Down by (%)	Power ratio
3	1.41	70.8	29.2	2
6	2.0	50	50	4
12	4.0	25	75	16
20	10	10	90	100
26	20	5	95	400
32	40	2.5	97.5	1600
36	63	1.6	98.4	4000
40	100	1.0	99.0	10^4
42	126	0.8	99.2	1.6×10^4
46	200	0.5	99.5	4×10^4
50	316	0.3	99.7	10^5
60	1000	0.1	99.9	10^6

[a]For example, if an amplifier or tape recorder response is down by 3 dB at a given frequency, its signal amplitude is at 70.8% of the peak-flat (or upper) value; or, its response is down by 29.2% at that frequency (Figs. 6-21, 8-9).
If an S/N ratio is 36 dB, the peak-to-peak signal voltage is 63 times the rms noise level, and the rms noise amplitude is 1.6% of the peak-to-peak signal amplitude (Fig. 6-23).
If the ratio of voltages in wave A to wave B is 26 dB, the amplitude of A is 20 times that of B. If the ratio of power is 26 dB, A contains 400 times more power than B.

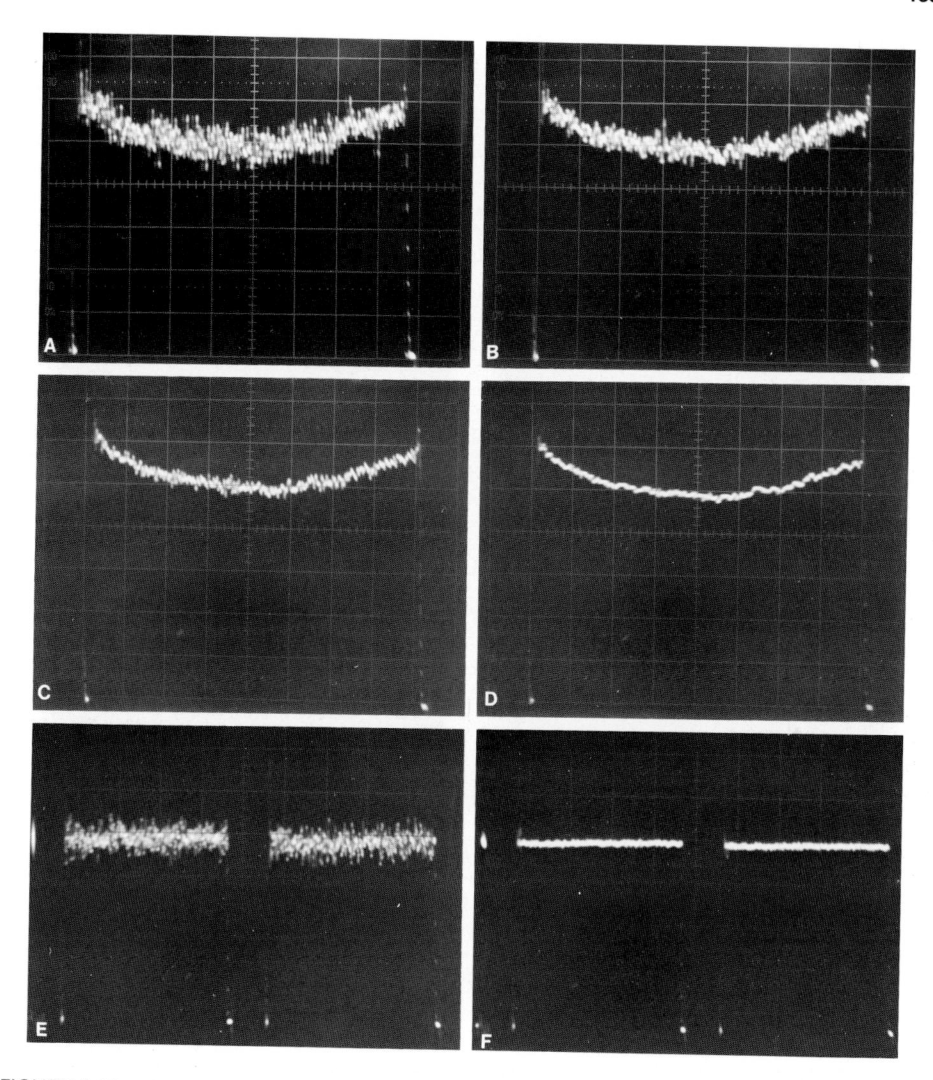

FIGURE 6-23. Cathode-ray oscilloscope traces (H-line scans). A noisy 1-H-scan video signal (A) from a SIT camera (of the subject in Fig. 6-24) has been averaged (with the Image-I processor) for 2 frames (B), 8 frames (C), and 256 frames (D). E and F are 2-H-scan traces with 4-frame-averaged background subtracted from 4-frame running-averaged live video (E), and 256-frame summed background subtracted from 256-frame summed incoming signal (F). The S/N ratio improves from 10 dB (A), to 20 dB (B), 26 dB (C), and 36 dB (D); and from 13 dB (E) to 40 dB (F). In traces D and F, the signal can actually be interpreted (compare with Fig. 6-24D). A through D: with shading; E and F: shading removed by background subtraction.

$1/\sqrt{256} = 6\%$. The precision can be improved by spatial, temporal, and other integrating and filtering techniques (Chapter 10, Appendix IV).

For example, if the measurement of intensity need not be made for every pixel, but could be averaged (with appropriate weighting factors) over several pixels, say in the 5×5 pixel area that surrounds each pixel, the precision of measurement would be improved by a factor of $\sqrt{25} = 5$. Likewise, scanning with a slit, instead of a small rectangular area, can provide better signal averaging when the image is scanned perpendicular to the direction of extension of the particular image feature of interest. In this case, we have sacrificed resolution along one axis, but gained it in another direction.

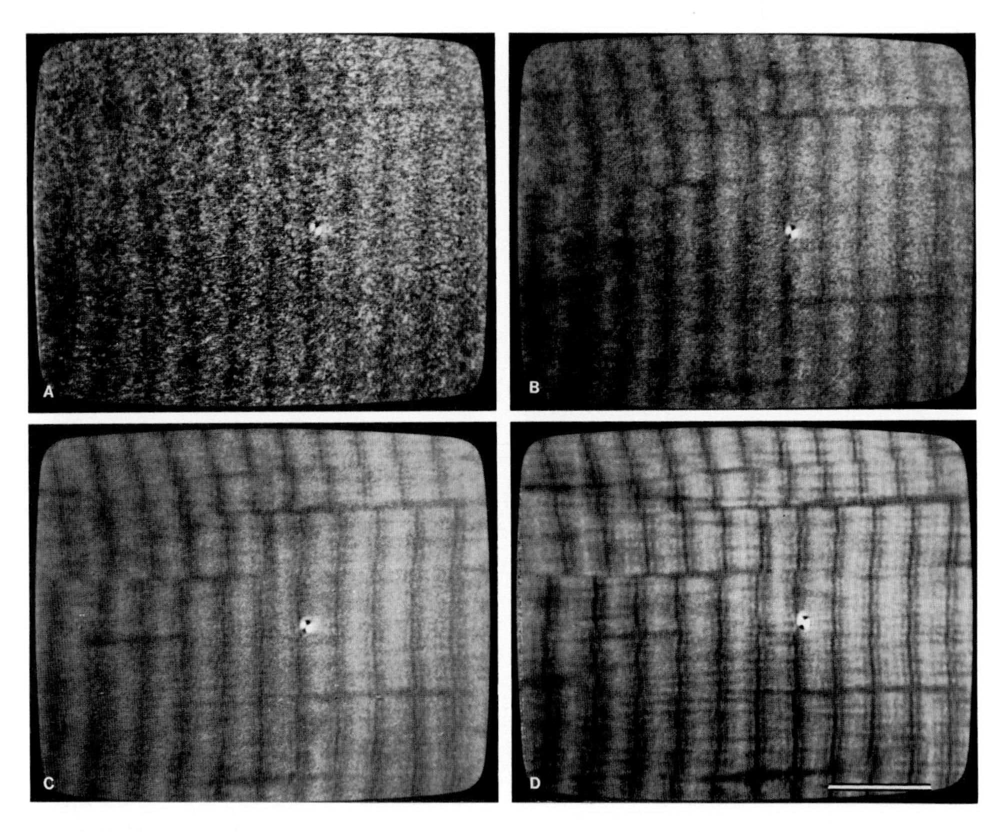

FIGURE 6-24. Monitor picture of a 360-nm-thick muscle section. Polarized-light images of this critical test specimen (Fig. 11-7) were captured at 6-nm compensation with a SIT camera (Dage–MTI Model 66). (A) 256-frame averaged background subtracted from live video (S/N = 10 dB); (B) 256-frame averaged background subtracted from 2-frame running-averaged live image (S/N = 14 dB); (C) 256-frame averaged background subtracted from 16-frame running-averaged living image (S/N = 30 dB); (D) 256-frame averaged background subtracted from 256-frame summed in-focus image (S/N = 40 dB). Cathode-ray oscilloscope traces in Fig. 6-23 correspond more or less to the video signals for these pictures. Digital processing by Image-I system. Scale = 5 μm.

The mathematical foundations and methods for extracting and enhancing selected attributes of the video signal have become a major science and art in electronic signal processing. Several texts on this topic have appeared in the last decade (e.g., Baxes, 1984; Castleman, 1979; Gonzalez and Wintz, 1977; Goodman, 1968; Oppenheim and Schafer, 1975; Pratt, 1978; see Chapter 10).

We will now leave the general discussion of video signal and noise, and survey how to hook up an array of video equipment so that the signals are compatible and synchronized.

6.6. SYNCHRONIZING THE VIDEO EQUIPMENT

Earlier in this chapter, we examined the nature of the picture signal and timing pulses that together make up the video signal, and noted the conventions used to standardize these signals. In Chapter 3, we surveyed the basic hookup and mechanics of operating the video equipment without exploring the detailed reasons for the suggested procedures. We shall now take a more detailed look at the rationale for the hookup, and the means for hooking up an array of video equipment.

Compatibilities of Scan Rates and Sync Pulses

As noted in Section 6.3, much common video equipment is broadcast compatible; (in the United States and Japan) with the 525/60 2 : 1 interlace NTSC color signal. However, compatibility does not imply identity either of scan rates or of the sync pulse formats. Nor does compatibility of equipment A and B with the NTSC signal necessarily mean that the signals from A to B, or from B to A, will always work properly. It may work in one direction but not the other way. Also, as we have noted, CCTV may use various scan rates and sync pulses that do not conform to the broadcast standard. Again, the use of nonstandard formats does not automatically imply incompatibility, as discussed next.

In simpler CCTV systems, e.g., with a single camera, VTR, and monitor(s), all with a common scan rate of 525/60, the sync pulses carried by the composite video signal can be expected to synchronize the video equipment, whether they were designed for monochrome or color.

In monochrome video, whether the camera provides a 2 : 1 or random interlace, a standard monitor can generally sync to the signal without trouble. In other words, a regular monochrome monitor can respond equally well to timing signals of the broadcast standard RS-170 or the various forms of the RS-330 types used only in CCTV. The only adjustments needed are usually the H and V HOLD controls on the monitor, so long as the incoming signal is properly terminated.*

With the low termination (or load) impedance of 75 ohms, the chance of picking up electrical noise is reasonably low. The coaxial cable itself also has a short time constant that prevents the degradation of high-frequency video signals for quite a long distance in the cable.†

Proper termination of the cable with the 75-ohm impedance provides a 1-volt peak-to-peak voltage of which the maximum picture signal takes up 0.7 volt, and the sync pulse takes up 0.3 volt. With unterminated inputs, the signal would have about twice this voltage, give rise to excessive signals, and distort the highlights in the picture. The signal can also reflect back from an unterminated end of the cable and introduce an interference pattern or **ghosts.**

NTSC color signals can also be carried along single coaxial cables with 75-ohm termination. The signal can synchronize 525/60 scan rate color (or monochrome) recorders and monitors. Naturally, color would not appear without the use of recorders and monitors designed with color capability.

Many recorders and processors, which can handle both color and monochrome signals, automatically check the incoming signal and pass it through optimized alternate circuits depending on whether the signal is for color or not.

Some equipment can be switch-selected for use in either the 525/60 or 625/50 scan mode, but commonly such selection for different *vertical* scan rates is not possible.

For *horizontal* scan rates that are not standard, monitors and VTRs can often lock on to the incoming signal if the departure of the H-scan rate is minor, say 250/60 to 280/60 1 : 1 field scan, or 512/60 2 : 1 interlace scan instead of the standard 525/60 scan rate.

With major differences in H-scan rates, one has to use special monitors with matching scan rates (some of which are switch-selectable for multiple scan rates). With monitors of higher scan rate, the bandwidth is also increased to handle the higher frequency needed to provide a proportionally higher horizontal resolution.

Sync Stripper

Within a monitor, VTR, or video processor, the composite video signal is separated into the sync pulses and the picture signal. The sync pulses and picture signal are amplified and modulated independently, and their waveforms are shaped separately through **processing amplifiers.** In these devices, an electronic circuit *strips* the sync pulses from the picture signal, which occupies a different voltage level.

A **sync stripper** can also be a separate piece of equipment that is used to separate out the H-

*With black and white or monochrome monitors, the termination is not so critical for obtaining sync, but with **color monitors** the wrong termination may prevent syncing altogether.

†However, a phase delay of 150 nsec or 0.002 H is induced by every 100-foot length of a 75-ohm coaxial cable (since the electromagnetic wave is traveling at two-thirds of its speed in vacuum). If, for example, more than one camera is used, and the cable on one camera is very much longer than the others, a *video delay circuit* may have to be inserted in the circuits of the other cameras to synchronize the output of all of the cameras.

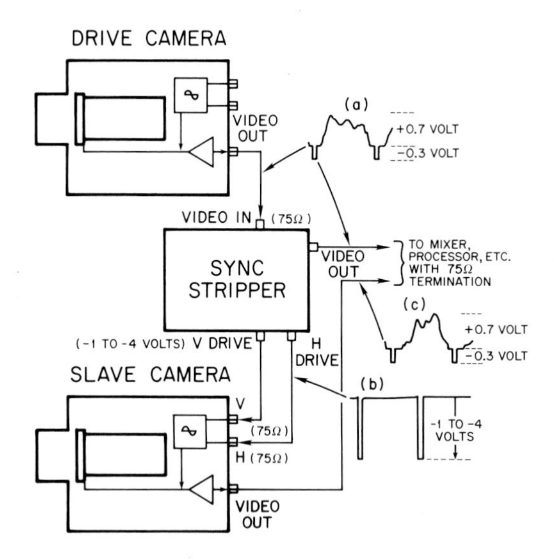

FIGURE 6-25. Use of a sync stripper for synchronizing cameras. A sync stripper can extract H- and V-timing pulses from the composite video output from one camera and use them to drive a second camera in synchrony. See text.

and V-sync pulses and which generates the H- and V-drives (Fig. 6-25). A *video mixer,* or *special effects generator,* usually contains a sync stripper and is equipped with terminals that provide the desired output pulses (Section 12.4). The isolated timing pulses are commonly negative-going, as are the sync pulses in the composite video signal. However, they have the higher amplitude of 1–4 volts, rather than 0.3 volt, as in the 1-volt peak-to-peak composite video signal (Fig. 6-25).

Sync pulses isolated by a *sync stripper,* or generated by a separate **sync source** such as a *master oscillator,* are used to drive several pieces of video equipment in syncrhony. For example, if we wish to take the output from two or more cameras and mix, process, record, or display them together, the two camera outputs must be scanning in sync. Otherwise, the uncoordinated sync pulses arising out of the two cameras would prevent a coherent video image or signal from forming. To synchronize the two cameras via a sync stripper, the H- and V-drive pulses derived from one of the two cameras (equipped with a high-quality oscillator) is made to drive the second camera. As shown in Fig. 6-25, the output (a) of the ''drive'' camera passes through the sync stripper. The timing pulses (b) from the sync stripper drive the H and V sweeps in the ''**slave**'' camera. The composite video signal (c) from the slave camera is now synchronous with (a), so that the two signals can be superimposed, share a field, be subtracted from each other, or presented in sequence without a **glitch**, etc., through a video mixer or special effects generator.

The oscillator in the drive camera is generally a *quartz crystal* (**xtal**) *oscillator* that provides a constant-frequency output. The oscillator output of the slave camera is synchronized by the drive signal through a *phase-locked loop* (PLL) *amplifier* that is present in every camera. The PLL amplifier is a device for locking the frequency and phase of an AC oscillation to an external AC signal.

Instead of using the H- and V-drive pulses generated by a sync stripper or a master oscillator, one can also use the composite video signal itself to drive some cameras. A camera equipped with a GENLOCK input derives the timing pulses internally by stripping a composite video signal (Fig. 6-26).

Synchronizing an Array of Video Equipment

The same principle used for synchronizing two cameras can be used to synchronize an array of video equipment.

The need for a common sync for an array of video equipment (e.g., two or more cameras) becomes evident if one reflects on the way the video image is formed. All of the video equipment receiving the video signal must scan the horizontal and vertical sweeps precisely in synchrony with the camera (or recorder) that generates the signal. In general, the more complex the array of video

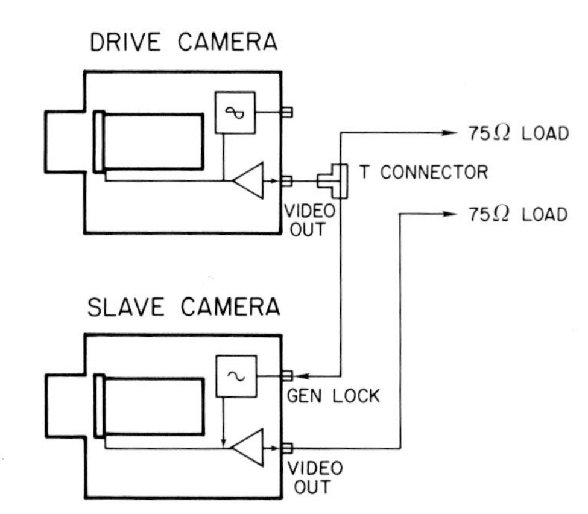

FIGURE 6-26. Use of genlock for synchronizing cameras. Multiple cameras can be driven in synchrony (without a separate sync stripper) by a built-in genlock circuit, which derives timing pulses from the composite video signal. See text.

equipment, the greater is the need to use specific common timing pulses generated by a single, reliable master oscillator, or sync generator. All of the component equipment must be driven synchronously.

Also, the need for accurate phase relationships between the H and V syncs tends to rise with the quality of the video equipment receiving the signal. In order to form proper images in many high-resolution monitors [which are equipped with automatic horizontal frequency control (AHFC) circuits with long time constants], the H and V syncs have to be more accurately in phase than in standard monitors with AHFCs that have shorter time constants. Also, more sophisticated VTRs and digital image processors tend to require more accurate H and V synchronization than do more tolerant (and generally less expensive) video equipment. In less-demanding equipment, the composite video signal itself generally provides adequate timing signals.

While cameras, processors, monitors, and VTRs during recording can usually be driven by external sync pulses, VTRs in playback generally cannot. During recording, VTRs accept the composite video signal and lay down synchronized magnetic records (Chapter 8). But once the recording is made, most VTRs can only use the recorded tape as the source of video (picture and sync) signals and cannot be synchronized or genlocked during playback by an outside timing signal.*

In **helical-scan** VTRs, the composite video signal is stripped and the V-timing pulses (control pulses) are laid down linearly along one edge of the magnetic tape. These control pulses are commonly placed several centimeters from the helical tracks that record the sequential fields (Chapter 8). Thus, in playback the composite video must be resynthesized by adding back, to each field scan, the V sync from parts of the tape that are physically separated from the helical scan that recorded the picture signal and H-sync pulses.

This mode of recording and playback imposes major demands on the mechanical parts and servo drive systems of the VTR, as well as on the dimensional stability of the magnetic tape itself. While the composite video arising out of VTR recorded and played back at standard speed is generally well synchronized, the timing of the H- and V-sync pulses can become discordant when we play back *time-lapse* recorded or copied tapes.

The pulse-cross pattern of the VTR output in Fig. 6-27A shows the discordance of the time base of the H and V syncs. Remarkably, even a composite video signal as unsynchronized as this can generate a regular H scan on the monitor with a short AHFC by the time the V scan descends to the level where the picture appears on the screen.

As described in Chapters 7 and 8, the speed of recovery from an out-of-phase H and V sync depends on the monitor circuitry, so that flagging or tearing of the upper region of the monitor image may or may not extend into the visible picture area. With a short AHFC, the sweep is brought back into synchrony by the PLL amplifier in the monitor before the V scan emerges from the overscanned region of the monitor.

*Most 1-inch studio-quality VTRs and some ¾- and ½-inch VTRs with editing or animation capabilities can, however, be synchronized during playback to an external reference timing signal.

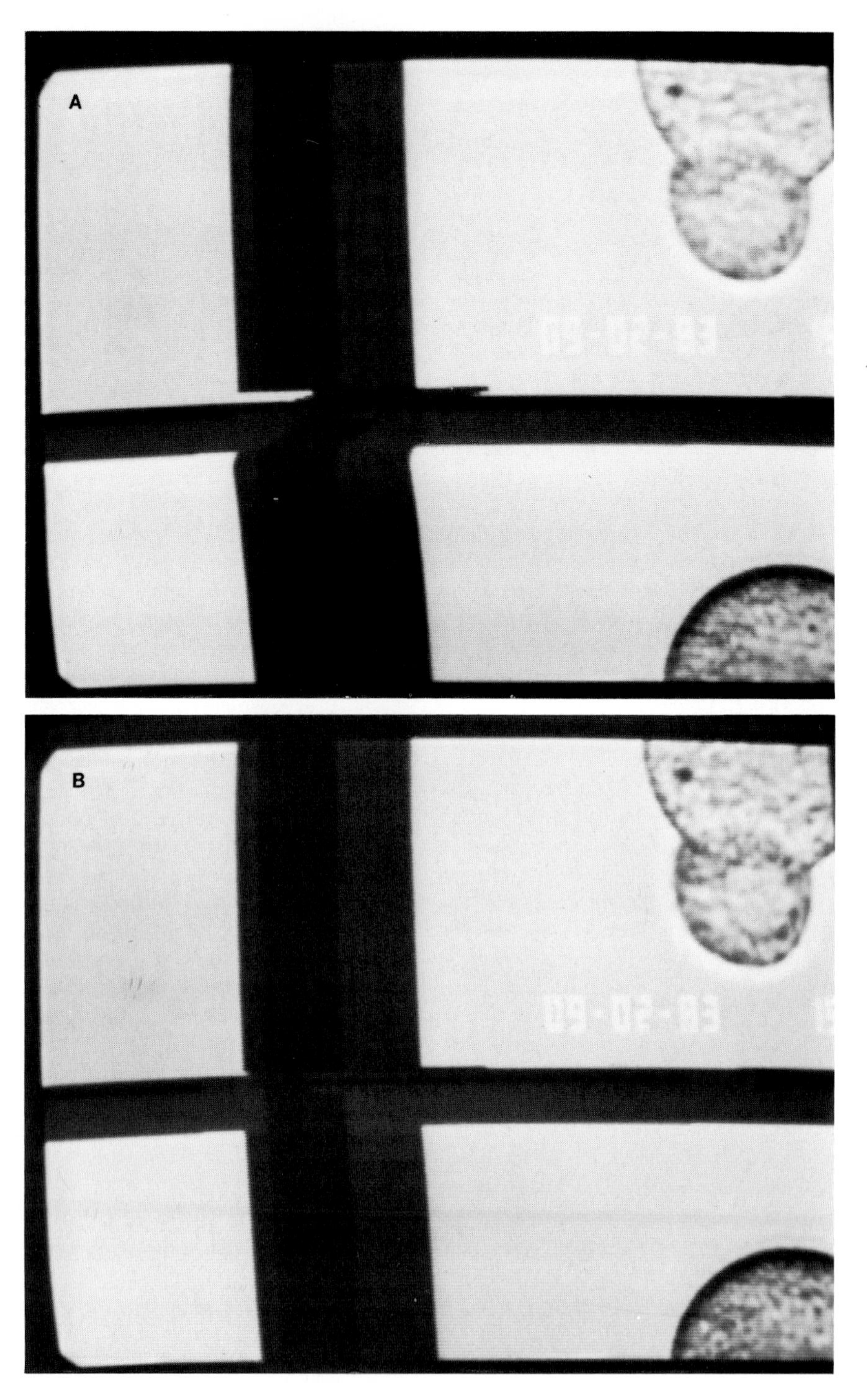

FIGURE 6-27. Pulse-cross pattern with and without a time-base corrector. (A) Pulse-cross pattern of the signal played back from a VTR shows the discordance of synchrony between the H and V scans. The timing error of the H-sync pulses at the beginning of the picture scan (just after the V-sync pulses) is seen as a distortion of the dark vertical stripe. (B) The same signal after passage through a time-base corrector, which corrects the time-base error of the scans. (See Figs. 6-13 and 6-14 for details of the pulse-cross pattern.)

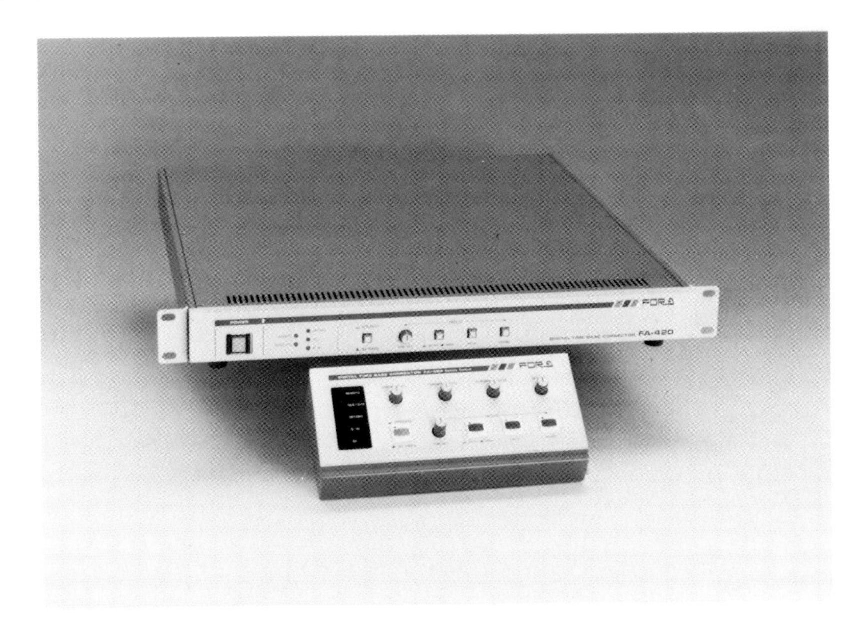

FIGURE 6-28. A modern commercial time-base corrector, a device for improving the precision of synchronization of H and V scans. High precision of synchrony between the H and V scans is required for applications involving digital storage or processing of the video signal. While still costing several thousand dollars, units such as this one, which are capable of correcting the time base for the entire frame (rather than for a few lines at a time), are much less expensive and considerably more compact than their predecessors. (Courtesy of For-A Corporation of America.)

When the timing of the H and V scans must be maintained with very high precision (such as for signals to be stored or processed digitally), their relationship *may* be improved by the use of a **time-base corrector** (Fig. 6-28). The time-base corrector has a (digital) memory that temporarily stores one to several lines of sync-stripped H-scan periods, and reapplies the corrected timing (H-sync) pulses. Figure 6-27B shows the pulse-cross pattern of the signal in Fig. 6-27A after passing a "2 H" time-base corrector.*

As seen in this example, a pulse-cross monitor provides a quick check of the state of the timing pulses in the composite video signal. Since monitors with pulse-cross capabilities can be obtained relatively inexpensively (for a few hundred dollars), it is advisable to use such a monitor routinely to check the quality of the video sync pulses.

Ultimately, however, the quality of synchronization that can be achieved among pieces of video equipment, especially a large array of equipment, has to be tested by actual hookup and operation. Several formats of synchronizing pulses are used for CCTV, and the H- and V-scan rates may deviate by different amounts from the standard.†

The degree of tolerance to deviation from expected sync pulse formats and H and V timing varies greatly with the type of video equipment, the tolerance generally being greater for the less expensive, consumer- or industrial-oriented equipment. Careful study of specifications and on-site testing are needed to establish the compatibility of sophisticated video equipment such as digital image processors and high quality VTRs with other video devices or signals from recorded tape.

*However, the time-base corrector is not a cure-all for video timing problems. In fact, for the output of time-lapse recorded video, the monitor picture stability may become considerably worse when a time-base corrector (TBC) is used; except for the TBC shown in Fig. 6-28, which appears to provide a happy exception.

†Strictly speaking, the H-scan rate for the 525/60 2:1 interlace monochrome standard is 15,750 Hz, while for color it is 15,734.26 Hz. The V-scan rates are 60.00 and 59.94 Hz, respectively (Ennes, 1979, Chapter 2). In practice, the H-oscillator frequency may deviate from these standards by one to several hertz, depending on the equipment.

7

Video Imaging Devices, Cameras, and Monitors

In this chapter we will examine some basic characteristics of video imaging devices, cameras, and monitors. The video imaging devices, or *sensors,* include the vidicons and related *camera tubes,* the highly sensitive *image-intensifier tubes,* and the compact *solid-state pickup devices.* The imaging devices are usually packaged as an integral part of the video camera. In some instrumentation cameras, one can choose from among several tube types.

In addition to examining the basic characteristics of video sensors and cameras, we will compare their relative merits for various modes of video microscopy. The characteristics of monitors will be considered somewhat more generally.

7.1. VIDEO IMAGING DEVICES

The video imaging device lies at the front end of the video camera. The device converts the brightness of each unit image area in the focused optical image into a corresponding electrical signal. Thus, the imaging device lies at the interface between the incoming optical image and the electrical video signal.

The balance of the video camera, as well as other components such as the processor, recorder, and monitor, also affect the quality of the video picture. However, since the focused optical image must first be captured by the imaging device before a video picture can be produced at all, the characteristics of the video sensor ultimately limit the sensitivity, wavelength dependence, and resolution of the whole video system.

Video Camera Tubes

Currently, the most commonly used video imaging devices are those of the *vidicon family of camera tubes.* The vidicon tubes are used in broadcast-style **three-tube color cameras** as well as in monochrome CCTV.

The vicidons are encased in a glass envelope, commonly some ⅔ to 1 inch in diameter, and 4 to 7 inches long. The interior of the tube is under high vacuum, and electrical connections except the signal electrode connection are made through multiple pins that pass through the glass envelope. The pins (including a short index pin) are arranged in a circle to fit a socket at the back end of the tube (Fig. 7-1).

At the front end of the vidicon tube, the optical image is focused onto the *target* through the *faceplate.* The faceplate is usually made of a flat plate of glass, but for extreme short- and long-wavelength applications, special tubes with faceplates made of quartz and other material are also available.

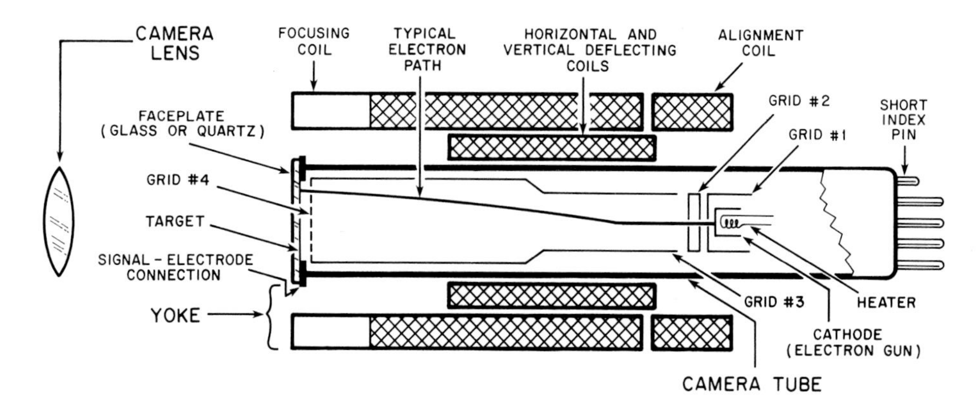

FIGURE 7-1. Diagram of a typical vidicon camera tube.

The target, which is deposited on the inner face of the faceplate, consists of three layers. The transparent *backplate electrode,* also called the *conductive layer,* or *target electrode,* is deposited directly onto the faceplate. The *photoconductive layer* is deposited on an intervening *dielectric layer* on top of the target electrode, and faces the *electron gun* located toward the back of the camera tube (Fig. 7-2).

The electron gun generates the electrons that are accelerated by a positive voltage toward the target. The trajectories of the electrons in the high vacuum are governed by the *focusing coil* (in the case of **magnetic focusing** tubes) and *electrodes* into a narrow pencil, or the *electron beam.* The electron beam is focused and forms a small spot, or the *aperture,* on the target. The aperture defines the size of the picture element.

The photoconductive layer in the target is the primary transducer in the vidicon family of camera tubes, and works in the following way.

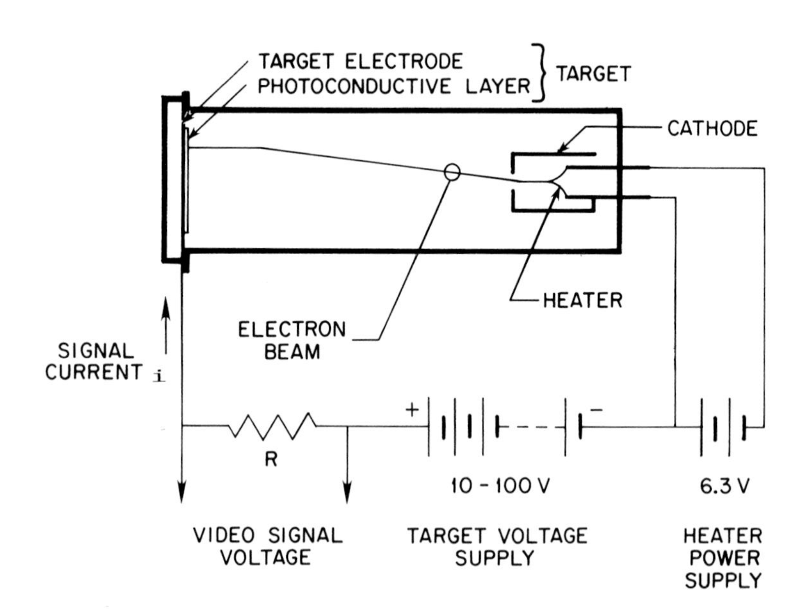

FIGURE 7-2. Schematic diagram of vidicon camera tube. The photoconductive layer on the target electrode is a transducer whose resistance is lowered as a function of the intensity of illumination falling on it. As the electron beam scans the target, a current flow is produced. The size of the current depends on the amount by which the resistance of the photoconductive layer has been lowered. The current flow is converted to the video output signal.

The electron beam sweeps the target and charges up its back surface with electrons. In the dark, the photoconductive layer is essentially an insulator, so that the electric charge remains on the surface of the target. When light strikes the photoconductive layer, the resistance of the layer drops. Electron-hole pairs are formed and stored in the region of the photoconductive layer illuminated. Within the *dynamic range* of most (but not all) types of vidicon tubes, the drop in resistance of the photoconductive layer is nearly proportional to the intensity of illumination.

The target electrode is set at a potential some 10 to 100 volts positive relative to the electron gun, or the *cathode*. When the electron beam lands on a particular target area that has been illuminated, the lowered resistance of the photoconductive layer transiently allows a current to flow (usually of a few to a few hundred nanoamperes). As shown in Fig. 7-2, the current (i) flows from an external positive voltage supply through a resistor (R), the target electrode, photoconductive layer, electron beam, cathode, and then back to the negative terminal of the target voltage supply. The i·R drop across the resistor gives rise to the *video signal voltage*, or the *output signal.*

In the camera tube, the electron beam is scanned electromagnetically or electrostatically by the *deflection coils* or *electrodes*. The electrodes are built into the evacuated camera tube, while the magnetic deflection coils are made as a separate unit known as the **yoke.** The tubular yoke slips over the middle of the camera tube (Fig. 7-1).

The deflectors constantly deflect the electron beam, which scans* the target in a raster (Section 6.2). In the 525/60 scan mode, the raster "lands" on a particular point on the target for a brief instant once every 1/30 sec. During most of the 1/30 sec that the beam does not land on a particular point on the target, the light hitting that point continues to produce electron-hole pairs. In other words, *the target is a storage device or integrator of light* (see **target integration**). The accumulated electron-hole pairs decrease the resistance of that region of the photoconductive layer.

Once every 1/30 sec, for the very brief moment that the electron beam does land on the particular target region, the signal current for that picture element flows. The current is inversely proportional to the resistance of the region, or proportional to the number of accumulated electron-hole pairs. Thus, within the dynamic range of the camera tube, the current for each picture element flows approximately proportionately to the brightness of the particular image point.

To recapitulate: in vidicon camera tubes, (1) the focused optical signal is stored in the target as electron-hole pairs, which lower the resistance for the corresponding picture elements in the target; (2) the sweeping raster generates a current through the successive picture elements in the target and the external resistor, giving rise to the picture signal; and (3) the electron beam recharges the target at the same time.

Many types of vidicon tubes are available today, the primary difference among the tubes being the composition of the target material (Table 7-1). Depending on the composition and character of the photoconductive layer, the vidicon tubes vary in **sensitivities,** respond differently to light of various wavelengths, and display different noise characteristics and resolving power. They also vary with respect to **geometrical distortions** and blemishes, as well as their **burn** and blooming characteristics when exposed to bright light. **Persistence** or **lag,** which shows up as a comet tail when a spot of light moves across the field, also varies with the target, or type of vidicon tube.

On the whole, the vidicon tubes are compact, reasonably rugged, and moderately priced (a few hundred to a few thousand dollars). They generally provide good image quality and moderately high horizontal resolution† (limiting resolution of 600 to 1000 TV lines PPH, depending on tube size and type). Several types of vidicons are also sufficiently sensitive to make them prime candidates for various forms of video microscopy as discussed in Sections 7.2 and 7.3.

However, for microscopy involving very low levels of image brightness, such as in darkfield, fluorescence, and high-extinction polarized-light microscopy of weakly retarding objects, the vidicons may not be sensitive enough. Image intensifiers or intensified camera tubes are then used instead.

*Should the deflectors fail, and the beam stays focused on a single spot on the target, the target is ruined almost instantaneously. Special circuits are built into the video camera to prevent such an accident.

†The resolution is a function of image brightness and contrast. The MTF or CTF curves (Section 7.2) express the relation between resolution and contrast.

TABLE 7-1. Characteristics of Some Standard 1-Inch Vidicon Camera Tubes[a]

	Chalnicon (Toshiba)	Newvicon (Panasonic)	Plumbicon (Phillips, Amperex), Vistacon (RCA)	Saticon (Hitachi)	Silicon (Hamamatsu)	Ultricon (RCA)	Vidicon, Sulfide vidicon (RCA)
Target material	CdSe	Cd·ZnTe	PbO	Se·As·Te	Silicon diode array	Silicon diode array	Sb_2S_3
Picture area (mm)	10 × 10	12.8 × 9.6					
Diagonal	14.1	16.0	16.0	16.0	16.0	16.0	16.0
Sensitivity (nA/lx)[b]	300[c]	450[c] 500	40[c]	30[c]	600[c]		20[c] (variable)
Responsivity (μA/lm per ft^2 at 2854°K)	2600		330/125[d]			5500/1030	250
Spectral response (nm)							
Min	400	400	400	400	400	300	400
Peak	700	750	500/480[d]	560	680	680/560[e]	550
Max	750	850	950/650[d]	700	1050	1100/ 750[e]	750
Dark current (nA)	0.8	1[c]	2	0.3 [f]	8	8.5	20
Gamma	0.95	~1.0	0.95	0.95[f]	1.0	1.0	0.65[g]
Limiting resolution	750/1000	800	1000	750[c]	550	700	1000/1100
Amplitude response (% at 400 TV lines)	45/55	45	60	45[f]	20	35~45	50/60
Third-field lag (%)	20	20	8/3[h]	4[f]	12	8	20
Shading (%)	10	10[c]	10[c]	10[c]	10		15

[a]From manufacturer's specification.
[b]1 lx = 0.0929 **fc** at 2856°K.
[c]Hamamatsu data sheet.
[d]With VG9 (1 mm) filter.
[e]With KG3 (5.5 mm) filter.
[f]RCA data.
[g]Low gamma, but dynamic range of 350 is 3–4 times wider than other vidicon tubes.
[h]Light biased for 8-nA peak current.

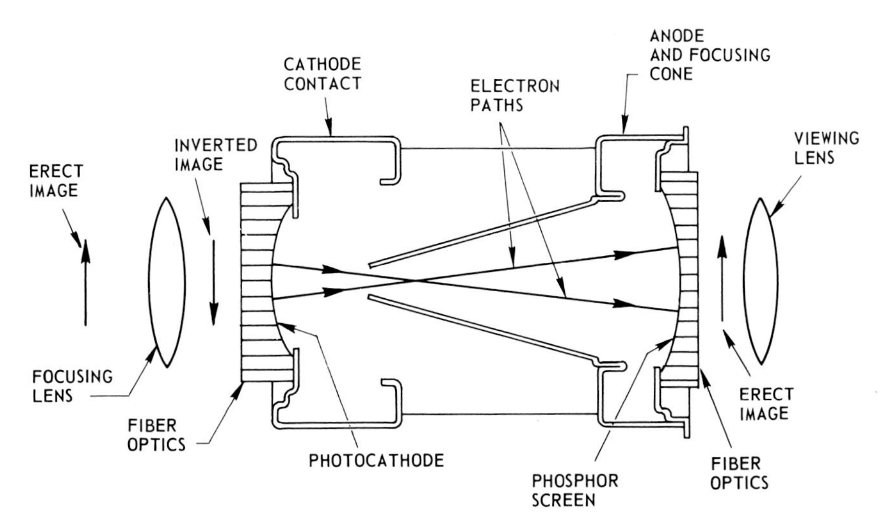

FIGURE 7-3. Schematic diagram of an electrostatic-type image intensifier tube (Gen I). (Courtesy of RCA, from RCA *Electro-Optics Handbook,* 1974, p. 174.)

Intensifier Tubes

At low levels of illumination, the signal current from a vidicon tube drops below the noise current of the tube or preamplifier, and the picture signal becomes buried in noise. The video picture can, however, be made visible by intensifying the dim image before it enters the target of the vidicon tube. An intensifying image tube, or *image intensifier,* can be inserted before the camera tube. Alternatively, an intensifier section can be incorporated in the same vacuum with the vidicon to make up an *intensifier camera tube.*

In an image intensifier, the photons, which make up the dim incoming optical image, eject photoelectrons from the photocathode (Fig. 7-3). The photoelectrons, released from the photocathode into the vacuum of the intensifier, are accelerated by an externally supplied high voltage. The accelerated electrons are imaged onto a phosphor by electrostatic or electromagnetic lenses (Fig. 7-3). In the phosphor, each impinging, energetic electron produces many photons, so that the image on the phosphor can be made 50–100 times brighter than the image that struck the photocathode. By cascading intensifiers, the brightness of an image can be multiplied 10^5 to 10^6 times (Fig. 7-4). These image intensifiers are known as **Gen I** or first-generation **intensifiers.**

The so-called **Gen II intensifiers** multiply their image signal by cascading secondary electrons much as in a photomultiplier. In a photomultiplier, each photoelectron, ejected from the photocathode by the incoming photon and accelerated toward a **dynode,** kicks out a few secondary electrons at the dynode. Each secondary electron in turn is accelerated to another dynode, where again three or four electrons are ejected for each electron entering the dynode at the proper angle. Thus, after ten or so stages of dynodes, the photoelectron is multiplied 10^4 to 10^5 times.

In the Gen II intensifier, a **microchannel plate** multiplies the electrons (Fig. 7-5). The microchannel plate is a thin wafer sliced across a parallel array of hollow glass cylinders. The inside of each cylinder is coated with material that emits secondary electrons. In vacuum, the accelerated electrons bounce along inside the cylinder, so that the cylinder acts similarly to the multiplier section of a photomultiplier tube. At the same time, the geometry of the electron image through the microchannel plate is maintained just as an optical image is maintained through a fiber-optic bundle.

In the multiple-stage Gen I intensifier, and especially in the Gen II intensifier, the amount of electron multiplication per photoelectron can fluctuate considerably. Nevertheless, the intensifier can raise the image brightness far above the noise level of the vidicon tube, so that an intensifier placed ahead of a regular vidicon tube can substantially improve its ability to detect and resolve features in a very dark image.

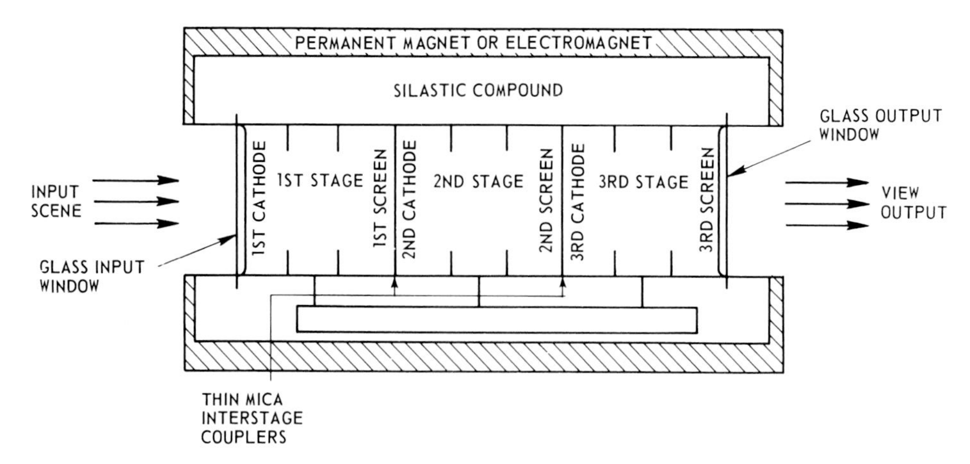

FIGURE 7-4. Schematic diagram of a typical three-stage magnetically focused image intensifier tube (Gen I). (Courtesy of RCA, from RCA *Electro-Optics Handbook,* 1974, p. 175.)

In an intensifier, the spontaneous emission of electrons from the photocathode can be quite low, so that the composition of the image can be limited primarily by photon statistics of the incoming low-light-level image [e.g., see Reynolds (1972) and Fig. 7-12].

Image intensifiers can be coupled optically with vidicon tubes through **fiber-optic** plates or via lens systems and placed in **intensifier vidicon** cameras. Alternatively, some video camera tubes are made with the intensifier built right into the tube. The *Silicon-Intensifier Target* (SIT) tube is a moderately inexpensive camera tube of this latter type.

In the SIT tube, a photocathode converts the incoming optical image into photoelectrons. The photoelectrons are accelerated by a high potential of several thousand volts and focused onto a silicon target (Fig. 7-6). The target is made up of an array of miniature **p-n junction** silicon **diodes.** Each accelerated electron impinging on the silicon target generates a very large number of electron-hole pairs. As the electron beam from the rear of the tube sweeps the target, the holes (collected on the p side of the diode) are neutralized and a signal current is generated as in a typical vidicon tube.

With this design, a SIT tube can give sensitivities that are several hundredfold greater than regular vidicon tubes, which do not have an intensifier section. The intensifier SIT (**ISIT**) is even more sensitive than the SIT. Both, however, tend to give noisy images as discussed later.

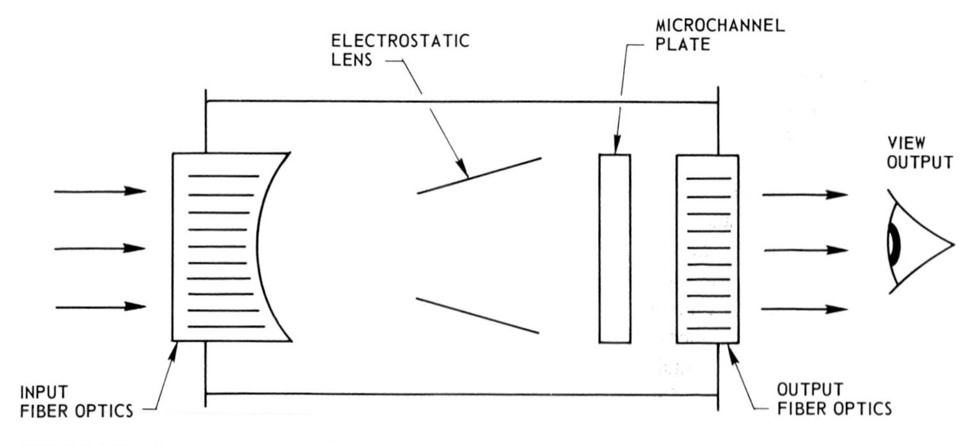

FIGURE 7-5. Schematic diagram of an electrostatic-focus-type image intensifier tube utilizing a microchannel plate (Gen II) to increase gain of the Gen I system. (Courtesy of RCA, from *RCA Electro-Optics Handbook,* 1974, p. 176.)

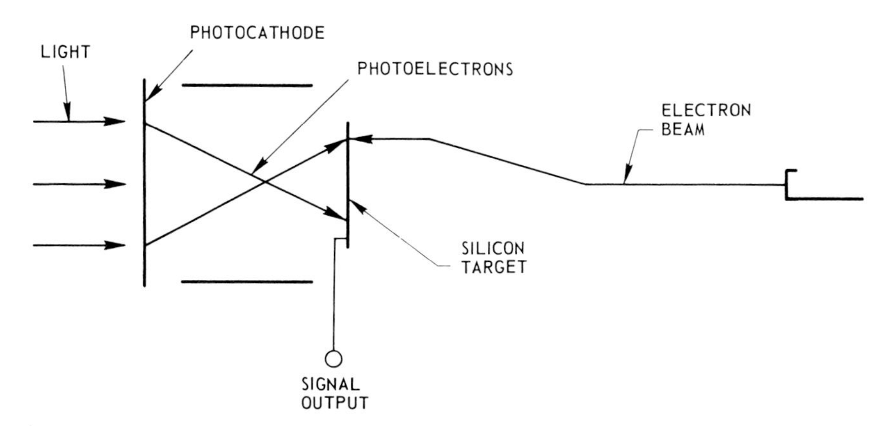

FIGURE 7-6. Schematic representation of SIT (silicon intensifier target) and **SEC** (secondary electron conduction) camera tube. (Courtesy of RCA, from *RCA Electro-Optics Handbook,* 1974, p. 183.)

The *image orthicon* and *image isocon* (Fig. 7-7) are intensifier video camera tubes with high gain and sensitivity.* They also incorporate an intensifier section ahead of the target, but the electron beam is scattered back from the target as the ''return beam.'' The return beam is further amplified within the camera tube through an electron multiplier that provides a large signal current (of up to several thousand nanoamperes).

The image isocon in particular gives a very high resolution (1100 TV lines), and having a low **dark current,** provides an exceptionally large intrascene dynamic range of 1000 : 1, compared to 100 : 1 or less for most vidicons. These tubes are used in broadcast, low-light-level color cameras, but require special care in their operation and are very expensive. Characteristics of representative intensifier camera tubes are compared with some vidicon tubes in Table 7-2.

Hamamatsu has recently introduced an intensifier camera for microscopy, consisting of a **photon counting** tube optically coupled through a relay lens to a low-lag Vidicon. The photon counting tube is an electrostatic-focus-type image intensifier that contains a two-stage microchannel plate. This camera has a very wide interscene dynamic range. The gain of the intensifier section can be varied continuously, allowing the camera to be used from the photon counting level to ordinary low light levels (10^0 to 10^9 photons/mm^2-sec).

Solid-State Sensors

Over the last 20 years, the vacuum electron tubes in oscillator, rectifier, and amplifier circuits have largely been replaced by solid-state semiconducting devices. The semiconducting circuit elements in turn are now built into *large-scale integrated circuits* on minute semiconductor chips.

Recently, some video imaging devices have also been fabricated on silicon and other wafers, or chips. These solid-state devices are very much more compact and consume miniscule power compared to their camera tube counterparts (Fig. 7-8). In addition, the solid-state detectors show low geometrical distortion since the array of light-sensing elements is permanently built into the wafer, and the location of the picture elements does not depend on scanning by an electron beam. Other advantages such as low lag, low burn, or ''sticking'' under bright illumination, immunity to mechanical vibration and shock, and long service life are touted. Some solid-state video sensors have also been shown to operate with very little noise, and a dynamic range of 50,000 : 1 above dark noise is considered possible [see Hall (1980) for an overview of solid-state pickup devices].

Applied to astronomy, a CCD (**charge-coupled device**) solid-state detector cooled to −100°C

*The image orthicon was often used to pick up the **luminance signal** in a four-tube color camera (Section 7.4). In modern cameras, the orthicon has been superseded by the more efficient and economical vidicon camera tubes.

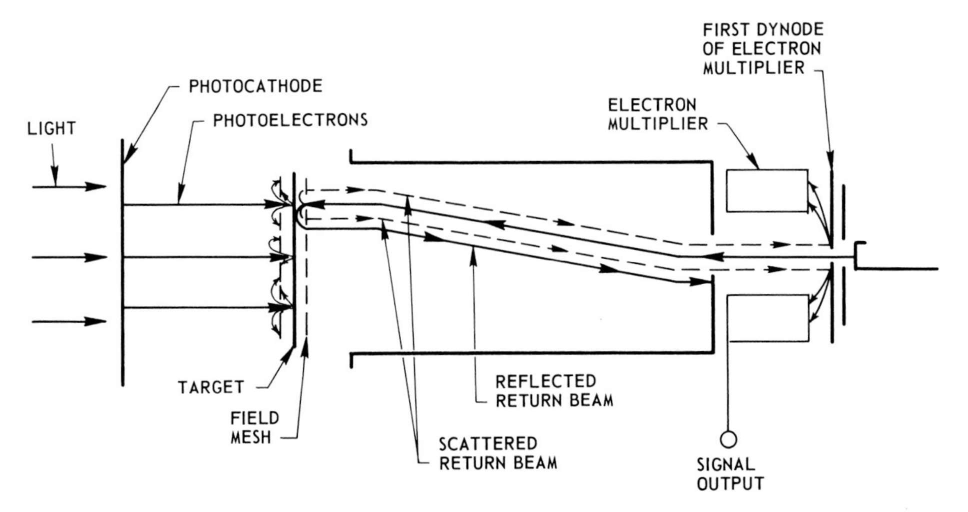

FIGURE 7-7. Schematic representation of image isocon, a return beam-type camera tube. (Courtesy of RCA, from *RCA Electro-Optics Handbook,* 1974, p. 183.)

or lower, was found to show exceptionally low noise (''the equivalent of 10 photons or less''), a dynamic range of about 5000 : 1, a quantum efficiency as high as 70%, and good linearity of response to light (Kristan and Blouke, 1982). Kristan and Blouke report being able to detect stellar objects of less ''than 25th magnitude: some 50 times fainter than the photographic limit'' with a Texas Instrument CCD (now made by Tektronix). Some of these **slow-scan** CCDs have a square, checkerboardlike array of sensors 800 pixels on a side. The sensor elements are just 15 μm on a side, and the whole **chip** occupies an area of only 17.8 by 17.8 mm.

CCDs respond to light by producing electron-hole pairs in the illuminated sensor elements. Each sensor element is a silicon photodiode built into the silicon chip, but isolated electrically from its neighbors. The photoelectrons produced in each photodiode immediately move into an adjoining, isolated potential well. The electrically positive potential well is formed in a metal oxide ''gate structure,'' or ''shift register,'' built into the CCD device in a layer adjoining the sensor elements.

The rows of electrons trapped in the checkerboard-arrayed gate structure are then all shifted by one row at the signal from a vertical shift register clock; the clock shifts the voltages on the alternate electrodes of the vertical gate and thereby moves the potential wells in the gate structure upwards by a notch. During this process of charge-coupled register shift, the top rows of electrons are shifted into an ''isolation transfer gate.''

In the isolation transfer gate, the packets of electrons representing each pixel in the row are sequentially shifted to the left. The electrons are shifted by charge coupling at the signal from a horizontal shift register clock. The horizontal shift is completed, and the transfer gate is emptied, before the vertical shift register clock moves the next row of trapped photoelectrons up into the isolation transfer gate.

The electrons shifted out of the isolation transfer gate enter an amplifier built into the CCD device. In the amplifier, the packets of electrons sequentially register the amount of photoelectrons produced by successive photodiodes, from left to right in one row, and then in the next row down. The output signal from the amplifier thus results in a raster scan of the photoelectrons that were generated in the two-dimensional array of sensors. (See the article by Kristan and Blouke for a pictorial explanation of CCD functions.)

Many solid-state video imaging devices related to the CCD, including **charge injection devices** (CIDs), **charging priming devices** (CPDs), and **metal oxide semiconductor** (MOS) pickup devices, have been developed since the CCD was invented in 1970. Compact monochrome and color TV cameras using these solid-state sensors are now appearing on the market with prices of about $1000 and less. While their resolution is somewhat limited by the limited number of sensor

TABLE 7-2. Important Performance Characteristics of Typical Camera Tubes and Intensifier Camera Tubes[a]

	Sb_2S_3 vidicon	PbO vidicon	CdSe vidicon	Silicon vidicon	SIT camera tube	SEC camera tube	Image orthicon	Isocon
Image diagonal (mm)	16	21	11	16	25	25	45	35
Typical faceplate illuminance (lx)[b]	20	3	1	0.5	2×10^{-3}	0.035	0.2	0.02
Responsivity (μA/1m)[c]	250	350	2750	4350	2.9×10^5	1.35×10^4	4×10^4	9.2×10^5
Typical signal level (nA)	300	300	200	300	300	150	8000	4000
Limiting center resolution, TV lines PPH (aspect ratio 4:3)	1100	600	700	700	750	600	750	1100
Lag, third field (%)[d]	18	3	12	8	10	3	3	3
Output dark current (nA)	20	0.5	1	8.5	10	0.001	—[e]	—[f]
Intrascene dynamic range[g]	350:1	60:1	60:1	50:1	50:1	50:1	100:1[h]	1000:1[h]
Photoconductor material, or spectral response	Sb_2S_3	PbO	CdSe	Si	S-20	S-20	S-10	S-20

Tube type	SIT	Isocon	SEC	ISIT	I-Isocon	I-SEC
			General Data			
Photocathode diameter (mm)	40	35	40	40	35	40
Responsivity (A/lm)	0.4	0.9	0.015	12	27	0.45
Intrascene dynamic range[i]	50:1	1000:1	100:1	—	—	—
Usable light range of illuminance[j]	10,000:1	2000:1	200:1	10^5:1	20,000:1	2000:1
Exposure damage limit (lx-sec)	10^5	10^6	10^2	—	—	—
			Typical performance at normal highlight			
Output current (nA)	500	3000	150	500	3000	150
Faceplate illuminance to obtain typical output current (lx)	2×10^{-3}	1.5×10^{-2}	3×10^{-2}	7×10^{-5}	5×10^{-4}	1×10^{-3}
Lag after 50 msec (%)	10	4	2	10	4	2
Limiting resolution, 100% contrast (TV lines)	1100	1100	600	1000	1000	500
400-line amplitude response (%)	56	75	20	41	56	13

(continued)

TABLE 7-2 (*Continued*)

Tube type	SIT	Isocon	SEC	ISIT	I-Isocon	I-SEC
	Performance at 10^{-4} lux faceplate illuminance					
Output (nA)	35	20	Not usable; signal is very low; picture would be very laggy and noisy	500	600	25
Limiting resolution (TV lines)						
100% contrast	700	250		900	900	300
30% contrast	600	175		550	600	250
Lag after 50 msec (%)	30	Not usable because of lag		10	13	18
S/N (4-MHz bandwidth)	3	2		3	3	3
	Performance at 10^{-5} lux faceplate illuminance					
Output current (nA)				100	60	
Limiting resolution (TV lines)						
100% contrast		Low signal	Very laggy	800	450	Not usable (low signal)
30% contrast				350	250	
Lag after 50 msec (%)		Very laggy		25	Not usable (very laggy)	Very laggy

[a]Courtesy of RCA, from *RCA Electro-Optics Handbook*, 1974, pp. 189, 207.
[b]1 lx = 0.0929 fc.
[c]Illumination from a tungsten lamp, 2856°K color temperature.
[d]Measured at the indicated typical signal level.
[e]Larger than the signal current and increases with excess beam current.
[f]Dark current is equivalent to 10% of the signal current. Dark current is uniform and independent of temperature.
[g]Illuminance range on the faceplate for a 50:1 (34 dB) signal range. Details may be observed in the total range.
[h]There is a useful compression of the signal above the knee for wide variations of scene luminance.
[i]Intrascene dynamic range is here defined as the range of illuminance in a single scene such that the output signal level has a range of 50:1.
[j]Usable light range is the total range of illluminance that can be accommodated by variation of gain within the tube as well as by the inherent intrascene dynamic range capability.

A

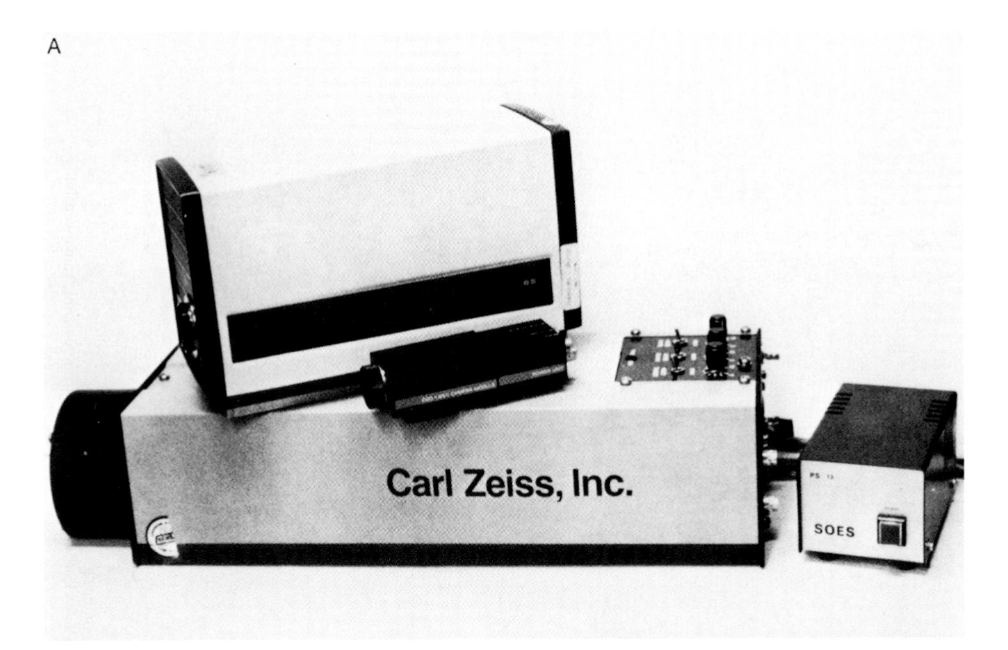

B

FIGURE 7-8. Photographs of tube and solid-state video cameras. Solid-state cameras are smaller, lighter, and require less power than the tube type. In A, a Pulnix–Sony CCD (charge-coupled device) camera, $5\frac{1}{4}$ inches long, with its separate power supply at the right, rests on a Zeiss–Venus three-stage intensifier tube camera, and in front of a Dage–MTI Model 65 series tube camera. (B) A hand-held General Electric CID (charge injection device) surveillance camera. (B: Courtesy of General Electric Co., from Publication LMC 7/82 15K.)

TABLE 7-3. Examples of RS-170 Compatible, Solid-State Pickup Device and Cameras[a]

		Cameras				
	Pickup device Charge-coupled device RCA SID 52501	Virtual-phase Charge-coupled device Texas Instruments	Interlying transfer charge-coupled device Sony XC-37[b]	Charge priming device Panasonic WD-CD100 color	Charge-coupled device Panasonic WV-CD50[b] monochrome[c]	Metal oxide semiconductor Hitachi KP-120
Dimensions (mm W × H × L)	30.8 × 20.3 × 2.67	21 × 25 × ?	43 × 29 × 73	69 × 60 × 196	41 × 41 × 55	62 × 48 × 131
Scan area (mm H × V)		10 mm (diagonal)	8.8 × 6.6	8.8 × 6.6		8.8 × 6.6
Number of elements (H × V)		328 × 490 (slow scan) 328 × 245 (for 525/60 TV rate)	384 × 491	404 × 256	510 × 492	320 × 244
Resolution (TVL PPH)	425 × 240	328 × 171	280 × 350	250 × 190	380 × 350	240 × 190
Min. illumination, (lx)	Sensitivity 65 mA/W with 2856°K source		400(f=4; with IR filter); 3(f=1.4, without IR filter)	100	5 (F=1.4 with)	10/5 (f=1.4)
Dark current (nA for image area)	4	5				
S/N ratio (dB peak signal/rms noise)	50			46	46	49 (random, 4.5 MHz) 52 (fixed pattern, 4.5 MHz) 46 (100 kHz to 4.5 MHz)
CTF (%, with IR filter)	67 (at 200 TVL)	50~60 (at 328 TVL)				
Gamma	1 (4~400 nA)	0.9~1.0				1
Dynamic range						1000~3000

[a]From manufacturer's specification sheets. See additional sources listed in Table 7-6 (p. 236); also, for concise explanations of CCD sensors, see "CCD: The Solid State Imaging Technology." For reviews, see Wilson (1984); Lewis (1985); and especially Hirschberg (1985). Note: Check for up-to-date listings. New devices with substantially improved characteristics are appearing at a rapid pace.

[b]These entries updated November 1985. See also Tektronix on p. xxiii.

[c]Gamma, AGC, and auto-aperture correction with ON-OFF switches.

elements (typically 300–400 horizontal by about 240 vertical TV lines; Table 7-3), their sensitivity at room temperature can be comparable to a Newvicon-type vidicon tube. Also, while earlier solid-state sensors are said to have suffered from missing pixels and other image blemishes, the image quality of some recent solid-state cameras is surprisingly good. In addition, the compact size, the possibility for low-noise and high-dynamic-range operation, and the potential for image integration, make the solid-state cameras interesting candidates for video microscopy.*

7.2. PERFORMANCE OF VIDEO IMAGING DEVICES

We shall next examine how the performance of various video imaging devices is compared, and consider which types are suitable for applications in video microscopy.

Resolution and Contrast

In microscopy, one strives for a well-resolved image with good contrast down to the finest image detail. The ability of video imaging devices to capture fine image details varies with the type of imaging device. Some camera tubes inherently have greater resolving ability albeit at a higher level of image brightness, while others have a lower resolution but can capture the image at a lower light level.

As with any imaging device, the contrast that the video camera tube can transfer from the incoming image to the output signal diminishes as the spatial frequency of the image detail is increased. In other words, the contrast for the finer detail in the image is reduced compared to the coarser features. Therefore, the modulation and contrast transfer functions, discussed in Section 5.7 for microscope lenses, are also used to specify the response of video camera tubes.

The *amplitude response curve* depicts the contrast that is transferred by a video imaging device as a function of number of TV lines PPH (Fig. 7-9). The **amplitude response** measures the percentage modulation of the signal amplitude, relative to the fully modulated input square wave (Fig. 7-10). In other words, the amplitude response gives the ratio of the amount (P) by which the video signal coming out of the actual imaging device is modulated, over the amount (Q) by which the signal from an ideal imaging device would be modulated for square waves with various spatial frequencies. The amplitude response curve is therefore a form of contrast transfer function curve for the video imaging device. Figure 7-11 shows the contrast transfer function for an SIT camera tube.

Since the human eye can just barely discern a contrast of about 3 to 5% for closely spaced stripes (Chapter 4), the image is said to have reached the *limiting resolution* when the image contrast reaches 3%. The resolution of the video imaging device also drops with reduced faceplate illuminance. Figure 7-12 illustrates the relation between typical limiting resolution and faceplate illuminance for some intensifier camera tubes.

The amplitude response curve and limiting resolution reflect the properties of the target and photocathode materials in the camera and image tubes. Nevertheless, these values can vary, even for the same tube, depending on a number of factors. The size of the sensor, the level of faceplate illumination, accelerating voltage of the electron beam, beam focus, etc. can all affect tube performance, so that the published figures and performance values illustrate "typical" values. In the camera tube specifications, the value for the amplitude response is commonly given as the percentage response to 400 TV lines (horizontally, for a distance equal to the picture height), under specified conditions of signal and dark currents (Tables 7-1 to 7-3).

Sensitivity and Wavelength Dependence

Surprisingly, the sensitivities of video imaging devices are not easy to compare. To start with, manufacturers use different units, and somewhat different standards, to express the measured sensitivity of their video imaging devices. Variation in sensitivity from tube to tube is sufficiently

*Many new solid-state video pickup devices and cameras have been introduced lately. Lewis (1985) provides a comprehensive review and list of vendors. Also, ITT and Hamamatsu have introduced solid-state cameras directly coupled to image intensifiers.

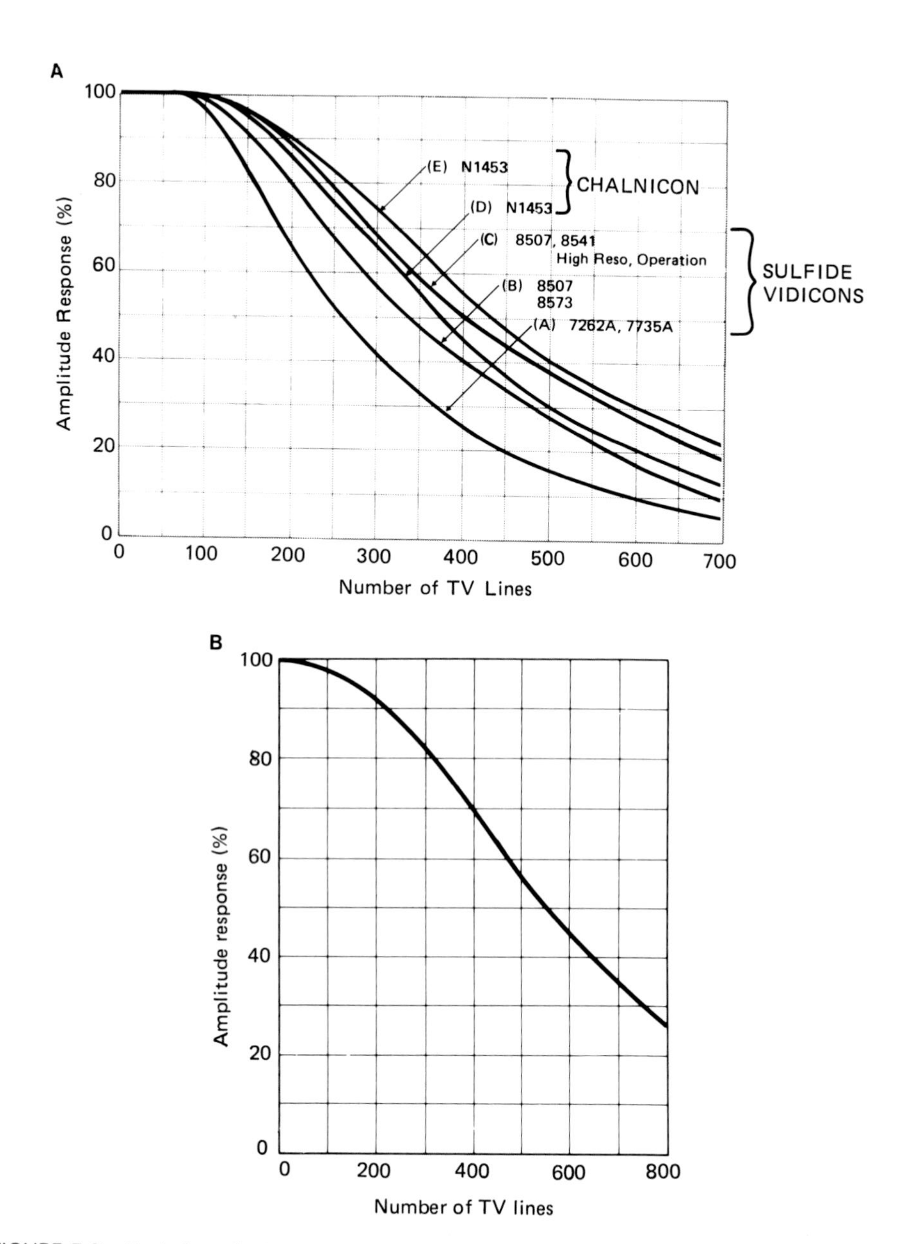

FIGURE 7-9. Typical amplitude response curves for selected vidicon tubes, under "standard" conditions (except where special conditions are indicated).

(A) Hamamatsu 1-inch vidicons. Sulfide Vidicons have Sb_2S_3 targets, which have lower sensitivity than those (CdSe). (From Hamamatsu Publication SC-5-3, with permission.)

(B) Hitachi 1-inch **Saticon** (H9362), made for TV viewing of X-ray fluorescent screens. Note the very high resolution of this tube. (From Hitachi publication CE-E4775, with permission.)

(C) Panasonic 1-inch Newvicon (S 4076). (Information courtesy of Matsushita Electronics Corp., Takatsuki, Japan.)

(D) Toshiba 1-inch Chalnicon (E 5063). Signal output current: 0.3 μA. (a) Beam current: 0.3 μA; standard operation. (b) Beam current: 0.6 μA; standard operation. (c) Beam current: 0.3 μA; high-voltage operation. (d) Beam current: 0.6 μA; high-voltage operation. (From Toshiba Publication EP-T100, with permission.)

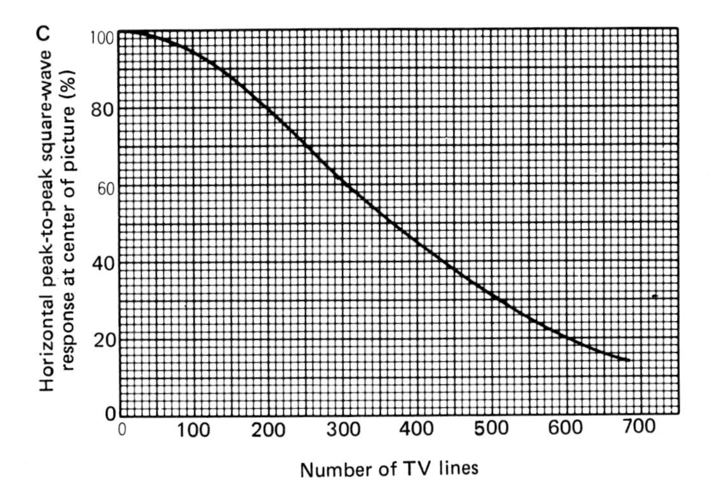

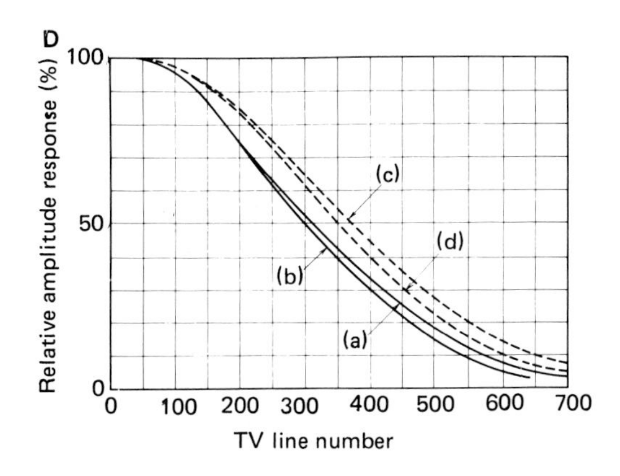

FIGURE 7-9 (*Continued*)

high (often of the order of ±50%), so that there is an understandable hesitance by the manufacturers to provide anything more than "typical" values for camera tube characteristics. The state of the art also changes, so that the specifications can improve over time.

More fundamentally, the *spectral sensitivity*, or the response of video imaging devices to different wavelengths of light, varies with the nature of the target, photocathode, or sensor, as well as the envelope or coating material. Since in most video applications the sensitivity of the imaging device relates to the visual sensation of image brightness, the sensitivities of the video imaging devices are commonly expressed with reference to the sensitometric characteristics of the human eye. Thus, the sensitivities of video imaging devices are often measured in photometric units or as a function of illuminance (which reflect the sensitivity curve for the photopic human eye; Chapter 4), rather than in radiometric units, which give the absolute energy flux (Tables 7-1 to 7-3).

Some manufacturers express the sensitivity of camera tubes as *responsivity* in units of microamperes of signal current per-**lumen** per square foot of illumination (μA/lm per ft^2). Others use *sensitivity* expressed as nanoamperes per **lux** (nA/lx). Still others express sensitivity in units of milliamperes per watt (mA/W) for example, to overcome the ambiguity imposed by the use of photometric units. The lumen and lux are photometric units, while the watt is a radiometric unit. One lux is the unit of illuminance resulting from the flux of 1 lm falling on 1 m^2, 1 lm being the luminous flux of light emitted from a standard 1-cd source into 1 steradian (for definitions and detailed discussions of detector characteristics, see, e.g., *RCA Electro-Optics Handbook,* 1974, Chapter 10).

Measurements of responsivity and sensitivity are standardized to the extent of using a standard light source, commonly a tungsten incandescent source operated at a color temperature of 2856 (or

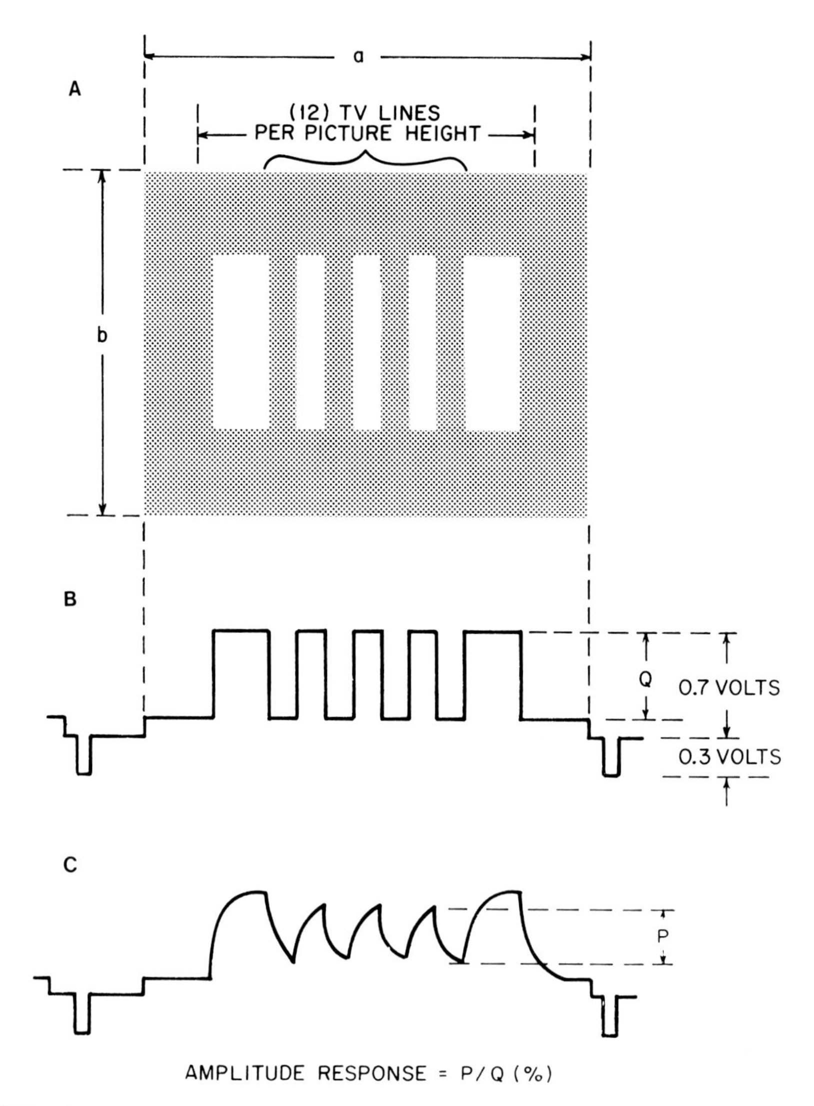

FIGURE 7-10. Definition of amplitude response. Ideally, the video signal (B) coming from an imaging device should correspond exactly to the distribution of luminance in the image (A). In actuality, the modulation is less (P) than ideal (Q). The amplitude response is defined as P/Q (%). It is the ordinate of the plots shown in Fig. 7-9. See also Fig. 7-39.

2854)°K. Since the major fraction of the emission from such a source lies in the far-red to IR region of the spectrum, and since video sensors vary dramatically as to their red, near-IR sensitivities (both on an absolute scale and relative to their green sensitivities), one needs to use special care in specifying or comparing sensor sensitivities. The problem becomes especially acute in comparing sensitivities of image tubes used for fluorescence microscopy, where, depending on the fluorochrome, the emission peak may lie in the red to far red, close to the upper end of the visual sensitivity curve. Fortunately, a "typical" **spectral response curve** is generally provided with the manufacturer's specifications of video imaging devices (Fig. 7-13).*

*Note: light in the far-red to IR region of the spectrum can give rise to additional flare and reduced contrast for several reasons (increased lens aberrations, failure of anti-reflection coating, increased specular reflection from "black" surfaces). The far-red and near-IR sensitivities of camera tubes such as Newvicon and **Ultricon,** or of CCD devices, can be reduced and their contrast improved by blocking out those wavelengths with suitable filters (e.g., HA-11 from Fish–Shurman Corp.).

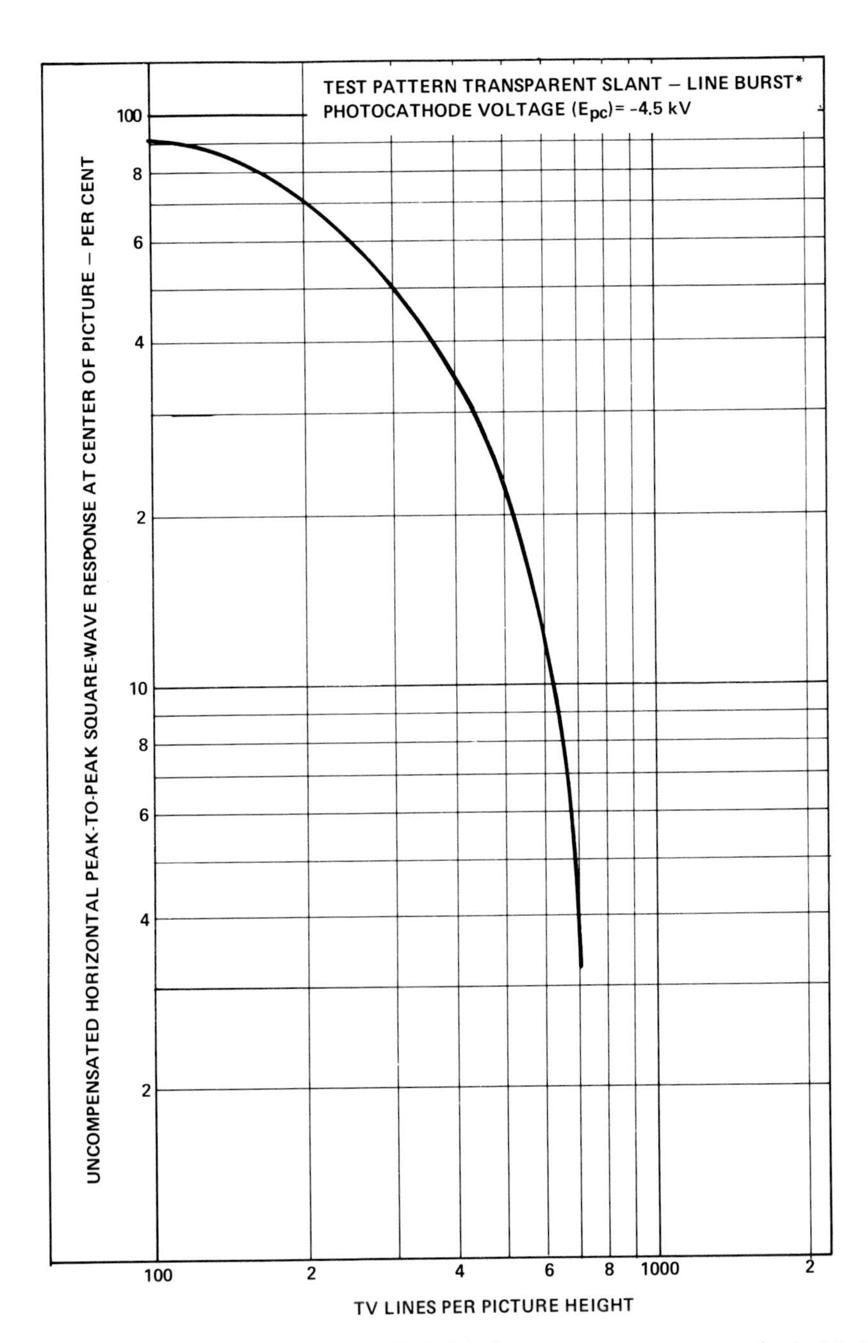

FIGURE 7-11. Square wave response curve. Typical horizontal square wave response (contrast transfer function) for RCA 4804/H series, 16-mm, fiber-optic faceplate-type SIT tube, operated under "standard" conditions. *Contrast transfer function is measured using the RCA P2000 slant-line burst pattern with horizontal center response balanced on the 400-line chevrons. (Courtesy of RCA, from RCA Publication 4804/H series.)

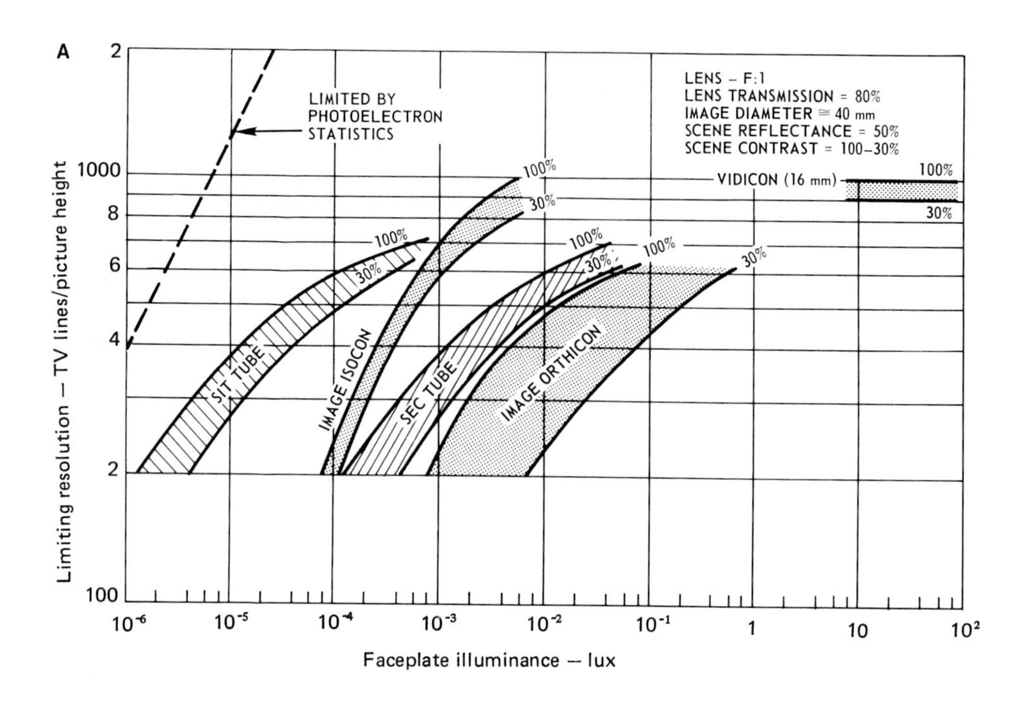

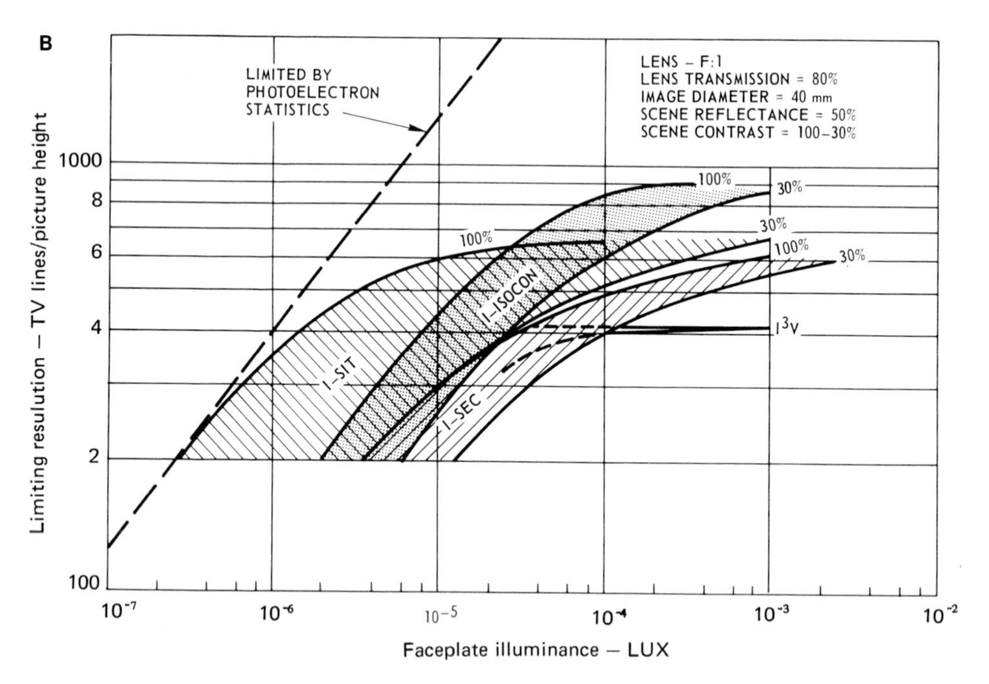

FIGURE 7-12. Resolution versus illuminance for several camera tubes (typical values). (A) Selected camera tubes. (B) Selected intensifier camera tubes. I^3V indicates a three-state image tube coupled to a vidicon. The values shown for each tube are for an image contrast range of 30–100%. Note that A and B cover different illuminance ranges. (Courtesy of RCA, from RCA *Electro-Optics Handbook*, 1974, pp. 205, 206.)

Light Transfer Characteristics

Acknowledging that the spectral characteristics vary with the type of video sensor, it is nevertheless important to know what typical levels of signal current are expected for different levels of sensor illumination. The plot log (signal current from a video imaging device) versus log (illuminance of the faceplate) is known as the **light transfer characteristics** curve.

Figure 7-14A–D shows light transfer characteristics curves of several imaging devices obtained from manufacturer's specification sheets. These curves are "typical" curves measured under electrode voltages, beam current and faceplate illuminance, etc. specified for the particular imaging device by the manufacturer. The signal current is normally measured with the faceplate illuminated by a standard incandescent light source with a color temperature of 2856°K. The size of the target, accelerating voltage, beam current, and the magnitude of the **bias light,** etc. affect the magnitude of signal current.

The light transfer characteristics curve and the sensitivity of the **sulfide vidicon** (antimony trisulfide vidicon, also just called Vidicon) can be shifted along the illuminance axis by altering the target voltage (Fig. 7-14B). Other camera tubes in the vidicon family, such as the Newvicon and Chalnicon, have to be operated at a fixed target voltage so that their light transfer characteristics curve cannot be shifted.

Gamma

The slope of the light transfer characteristics curve, or the exponent that defines the curve approximating the output versus input relationship, is known as the **gamma** of the video imaging device. The gamma of the imaging device is determined primarily by the nature of the target material. For several types of camera tubes in the vidicon family and for solid-state pickup devices, the gamma is nearly 1.0, while for the sulfide vidicon and some intensifier camera tubes, the gamma is less than 1.0 (Tables 7-1, 7-3, Fig. 7-14A–D).

The sulfide vidicon with its low gamma of 0.65 responds to a broader range of illuminance, or has a greater dynamic range, than other tubes of the vidicon family. The broader dynamic range is an advantage for commercial and other color video, but the lower gamma is a drawback for video microscopy in monochrome, especially when quantitation of image brightness is desired. For photometry, one needs a video camera tube that provides a gamma of 1.0, or at least a video camera that compensates for the sensor and provides a camera gamma of 1.0. The need for unity gamma can be seen by examining the nature of the light transfer characteristics curve as follows.

The light transfer characteristics curve is plotted as log signal output current versus log illuminance of faceplate (Fig. 7-14A–D), rather than linearly as signal output current versus illuminance of faceplate. The gamma, or slope (k), of the curve therefore provides the exponent relating illuminance to the signal output current, or

$$i/i_D = (I/I_D)^k \tag{7-1}$$

The illuminance I gives rise to the signal output current i; i_D is the dark current and I_D the level of illuminance that (under the specific conditions of measurement) would give rise to a signal current equal to the dark current.

In other words, the signal output current generally rises with the kth power of the illuminance. When k is less than 1.0, the sensitivity of the imaging device decreases with rising illuminance. Only when k = 1.0, or the gamma is unity, is the signal output current proportional to the illuminance of the faceplate. For the camera tube whose gamma is not exactly 1.0, some cameras provide built-in **gamma compensation** circuits to provide an effective gamma of 1.0 for the camera output signal. In color video, gamma compensation may be used to reduce the gamma of the camera (Fig. 7-34).

In some camera tubes such as the image isocon and orthicon, the gamma itself decreases at higher illuminance, and the light transfer characteristics curve displays a "knee" (Fig. 7-14D).

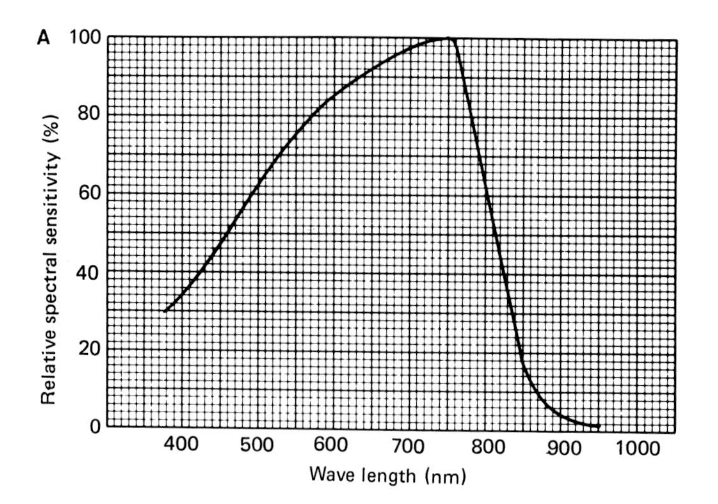
A
100
80
60
40
20
0
Relative spectral sensitivity (%)
400 500 600 700 800 900 1000
Wave length (nm)

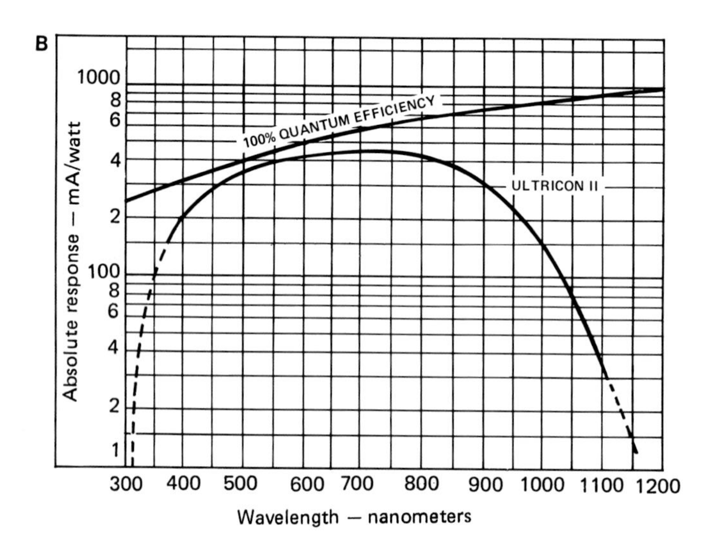
B
1000
8
6
4
2
100
8
6
4
2
1
Absolute response — mA/watt
100% QUANTUM EFFICIENCY
ULTRICON II
300 400 500 600 700 800 900 1000 1100 1200
Wavelength — nanometers

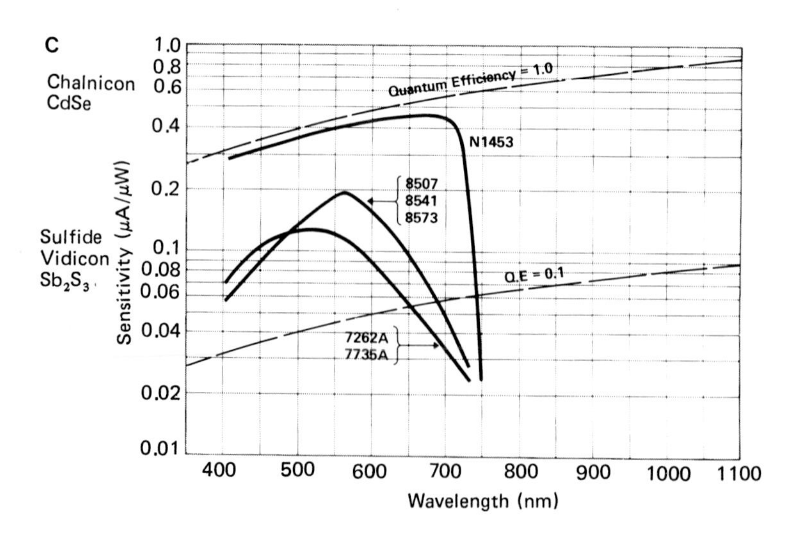
C
Chalnicon
CdSe
Sulfide
Vidicon
Sb_2S_3
1.0
0.8
0.6
0.4
0.2
0.1
0.08
0.06
0.04
0.02
0.01
Sensitivity (μA/μW)
Quantum Efficiency = 1.0
N1453
8507
8541
8573
Q.E = 0.1
7262A
7735A
400 500 600 700 800 900 1000 1100
Wavelength (nm)

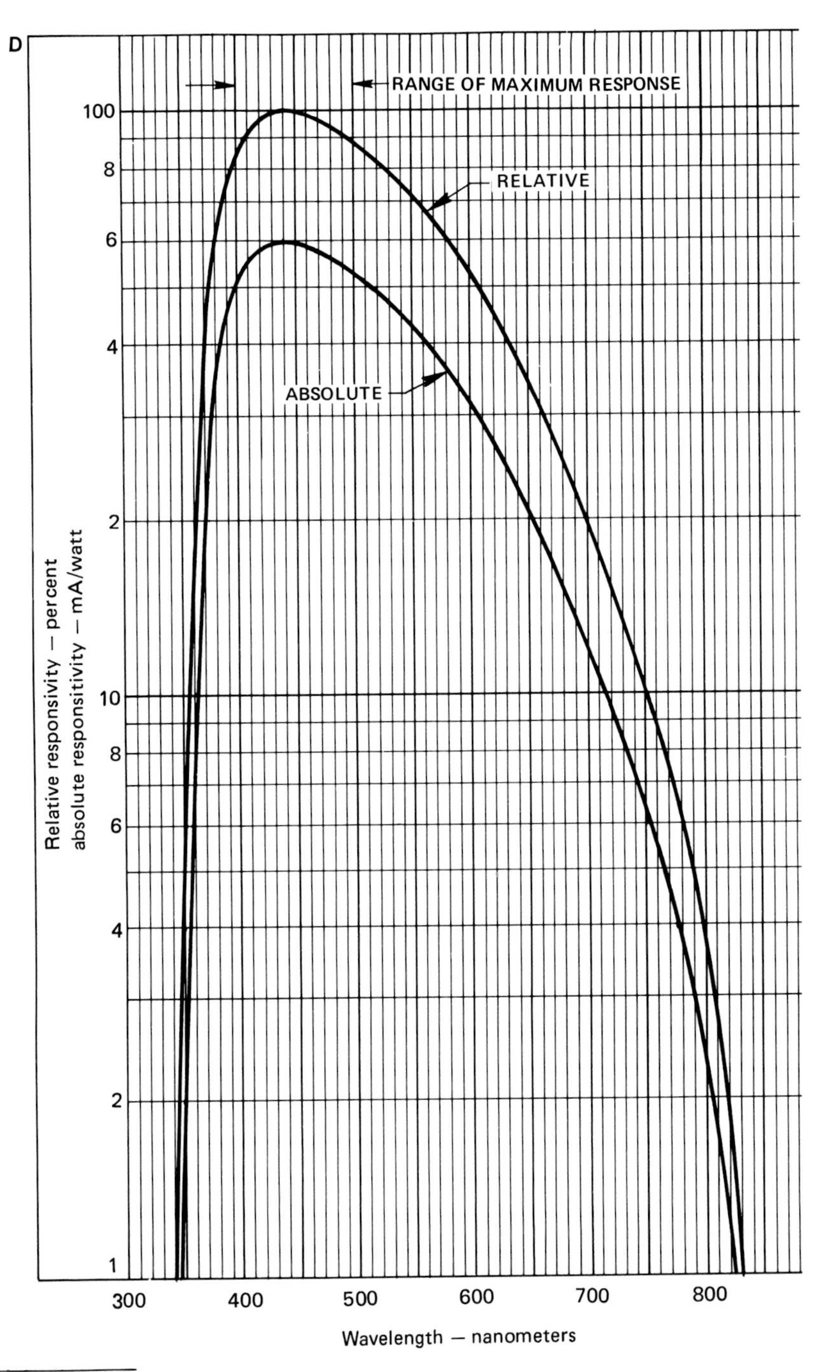

FIGURE 7-13. Spectral response curves for selected video imaging devices. Note that manufacturers may use different units for the ordinates of such plots. All are standardized to a 2856°K tungsten light source. The curves show typical values for "standard" operating conditions.

(A) Panasonic 1-inch Newvicon (S 4076). (Information courtesy of Matsushita Electronics Corp., Takatsuki, Japan.)

(B) RCA Ultricon II, a silicon target vidicon (Courtesy of RCA, from RCA Publication 4532/U, 4833/U, 4875/U series.)

(C) Hamamatsu 1-inch vidicons. (From Hamamatsu Publication SC-5-3, with permission.)

(D) RCA. Typical photocathode responsivity of 16-mm SIT-Multialkali (NaKCsSb) as modified by fiber-optic window. (Courtesy of RCA, from RCA Publication 4804/H series.)

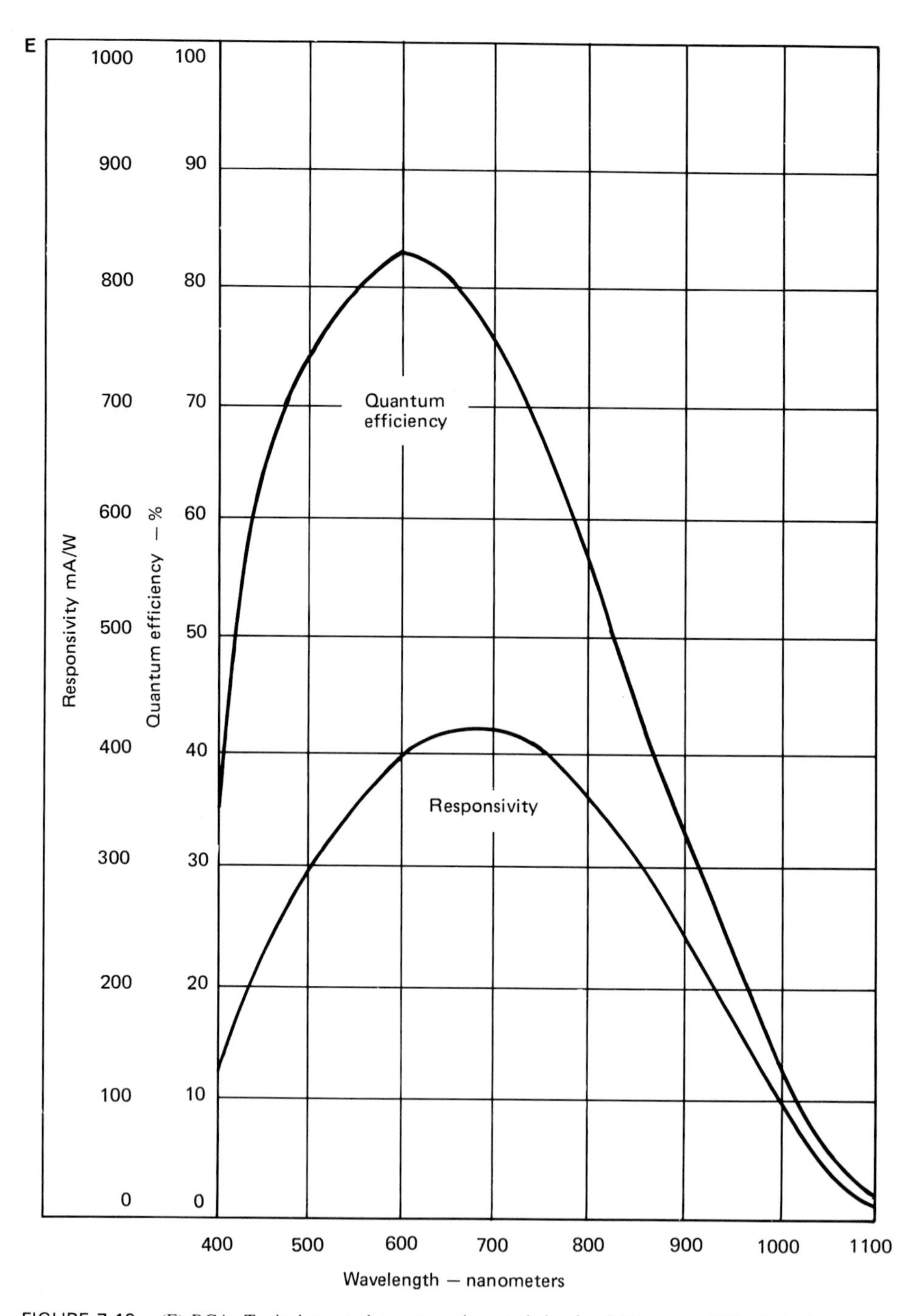

FIGURE 7-13. (E) RCA. Typical spectral response characteristics for CCD sensor (SID 504). Courtesy of RCA, from RCA Publication SID 504.)

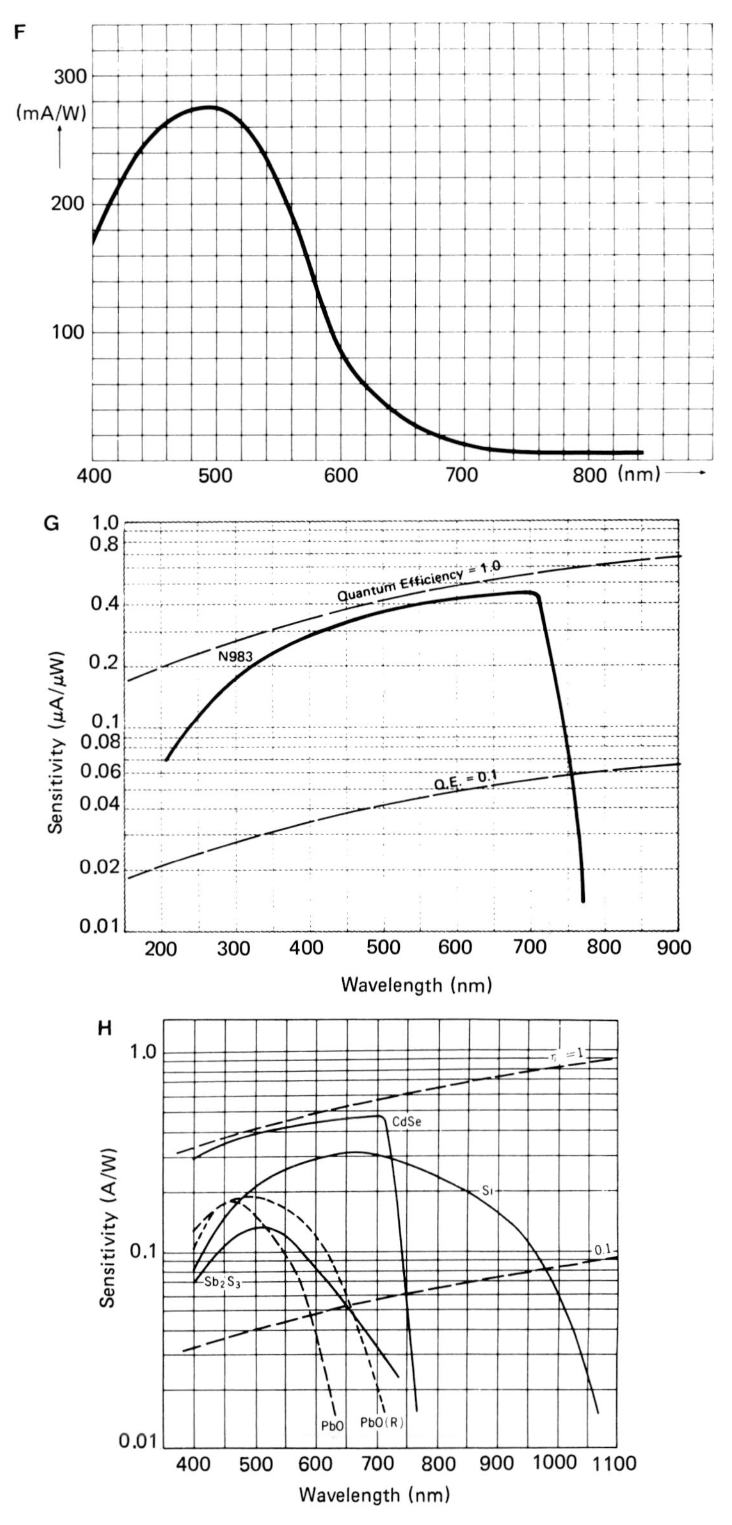

FIGURE 7-13. (F) Amperex. Typical response curve for 1-inch Diode Gun Plumbicon (XQ2072/02 and 03). (From Amperex Publication XQ2072/02, XQ2072/03, with permission.)

(G) Hamamatsu 1-inch ultraviolet-visible (CdSe) vidicon. (From Hamamatsu Publication SC-5-3, with permission.)

(H) Comparison of photocathode sensitivities on an absolute scale. (Courtesy of Hamamatsu Corp.)

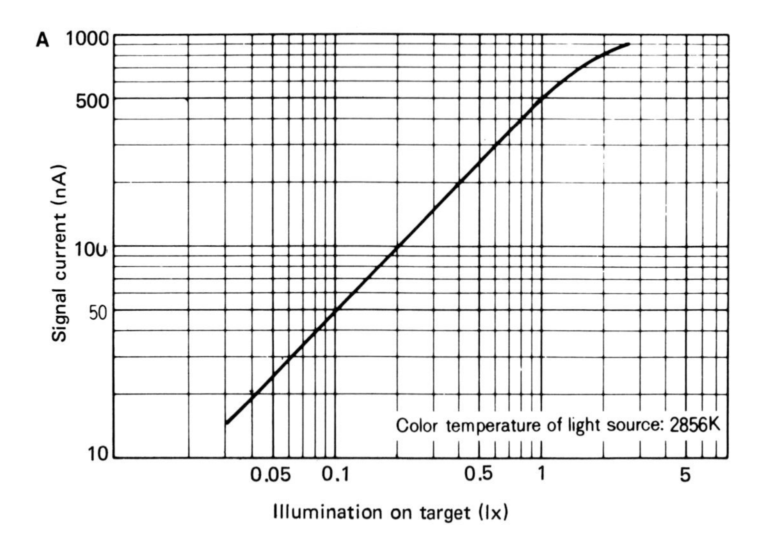

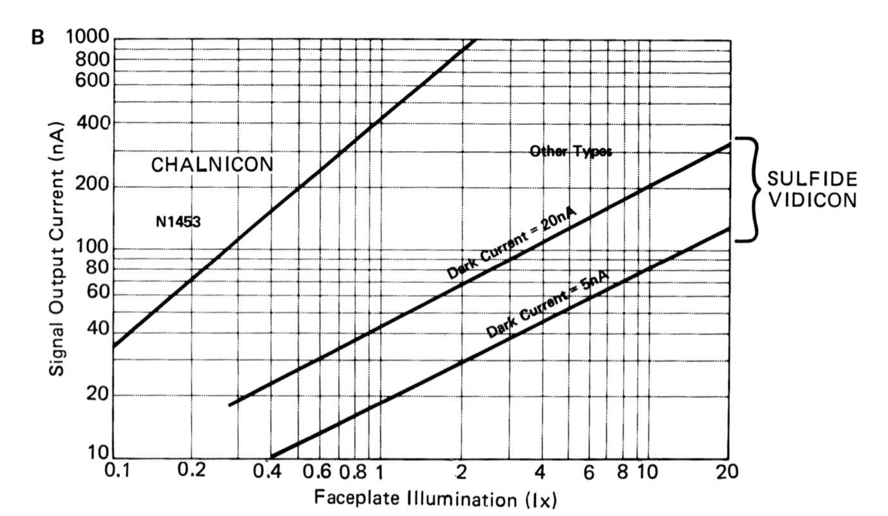

FIGURE 7-14. Light transfer characteristics curves for several camera tubes. Curves show typical values under "standard" operating conditions.

(A) Panasonic 1-inch Newvicon (S 4076). (Information courtesy of Matsushita Electronics Corp. Takatsuki, Japan.)

(B) Hamamatsu 1-inch vidicons. (From Hamamatsu Publication SC-5-3, with permission.)

(C) Hitachi 1-inch Saticon (H9362). (From Hitachi Publication CE-E4775, with permission.)

(D) Toshiba 1-inch Chalnicon (E 5063). (From Toshiba Publication EP-T100, with permission.)

(E) RCA. Selected imaging devices. (Courtesy of RCA, from Publication IMD 101.)

Dynamic Range

The dynamic range of a video imaging device expresses the range of faceplate illuminances over which the device responds meaningfully. Below the dynamic range, the signal is buried in noise, and above the dynamic range, the tube "saturates," and an increase in illuminance no longer gives rise to a significant increase in signal current.

Two types of dynamic ranges may be given for the same imaging device, the *intrascene*

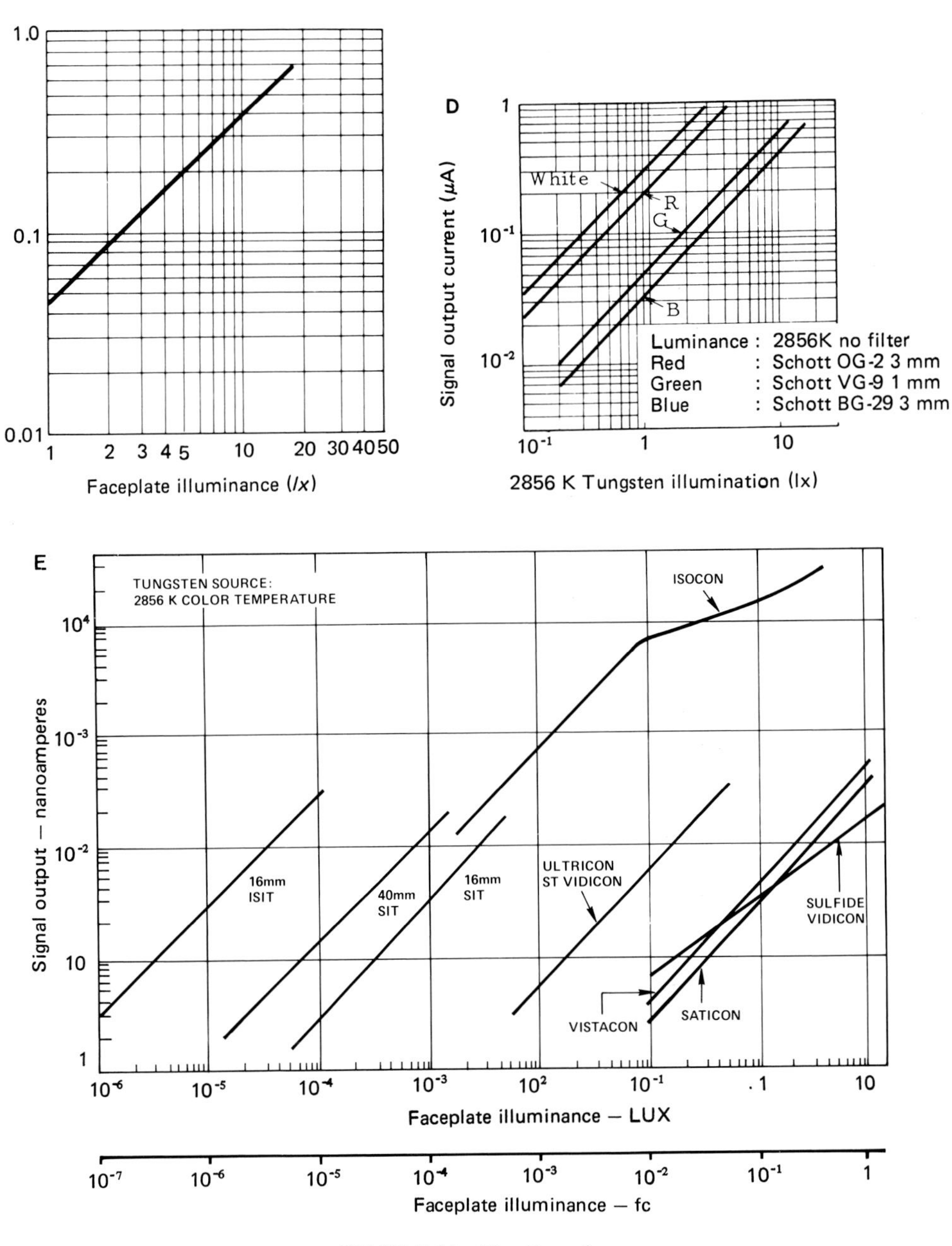

FIGURE 7-14 (*Continued*)

dynamic range and the **usable light range.** Both are expressed numerically as the ratio of maximum and minimum useful illuminances.

The intrascene dynamic range for the vidicon family of camera tubes is typically 70 : 1 to 100 : 1, except in the sulfide vidicons, for which the dynamic range is about 350 : 1. For the vidicon family of camera tubes, the lower end of the intrascene dynamic range is limited by amplifier noise, and the upper end by tube saturation. For intensifier camera tubes, the lower end is limited by tube noise.

The useful light range of a video imaging device may be considerably greater than the intrascene dynamic range for the same device. As already mentioned, the target voltage can be adjusted to alter the sensitivity in sulfide vidicons. Also, the high voltage supply for the intensifier can be adjusted to control the intensifier gain over a wide range. Thus, while an intrascene dynamic range of 50 : 1 is given for a SIT tube (for a signal range of 50 : 1), its usable light range is listed as 10,000 : 1 (see footnotes to the lower part of Table 7-2).

Video cameras commonly incorporate manual or automatic circuits to adjust the *target voltage*

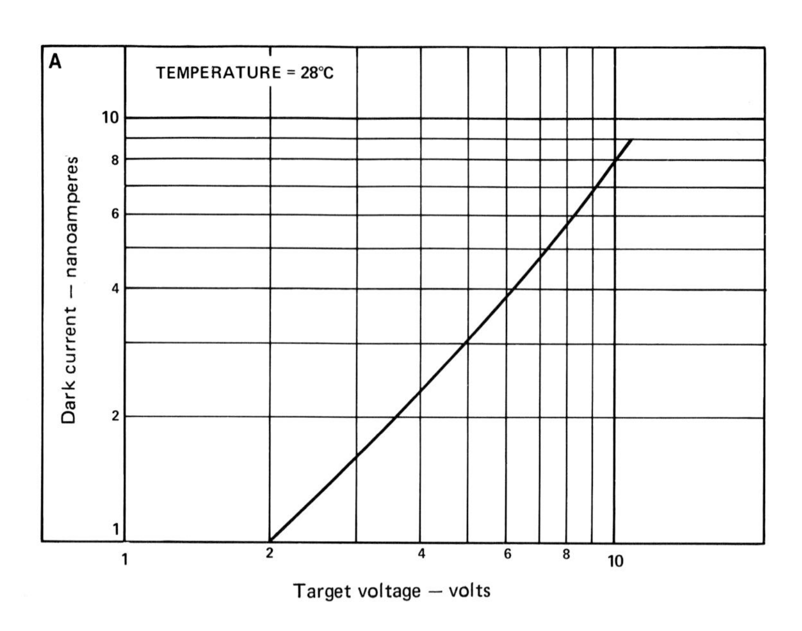

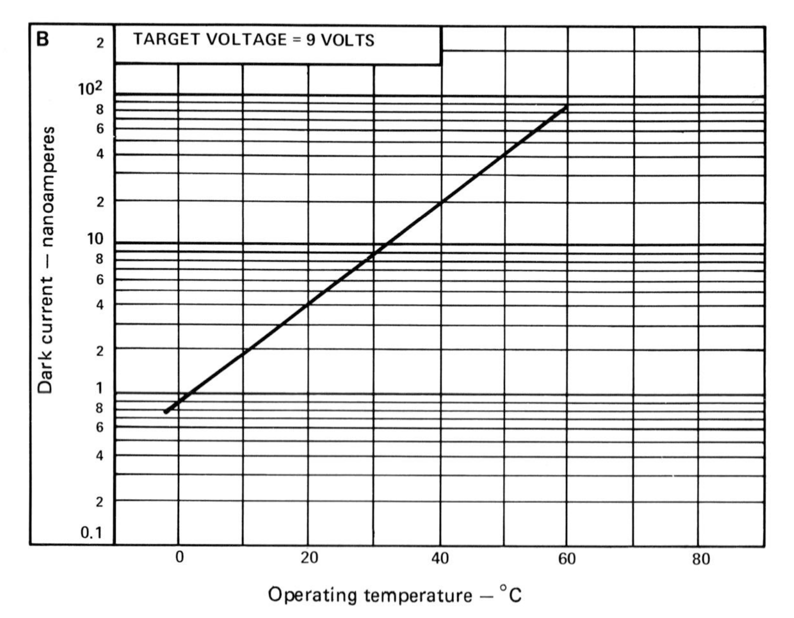

FIGURE 7-15. Dark current as a function of target voltage and operating temperature. Typical dark current as a function of (A) target voltage and (B) operating temperature, for RCA 4804/H series, 16-mm, fiber-optic faceplate-type SIT, under "standard" operating conditions. (Courtesy of RCA, from RCA Publication 4804/H series.)

when the camera tube permits such adjustments. When an intensifier is present, its *high voltage* supply, and hence intensifier gain, is likewise regulated. In addition, most cameras incorporate an **AGC** (automatic gain control) circuit, which keeps the amplifier gain to an operating level appropriate for the output signal from the imaging device. Normally, such controls provide an optimally modulated signal output from the camera. However, these same functions can also interfere with quantitation in video microscopy, if the image brightness, or range of brightnesses, varies between scenes.

Dark Current and Noise

The amount of output current that flows through a video imaging device when the device is placed in the dark is known as its *dark current*. The level of dark current typically varies from 1 nA or less to over 10 times that value, depending on the imaging device (Tables 7-1 to 7-3). For the same target, the level of dark current varies with the target voltage, the size of the signal current, and the temperature at which the device is operated. The target voltage and dark current can be varied over a wide range in sulfide vidicons. The sensitivity of the sulfide vidicons rises with greater target voltage and dark current (Fig. 7-14B). Figure 7-15 shows typical dark currents of an SIT tube (16-mm image diagonal, RCA 4804/H series) as functions of target voltage and operating temperature.

Depending on the type of imaging device, the dark current may or may not limit the level of noise introduced into the video signal. For example, in the vidicon family of camera tubes, the amplifier noise current becomes dominant at threshold levels of faceplate illumination. The video scene displayed on the monitor becomes covered with *snow* originating from the "white" noise introduced by the input stage of the amplifier rather than by the dark current of the vidicon itself. Thus, "vidicons having the highest responsivity have the highest signal-to-noise ratio at a given light level" (*RCA Electro-Optics Handbook,* 1974, p. 190).

On the other hand, the noise level in some intensifier camera tubes and avalanche solid-state devices can themselves determine the threshold illuminance. Figure 7-16 shows typical S/N ratios to be expected at various levels of faceplate illuminances from the types of camera tubes listed in Table 7-2 and Figure 7-14E.

Not only the amount of dark current but the *character* of dark current and noise varies with the type of imaging device. The noise arising in the dark from some devices is nearly random white noise appearing as a fine snow on the video monitor, while the noise from some intensifier tubes

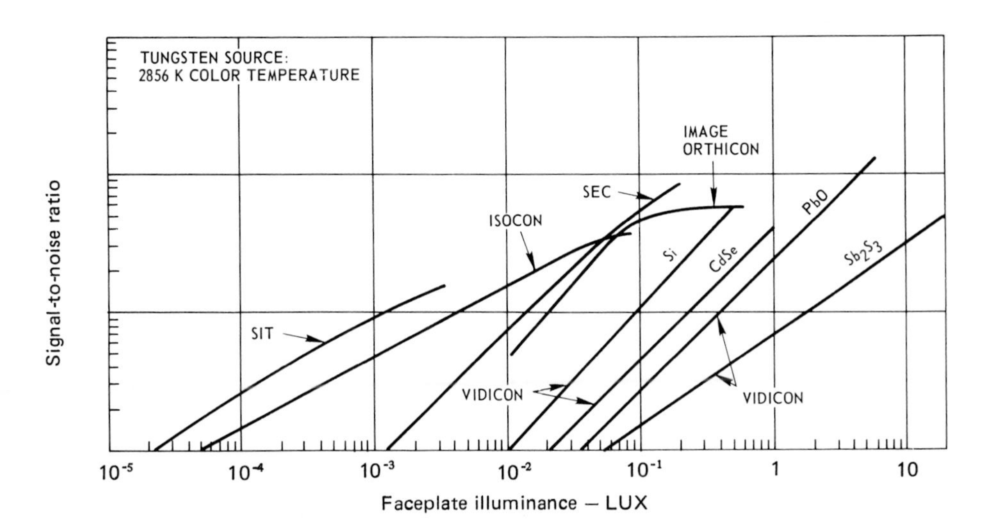

FIGURE 7-16. Signal-to-noise ratio characteristics of selected camera tubes. Typical S/N ratios are plotted as functions of faceplate illuminance. (Courtesy of RCA, from *RCA Electro-Optics Handbook,* 1974, p. 191.)

appears as discrete random bursts of bright spots of light. The frequency of occurrence, geometrical distribution, and contrast of these noise patterns vary considerably depending on intensifier tube type, operating conditions, and *camera design.*

At extremely low levels of sensor illumination, statistical fluctuation of the low number of photons that make up the image also introduces random bright spots, which are sometimes difficult to distinguish from the noise from the dark current originating in the video camera and image tubes. We shall return to a discussion of video cameras suitable for low-light-level microscopy later.

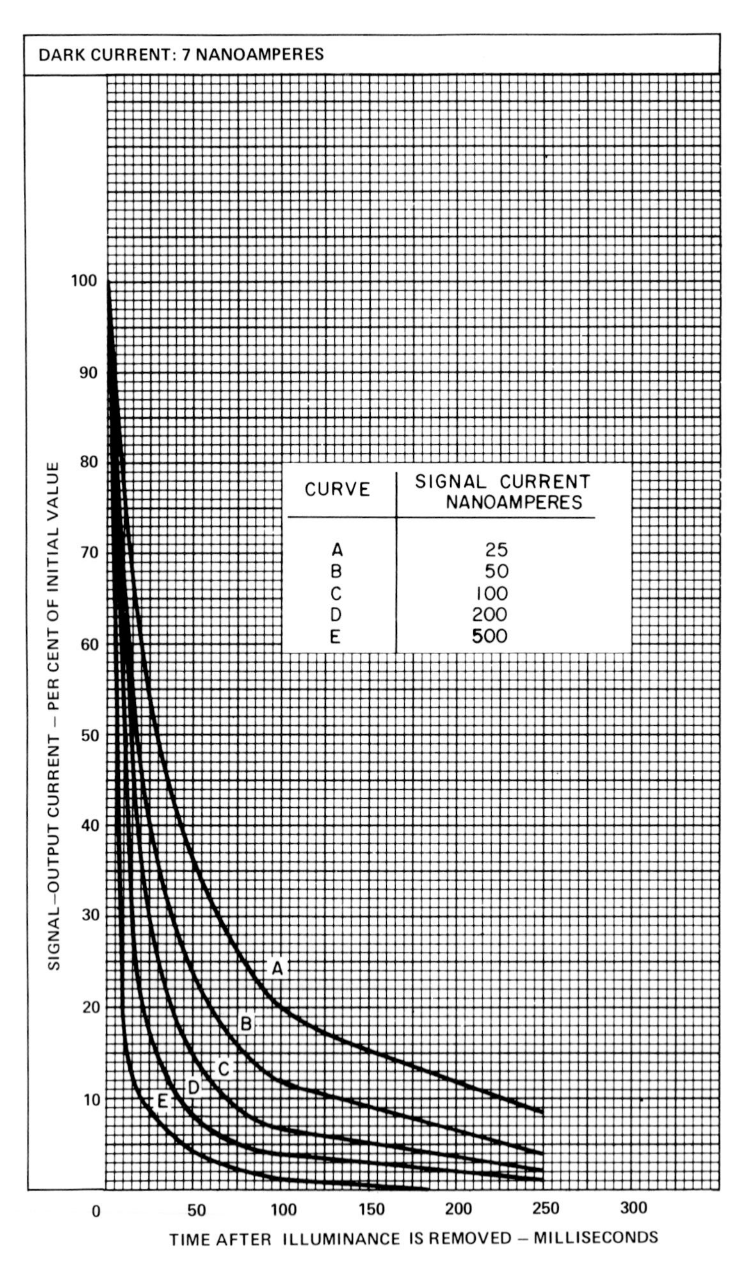

FIGURE 7-17. Typical persistence characteristics for SIT. RCA 4804/H series, 16-mm, fiber-optic faceplate operated under "standard" conditions. (Courtesy of RCA, from RCA Publication 4804/H series.)

For some applications, the level of image tube dark current is raised intentionally by applying a bias illumination. The bias illumination reduces lag, as described below.

Lag

The target in the vidicon family of camera tubes, and the phosphor excited by the accelerated electrons in the image intensifiers, both exhibit a finite decay time. Such decay gives rise to a *smear* or a white *comet tail* when a bright object moves in the video scene. Solid-state pickup devices can also exhibit a *vertical white stripe* that smears the parts of the image that are very bright (Fig. 7-25). The phenomena that give rise to these image defects are collectively known as the *lag*, or *decay lag*, of the video imaging device.

The decay lag or *persistence* is represented in a graph as a function of number of TV fields, or time, after the illumination is removed. Numerically, lag is also expressed as the percentage residual signal in the third field, or 50 msec after the illumination is removed. As shown in Figs. 7-17 and 7-18 and Tables 7-1 and 7-2, lag is a function of the tube type and faceplate illuminance. A *third-field lag* of greater than 10% is said to be objectionable for broadcast TV.

In SIT cameras, the comet tailing caused by lag can become prominent at low light levels, and limit their use in video microscopy. For example, Fig. 7-19 illustrates a swimming ciliated sea urchin embryo, some of whose cells were labeled with a fluorescent dye. The comet tailing grossly deforms the outline of the cells in the swimming embryo.

Among the vidicon family of camera tubes, the Newvicon and Chalnicon, which are otherwise well suited for a wide range of applications in video microscopy, tend to show significant lag even with moderate illumination. Figure 7-20 illustrates the appearance of sea urchin sperm swimming in

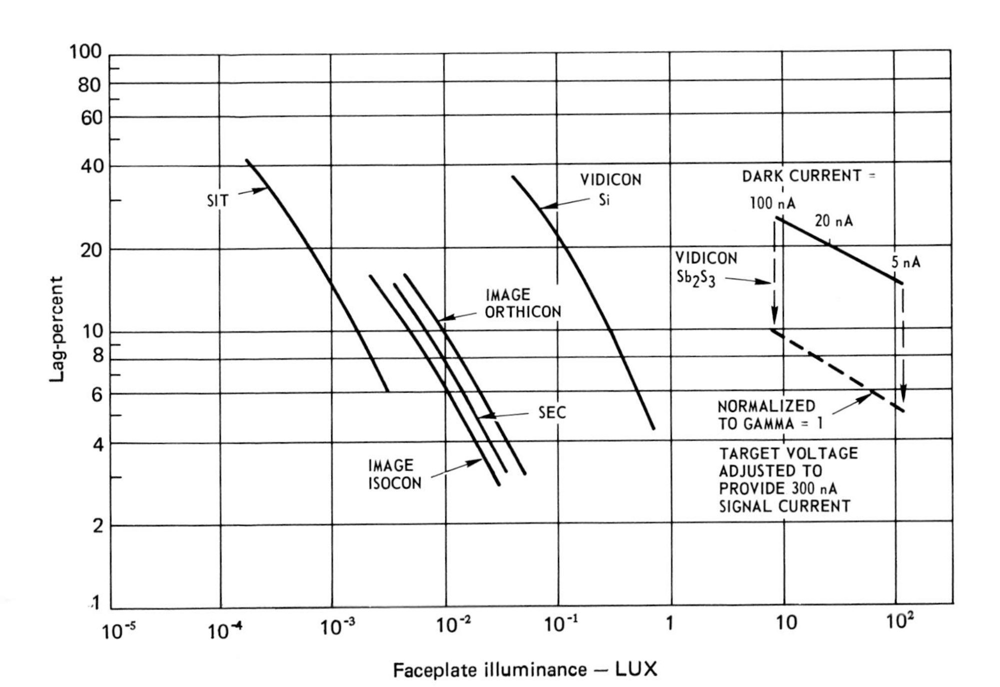

FIGURE 7-18. Typical third-field lag of selected camera tubes. Lag percent is plotted on a log scale as a function of faceplate illuminance. In the case of the Sulfide Vidicon (Sb_2S_3), which has a gamma of about 0.65, corrections have been made to show the lag percent in terms of the light level. This normalization (to gamma = 1), which allows direct comparison with the other curves, lowers the lag values by a substantial amount, as shown. (Courtesy of RCA, from *RCA Electro-Optics Handbook,* 1974, p. 192.)

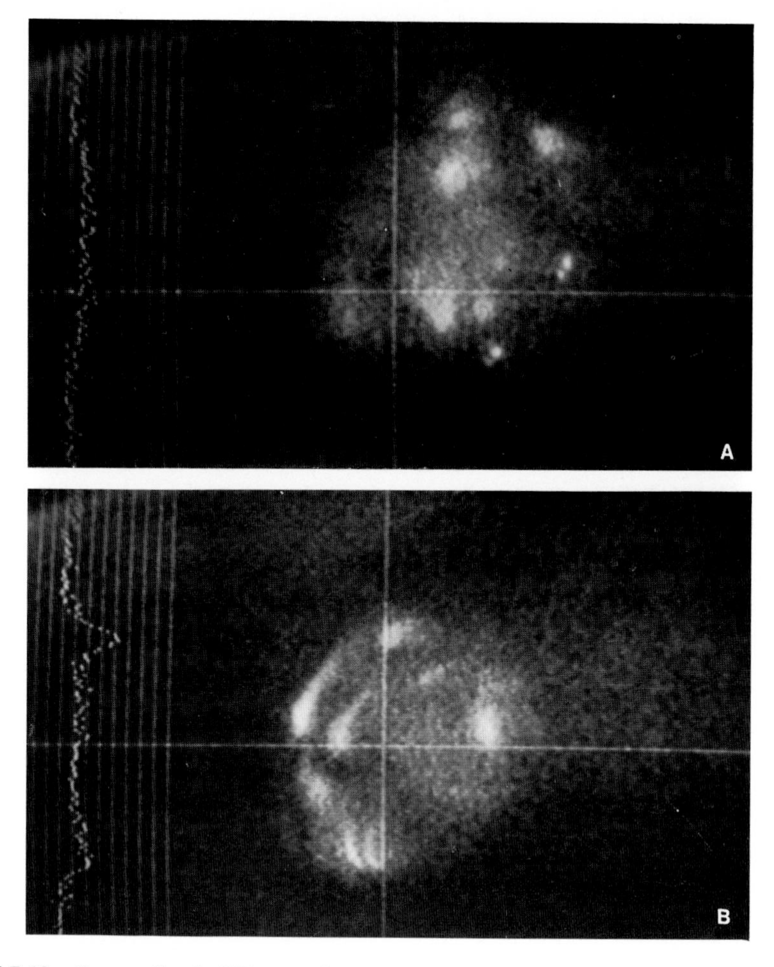

FIGURE 7-19. Comet tailing in SIT camera image. These are very-low-light-level fluorescence microscope images of a living sea urchin embryo. Certain cells in the embryo (descendants of a single micromere injected with lucifer yellow at the 16-cell stage) show as distinctly fluorescent bodies. The fluorescence was too weak to be seen with the eye. In A, the embryo is stationary. In B, it is swimming, and the fluorescent cells show comet tails characteristic of low-light-level SIT camera images.

a helical path. The corkscrew appearance of the head of the swimming, but not the stationary, sperm is a video artifact introduced by the lag in the Newvicon tube used to record this scene.

Decay lag can be reduced in some camera tubes by raising the level of faceplate illuminance by providing *bias illumination*. Figure 7-21 shows typical effects of bias illumination on the decay lag of **Plumbicon** and Chalnicon tubes.

In addition to decay lag, some camera tubes also have a *buildup lag*. The buildup lag in 1-inch Plumbicon and Chalnicon tubes is shown in Fig. 7-22 as a function of bias illumination. As the figure shows, the conspicuous buildup lag of low-light-level scenes displayed by Chalnicon tubes can be reduced by bias illumination.

SIT tubes and some other intensifier camera tubes rely on the lag of the vidicon targets and the intensifier phosphor to smooth out the noise that arises in the intensifier section. Some of these intensifier tubes can also show a very prolonged buildup lag. While I have not been able to find published data on the buildup lag for these tubes, we have noticed that the image contrast in extremely low-light-level images tends to build up and improve slowly over a period of several

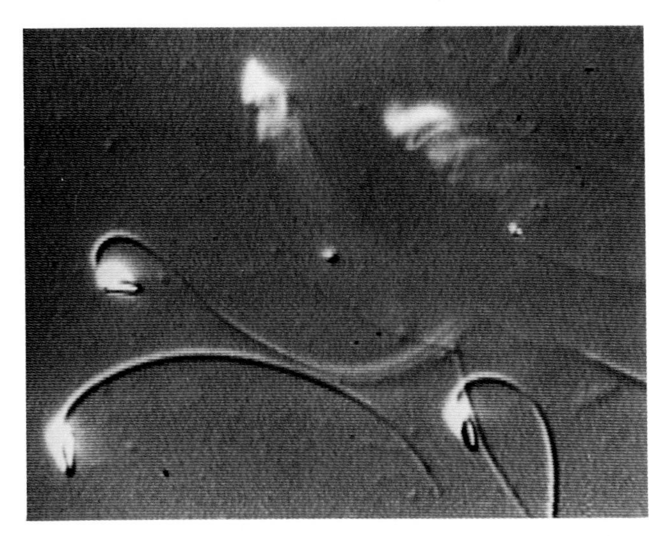

FIGURE 7-20. Frozen-field image of swimming and quiescent sperm. This single frozen-field (1/60 sec) image of *Arbacia* (sea urchin) sperm was produced by DIC optics and captured with a 1-inch Newvicon camera. The lag in the Newvicon tube results in the corkscrew-shaped track of the swimming sperm (top of figure).

seconds. Curve (a) in Fig. 7-22C suggests that an extended buildup lag could contribute to the observed slow buildup of the image, although slow improvement of contrast by image summation and noise reduction in the vidicon target or intensifier phosphor may also be responsible.

Blooming

The upper end of the dynamic range of a video imaging device is generally limited by saturation as described earlier. In some camera tubes, the signal also starts to spill out into adjacent pixels, so that the area of the highlight expands. The spreading of the highlight is not limited to a narrow region bordering the highlight, but can cover a large area (Fig. 7-23). Such spreading of the highlight is known as *blooming.**

Blooming is especially noticeable in the **silicon vidicons,** and also to some extent in the SIT tubes (Fig. 7-24). A new silicon target vidicon from RCA, the Ultricon, shows less blooming than the standard silicon vidicons (Fig. 7-24A). The Ultricon provides a very clean, sharp image with low lag at moderately low light levels (Table 7-1). With its unity gamma, relatively broad dynamic range, and extended red sensitivity, the Ultricon is a very attractive candidate for video microscopy. Unfortunately, even with the improvements, we have been unable to use the Ultricon for polarized-light and DIC microscopy. The image bloomed uncontrollably in the highlights of images that contained a high range of intensities.

In some solid-state pickup devices, especially the CCDs, white vertical stripes may appear above and below the image highlights (Fig. 7-25). This form of blooming is caused by incomplete transfer of the electrical charges in the CCDs. The design of solid-state pickup devices is, however, improving rapidly and many of the new types of devices suffer much less, or hardly at all, from blooming.

Burn, Sticking

Some camera tubes, especially the highly sensitive image isocon, SIT, and other intensifier camera tubes, can be damaged, or their lives shortened, by exposure to excessively bright light. The

*Blooming can also be introduced by video monitors.

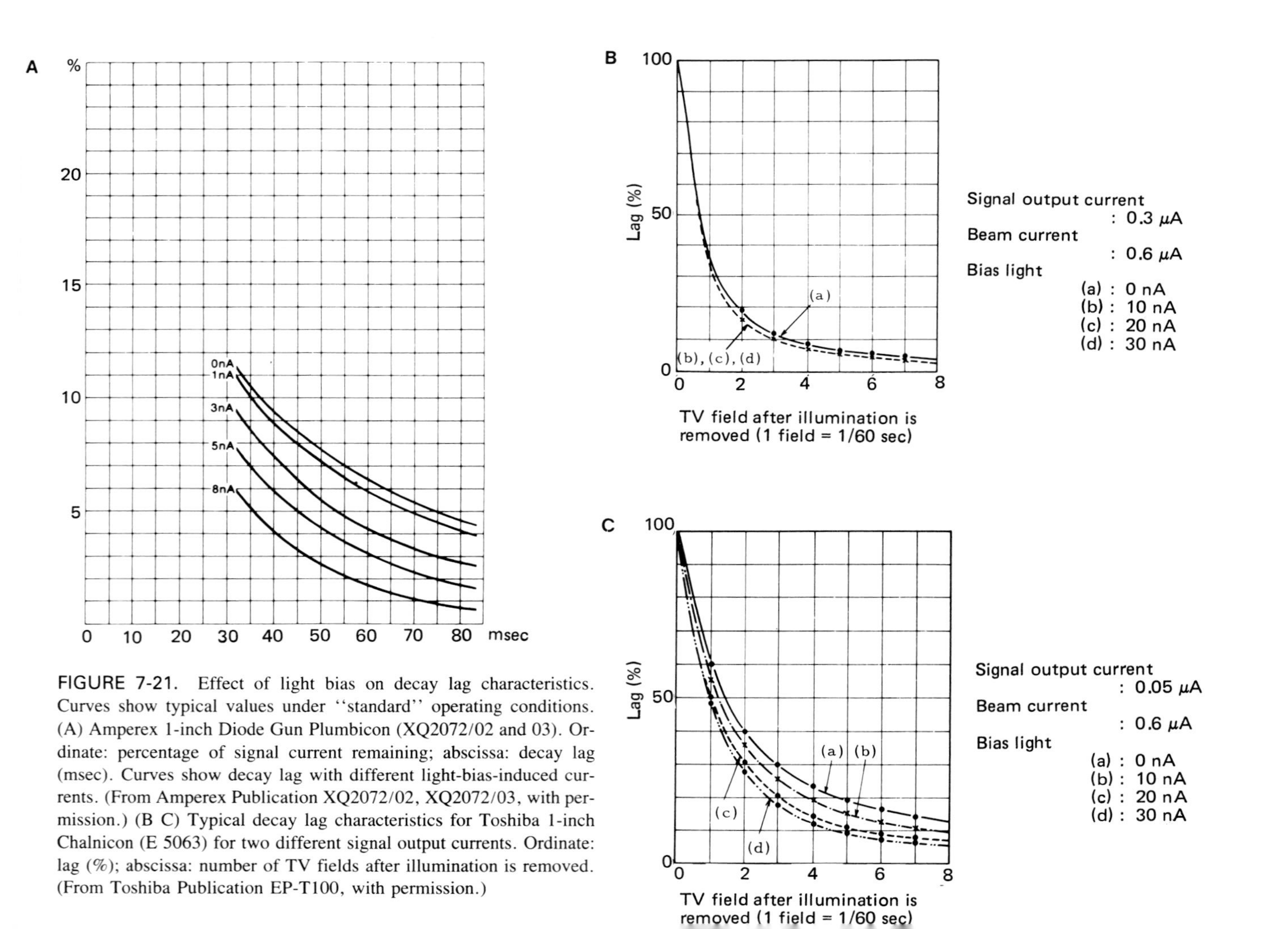

FIGURE 7-21. Effect of light bias on decay lag characteristics. Curves show typical values under ''standard'' operating conditions. (A) Amperex 1-inch Diode Gun Plumbicon (XQ2072/02 and 03). Ordinate: percentage of signal current remaining; abscissa: decay lag (msec). Curves show decay lag with different light-bias-induced currents. (From Amperex Publication XQ2072/02, XQ2072/03, with permission.) (B C) Typical decay lag characteristics for Toshiba 1-inch Chalnicon (E 5063) for two different signal output currents. Ordinate: lag (%); abscissa: number of TV fields after illumination is removed. (From Toshiba Publication EP-T100, with permission.)

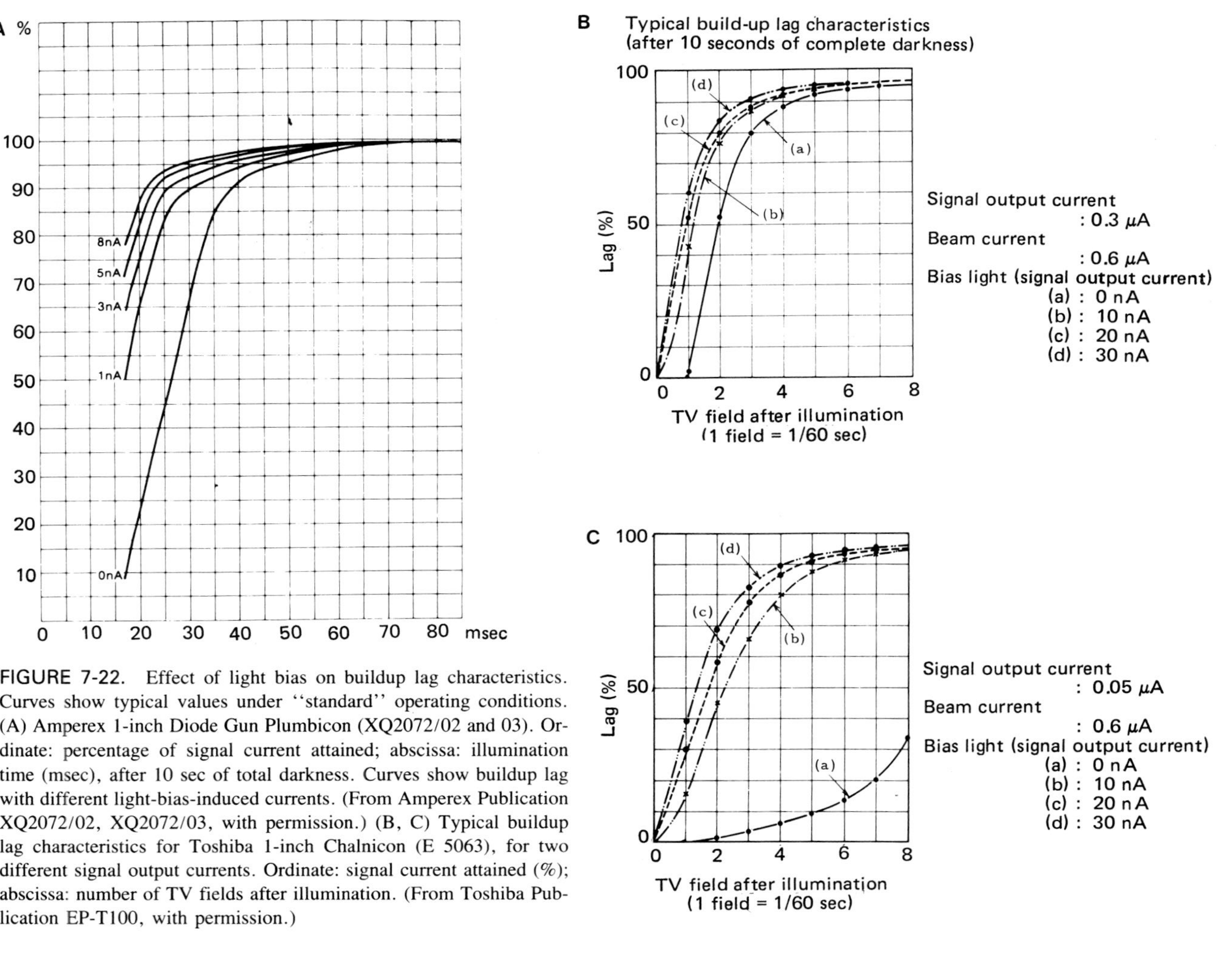

FIGURE 7-22. Effect of light bias on buildup lag characteristics. Curves show typical values under "standard" operating conditions. (A) Amperex 1-inch Diode Gun Plumbicon (XQ2072/02 and 03). Ordinate: percentage of signal current attained; abscissa: illumination time (msec), after 10 sec of total darkness. Curves show buildup lag with different light-bias-induced currents. (From Amperex Publication XQ2072/02, XQ2072/03, with permission.) (B, C) Typical buildup lag characteristics for Toshiba 1-inch Chalnicon (E 5063), for two different signal output currents. Ordinate: signal current attained (%); abscissa: number of TV fields after illumination. (From Toshiba Publication EP-T100, with permission.)

FIGURE 7-23. Blooming. Montaged picture of the bright rulings in a precision-ruled, metal-coated stage micrometer (Bausch & Lomb precision micrometer). The top half of the monitor picture shows a normal image. In the bottom half, the excessively bright lines induced blooming, which has expanded the bright image areas into adjacent areas in the dark background. Also evident is a reversed-polarity smear, the dark shadows trailing (i.e., appearing to the right of) the very bright lines.

damage is generally beam current dependent, so that the automatic protection circuit in the camera can prevent permanent damage up to a point.

Without reaching the level of illumination that actually damages the tube, the highlight of the image may still be temporarily "burned in" or "stick" in the camera tube. The tendency to *burn* or show *sticking* is a characteristic that varies according to the type of camera tube. Solid-state pickup devices are reported to show little burn. Compared to other tubes in the vidicon family, some Chalnicons and sulfide vidicons show a greater tendency to burn, especially at high dark currents. Silicon vidicons tend to burn less than other vidicon types, but the SIT tube is quite susceptible to burn. When the image does stick in a camera tube, exposure to a uniform, moderately bright illumination can speed up its dissipation.

Shading

An image that gives the appearance of suffering from nonuniform illumination is said to be suffering from shading. For such an image, one or more of its sides, or the middle of the field, appears brighter or darker in a continuously varying shade.

In video microscopy, shading can arise from nonuniform illumination, vignetting in the optical system, or by departing from Koehler illumination. Alternatively, shading can arise in the video system itself.

In the video camera or imaging device, shading is defined as the variation in amplitude of the signal output current when the sensor is illuminated uniformly. The envelope of the camera output, visualized on an oscilloscope, appears tilted or bowed despite the uniform illumination (see Section 9-3, 10.10).

Shading in video imaging devices can arise from nonuniform sensitivity of the vidicon target, intensifier photocathode and phosphor, or of the solid-state sensor elements. Shading in the camera tube can also arise from the geometry of the electron optics, or the landing angle and scanning

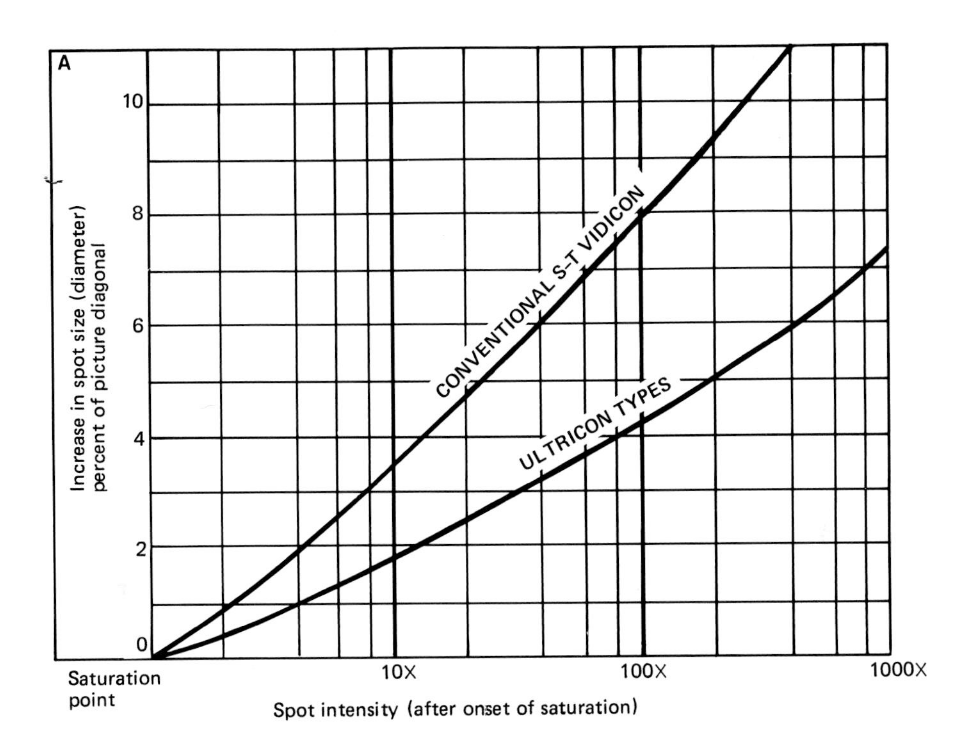

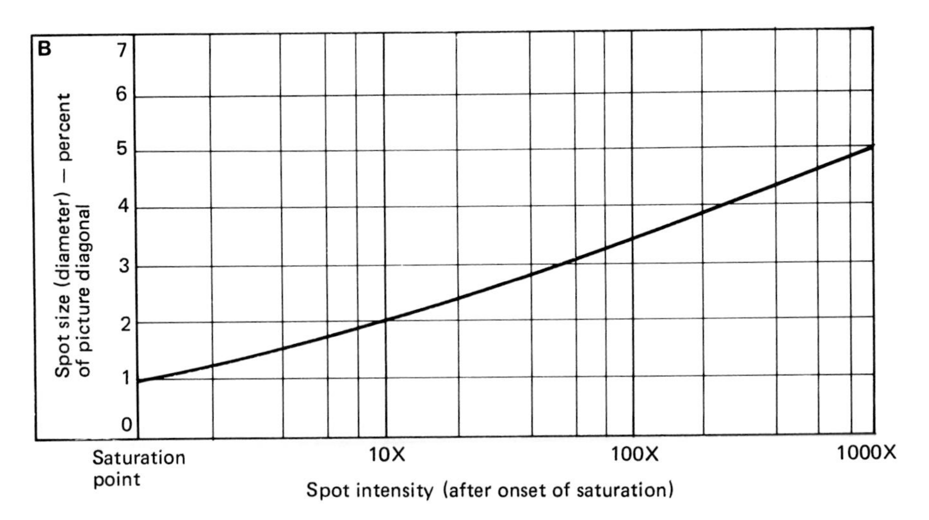

FIGURE 7-24. Blooming characteristics. Curves show typical values under "standard" operating conditions. (A) Comparison between low-blooming Ultricon types and conventional silicon target vidicons (not to be confused with SIT tubes). (Courtesy of RCA, from RCA Publication 4532/U, 4833/U, 4875/U series.) (B) Typical blooming characteristics for RCA 4804/H series, 16-mm, fiber-optic faceplate-type SIT. (Courtesy of RCA, from Publication 4804/H series).

GE TN2505 Solid State CID Camera
Non-Blooming
F 5.6 100 mm Lens

Competitive Solid State Camera
Blooming
F 5.6 100 mm Lens

FIGURE 7-25. Blooming in a solid-state imaging device. In solid-state sensors of the CCD type, blooming can cause white vertical stripes flanking bright highlights in the image. Some other types of solid-state sensors, such as CID, are less prone to blooming. (Courtesy of General Electric Co., from Publication LMC 7/82 15K.)

pattern of the scanning beam. The electron optics and landing pattern of the image-forming electrons in the intensifiers can also give rise to shading.*

Shading can be corrected in the video camera by dynamically regulating the beam current in the camera tube. A slowly rising or dropping "sawtooth" or parabolic voltage (or a combination of the two) added to the blanking signals modulates the cathode potential. The cathode potential controls the beam current in synchrony with the horizontal and vertical sweep, and provides the opportunity for dynamics shading correction (as in Fig. 9-6; see also Chapter 10).

With **shading compensation,** the signal output current becomes uniform over the field when the sensor is provided with a uniform, moderate level of illumination. For critical photometry, one may additionally need to test for uniform light transfer characteristics across the picture area (for the particular wavelengths involved), that is, the absence of shading at all relevant levels, and not just at a particular level of illuminance.

Geometrical Distortion

The image of an orthogonal grid pattern captured by the video imaging device may be distorted on the monitor or not maintain the relative spacings of the grid. Likewise, a circle may appear as an ellipse or become egg shaped (Fig. 7-36). Such geometrical distortions in video can arise from imperfections or maladjustments of the display monitor, the amplifiers and other video electronics, or the video imaging device itself.

Among the video imaging devices, the Gen I intensifiers can give rise to significant amounts of distortion. The foreshortened electron optics in the intensifier give rise to **pincushion distortion,** with the peripheral spacings of a grid pattern expanding away from the orthogonal axes of the scene. The pincushion distortion in the intensifier camera tubes can amount to 3–4% or greater at the circumference of a centrally located circle whose diameter equals the height of the picture.

In the vidicon family of camera tubes, geometrical distortion is generally less than 1% (except for a maximum value of 2% for the Plumbicon, as listed in "Imaging Device Performance in the C-1000 Camera," Hamamatsu Systems, Inc.). Distortion in vidicons is minimized by proper design of the electron optics and selection of the accelerating voltages, but can become larger if the horizontal or vertical sweep becomes *nonlinear*. Nonlinear sweep converts a circle into an egg shape. As described earlier, nonlinearity of sweep can arise from improper design or operation of the deflection circuit that includes the high-impedance deflection coils.

Geometrical distortion in the vidicon camera tubes can worsen if the sweep of the electron beam is misaligned relative to the center of the target. The adjustable *deflection magnets,* which are built into the yoke assembly to align the beam, can be used to compensate, for example, for strong external magnetic fields. However, if not properly positioned, these alignment magnets can offset the beam and introduce additional geometrical image distortion.

Geometrical distortion, focusing errors, and shading in vidicon camera tubes are sometimes reduced by **dynamic focusing.** Dynamic focusing automatically regulates the deflection, focusing, and landing angle, of the electron beam as the beam is scanned. Thus, dynamic focusing can correct for some of the imaging problems that are difficult to compensate with static electron optics.

In solid-state pickup devices, the location of the picture elements is fixed by the orthogonal array of photodiodes. Thus, geometrical image distortion can be very low. In addition, the distortion is less susceptible to external fields, since there is no scanning beam or electron optics to be perturbed.

Solid-state sensors arranged in a single row, or *linear imaging devices* or *linear position sensors,* can provide very high resolution and positional accuracy along a single line. A well-resolved, large picture free of geometrical distortion is obtained by scanning the image with a linear image sensor containing several thousand pickup elements. Such devices are produced by a number of manufacturers,† and are used in applications such as facsimile, computerized tomography, and optical character recognition.

*In addition to the video imaging device itself, the electronic circuits in the video camera and other components, and especially the properties of the monitor, can be responsible for shading.

†For example, Fairchild, Hamamatsu, E G & G Reticon, Texas Instruments.

Blemishes

Discrete permanent defects in the video image that appear independently of the scene are known as *picture blemishes*. Commonly, blemishes arise from imperfections in the video imaging device. Blemishes in the imaging device can arise in the vidicon target, intensifier photocathode or phosphor, sensors and gate structures in the solid-state sensors, or in the faceplates, including the fiber-optic plates.

The most obvious of the common blemishes are the bright *spots* originating from hot spots on the conductive target in the vidicon tubes, or by excessively emitting electron sources in the cathode of the intensifier. Less obvious by casual inspection are those resulting from missing pixels, or defects that reflect discrete sensor areas with low sensitivity. In the Gen II intensifiers and solid-state pickup devices, these defects can be particularly prominent. Blemishes can also take the form of regular patterns (e.g., from the fiber-optic plate), smudges, streaks, and mottled or grainy background.

Since picture blemishes noticeably detract from the quality of the video image, manufacturers carefully specify the amount of blemishes allowed for each tube type. The number, and size of spots (commonly expressed as number of scan-line widths), with greater than some defined contrast allowable in different zones of the image, are tabulated in most camera tube specifications.

In video microscopy, a variety of sources in the microscope optical train other than the video sensor can introduce prominent picture blemishes. The means for detecting, isolating, and, where possible, removing the source of the blemishes in a video microscope are detailed in Sections 3.9, 3.10, and 5.10.

7.3. CHOICE OF VIDEO CAMERAS

General Considerations

As discussed in the previous section, the video image pickup device ultimately limits the resolution and sensitivity of the video system. We should therefore choose the type of pickup device that meets our particular needs in video microscopy. However, as an end user, we seldom purchase the pickup device by itself. Instead, we purchase a *video camera*, which includes the camera housing with lens and camera mounts, the pickup device, yoke, power supply, clock and drive circuits, video amplifiers, various signal-adjusting circuits, and the signal output (and sync) connector(s).

Cameras may be designed as single, self-contained boxes to which the lenses, microscope, AC power,* and video output cables are connected. Other cameras are designed in two units with a *camera head* separated from a *control box* that also includes the power supply. The two units are connected with a flexible, shielded multiple cable. The two-unit design provides several conveniences such as a smaller and considerably lighter head, remote control of the video camera functions, control of camera functions without risk of vibrating the microscope, etc. On the other hand, there is some increased risk of picking up ground-loop or RF interference in the circuit and cable that link the two units.

Many factors enter into the choice of video cameras for microscopy: monochrome or color; desired sensitivity, resolution, contrast, and gray shades; acceptable degree of lag, geometrical distortion, picture blemishes, and noise; dynamic range and safety (shutoff) features; automatic and manually adjustable signal-processing features; scan rates, interlace, and clock stability; external sync capability; type of power supply and environmental requirements; mechanical construction, including type and quality of lens mount and camera mounting screws; dimensions and weight; ruggedness, durability; freedom from electromagnetic interference; service reliability; special attributes and convenience features; potential for use other than with the microscope; cost.

*Many cameras, especially color cameras, are DC operated by batteries to provide mobility and permit field work. An AC power adapter is provided with most such cameras.

The criteria for selection of some of these factors follow from our earlier discussion. Others require further elaboration, as follows.

Monochrome or Color

Monochrome and **color cameras** differ from each other in more ways than just the ability of the latter to handle color. The characteristics required of the pickup tubes are different, as are the nature of the output signal, and the optical arrangement and camera circuitry necessary to produce the color signal. We shall defer a discussion on color cameras to a later section, but it is important to realize that even an expensive color camera cannot necessarily substitute for a good monochrome camera. In a color camera, much design effort goes into the generation of a good color signal with its limited bandwidth, while with a monochrome camera more emphasis can be placed on the achievement of high sensitivity, high resolution, high or variable contrast, and good gray scale. The counterpart of this argument is that color cameras tend to be limited in sensitivity, resolution, achievable luminance contrast, and luminance gray scale, compared to monochrome cameras designed to excel in these qualities.

Monochrome cameras with similar sensitivity and image quality also tend to be considerably less expensive than color cameras. Color cameras clearly have their place in microscopy, but should not be purchased or used as a substitute for monochrome cameras.

Sensitivity

The sensitivity of the video camera is almost exclusively determined by the sensitivity or responsivity of the video imaging device. Therefore, one should select a camera equipped with the imaging device that provides the needed sensitivity. However, one should bear in mind that the camera sensitivity required for different modes of microscopy may or may not be proportional to the luminance of the image commonly experienced in viewing or photographing the microscope image. As already mentioned, the sensitometric characteristic of a camera tube can be quite different from the radiance response of our eye. Also, photographic emulsions tend to be more sensitive in the blue than toward the red end of the visible spectrum.

In addition, some video sensors and cameras respond nearly linearly to varying levels of illumination (i.e., with a gamma of 1.0), while our eye and photographic plates respond logarithmically. These differences in response provide an interesting advantage for video as applied to microscopy in polarized light, interference contrast, DIC, fluorescence, darkfield, etc., as extensively discussed elsewhere.

Without entering into a detailed discussion on these topics, we shall classify cameras and imaging devices according to the sensitivities generally needed for different modes of microscopy. Table 7-4 lists a representative series of (1-inch) imaging devices in approximately descending order of luminous sensitivity.* The listed applications in microscopy indicate typical uses as approximate guidelines.

Resolution

As discussed earlier, the video imaging device characterizes the limiting resolution, amplitude response and their dependence on faceplate illuminance, of the video camera. Nevertheless, the balance of the camera also plays an important role in determining whether these sensor characteristics are realized. The camera supplies the regulated power, electrode potentials, the sync and blanking signals, the raster drive, and other voltages needed to operate the imaging device. Accurate focusing of the beam is needed to provide the small, regularly shaped aperture that scans the image. Highly stable, and strictly linear, horizontal and vertical sweeps free from AC line interference, as well as precisely timed sync pulses, are needed to accurately position the scanning aperture. Thus,

*The cameras were tested in our Laboratory, many in collaboration with the staff and students of the Analytical and Quantitative Light Microscopy Course, Marine Biological Laboratory, Woods Hole, in 1981 and 1983.

TABLE 7-4. Sensitivity of Selected 1-Inch Video Cameras

Imaging device (camera source)	Character of noise[a]	Sensitivity class	Typical applications in microscopy
I³-Vidicon (Zeiss–Venus)[e]	Snowstorm		
ISIT (RCA,[b] Sierra,[b] Dage–MTI)	Dancing bright spots	Quantum limited, to extremely sensitive	Extremely low-light-level chemiluminescence, phosphorescence, fluorescence
I-Isocon (RCA, Cohu)[c]			
I²-Newvicon (Cohu)	Dancing bright spots		
Isocon (RCA, Cohu)[b]			Fluorescence; darkfield; high-extinction polarization
I²-Vidicon (Zeiss–Venus)	Fine-grain salt and pepper	Very sensitive	
SIT (Dage–MTI)	Finer-grain salt and pepper		
MOS (Hitachi)[d]			Polarization with moderately low retardation; DIC interference; modulation contrast; phase contrast; brightfield
Newvicon (Dage–MTI)			
Ultricon (Dage–MTI)	Very fine white snow	Sensitive	
Chalnicon (Hamamatsu)			
CID (Panasonic)[d]			
Plumbicon (Amperex, Hamamatsu)[c]		Better than average sensitivity	Phase contrast; brightfield; color
Saticon (JVC)			
Sulfide vidicon (Sony)[d]		Average sensitivity	

[a]At limiting light levels.
[b]Personal communication and catalog description only.
[c]Catalog description only.
[d]2/3-inch tube or format.
[e]See also Photon Counting Camera (Hamamatsu), and intensified CID video camera (Xybion).

the quality of the power supply, sync-pulse and sweep generators, as well as the mechanical support for the video camera tube, yoke, and the camera body relative to the microscope optical train, all affect the resolution of the image. Most important are the precision of the master oscillator, and the abilities of the video amplifiers to properly shape and faithfully propagate the video signal arising from the imaging device. As discussed in Chapter 6, adequate bandwidth and proper phase-handling qualities of the amplifiers and other video circuits are essential for handling the high-frequency video signal without distortion and for providing an image with high resolution. Some high-resolution cameras incorporate an "aperture correction" circuit to maximize resolution.

Most instrumentation cameras are designed to deliver a horizontal resolution limited by the video imaging device itself. The vertical resolution is limited by the number of scan lines, vertical blanking interval, and the Kell factor. Some cameras can be used with greater than 525 scan lines, and given an appropriately reduced aperture size and the needed scanning precision and bandwidth, can provide resolutions that are limited by the target in the camera tube, both horizontally and vertically.

In video microscopy, the resolution of the video camera should, if possible, be somewhat above that of the microscope. Where possible, the limiting resolution of the camera should approximately double that of the microscope (Section 10.4, Appendix II), but the available light level or the need for a wider field of view may preclude an optimum match. Generally, a camera resolution of no less than 600 horizontal by 340 vertical TV lines is desirable for obtaining a "presentable" picture. As discussed elsewhere, even lower-resolution images can be acceptable at times, especially for high-contrast, dynamic scenes. Nevertheless, since the product of the MTFs of all video components determines the MTF of the whole video train, cameras with higher resolution and amplitude response tend to yield a better-resolved, crisper image. Cameras yielding above 800 TV lines of horizontal resolution in fact yield a noticeably sharper image than cameras yielding 600 horizontal TV lines PPH, even when the tape recorder, monitor, etc. limit the overall resolution of the video train.

Sensor Size

The sensitivity and resolution of a video imaging device improve with increasing sensor size. Thus, if cost, size, and maintenance were of no concern, one would probably choose cameras with the largest sensor area for video microscopy. In practice, the image format used in the "1-inch" tube seems to provide a reasonable compromise between sensitivity and resolution versus price and availability. Cameras having a ⅔-inch tube are less expensive and are used extensively in small video cameras, but except for color work, are not recommended as the main camera in video microscopy. One-inch tubes are widely used in instrumentation cameras and provide a significant improvement in sensitivity and resolution, as well as generally improved image characteristics, over the ⅔-inch tubes.* For video microscopy in extremely low light levels, intensifier tubes with greater than the standard 1-inch format could provide significant improvements in resolution, image quality, and sensitivity. Several types of intensified tubes with greater than 16-mm image diagonal are listed in Table 7-2 (e.g., see "Imaging Devices," IMD 101, RCA). Table 7-5 lists nominal camera tube sizes and standard active sensor areas.

Contrast

By using video, we can boost or otherwise adjust the image contrast, features that are of major advantage to microscopy. Contrast can be adjusted by analog and digital means after the video signal leaves the camera (Chapters 9, 10), but the camera itself can also be equipped with circuits that substantially boost or allow adjustment of image contrast.

*The same microscope field projected on a ⅔-inch camera tube compared to a 1-inch tube would provide over two times the light per unit area of the sensor, since the same amount of light would be concentrated in an area only half as large [$(11\ \mathrm{mm}/16\ \mathrm{mm})^2$, Table 7-5]. Nevertheless, for reasons that are not obvious, 1-inch camera tubes actually perform better than ⅔-inch camera tubes with the same type of target material.

TABLE 7-5. Standard Video Sensor Dimensions

Nominal tube[a] or device diameter		Image diagonal		Active area of sensor (width × height)	
mm	inch	mm	inch	mm × mm	inch × inch
18	2/3	11	0.43	8.8 × 6.6	0.346 × 0.260
25	1	16	0.63	12.8 × 9.6	1/2 × 3/8
		16	0.63	14.1 × 14.1[c]	
		25	1.0[b]		
		40	1.6[b]		
38	1.5	25.4	1.0		
		40	1.6[b]		
		80	3.15[b]		
51	2.0	33	1.3		
		45	1.8[b]		
81	3	41	1.6		

[a]For intensifier tubes, diameter of vidicon section.
[b]Maximum image diagonal for intensifier (from RCA "Imaging Devices," Publication IMD 101).
[c]Toshiba Chalnicon.

Before considering the contrast-boosting or -adjusting functions of the camera, we should briefly recall the nature of image contrast. As discussed in Chapter 4, contrast is a measure of the difference in luminosity of adjoining areas that gives rise to a visual sensation of luminance differences. Numerically, we adopted the definition for contrast of $\Delta I/I$, or the difference in intensity (ΔI) over the intensity (I) of the background. *Contrast is thus increased either by raising* ΔI *or by lowering* I. In other words, the gain of the camera need not necessarily be changed, but the contrast of the same scene can be increased by raising the threshold of the camera, or the level of light that the camera accepts as being the black level. Reducing the background (I) without changing ΔI effectively raises the contrast. The **auto black,** or **automatic pedestal control** circuits provided in some cameras, very effectively increase image contrast in this fashion. The same features can also be used to markedly improve the S/N ratio, and thus contrast, in intensifier cameras working at the lower end of their sensitivities.

In addition to using the black level adjustment, contrast can also be boosted or adjusted with the **video gain** control. While maintaining the gamma at unity, so that the video output signal is still proportional to faceplate illuminance, the gain of the video amplifier can be raised to give a greater signal for the same increment of faceplate illuminance. The curve relating camera output to faceplate illuminance remains a straight line (within the dynamic range of the camera), but the steepness of its slope is raised.

In applying video to microscopy, *adjustable black level and gain control are essential features.* The black level and gain are adjusted *automatically* in some cameras to optimize contrast of the scene; other cameras provide the option of switching to *manual control.* The latter provision combined with a gamma of unity is needed for applications involving photometry. Manual control also permits more precise control of image contrast by the operator, a feature especially useful for intensifier cameras. Contrast enhancement in video microscopy is discussed in Sections 3.5, 3.7–3.9, 4.5, 7.5, 8.4, 9.3, 10.6, and 11.3.

Gray Scale

The light transfer characteristics curve, noise characteristics, shading, and blooming, all affect the number of shades of gray, or **gray values,** that the video camera can capture. For visual observation, subtle image details can be brought out as the number of gray values is raised, up to a point. Even more gray values are needed for digital image processing and photometry (Chapter 10).

Camera manufacturers often specify the number of steps of gray scale rendered by the camera. Gray scales used as test targets are commonly made with discrete steps of $\sqrt{2}$ of reflectance or transmittance, so that the number of steps presented in the camera specifications does not have the same meaning as the number of gray values that can be represented by the signal from that camera. Since the number of gray values that is, in fact, meaningful depends on the size and shape of the specimen, as well as on the shading, blooming, and the character and magnitude of the noise signal, careful evaluation is needed to select the camera that is optimal for any particular type of application, especially in video microscopy involving photometry.

Lag, Geometrical Distortion, and Other Picture Defects

Lag, bloom, burn, picture blemishes, shading, and geometrical distortion are introduced primarily by the imaging device. As discussed earlier, lag can be reduced in some tubes with bias illumination, and shading can be corrected in the cameras. Most of the other picture defects cannot be easily corrected, or would require expensive devices to correct, so that the limiting characteristics of the imaging device should be kept in mind when choosing the video camera.

Given the same camera tube, shading and geometrical distortion can be worse in less expensive cameras, since their power supplies tend to be poorly regulated and the horizontal and vertical drives tend to depart from linearity.

Noise

The level and character of the noise varies with the video imaging device, as already discussed (Tables 7-1 to 7-4, Section 7.2). The sensitivity levels, bandwidths, and filtering and other electronic circuit characteristics chosen for the camera, also affect the noise in the camera output signal.*

The camera designer not only selects the type of imaging device, but the means for optimizing its performance. Should the camera be made more sensitive but allow additional noise, or vice versa? How much can one sacrifice in bandwidth, and hence resolution, to keep the noise down? What kind of filters will sharpen the picture detail, or conversely, portray uniform gray areas without introducing ringing or unsharp edges? How much does one shield against noise generated by AC-line and other electromagnetic interference and not lose circuit flexibility and compact size of the camera? The camera design engineer makes many such adjustments and compromises among conflicting needs, and with the constraint of cost. But noise tends to creep in at all levels in the camera and elsewhere in the video system.

In general, one needs to choose a camera with adequate sensitivity, but not much greater than that needed for the intended type of microscopy. Not only will the camera be more costly, but the level of noise from many sources is likely to rise.

Dynamic Range and Safety Shutoff

For video microscopy, both the intrascene dynamic range and the useful light range can influence the utility of the camera. A camera with a broad useful light range coupled with *automatic gain control*, or in some cases **automatic target control,** would adapt to scenes with varied brightnesses. This is especially useful with low-light-level microscopy, where occasionally a much brighter image may suddenly enter the scene.†

*In fact, one should be aware that some cameras are equipped with **automatic bandwidth suppression** circuits. These circuits reduce the passband at low signal levels, reducing the apparent amount of noise that is seen at low light levels.

†The broad automatic light range listed for many cameras includes automatic lens iris control, a feature not directly adaptable to video microscopy.

Most cameras have a built-in AGC circuit to adjust the amplifier gain to the level of the incoming signal. Some vidicon and solid-state cameras have a HIGH–LOW GAIN switch that adjusts camera sensitivity, while intensifier cameras may be equipped with a high-voltage adjusting knob for controlling the intensifier gain over a wide range, and a target control to adjust the vidicon gain.

In addition to these manual (or automatic) controls for camera sensitivity and image contrast, some high-gain cameras are equipped with *automatic shutoff* circuits that protect the camera in case of excessive exposure of the tube to light. Without the fast response provided by the automatic protection circuit, the camera tubes and circuit in an intensifier camera might be damaged before the exposure could be interrupted manually. The importance of the automatic shutoff feature increases with the sensitivity (and cost) of the camera.

In video microscopy, cameras equipped with Newvicon, Chalnicon, and related types of vidicon tubes may encounter luminance ranges in the scene that are considerably greater than the dynamic range of the camera tube. In order to prevent an undesirably high white signal (which can introduce **streaking** in the monitor and otherwise distort the image), some cameras are equipped with a **white clip** or **white punch** circuit. The white clip or punch circuit also allows a higher contrast to develop in the midbrightness range of the image. With the white clip circuit, the part of the image above a certain brightness level no longer gives rise to an increased signal. With white punch, the signal for the highlight above a certain brightness is flipped over in sign, so that in that range greater intensity is represented as decreasing brightness on the monitor image; hence, very bright peaks are "punched out" but still traceable.

Signal-Processing Features

Camera features that were listed above, under sensitivity and contrast, allow us to regulate image contrast and brightness. Also, the filtering circuits built into the camera can control the noise and enhance edges and fine details in the image. Therefore, the camera circuitry itself provides or allows varying degrees of image processing.

The sharpness, and overall excellence of the image quality provided by modern video equipment, depend to a significant degree on these features designed into the camera or camera control box circuitry. For video microscopy in particular, these features, coupled with the moderately high sensitivity and desirable image qualities of the Newvicon and Chalnicon tubes, have made these instrumentation cameras (and sometimes even some surveillance cameras) most useful. The image can also be processed in several ways after the video signal leaves the camera. The equipment and principles of analog and digital signal processing are described in Chapters 9 and 10.

Scan Rates, Interlace, and Clock Stability

The relations of scan rates and interlace to image resolution were discussed extensively in Section 6.4. Special high-resolution cameras use more than 525 scan lines/frame or nonstandard interlace to achieve the greater resolution. A high-resolution video image is made up of a greater number of picture elements and clearly has desirable features. Also, some cameras are driven with nonstandard V- and H-scan rates to improve compatibility of the output signal with computer interfacing (e.g., Hamamatsu). These advantages are sometimes offset by incompatibility with other commonly available video equipment including VTRs, processors, and standard computers.

The need for a stable clock, or master oscillator, and precisely linear horizontal and vertical scan for producing an undistorted image, has been emphasized repeatedly. Without these qualities and a well-focused small scanning aperture, the camera could not produce a signal that yields an excellent image. While a well-balanced design is important for camera and other video equipment, the quality of the clock and timing circuits takes on an importance next only to the imaging device itself. High-quality cameras contain a clock governed by a quartz crystal oscillator, or are driven by an external timing generator.

External Sync Capability

In order to maintain the camera timing circuits under the control of an external sync source, many cameras are equipped with *external sync* capability. The external source can be a sync generator that runs many video components in strict synchrony, or can be another camera, VTR, etc. with whose output one wishes to superimpose or montage the camera output (Chapter 12).

The external sync on the camera may either use horizontal and vertical external sync pulses (generally at −4 volts), or take the form of a *genlock*. For genlock, the composite video output (generally 1 volt peak-to-peak) of the other camera, VTR, etc. is simply connected to the GENLOCK terminal. The camera strips and uses the sync pulses provided by the external composite video signal to drive its timing circuits. With some cameras the external sync capability is a standard feature, while with others it is **optional.**

Other Considerations

I mentioned the pros and cons of the single- versus two-unit design of the video camera at the beginning of this section. As implied there, some cameras are more susceptible to ground-loop or common-mode *AC interference* than others. This type of noise can interfere with the displayed image and also disturb the sync. Ground-loop interference can make the vertical sync unstable, or can generate intrusive noise including horizontal lines or bars at fixed heights across the monitor screen. Common-mode interference can result in regular, intrusive geometrical patterns that are spaced across the screen.

Whenever possible, one should first try out a camera on the actual microscope and site. Without such trial, it is often difficult to predict camera susceptibility to electromagnetic interference or ground-loop problems in a particular environment. Unforeseen incompatibility with the intended application may also show up.

While the electronics of video cameras are generally well engineered, the mechanical design does not always match the quality of the electronics. The quality of design and machining of the *lens mount* should be observed closely, and the fit of the camera lens mount with the lens or coupler should be tested. The fit of the threads should be neither too loose nor too tight, and the lens or coupler should sit snugly, without play, against the shoulder on the camera. The lens mount and/or the *camera support* generally define the relation between the positions of the microscope image and the video imaging device. Reliable and secure alignment of the two are therefore very important.

Likewise, the convenience and reliability of the device for *focusing* the camera tube inside the camera is important. Inside the camera, the pickup tube should be held firmly in place by an adjustable clamp. While the clamp must securely position the tube, the clamp must also be releasable for tube replacement. The tube also needs to be released for rotational alignment of the tube relative to the yoke; with magnetic deflection camera tubes, tube orientation can affect ringing in the image.*

U.S. Sources of Instrumentation and Special-Purpose Cameras

Several sources of *instrumentation cameras* in the United States are listed in Table 7-6, which also includes some sources of *special-purpose cameras* described below.

Video sensor noise commonly rises with temperature, and can be reduced in some cameras by lowering their temperature. CCD cameras are especially suitable for noise reduction by chilling. As mentioned earlier, the low noise in *chilled CCD cameras* can allow photon integration for very long periods, thus yielding very high sensitivity as well as an expanded linear dynamic range. These are characteristics desirable for *precision photometry*, provided the long integration time (and loss of temporal resolution) can be tolerated. CCD cameras cooled thermoelectrically and with liquid nitrogen are now commercially available (e.g., Photometrics Ltd.).

*The short index pin is intended to indicate the direction of the horizontal raster. Fine-tuning of tube orientation is sometimes needed to eliminate camera ringing.

TABLE 7-6. Selected U.S. Sources of Monochrome Instrumentation Video Cameras

High resolution	Low light level	Solid-state	Special scan rate or imaging rate		
√	√	√	√	Cohu Electronics Division	5725 Kearny Villa Road, San Diego, CA 92138
√	√		√	Dage–MTI	208 Wabash Street, Michigan City, IN 46360
		√	√	Data Copy Corporation	1070 E. Meadow Circle, Palo Alto, CA 94303
		√	√	E G & G Reticon	345 Potrero Avenue, Sunnyvale, CA 94086
		√	√	Fairchild CCD Imaging	3440 Hillview Avenue, Palo Alto, CA 94304
		√		General Electric Company	890 Seventh North Street, Liverpool, NY 13088
√	√	√	√	Hamamatsu	332 Second Avenue, Waltham, MA 02154
		√		Hitachi America, Ltd.	500 Park Boulevard, Suite 805, Itasca, IL 60143
√	√	√		Ikegami	37 Brook Avenue, Maywood, NJ 07607
√	√	√		ITT, Electro-Optical Products Division	3700 E. Pontiac Street, Fort Wayne, IN 46801
	√	√		Javelin Electronics	P.O. Box 2033, Torrance, CA 90510
		√	√	NAC	Instrumentation Marketing Corporation 820 S. Mariposa Street, Burbank, CA 91506
	√	√		Panasonic, Matsushita Electric	One Panasonic Way, Secaucus, NJ 07094
			√	Photmetrics	1735 E. Ft. Lowell Road, Tucson, AZ 85719
	√	√		RCA, Electro-Optics and Devices	New Holland Avenue, Lancaster, PA 17604
√	√	√	√	Sierra Scientific	2189 Leghorn Street, Mountain View, CA 94043
		√	√	Sony Video Communications	5 Essex Road, Paramus, NJ 07652
		√	√	Spin Physics, Kodak	3099 Science Park Road, San Diego, CA 92121
		√		Texas Instruments, Optoelectronics	P.O. Box 5012-M/S34, Dallas, TX 75222
		√	√	TriTronics, Video Products Division	2921 W. Alameda Avenue, Burbank, CA 91505
√	√			Westinghouse	Westinghouse Circle, Horseheads, NY 14845
	√	√	√	Xybion	7750-A Convoy Court, San Diego, CA 92111
	√			Zeiss–Venus	One Zeiss Drive, Thornwood, NY 10594

Image-scanning cameras with linear solid-state sensor arrays (**image dissector**) are used to obtain images with high resolution and *low geometrical distortion* (e.g., Data Copy Corp.). Some solid-state sensors are also used for *high-speed* video imaging (e.g., Spin Physics Motion Analysis System). Other cameras use vidicon-type camera tubes and high-speed shutters and stroboscopic flash illumination to *freeze* the images of rapidly moving objects (e.g., Sony SVM 1010 Video Motion Analyzer). Still other cameras use special optical scanners or shutters to expose segments of the video camera tube to obtain recordings at multiple TV rates (e.g., TriTronics Stop Motion High Speed Monochrome Camera; see Section 8.6).

7.4. COLOR CAMERAS AND THE ENCODED NTSC SIGNAL

The majority of video cameras available today are made to reproduce color rather than monochrome images. In principle, color cameras carry out functions of four monochrome cameras, one each for the three primary colors, red, green, and blue, and one for the *luminance,* or levels of luminosity, of white and gray shades. Color cameras are thus considerably more complex than monochrome cameras, both optically and electronically. Their sensitivity is lower, contrast for monochrome scenes is limited, and they cost more than monochrome cameras. Color cameras do have a place in video microscopy as described below, but should not be used in place of a monochrome camera.

Despite these complexities, modern color cameras are efficiently engineered, and their sensitivity, image quality, and color have been sufficiently improved, so that late-model color cameras (especially those using Saticon, Newvicon, etc., type vidicon tubes) can produce very respectable images of colored scenes seen through the microscope. Compact color cameras or camera heads are available, both in the tube type and with solid-state pickup devices.

Color cameras are needed to capture the colors in the scene, but not for color-coded or "pseudocolor" display of scene features. For the latter, only the monitor needs to be capable of handling color; colors are assigned after the fact to features of the image captured by one or more monochrome cameras.

Color cameras are generally of two types, the three- and four-tube types, or the single-tube (or pickup device) type. In the three-tube type, the incoming optical image is split into three images with a set of beam-splitting prisms or mirrors that, combined with "trim filters," produce the red, green, and blue images (Fig. 7-26). The three images are focused onto the faceplates of three, matched camera tubes driven by the same clock, so that the red, green, and blue images produce separate, but synchronized, R, G, and B signals. In the four-tube type, a fourth camera tube captures the luminance distribution in the image. In the single-tube type, the camera tube itself splits the image into its three primary color components and produces the RGB signal.

The signal coming out of the color camera in principle can be of the *RGB* type containing the three synchronized (composite) video signals for the red, green, and blue channels (Fig. 7-26A). Or it can be of the **YRGB** type, with the fourth signal (**Y**) for luminance, or gray shades, provided separately. Commonly, the RGB or YRGB signals are *encoded* (by a **color encoder**) in the camera and combined into a single *NTSC color signal* (Fig. 7-26B).

The RGB signal retains the resolution of each of the three primary color pickup tubes, but can only be viewed in color through an RGB monitor. The signals to the RGB monitor have to be carried over three separate coaxial cables, and the RGB signal cannot be recorded with a standard VTR.

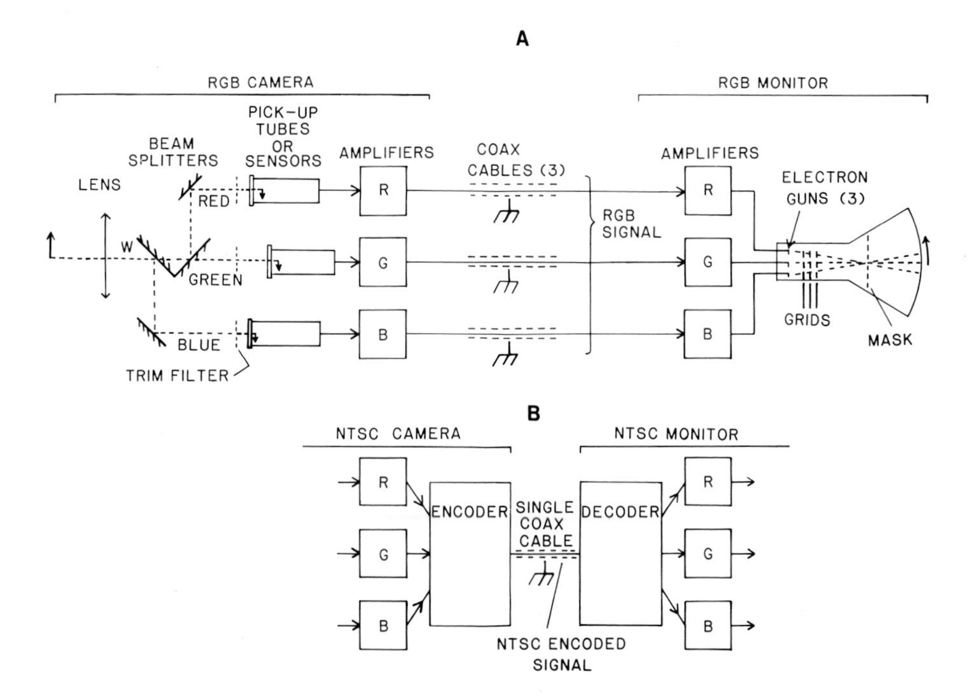

FIGURE 7-26. Schematic diagrams of RGB and NTSC formats for handling color video signals. See text.

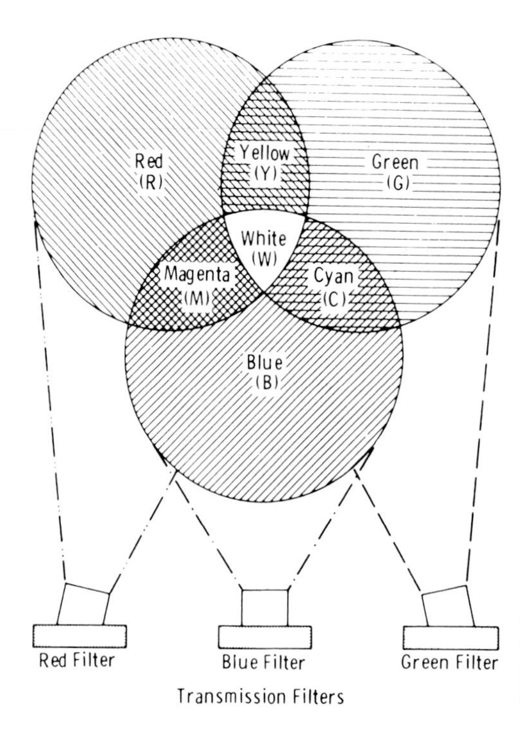

FIGURE 7-27. Combinations of light of primary colors. Light (not pigment) of the three primary colors (red, green, and blue), when combined in appropriate ratios, can give rise to the sensation of all the colors. (From Ennes, 1979.)

In contrast, the encoded NTSC signal can be carried over a single cable or channel, can be recorded by a VTR, and can be viewed on a standard color monitor that contains a **decoder.** The NTSC signal can also be viewed without color on a monochrome, or ''black and white,'' monitor with the colors represented by naturally appearing shades of gray.

By encoding and merging the three RGB signals or the four YRGB signals into a single NTSC signal (or PAL* or SECAM† signal in the European and other formats), we gain the convenience of simplified transmission and a brandwidth no greater than that required for monochrome TV transmission. The encoding into a single signal with restricted bandwidths is necessary for broadcast and other production applications and for VTR recording. However, we pay for the convenience by the major loss of resolution in the images formed by the three primary colors (the pixel-to-pixel color resolution is lost), as well as some compression of the amplitudes of the color signals. The rationale for the compromise used in *color video encoding* follows.

To an average human observer, light of the three primary colors superimposed in various ratios gives rise to the sensation of all colors, including their *hue* and *saturation* (Figs. 4.19, 7-27, 7-28). [Red, orange, green, blue, etc. are hues; an unsaturated red is pink, or if totally unsaturated, it is white. The degree of saturation measures how much white is mixed with the particular hue (Section 4.7).] When light of the three primary colors, red, green, and blue, are added together in an appropriate ratio, the sum appears to an average observer as white or gray depending on the luminance level.

In a colored scene, we perceive the image *detail* primarily through our gray shade, or luminance, sensation. Color does not contribute appreciably to the perception of image detail (narrow colored stripes are perceived as shades of gray), and color needs only to be ''painted in'' with a much lower resolution. NTSC and other encoded color video systems take advantage of this property of our visual system (Bedford, 1950).

In color encoded video, the *luminance signal,* or *Y signal,* either is provided by a separate

*Phase-alternating line system (Telefunken).
†Systéme Electronique Couleur Avec Mémoire (Henri de France).

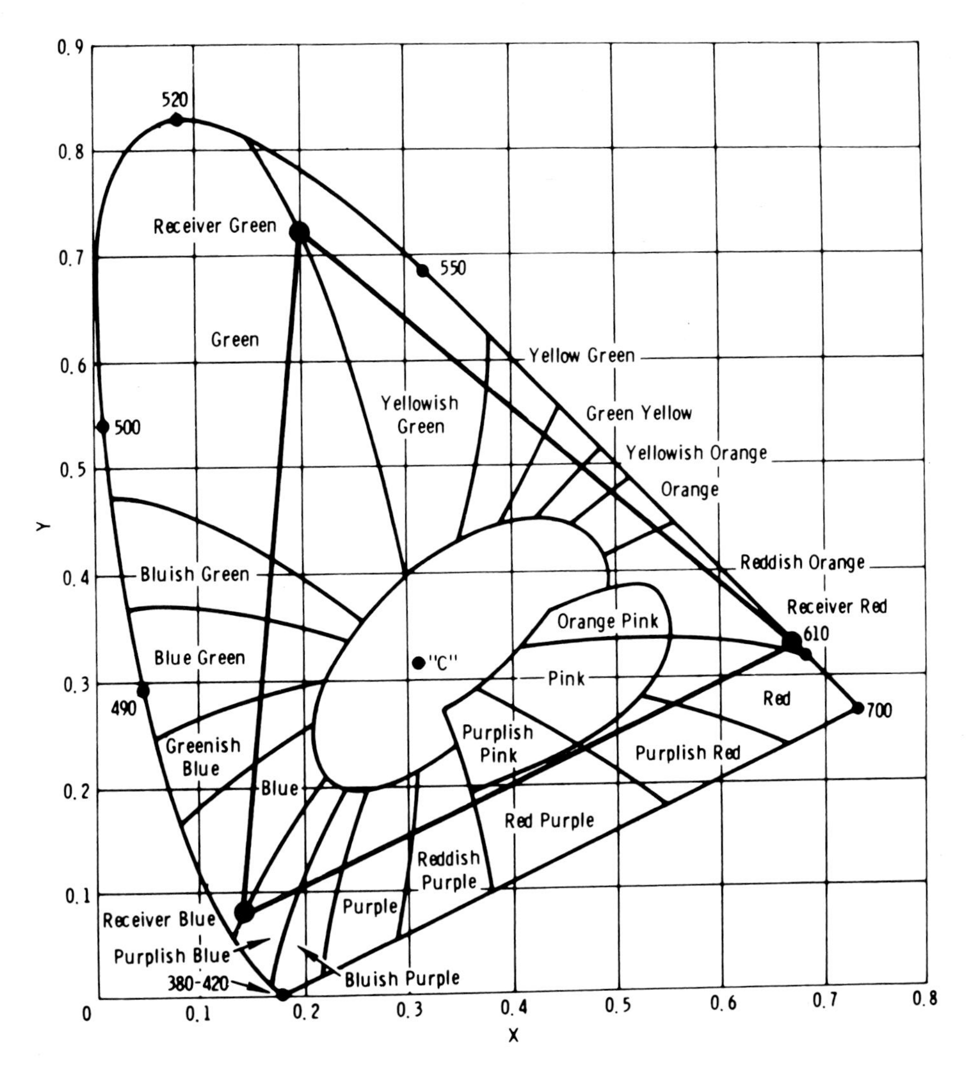

FIGURE 7-28. CIE chromaticity diagram. Colors at the periphery of the curve are saturated. Toward the center, they are less and less saturated, until they become gray or white at point "C." See text and Figs. 4-18 and 4-19 (color). (From Ennes, 1979.)

camera tube or is derived by adding 0.3R + 0.59G + 0.11B. R, G, and B are the corrected signal outputs from the red, green, and blue camera tubes. Given these ratios, the R, G, and B signals add up to a white or gray Y signal, and hence provide the luminance for each image point. The Y signal is transmitted with a bandwidth of 4.2 MHz, the maximum bandwidth available for the video signal per TV channel. The Y signal contains the high-frequency information needed to convey the image detail and is therefore assigned this maximum bandwidth.

The red, green, and blue information for the NTSC encoded signal is converted into three *difference* signals (R − Y), (G − Y), and (B − Y). But since 0.3R + 0.59G + 0.11B = Y, only two of the three difference signals together with the Y signal are needed for complete color encoding. The (R − Y), (G − Y), and (B − Y) signals, which have properties of vectors, are transformed, or matrixed, into mutually orthogonal vectors known as the **I and Q vectors.** I stands for in-phase, and Q for quadrature, Q lagging in phase by 90° over I. The I vector is oriented along the orange–cyan axis and is assigned a wide bandwidth of ±1.3 MHz, since we perceive the greatest image detail with the orange–cyan hues. The Q vector is oriented along the magenta–green axis and is assigned a narrower bandwidth of 0.5 MHz because of our relative inability to resolve magenta and green details.

The I and Q vectors modulate the amplitude and phase of a 3.58-MHz **color subcarrier,** or the *chrominance reference wave.* The wave generated by taking the difference between the modulated wave and original **subcarrier** is known as the **chrominance signal.** The chrominance signal added onto the Y signal produces the NTSC signal (Fig. 7-29A). Thus, the color encoded signal is a sum of the luminance and chrominance signals. The amplitude of the chrominance signal (above and below the level of the luminance signal) provides the saturation of the color. The phase of the chrominance signal, relative to the color reference signal, provides the hue. The amplitude of the luminance signal provides the luminance of that image area. The chrominance signal becomes zero for scenes that are made up exclusively of shades of white and gray, namely for monochrome scenes (Fig. 7-29B). This is the reason for using the difference signals (R − Y), (G − Y), (B − Y), rather than the R, G, and B signals directly, to produce the I and Q vectors. These vectors modulate the 3.58-MHz subcarrier and produce the chrominance signal.

In the NTSC signal, nine cycles of the 3.58-MHz chrominance reference wave are added to the back porch of each horizontal blanking cycle (Fig. 7-29A,C). This reference wave, or the *color burst,* provides a standard frequency and phase for the chrominance signal in each horizontal sweep, thus allowing the chrominance signal to be decoded in the receiver or monitor. The decoded chrominance signal, combined again with the luminance signal,* gives back the RGB signals that are displayed on the color monitor.

As this very abbreviated description of NTSC color encoding shows, the color signal is complex. Nevertheless, the NTSC output from a color camera can be handled just as simply as the output from a monochrome video camera. So long as the monitor, recorder, mixer, etc. are provided with NTSC capabilities, the signal can be handled through single, standard coaxial cables. No special knowledge is required to handle the color video camera, except for a few *preliminary adjustments.*

For the three (or four)-tube color cameras, the raster scans in the pickup tubes need to be aligned with the red, green, and blue (and the monochrome luminance) images or placed in **register,** before the camera is used. Cameras are provided with test charts and fairly simple adjustments for electronically placing the YRGB signals of the test chart in register. With single-tube cameras, the separation of the three primary color images takes place within the tube itself so that the registration problem is simplified.

The other main adjustment needed is the **white balance.** The camera is exposed to a reference white scene for the particular illumination being used, and the WHITE BALANCE button or switch on the camera is depressed. This circuit automatically adjusts the gains of the R, G, and B channels to provide the relation $Y = 0.3R + 0.59G + 0.11B$ for the particular illumination.†

The dynamic range, or usable light range, of the color camera is generally achieved by the use of an *auto-iris* circuit, which drives a motorized iris and optimizes the amount of light falling on the sensors. In addition, a HIGH–LOW GAIN switch allows the adjustment of the amplifier gain to match the level of available light.

For video microscopy, a conventional camera lens and iris cannot be used because of vignet-

*For example, (R − Y) + Y = R. For details of color encoding, see, e.g., Ennes (1979), *Raster Graphics Handbook* (1980), Sony Color Signal Processing: Video Recording Course Booklet. The complete waveform and timing characteristics of the NTSC composite color video were specified in the November 1977 EIA tentative standard as RS-170A. As shown in Fig. 6-10, RS-170A is similar to RS-170 but uses two successive color frames (A and B) with a total of four color fields (I and II for frame A, III and IV for B). The color reference subcarrier is specified at 3.579545 MHz ± 10 Hz. The phases of the "3.58"-MHz subcarrier are alternated 180° in fields III and IV relative to I and II. For monochrome images, the burst is omitted, and the timing pulses for fields III and IV are identical to I and II, respectively. The alternation of phases in the subcarrier produces less variations in the color of the image reproduced through a color receiver, especially for signals derived from high-quality color recording such as on a 1-inch studio recorder. The field rate, or vertical frequency, for color signals in the NTSC format is actually 59.94 Hz rather than 60 Hz as in monochrome video (see Ennes, 1979, Section 2-10).

†For some light sources, white-balancing color filters have to be used.

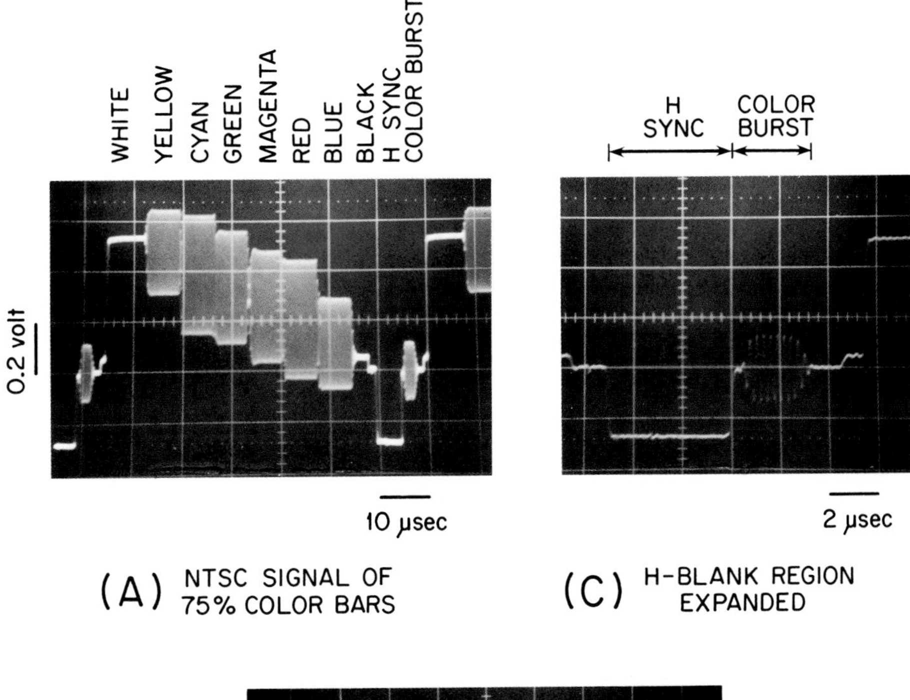

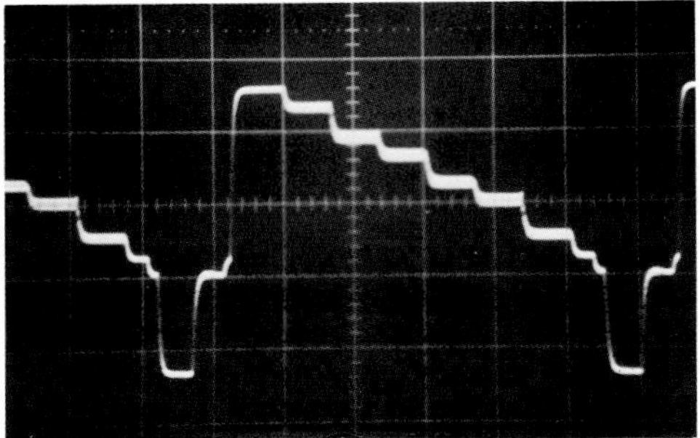

(B) CHROMINANCE SIGNAL AND COLOR BURST ARE ELIMINATED FROM (A) BY SHUNTING THE OSCILLOSCOPE INPUT BY A 0.05 μF CAPACITOR. THE LUMINANCE SIGNAL PERSISTS AS A GRAY SCALE.

FIGURE 7-29. Oscilloscope displays of the H-sweep of an NTSC color signal (A) of 75% **color bars.** (B) The chrominance signal and color burst are eliminated from A by shunting the oscilloscope input by a 0.05-μF capacitor. The luminance signal persists as a gray scale, but the colors become completely unsaturated, so that the display turns into a monochrome picture. (C) H-blank region from A, expanded, showing the color burst.

ting, so that the auto iris is of no help and the dynamic range of the color camera becomes quite limited. The image brightness of the microscope must therefore be regulated by the use of neutral-density filters, an attenuating diaphragm preceding an optical scrambler (Fig. III-21), or with a brighter light source to match the narrow level of illumination acceptable to the color camera.

The color camera is inevitably less sensitive than monochrome cameras using the same type of video imaging device, since the image has to be shared among the sensors responding to the separate primary colors. The resolution in the encoded color signal is also limited, especially for the nonluminance channels, for reasons discussed above. However, the amount of image information, if not spatial resolution, is very great in a color image (Chapter 10).

For video microscopy in color, one may wish to use the RGB signals from a color camera, rather than the NTSC signal, in order to maximize the image resolution in each color. Even though color video cameras in principle contain the RGB signals, the signal output provided by most color cameras is already encoded into a single NTSC (or PAL, SECAM) composite video signal. To access the RGB signal with its higher resolution, one needs a camera that provides a separate set of outlets for the R, G, and B composite video signals. Some three-tube cameras provide such outputs. For single-tube color cameras, the RGB signals are generally not available. With the compact and complex design of the single-color pickup device, one cannot expect from it the resolution obtainable with three-tube cameras, although the **Trinicon** tube, for example, can give rise to remarkably high-quality color images.

In VTRs, monitors, and image processors capable of handling encoded color and monochrome signals, the circuitry and equipment can operate in different modes depending on the incoming signal. Thus, in equipment having *automatic color detection,* the signal is processed as encoded color video if the color burst is present, whereas the signal is processed as a monochrome video signal if no color burst is detected by the **chroma detector.** For the monochrome signal, there is no need to filter out the color subcarrier, so that the circuitry can be devoted more efficiently to handling the luminance signals alone. The bandwidth can be fully utilized to provide a higher-resolution luminance image with crisper boundaries and image detail, and with less artifact and distortion of areas with differing shades of grays.

7.5. VIDEO MONITORS

The last link in the video chain, the *display device,* reverses the process that took place in the video camera and converts the electronic signal back into a *picture* that we can see.

Commonly, we view the picture on a *cathode-ray tube* (CRT), or the *picture tube,* in the *video monitor.* Alternatively, we can view a picture projected with a *video* or **light valve** *projector* (Section 12.5). Lately, solid-state and *liquid-crystal devices* as well as printout-type display devices are becoming more common, especially for the display of digitized information.*

It is also possible to use a TV set instead of a monitor to display the video picture. The video signal is first converted by an **RF converter** to a broadcast-frequency **FM** signal, which is then connected to the antenna terminal of an appropriately tuned TV set. While it is possible to use a TV set (and this may, in the absence of an NTSC color monitor, be the only way to display a color picture played back from a VTR, for example), a video monitor or **receiver–monitor** is preferred for CCTV, and especially for video microscopy. A monitor bypasses the need to use an RF signal that by **FCC** rules must limit the bandwidth of the signal. By using a monitor rather than a TV set, less noise from electromagnetic interference is picked up, there is less opportunity for signal degradation, and, most importantly, one need not lose resolution by cutting down the bandwidth.

Some display devices take several seconds or minutes to form or display a complete picture. In a standard video monitor or projector, the picture is refreshed at the same rate as in the camera, so that the picture follows the changing scene captured by the video camera in real time.

The Picture Tube and Basic Monitor Circuitry

The CRT or *picture tube* is essentially a large, evacuated glass flask enclosing an electron gun, electrodes, and phosphor screen (Fig. 7-30). The heated *cathode* in the electron gun, located at the neck of the flask, emits a beam of electrons. The **ultor** and high-voltage *anode* accelerates the electrons to many kilovolts. The focusing coil and electrodes focus the electron beam onto the *phosphor screen* located at the base of the flask, which is the *faceplate* or the front end, of the picture tube.

*For a survey of raster display devices and excellent discussions on monitors, see *Raster Graphics Handbook* (1980). See also Aldersey-Williams (1984), Graff (1984), and Hirshon (1984a) for reports on flat-panel display devices.

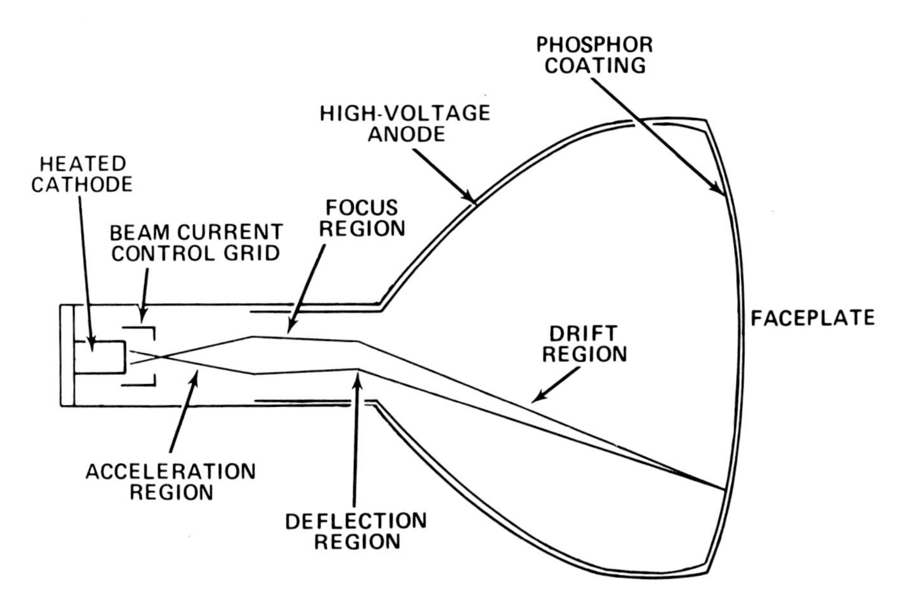

FIGURE 7-30. Basic design of a conventional cathode-ray tube. Focus and deflection mechanisms can be electrostatic or magnetic. See text. (From *Raster Graphics Handbook,* 1980, p. 2-59, with permission.)

The accelerated electrons landing on the screen excite the phosphor. The excited phosphor emits light from each *spot* whose brightness varies with the accelerating voltage and the current of the electron beam. The beam current is modulated by the amplitude of the video signal imposed on the *grid* as the beam traces the raster. Thus, each spot on the phosphor is excited according to the amplitude of the video signal and corresponding intensity of the *landing beam* (Fig. 7-31).

In the monitor, a nonlinear **processing amplifier** processes the composite video signal. Additional circuits strip the sync signal and clamp the black level of the processed signal (to the back porch of the blanking intervals). Once stripped, the V- and H-sync signal are separated from each other. A low-frequency pass filter passes the 60-Hz V-sync pulses but blocks the 15,750-Hz H-sync pulses. A high-pass filter passes the high-frequency H-sync pulses but blocks the V-sync pulses.

In the monitor, the separated sync pulses synchronize the outputs of two free-running oscillators that provide the timing triggers for the vertical and horizontal deflection circuits (Fig. 7-31). The outputs of the oscillators trigger the onset of the V and H flybacks, which are followed by the vertical and horizontal scans, so that the raster is synchronized with the composite video signal reaching the monitor (Chapter 6). In the monitor, discrete blanking pulses are added to the picture signal to ensure complete blanking of the retrace beam during flyback. The free-running oscillators continue to deflect the electron beam in a raster whether or not the monitor receives any signal.

Picture Size

The electron beam in the picture tube scans an area of the phosphor considerably larger than the target area in the camera tube. Nevertheless, so long as the H and V deflections scan the raster linearly, with the proper aspect ratio, and in good synchrony with the video signal, a geometrically undistorted, proportionally magnified picture of the camera tube image should appear on the video screen. The aspect ratio is commonly 4 H : 3 V, although some cameras use a 1 : 1 or other ratio.

Monitors are available with picture diagonals ranging from an inch or two to 7, 9, up to 25 inches and larger (Fig. 7-32). With small picture tubes, the raster may be so fine that the picture may appear to be well resolved. With larger picture tubes, the raster is proportionately coarser, so that the picture may appear to be poorly resolved if viewed from too close a distance.

Viewed from distances proportional to the size of the picture, the resolution of the monitor

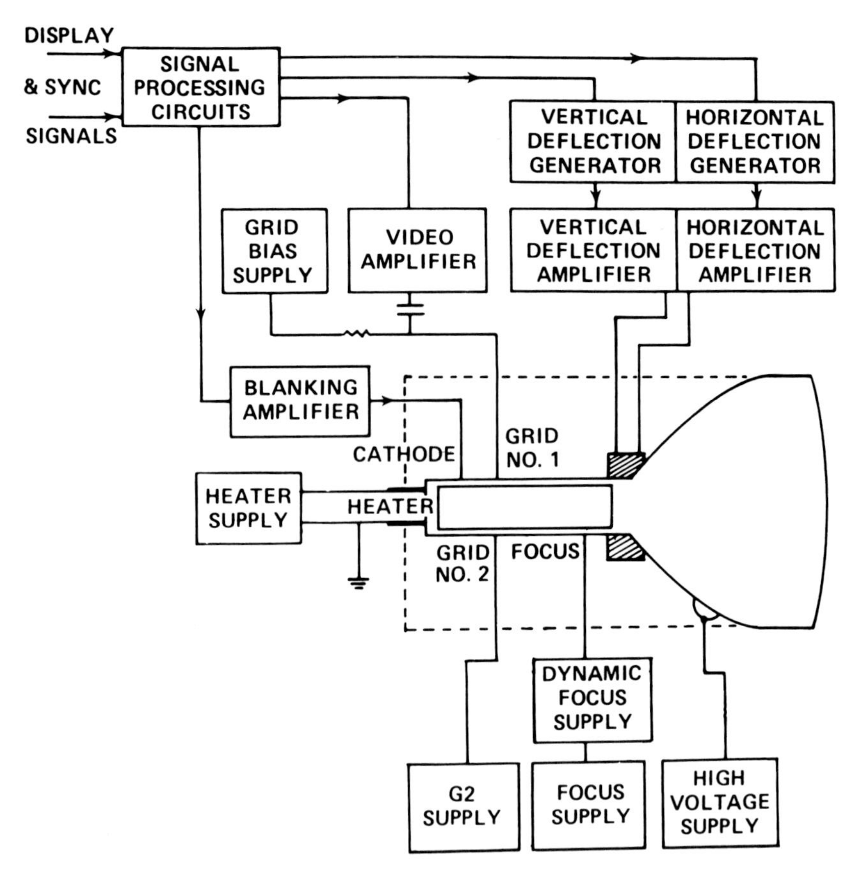

FIGURE 7-31. Block diagram of monochrome-monitor signal-processing circuits. See text. (From *Raster Graphics Handbook,* 1980, p. 9-15, with permission.)

should in principle be independent of picture size. In fact, however, because the spot that makes up the picture is usually 0.3 mm or larger in diameter, very small monitors cannot yield resolutions as high as larger monitors. Generally, at least a 5- to 6-inch monitor is needed to resolve 512 TV lines and a 13-inch monitor to resolve 1100 TV lines, horizontally per picture height.

Resolution of the Monitor

The size and shape of the landing beam, the luminance distribution within the phosphorescent spot, the number of scan lines, and the bandwidth of the circuit limit the ability of the monitor to display image detail.

The landing beam produces an image of the crossover of the electron beam, focused onto the phosphor screen by electrostatic or electromagnetic lenses (Fig. 7-30). The structure of the electron gun, the beam current, acceleration, the geometry of the electron optics and the quality of their voltage and current supplies, all govern the size and shape of the landing beam.

Given the finite size and shape of the landing beam and the light scattering in the phosphor, the excited spot is neither infinitely small nor uniformly luminous. Rather, the spot is a small disk with a Gaussian distribution of luminosity. As the beam scans the phosphor, the picture is formed by overlapping Gaussians (Fig. 7-33). Therefore, the H scan lines, for example, are not luminous lines separated vertically by discrete dark lines, as may appear by casual observation of the monitor. Instead, the dark lines are merely less bright regions where the skirts of the Gaussians of the

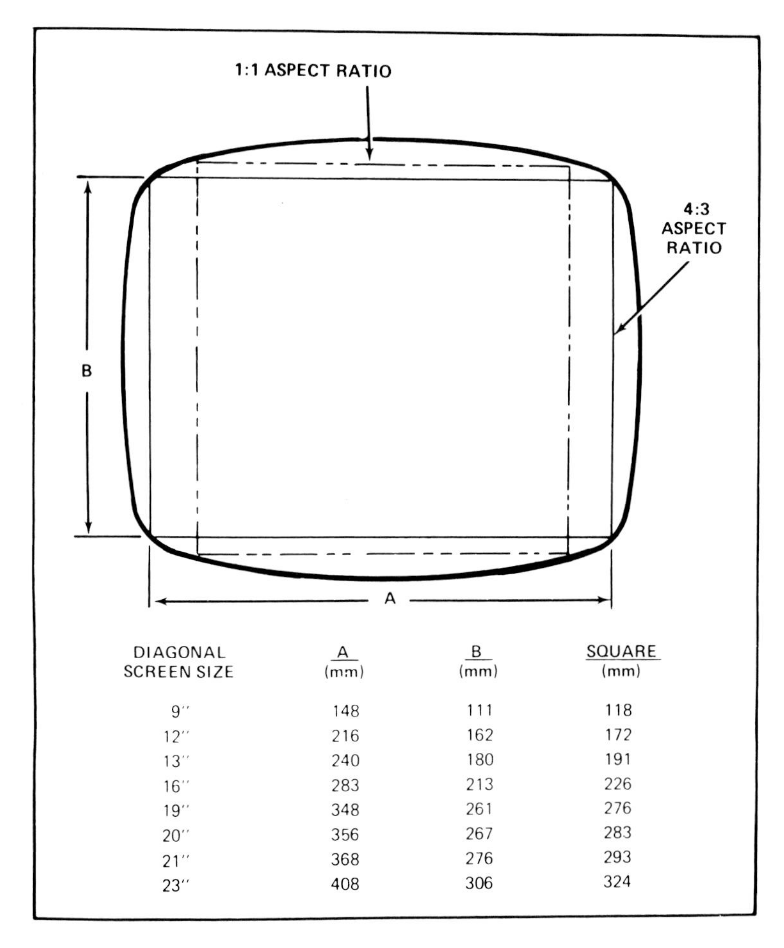

DIAGONAL SCREEN SIZE	A (mm)	B (mm)	SQUARE (mm)
9″	148	111	118
12″	216	162	172
13″	240	180	191
16″	283	213	226
19″	348	261	276
20″	356	267	283
21″	368	276	293
23″	408	306	324

FIGURE 7-32. Raster dimensions as functions of monitor screen size. Raster dimensions are shown for monitors of various diagonal sizes, using aspect ratios of both 4:3 and 1:1. (From *Raster Graphics Handbook*, 1980, p. 9-5, with permission).

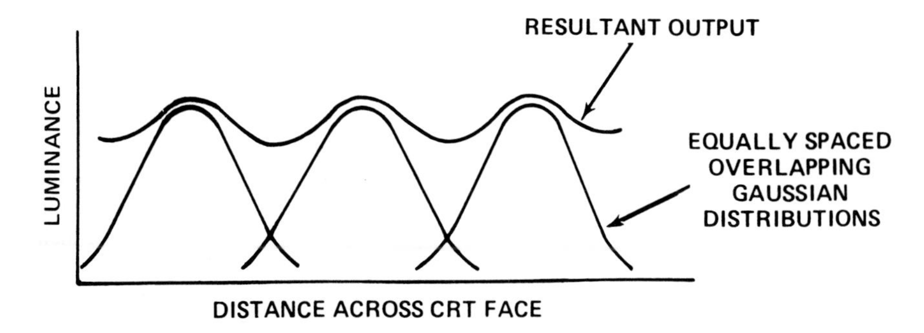

FIGURE 7-33. Effect of overlapping raster lines or pixel images. Each excited spot on the monitor phosphor is not uniformly bright, but rather has a Gaussian distribution. The resultant output curve is the sum of the brightness of the adjacent scan lines or pixels. (From *Raster Graphics Handbook*, 1980, p. 9-8, with permission. Source: Tannas, 1981.)

adjacent lines overlap with one another. The situation is similar to the microscope image of many point sources that is made up of overlapping Airy diffraction patterns.

The landing beam in the picture tube must be focused to its minimum dimeter, and be free of astigmatism. The beam should also remain in focus as it scans the large area of the phosphor screen. Higher-quality monitors incorporate stable power supplies that maintain good focus and deflections, and a *dynamic focusing* circuit to keep the beam in focus as it is deflected over wide angles.

Once the landing beam is properly focused, the number of scan lines govern the *vertical resolution* of the monitor just as in the video camera (see Section 6.4). The bandwidth of the signal and frequency response of the circuit ultimately limit the *horizontal resolution* of the monitor. As in the video camera, a minimum of 1-MHz bandwidth is required for every 80 TV lines of resolution (for a horizontal distance equal to the picture height). Thus, a circuit with a good frequency response up to 10 MHz is required in order to gain a horizontal resolution of 800 TV lines, or 400 line pairs, per distance equal to the picture height.

The electronic circuit in a video monitor must respond well to a broad range of signal frequencies ranging from about 30 Hz to several megahertz. We need good low-frequency response in order to display larger, uniformly lit areas without appreciable falloff or shading. We need high-frequency responses to provide good image resolution and proper representation of vertical edges, since both involve rapid electronic transitions. In less-expensive monitors, resonating and filtering circuits are used to economically enhance the appearance of image details and to crispen edges.

In general, the time constants of video amplifiers determine the shape, or the ''aperture,'' produced by a sharp electronic pulse. **Aperture correction** circuits, incorporated in high-quality monitors, modify the horizontal and vertical transitions and sharpen the edges of each spot electronically.* High- and low as well as mid-frequency responses must be properly balanced to minimize extraneous edge structures and trailing shades. In all, the monitor circuit is designed to compensate and optimize the overall MTF of the video system.

The resolution and contrast of the picture displayed on the monitor are also affected by adjustments made by the user.

Gamma, Contrast, and Picture Brightness

In video monitors, the brightness of the picture is generally not linearly proportional to the strength (voltage) of the incoming video signal. Rather, the brightness (I) increases as an exponential function of the signal (S), with the realtionship $I = S^G$ (Fig. 7-34). G, the *gamma* of the monitor, usually is between 2.2 and 2.8 (*Raster Graphics Handbook*, 1980). This high gamma compensates for the low gamma of some vidicon cameras, such as those using sulfide vidicons, a condition that is especially important for rendering proper balance of hue in color video systems.

For video microscopy in monochrome, the nonlinear response, or the high gamma, of the monitor is often a blessing rather than a flaw to be corrected. For example, one can display the incoming signal shown in Fig. 7-35A as though it had been the signal in Fig. 7-35C. The amplitude of the signal is first boosted as in Fig. 7-35B by increasing the amplifier gain (by turning up the CONTRAST knob). Then the picture is darkened by reducing the beam current (by turning down the BRIGHTNESS knob). The effect (Fig. 7-35C) is as though the signal for the highlights were boosted exponentially with the background suppressed. In a darkened room, the picture can be made even darker, so that fainter details can be brought out.

These adjustments enhance the details in the picture highlights and suppress the background, including the optical and electronic noise encountered in low-light-level microscopy, where the darker background often contains much of the undesirable noise. Thus, *monitors with a good range of GAIN (CONTRAST) and BRIGHTNESS controls can be very helpful for video microscopy.*

While these adjustments provide useful pictures for viewing the monitor, one must also

*For the H scan, aperture correction is achieved by raising the high-frequency response of the amplifier (Ennes, 1979, Section 4.9). For correcting the scanning aperture vertically, a one-line delay circuit stores the signal for a 1-H interval, so that the signal can be compared with the previous line. See Ennes (1979, Chapters 3–5).

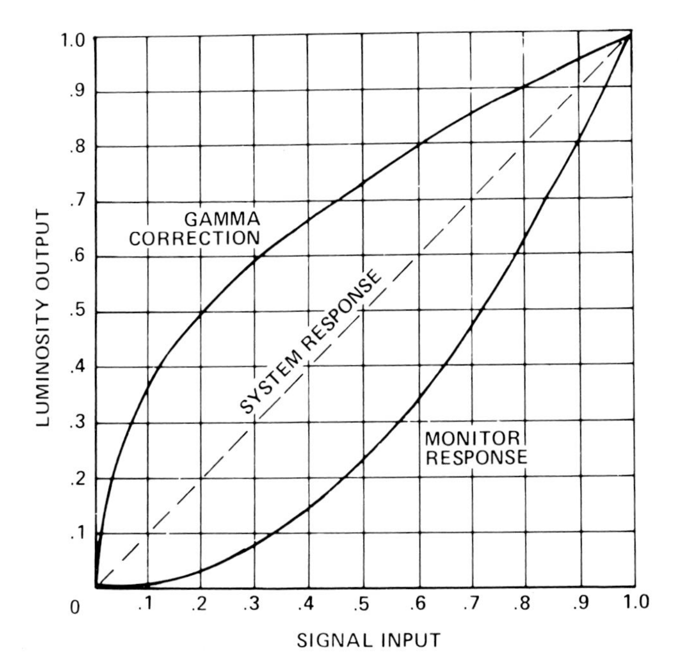

FIGURE 7-34. Relation among camera, monitor, and system response. Monitors typically respond with a gamma of about 2.5. In order to have the entire video system respond linearly to the signal (gamma = 1.0), a gamma correction is introduced in the camera circuit to compensate for the monitor response. This correction is essential for color reproduction; however, it is not necessarily desirable for video microscopy in monochrome. (From *Raster Graphics Handbook,* 1980, p. 8-18, with permission.)

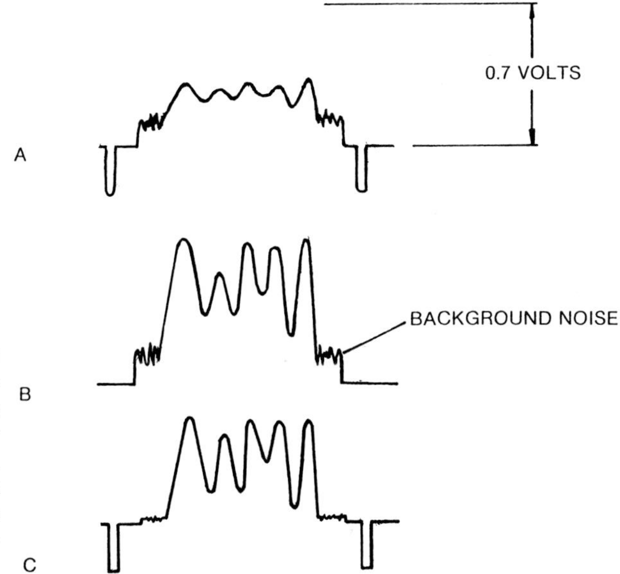

FIGURE 7-35. Use of monitor CONTRAST and BRIGHTNESS controls to optimize the picture. (A) Incoming signal. (B) Turning up the CONTRAST knob expands the signal amplitude, nonlinearly, because the monitor has a high gamma. (C) Then, the BRIGHTNESS knob is turned down to suppress background noise.

remember that a signal of the type shown in Fig. 7-35A cannot be recorded and played back as well from a VTR as a signal that was already optimized as in Fig. 7-35C. Whenever one records the signal on a VTR, especially if the recording is to be viewed on another monitor or projected, one should make every attempt to first enhance the signal to the proper level. We can enhance the signal by selecting cameras with appropriate controls (Section 7.3) and by processing the signal with analog or digital devices (Chapters 9, 10). A standard monitor with the controls set to their home position (Section 3.5), or a cathode-ray oscilloscope display, helps determine whether the picture signal properly spans the needed 0.7 volt level.

The contrast and brightness of the video monitor are adjustable over a wide range, but the maximum contrast and brightness attainable, as well as the gamma and color of the picture, are ultimately determined by the type of picture tube and especially its phosphor.

Types of Phosphors, Hue, Persistence

In CCTV, we view or photograph the picture on the monitor, or may copy the picture with another video camera (Chapter 12). Depending on the type of sensor (the eye, photographic emulsion, or target in the video camera tube), the nature of the image, and the viewing environment, we may adjust the monitor differently or select different types of monitors.

In CRTs, phosphors with a variety of *color* or *tints* are used depending on their primary function (Table 7-7). For visual observation, we may prefer a phosphor with a peak emission lying close to 550 nm, matching the wavelength of peak sensitivity of the light-adapted human eye. A bluish-tinted phosphor gives a colder (subjectively greater contrast), and a yellowish phosphor a warmer (less harsh), feel. Some CRTs use mixed phosphors to provide neutral grays and whites, or to provide selected colors for color display. Others use an orange tint or green phosphor to minimize eyestrain, e.g., in word processing. For photography, we may choose a bluer phosphor (or blue phosphor mixed with others) to take advantage of the greater blue sensitivity of orthochromatic emulsions.

In addition to color, the *persistence* of phosphors varies widely (see ''Decay to 10% Brightness'' in Table 7-7). Long-persistence phosphors such as those used in radar display might be useful for image integration and noise reduction in video microscopy. But long-persistence phosphors would be inappropriate for most applications where the specimen is moving or changing rapidly. The picture of moving objects would be blurred, slowly moving bright objects would show pronounced comet tails, and it would be difficult to find the proper focus rapidly. On the other hand, extremely short-persistence phosphors of the type used in some oscilloscopes and flying spot scanners tend to have poor radiant efficiency and are not normally appropriate for video microscopy.

The persistence of the CRT in standard video monitors appears to be reasonably short compared to most video cameras, so that problems relating to long persistence in video microscopy are more likely to reside in the camera than the monitor.

Geometrical Distortions

Some camera tubes, especially intensified camera tubes, can distort the video picture by 3% or more. Others distort the picture less than 1% at the edge of a central circle whose diameter equals the picture height. Geometrical distortions in the monitor commonly fall between these values. Test charts or patterns (Figs. 7-36, 7-40) are used to assess distortion.

With some monitors, an undistorted signal of proper amplitude, e.g., a properly terminated signal produced by a camera with low distortion or by a test signal generator, still produces a distorted picture. The picture of a uniformly spaced rectangular grid might expand away from the central X–Y axes by several percent of the picture height toward the edges of the screen, or show *pincushion distortion* (Fig. 7-36D). Or the spacing of the grid may not be uniform from left to right or top to bottom, *nonlinear* conditions that can turn circles into egg shapes (Fig. 7-36A,B).

For high-quality video monitors, several manufacturers specify distortions and nonlinearities not to exceed 1–2% at the edge of a central circle whose diameter equals the picture height. If the

TABLE 7-7. Phosphor Characteristics[a]
Characteristics of Phosphors Used in Cathode-Ray Tubes
(Complete Data Are Available in JEDEC Publication No. 16-C*)

Type	Fluorescent Color (Kelly Chart)[1]	Spectral Peak (nm)	Spectral Output Wavelength (nm) 300 350 400 450 500 550 600 650 700	CIE Coordinates x	CIE Coordinates y	Persistance Class[2]	Decay to 10% Brightness	Applications
P1	Yellow-green	525	P1	0.218	0.712	Medium	24 msec	General-purpose visual display
P2	Yellow-green	535	P2	0.279	0.534	Medium short	25-100 μsec[6]	Used in cathode-ray oscillographs
P4	White (sulfide)	460 560	P4 Sulfide	0.270	0.300	Medium short	20 μsec 60 μsec	Used in direct-view television
P4	White (silicate-sulfide)	450 540	P4 Silicate Sulfide	0.317	0.331	Medium short	20 μsec 60 μsec	Used in theater television projection tubes
P7	Purple blue	440	P7	0.151	0.032	Medium short	~50 μsec[6]	Used for radar and oscillography
	Yellow-green	555		0.357	0.537	Long	0.35 sec	
P11	Blue	460	P11	0.139	0.148	Medium short	~50 μsec[6]	Photographic recording
P12	Orange		P12	0.557	0.442	Long	200 msec	Low frame rate displays and radar
P14	Purple-blue	440	P14	0.150	0.093	Medium short	25 μsec	Used for radar
	Yellow-orange	600		0.504	0.443	Medium	5 msce	
P16	Violet[4]	380	P16	0.199	0.016	Very short	0.12 μsec	Flying spot scanners and photographic applications
P19	Orange	590	P19	0.572	0.422	Long	220 msec	Used in radar
P20	Yellow-green	560	P20	0.426	0.546	Medium short	~.2 msec[6]	Bright visual displays
P22[3]	Violet	450	P22[3]	0.155	0.060			Three-color phosphor screen used for color television
	Green All sulfide	515		0.285	0.600	Medium short	—	
	Red	680		0.663	0.337			
P24	Green	510	P24	0.245	0.441	Short	—	Used in flying spot scanners
P25	Orange	610	P25	0.569	0.429	Medium	45 msec	Used in radar
P26	Orange	595	P26	0.573	0.416	Very long	~10 sec[6]	Used in radar
P28	Yellow-green	550	P28	0.370	0.540	Long	.5 sec	Used in radar
P31	Green	520	P31[8]	0.226	0.529[9]	Medium short	38 μsec	Oscilloscope tubes
				0.193	0.420			
P33	Orange	587	P33	0.559	0.440	Very long	7 sec	Used in radar
P36	Yellow-green	550	P36	0.400	0.543	Very short	.25 μsec	Flying spot scanners
P37	Green-blue	470	P37	0.143	0.208	Very short	.15 μsec	Flying spot scanners and photographic applications
P38	Orange	600	P38	0.591	0.407	Very long	1 sec	Low frame rate displays and radar
P39	Yellow-green	525	P39	0.223	0.698	Long	150 msec	Medium frame rate displays and radar
P40	White	440	P40	0.276	0.312	Medium short	150 μsec	Longer persistence version of P4
	Yellow-green	550				Long	~.5 sec	
P41	Orange-yellow	590	P41	0.541	0.456	Long	200 msec	Low frame rate displays with light pen
		380				Very short	0.12 μsec	

(*continued*)

TABLE 7-7. (*Continued*)

Type	Fluorescent Color (Kelly Chart)[1]	Spectral Peak (nm)	Spectral Output Wavelength (nm) 300 350 400 450 500 550 600 650 700	CIE Coordinates x	CIE Coordinates y	Persistance Class[2]	Decay to 10% Brightness	Applications
P42	Yellow-green	520	P42	0.238	0.568	Medium	10 msec	Integrating phosphor for low repetition rate displays at high brightness
P43	Yellow-green	544	P43 Very Sharp Distribution Curve	0.333	0.556	Medium	1.2 msec	Bright high contrast displays[7]
P44	Yellow-green	544	P44 Very Sharp Distribution Curve	0.300	0.596	Medium	1.2 msec	Bright high contrast displays[7]
P45	White	—[5]	P45 Very Sharp Distribution Curve with Multiple Peaks	0.253	0.312	Medium	1.7 msec	Alternative to P4 for high brightness white displays
P46	Yellow-green	530	P46	0.365	0.595	Very short	0.16 μsec	Flying spot scanners
P47	Purple-blue	400	P47	0.166	0.101	Very short	0.08 μsec	Flying spot scanners and photographic applications
P48	Yellow-green	400 530	P48	0.365	0.474	Very short	0.12 μsec	Flying spot scanners. Blend of P46 and P47 phosphorus. Proportions may vary with system requirements.
P49	Green (10 kV)	523	P49 17kV Green	0.315	0.615	Medium	1.2 msec	Multicolor displays. Color is dependent on voltage.
	Red (17 kV)	615	10kV Red	0.672	0.372	Medium	20 msec	
P50	Yellow-green (15 kV)	537	P50 15 kV Yellow-Green	0.398	0.546	Medium short	20 μsec	Multicolor displays. Color is dependent on voltage.
	Red (8 kV)	—[5]	Has Multiple Peaks 8 kV Red	0.655	0.340	Medium	5 msec	
P51	Yellow-green (12 kV)	545	P51 12 kV Yellow-green	0.414	0.514	Medium short	31 μsec	Multicolor displays. Color is dependent on voltage.
	Red (6 kV)	620	6 kV Red	0.675	0.325	Medium	2.2 msec	

Notes:

1. See Fig. 7-28 (p. 239), for Kelly Chart.
2. Joint Electron Device Engineering Council persistence classifications (based on time for phosphorescence to decay to 10% of initial brightness)

Very long	1 sec and over
Long	100 msec to 1 sec
Medium	1 msec to 100 msec
Medium short	10 μsec to 1 msec
Short	1 μsec to 10 μsec
Very short	Less than 1 μsec

3. All sulfide phosphors listed here. Other phosphor compositions are available also.
4. Emits principally in the ultra-violet region
5. Multiple peaks
6. Depending on anode current
7. Emits a high proportion of its radiant energy in a band between 540 and 545 nm, making it suitable for applications employing optical filtering techniques.
8. LO-current: 425-600 nm. HI-current: 405-600 nm. Both settings produce a peak at about 520 nm.
9. Top row: LO-current. Bottom row: HI-current.

* Available from: Joint Electron Device Engineering Council (JEDEC) c/o Electronic Industries Association, 2001 Eye Street N.W., Washington, D.C. 20006.

[a]Courtesy of Eastman Kodak Co., from "Kodak Films for Cathode-Ray Tube Recording," Publication P-37.

distortions are much greater, the monitor controls may require adjustment, the monitor may be overdriven, or it may simply be of rather poor quality or defective. One can check out the last possibility most expediently by substituting another monitor.

If one fails to adjust the aspect ratio of the monitor to 4 : 3 (H : V), the horizontal and vertical grid spacings would not conform to those in the standard incoming signal, and circles would appear as ellipses. Most monitors provide adjustments for picture height, and on some monitors, also picture width. These adjustments provide the proper aspect ratio and H- and V-grid line separations. In addition to the picture height and width controls, many monitors also provide vertical and horizontal linearity controls. The controls for picture size and linearity tend to interact with each other (in much the same way the controls for contrast and brightness interact), so that the two are trimmed in sequence several times.

One can also distort the picture by setting the CONTRAST and BRIGHTNESS controls too high. Overdriving the monitor with too high an input signal (which can be caused by not properly terminating the last monitor or video equipment) can also distort the picture. When we overdrive the monitor, its power supply becomes overloaded and the beam focus as well as scanning pattern can be altered, so that the spot and geometry of the picture are distorted. As described earlier, higher-quality monitors contain better-regulated, greater capacity power supplies which thus present less opportunity for picture distortion.

Geometrical distortions in the monitor can produce improperly shaped images or nonuniform magnification of the video microscope scene and lead to erroneous measurement of distances and angles at the face of the monitor. However, as long as we do not overdrive the monitor, it is possible to calibrate and measure absolute distances in the specimen quite accurately in spite of some distortion in the monitor. As discussed in Chapters 9 and 12, we can electronically introduce fiducial marks before the signal enters the monitor so that the monitor distortions do not seriously affect the accuracy of calibrations.

In applications where the geometrical distortions must be minimized, one uses special low-distortion monitors with flat faceplates. Such tubes are used in instrumentation cathode-ray oscilloscopes, for photographing the pictures produced by a scanning electron microscope, and in computer graphics terminals, etc.*

Shading, Streaking, Ringing, and Other Picture Defects

In addition to overall geometrical distortions, one may find localized defects in the picture on the monitor. Some local defects are obvious, while others may subtly affect important details of the image and lead to misinterpretations.

Shading in the camera was discussed in Section 7.2. As in the camera tube, the monitor can also introduce shading (Fig. 7-36D). Shading in the monitor becomes especially pronounced when one raises the monitor contrast and reduces the brightness. Since it is not easy to correct shading that originates in the monitor, one should: (1) select monitors with minimum shading under the conditions of the intended use; (2) be aware of the shading that is characteristic of the particular monitor(s) that is used; (3) keep in mind that a monitor may provide a uniform field at normal brightness, but show extreme shading when the brightness is reduced.

Discrete bright areas of the picture can sometimes be trailed (to the right) by dark lines (Fig. 7-37). In fact, the dark lines can persist over the whole period of the horizontal scan, so that a bright area can be surrounded to the left and right by a dark band. Such streaking, or "reversed-polarity *smear*," sometimes arises in the camera but can also be introduced by too large an input voltage or from inadequate low-frequency response of the monitor circuit (Ennes, 1979, Chapter 3). When the monitor GAIN is turned up too high, the highlights may also start *blooming* (spill out and spread into adjacent areas). The streaking of the extreme highlight as well as its blooming can be eliminated by lowering the monitor CONTRAST or BRIGHTNESS controls. Otherwise, the level of the signal coming into the monitor may be too high or the defects may have been produced in the camera.

*"Display Devices," a publication by RCA, lists CRTs appropriate for various applications together with several characteristics of the phosphor screens.

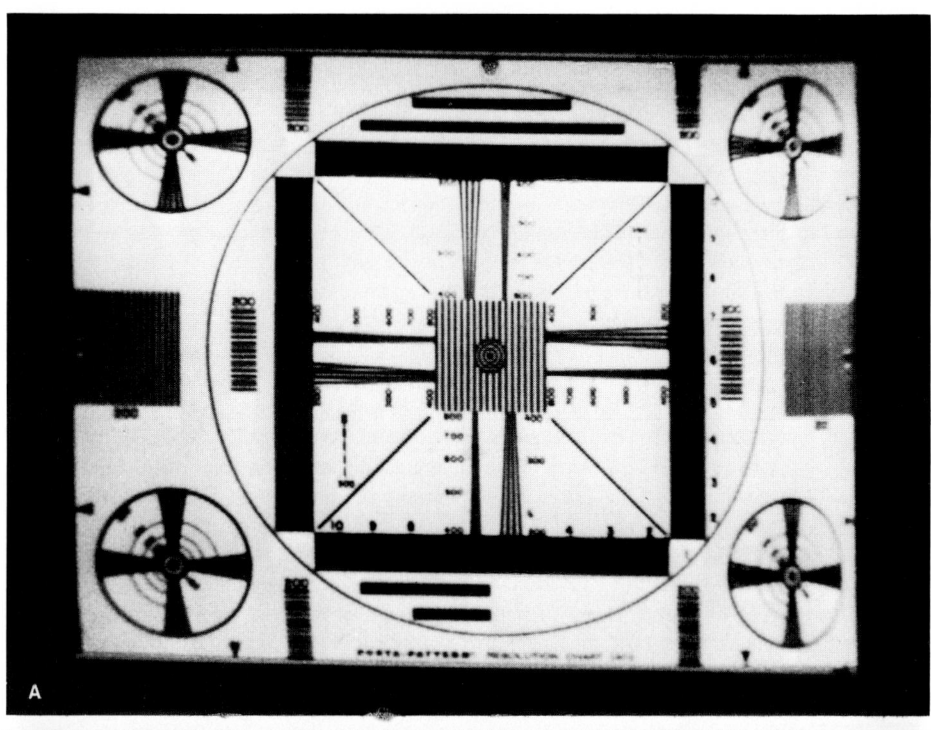

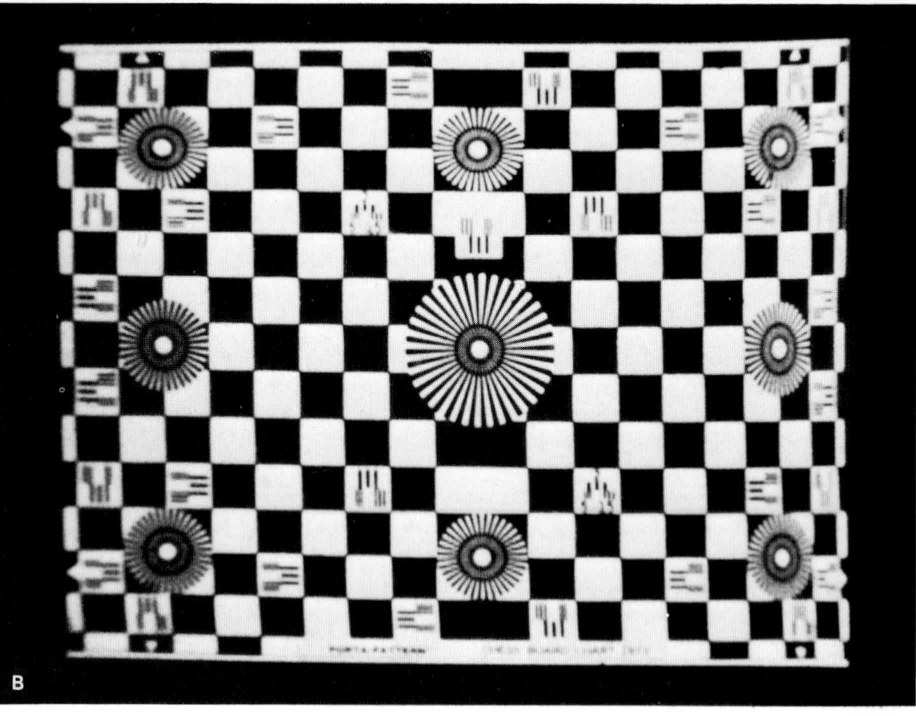

FIGURE 7-36. Distortions of the video picture. (A, B) Test patterns photographed off a monitor that shows nonlinearity. Notice that the circles on the left are elongated horizontally, and those on the right are compressed. The original test chart for A is shown in C. Such test charts are available, for example, from Porta-Pattern.

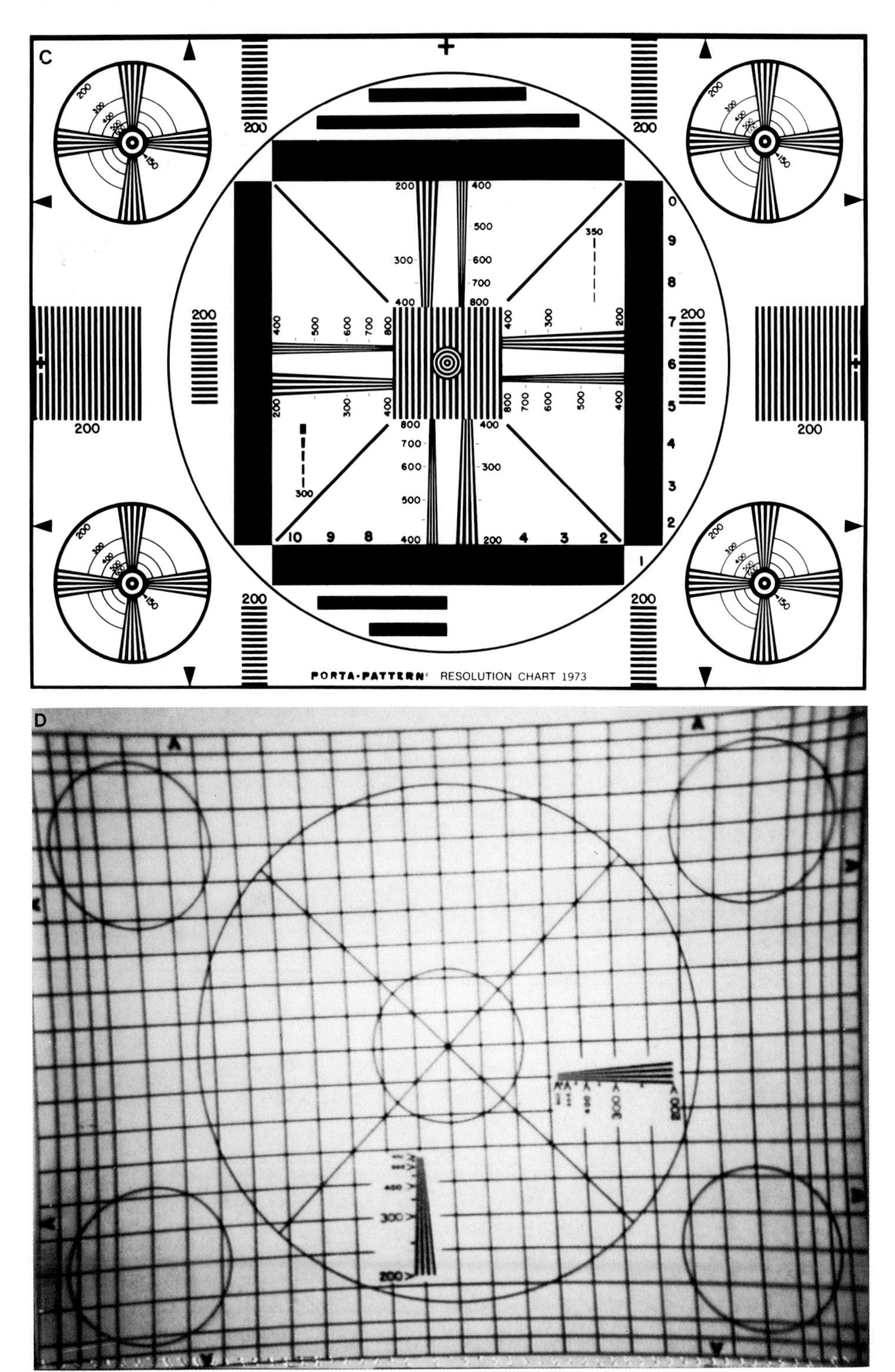

(D) Monitor picture of an alignment test chart showing both pincushion distortion and shading (luminance fall-off) produced by a 16-mm SIT camera.

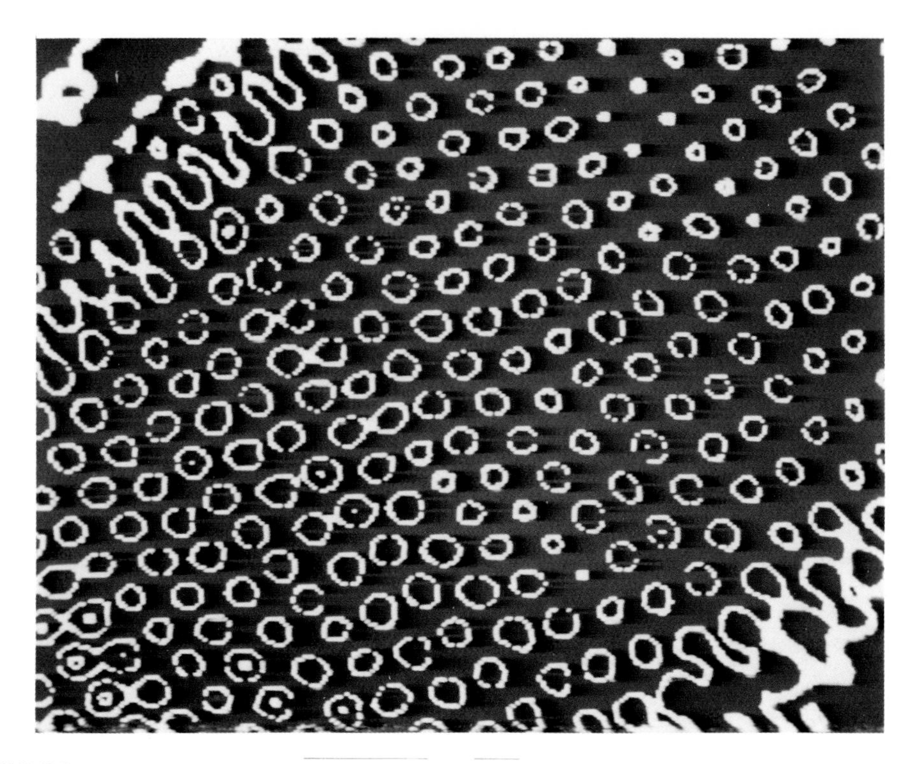

FIGURE 7-37. Reversed-polarity smear. The reversed-polarity smear in the monitor picture appears as dark "shadows" trailing to the right of each of the bright rings. The signal amplitude for the bright rings was excessively high. The signal was generated by digitally thresholding the zoomed image of the diatom shown in Fig. 8-19. See also Fig. 7-23 and text.

Even without such high levels of video signal, one may find spurious fine fringes at the transitions from a bright to dark area or vice versa. Such fringes commonly follow (appear to the right of) the transition, especially at sharp vertical edges or lines. These fringes appear when the circuit rings after a sudden fall or rise in the video signal. The circuit rings when the high-frequency response is boosted by a resonating circuit without adequate damping. Since *ringing* reflects the design of the monitor circuit (an attempt to increase the high-frequency response to crispen edges without the use of an exceptionally wideband amplifier), some degree of ringing is common in many video monitors.

In video microscopy, the user can distinguish ringing introduced by the video system from, for example, diffraction fringes introduced by the microscope optics. One turns the specimen or camera relative to each other to change the angle or direction that the suspected edges make with the video scan direction. Ringing originating in the video system would continue to appear on the same side of the monitor, while patterns residing in the microscope image would turn with the image.

Besides the fine fringes at the edges of vertical transitions, stationary or wandering repetitive fringes may appear superimposed on the monitor picture. These fringes may run vertically or diagonally, or may make vertical lines in the picture appear wavy. Such periodic patterns reflect a parasitic *oscillation* in some part of the video system, commonly in the camera or processor, but also sometimes in the monitor. The feedback that leads to such oscillation could also take place, inductively or capacitatively, between video components through the connecting cables or via the power line. Conversely, a high-frequency noise signal may enter through the AC power line, or by radiation from some nearby equipment (e.g., see Fig. 3-13). Careful grounding of the video equipment, especially the light baffle or lens mount, may eliminate these oscillations, but it may be necessary to call in an electronics expert to overcome these problems.

We will next consider how to critically adjust the monitor and survey the testing devices that are used to assess the quality and performance of the monitor.

Critical Adjustment and Testing of the Monitor

We discussed the general adjustments for the CONTRAST, BRIGHTNESS, and SYNC controls on the monitor in Section 3.5 and above. To critically adjust the H and V HOLD controls, we can take advantage of the pulse cross (Section 6.3) or the underscan modes, since in these modes the monitor displays the first few scan lines that follow the vertical blanking period. (In the standard mode, most monitors overscan and mask the edges of the raster including the first several lines.) With these lines showing, trim the H HOLD control to minimize the flagging or hooking at the very top of the raster.

For fine-tuning the V HOLD, examine the scan lines closely with a low-power magnifier, or an inverted microscope oscular (Section 5.10). First, measure the number of scan lines per unit height and determine the number of scan lines for the full height of the picture. If the V HOLD is set properly, you should find approximately 480 lines when all of the scan lines are visible, i.e., on an underscanned monitor. If the V HOLD had not been adjusted correctly and the monitor was syncing to

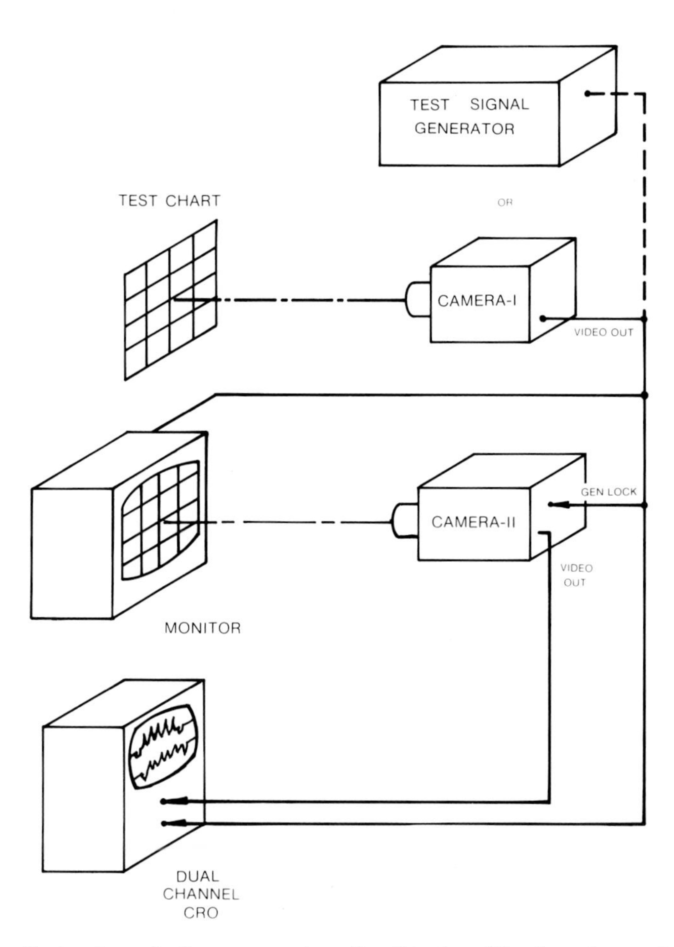

FIGURE 7-38. Hookup for evaluating camera and monitor distortions. Waveforms from various components in the video train are compared, on the dual-channel oscilloscope, with waveforms produced by a test signal generator, or with a camera of known quality. See "Test Procedure for Precision Raster Scan Displays" (Visual Information Institute, Inc.).

every field rather than to each frame (made up of two interlaced fields), only half the number of scan lines would be present. The vertical resolution then drops to about half of the maximum achievable.

Once the proper number of scan lines are present, exactly interlace the odd and even scan lines in the following manner. Again with the aid of a magnifier, fine-tune the V HOLD until the alternate scan lines mesh exactly in between each other.* As discussed in Section 6.4, proper interlace is necessary in order to obtain maximum vertical resolution. Naturally, one can only achieve 2 : 1 interlace with the signal from a 2 : 1 interlace camera, not from a random interlace camera.

Picture defects such as distortion, nonlinearity, shading, streaking, ringing, etc. can arise in the monitor or elsewhere in the video train as discussed earlier. For video microscopy, it is helpful to know the source of these imperfections even if they cannot be eliminated altogether.

To examine the performance of the monitor as well as the quality of the video signal entering the monitor, we can display the signal waveform on a cathode-ray oscilloscope (CRO) (Fig. 7-38).† For example, Fig. 7-39C shows an oscilloscope trace of an H sweep through a resolution *test chart*. The signal is moderately free from shading and distortion, but shows a reduction of signal amplitude (hence in the CTF) for the high-frequency components (Fig. 7-39B). Figure 7-39A shows the original test chart.

One can examine the performance of the monitor by inspecting the picture and comparing it with the waveform of the incoming signal displayed on the CRO. Alternatively, one can superimpose the reflected image of a Ball Chart (EIA Linearity Chart, Fig. 7-40) with the picture of a standardized cross grid displayed on the monitor to determine the amount of distortion introduced by the monitor. One can also copy the monitor display optically with a high-quality video camera and compare its ouput on a two-channel CRO with the signal entering the monitor. In this way, we can measure the distortions and determine the CTF of the monitor.

As mentioned above, we can generate a test signal by focusing a standard test chart with a high-quality camera. For more stringent tests, one uses a *test signal generator* that produces calibrated video signals.‡ The test signal generator produces standardized video signals that give rise to accurately spaced vertical and horizontal grid lines, multiburst (a series of regularly spaced black and white bars provided with discrete spatial frequencies), dots in a grid pattern, color bars, gray scale, flat field, window (a white square about a third of the picture width on a black background or vice versa), sine-squared pulse (used together with the window and flat field to calibrate the high-, medium-, and low-frequency responses of the video system or components), etc. With these standard signals, one can test the performance and measure the CTF or MTF of monitors as well as of video recorders, processors, etc.

Monitors with Special Features

As discussed earlier, monitors can accept incoming signals with scan rates, interlace, and timing pulses that differ substantially from the standard. H-scan rates can vary by as much as 10%, and the timing pulses in the V-blank interval can follow a variety of formats. Nevertheless, if the *scan rate* or, more importantly, *video format* (Section 6.3) departs too far from that intended for the monitor, the monitor may be unable to sync to the signal.

Generally, a 525/60, 2 : 1 interlace standard monitor can accept 260/60, 1 : 1 interlace signals as well. One can also drive a U.S. or Japanese standard monochrome video monitor either with an RS-170 or -330 signal or with RS-170A, the composite NTSC standard for color signals. The monochrome RS-170 or -330 signal can also drive a 525/60 NTSC color monitor.

To display the signal generated by a high-resolution video camera (with large numbers of scan

*To see the scan lines in greater detail, turn your head sidewise and take advantage of the greater horizontal resolving capability of the eye (suggestion made by Ed Horn).

†For a guide to the use of the oscilloscope, see, e.g., Middleton (1981), as well as the user's manual for your particular oscilloscope.

‡Standard charts are available, e.g., from Video Information Institute, Inc., or, as 2 × 2 transparencies (Porta-Pattern), from Comprehensive Video Supply Corp. Test signal generators are available from these and other sources. For details of testing equipment and procedures, see, e.g., Applications Guide for VII Test Instruments (Visual Information Institute), and Chapters 3, 4, 6 in Ennes (1979) andChapter 10 in Showalter (1969).

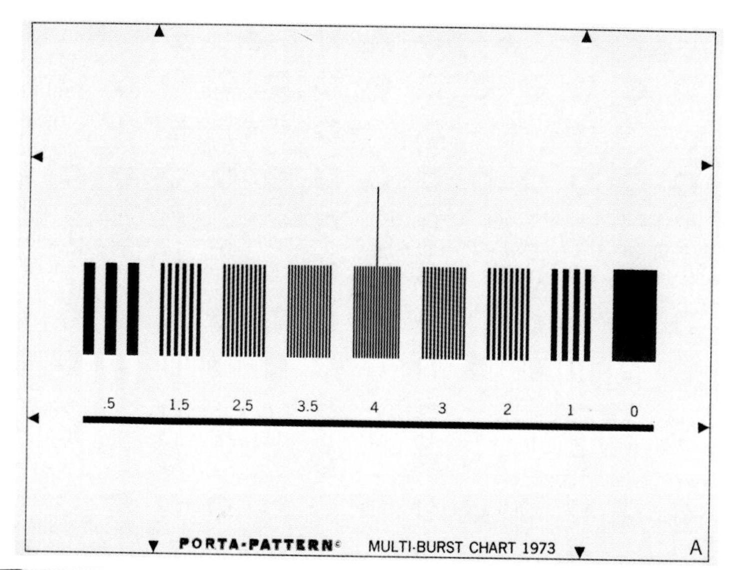

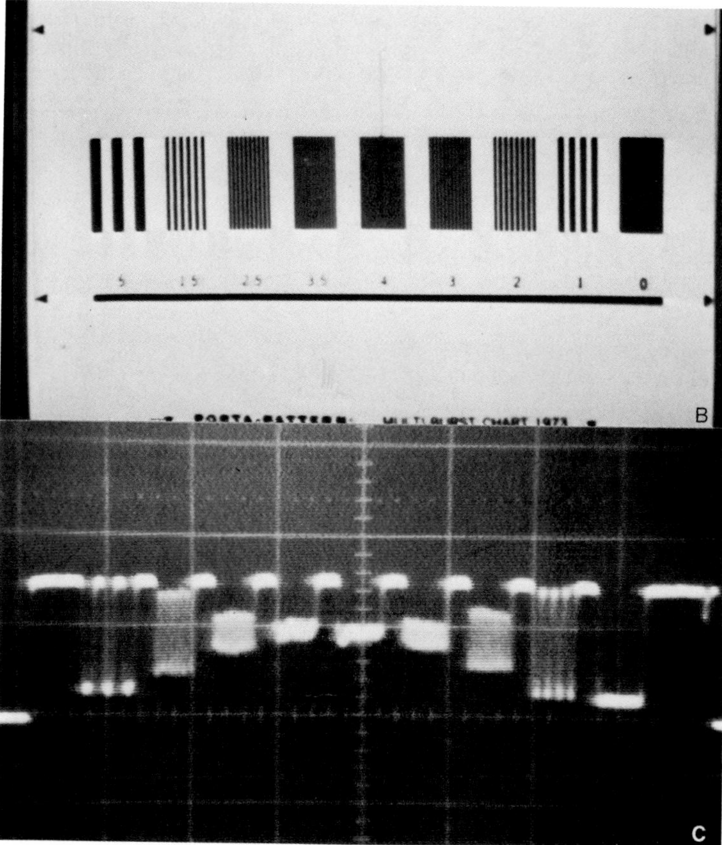

FIGURE 7-39. Loss of amplitude response at high frequency. (A) Original test chart, a "multiburst" pattern from Porta-Pattern. (B) The same test chart seen through a camera and a monitor. (C) CRO trace of an H scan through the middle of B.

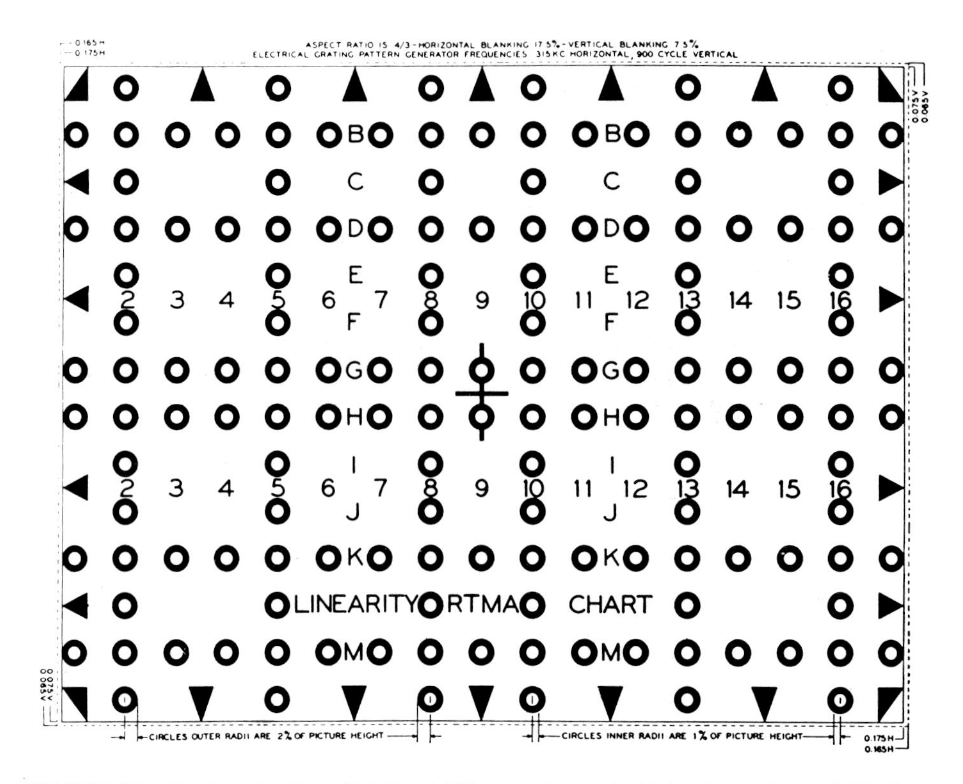

FIGURE 7-40. EIA Linearity Chart (Ball Chart). When superimposed with the picture of a standardized cross grid displayed on the monitor, distortion introduced by the monitor can be measured. Inner radii of circles are 1%, and outer radii 2%, of picture height. Tips of the triangles indicate boundary of active picture area. Aspect ratio, 4:3; horizontal blanking, 17.5%; vertical blanking, 7.5%. Test pattern generator frequencies: 315 kHz, H; 900 Hz, V. [From IEEE Publication Std 202-1954, reaffirmed 1972, with permission; copyright 1934, IRE (now IEEE).]

lines such as 1025, 1225, etc.), one needs a *high-resolution monitor* that can scan at the designated rate and which has a bandwidth commensurate with the resolution (e.g., up to 30–40 MHz). Some high-resolution monitors are versatile, and, for example, the Conrac QQA color monitor can be switch selected to three preset H-scan rates between 15 and 37 kHz (500 to 1225 lines at 60-Hz fields) and lock onto field rates of from 15 to 60 Hz.*

Most video monitors are overscanned by several percent to hide the (imperfect) borders of the picture area. Some monitors are provided with an UNDERSCAN switch or button that allows one to view the whole active picture area. With that option, one can check the first few scan lines that follow the V-blanking period and adjust the H HOLD critically. The underscan option is also useful for recovering parts of the image that may be hidden at the edges of a picture in an overscanned monitor. In a tape-recorded scene, an important feature of the specimen may lie just outside of the area visible with a standard monitor, but could be recovered by underscanning.

We have already seen the significance and utility of the *pulse-cross display* (Section 6.3). For video microscopy, it is useful to have available at least one monitor with this capability, so that the state of the V- and H-sync pulses, as well as the timing pulses during the V-blanking period, can be examined readily.

Depending on the monitor, we can either derive the sync pulses internally from the composite

*High-resolution monitors are also available, e.g., from: Dage–MTI, Ikegami (monochrome); Sony, Videotek (Mitsubishi) (color). High-resolution monitors are generally equipped with the special circuits for dynamic focusing, aperture correction, shading correction, etc., as well as more stable sync generators and power supplies.

video signal, or externally from separate sync pulses using a noncomposite or composite video signals. The INT(ernal) EXT(ernal) SYNC SELECT switch connects the monitor sync circuit either to the −0.3-volt sync pulses stripped from the composite video signal in the monitor, or to the −4-volt sync signals supplied separately to the monitor. By using stronger, separate sync signals, these monitors can generate a raster that is highly stable.

For some applications, especially for the display of color signals, one may use a TV receiver instead of a video monitor. In such cases, an *RF converter* is used to convert the NTSC signal into a broadcast-band TV signal. Some VTRs and many home computers contain RF converters. The receiver demodulates the TV signal and extracts out the NTSC signal. TV receivers are also used for multiple display of video, but with the higher frequencies and the extra steps involved in RF conversion and demodulation, the image contrast and resolution on a TV receiver generally do not match the display made directly on a video monitor. Some video monitors are equipped with an RF INPUT jack so that the monitor can accept either an RF modulated input or a regular video input.

V and H HOLD, and CONTRAST and BRIGHTNESS CONTROLS are standard features on all video monitors, usually accessible from the front panel. Many monitors are also equipped with V HEIGHT and LINEARITY controls as well as FOCUS controls, generally as screwdriver-adjustable potentiometers. In some monitors, the *H width* and *linearity,* as well as V and H *positions* and beam *astigmatism,* can also be corrected using screwdriver slots accessible from outside the monitor.*

As discussed elsewhere, the design of the *AHFC* (automatic horizontal frequency control) can vary considerably depending on the monitor. A slow AHFC is desirable for reducing picture glitches originating from external electrical noise (e.g., switching noises), but the monitor needs a fast AHFC to sync to incoming signals with irregular sync pulses (as in broadcast reception). The sync pulses coming out of a VTR, especially a **time-lapse VTR,** can be quite irregular. Without a very fast AHFC, such signals tear or flag at the upper parts of the video picture (Section 8.5; Fig. 8-16). Some monitors are equipped with a switch for selecting between fast and slow AHFC. Even on the fast setting, the monitor may not respond rapidly enough to the time-lapse VTR output and the top of the picture may flag. Some of the least expensive monitors (e.g., Sanyo Model 4290) seem to synchronize especially well to the signal from a time-lapse VTR.

Most monitors are equipped with a switch that allows the choice of using or bypassing the *DC restoration* or *black reset.* With the switch on, these circuits automatically bring the picture signal to an appropriate **average picture level** (APL). The DC restoration is helpful for viewing pictures where the APL, or the fractions of the picture signals representing the highlights, changes frequently. On the other hand, one may wish to bypass the DC restoration circuit and display the absolute brightness of the picture regardless of how much the average brightness of the scene changes. In video microscopy, one may wish to keep track of the absolute level of the signal to observe intensity changes of a particular feature in the scene or to monitor the incoming signal level, which affects the quality of subsequent processing and recording.

The *power supply* in the video monitor, as in the camera, importantly affects the performance of the monitor. If the power supply in the monitor is overloaded, as the beam current is increased, the picture on the monitor expands, becomes distorted, or may bloom or streak, etc. *High-quality monitors* have power supplies that are well regulated and with greater capacity. They should in principle provide wider ranges of contrast and brightness controls as well as better S/N ratio. However, one feature that is lacking in many, if not most, higher-priced monitors is a fast-enough AHFC that can accept and sync to the signal from a time-lapse VTR without flagging.

The distortion and nonlinearity of the raster may not exceed 1–1.5% in monitors of good quality. However, the faceplate of the picture tubes may still be quite curved, so that a photograph or projected image of the picture can be distorted beyond acceptable levels. Such distortions can limit the use of video monitors for photography or morphometry.† Special flat-faced, or *flat screen,* picture tubes are made to avoid the distortion arising from curvature of the picture tube faceplate.

*Inside the monitor chassis, the power supply for the picture tube reaches many kilovolts and is quite dangerous. Only qualified technicians should attempt to make internal adjustments or repairs to the monitor.

†But see the subsection Geometrical Distortions earlier in this section for means of getting around the limits imposed by distortions in the monitor.

They are often incorporated into monitors that are used for taking photographic records of the video picture.

Color Monitors

We surveyed the general principles of color video including the functions of the camera and the nature of the YRGB and encoded (NTSC) video signals in Section 7.4. In the color monitor, the process in the color camera is reversed and the RGB or encoded (NTSC) signal is made to generate a color picture.

In the color picture tube, fine dots, or stripes, of red, green, and blue phosphors are deposited in an interlacing pattern on the glass faceplate. Each type of phosphor is excited by the electron beam emitted by one of three electron guns. Figure 7-41 shows two of the several geometrical configurations used for the electron guns, a metal (**shadow**) **mask** placed near the faceplate, and phosphors in a color picture tube. The geometry of the guns, the distribution of perforations or slits in the metal mask, and the pattern of phosphor deposition ensure that the beam from each gun lands only on its corresponding phosphor. The other two phosphors lie in the shadow of the metal mask. In these picture tubes, the three electron beams, each modulated by the R, G, and B signals, are scanned together to generate the raster and display a color picture.

In essence, a color picture tube carries out the function of three monochrome picture tubes. The color monitor also contains three circuits the equivalent of three monochrome monitors, one each for the red, green, and blue channels. Only the raster-generating circuit and, if present, the decoding circuit are shared.

Some color monitors accept three separate signals for the red, green, and blue channels.* The *RGB monitor* processes each of these three signals much as in a monochrome monitor and feeds them to the corresponding input of the picture tube. As in monochrome monitors, the *resolution* of the RGB monitor is limited by the quality of the picture tube and the bandwidth of the monitor circuits but *not by encoding* of the color video signal.

Other color monitors accept only encoded signals such as the NTSC. In those monitors, the single NTSC encoded signal is first separated into its luminance and chrominance signals. From them, the RGB signals are decoded by reversing the arithmetic process used in the camera (Section 7.4). The R, G, and B signals are then each processed through their circuits and sent to the picture tube as in an RGB monitor.

Picture resolution is rather limited in monitors using only the encoded signal, since the bandwidth assigned to an encoded video signal is limited, e.g., to 4.2 MHz in NTSC. Furthermore, the 3.58-MHz color carrier frequency has to be removed from the NTSC signal, since otherwise the whole picture would be covered with a regularly spaced pattern numbering about 190 per H scan. But if the color carrier is removed with a standard **notch filter,** a fairly broad band of frequencies around the color carrier is also eliminated from the luminance signal. That means a considerable loss of resolution as one sees on conventional color TV receivers.

In *high-resolution* color monitors, a special **comb filter**† is used instead of a notch filter to more selectively remove the color carrier with minimal loss of its neighboring frequencies. The higher-frequency luminance signal, and the corresponding high spatial frequency in the picture, is thus preserved. Some high-resolution NTSC color monitors have horizontal resolutions up to 550 TV lines PPH or greater, although the high-frequency details are produced by the luminance signal only. Some RGB monitors have horizontal resolutions exceeding 1000 TV lines.

Some high-resolution color monitors accept both an RGB and an NTSC (or PAL, SECAM) input, but most monitors accept one or the other.‡ The input for the RGB monitor can be three

*Some RGB monitors also require a separate sync signal.

†For a concise explanation of the comb filter (which in a color monitor is in fact a 1-H delay line), as well as other practical aspects of NTSC signal processings, see, e.g., "Sony Basic Video Recording Course" Booklet 8, Color Signal Processing.

‡Some expensive color monitors made for the international market are highly versatile, being equipped with switch-selectable circuits that accept 525/60 or 625/50 scan rates in RGB or in the NTSC, SECAM, or PAL formats.

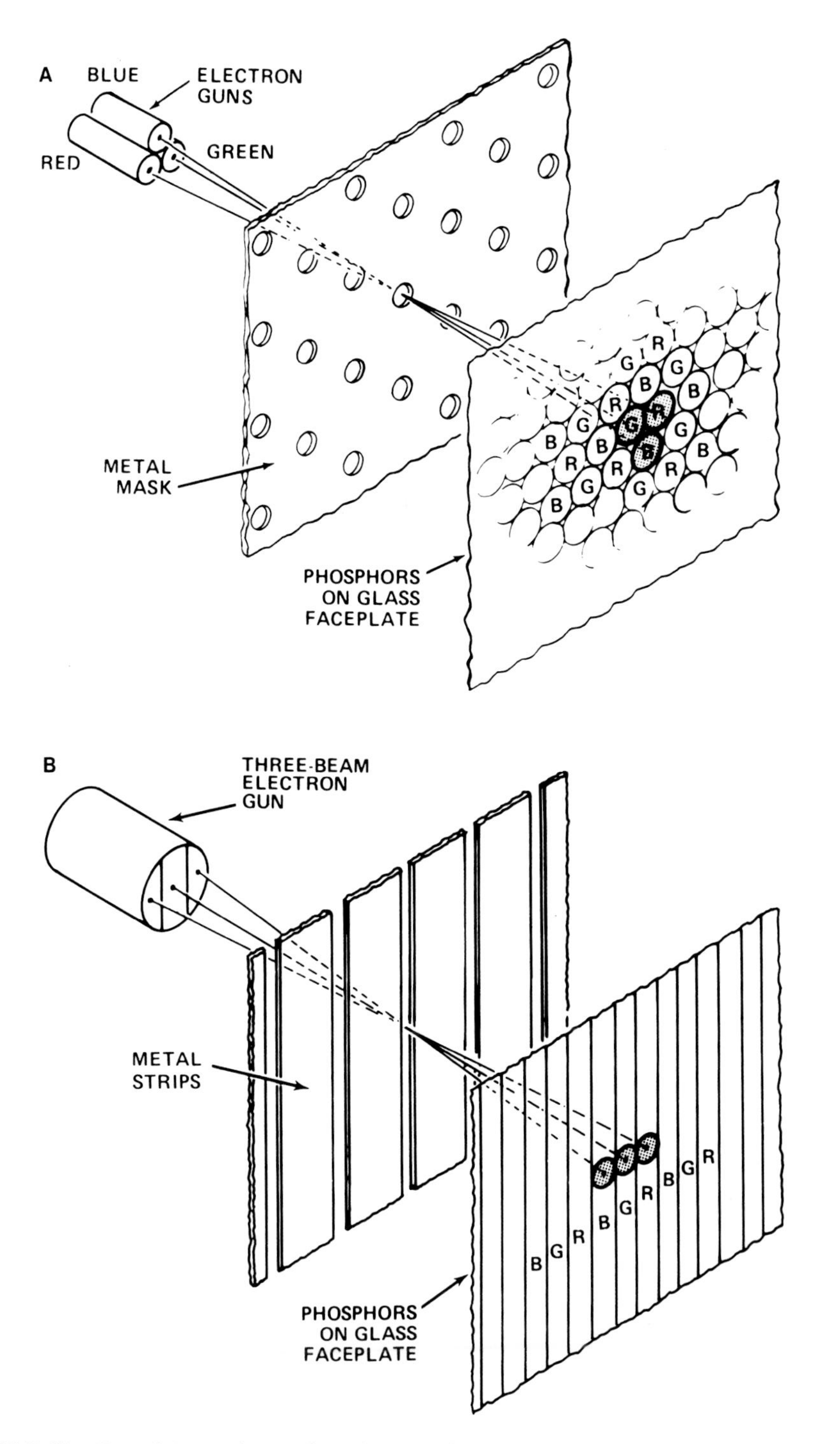

FIGURE 7-41. Two of the monitor configurations used in color picture tubes. See text. (A) Delta–delta configuration. (B) Single-lens, three-beam electron gun with vertical masks and phosphor stripes. (From *Raster Graphics Handbook,* 1980, pp. 9-10, 9-13, with permission.)

composite video signals, each with a 1.0-**Vpp** amplitude. Conversely, the RGB signals can contain just the 0.7-Vpp picture signals, accompanied by a separate −4-V sync signal. NTSC monitors accept a standard 1.0-Vpp, composite, encoded video signal. The composite video signal, as well as the separate picture and sync signals, each require a 75-ohm termination.

Color monitors require some adjustments that we do not encounter in monochrome monitors. In color picture tubes with three guns, the **convergence** of the three beams and their focus and astigmatism require occasional adjustment. Color monitors are intermittently **degaussed** in order to discharge the residual magnetism or charge that may affect proper landing of the beam and proper rendering of color.

Just as one adjusts the contrast and brightness on a monochrome monitor, one adjusts a color monitor to display pictures with properly balanced hue and saturation. We can conveniently make these adjustments with the aid of *color bars*. Vertical stripes, from left to right, of the following color, make up the color bars: white, yellow, cyan (pale sky blue), green, magenta (reddish purple), red, blue, and black. Color bar generators and some color video cameras generate the color bar signal (Fig. 7-29). Using a standard color bar signal, one adjusts the HUE and COLOR (and sometimes also the BRIGHTNESS) *control* knobs on the monitor to obtain the proper color as well as the balance of colors that provide neutral white and grays. The AUTO(matic) color balancing buttons on some video monitors and TV receivers provide balanced hue and saturation for standard incoming signals.

8

Video Recorders

8.1. INTRODUCTION

The video signal can be recorded in many ways. For transient storage and processing, the signal can be digitized and placed in solid-state chips in a frame buffer. For long-term storage of digital computer-processed video images, floppy or hard magnetic **disks** can be used. These magnetic disks, while too slow to capture (fully-resolved) dynamic scenes in real time, can in principle store digitized images without loss of spatial resolution. Video pictures can also be recorded by photographing the video monitor to produce sequences of snapshots or copied with a **kinescope** to produce a motion picture (Sections 12.1, 12.6).

More conveniently, we can record the video signal directly onto magnetic or optical media, commonly onto magnetic tape. A *video tape recorder* (VTR) allows instant recording and immediate playback of the video and audio signals. Once recorded, we can examine the scenes, with the accompanying sound, repeatedly, at natural speed, or, depending on the recorder, scrutinize the events slowed down, sped up, or frame-by-frame.

Some magnetic disks built into video recorders can capture one or more video frames (or fields) in real time for subsequent slow-scan transmission or for motion analysis. Laser disks, used in the **optical memory disk recorders** (OMDR), allow video-speed recording of over 10,000 single video frames, with the option of very rapid search and retrieval of any designated frame (Section 8.9).

In this chapter, we will consider the function, choice, and use of VTRs, including time-lapse VTRs. We will also examine some magnetic and optical disk recorders that are capable of instant recording and playback. The mechanisms for copying and editing videotape are detailed in Section 12.3.

8.2. HOW VTRs WORK

Video Tape Recording

Stripped to its bare elements, the composite video signal is made up of a linear sequence of voltages and time delays that specify the brightness and location of every image point, and of electrical impulses that signal the start of the horizontal and vertical sweeps (Sections 6.2, 6.3). In a VTR, these signals are recorded on tape as a series of magnetic fields in much the same way that audio signals are recorded in an audiocassette recorder. However, as seen in Chapter 6, the video signal is an electrical wave with very high frequency. The video signal requires a bandwidth of 1 MHz for every 80 black and white vertical lines that are to be resolved horizontally (i.e., along the scan lines) for a distance equal to the height of the picture (Sections 6.4, 6.5). Therefore, in order to

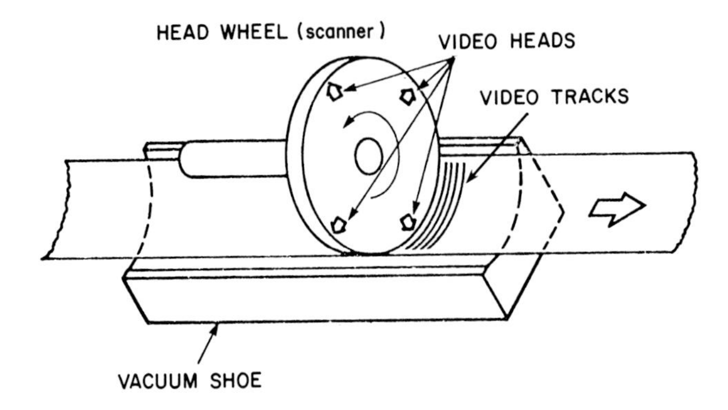

FIGURE 8-1. Segmented recording. A head wheel (scanner) rotates across the tape at a high constant speed as the tape moves at moderate speed in the direction indicated by the arrow. The video recording heads impart the magnetic imprints, or video tracks, on the tape. Since many separate, sequential tracks are required to record the signal from a single video field, this recording arrangement is said to be "segmented." (From "Sony Basic Video Recording Course," Booklet 3, with permission.)

just match the horizontal resolution to the vertical, or to resolve 340 TV lines horizontally per picture height (PPH), one would need a video bandwidth of 4.25 MHz (1 MHz × 340/80).*

To record a video signal with 340 TV lines of horizontal resolution PPH, one would also need a recording frequency of no less than 4.2 MHz.† Therefore, in order to attain even a modest horizontal resolution that just matches the vertical resolution, we must deal with frequencies that are over a thousand times greater than the frequency encountered in high-fidelity audio recording!

If we were to record a signal with this frequency along the length of a magnetic tape as in audio recording, the tape would have to rush past the stationary *recording* **head** some 1000 times faster than in audio, or at a velocity of 6–10 m/sec. That would present quite a mechanical design problem, on top of which the whole reel of tape would be used up in no time at all.‡

In practice, the video signal need not be recorded along the length of the magnetic tape with a stationary head as was the case in the early prototypes in the 1950s. Rather, the magnetic recording heads can be mounted on a drum, the *scanner,* which is spun across the tape at a high constant speed. As the tape travels at a moderate speed past the spinning scanner, the magnetic imprints, or the *video tracks,* are laid across the width of the tape (Fig. 8-1). With this arrangement, the *speed of tape transport* can be considerably reduced while the *writing speed* (the speed at which the head traverses along the video tracks) remains at the high rate that is required to provide the needed recording frequency.

While the tape transport speed is reduced to a reasonable level with this scheme, the signal from a single video field cannot be fitted into a single track, even across a 2-inch-wide videotape. Several sequential tracks have to be laid down before the signal for a full field is recorded, e.g., 16 tracks for a 2-inch-wide tape. Since the track for each video field is thus segmented, magnetic recording of this type is known as **segmented recording.** Segmented recording has been used widely in broadcast studio recorders.

*In contrast to the horizontal resolution, which is set by the bandwidth of the signal, the vertical resolution is set by the number of scan lines per frame. For standard 525/60 video, the vertical resolution is approximately 485 (the number of active video scan lines) × 0.7 (the Kell factor) = 340 TV lines (Section 6.4).

†Actually, one needs a greater bandwidth than this because of the additional frequency required by the sidebands for FM recording, as discussed later.

‡I am told that such a machine was actually designed and abandoned due to the problems described here. However, currently there is a renewed interest in a form of longitudinal recording, and at least the Kodak Spin Physics high-speed VTR uses a form of longitudinal recording as described in Section 8.6.

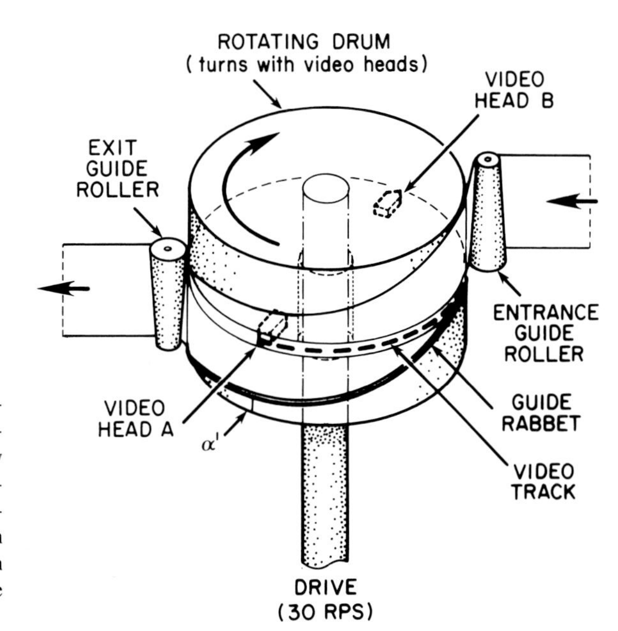

FIGURE 8-2. Helical-scan recording. In this method, used in most formats today, the tape is guided helically around the rotating scanner. This results in the video head taking a diagonal swipe on the tape. Generally, each track registers a complete field, in which case the recording is said to be nonsegmented.

In an alternative arrangement, the tape is made to run helically around the spinning scanner (Fig..8-2). With this arrangement, the video tracks are laid down diagonally at a shallow angle (α) to the length of the tape (Figs. 8-2, 8-3). Each track can now be made long enough to register a complete video field. This mode of recording, which can be **nonsegmented,** is known as *helical-scan recording*. Nonsegmented helical-scan recording is used on all small-format VTRs such as the ½-inch *videocassette recorders* (VCRs) commonly used for home video (i.e., consumer products) and the ¾-inch tape machines commonly used for industrial applications. Each track in these

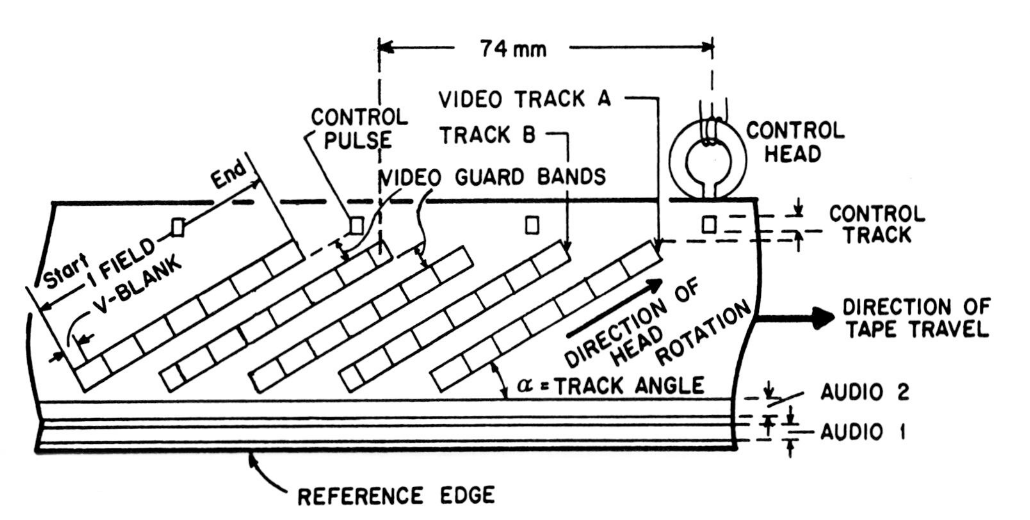

FIGURE 8-3. Schematic of U format. Each video track is laid down at the track angle, α, to the length of the tape. Each track represents a complete video field, as indicated at the left. Alternate tracks A and B represent the odd and even fields. A control (timing) pulse is laid down (along the control track) every two video tracks, signaling the end of a complete frame. On the side opposite the control track, adjacent to the reference edge, audio tracks 1 and 2 are laid down. All tracks are separated from one another by guard bands to prevent magnetic interference, or "cross talk."

recordings starts a fixed number of H intervals before the onset of the V-sync pulse and ends during the V-blanking interval.

With modern helical-scan recorders, the magnetic record corresponding to each signal impulse is laid down at a very high density; in fact, the recording gap in the read/write head is only 1 μm or less, and the width of the track is reduced to the order of 100 μm.

Regardless of the size of the magnetic record, the fields from adjoining tracks must not interfere with each other. Interference from adjoining tracks would result in **cross talk** and loss of picture quality. In order to prevent cross talk, the adjoining tracks are separated by a finite space known as the *video guard band* (Fig. 8-3). In the smaller high-density recording formats such as the ½-inch **beta** and **VHS,** the width of the **guard band** may be reduced to zero, or the adjacent video tracks may actually overlap. In these formats, the heads that record the alternate tracks are turned so that their gaps make an angle (2β) relative to each other (Fig. 8-4). The 2β° azimuth angle between

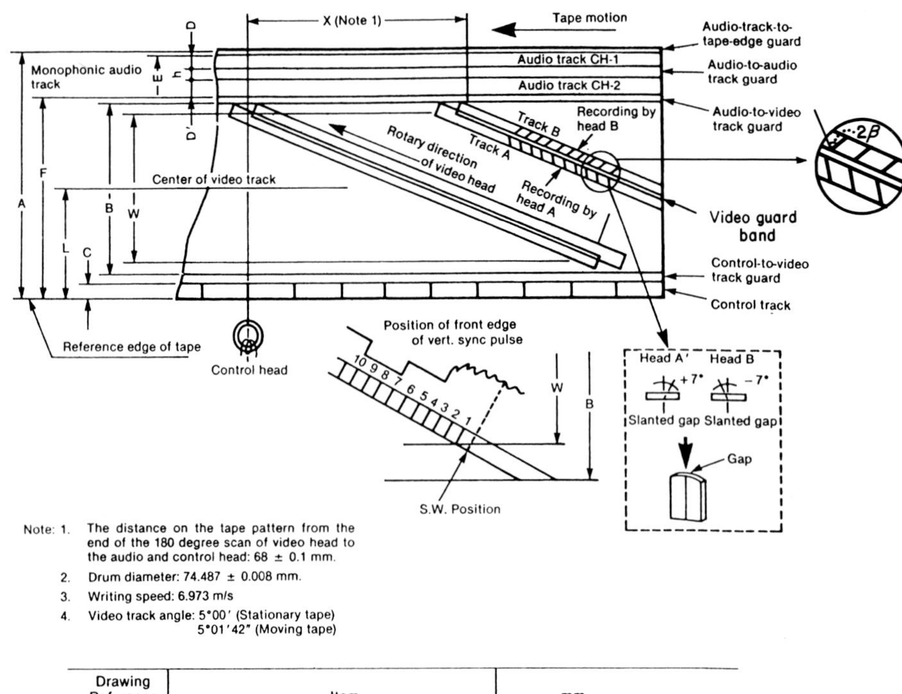

Drawing Reference	Item	mm	
1 (A)	Tape width	$12.67\ -{}^{0}_{0.04}$	
2 (B)	Video recording zone width	10.60	
3 (C)	Control track width	0.6	
4 (D)	Audio track width (stereophonic)	0.35	Ch-1 available for dubbing
5 (D′)	Audio track width (stereophonic)	0.35	Ch-2
6 (E)	Audio track width (monophonic)	1.05 ±0.05	
7 (F)	Audio track reference	11.51	
8 (H)	Audio-to-audio track guard width	0.35	Reference value
9 (P)	Video track pitch	0.0292	
10 (W)	Video recording zone effective	10.2	
11 (L)	Video track center (from reference edge)	6.01	

Betamax format (industrial).

FIGURE 8-4. Schematic of Betamax (industrial) format. In ½-inch formats, adjacent video tracks may overlap, as shown, in order to achieve a very high density of recording. Cross talk is minimized by recording the magnetic fields so that they are angled with respect to those on adjacent tracks (see insets). The angle, 2β, between the azimuths of the magnetic fields in adjacent video tracks is 14° in Betamax and 12° in VHS. See Section 8.5. (After "Sony Basic Video Recording Course," Booklet 4, with permission.)

the magnetic fields in the adjacent tracks reduces the cross talk to an unobtrusive level. Further specifics of the various recording formats are described in Section 8.3.

In addition to the diagonal video tracks, **audio tracks** and a **control track** are laid down near the edge of the tape, e.g., as shown in Figs. 8-3 and 8-4. The two audio tracks can be used for stereo recording, or one track can be used for recording simultaneously with video while the other is used for recording time codes and other special signals or for audio dubbing at a later time.

On the control track, the *control head* records a single magnetic impulse, derived from the vertical-sync pulses, once every video frame (i.e., at 30 Hz for 525/60 video). In playback, the control pulses picked up by the control head signal the tape position and tape speed to the servo systems.

The Scanner

As described above, the **video heads** on the scanner sweep across the moving videotape and lay down the video signal as a series of magnetic tracks. In many VTRs, the same heads are used to both record (write) and recover (read) the magnetic information, i.e., as *read/write heads*.

In nonsegmented helical-scan recorders, the scanner commonly consists of a pair of video read/write heads mounted exactly 180° apart on the lower face of a precision revolving *drum* (Fig. 8-2).* For 525/60 scan-rate video, the drum motor, or belt, drives the drum at 30 revolutions/sec (1800 rpm). Each head thus makes a complete sweep of the track, representing a full video field, in the time the drum makes a 180° rotation.† The speed of the scanner rotation, as well as its phase relative to the incoming video (during recording) and the video tracks (during playback), are closely regulated by servo mechanisms as described later. Two small magnets, mounted on the lower face of the scanner, generate electrical pulses (in a fixed pickup coil) that signal the speed and orientation of the rotating scanner itself.

Below the rotating upper drum, there is a lower *stationary drum*. The drive shaft for the upper drum passes through an opening in the stationary drum. The outer cylindrical surface in the upper portions of the stationary drum is machined and finished to a very close tolerance to match the diameter of the upper drum. The cylinder surface of the lower portions of the stationary drum is made somewhat larger. A discrete step, describing a helix along the surface of the drum, lies at the interface of the upper, thinner and the lower, fatter portions of the stationary drum. The videotape is guided down around the drum on this helical step, or the *guide rabbet*. The rabbet, together with the tapered entrance and exit *guide rollers*, define the helical path of the tape around the drums (Fig. 8-2).

The tape, driven at a set slow speed by the *transport mechanism*, slides over the rapidly spinning top drum down onto the bottom stationary drum. The video heads, mounted on the bottom of the spinning top drum, thus describe diagonal paths along the inner surface of the tape, which is coated with the thin layer of magnetic material that records the signals. Actually, the tape does not slide directly over the surface of the highly polished drums. Rather, the tape glides on a thin film of air that is trapped (or introduced) between the tape and the drums. This film of air keeps the tape lifted very slightly above the surface of the drums, reducing friction and protecting the tape.‡

The diagonal path of the recording head relative to the lower edge of the tape (the *reference edge*), which rides on the guide rabbet, gives rise to the *track angle* (α, Fig. 8-3). In principle, the track angle is determined by the helical pitch (α') of the rabbet, the diameter and rotational speed of the drum, and the speed of tape transport. Figure 8-5 shows how the track angle increases with tape

*The description of the scanner and several sections that follow illustrate the arrangements used in some **U-matic** systems. The details vary with the tape format and manufacturer, as well as vintage. For further details, see, e.g., Hobbs (1982) and "Sony Basic Video Recording Course."

†In 525/60 scan-rate video, there are 525 scan lines/field, 60 fields/sec, and 2 fields/frame (Section 6.3). At 30 revolutions/sec, the two heads in the configuration described here record a track each for the odd and even fields, in tandem for a total of 60 fields/sec.

‡However, the read/write heads do usually contact the tape and even press into it. Therefore, for high-quality recording, especially for time-lapse work and for data that are frequently inspected in **still frame** (or still field), we need special high-quality tape with a binder that does not readily allow the metal oxide magnetic material to be shed and create dropouts (Section 8.8).

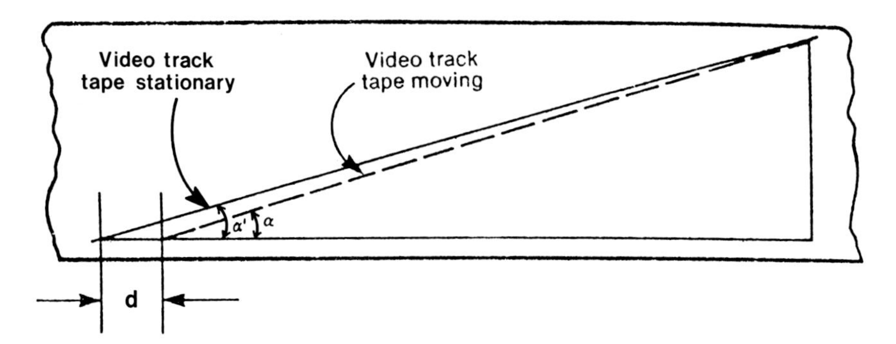

FIGURE 8-5. Video track angle. The video track angle (α) for *moving* tape is not the same as the rise angle (α') for the guide rabbet shown in Fig. 8-2; α' provides the track angle for stationary tape. If the tape moves a distance, *d*, during the recording of a single video field, and the scanner is rotating in the same direction as the tape transport (as in U format), then α is greater than α', as shown here. For standard-speed, U-format recording, $d = 1.588$ mm and $\alpha = 4°57'32.2''$; for time-lapse recording, the tape transport speed is slowed, resulting in a decrease in *d* and in α. Because adjacent video tracks are very narrow and closely spaced, α must be maintained to a very close tolerance, in order to avoid mistracking during playback. (After "Sony Basic Video Recording Course," Booklet 3, with permission.)

movement (when the tape is transported in the same direction as the head rotation).* In practice, tape tension, stiffness, deformation, and other factors can also affect the track angle. Humidity and temperature can thus affect VTR function in many ways.

Tape Path and Transport

Figure 8-6 shows a schematic example of the tape path in a ¾-inch VCR.† The tape leaving the *supply spool* of the *cassette* runs past the first *optical sensor* and then over the **erase heads.** These heads erase the audio and video tracks when the VTR is in the recording mode. The tape is then twisted slightly by the *entrance guide* (Fig. 8-2) and passes helically down the scanner as described above. Leaving the scanner after wrapping around the drums for somewhat over 180°, the tape passes over the *audio and control head stack* where the audio and control signals are recorded onto or read off from the tape. From there, the tape passes between the **capstan** and *pinch roller* where the *capstan motor* drives the tape at its proper transport speed. After encountering the *tension roller* and the second optical sensor, the tape is finally wound back into the cassette onto the *take-up spool.*

In a VCR, the tape is not manually threaded through the path just described as in a *reel-to-reel* VTR. Instead, the tape is threaded out of the cassette and guided through its path automatically when the record or playback modes of the recorder are activated. *Automatic threading arms* draw the tape out of the cassette, and a *threading ring* (on which the pinch roller and several other rollers are mounted) guides the tape around so that it is pressed between the pinch roller and stationary capstan, and wrapped around the scanner. When the *stop* or **eject** (or depending on the VCR, the *fast forward* or *rewind*) sequence is activated, the reverse process takes place and the tape is wound back into the cassette. The tape is thus protected in the cassette and does not come in contact with the operator's fingers or other sources of contamination.

Whether on a VCR or a reel-to-reel machine, the supply and take-up reels are each driven or braked to maintain appropriate *tension* on the tape. Additionally, a spring-mounted, or electronically controlled, tension roller fine-tunes the tape tension. If the tape tension or slack becomes excessive, a switch on the tension bar shuts off the drive motors.

The clear regions devoid of magnetic coating, at the two ends of the tape, signal the start and

*Depending on the specific recording format, the scanner spins in the same direction as the tape transport or in the opposite direction.

†See footnote * on p. 267.

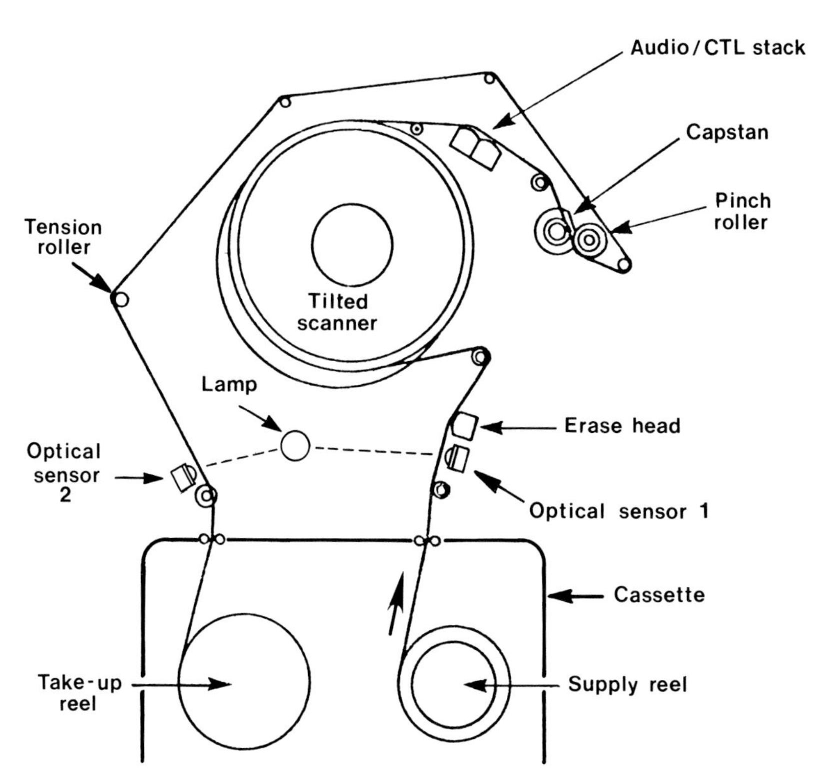

FIGURE 8-6. U-matic type II tape path. (After "Sony Basic Video Recording Course," Booklet 5, with permission.) See text.

end of the tape. These regions are detected by the optical sensors, which stop the drive motors or otherwise activate a series of appropriate steps. Breaks in the tape, as well as scratches that are large enough to let light through the otherwise dark magnetic layer, also trigger the optical sensors to shut off the motors. Most VTRs incorporate a mechanism for shutting down the drive if *dew* is sensed on the drum. The servo-drive motors on some VTRs are also inactivated if a proper *input video signal* is missing while the VTR is in the record mode.

Servo Controls

As discussed earlier, the video tracks must be recorded at a precise angle (α) relative to the bottom edge, or the reference edge, of the tape. Moreover, recording at the beginning of each track has to start at a precisely specified point of the video cycle (e.g., 6.5 H before the leading edge of the V-sync pulse for the U-matic system).

To satisfy these conditions, the scanner must rotate, and the tape must be transported over the scanner, at exactly specified speeds. Moreover, the positions of the revolving heads must be precisely *locked in phase* with the incoming video signal so that the A and B heads alternately come in contact with the tape just before the corresponding V-blanking periods.

To achieve these conditions, the VTR incorporates a series of *servo controls*. During recording, the speed and phase of scanner rotation are detected by electrical pulses that are generated in a stationary pickup coil by two asymmetrically placed magnets spinning with the scanner. The servo motors, driving the scanner and the capstan, are governed by comparing these pulses with a standard 30-Hz signal. The servo motors are also phase locked with the 30-Hz signal derived from the alternate V-sync pulses of the incoming video signal. The speeds and phases of rotation of the scanner drum and the capstan are thus controlled through a number of servo loops.

During playback, the read/write heads have to follow exactly along the video tracks laid down during recording, without running out into the guard bands or crossing over into another track. As in recording, the heads must also meet the beginning of the tracks at just the right time. Furthermore, the signals picked up from the successive tracks by the A and B heads must be precisely synchronized, so that they can be switched and recombined to produce the properly timed odd and even fields in the outgoing video signal.

To fulfill these conditions during playback requires even more precise control than in recording. Not only must the relative speed and phase of scanner rotation and tape transport reproduce those encountered in recording, but any imperfection of recording speed or phase, as well as changes brought about on the tape after the recording, must be compensated. The pulses read off from the control track provide the additional frame of reference needed to synchronize the tape transport and head rotation during playback.

Even by adding the control pulses to the servo input, a variety of mismatches can appear during playback of the tape. For one thing, the video track and the corresponding control pulse are separated by many centimeters on the tape. Repeated playback, or changes in humidity or temperature, could alter tape dimensions. Any change in tape dimension, or tension, would introduce *timing errors* by which proper intervals are no longer maintained between the re-added V-sync pulses and the H-sync pulses and picture signals.

Slight differences in tape tension between recording and playback could also introduce **skew**; the picture becomes unstable or the top of the picture skews to the left or right. Also, distortion of the tape, or error in tape speed relative to the scanner, could cause the heads to *mistrack* and wander into the guard band or even cross a guard band into an adjacent track. Crossing the guard band gives rise to noise (**guard-band noise**), which is displayed as an unsightly **noise bar** that bisects the upper and lower portions of the picture displayed (Fig. 8-7).

Despite these potential sources of trouble, most modern VTRs work with remarkably little trouble given some simple, appropriate care. Later in this chapter, we will go over the basic care that is required for proper use and maintenance of the VTR.

Frequency Response and Signal Processing in Magnetic Tape Recording

We have seen how an interlocking array of servo control circuits is used to regulate the motors that drive the head and tape in the VTR. An additional series of circuits is needed in the VTR to *process* the video signal before and after magnetic tape recording. These signal-processing circuits will be described after a discussion of the *frequency responses* encountered in magnetic recording.

Here we will emphasize monochrome recording, or recording of the luminance signal only. We will not dwell on the circuits that deal specifically with color recording. In fact, a considerable portion of the circuitry in modern video recorders, and ancillary devices such as time-base correctors, are designed to meet the special needs of color recording. Although such circuits are needed for handling color signals, we pay a considerable penalty in resolution and picture definition when we have to use these circuits for monochrome recording.

In monochrome video recording, we deal only with the luminance signal and need not worry about the chroma signal or the (3.58 MHz) color subcarrier. Even so, we are dealing with signals that range from a low subsonic frequency of around 10 Hz to a high in the radio frequency range of 4 to 10 MHz, or greater, depending on the maximum frequency that can be recorded.

The efficiency of playback of magnetic recording is highly dependent on the frequency of the recorded signal. At a low signal frequency (long wavelength of the magnetic record*), the change in magnetic field seen by the pickup head per unit of time (i.e., per unit distance of tape movement) is small (''A'' in Fig. 8-8). Since the current induced in the coil of the pickup head is proportional to the *rate of change of the magnetic field with time,* the magnitude of the signal generated in the coil is low at low frequencies. At higher recorded frequencies, the induced magnetic field changes faster and a stronger signal is generated in the coil.

Quantitatively, each time the signal frequency is reduced to a half (lowered by one octave), the

*The wavelength λ of the magnetic record is given by $\lambda = V/f$, where V is the head velocity relative to the tape and f is the frequency of the signal.

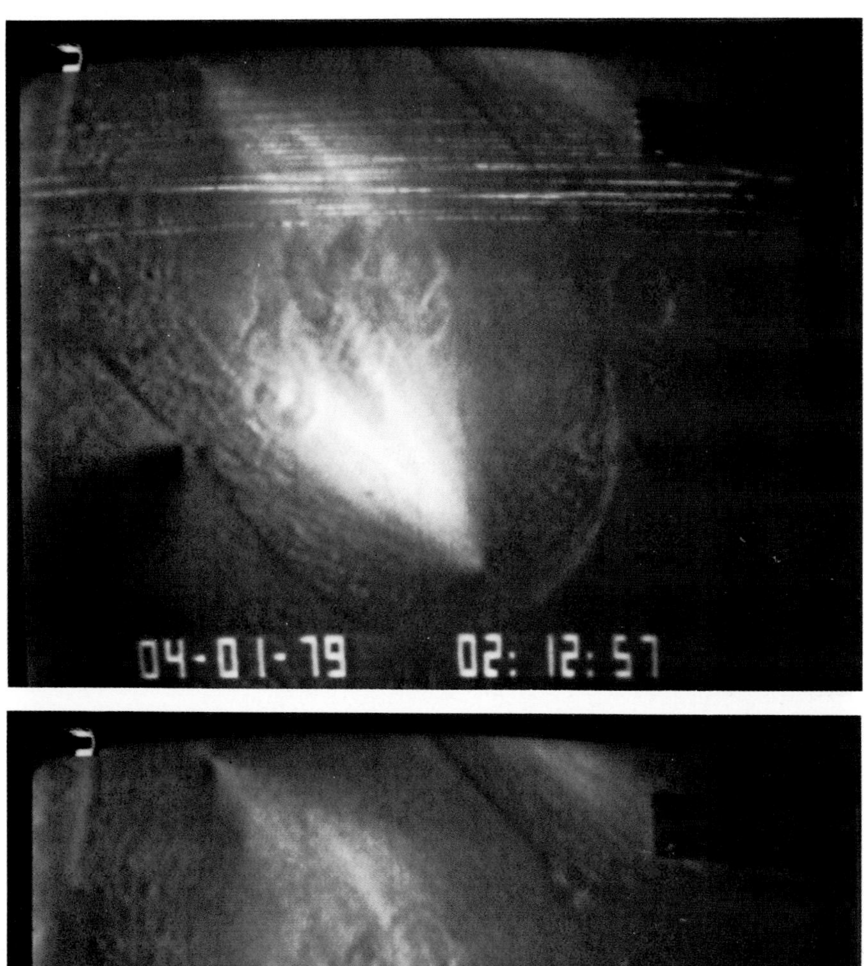

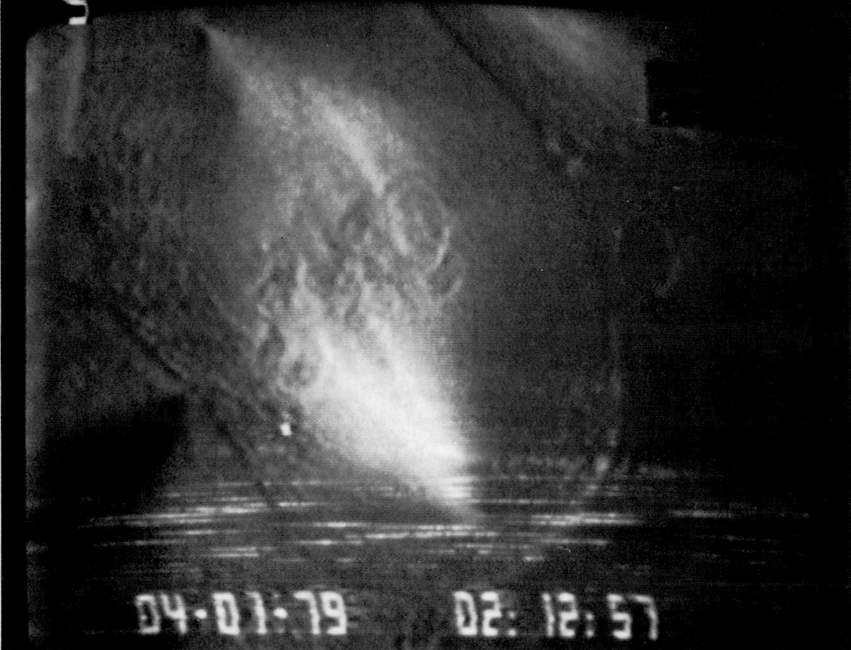

FIGURE 8-7. Noise bar. Noise bars (guard band noise) result from mistracking of the recording heads onto or across guard bands during playback. Mistracking often is caused by playing a time-lapse recorded tape at the wrong speed, as shown here.

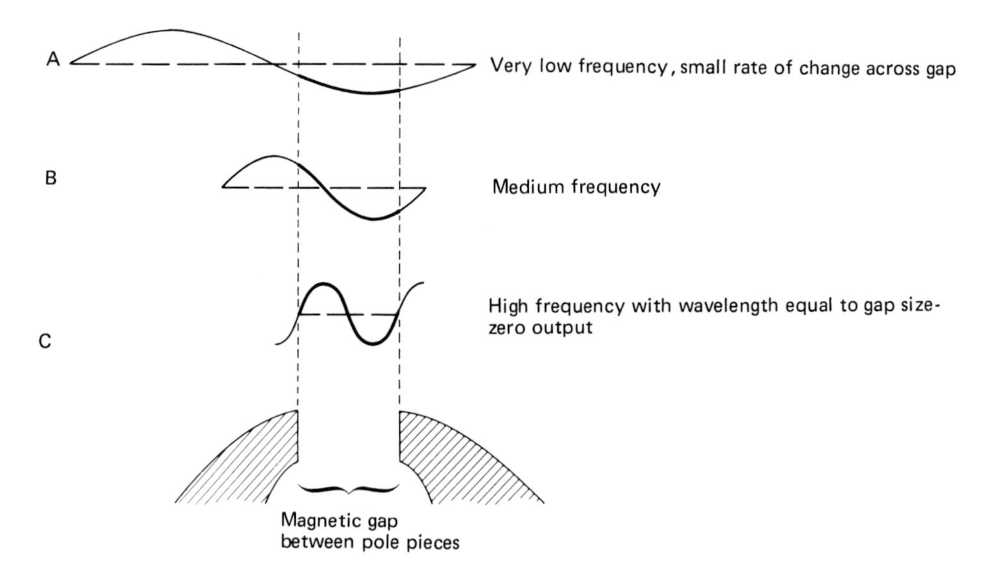

FIGURE 8-8. Dependence of signal strength on head-gap size and frequency. The strength of the signal induced in the coil of the pickup (playback) head is determined by the rate of change of its magnetic field. This change is produced by the magnetic tape, which moves past the head at constant speed. At low frequency (long wavelength) of the magnetic record, as in (A), the rate of change with time is small. At higher frequencies, the rate of change with time increases, for a given amplitude, until the wavelength (indicated by dashed lines) is twice the gap size of the pickup head (B). At still higher frequencies, the signal decreases until it becomes zero, when the wavelength equals the head-gap size (C). (From Ennes, 1979, p. 348.)

signal strength also drops to a half (6 dB down). This change in signal strength as a function of signal frequency in magnetic tape reproduction is known as the *6 dB/octave rule* (''a'' in Fig. 8-9A). At the lower frequencies, the miniscule voltage generated in the coil of the pickup head falls below the noise level and cannot be detected. The signal increases with frequency to a limit (''b'' in Fig. 8-9A), where the wavelength approaches twice the head-gap size. At higher frequencies, energy cancellation within the head causes a sharp decrease in response, until it reaches zero, when the wavelength equals the head-gap size.

The finite thickness of the magnetic coating on the tape also results in a loss of magnetization as the frequency rises. Furthermore, the closely spaced adjacent magnetic fields work against each other and reduce the field strength. These many factors rapidly diminish the high-frequency response of magnetic tape during playback (Fig. 8-9A,B).

Over the years, the high-frequency response of magnetic tape recording has been raised by improvements of the magnetic coating material and design of the read/write head.* At the same time, the push for more recording time on smaller-format tapes has raised the recording density and mitigated against raising the high-frequency response needed for greater resolution.

In the VTR, the signal is processed electronically to reduce the falloff of frequency response at both the low- and high-frequency ends. First, the total *octave range* used in recording is substantially reduced by *frequency modulation* (FM). For example, a 10-Hz to 4-MHz video signal, modulating a 5-MHz carrier frequency, is converted to a 1-MHz to 7-MHz FM signal as explained below.

In FM recording, the amplitudes, or voltages, of the **amplitude modulated (AM)** video signal are converted to frequencies; e.g., peak white by 6.8 MHz, blanking level by 5 MHz, and the tips of the sync pulses by 4.3 MHz. The 10-Hz to 4-MHz video frequency signal represented (asymmetrically) in the sideband extends the total frequency so that the FM wave spans 1 to 7 MHz in this

*A particularly hopeful approach is the use of **vertical field** (as opposed to lateral) **magnetic recording.** So far, this approach does not seem to have been used for VTRs.

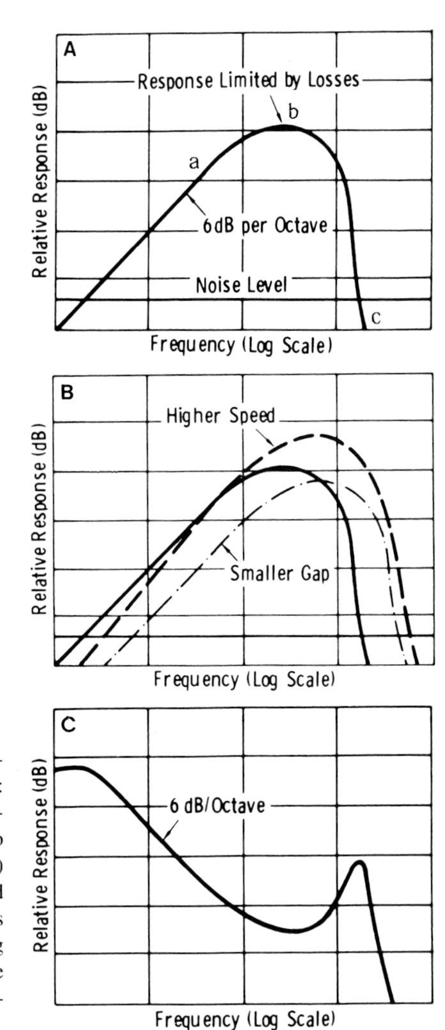

FIGURE 8-9. Frequency response in magnetic tape recording. (A) Uncompensated frequency response curve. a: region exhibiting 6 dB/octave response; b: maximum response. At higher frequencies, signal drops rapidly to zero (c). (B) Changing the tape speed or gap size (dashed lines) shifts the response curves relative to that shown in A (solid line). (C) Equalization curve, which electronically corrects the magnetic recording characteristic of A and B. The rising response at the high-frequency end is to correct for the rolloff caused by gap size and tape magnetic coating factors. See text. (After Ennes, 1979, p. 349).

example. By using FM, the range of the signal is now compressed from over 18 octaves for the original 10-Hz to 4-MHz wave down to less than 3 octaves for the FM signal.* By limiting the frequency to a *narrower octave range,* the response curve that follows the 6 dB/octave rule no longer dips way below the noise level in the lower-frequency range. Thus, the S/N ratio is vastly improved. In addition, by using FM, the various signal voltage levels, such as the peak white, blanking level, and sync tip, are unambiguously represented by discrete frequencies rather than by ambiguous amplitudes.

To take advantage of these features of FM recording, the signal entering the VTR is processed in the following manner. The color subcarrier and the chroma signal, if present, are stripped and sent to a separate circuit for color signal processing. Once the luminance signal and the sync pulses are freed from the chroma, the signal levels are adjusted by an *automatic gain control* (AGC) circuit; the sync tips and peak white are clipped to their standard levels; and the signal is preemphasized (Fig. 8-9C). The *preemphasis* helps to offset the dropoff in the low and high ends of the frequency response of magnetic recording that we have just discussed. The signal is then modulated into an FM signal. In some recorders made for use in both monochrome and color, the color circuit and certain passband filters are shut off to provide a wider passband for the monochrome signal

*Recall that the frequency doubles for each rise in octave.

when the VTR senses the absence of a color subcarrier. This automatic switching circuit is known as the **color killer.** Automatic switching provides optimized circuits for both color and monochrome recording, yielding higher-resolution recordings in the monochrome mode free from interference from the color subcarrier.

In *playback,* the small FM signals from the A and B pickup coils are first amplified through their respective preamplifiers. The signals from the two recording heads are then *switched* alternately between every field and *balanced* to the right levels.* Joined together, they reconstitute a single video signal with two fields per frame.

Several other operations take place in the playback circuits. Before *FM demodulation,* the playback signal levels are adjusted by another *AGC* circuit, their extreme levels are limited, and, in addition, *dropouts* are compensated, as follows.

Unless the magnetic recording is perfect, some parts of the signal can be expected to be missing in playback. For example, a dust particle on the tape or a minute hole in the magnetic coating would lead to a brief missing signal, or dropout. In many VTRs, if the dropout is within a single H-scan line, the dropout is compensated by substituting the whole (or a part of the) defective line with the previous scan line. To accomplish **dropout compensation,** the VTR continuously stores a single horizontal interval of the frequency-modulated video signal. The FM signal is ordinarily played back through a *1-H delay line.* When the envelope of the FM signal shrinks, signaling a dropout, the stored, previous line is automatically made to replace the current, defective line. Thus, the dropout is compensated by line substitution of the luminance signal.†

After the 1-H delay line and dropout compensation circuit, the FM signal is demodulated; the signal that was emphasized during recording is *deemphasized;* the *high-frequency noise* in the luminance signal is detected and canceled or otherwise "cored out"; and the signal levels are readjusted to the *standard video levels* before the chrominance information is added back.

Even this brief description of the signal-processing circuits (for the luminance portion of the video signal) shows the complex steps that are required to record and recover a video signal in a VTR. Altogether, the signal-processing circuits, as well as the recording medium, the read/write heads, and the servo drives, in a VTR are impressively well designed to collectively provide the specified performance. Nevertheless, it is well to keep in mind that the signal played back from a VTR has gone through many transitions. For quantitation of image parameters using video tape recording, appropriate calibrations are clearly in order.

As also should be apparent from these descriptions, the mechanical and electronic parts of a VTR, and especially of a VCR, all function interdependently. In addition, some of the components are delicate, and they are packed with very high density in several layers. While the routine care and maintenance of the VTR described in Section 8.7 should be attended to by the user, repair and interior adjustments of a VTR must be delegated to *experienced* service personnel.

8.3. VIDEOTAPE FORMATS

Magnetic tape of several different sizes are used for video recording. One- and two-inch-wide tapes, mostly used in reel-to-reel machines have long been the standard in broadcast studios. Three-quarter-inch tapes supplied in cassettes are used for industrial, small studio, and professional portable recording. Half-inch tapes are widely used in reel-to-reel form for security and laboratory recording and in various cassette formats for home entertainment and educational purposes. Half-inch tapes run at high speed are also used in compact *"Recams"* for electronic news gathering (ENG), and even narrower (8 mm) tapes have been introduced for home video *"camcorders."* In both of these systems, the VCR sits piggyback on the camera (Fig. 8-10).‡

*Recall that the A and B fields were recorded on successive, separate tracks.

†Some modern broadcast series VTRs store a whole video line, rather than the FM luminance signal alone. This allows dropout compensation for the full color information.

‡A Recam is a broadcast-quality camera/VCR that is combined into a single physical unit. Using the **M-format** ½-inch tape, it achieves broadcast-quality color exceeding the performance of U-format ¾-inch VCRs (e.g., see Renwanz, 1983). Camcorders are similarly packaged, lightweight, compact consumer products that record on VHS, beta, or the new ¼-inch tape (e.g., see "All-in-One Camcorders," 1983).

M-format

Broadcast Quality Picture with 1/2″ Tape

The RECAM's professional quality results are made partly possible by a special recording process which records Y (3.8MHz) and I/Q (1.0MHz) information on separate video tracks. This means that Y and I/Q signals are delivered directly from the camera to the recorder. In addition, the I/Q signals are multiplexed for even greater freedom from intermodulation. Also contributing to the outstanding picture quality are the 4 video recording heads and the high tape speed of about 8 ips with a recording wavelength of 27,000 cycles per second. Although compact VHS tape is only 1/2″ wide, the M-format recording system delivers broadcast quality video, 2 high quality audio tracks, a control track, and a time code track all on this space saving, easily available tape.

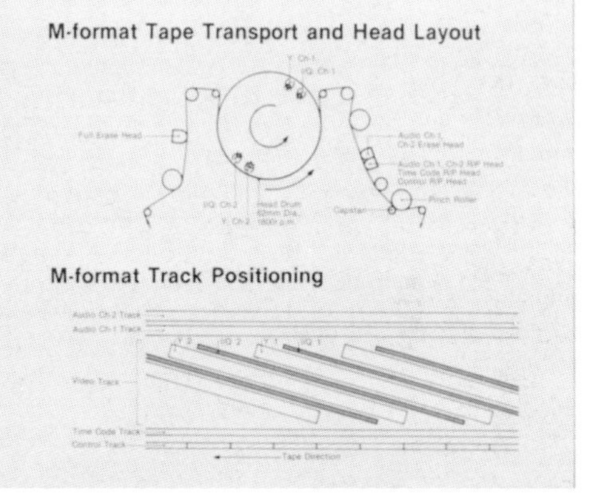

FIGURE 8-10. Recam using high-performance M-format. The new ½-inch M-format, used in Recams such as this one (Panasonic B-100), exceeds the performance achievable with the ¾-inch U-format VCRs. (From Panasonic Publication B-100; information courtesy of Matsushita Electronics Corp., Takatsuki, Japan.)

The tape width and type of tape transport, whether reel-to-reel or cassette, are fixed for a given VTR. The different recording formats including the basic speed(s) of tape transport are not interchangeable so that differently formatted tapes can be played back only on the intended types of VTR.

In choosing the videotape format for a particular application, ones needs to consider the image resolution and fidelity, the range of equipment available, and the price of the VTR and tape. Since high-frequency magnetic recording is essential for high resolution and fidelity, the widest-format tapes might be the format of choice if cost were of no concern. In contrast, ½-inch cassette formats are both convenient and economical, but their recording quality may not be adequate for many applications in microscopy. We shall now examine some representative videotape formats.

One- and Two-Inch Formats

One- and two-inch videotapes are used in various formats. The video tracks may lie at a relatively large angle to the length of the tape and be segmented (as in the BCN-format). Others may be nonsegmented and use helical scan as in the narrower-format tapes. There, a single track recorded at a shallower angle registers the signal from a whole video field.

The so-called **C-format** for the 1-inch tape uses a nonsegmented, helical scan. However, information for the V-blank interval is recorded on a separate, shorter helical track. In the C-format, there are three audio channels to the side of the video channel, one of the tracks being commonly used to store the **SMPTE** (Society of Motion Picture and Television Engineers) **time code**. The width of the video and sync tracks is between 0.125 and 0.135 mm. Each video track, tilted 2°34′ to the tape edge, runs for slightly over 410 mm. Depending on reel size and equipment, some C-format tapes can provide up to 3 hr of recording.

Three-quarter-Inch U format

In contrast to the 1-inch format, the video tracks in the smaller tape formats are considerably narrower and shorter per field recorded. We will examine the ¾-inch **U format** in some detail.

During forward motion of a U-matic VCR, the tape unwinds out of the right reel of the cassette and is taken up by the left reel. The tape is threaded sequentially around the erase head, the scanner, the control and **audio heads,** and the capstan, as shown schematically in Figs. 8-2 and 8-6. As these figures illustrate, the two video heads on the scanner spin in the same direction as the tape is transported. Since the tape travels helically *down* the drums as the scanner spins at a constant height, the video track in the U-matic format runs from the lower left to the upper right on the magnetic tape (Fig. 8-3).*

These diagonal video tracks use the central 15.5 mm of the ¾-inch (19.00 ± 0.03 mm)-wide magnetic tape. Since the A and B heads scan a little over 180° on the drum, the tracks each cover slightly more than a full video field; the effective width occupied by the video tracks is 14.8 mm.

The pitch of the video track, or the distance between adjacent video tracks measured at a right angle to the track, is 0.137 mm. Of this distance, 0.052 mm is used as the video guard band so that the actual video track is only 0.085 ± 0.007 mm wide. The video track is 171.22 mm long and makes an angle of 4°57′33.2″ with the reference (bottom) edge of the tape. The tape travels at a speed of 95.3 mm/sec, and the effective writing speed for the U format is 10.261 m/sec. The maximum recording time per cassette varies with tape length and cassette sizes, but is commonly 20 or 60 min.

In the U format, the control track occupies the top 0.60 mm of the tape, below which a guard band separates it from the video tracks (Fig. 8-3). In the control track, the *control pulses* that regulate the speed of the tape and scanner servo-drive appear once every two video fields, or once per video frame. They are spaced 3.18 mm apart along the length of the tape. The two audio tracks situated near the bottom of the tape, or next to the reference edge, are each 0.80 ± 0.05 mm wide and protected with 0.2-mm-wide guard bands. Measured along the tape, the read/write heads for the control and audio tracks are placed 74.0 mm ahead of the effective end of the corresponding video track (Fig. 8-3).

Half-Inch Formats

For the ½-inch *Beta* and *VHS* formats, the supply reel is on the left of the cassette (Fig. 8-11). Therefore, the video tracks are tilted oppositely to the U format. More importantly, the A and B video tracks in the high-density formats actually overlap with each other. Even with the overlap, the pitch for the video track in the Beta format is only 0.029 mm, or 29.2 μm, and the head gap is a scant 0.60 μm! In the VHS format, the video track is 59 μm wide.

In order to avoid cross talk between the A and B tracks in these high-density formats, the azimuth directions of the A and B head gaps are tilted relative to each other. The magnetic fields make an angle (2β) of 14° to each other in the Beta format, and 12° in the VHS format (Fig. 8-4).

The audio channels are placed at the top of the video tracks in the Beta format and at the bottom in the VHS, with the control tracks on the opposite edges. Tape speeds are 40 mm/sec for the 1-hr industrial Beta, 20 mm/sec for the 2-hr version (Beta II), 33.4 mm/sec for the 2-hr VHS (SP), and half of that speed for the 4-hr version (LP). SLP cassettes run for 6 hr.

M-format

A ½-inch format capable of exceptionally high-fidelity video and audio recording has recently been introduced. Used in "Recams," the M-format uses a VHS T-120HG cassette at 8 **ips** to give a total recording time of 20 mins. The M-format is said to provide a bandwidth of 3.8 MHz for the

*The tape is viewed from the side with the magnetic coating.

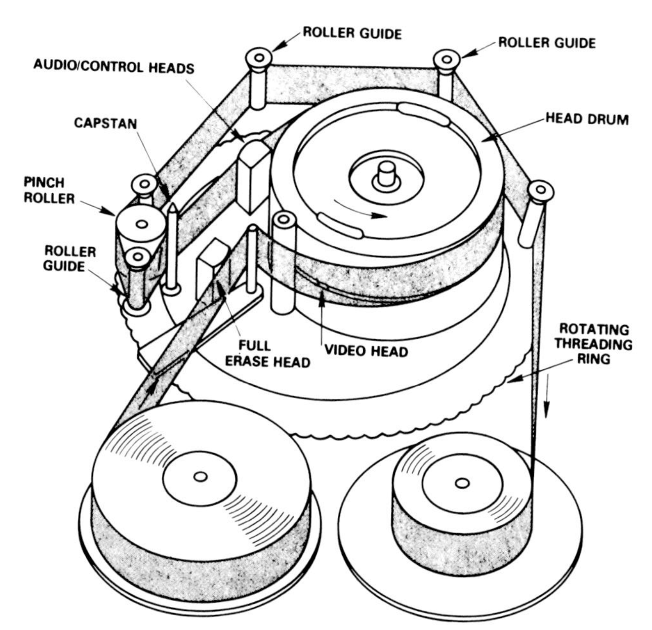

FIGURE 8-11. Conventional beta tape path. In the Betamax and VHS ½-inch formats, the supply reel is on the left of the cassette, rather than the right, as in the U format (Fig. 8-6). As a result, the video tracks are tilted oppositely to those in U format (cf., Figs. 8-4, 8-3, and 8-6). (Courtesy of Sony Corp.)

luminance signal, with a video S/N ratio of 47 dB. Unlike standard *color-under* recording formats, the M-format uses a four-head scanner to record a wide-bandwidth luminance signal on its own track, separately from the I and Q color signals (Fig. 8-10). The separation of the luminance and I and Q tracks prevents interference and the consequent formation of herringbone patterns. This eliminates the need to filter out the 3.58-MHz color subcarrier from the luminance signal as is done using a comb or notch filter in the color-under scheme.*

While the vastly improved color quality and exceptionally high S/N ratio achieved by the M-format are attractive, the Recam systems are expensive, and their design does not appear immediately suited for application in video microscopy.

8.4. CHOICE OF VTRs FOR VIDEO MICROSCOPY

Introduction

As described above, there are several sizes and formats of videotape. There is an even larger assortment of VTRs available today, ranging from large-format studio recorders to quarter-inch-format home consumer products. Most of the VTRs (sold in North America and Japan) are made to record and play back standard NTSC color video and accept EIA RS-170, RS-170A, and RS-330 signals. A few are made for monochrome only, and some high-resolution VTRs only accept nonstandard video signals.

Most VTRs both record and play back video and audio signals, although some portable recorders attached to the camera (Recam and camcorders) require a separate playback deck, and some decks (designated *tape players*) are made for playback only. Many VTRs are equipped with **pause** or *still frame* capability, while some also allow *slow scan* and *speed search*. In these modes, many, but not all, VTRs show some degree of picture breakup caused by the intruding guard band noise.

Special VTRs, known as **editing decks,** are made specifically for videotape editing, although some standard VTRs can also be used for **assemble editing** as described later.

*See, e.g., Faroudja and Roizen (1981) for an explanation of "color-under," or "heterodyne," recording.

In general, VTRs must run at a closely regulated fixed speed. However, many ½-inch VCRs can also be used in the *extended play* modes, at a sacrifice of some fidelity, by running the tape at defined lower speeds of a half, quarter, sixth, eight, etc. of the standard speed.

As mentioned earlier, *1-inch and larger-format VTRs* are designed primarily for studio or production TV use. They are expensive, quite bulky and heavy (although some portable units are available), machined to exceptionally high tolerances, and are expected to be regularly maintained by professional service personnel (Fig. 8-12). They also require input signals with well-corrected time bases. Time-lapse capabilities are generally not available (but see Sony BVH-2500 in Table 8-1), although capabilities for noiseless still-frame display and some degree of speed-up and noise-free slow playback are provided in order to accomplish, for example, synchronized dubbing. Also, some 1-inch VTRs provide animation capabilities. These large-format VTRs provide the highest resolution, picture fidelity, and stability (Table 8-1).

FIGURE 8-12. One-inch C-format reel-to-reel VTR. Large-format recorders such as this Ampex VPR-80/TBC-80 are capable of providing the highest resolution and picture fidelity, but are expensive, bulky, and require expert maintenance. Several 1-inch VTRs are compared in Table 8-1. (Courtesy of Ampex Corp.)

TABLE 8-1. Partial Specifications for Some 1-Inch Reel-to-Reel VTRs

Make	Ampex–Nagra	Ampex	Hitachi	General Electric	Sony
Model	VOR-5 Portable VTR	VPR/TBC-3 Basic Animation VTR	HR 230-2 1-inch Production VTR	HPR-50[a] Very High Resolution Monochrome VTR	BVH-2500 Delta T, Animation, Time-lapse VTR
Tape format	C-format	C-format	C-format	C-format, 800-line resolution	C-format
Record heads	3 (Erase, Record, Play)	3 (Record, AST, Erase) or 6 (with sync option)			2 Super Bimorph (Record/Play, Erase), 1 Video Confidence Special Head Position Detector
Tape speeds	9.6 ips nominal	9.606 ± 0.02 ips	9.606 ± 0.02 ips	9.606 ± 0.2 ips (525/60) 8.87 ± 0.2 ips (1023/60)	9.6 ips, plus 1/60, 1/30, 2/30, 3/30, 4/30 of standard speed
Video write speed	1009 ips nominal	1009 ips		1009 ips nominal	1007 ips
Record time	20 min (5.5-inch reel)	190 min (14-inch reel)	3 hr (14-inch reel)	92 min (10.5-inch reel)	200,000 frames/400,000 fields (11.75-inch reel)
Video	NTSC 525/60	NTSC/PAL-M 525/60		525/1023 switch-selectable	
Bandwidth	5.0 MHz (−3 dB)	Flat to 4.2 MHz ± 0.5 dB −3 dB at 5.0 MHz	4.7 MHz (−3 dB)	10 MHz (−3 dB)	Flat to 4.2 MHz ± 0.5 dB −3.0 dB at 4.5 MHz
S/N ratio	46 dB (p–p/rms)	46 dB p–p/rms noise	48 dB (p–p/rms)	40 dB	48 dB p–p/rms noise 47 dB (R&S unweighted)
Audio	50 Hz–15 kHz (−2 dB)	± 1 dB 200 Hz–12 kHz ± 2 dB 50 Hz–18kHz	50Hz–15 kHz (−3 dB)	50 Hz–15 kHz (± 3 dB)	50 Hz–15 kHz (+ 1.5–3 dB) 200 Hz–7.5 kHz (± 1.0 dB)
S/N ratio	56 dB (audio 1, 2) 50 dB (audio 3)	56 dB (audio 1, 2) 54 dB (audio 3, wide band)	56 dB (audio 1, 2) 50 dB (audio 3)	46 dB Time code wide band (audio 3)	56 dB (audio 1, 2) 50 dB (audio 3)
Power	10–18 V DC 40 W nominal	50/60 Hz 95–270 V AC 2.6 kW peak surge	AC 110/117/220/240 50/60 Hz 600 V AC	100/110/120/130 V AC ± 10% 200/220/240/260 V AC ± 10% 5.5 A (nominal 110 V)	48–64 Hz 110–120 V AC 550 W
W × H × D (inches)	17.5 × 5.5 × 8.5	30 × 75.5 × 32	19 × 31.3 × 22.9	19 × 24.5 × 15	Rack mount: 22.5 × 26.8 × 23.2
Weight	15 lb with 20-min reel & battery	650 lb	187 lb	120 lb	155 lb
List price (12/84)		\$89,000[b]	\$78,000[c]	\$45,000	\$75,000[d]
Features	Very compact, portable	Pinchless (vacuum); auto scan track; clean still to 3× forward, 0–1× backward; shuttle up to 500 ips, forward/reverse; animation, min. 1 field/cycle, min. cycle ca. 3 sec	Air-tension arm posts; noncontact head drum (in standby); 1× reverse to 3× forward without noise bar; slow motion sequence store	Variable play, 0–1 1/2x speed; auto scan tracking; dynamic recording (not frame-by-frame); search to cue	Single field or frame; time-lapse recording without preroll; dynamic tracking head at 0.2–0.6 sec/cycle; still frame recording; 8-bit 4FSC A/D line-by-line video expansion; cue accuracy ± 0 frame

[a]Modified Ampex (see Fig. 8-12).
[b]With TBC and animation controller.
[c]With TBC and slow advance control.
[d]Without TBC.

FIGURE 8-13. Three-quarter-inch front-loading U-matic VCR. The ¾-inch format often is a good choice for video microscopy; it provides good-quality recording and playback, reasonably high resolution, and convenience in use, at a cost very much less than 1-inch formats. The Sony VO-5800 H, shown here, and the Panasonic NV-9240 XD are high-resolution versions, especially suited to video microscopy (Tables 8-2, 8-4).

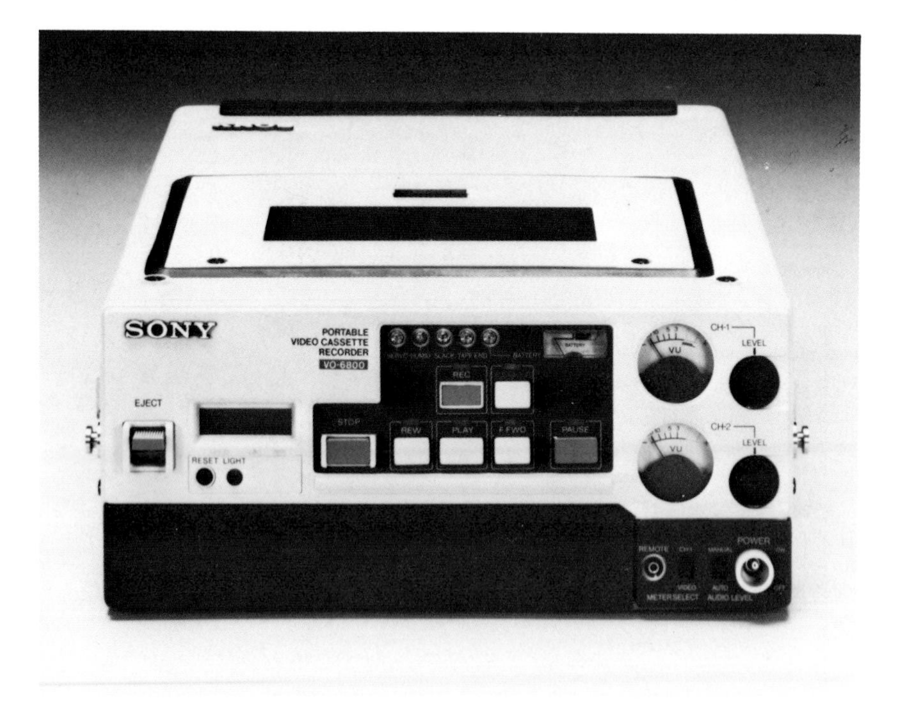

FIGURE 8-14. Portable U-matic VCR. This Sony Model VO-6800 weighs under 13 pounds and is less than 14 inches long (one-quarter the weight and one-half the size of the model 5800 H in Fig. 8-13), making it easily transported and used in the field. (See Table 8-2.) (Courtesy of Sony Corp.)

The *¾-inch U-format VCRs* provide a very wide choice of instrument types and capabilities (Figs. 8-13, 8-14). They are widely used for industrial, educational, scientific, and medical application as well as for electronic news gathering. The ¾-inch VCRs provide a good-quality, medium- to moderately high-resolution image (Table 8-2). At normal speed, many U-format VCRs record up to 60 min, and the cassettes are each provided with protective plastic cases that fit neatly onto a standard bookshelf. While the ¾-inch VCRs are professional-quality equipment, generally costing several times more than the smaller, consumer products, most U-format VCRs are designed for simple hookup and convenient operation, and should not require frequent maintenance by technical service personnel.

Some *½-inch Beta- and VHS-format VTRs* are made for industrial, scientific, and security applications, while other consumer-oriented products are popular for home video recording and playback. These VCRs are often available at a modest price, are reasonably compact, convenient to operate in conjunction with a regular TV set, and can even be programmed to record at preset hours (Table 8-3). The very clever ½-inch recording formats (Section 8.3) now use cassettes (which can hold several hours of recording) that are smaller than regular-size paperback books.

Among the many VTRs available, which should one choose for video microscopy? In principle, most any VTR could be used for general recording and playback of video microscope scenes at normal speed. However, the following considerations may narrow the choice.

1. Do you need to make time-lapse recordings? That is, do some of the events occur so slowly (such as cells dividing, embryos developing, etc.) that you need to speed up the events in playback more than a few times the normal speed? If yes, you will need a *time-lapse* VTR (Section 8.5).

2. Do the events to be recorded take place so fast that an exposure of 1/30 or 1/60 sec would blur the image intolerably? If so, you may have to consider a special *high-speed* camera and recorder (Section 8.6).

3. Do you need playback resolution of considerably higher than 400–500 lines PPH in monochrome? Can you use nonstandard-format video, and afford a budget of tens of thousands of dollars

TABLE 8-2. Some 3/4-Inch Standard-Speed VCRs

Make	JVC	Sony	Sony
Model	CR-4900 U Portable	VO-5800 19-inch rack/bench	VO-6800 Portable
Format	U format	U-matic	U-matic
Record heads	Rotary 2 R/W 2 confidence	Rotary 2 R/W	Rotary 2 R/W 2 confidence
Tape speed	9.53 cm/sec	9.53 cm/sec	9.53 cm/sec
Record time	20 min (KCS-20)	60 min (KCA-60)	20 min (KCS-20)
Video	NTSC	NTSC	NTSC/PAL/SECAM
H resolution (S/N ratio)	260 lines (46 dB)	340 (48 dB) B/W 260 (46 dB) color	260 (NTSC) 250 (45 dB) (PS)
Audio	50 Hz–15 kHz (48 dB)	50 Hz–15 kHz (48 dB)	50 Hz–15 kHz (50 dB)
Operating position	Horiz/vert	Horiz (top load)	Horiz/vert
Power	12 V AC/DC 12 W (Rec) 15 W (Mon) 19.3 W max	100–120 V ± 10% AC 50/60 Hz ±10% 55 W	12 V DC 15.5 W (NTSC)
W × H × D (inches)	13.8 × 5.5 × 13.9	17.6 × 9.4 × 20.5	10.3 × 4.6 × 13.4
Weight	19.6 lb incl battery	53 lb	12 lb 20 oz
List price (12/84)	$4250	$4590	$3800
Features	Video confidence SMPTE time-code capacity Backspace frame-servo editing 10× fast search forward/reverse	Video 1 & 2 sync Elapsed time in min/sec **LED** Frame-servo editing search: forward/reverse at 1/30 to 5× standard speed	Video confidence Remaining time indicator Frame-servo editing 10× fast search forward/reverse

TABLE 8-3. Some 1/2-Inch Standard-Speed VCRs[a]

Make	JVC	Panasonic	Panasonic	Sony	Pansonic
Model	BR-6200 U	NV-8420	NV-8950	SLO-323	AG-2400
	Portable	Portable	Dynamic Tracking		Portable
Format	VHS	VHS	VHS	Beta	VHS
	NTSC color B/W	NTSC color	NTSC color	NTSC color B/W	NTSC
Tape speed	1.31 ips SP	1.31 ips	1.31 ips SP	Beta I (rec/play) 40.0 m/sec	SP mode 1-5/16 ips
	0.43 ips EP		0.44 ips SLP	Beta II (play) 20.0 mm/sec	LP mode 21/32 ips
				Beta III (play) 13.3 mm/sec	SLP mode 7/16 ips
Record time		120 min (NV-T120)	360 min (NV-T120) SLP	60 min (L-500 in Beta I)	120 min (NV-T120) SP
H resolution	240 (SP)	240 color	240 color	300 B/W, 240 color	240 color
S/N ratio	46 dB	45 dB	45 dB	45 dB	45 dB
Heads	2 for SP	2 rot. heads	4 rot. heads	2: Beta I	2: conventional R/W
	2 for EP	1 audio erase	2 piezo actuated	2: Beta II & III	2: still, frame adv. slow playback
Audio	100 Hz–10 kHz (± 6 dB)	50 Hz–10 kHz	50 Hz–10 kHz (SP)	50 Hz–10 kHz (50 dB)	50 Hz–10 kHz (SP)
	46 dB (Dolby)	43 dB	48 dB	50 Hz–12.5 kHz (45 dB)	43 dB
Power	DC 12 V 9 W	DC 12 V 7 W (play)	AC 120 V 60 Hz 70 W	120 V 50/60 Hz 70 W	DC 12 V 6 W (record)
W × H × D (inches)	10.7 × 4.1 × 12.4	9.4 × 3.7 × 9.6	17.8 × 6.6 × 16.8	15.5 × 5.8 × 13.9	8.4 × 2.8 × 10.4
Weight	13 lb	8.3 lb	36.3 lb	27 lb 2 oz	4.8 lb plus battery
					5.72 lb incl battery
List price (12/84)	$1185	$995	$1995	$1650	$950
Features	Still frame	Still frame	Noise-free in any mode	Noiseless still	Clean still and slow frame advance
	Frame advance	Frame advance	Still frame, 1/3–1/10		
	3×EP 1/2× play	Fixed noise bar	2–8× search	Frame-by-frame advance	Record 2/6 hr
	7×SP 21×EP scan	1/4–1/30× play	Blank area search	1/10–2× forward	Playback 2/4/6 hr
	Insert edit	Speed search	IR remote control	Search 6–15× forward	Multifunction display
	IR remote control	12-function remote		5–10× reverse	Insert edit, IR remote

[a]For comparison charts of consumer-product Beta and VHS 1/2-inch VCRs, see, e.g., *Video* Nov. 1983, pp. 60–65 (table models), and May 1984, pp. 69–75 (portable VCRs).

for the VTR alone? If the answers to all of these questions are yes, you may wish to consider a *nonstandard* (e.g., 1023 scan line), *ultra-high-resolution* video system, including special high-resolution cameras and monitors in addition to the VTR.

Aside from these special requirements, the following considerations also enter into the selection of VTRs for video microscopy: color capability or monochrome only; resolution; contrast and picture *definition;* still frame and variable speed display; editing capability; portability and RF conversion; compatibility; ease of use and convenience features; reliability of equipment and availability of service; cost of equipment, maintenance, and supplies. We will consider these conditions in turn, coupled with my evaluation of some VTRs that appear especially suited for video microscopy.

Color Capability or Monochrome Only

Most VTRs used in North America and Japan today, except for those designed for special applications, are made to record and play back standard NTSC color and monochrome video signals.* In microscopy, *color* can be useful for recording video images at low power, such as when observing live specimens or performing operations under a dissecting microscope or for examining conoscopic interference patterns in polarized light. For such applications, or for transmitted-light observations of stained or highly colored specimens, modern color cameras can provide adequate sensitivity even at high magnifications and their NTSC output can be directly recorded with a standard VTR. However, for many other forms of microscopy, NTSC color cameras may not provide the sensitivity, contrast, or resolution needed to capture the image of interest.

Color VTRs can also be used to record the false color or pseudocolor output of a digitally processed microscope image, provided the resolution and picture quality given by the NTSC signal are adequate. Note that *an RGB signal cannot be directly recorded on a VTR;* the three-channel signals of RGB must be converted to a single-channel signal such as NTSC (with accompanying loss of resolution and luminance contrast) for a standard VTR to be able to record the signal.

Color video requires considerably more complex electronic circuitry than monochrome, and the bandwidth for color video recording is limited. VTRs made specifically for handling *monochrome* (and to some extent those with color-killer circuits) are not limited by these constraints. Thus, they tend to provide higher resolution, greater contrast, better picture definition, and greater S/N ratio for monochrome recording than those that are made for general, color and monochrome recording.

Some VTRs made for medical, scientific, and engineering applications provide the maximum advantage of recording and playback in monochrome as discussed in the next section.

Resolution

The horizontal resolution of a VTR increases with head speed, bandwidth, and reduction in gap size of the read/write head (Section 8.2). With standard-format VTRs, the resolution generally increases as the width of the tape increases. Tables 8-1 through 8-3 list resolutions (or bandwidths) and other characteristics for representative, *standard-format* VTRs available today.

The horizontal resolutions provided by these VTRs in the ½- and ¾-inch formats are clearly way below the (700–800 lines and greater) resolution provided by a standard monochrome camera or monitor. But despite the limited resolution, the moderately priced ¾-inch and some of the ½-inch VCRs have served video microscopists well for the past several years. One-half-inch VCRs with monochrome resolution of no more than 280–300 TV lines still give a reasonably good first-generation tape, and ¾-inch VCRs with 320- to 340-line resolution and moderately high S/N ratios even give quite respectable second- or third-generation, or edited, tapes.

Nevertheless, the limited resolution of these moderately priced VCRs imposed quite a constraint on the use of video in microscopy. Given the limited resolution of the video system set by the VTR, the optical image had to be magnified higher in the microscope (than would have been

*See Section 6.3 for descriptions of video standards used in other countries.

TABLE 8-4. High-Resolution VTRs[a]

Make	IVC	Panasonic	Sony
Model	IVC-1010[b]	NV-9240 XD	VO-5800 H
Tape format	1-inch reel-to-reel	3/4-inch U-cassette	3/4-inch U-cassette
Tape speed	6.91 ips	3.75 ips	3.75 ips
Record time	1 hr max (8-inch reel)	60 min (KCA 60)	60 min (KCA 60)
Video in	525, 625, 875, 1023 lines/60 Hz	525/60 2:1 or random interlace	525–1125 lines, 60 fields/sec
H resolution	800 lines B/W 10 MHz (4 dB down)	500 lines B/W 280 lines color	430 lines B/W (at 525/60)
S/N ratio	(43 dB) 40 dB (R&S)	48 dB (R&S)	43 dB
Audio	7.5 Hz–10 kHz (40 dB) 250 Hz–7.5 kHz	2 track 50 Hz–15 kHz 48 dB	2 track 50 Hz–15 kHz 48 dB
Power	115 V ± 10% 60 ± 5 Hz 230 V ± 10% 50 ± 5 Hz 450 W max	AC 100/120/220/240 V 50/60 Hz 130 W	AC 100–120 V 50/60 Hz 75 W
W × H × D (inches)	22.5 × 13.5 × 14	26.1 × 9.4 × 18.8	17.6 × 9.4 × 20.5
Weight	80 lb	77 lb	53 lb
List price (12/84)	$18,000	$4500	$6000
Features	Var. slow motion Freeze field Superhigh resol. Min, 0.1-min timer	Head amp freq adj Drop out comp on/off, Sync, **SC IN** SMPTE time code on channel 1	1/30–5× search forward/reverse V-interval switching between two video inputs LED min, sec timer Backspace edit

[a]See also General Electric HBR-50 (Table 8-1)
[b]Discontinued.

necessary without the VTR) to match the video resolution to that of the microscope (see Appendix II and Inoué, 1981b). That meant a loss in the size of the field of view, as well as a drop in the illuminance on the video camera faceplate. To compensate for the latter, the specimen had to be illuminated more intensely (increasing the likelihood of adverse effects on living cells) or the camera sensitivity had to be raised (resulting in a lower S/N ratio).

Fortunately, at least two *high-resolution* ¾-inch VCRs became available in 1984 (Sony VO-5800 H, Fig. 8-13; Panasonic NV-9240 XD, Table 8-4). As shown in Table 8-4, these U-format VCRs provide horizontal resolution of 430 to 500 TV lines, with S/N ratios of 43 dB or better for 525/60, 2 : 1 interlace monochrome video. The resolution achieved by these VCRs is considerably greater than with other ¾-inch VCRs made for standard color and monochrome video recording. Provided the features that are not available on these VCRs do not detract from the intended application, these VCRs would be my first choice for standard-speed recording in video microscopy today.

In contrast to these two U-format high-resolution VCRs, a 10-MHz-bandwidth, 1-inch VTR provides an *ultrahigh resolution* of 800 TV lines horizontally. This reel-to-reel VTR, the IVC-1010 (as well as the General Electric HBR-50, Table 8-1), can accept 60-field, 525-line EIA RS-170 signals or signals with higher numbers of scan lines up to 1023 to provide increased vertical resolution (Tables 8-1, 8-4). Naturally, the camera and monitor must also have the high bandwidths and be capable of working at these nonstandard sweep rates for the video to be recorded and played back at the intended high resolution.

Contrast and Picture Definition

In a VTR, the video signal is processed in a most complex fashion as described in Section 8.2. It is therefore not surprising that the quality of the signal played back from a VTR depends on many factors that are not apparent from the simple technical specifications usually available. Nevertheless, one measure, the *S/N ratio,* is a good indicator of the overall quality of the picture that can be recovered from a VTR. With S/N ratios of less than about 40 dB (rms noise = 1% of peak-to-peak signal amplitude), snow in the picture as well as the lack of picture detail can be quite detracting. The lower-contrast, finer details in the picture are either lost in the snow or can be lost if one uses a filtering circuit to suppress the higher-frequency noise that gives rise to the snow.

The sharpness of the picture, or *picture definition,* is also affected by the low- and high-frequency responses of the VTR and its phase characteristics. Edges may either be blurred or otherwise be bounded by excess lines, and highlight or shadow details may be lost in larger or smaller image features depending on the design characteristics of the VTR.

Still-Frame and Variable-Speed Display

One often needs to analyze, frame-by-frame, events or motion recorded by video. Some VTRs enable *still-frame* or *still-field* display coupled with *variable-speed* forward and reverse playback. These are not only standard features on many 1-inch-format studio VTRs, but are also available on a number of ½-inch VCRs as well.

The Panasonic NV-8950 **dynamic tracking** VCR is a notable example of a ½-inch VHS-format VCR providing *clean,* variable-speed playback as well as still-frame display. The playback, free of noise bars at nonstandard speeds, is achieved in this four-head VCR by swinging a pair of video heads with a piezoelectric actuator during the playback. In this way, each *floating head* swings during drum rotation, adjusting to the speed of drum rotation and tape transport. The two floating heads thus continually follow the tracks precisely. The heads do not cross the guard bands, so that the picture remains free of noise bars at various playback speeds. While the horizontal resolution of this VHS color recorder is limited to 240 lines (with an S/N ratio of 45 dB), the picture can be frozen a single frame at a time.* The forward speed can be adjusted between 0 and 2×, and the reverse slow-scan speed can be adjusted between ⅓ and ⅒, continuously. This industrial-grade VCR, presently listed at about $2000, could prove quite handy as a motion analyzer where its limited resolution is not a drawback. Some ½-inch consumer-product VCRs also provide clean still-field and fast- and slow-scan features (see footnote to Table 8-3).

Some of the ¾-inch VCRs allow variable-speed playback and still-field display, although not totally free from noise bars as in the floating-head ½-inch VCRs and 1-inch studio recorders. An example of well-built VCRs with such features are the Sony VO-5800 series U-matic VCRs, including the high-resolution monochrome VO-5800 H already mentioned (Fig. 8-13).

A portable, *magnetic disk recorder* and monitor designed specifically for *motion analysis* holds 10 sec of NTSC video signal (Sony SVM-1010). The picture can be advanced field-by-field (**field-sequential playback**), forward or backward, or played back at two preset slow speeds. This convenient motion analyzer has unfortunately been discontinued, but the NV-8950 and some other ½-inch VCRs providing clean still field and slow scan may be used to assume some of its functions (see also *optical memory disk recorders,* Section 8.9).

Editing Capability

As mentioned earlier in this section, special VTRs are made for editing videotape. The *editing decks* that are available for the various 1-, ¾-, and ½-inch formats are used together with a compatible tape player and editing controller for professional-quality editing. Editing decks have special capabilities including **insert editing,** locking to external sync during playback, variable slow- and fast-speed viewing, and capabilities for adding SMPTE time codes and cues that are used for fast search and for locating specific frames.

Insert editing capabilities allow the insertion of video in sync with the portions recorded ahead and behind the insert without leaving noisy frames at the insert. In conventional recorders, the tape

*Note, however, that except for some (special) VTRs with animation capabilities, the tape in these VCRs does not strictly advance one frame (or field) at a time. Instead, it may skip a few frames.

is erased by the stationary erase head several frames ahead of recording, so that the front end of a recorded segment is erased. This gives rise to noisy frames if insert editing is attempted. In contrast, editing decks are equipped with a pair of "flying erase heads" on the scanner, so that the tracks are erased just ahead of the recording. While insert editing requires a VTR with flying erase heads, *assemble editing* can be accomplished with several nonediting decks equipped with appropriate *pause* functions as described later.

Portability, RF Conversion

Most of the standard 1- and ¾-inch VTRs are quite heavy and cannot be carried around very conveniently. Portable units, developed primarily for ENG (electronic news gathering, for broadcasting) work, are available in these formats (Tables 8-1 to 8-3). These relatively compact, battery-driven VTRs can also be run with an AC power supply unit.

The cassette capacity of the portable VCR may be limited (e.g., to 20 min instead of 60 min for the U-format VCR). However, a portable unit provides the convenience of recording or playback at a site away from the main facilities, serves as a backup or second VCR, and allows *reliable playback* of one's tape on a known VCR (not all VCRs with the same tape format can be counted on to do so; see next section on compatibility). We have also used our portable U-format VCR (Sony VO-4800) to obtain clean *assemble editing* for presentation of video data at scientific meetings (Section 12.3).

A number of the portable VCRs are equipped with *RF converters* to allow direct playback or recording to and from regular TV sets. Most consumer-product VCRs are equipped with RF converters, but some may not provide standard video signal outputs that allow the video to be displayed directly on a video monitor. While RF conversion provides the convenience of using TV sets for display, some image degradation is inevitable in the RF converted system.

A new ¾-inch U-matic portable VCR (Sony VO-6800, Fig. 8-14), which is significantly smaller and lighter than the VO-4800, plays back impressively sharp pictures of video microscope scenes. Other manufacturers have also introduced a new generation of portable U-format VCRs with improved features and performance.

Compatibility

Among the types of video equipment commonly encountered in video microscopy, the VTR is the one most likely to cause problems of (in)compatibility.

First, a VTR will only play tape made for that VTR's particular format. For example, Beta-format tapes cannot be played back on VCRs made for VHS format and vice versa. In these cases, the cassettes themselves are different. But even among VCRs that use the same cassette, the format or speed of tape transport may be different. For example, the ¾-inch time-lapse VCRs run "60-min" cassettes for 72 min at "normal speed" instead of for 60 min. Clearly, such a tape could not be made to play back on a standard U-format VCR.

Even tape recorded on a VCR of the same format will not always play back well on another VCR, or even on the same VCR used for recording, if extreme change in temperature, humidity, mechanical adjustments, or other conditions arise that change either the tape tension or **tracking.** Subtle differences in tape tension can arise from variations in VCR design, operation, or maintenance, or from environmental factors, and introduce an error known as *skew*. The tracking angle can change with tape stretching or variation in tape transport speed. A number of VTRs are equipped with SKEW and TRACKING controls to compensate for these variations.*

Ease of Use, Convenience Features

Most VCRs are designed for easy and convenient use. Functions such as play, pause, stop, rewind, fast forward, and record are commonly operated by push buttons. In some VCRs, relay-

*Skew results in a picture whose top is bent to the left or right, while tracking errors result in wavy vertical lines or streaks and noise bars in the video picture. See Section 8.5 for further discussions.

operated programs allow the user to go directly from one function to another, e.g., from rewind, fast forward, or record to play or eject, without having to first press the STOP button.

Some VCRs have manually operated levers for loading and ejecting the cassette, while in other VCRs these functions are motorized. The motorized cassette loaders reduce the risk of tape damage and cassette jamming that can take place on the manually loading VCRs.

Reel-to-reel VTRs clearly require more knowledge and practice in their handling than VCRs. The tape must be threaded carefully, exercising special care to protect the delicate read/write heads.

In some VTRs, a number of indicator or warning lights are provided: *dew* or *auto off* (automatic shut down when condensation of moisture is sensed on the drum, or when the tape becomes slack or broken), *servo* (lack of proper sync pulse in the incoming video in the record mode), *end of tape* (approaching end of tape length available for recording), etc. **Standby,*** *record, audio dub,* and other signals are also provided by indicator lights.

Some VTRs provide *auto-rewind,* a function that makes the tape wind back automatically when the (clear) end of the tape is reached. The same VTRs may incorporate a RESET button on the tape counter that makes the tape rewind only to the zero mark indicated on the counter, which can be reset. By using this and other options, some VCRs can be made to play back a selected portion of the tape repeatedly.

Cues, or SMPTE *time codes,* can be registered on the tape in some VTRs, allowing one to search and locate those particular spots automatically. As already mentioned, some VCRs allow fast and slow scan, features that are particularly helpful for locating specific scenes or events.

Many VTRs provide audio dub capability. When this button is depressed together with the PLAY button, one can dub (add) sound onto one of the audio channels (usually channel 1), without erasing the previously recorded video and audio signal.

Reliability, Availability of Service

On the whole, modern VTRs work remarkably reliably. Nevertheless, failures or malfunctions can occur, sometimes introduced by operator error and other times by failure of a component. Considering the mechanical and electronic complexity of VCRs, I have been impressed by the limited number of problems I have experienced with VCRs. However, the frequency of trouble will naturally be affected by the use and care of a VTR. These topics are covered in a later section.

In addition to selecting VTRs with a reputation for reliability, it is prudent to investigate the availability and location of a manufacturer's authorized service facility before choosing a particular VTR.

Cost of Equipment, Maintenance, and Supplies

In selecting a VTR, one should consider the cost of maintenance and tape in addition to the initial cost of the VTR itself. Representative prices for standard-speed VTRs are shown in Tables 8-1 to 8-4.

As mentioned earlier, trained personnel may be required to routinely maintain a large-format studio recorder, while the ¾- and ½-inch VCRs require relatively infrequent routine maintenance.

One-inch tape costs up to $200 for an 8-inch reel (1-hr recording time), while a 60-min ¾-inch cassette costs from $20 to $30. These usually must be ordered from professional video supply outlets. Half-inch cassettes are commonly available in video and record stores for well under $10 for a cassette that records several hours of video.

Thus, the initial equipment, as well as the cost of maintenance and supplies, rises exponentially as the tape format increases. As implied earlier, ¾-inch and a few ½-inch VCRs probably offer the best compromise in terms of cost and performance for most applications in video microscopy today.

*Standby is actually a *warning* signal, cautioning the operator not to attempt additional operations when this indicator is lit.

8.5. TIME-LAPSE VTRs, TIME-BASE ERROR

Time-Lapse VTRs

Some standard-speed VTRs allow sped-up and slow-scan playback within limited ranges (Section 8.4). However, when we must speed up very slow events, we need to use a *time-lapse recorder*.

Time-lapse VTRs are designed to record at normal speed (''real time'') and also, e.g., at 1/10th, 1/20th, . . . , 1/80th, etc. times the normal speed. During time-lapse recording, the tape is transported at 1/10th, 1/20th, etc. of the normal speed, while the speed of rotation of the heads is maintained at its normal rate. A track containing one video field is recorded every 10th, 20th, etc. fields. Thus, the VTR achieves time-lapse, or **skip-field,** recording by writing a single track every N fields on a tape moving at $1/N$ times the normal speed. When the tape is played back at standard speed, the scene speeds up N times. The elapsed time interval (t) between each recorded field is given by: $t = N/60$ (for 60 field/sec format video).

From the standpoint of the user, it is very simple to achieve time-lapse recording and sped-up playback that compresses time. One simply sends a standard video signal from a video camera (or its output through a monitor) to the input terminal of a time-lapse VTR operating in the record–play mode.* When any of the time-lapse settings are selected, the recorder switches directly to the desired time-lapse recording rate. Since the recorder is merely sampling every $1/N$th field out of the standard video signal coming from the camera anyhow, no change is required in the camera setting or the microscope illumination when one goes from one recording speed to another. Also, a time–date generator, hooked up in series with the camera and the recorder input, automatically inserts the exact time that each video field was recorded (Figs. 1-6, 3-11). To review the tape speeded up N times, or with the time compressed N times, one simply plays back the tape on the same VTR at normal speed. In addition, playback is usually field sequential in time-lapse recorders, so that the time resolution is twice that given by standard recorders that sequentially play back whole frames made up of two interlaced fields.

In contrast to the apparent simplicity of its use, the mechanism inside a time-lapse VTR is particularly complex. First of all, special designs are needed to lay down tracks that can be exactly followed during playback. A problem arises because, without the special design, the track angle would change between recording and playback as explained below.

During both recording and playback, the head drum must rotate at a constant standard speed in order to accept and play back video signals conforming to the standard 525/60 format. On the other hand, the tape transport is slowed down to $1/N$th its normal speed during recording, while it runs at normal speed during playback. Therefore, the track angle would change between recording and playback (Fig. 8-5). Moreover, the magnitude of change in the track angle would vary with the rate of time lapse.

Without a special arrangement, the read/write head would run off the recorded track, cross a guard band, and enter the next track during each rotation in playback.† Such *mistracking* would result in the picture being traversed by a broad noise bar (Fig. 8-7). In addition, the head gap would not lie at quite the proper angle relative to the magnetic recording on the track, so that the resolution and signal strength would drop and noise could rise in the picture area.

Manufacturers have used different designs to overcome the mistracking and other difficult problems associated with time-lapse video recording. For example, the head drum is canted to different angles depending on the extent of time lapse in the Sony TVO-9000 VCR; the path of the tape transport is elevated or dropped in the GYYR TLC-2051.

Another problem that is exaggerated in time-lapse recording is *time-base error*. With the tape being transported at very different speeds during recording and playback, and the control pulse and video tracks being separated by many centimeters, it becomes especially difficult to hold the tape

*The VTR does not start recording unless both the RECORD and PLAY buttons are pressed (Section 12.3).

†The center-to-center distance of the tracks measured along the length of the tape recorded at high rates of time lapse (very slow tape transport) is (172.15 mm/sec) × (60 fields/sec)$^{-1}$ = 2.87 mm/field. The longitudinal displacement between these tracks and the tracks at normal tape speed is 1.59 mm/field (Fig. 8-5).

dimension and speed so that during playback proper timing relationships are maintained between the V-sync pulse and the H-sync pulses and picture signals. Adjusting the TRACKING control in playback can help to a limited extent, but the fluctuating time-base errors encountered in time-lapse recorded tape are very difficult to correct. We will further discuss time-base errors and corrections a little later.

Commercially Available Time-Lapse VTRs

While these difficult problems have not been totally overcome, time-lapse VTRs with varying degrees of sophistication have appeared on the market in the past several years. Some aim primarily at the security market, recording at intervals of many seconds to minutes, except when an intrusion alarm automatically triggers real-time recording for a specified duration. Others aim at a broader range of applications and provide images that are substantially better than those required for intrusion surveillance alone.

Today, only a few time-lapse VTRs appear to be available, most using ½-inch VHS cassette tapes.* Among these are the GYYR TLC-2051 and the Panasonic NV-8050 and AG-6010 (Table 8-5). Each of these VCRs has its advantages and limitations. The GYYR, which was designed primarily for surveillance applications, can provide a still field free from noise bars automatically, but the resolution and quality of the picture played back leave much to be desired. The Panasonic NV-8050 provides a better-quality picture, but the noise bar cannot be completely eliminated; it can only be chased high or low in the picture so that it intrudes less into the area of interest. The Panasonic AG-6010 is said to provide clear still and slow playbacks but provides only a total of four speeds.

Two ¾-inch time-lapse VCRs were marketed for a few years but have since been discontinued (Table 8-5). This is most regrettable, since the Sony TVO-9000 in particular gave good-quality monochrome time-lapse recording, as well as clean still fields (Figs. 1-5, 1-6, and 9-11 are examples of still-field pictures photographed off playback from this VCR). In addition, after normal-speed or time-lapse recording on the TVO-9000, the picture played back in the 96-hr mode could be used to observe or analyze successive fields, interrupted only briefly by guard band noise.

Two alternatives to time-lapse VCRs are available today, albeit at considerably greater cost. Several 1-inch C-format VTRs have animation recording capabilities, and the Sony BVH-2500 incorporates a time-lapse and animation feature (Table 8-1). The other alternative is to take advantage of the capabilities of the new OMDRs (see Section 8.9, Table 8-7). Unlike with C-format VTRs, playback speed with an OMDR may not quite reach video rate (depending on the model), but scenes recorded in the time-lapse or animation mode can be played back at a wide range of speeds. The quality of the pictures is excellent and one has essentially instant access to any of the frames recorded on the disk; they can be examined for any length of time without concern for tape wear. The cost of the recording disks and quality of image compare very favorably with 16-mm motion picture film (see Section 8.9). Since the prices of OMDRs have dropped significantly in the short time since their introduction, OMDRs eventually may replace time-lapse VCRs for many applications in video microscopy.

The absence of a high-resolution time-lapse VCR appropriate for video microscopy is most unfortunate. Time-lapse VCRs are crucial for analyzing the physiology and pharmacology of cell division, cell motility and phagocytosis, embryonic growth, tissue cell differentiation, and many other basic biological events, as well as for monitoring the growth, erosion, and other slow changes in crystals, emulsions, and other structures of interest to industrial and material scientists. We desperately await the production of a *reliable, high-resolution time-lapse VCR with clean still-field frame, forward and reverse single-field frame advance, and clean slow- and fast-scan capabilities.*

*Some companies provide custom modifications to VTRs, e.g., for animation or very-high-resolution recording. They include: EOS Electronics AV Ltd., who modify the Sony 5850, ¾-inch U-matic editing deck for animation (which is a type of time lapse); and Merlin Engineering Works, who refurbish or custom-modify 1-inch VTRs for high-resolution frame-by-frame recording. The EOS-modified Sony 5850 is used for time-lapse recording and frame-by-frame analysis of motion by a number of cell biologists today.

TABLE 8-5. Time-Lapse VCRs[a]

Make	GYYR	Panasonic	Sony
Model	TLC-2051	NV-8050	TVO-9000[b]
Cassette	1/2-inch VHS	1/2-inch VHS	3/4-inch U
Video in	525/60 B/W NTSC color opt.	NTSC color B/W RF	NTSC color B/W RF
H resolution	300 B/W (240 color)	300 B/W 240 color	320 B/W 240 color
S/N ratio	40 dB	45 dB	45 dB
Record time	2 hr (real time) 12, 18, 24, 48, 72, 120, 240, 50–999 hr programmable	2 hr (real time) 12, 24, 48, 72, 120, 144, 240 hr One shot	1.2 hr (real time) 12, 24, 48, 72, 96 hr
Playback	3× scan forward/reverse	Normal speed, 2×	Same, pause
Audio	100 Hz–10 kHz (2 hr) 1 channel	50 Hz–10 kHz (2 hr) 40 dB, 1 channel	60 Hz–12.5 kHz (1.2 hr) 43 dB, 2 channels
Power	117 ± 10 V AC 60 Hz 40 W (run) 25 W (stop)	100/120/220/240 V AC 50/60 Hz 70 W	120 V AC ± 10% 60 Hz ± 0.5% 160 W
W × H × D (inches)	17.1 × 5 × 13.4	16.9 × 7.9 × 17.4	25.4 × 8.8 × 18.3
Weight	18 lb	40.7 lb	80 lb 7 oz
List price (12/84)	$3595	$2795	($7500)
Features	Time–date generator Step single field 3× search Alarm recall 1–99 fields/pulse	1/2H compensation Slow advance Fast search Alarm view One-shot operation	Pause Still adjust Memory rewind Video alert

[a]See also Sony BVH-2500 (Table 8-1), and footnote p. 289.
[b]Discontinued.

Some Practical Considerations

As pointed out earlier, one cannot take a ¾- or ½-inch cassette recorded on a time-lapse VCR and successfully play it back on a standard-format VCR. Even though the cassettes are identical, the tape speed and recording format generally are not.*

While the differently formatted tapes themselves are not interchangeable, the signal played back through the original VCR is usually compatible with the input of another *analog* video device. Thus, one can often simply play back the tape on the original VCR and transfer its output onto a second VCR that has the desired new format or speed.†

In some cases, even this approach will not work, especially if the primary recording has come from a random interlace camera, or if the second VTR (is an **editor** that) distinguishes and requires the presence of proper sync pulses for the odd and even fields. For such cases, one would have to copy a time-lapsed scene "optically" through a video camera. The camera records off a monitor that displays the output of the time-lapse VTR to which the camera is made to synchronize (see Section 12.2).

*Scenes recorded on a ¾-inch time-lapse VCR can, however, be played back on some U-format VCRs (such as the Sony VO-5800) in the *fast search* mode. The picture displayed is not perfect, but this trick saves a great deal of time and anguish when one is trying to quickly review time-lapsed records or is trying to locate a particular scene.

†During the copying process, we can even time lapse an already time-lapsed tape (or its copy) and achieve extreme compression of time. Or we can make a time-lapsed copy of a tape originally recorded at normal speed (see Section 12.3 for hookup and copying procedures). Note, however, that the output from time-lapse VTRs is not always compatible with other VTRs.

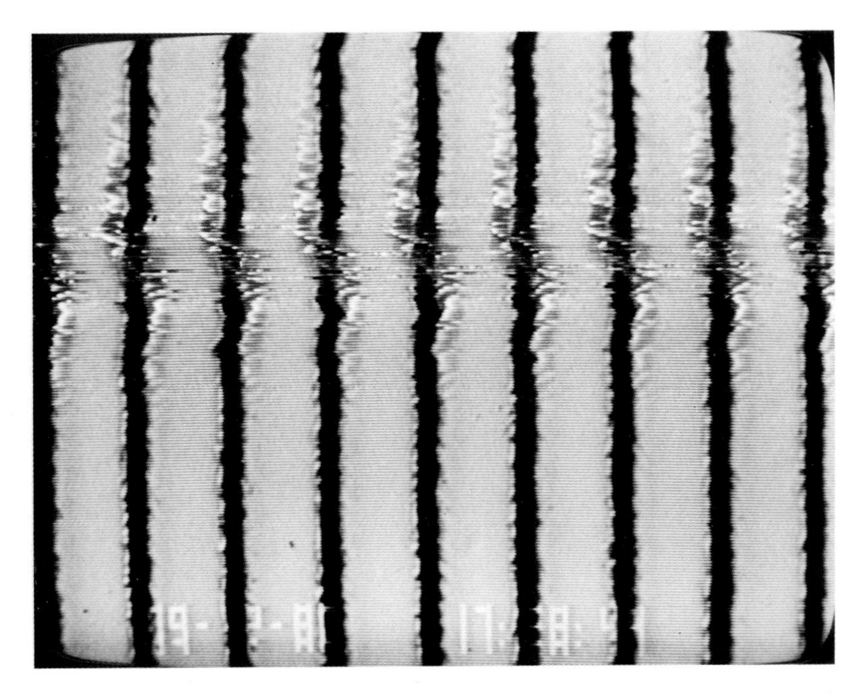

FIGURE 8-15. Improperly adjusted TRACKING control. Slight misadjustment of the TRACKING control during playback can produce effects such as the horizontal shimmer shown in this monitor picture. If the TRACKING control is farther out of adjustment, the entire picture breaks up.

Time-Base Error

While the output of the signal played back from a time-lapse VTR more or less conforms to the EIA RS-170 or -330 format, some timing pulses may be absent and the time-base errors can be quite large. A piece of video equipment receiving such a signal may or may not be able to *sync* with the incoming signal, depending on the severity of the timing problems and the type of equipment involved.

Thus, when a time-lapsed scene is played back and viewed on a monitor or video projector, the picture may break up for several reasons. With the V and H HOLD controls of the display device carefully adjusted, the picture may still contain a noise bar or the vertical lines in the picture may shimmer horizontally (Fig. 8-15). Some of these problems can be cured by adjusting the TRACKING control on the VCR.

Despite all these adjustments, the top of the picture may *flag,* or tear and flutter, when a time-lapsed scene is played back (Fig. 8-16). Flagging appears when the H-sweep oscillator in the display device is unable to latch onto (synchronize to) the irregular H-sync pulses for several scan-line intervals after vertical retrace.

Monitors and projectors with slow *automatic horizontal frequency control* (AHFC) are immune to power-line noise, but have a more difficult time latching onto a signal with large time-base errors. Thus, in a monitor with a slow AHFC, the flagging intrudes way down toward the middle of the picture. Display devices with fast AHFC are less immune to power-line noise, but can more quickly synchronize to the errant H-sync pulses.

Therefore, one should choose monitors with fast AHFCs for displaying time-lapsed records. Ironically, some of the least expensive monitors have the fastest AHFC and display the least flagging. Some monitors and projectors are equipped with switches for selecting between slow and fast AHFC.

To minimize the time-base error during time-lapse recording, one should use the signal from a 2 : 1 interlace camera rather than a random interlace camera. Also, one should take precautions to

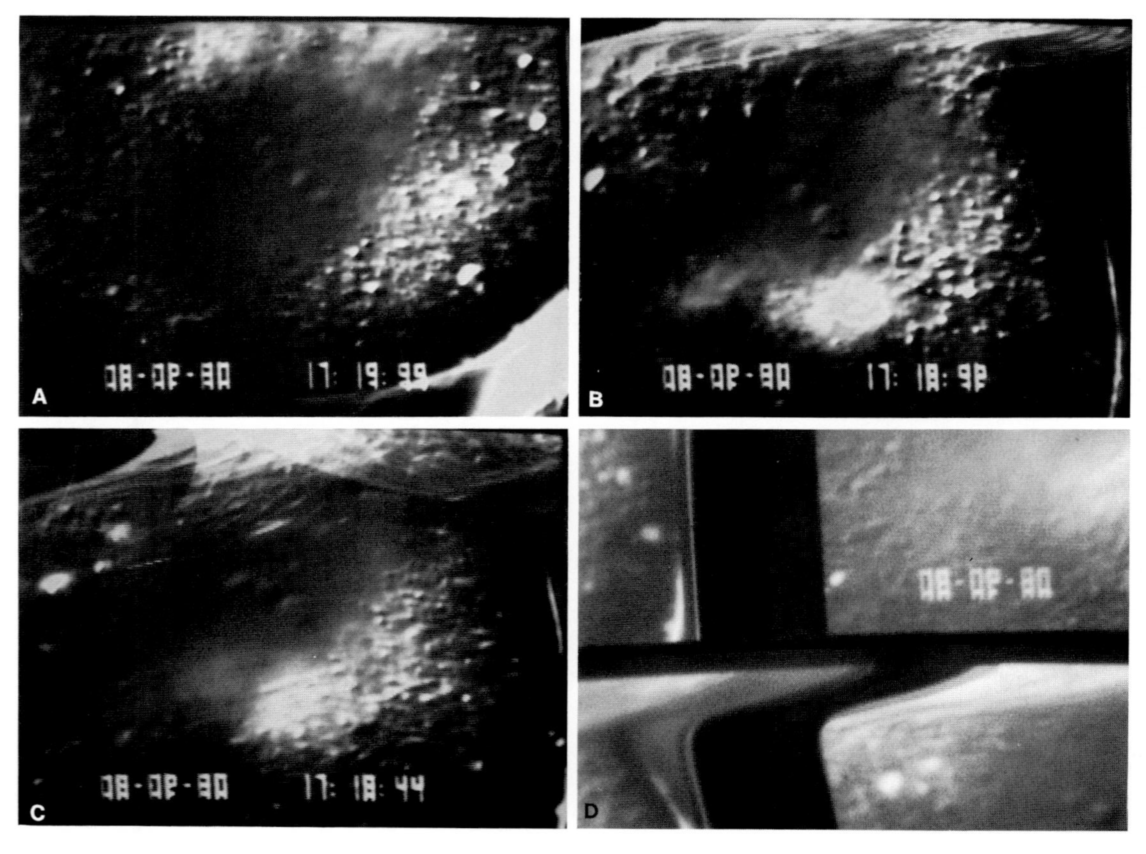

FIGURE 8-16. Flagging. The monitor picture in A shows a scene recorded at standard speed. B and C are time-lapse scenes exhibiting flagging to different degrees. D is a pulse-cross presentation of B. The pictures are 1/30-sec frozen frames of images produced by DIC microscopy of a meiotic spindle in an oocyte of a marine worm (*Chaetopterus pergamentaceous*).

minimize electrical and mechanical disturbances during time-lapse recording. Any disturbance that alters the head or tape speeds, or tape tension, during the recording can permanently affect the time base of the recorded material. Thus, it is important to avoid AC power-line circuits that fluctuate frequently, as well as to protect the VCR from jarring or bumping while it is recording in time lapse.

Time-base error caused by tape stretching rather than erratic VCR behavior can be minimized by using broadcast-quality tape (designated BR) and also by "prestretching" or adapting the tape to the VCR environment. This is accomplished by running the tape through the VCR in the fast forward mode, then rewinding the tape back to the source spool *before recording onto the tape*.

In comparison to analog video equipment, *digital* equipment requires input signals with better-corrected time bases. Modern *time-base correctors* (TBCs) are digital memory devices used to correct the time base of signals played back from standard-speed VTRs. These TBCs are used with virtually every VTR in broadcasting. They are needed to correct the luminance time-base errors, and bring the color burst and the chrominance signals back into phase, when a tape is played back, even from a broadcast-quality VTR.*

TBCs appear unable to synchronize to the output of time-lapse VCRs. Apparently, the timing error is too large or erratic to be corrected by the TBC. The picture actually jumps worse when a TBC is used at the output of a VCR playing back time-lapsed records (but see footnote * on p. 189).

Digital image processors (Chapter 10) also require input with good time-base correction. While the output from a standard-speed VCR can be accepted by many digital processors, and even more stably with the help of a TBC, the output of a time-lapse VCR usually generates nothing but confusion.

With time-lapse recording, we compress and alter the time frame for observing or measuring an event. We therefore need a frame of reference for the time at which each scene was taken. Some time-lapse recorders are equipped with an internal *time–date generator*. In others, time–date generators can be installed as options. In any event, a separate time–date generator, *elapsed-time meter*, or *field/frame counter* can simply be looped into the video circuit between the camera and the VTR (Fig. 3-11). The size and contrast of the displayed digits can be selected, and they can be placed and recorded onto the desired region of the scene. Once the signals containing these alphanumeric characters are introduced into the video signal, the time of the event and other identifying marks are conveniently displayed, superimposed on the original image, whether the tape is copied or played back at standard speed or field-by-field.

8.6. HIGH-SPEED RECORDING

Several problems can arise when we try to make standard-video-rate recordings of scenes involving rapidly moving objects. The motion may simply be so fast that nothing but a blur is registered. With motions that are not quite as fast, the object can leave a comet tail, such as those seen in the freeze-field image of swimming sea urchin sperm in Fig. 7-20. The comet tail arises from the decay lag in the camera tube (Section 7.2). The motion may be even slower, so that the image is not very blurred at the standard video rate of 60 fields/sec. Yet the object could have moved enough to displace its image between the odd and even fields. In such a case, a still-frame picture jitters in the region where the image had moved between the two fields.

Two approaches are used in video to capture sharp, frozen images of moving objects onto tape. In the first approach, the camera is equipped with a *rapid (mechanical) shutter* that is synchronized to the V-sync pulses. The shutter opens for a brief instant every 1/60 sec during V blanking and exposes the video pickup device to the image for a *brief fraction of the field time*. Even though the target of the camera tube or the solid-state pickup device is only exposed to a brief flash of the image, the pattern of resulting electrical charge is stored there until it is swept at the standard scan rate (Section 7.1). Thus, the pickup device is scanned at the standard video rate to produce a 60 field/sec signal.

*For a more detailed explanation of TBCs and a list of manufacturers, see, e.g., Acker (1983); see also Noronha (1981).

TABLE 8-6. High-speed Recording Systems[a]

Make	NAC	Spin Physics
Model	HVRB-200SS Field-sequential high-speed video system	SP-2000 Motion analysis system
Cassette	1/2-inch VHS T-120	1/2-inch high density 1,000 ft
Camera	320 H × 244 V 2/3-inch MOS	192 H × 240 V MOS
"Frame rates"	200, 60 fields/sec	2,000, 1,000, 500, 200, 60 fps at full frame
Shutter	At 200 fields/sec: 1/200, 1/1,000, 1/2,500, 1/5,000, 1/10,000 sec	Fully open
Frame Format	Full frame	1/2, 1/3, 1/6 partial frame At 2000 fps, provides 4,000, 6,000, 12,000 pictures/sec at 1/6 frame
Record time	V-32 VTR: 36 min (200 fields/sec) on VHS T-120 cassette	34-track recorder: 25 min (60 fps) 45 sec (2000 fps)
Playback	V-31 VTR: 200 fields/sec H-resolution: 200 TV lines 1-15 fields/sec Slow 10 fields/sec Reverse Fast forward, still field, single field advance	34 track recorder: 60 fps, slow motion freeze frame (from buffer memory)
Monitor	Special 100 fields/sec	NTSC standard 525-line monitor 220 line B/W
System price (circa December 1984)	\$32,000–\$35,000 (monochrome) \$40,000 (color)	\$110,000–\$150,000

[a]See also Xybion SVC-09, and TriTronics, PCSM 6500.

In some cases, a strobe flash source, syncrhonized to the shutter, is used to provide an exposure much shorter than achievable with the mechanical shutter alone. In other cases, the image is optically dissected within the field time into horizontal segments, so that the camera tube picks up a sequence of scenes that are montaged vertically. With the camera output remaining at the standard video rate, such a camera can capture events that are separated by time intervals a fraction of the video field rate.*

In these cases, the video recorders themselves can be standard VTRs recording at normal speed. The scenes are slowed down and analyzed in the field-by-field or slow-play mode.

In the second approach, the recording itself is speeded up. A special *high-speed VTR* is connected to the (shuttered) camera. The synchronized camera and VTR record the scene at many times the normal speed. The VTR played back at standard, or slower, video rates provides a slow replay of the rapid event. Examples of this group of high-speed recording systems are the Spin Physics SP-2000 from Kodak and the NAC HSV-200 system handled by Instrumentation Marketing Corp.

The NAC HVRB-200SS system uses 2-hr, ½-inch VHS cassette tape, transported in a special recorder at 3.3 times the normal speed. The camera operates at 200 or 60 fields/sec (Table 8-6). Depending on the shutter speed, the exposure can be as low as 1/10,000 sec. The scenes are played

*Several companies manufacture high-speed shuttered video cameras, e.g., NAC, Sony, TriTronics, Xybion. The Sony RSC-1010 shuttered vidicon camera, coupled to the video motion analyzer SVM-1010 referred to earlier, is a moderately priced, conveniently portable unit. Unfortunately, this item too has been discontinued.

Most high-speed camera manufacturers use solid-state pickup devices in their cameras today. The solid-state pickup devices are very much more compact than the vidicon tubes, but more importantly, they show considerably less lag, burn, and geometrical distortion.

back on separate VTRs at either 200 or 60 sec. Both players allow variable slow or still-field display. The 200 fields/sec playback VTR provides field-sequential playback of standard NTSC format VHS recordings so that it can be used for motion analysis on VHS tape taken with standard 30 frames/sec, 2 : 1 interlaced cameras.

The Spin Physics SP-2000 unit incorporates a sophisticated 600 MHz magnetic tape recorder that records 32 parallel tracks that run along the length of the tape. Two stacks of 17 read/write heads record the alternate tracks to produce *32 linear video tracks* and 2 control tracks. The ½-inch magnetic tape runs at speeds of up to 250 inches/sec. In playback, the parallel FM signals from the tape are demodulated and stored in a digitized frame buffer, one frame at a time. The signal is retrieved from the frame buffer at the desired rate in NTSC, standard video format. The operation of the Spin Physics unit, which speeds up recorded events by 33 to 200 times, is controlled by a built-in programmable microcomputer.

When any of these high-speed cameras are used, the illuminance on the pickup device drops as the exposure time is reduced. Thus, unlike regular skip-field or skip-frame recording, image brightness must be increased proportionately with the shutter speed, when high-speed recording is applied to video microscopy.

One high-speed, color video system has recently attracted worldwide attention. This is the *Super Slo-Mo* system developed by Sony for ABC and used at the 1984 Summer Olympic Games. The Super Slo-Mo system uses a special high-definition, low-lag camera that shoots the scene at 90 frames/sec. The 525-line/180-field, 2 : 1 interlace R, G, and B signals each have a 20-MHz bandwidth. These are sent from the remote camera to the camera-control unit by *fiber-optic links*. At the camera-control unit, the R, G, and B signals are digitized by high-speed **A/D converter** chips. These signals are then coupled and digitally converted into three successive sequences of standard-rate (525/60) NTSC signals. The signals are recorded on *three sequential tracks,* in the standard C-format, on a modified 1-inch VTR. The VTR played back at standard C-format speed, then gives rise to full-sized pictures that are slowed down three times. Despite the high-speed recording and slow playback, the system provides pictures with the full resolution of a normal-speed 525/60 camera and recorder.*

While systems such as the Super Slo-Mo and the High Definition TV may not be generally accessible for some time, some of the components developed for such systems (e.g., the high-speed A/D converter chip) could become available sooner and open up new opportunities for achieving greater speed and resolution in video microscopy, and for image improvement through digital image processing at a faster rate.

8.7. CARE AND MAINTENANCE OF THE VCR

As we saw in Section 8.2, a VCR is a complex piece of equipment that incorporates sophisticated mechanical, electronic, and electromechanical designs. Signals are recorded and played back from compact magnetic tape at very high precision and density. Electronic feedback loops and servo controls continually correct the system's performance at various levels. Special circuits process the signal before and after recording to compensate for the limited capabilities of the tape and read/write heads. In fact, successful design of the VCR, including the videotape, required the convergence of such a broad range of technology that the VTR was chosen as one of three sample industrial products studied by a panel of the National Science Foundation to trace the diverse scientific research and technological innovations that were needed for their successful developments.†

General Precautions

While the interior of a VCR is indeed complex and requires delicate adjustments by trained personnel, its external appearance and use are remarkably simple and *almost* foolproof. Nevertheless, it is possible to misuse and damage the equipment.

To prevent such a mishap, be sure to *first read the owner's manual* before using a VCR. *Heed*

*See Spotlight: The Super Slo-Mo System (1984).

†"Technology in Retrospect and Critical Events in Science," National Science Foundation.

the warnings in the manual. If you follow the joke about reading a manual only if all else fails, costly repair bills and unwarranted downtime and grief may result.

To reiterate, a VCR is deceptively simple on the outside, but extremely complicated and delicate inside. Above all, do not abuse. Keep liquids, dirt, dust, and smoke away from VCRs. Keep the tape away from magnets. Make sure that *all* users read the manual before operating, especially a time-lapse VTR, for the first time. More detailed practical pointers in the use and care of the VCR now follow.

When the small *standby light* is on, do not attempt another operation until the light goes off. Always wait until the standby light goes off and the operations (tape threading/unthreading, or cassette loading/unloading, etc.) are completed. Do *not* turn off the VCR while the standby light is on; the VCR may hang up in the middle of an operation and become jammed.

Do not attempt to use the VCR if the *dew* or *tape slack* indicator lights are on. The problem must first be corrected before the VCR can be made to work. Avoid extremes of temperature and sudden temperature change. If you need to bring the VCR from a cold area to a warm room, keep it sealed in a plastic bag until the temperature of the equipment rises. If dew does appear, you will have to wait until it disappears or place the VCR in a drier area.

Do not stack equipment and block the ventilating ports or grills. Adequate ventilation is needed to prevent heat buildup.

Keep liquids, dirt, dust, and other potential sources of contamination away from the VCR. No smoking should be permitted in a room containing a VCR (or with expensive microscope optics).

Do not abuse, and especially do not move, bump, or jar a VCR, particularly during time-lapse recording.

Select and use an electrical outlet with minimum line-voltage fluctuation. Put the VCR on a separate line from equipment, such as air conditioners, that draw heavy currents and introduce "brush noise." Turn off the VCR when RF spikes are expected, e.g., from igniting an arc lamp nearby.

Playback

Before playing back any tape previously recorded with important information, always remove the red WRITE-PROTECT button on a ¾-inch cassette (Fig. 8-17), or break the tab at the back end on a ½-inch cassette to activate the **write-protect** function of the VCR.* If it becomes necessary to record again, the button can easily be reinserted, or the tab function can be restored by inserting a small block of wood or eraser to replace the broken tab. But once the data are erased, they are gone forever.

If the picture played back is covered with noise, stop the VCR immediately and check out the source of the noise.† A noise bar may indicate improper adjustments of the SKEW or TRACKING control. Torn, flipping, or rolling pictures may indicate the use of incompatibly formatted tape,‡ improper adjustment of the SKEW control, or of the monitor V and H HOLD. A picture covered with short horizontal hash or snow could indicate a clogged head, a dirty tape, a tape with the wrong format, or electronic trouble.

Continued operation with a *clogged head* is risky. It can irrevocably ruin the tape and possibly even harm the read/write head or even the drum cylinder. A *dirty tape* can likewise abrade and ruin the tape, head, and cylinder.

Scratches on videotape can be seen by examining the oxide surface in reflected light. Open the tape cassette cover by depressing the COVER RELEASE (Fig. 8-17). Abrasion on the tape appears as light-scattering streaks running diagonally at the track angle. A tape gouged by a clogged or damaged head shows a prominent diagonal line(s) where the oxide has been scraped off (Fig. 8-17).

One can often distinguish the source of trouble by trying out another tape of known quality

*A VCR loaded with a cassette that is write-protected cannot be made to enter the record mode.

†This comment obviously does not apply to the blank part of the tape. In those parts, there would be no picture, and the monitor would display nothing but noise on a black background.

‡Recall that the recording formats for time-lapse VCRs are different from standard-speed VCRs even though they use the same cassettes.

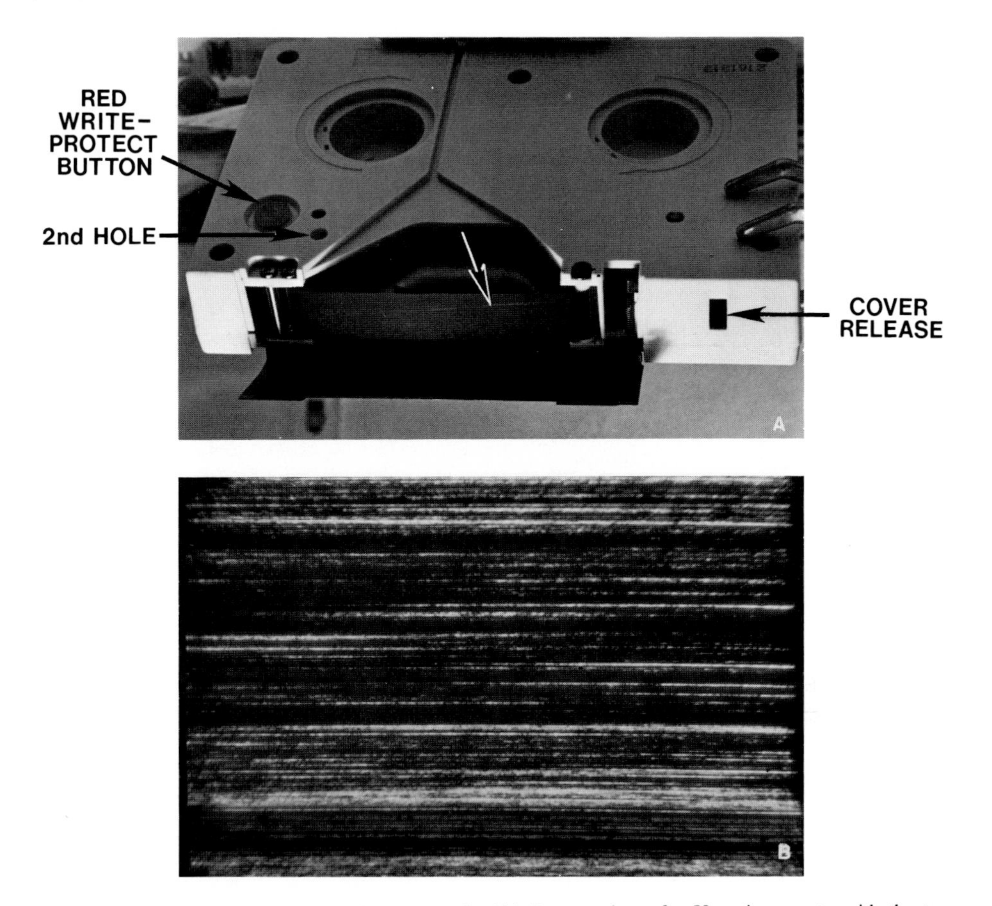

FIGURE 8-17. U-matic cassette and tape scratch. (A) Bottom view of a U-matic cassette with the tape exposed, revealing: (1) the red WRITE-PROTECT button, which, when removed, prevents inadvertent erasure of recordings already on the tape; (2) the "second hole" near the red button indicates that the cassette contains the KCA- or KCS-type tape necessary for time-lapse recording; (3) the COVER RELEASE, which, when depressed, opens the cover and exposes the recording surface of the tape; (4) the diagonal scratch (arrow) that was produced on the tape when it passed through a clogged head. The scratch accurately indicates the track angle on the tape. (B) Monitor picture of playback of the scratched portion of the tape shown in A.

(with recordings that are not too important), although one needs to keep in mind that a dirty tape can also clog the head. If the noise persists after cleaning the head and using good, compatibly formatted tape, the trouble is most likely in the electronics. Check out the coaxial cable linking the VCR and the monitor; test the conductivities of the axial wire and the braided ground shield between the two connectors. (In fact, this is one of the first things to check; try substituting with another cable.) If the cable and terminals are not the source of the noise, and the monitor is performing without trouble, the VCR needs to be repaired. Unless the problem is a clogged head, which can be cleaned as described later, the VCR should immediately be sent in for service before more damage is done.

Pressing the PAUSE button in the playback mode produces a still-field picture. The paused picture in the still-field mode is noisier than when the tape is played back without pausing. When the tape is running, the random noise, which varies from frame to frame, becomes averaged out in our perception or is integrated as it is photographed. The paused picture captures the event *frozen* in a 1/60 sec time interval.

While the pause function is useful and important for video microscopy, the tape can be damaged if held in the pause mode for too long. To prevent such damage, most recorders automatically skip out of pause after 8–10 min. In the case of the GYYR TLC-2051, the transport motor in the pause mode steps the tape onto the next field about once every minute for the same reason.

Recording Operation

Before inserting a tape cassette for recording, make sure you have *labeled* and identified the cassette and its container. At the very minimum, write the date on the labels. If more than one tape is used on the same day, also enter the sequence number. Always enter VCR recording on a *master log*. While details should be entered in a laboratory notebook, the date, starting and ending counter numbers, name or initials of persons involved, and a few words identifying the subject matter should be entered into the master log for quick reference. A convenient arrangement is to include this information together with the microscope lamp log.

Before recording with a VCR, make sure all of the video and audio connections have been made correctly. Make sure the microphone is working and connected to the proper MIKE jack, not to the AUDIO IN jack unless the microphone is followed by an audio amplifier. If copying from VCR to VCR, make sure that the already amplified signal from the AUDIO OUT terminal is connected to AUDIO IN and not to MIKE. Double-check the termination and the channel select switches.

Make a habit of recording observations and comments with audio together with the video. While one should also keep some notes and sketches in a notebook, *voice records* on the videotape can yield valuable information easily lost in the rush of an exciting observation. Use *audio channel 2* for concurrent recording with video; that leaves the audio-dubbing channel 1 available for adding audio cues or comments later.

Always use a *time–date generator* to record the actual time and date of observation onto the videotape. That allows coordination with other notes taken. Also, the time–date record, coupled with the 30-Hz-frame and 60-Hz-field rates, provide important time frames for analyzing the video data.

Before recording a scene, make sure that you are not recording over, and erasing, previously recorded material that should have been kept. The time spent in brief playback of the next segment of the tape is always good insurance.

In the recording operation, do not forget that pressing the RECORD button, or RECORD after the PAUSE button, does not start the record mode. The live video and audio become connected to the output terminals as though we were recording, but no recording actually takes place until the RECORD and PLAY buttons are pressed together. In the pause mode, the PAUSE button must be pressed once more to leave the pause mode. In fact, during recording, it is all too easy to forget that one had paused and interrupted the recording. I have missed precious footage of recording on several occasions by not paying close attention to the *pause indicator light.*

The pause operation in several VCRs otherwise serves a useful function in recording. It lets us record sequences of scenes with smooth transitions free from glitches at the junction of the scenes. By ending the recording of a scene by pressing the PAUSE rather than the STOP button, the RECORD and PLAY buttons remain depressed so that the VCR remains ready to continue recording. When the PAUSE button is pressed again, the VCR locks onto the sync pulses of the incoming video signal so that the previous recording, as well as the heads and tape drive, quickly become synchronized with the new signal. Thus, the new sequence is recorded, abutted to the sequence recorded before the pause, without intervening, noisy blank frames and without loss of sync.*

In fact, the transition using pause from one scene to the next can be so smooth that without the jump in time indicated by the time–date signal, or without explicit voice notation, one can easily overlook the fact that a transition has taken place.

To achieve good recording, set the TRACKING control to its neutral position. During playback, the tracking control *electronically* adjusts the time delay, which occurs after the control pulse is read, to the time when the video head starts reading the video track. Although this control normally does not function in the record mode, it is still safer to reset the TRACKING knob to its neutral position before you start recording. The SKEW control, which is a device that *mechanically* adjusts the tension on the tape, is automatically returned to its neutral position when the RECORD button is pressed. Also make sure that the incoming signal is at the correct level (0.7 volt for the picture

*This application of the pause function is not only useful for making sequential original recordings from the camera, but also for assemble editing as described in Section 12.3.

signal, −0.3 volt for the sync pulses). This precaution is especially important when copying from VCR to VCR or recording titles from title generators or computers. The signals from these sources are often too high, badly distorting the highlight regions in the recorded tape. The consequences may not become obvious until the tape is played back.*

Noise and *blemishes* introduced into the original scene become part of the permanent record. Some degree of random noise can be averaged out later, but shading, hot spots, and other blemishes cannot easily be removed later. Unless the background to be subtracted is recorded on a separate segment of the tape, even digital image processing cannot easily remove these picture defects. Therefore, at the time of initial recording, pay close attention to cleanliness of the camera tube, proper illumnination, out-of-focus images, and the cleanliness and alignment of the microscope described in Chapters 3 and 5.

Troubleshooting and Repair

What should one do if the VCR will not record, or only plays back noisy pictures, or no picture is seen on the monitor in playback, or the VCR simply will not turn on? A brief list of likely problems and a guide for troubleshooting is usually provided in the *owner's (or operator's) manual* that comes with the VCR.

Follow the troubleshooting guide in the owner's manual. But first also make sure: (1) that all cables are correctly hooked up and properly terminated (even experienced persons can make mistakes); (2) that all selector switches are properly positioned; (3) that there is a good signal from the camera (make sure by connecting the camera cable directly to the monitor); (4) that the video components hooked into the system are all powered up (that includes the VCR, and if hooked up in the system, the video processor, switcher, and special effects generator; these units will not relay the video signal unless they are powered up); and (5) that the cassette is not the wrong type (you can only record on certain types of tape with a time-lapse VCR) or write protected [the VCR will not enter the RECORD mode; see Section 12.3 and items (1) and (2) in Fig. 8-17].

If the problem still persists, contact the original sales agent, an authorized service center, or your audiovisual department. However, *in no case should personnel not trained in the service of VCRs, or unfamiliar with the particular type of instrument, attempt to fix structural and electronic problems inside the VCR.* Read/write heads can easily be damaged, the drum cylinder can be scratched, guide pins can be bent, and the electronics can be hopelessly misadjusted. Special tools and jigs and **alignment tape** as well as detailed specification of tolerances, are needed to adjust each VCR to factory specification after it has been serviced.†

Head Cleaning and Replacement

Depending on the quality of the magnetic tape used, and the care given to maintaining the tape and VCR, the read/write heads have to be cleaned at infrequent or shorter intervals.‡

The heads should not be cleaned unless it is time for scheduled maintenance service or some symptom suggests the need. So long as there is no noise in the playback picture attributable to the VCR head, and the tape shows no sign of abrasion, probably the safest approach is to have the manufacturer's designated service center check out and clean the VCR once every 500 to 1000 hr of actual operation. The service interval varies with the type of equipment and the conditions and environment of use. Check with the manufacturer's representative.

In cleaning the head, it seems inadvisable to use the so-called "head-cleaning tape," especially those that are mildly abrasive. If you decide to clean the heads yourself, be sure to assemble

*See Section 12.3 for methods of monitoring and adjusting the signal voltages.

†See chapters on trouble-shooting and repair, e.g., in Hobbs (1982); the service manual for the particular VTR; or the "Sony Basic Video Recording Course."

‡By exclusively using the KCA- or KCS-type tapes specified by the VCR manufacturer (Sony), and by taking good care of the VCR and the tapes, I have not had to clean the heads on our time-lapse and portable VCRs for over 1000 hr of actual running time. However, the heads did clog and necessitated cleaning when we ran a substitute brand tape for just a few hours.

the proper tools and receive instructions from a qualified individual *before* making your first attempt.

For cleaning read/write heads, drum cylinders, and parts in the tape path, a special chamois-tipped cleaning tool is moistened with isopropyl alcohol or specially formulated solvents. First remove the top cover of the VCR *with the power cable unplugged.* Then apply the moistened face of the tools, so that the tool face lies tangentially to the head drum, with very mild pressure.* The head drum is turned gently with the fingers for a few turns, while the cleaning tool is held in such a way that *no pressure is exerted on the heads up or down, i.e., along the height of the cylinder.* The head is easily damaged if pressure is applied in that direction. Both drum cylinders, as well as the erase, audio, and control heads, and the guide rollers and drive capstan are cleaned with the same type of tool. Keep the tip of the tool clean and moist (not excessively wet), and do not forget that the heads, rollers, and other parts inside the VCR can be quite delicate.

Even under the best conditions, the read/write heads in the VCR are not built to last indefinitely. In fact, the heads (together with the head drum) on a ¾-inch VCR are *meant to be replaced* after about 2000 hr of use. Larger-format VTRs require more frequent, and smaller-format VCRs require less frequent, replacement of heads. Some VCRs have elapsed-time meters that are visible to the user; others have indicators that are accessible only during service.

When Not in Use

During use, the vent grills of a VCR should be fully exposed so that air circulation is not impeded and the interior is adequately cooled. However, when the VCR is not in use overnight or longer, turn off the power, and cover and protect the VCR from dust and other contaminants. Also, rewind the tape onto the supply spool rather than leaving it threaded at some other point.

The VCR should be turned off before being covered so that heat does not build up. Also, when any piece of video equipment is not use for several days, it is wise to unplug the power cords to prevent a high-voltage surge from entering and damaging the equipment. (I have not had such problems with a VCR, but have with time–date, and time-code, generators).

8.8. CHOICE AND CARE OF THE VIDEOTAPE

The complex machinery and electronic circuitry in a VCR are all geared to providing the best possible recording and playback from the magnetic recording tape. The tape that is used to store the video and audio information thus plays a crucial role in the process of video recording. Minute magnetic fields must be registered in crowded but orderly rows, and retained without loss of strength for many years. The binder, which supports the fine ferric oxide or chrome oxide magnetic particles, must form a uniform micrometer-thick layer, and retain the particles firmly in place withstanding the abrasion during many tortuous passes through the tape path. The mil-thick plastic base that supports the magnetic layer must retain precise dimensions as it repeatedly twists around the guide rollers and contorts down the head drum, whirling at high speeds.

The superior quality of the magnetic tape and the care given its use and storage are thus clearly vital to the quality of the magnetic recording and its preservation.

Choice of Tape

Among the videotapes available, properties can be quite varied. Some record well up into the higher frequencies, some have better S/N ratios, some have fewer dropouts, some retain their magnetism better, and others hold their dimensions more stably.

Of the many types of tape available, those used for time-lapse recording must meet particularly critical requirements. We need dimensionally stable tape capable of high-density recording. Such

*For illustrations on the use of the head-cleaning tools, see, e.g., Hobbs (1982) and "Sony Basic Video Recording Course." Be especially careful not to substitute rubbing alcohol, which may contain oils or emollients.

tape for *¾-inch time-lapse recording* is sold in cassettes identified with the prefixes *KCA* and *KCS*. The tape in cassettes so labeled also withstands the heavy abrasion wear encountered in still-frame operation, without shedding oxide and clogging the heads. The number following the prefix designates the number of minutes the tape would run at standard speed. Those with suffixes of *BR* (broadcast application) are of even higher grades and yield improved time-lapse records.

The KCA and KCS cassettes can be identified by the two small holes, instead of the usual single hole, that are present on the bottom of the cassette, next to the red WRITE-PROTECT button (Fig. 8-17). A small probe in the VCR, not encountering resistance at the outer of the two holes, identifies the cassette as being of the proper type. Conventional U-matic cassettes such as the KC-30 and KC-60 types do not have the outer small hole and are rejected by ¾-inch time-lapse VCRs.

Many brands and grades of ½-inch video cassettes are available for the Beta and VHS formats. Judging from the survey made by *Video* magazine, no single brand of tape achieves the best marks in all the tests performed.* For monochrome video microscopy, where good performance is desirable in luminance S/N ratio, response at high frequency, low dropout rate, and stability of the record, the TDK Pro T-120 and TDK Hi-Fi T-120 appear to be preferable, among the VHS blank tapes, to the others tested.

Some extended-play cassettes contain tape with an exceptionally thin tape base. The thinner tape is more likely to stretch and be dimensionally unstable to heat and humidity. Also, they are more likely to shed their magnetic coating, and the thin tape can slip between the drive capstan and pinch roller on a VCR designed to use thicker tape. Again, one should use tape of the type specified by the VCR's manufacturer.

Storage of Recorded Tape

Some of the precautions to be followed in storing recorded tape are obvious ones. The cassettes should be stored away from extremes of temperature† and humidity, well protected from dirt, dust, and smoke.

After recording, the cassette should be write-protected and returned promptly to its properly labeled case. The cassette must be kept away from equipment that generates strong magnetic fields, including loudspeakers and telephones.

Less obvious are the precautions that help maintain tape dimension and shape. After recording, and especially after repeated playback of a particular scene, fast forward the tape to the end and then rewind it completely so that the tape is neatly wound back on the supply spool. An irregularly wound tape has edges protruding against the inner walls of the cassette. Also, do not store cassettes lying flat on their sides; rather, stand them in a cassette case on edge, the way books are usually arranged in a bookshelf.

Both of these precautions minimize the lateral compression of the tape during storage. When we recall that the closely spaced, microscopic, video tracks are located only in reference to the tape's edge that rides on the rabbet, it becomes clear why we need to take such special care to protect the edge of the tape.

8.9. MAGNETIC AND LASER DISK RECORDING

Magnetic Disk Recording

Computer programs and data are frequently stored on "floppy" or "hard" *magnetic disks*. In a *disk drive*, concentric tracks of magnetic records holding anywhere from 100 kilobytes to many megabytes of characters (of 8-**bit** information) can be laid down on the disk depending on its type. The information is recorded to, or read back from, the disk according to instructions given by the computer.

*Braithwaite (1985); see also the March 1985 issue of *Video* for tests on Beta-format blank tapes.

†At elevated temperatures, the rate of magnetic print-through from one tape layer to another increases, the rate roughly doubling when the temperature rises from 20°C to 40°C (van Wezel, 1981).

Floppy disks are circular, thin plastic disks, usually with diameters of 3½, 5¼, or 8 inches. They are coated with magnetic material similar to that used on audio- and videotapes. The disks, permanently protected in plastic or paper sleeves, are inserted as needed into the disk drive. In the disk drive, the magnetic read/write head, which travels along the radius of the disk, is placed less than 1 μm above the surface of the disk through an elongated opening in the sleeve. As the drive spins the disk at a closely regulated rate inside the stationary protective sleeve, computer instructions position the read/write head onto appropriate tracks on the disk to record or read back the digital signals.

The floppy disks can be interchangeably inserted or removed from the disk drives as needed, but they are somewhat fragile. They can be damaged by mishandling or by contamination, even from very fine particles, such as the oil droplets contained in cigarette smoke. Magnetic disks can also be erased accidentally by moderately strong magnetic fields that may be present around video monitors, speakers, and ringing telephones.

Hard, or Winchester, disk drives operate similarly. But usually, the hard disk is permanently built into the drive, which is equipped with fine-pored air filters. A hard disk can hold many more times information (5 up to 10s of megabytes) compared to the 100 kilo- to 1 megabyte for the floppy disk. Hard disks can also transfer information at rates very much faster than with floppy disks.

Magnetic disk recorders are also used to store video images that have been digitally processed (Chapter 10). For a video frame digitized to 512 × 512 (or 640 × 480) pixels with 256 (8 bits or 1 byte) shades of gray, an 8-inch, 0.5-megabyte floppy disk would just barely hold two video frames. A 20-megabyte hard disk would hold 80 frames of similarly digitized video.

The information on a magnetic disk can be **randomly accessed** by addressing any sector on any track. But the read/write rate is limited by the speed of digital information transfer between the disk drive and computer; reading or writing a single video frame may take seconds or minutes.* Nevertheless, magnetic disks hold digital information with very high precision; the error rate per bit is exceedingly low.

Magnetic Disk versus VTR Recording

Compared to magnetic disks, the tape in a VTR stores vastly greater amounts of information. The nature of the video signal played back from a VTR and its intended uses, however, are quite different from those used in a digital computer. For a digital computer, the error rate of memory must be kept very small, and each sector on the disk must be quickly accessible. For video, small errors or dropouts are less significant. The read/write speed of a VTR is very much faster than for magnetic disk recorders, but it may take minutes to spool and locate a particular frame on a VTR.

Laser Disks

A new technology using *laser disks* instead of magnetic tape or disk has emerged recently. Despite the youth of this technology, it has proved to be very powerful, practical, and appealing. Each laser disk holds a vast amount of memory (of the order of a gigabyte = 10^{12} bytes) and provides 30–60 min of video and audio playback with very high fidelity.† Also, with laser disks, any video frame on a disk can be *randomly accessed,* or recalled and played back virtually instantly. The sale of laser video and audio disk players is rapidly rising even in the consumer market, and *interactive video disk systems* are beginning to be used for display, entertainment, instruction, and training.

The laser disk is an encapsulated plastic or glass disk, a few millimeters thick and ranging in diameter from 10 to 30 cm. The surface of the disk is engraved with a series of concentric or helical fine grooves. The grooves are less than 1 μm wide and spaced apart from each other by exact

*However, the *analog* motion analyzer, Sony SVM-1010, for example, does read and write on a magnetic disk at video field rates (Section 8.4). See also Video (Magnetic) Disk Recorder/Reproducer systems made by Precision Echo.

†*Laser audio disks* with exceedingly low distortion and S/N ratios of 90 dB are already available.

distances of 1 to 3 μm (depending on the format). The regularly spaced grooves act as a reflection grating and the disk gives rise to beautiful interference colors. An FM video signal is recorded in the grooves of the disk as a series of discrete submicrometer depressions (the "holes"). Each concentric track, which is coded with an address, registers a single video frame.

In a *laser disk player,* the beam from a low-power laser (usually a 2- to 3-mW diode laser) focuses a minute spot (< 1 μm) of light onto the concentric or spiral track. As the disk spins at its regulated, constant speed, the laser spot senses the holes as regions with lower reflectivity. Light reflected from the tracks on the spinning disk is thus modulated and picked up by a photodiode through the same lens that focused the laser beam. The photodiode converts the modulated light beam into an FM, video playback signal. The beam-focusing/pickup lens moves along the radius of the disk similarly to the read/write head in a magnetic disk drive. Being controlled by a built-in **central processing unit,** the lens can be positioned very quickly to read off any track addressed.

Because the signal is recorded as discrete submicrometer holes on the video disk surface, the recording density is high and the fidelity and S/N ratio in playback are excellent. The laser disk is also very much more durable, and less susceptible to contamination and erasure, than magnetic disk or tape. The high-NA lens that records and reads the laser beam need only focus the beam on the hole. The lens does not have to lie within 1 μm of the disk as with magnetic recording, so the probability of abrasion and "head crash" (dreaded by magnetic disk users) is virtually eliminated.

Each concentric track, which records a full video frame, has a track label, so that any track can be accessed from the keypad using the built-in microprocessor or by instruction from an external microcomputer via an RS-232C interface. Any one of the 50,000 to 100,000 frames recorded on a video disk can be accessed in about 1 sec. A single frame picture played back is rock steady, and one can play back the same frame indefinitely without concern for disk wear. Video disks can commonly be played back together with the audio record with the frames sequenced at standard video rates. The journal *Cell Motility* now distributes a video disk supplement of video microscope recordings.

Optical Memory Disk Recorders

Initially, video disks were mass-produced with prerecorded programs that could only be played back. In 1982, Panasonic introduced a series of **video disk recorder**/players, or *optical memory disk recorders* (OMDRs; Table 8-7).

The disk used in the OMDR contains a thin recording layer of heat-sensitive material (Fig. 8-18). This may be a silver emulsion in an organic gel matrix (Drexon Optical Memory Discs), or be made, e.g., of tellurium suboxide (Panasonic Optical Memory Disks). A laser beam focused by a high-NA lens melts or ablates a hole less than 2 μm in diameter into the heat-sensitive layer.* As the disk spins, short pulses of the laser beam, modulated by the FM video signal, record a series of holes at rates up to 10 megabits/sec.

The Panasonic OMDRs (Fig. 8-18), whose size is in-between that of a full-size ¾-inch cassette deck and a portable U-matic VCR, can record or play back full video frames, a single frame at a time (or sequentially at video rates in some models). Recording of a single frame is at video rate, and one can immediately play back the recorded frame without any processing (which is needed in photography) or waiting for tape to rewind (as in a VTR). Ten thousand to twenty-five thousand frames can be recorded on each 20-cm-diameter disk, depending on the model, and any track can be accessed within about 0.5 sec, before or after recording. Either color or monochrome recording is possible, but the high-resolution monochrome unit (TQ-2021FBC) is of particular interest for video microscopy.† This unit provides a recording capacity of 10,000 frames, each with a horizontal resolution of 450 lines with an S/N ratio of 45 dB. With this magnitude of resolution on a single still frame, the picture does contain some very-fine-grain random noise. The noise from the single frame

*The output power of the recording laser may be as low as 20 mW.

†In mid-1985, Panasonic has introduced a high-resolution, monochrome, *spiral* track OMDR. The new model, TQ-2025 F can record 16.000 frames continously in real time or as individual frames at video rate. It is also equipped with sound tracks.

TABLE 8-7. Optical Memory Disk Recorders

Make	Panasonic	Panasonic	Panasonic
Model	TQ-2021FBC[a] High Resolution B/W Still Video Recorder	TQ-2022FC Color Still Video Recorder with Motion Playback	TQ-2023F[b] Motion Video/Audio/Still Video Optical Disk Recorder
Video in	525/60 RS-170 B/W	NTSC	NTSC color
H resolution	450 B/W	340 B/W, 300 color	300 TV lines
S/N ratio	45 dB	45 dB	45 dB
Recording	Step single frame 1/30 sec	Step single frame 1/30 sec	13.3 min video rate Single frame 1/30 sec
No. of frames	10,000	15,400	24,000
Access time	1/2 sec average	1/2 sec average	1/2 sec maximum
Disk diameter, speed	8 inches, 1800 rpm	8 inches, 1800 rpm	8 inches, 1800 rpm
Tracks	Concentric	Concentric	Spiral (Auto adjust for Concentric disks)
Playback	Step single frame Variable note 20-0.5 frame/sec step frame	Motion Playback 8 min total, Step single frame	Video rate, Variable rate still/frame advance, Variable speed slow motion, Variable fast speed
Audio	None	None	20 Hz–20 kHz (± 3 dB) 70 dB 2 channels
Power	AC 120 V 60 Hz 120 W	AC 120 V 60 Hz 120 W	AC 120 V ± 10% 50/60 Hz 120 W
W × H × D (inches)	20.8 × 7.7 × 18.3	20.8 × 7.7 × 18.3	20.9 × 7.2 × 19.0
Weight	57.2 lb	57.2 lb	59.1 lb
List price (12/84)	$18,000	$27,000	$34,900

[a]Replaced in 1985 by TQ-2025F High Resolution B/W Motion/Still (16,000 frame spiral track) Video Optical Disk Recorder/Player (10/85 list price $23,700).
[b]Disks can also be played back on TQ-2024F Motion Video-Audio/Still Video Optical Disk Player (12/84 list price $3,975).

disappears when the signal is integrated digitally, or summed photographically, during the ¼ to ⅛ sec required for exposure. As shown in Figs. 8-19 and 8-20, the quality of the still-frame or freeze-frame video picture can be truly impressive.

This high-resolution monochrome OMDR can serve as an effective time-lapse or animation recorder, or as a frame-by-frame motion analyzer. Consecutive recordings can be made at intervals as short as 0.5 sec (clocked by an external intervalometer), and the internal computer logic can be programmed to play back at between 10 frames/sec and 2 sec/frame. For example, for scenes recorded once every 30 sec, over a 2-hr period, the number of frames required would be (2/min × 120 min) = 240 frames. Forty such sequences could be recorded on a disk costing $250 (Dec. 1984 price) for about $6 per sequence. The sequence could be played back, immediately after recording, at rates slowed to between 15 and 300 times actual speed, with any specified frame being retrievable within 0.5 sec. As we just saw (Figs. 8-19, 8-20), the quality of the single-frame image is excellent; it is superior to that obtainable from a single frame of 16-mm motion picture (Plus-X) film.

The signal played back from an OMDR is exceptionally stable. Not only are flagging and other problems associated with time-base errors completely absent, but the signal is so stable that it can readily be accepted by a digital image processor (although a 2-H time-base corrector does help to stabilize the picture further).

The whole technology of laser video recording is developing rapidly, and preliminary specifications have been released for *erasable* read/write video disks. The OMDRs are still expensive, costing several times the price of a good-quality U-matic VCR. Each disk also costs about 10 times more than a 60-min U-matic cassette. However, in the two and a half years since the OMDRs were introduced, their prices have already dropped very significantly. Undoubtedly, these superior recorders will soon find many applications in video microscopy (e.g., see Edelhart, 1981; Currier, 1983; Hecht, 1984, Hirshon, 1984b).

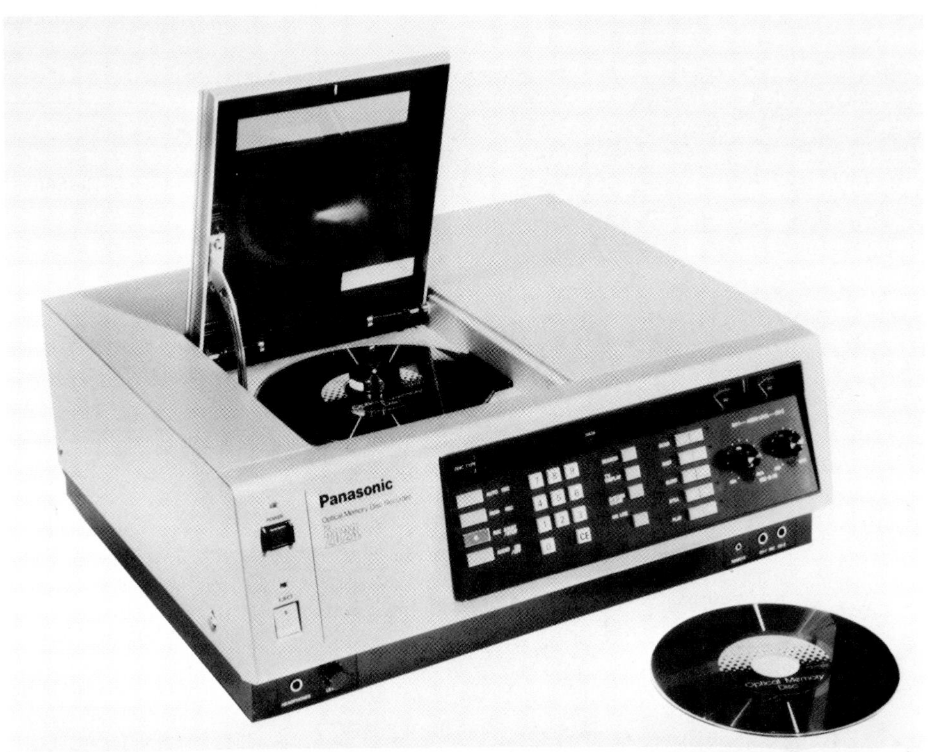

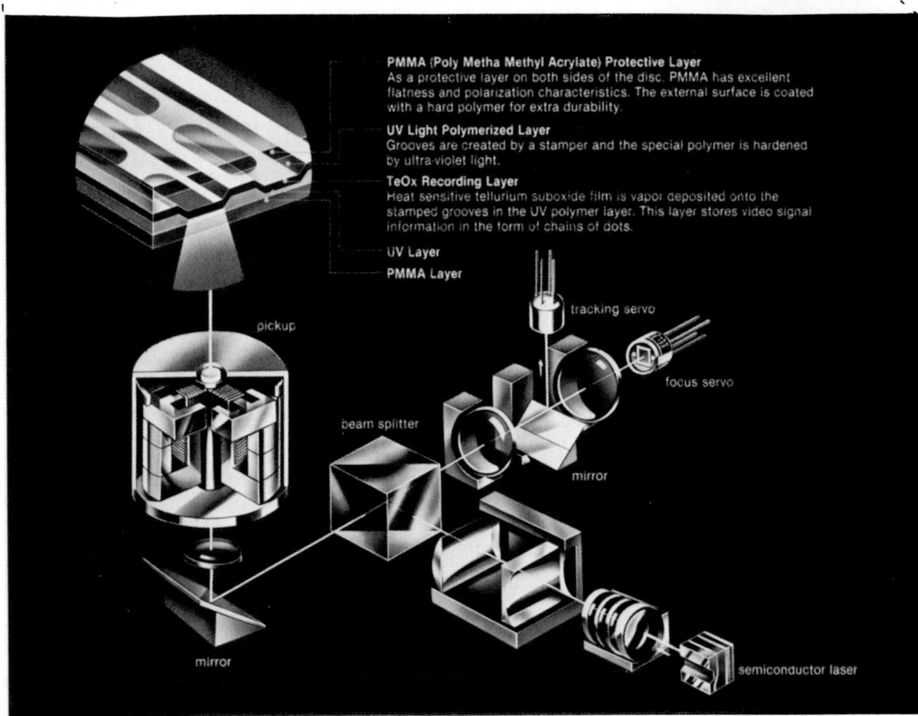

FIGURE 8-18. Optical memory disk recorder. Top panel: Appearance of Panasonic model TQ-2023F. Bottom Panel: Schematic of the optical system. The optical memory disk recorder shown here uses a solid-state laser to store and playback onto the removable video disk (top panel). Each frame is recorded in 33 msec and is not erasable. The microscopic record is produced in a heat-sensitive recording layer (inset) by a narrowly-focused laser beam (bottom panel). A single disk (8 inch-diameter), such as that in the foreground, can store as many as 25,000 video frames (10,000 frames in model TQ-2021 FBC; see Table 8-7). (Courtesy of Matsushita Electronics Corp., Takatsuki, Japan.)

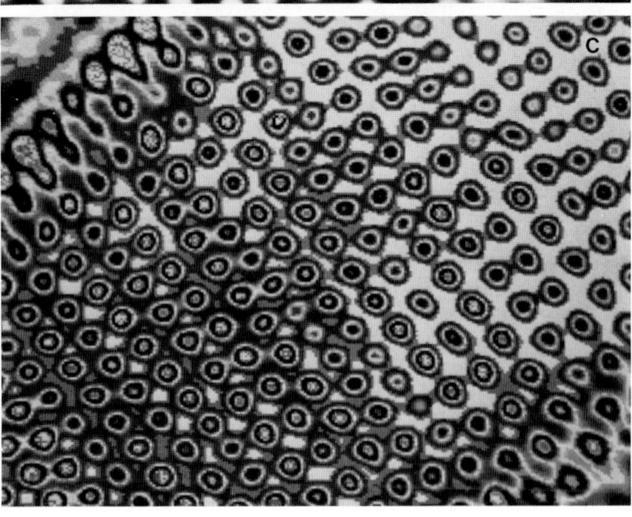

5 μm

FIGURE 8-19. High-resolution pictures played back from an OMDR. Diatom frustule seen in high-extinction DIC (A) and polarized-light (B) microscopy. For both DIC and polarized-light microscopy, the same 100/1.35 Plan Apo oil immersion objective (Nikon) was combined with an NA 1.35 rectified condenser (Nikon), oil contacted to the slide, which provides full NA. The pictures were taken through the author's inverted optical bench microscope (Section 5.11), using 546-nm monochromatic illumination. (A) Single frozen frame (1/30 sec) from the Newvicon camera (Dage–MTI Model 65) was recorded on a monochrome high-resolution OMDR (Panasonic Model TQ-2021FBC). The picture played back from the OMDR was photographed off a 13-inch high-resolution monochrome monitor (Panasonic WV-5410) (× 3300). (B) An image of the same specimen was electronically zoomed and processed to smooth the pixels and raise contrast. The image, processed with a real-time digital image processor (essentially identical to the Interactive Video System Image-1), was then recorded on the OMDR. (C) Isointensity contours of the frustule pores generated with the Image-I processor (the original was in pseudo color). The center-to-center spacing of the frustule pores in this diatom is 0.4 μm. The 5-μm scale is for B and C.

FIGURE 8-20. Test pattern. The test pattern stored on an OMDR (Panasonic TQ-2021FBC) was photographed off a 13-inch high-resolution monochrome monitor (Panasonic WV-5410). The photographs in Figs. 8-19 and 8-20 were taken on 35-mm film (Kodak Plus-X) with a 55-mm f/2.8 Macro lens (Nikon). The V HOLD control on the monitor was carefully adjusted to yield exact 2:1 interlace; a Ronchi grating was not used (see Section 12.1). In properly interlaced frozen-*frame* pictures, the scan lines are almost imperceptible at this enlargement. Compare with the frozen-*field* pictures shown in Figs. 1-6 and 9-11. The geometrical distortion is due to the monitor.

9

Analog Video Processing and Analysis

9.1. INTRODUCTION

In many ways, the light microscope is a highly perfected instrument. It produces an image resolved to a fraction of the wavelength of light, with excellent contrast, and with aberrations reduced to a bare minimum. It portrays images in many contrast modes and allows measurements of various optical quantities nondestructively. Still, practitioners of modern microscopy are continually striving to register fainter images, explore finer details, and record subtler contrast. We strive to detect weaker fluorescence, measure lower birefringence, and reveal miniscule absorbances and optical path differences, all with an effort to uncover dynamic physical and chemical details in finer and finer domains of the specimen.

Photography has long been a supportive companion for microscopists. With photographs, we gain permanent records of the scene, can detect change, and analyze the recorded picture in detail. We can calibrate the photographs so as to calculate numbers; establish density and distribution; determine topological relationships; measure distances, lengths, angles, and areas; and so on. By using a series of photographs taken at selected time intervals, we can follow the time course of changes, track movement, and measure velocity. With photographic photometry, we can also measure light intensity, absorption, spectral characteristics, optical path difference, dry mass, birefringence, **dichroism,** etc.

With photography, we can also enhance contrast, correct shading, and apply optical filtration to bring forth subtle, hidden features of the microscope image.

While we use photography to boost the performance of the microscope in many ways, photographic processing tends to take time, requires superior skills, and does not provide instant feedback. Also, analysis by photography demands careful calibration and tedious routines of measurements. In contrast, *video processing* tends to be simple, can be carried out *on-line,* and often in *real time.*

Video, coupled with signal processing, can improve temporal and spatial resolution, reduce noise, and raise the sensitivity for detecting optical signals. Video processing enhances the microscope image, and selectively corrects or improves the image in a manner not easily accomplished by optical means alone. With video image processing, we gain immediate display of the results, saving time and permitting on-line experiments with image manipulation. We can thus capture image features that are difficult or impossible to see without the immediate feedback.

Image analysis by video can improve precision, provide shortcuts, and save time, especially if a computer is linked to gather and analyze the data.

Thus, with video we can process and analyze the image as in photography, but with greater precision, speed, and directly on-line.

The video image, or, more appropriately, the video signal, can be processed and analyzed by analog or digital means. In *analog processing and analysis,* one treats the strength of the video signal as a continuously varying function. Also, the time interval in the signal that determines the location of an image point is treated as a continuum. In *digital processing and analysis,* the strength of the video signal as well as the location of the image point, the **pixel,** are treated as multiples of discrete units, commonly expressed in binary form (Chapter 10).

By digitizing the pixel location and intensity, low levels of noise or minor fluctuations in pixel locations and signal strengths no longer perturb the signal. While the digitized pixel locations and intensities cannot be dissected into infinitely small steps,* the pixel location and intensity are defined unambiguously. The message can thus be portrayed exactly and reproducibly, and the signal can be stored, transmitted, processed, and quantitated quite free from noise. This is the crux of digital electronics, including digital computers.

An example will serve to show just how immune to noise *digital transmission* can be. We transmitted a digitized video signal of a microscope image in color over the regular telephone line from Woods Hole, Massachusetts, to Boulder, Colorado, some 2000 miles away. There the signal was recorded and transmitted back again. The signal received over the phone line and played back in Woods Hole showed absolutely no noise or glitches. The picture was so perfect as to be indistinguishable from the original in every detail and in subtle shades of color! I was most impressed that the noise so common to the regular telephone line, in no way interfered with the two-way digitized transmission over such a distance and, being in the ''slow scan'' mode, spanning periods of over 2 min each way.† The same principle is, of course, used to span millions of miles of space and, coupled with digital processing, has provided incredible insights into features of distant planets.

Once digitized, the video signal can be processed or analyzed with a digital computer. Vast opportunities then arise for signal and image manipulation, geometrical corrections, noise reduction, **feature extraction,** and quantitation of various image aspects. With the explosive rise in the efficiency of digital memory, computer **hardware** and **software,** and the concomitant rapid drop in their prices, we may expect much of video technology, especially processing and analyzing techniques, to become digitized over the next decade.‡

At present, the majority of current video cameras, recorders, and monitors operate on analog principles. Compact *analog processors and analyzers* can be coupled directly to them in order to rapidly process and analyze the video signal and image. The operation of analog video processors and analyzers requires little special training and, although somewhat limited in flexibility, they tend to be compact, are easy to use, and today cost a small fraction of the price of digital devices.

In this chapter we discuss the principles and applications of analog video processing and analysis, and survey some devices available on the market. Digitized image processing and analysis are discussed extensively in the next chapter, by Walter and Berns (see also Appendix IV).

9.2. GENERAL PRINCIPLES OF ANALOG PROCESSING AND ANALYSIS

As discussed in earlier chapters, the video camera, monitor, and recorder all contain (analog) signal-processing circuits. Together, these devices are designed to economically yield video pictures with reasonably high MTF, low distortion, and noise.

In video microscopy, the microscope itself is adjusted to provide the camera with an image that works optimally with the video train. Furthermore, the microscope can in many ways be considered an optical analog of video. The microscope is used not only to generate a magnified image with

*But in fact, neither can we meaningfully dissect the analog signal into infinitely small steps. Along both the time axis and intensity axis, noise introduces ambiguities.

†This demonstration was made during our course, ''Analytical and Quantitative Light Microscopy for Biology, Medicine, and the Materials Sciences,'' on May 8, 1983. The signal was sent both ways in the slow scan video mode at 4800 **baud** (bits per second) with a Colorado Video Model 285C Digital Color Transceiver coupled with a Universal Data Model 208, 4800-baud **modem.** By using the slow scan mode, the video bandwidth is compressed and also contributes to a considerably improved S/N ratio (Section 9.4).

‡For a concise, lucid discussion of digital signals and devices, see Keyes, R. W. (1985) What makes a good computer device? *Science* **230:**138–144.

specific resolution and contrast, but to selectively filter, or effectively convolute, selected parts of the Fourier harmonics that make up the image (Section 10.12, Appendix I).

In addition to the processing that takes place in the microscope, video camera, recorder, and monitor, we can add separate analog processing devices to further enhance the image. With these devices, we modify the electronic video signal that produces the picture. Likewise, we use analog video analyzers to extract image features and inject electronic signals that appear together with the original picture to allow measurements unaffected by monitor distortion.

Before discussing the actual methods of analog processing and analysis, we need to recall the relationship between (1) the two-dimensional raster that scans the image in the camera and draws the picture on the monitor, and (2) the one-dimensional video signal that links the two.

To briefly recapitulate the discussion in Chapter 6, coordinates in space along the horizontal (X) and vertical (Y) axes of the camera and monitor are traced out by a descending row of left-to-right horizontal scans, represented as sequential intervals of time in the video signal. $X = 0$ appears at the left edge of the active picture area at the end of the H-blanking period. $Y = 0$ appears at the top of the picture at the end of the V-blanking period. The picture has a standard aspect ratio, usually with a width-to-height ratio of 4 : 3 or 1 : 1. For CCTV in the United States, the H blanking hides no less than 14% of the H period, and the V blanking no less than 7% of the V period (Tables 6-1, 6-2). The intensity for each image point is represented by the positive voltage (usually spanning from 0.05 to 0.7 volt) above the blanking level, which in turn is provided by the voltage of the back porch of the H-blanking signal (Figs. 6-7, 6-8). The H- and V-sync pulses occupy the lower 0.3 volt of the composite video signal, with the negative-going voltages at the beginning of the sync pulses signaling the start of the H and V scans, respectively.

To process or analyze the video signal, the sync signal is first *stripped* from the composite video signal. By separating the sync and picture signals, the picture signal can be amplified linearly or nonlinearly and then recombined at the desired pedestal level with the *reshaped* sync pulses. Oscillators and analog delay lines are used to advance or delay the picture signal relative to the sync pulses, and to correct *phase* relationships. In addition to enhancing the signal, these circuits are also used to add markers and extract signals at various specified locations on the video signal, namely with a (variable) specified delay relative to the V- and H-sync pulses.*

In analog processing and analysis, appropriate amplifiers, mixing circuits, filters, oscillators, and delay lines are used to modify the gamma of the system, alter contrast, adjust the background, threshold the picture, draw contours, position the picture or other signals to be superimposed, extract signals, etc. Thus, we can manipulate the picture and extract information in a variety of ways with appropriate analog video circuits.

9.3. GRAY LEVEL MANIPULATION

Once the sync pulses are stripped and the picture signal is isolated, one can use amplifiers with various characteristics to transform the picture signal. For example, an amplifier can be *linear* as in Fig. 9-1B, and provide output signals (C) proportional to the incoming signal (A). As the *gain* of the amplifier is changed, the slope (S) relating the output to the incoming signal is varied, so that the amplitude of the outgoing signal (contrast in the picture) is proportionally increased (a) or decreased (c).

Alternatively, the amplifier can be *nonlinear* as in Fig. 9-2B, and selectively expand the highlights (a), or shadow (c), depending on the *gamma* (γ) of the amplifier. Linear and nonlinear amplifiers can be placed in sequence or stacked to add or subtract their outputs and generate complex transfer functions.

By adding constant positive or negative voltages to the amplifier input, the baseline of the signal can be raised or lowered. Then, by saturating an amplifier or diode, the output voltage above

*I have intentionally avoided using circuit diagrams in this book. Some of the basic video circuits are described in Showalter (1969), Lancaster (1976), and van Wezel (1981). Some issues of popular magazines such as *Computers and Electronics*, *Byte*, *Creative Computing*, etc. carry articles on practical circuitry for video. The detailed circuits are generally provided in the service manual for the specific piece of equipment.

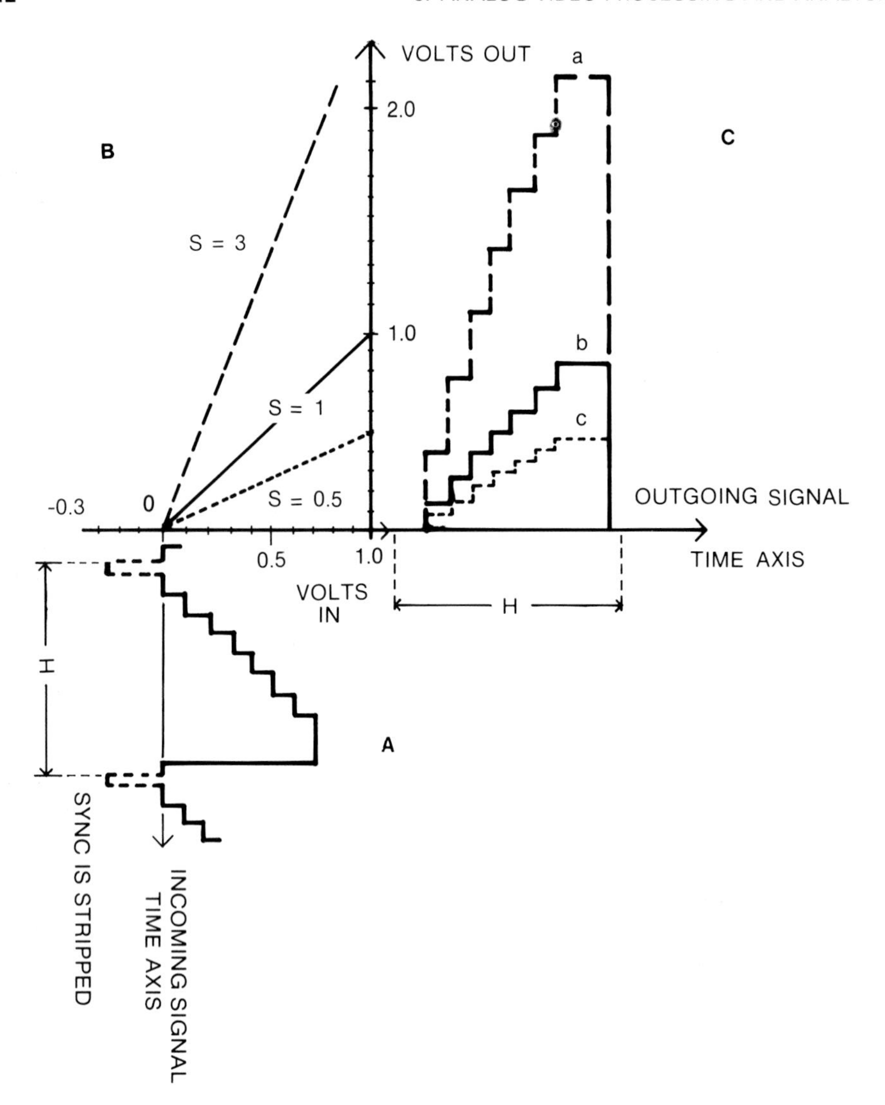

FIGURE 9-1. Linear amplifier. A linear amplifier (B) with gain, S, processes the incoming signal (A) to various amplitudes (C). See text.

a preset level (or below if the signal is inverted) can be **compressed** or clipped off. The black level, or *pedestal,* and the **white level** of the processed signal can thus be adjusted.

In this manner, the incoming signal of the type shown in A, Figs. 9-1 through 9-4, can be converted into variously modified signals as shown in C, Figs. 9-1 and 9-2 B through F in Fig. 9-3, and B through D in Fig. 9-4. Photographs of a scene (Fig. 9-5A) treated with an analog processor are shown in Fig. 9-5B–D. Thus, analog processing can be used to *expand* highlight, midrange, or shadow details, or to *threshold* (**image thresholding**) and provide *contoured* pictures.

With an analog device we can also correct *shading*. A sawtooth and/or parabolic wave can be added to the H or V blanking and correct for scenes shaded in a variety of ways (Fig. 9-6).

In an analog processor, the amplifier gain, gamma, pedestal, and white levels can be altered by simple turns of **potentiometers,** while the effect of the manipulation is directly observed on the monitor or with an oscilloscope. Likewise, various waveforms can be generated and added to the signal while observing the results. In this way, the gray level of the active scene can be manipulated, the image can be thresholded, and the shading corrected, quite simply and effectively by using analog devices.

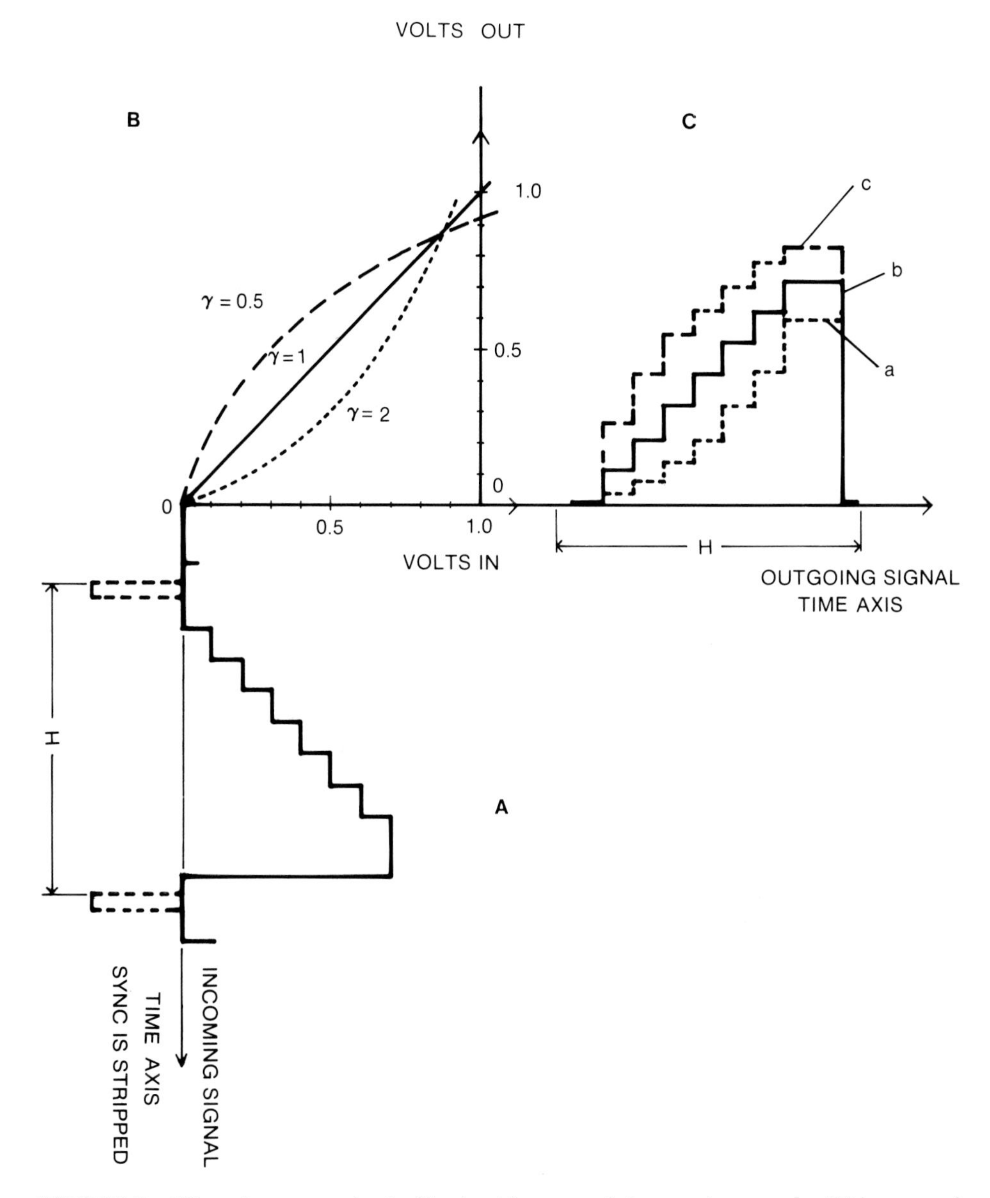

FIGURE 9-2. Effect of gamma on signals. The signal from a regularly stepped **gray wedge** (A) is processed nonlinearly [as a function of (volts in)$^{\gamma}$] if the amplifier gamma (γ) is different from 1.0 (B). The outgoing signal (C) has expanded highlights (a) if $\gamma > 1$, or expanded shadow (c) if $\gamma < 1$.

An extreme form of thresholding is used to **key** certain areas of the picture. The keying blanks out selected areas of a picture according to a mask (Section 12.4). The mask is either introduced from a high-contrast image focused by another camera, derived from the thresholded signal itself, or generated in a signal generator.

By keying, one can *mix* or *split the frame* (Fig. 12-14) with two synchronized video signals and provide various types of stationary or *moving* boundaries. Thus, we can *montage* or *superimpose* halftone pictures, drawings, titles, and pointers on each other; we can make one **fade** into another or **wipe** one in and another out, etc. with a special effects generator (video mixer). We can also use analog devices to superimpose video line traces, fiducial marks, alphanumeric characters, etc. onto the video picture. *Video analyzers, video pointers, time–date generators,* etc. are used in the video

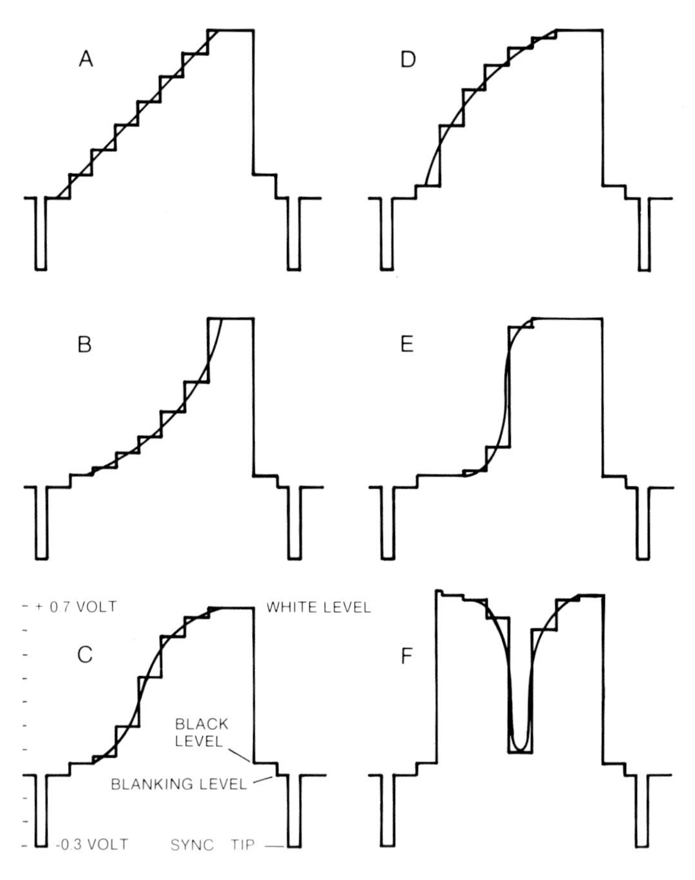

FIGURE 9-3. Modified output signals. (A) Incoming, or unmodified output signal from a regularly stepped gray wedge, shown modified in B through F. (B) Highlights expanded. (C) Midrange contrast expanded. (D) Shadow detail expanded. (E) Thresholded. (F) Contoured. The reference parameters in C apply to all of the panels.

circuits to accomplish these purposes. We can also display limited *pseudocolor* by using thresholded analog signals.

In addition to altering the gray scale, keying, and thresholding, we can use analog processing selectively to enhance *contrast at boundaries* in a video image. The signal for the edges can be selectively manipulated by differentiating the signal (with an inductive circuit). For example, one can *crispen edges* by using a **sharpening filter** or by taking the second derivative of a signal, inverting it, and adding the inverted second derivative back to the original signal (Fig. 9-7). One can also obtain *differential contrast,* producing microscope images similar to differential interference contrast (Nomarski), by adding together signals V and a in Fig. 9-7.

Alternatively, one can achieve a similar effect by adding a slightly delayed inverted signal to the original signal. O'Kane *et al.* (1980) have achieved such differential **edge enhancement** of the microscope image by using an analog delay line to shift the signal and a video mixer to superimpose the inverted signal.

9.4. ANALOG FILTERING AND NOISE REDUCTION

In the previous section, we considered primarily the transfer characteristics of the video amplifiers rather than their time constants. In fact, even a "simple" square wave is made up of

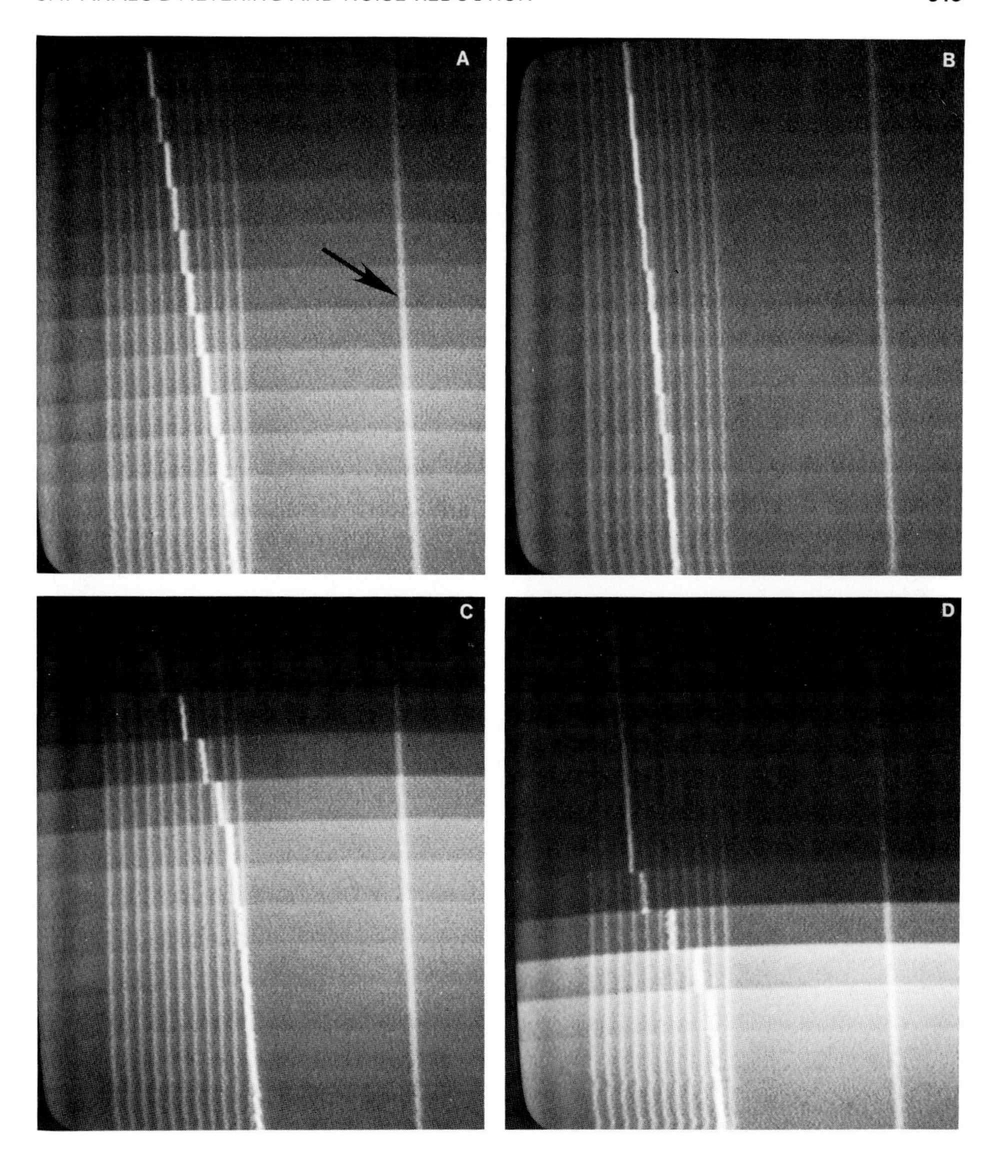

FIGURE 9-4. Monitor pictures of gray wedge and vertical **line scans.** The line scan on the left of each panel shows the luminance of the gray wedge, measured along the fiducial line (arrow in A). The line scan was generated by an image analyzer (Colorado Video Model 321, Fig. 9-10) and the signals were processed with an analog video processor (Colorado Video Model 604). (A) Unmodified signal; gamma = 1.0. (B) Gamma = 1.0, but gain is reduced. (C, D) Signal threshold at different **gray levels.**

many Fourier components (Section 10.12), and faithful reproduction of the square wave demands uniform amplification of the various harmonics and the retention of proper phase relationships among them. In video, we need extremely broadband amplifiers that respond to AC signals with frequencies ranging from about 30 Hz to several megahertz, and with appropriate boosting circuits. However, in any electric circuit, *noise* rises with the square root of the bandwidth, the absolute temperature, and the gain. Attempts to obtain high amplification in video together with high resolution and high fidelity can be frustrated by the increased noise.

In order to obtain video signals that are processed to bring out image detail, we need to start off with high-resolution cameras and other video equipment providing high S/N ratios. On the other

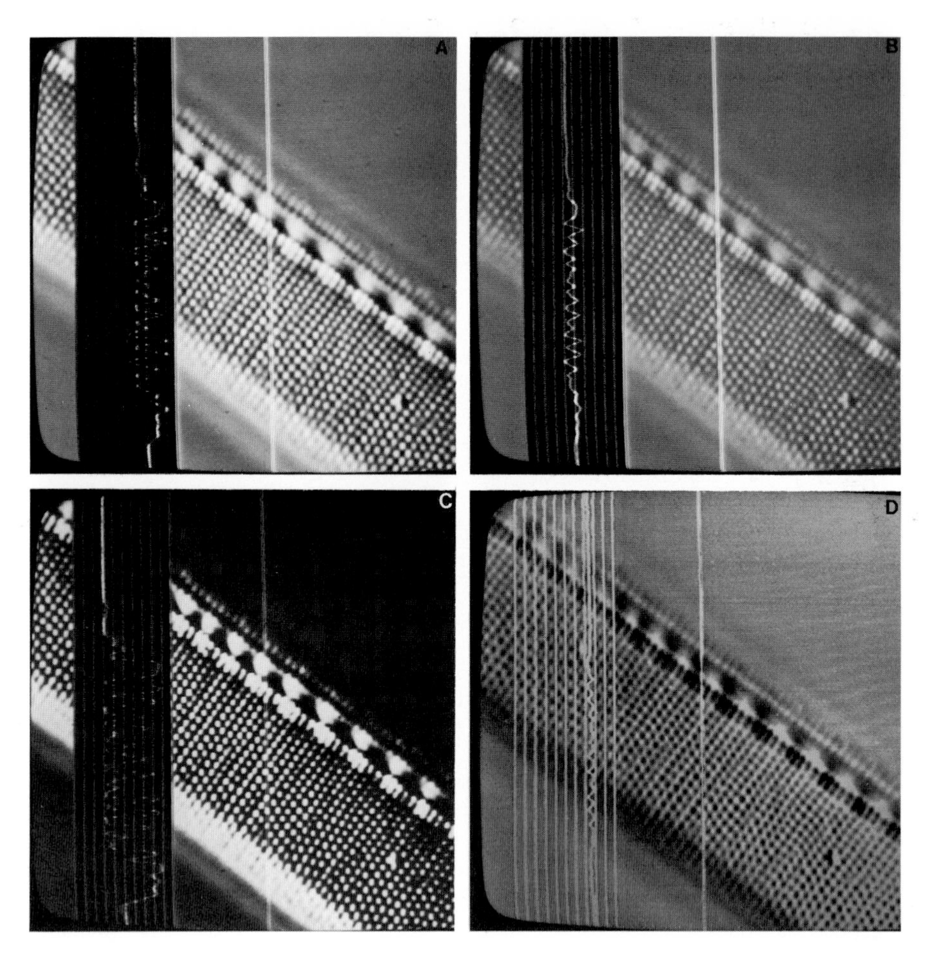

FIGURE 9-5. Examples of analog image processing. (A) Unprocessed. (B) Gain reduced. (C) Thresholded. (D) Inverted contrast from A. The line scan on the left shows the luminance of the image measured along the white vertical line near the middle of each panel.

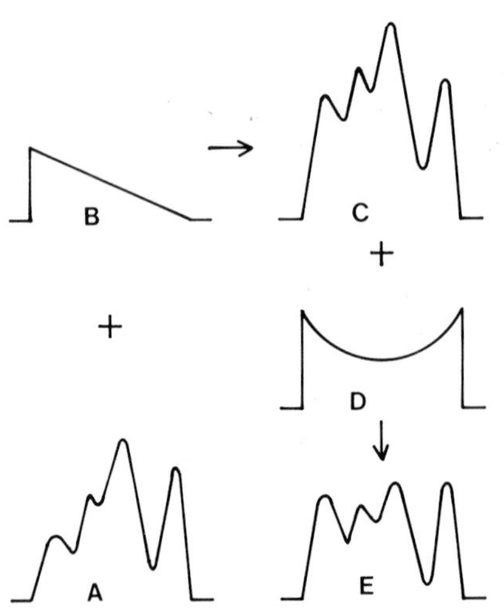

FIGURE 9.6. Shading correction. The signal in E is a faithful representation of the image. However, shading produced the distorted signal A. By adding B to A, the right-to-left signal fall off is compensated, resulting in C. The center-to-edge signal falloff is compensated by adding D. The final result is E.

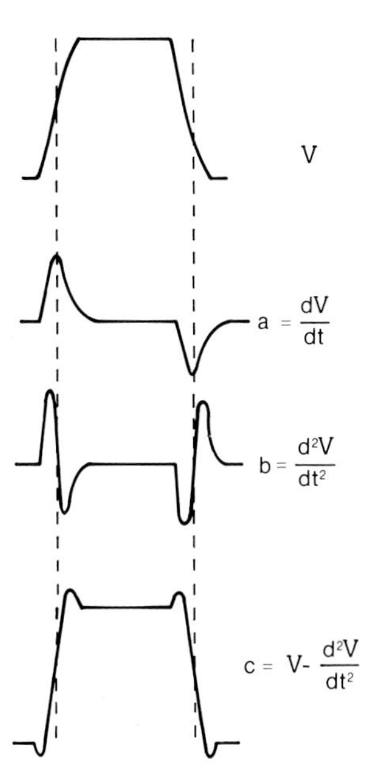

FIGURE 9-7. Edge sharpening. Analog signal processing can be used to sharpen the edges of an object displayed in video. The second derivative (b) of the unmodified signal (V) is inverted and added to produce c. (From van Wezel, 1981, Fig. 2.79.)

hand, the S/N ratio in a video camera varies with faceplate illumination. In video microscopy, the brightness of the image may be well below that optimum for a particular camera. The light level may be so low that the signal from even a highly sensitive camera is limited by noise generated in the camera or from photon statistics.

Noise can be reduced in analog processing by *compressing* the video *bandwidth*. **Smoothing filters** are used to trim the higher-frequency range. Naturally, the higher-frequency signal would also be lost together with the noise, and sharp transitions become softened, but without filtering the noise may have drowned out the picture altogether. By sacrificing some of the finer details in the picture, one can recover a less noisy, if somewhat less well resolved, picture. Figure 9-8 illustrates the effect of filtering on the video picture and signal.

The video bandwidth can also be reduced by *slow scan* if the scene remains unchanged over time, or if temporal resolution can be sacrificed. In the slow scan mode, each video frame is not scanned in 1/30 sec, but over a period of several seconds to minutes. The pulse intervals representing a given spatial frequency or spatial resolution then become proportionately expanded in the video signal. The signal bandwidth and video frequency response, required to provide a given spatial resolution in the image, are thereby reduced in inverse proportion to the scanning duration per frame.

With the frame scanned several hundred to several thousand times slower, the video bandwidth can be similarly compressed and the S/N ratio improved by 10- to 100-fold (i.e., by the square root of the ratio of the bandwidths). I have already described a striking example of noise reduction by slow scan transmission of a digitized micrograph.

Besides filtering the video signal or using a slow-scan circuit, one can also use other forms of analog processing to reduce random "noise." The video image can be *summed or averaged* to recover a less noisy picture. For example, very dim video images can be stored and *integrated* on the target for varying durations, depending on the characteristics of the target in the camera tube. Thus, Reynolds and co-workers (Gruner *et al.*, 1982) have designed a camera* for integrating the low-light-level signal in a chilled SIT-vidicon tube by *holding off the beam sweep,* at times by as

*The analog integrating camera (with the slow-scan digital readout) was designed for use with low levels of X-rays. However, the same design, with the phosphor scintillator removed, would be applicable to low-light-level video microscopy.

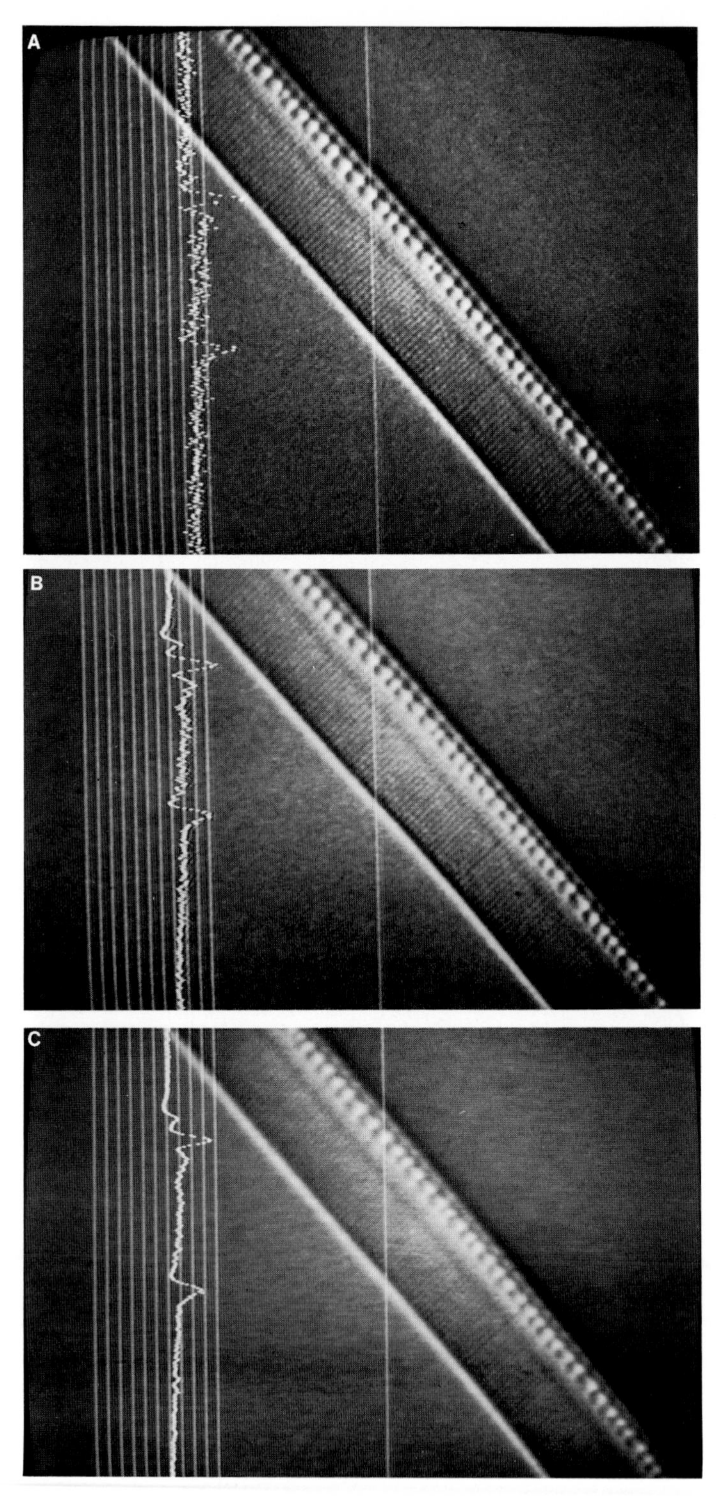

FIGURE 9.8 Noise reduction by bandwidth compression. Monitor pictures and vertical line scans of a diatom taken with DIC optics. (A) The full-bandwidth (20 MHz), low-light-level SIT camera image is noisy. In B and C, the noise is filtered by bandwidth reduction. The noise is reduced, but at some sacrifice of resolution and contrast. B bandwidth = 2.5 MHz; C bandwidth = 1.5 MHz.

much as several minutes. In addition to holding off the beam sweep to reduce the signal fluctuation introduced by photon statistics, Gruner *et al.* further reduced target leakage and camera noise by chilling their SIT-vidicon tube to −55°C and by using slow-scan video.*

Even in a Newvicon camera the noise in a low-light-level signal can be averaged out to some extent by holding off the beam. In an exploratory experiment, the image on a 1-inch Newvicon tube was successfully integrated for several seconds at room temperature before the picture became washed out, presumably by leakage of the target.† Noise can also be averaged out photographically by extended exposure of the monitor picture (Chapter 12).

9.5. IMAGE ANALYSIS

In video microscopy, *distances* in the specimen can be measured simply by placing a calibrated scale against the faceplate of the monitor displaying the picture. The scale is calibrated with a stage micrometer whose image is formed through the same microscope optics and video system as the specimen.

While this approach is simple, and is adequate for determining approximate distances and picture magnification, it is not very precise. A considerable amount of parallax can exist between the scale placed on the outer face of the picture tube and the picture itself; the picture is formed on the phosphor lying on the inner face of the thick, implosion-proof glass wall of the cathode-ray tube. Also, the picture can be distorted geometrically and may be nonlinear due to imperfections in the monitor and/or the camera. Furthermore, the aspect ratio of the monitor may not match that in the camera. In all, without adequate precautions, one could easily encounter errors of several to greater than 10%.

Instead of placing a scale on the monitor, one can measure distances (and hence also areas, angles, and velocities) with considerably greater precision by using a *video analyzer.* An analog video analyzer (Fig. 9-10), coupled to the video train as shown in Fig. 9-9, inserts *fiducial marks* (together with line traces in some analyzers) on the monitor, superimposed with the picture of the specimen. By adjusting a pair of precision potentiometers or a joystick, the fiducial marker(s) can be moved around and placed on points of interest to measure absolute distances in the picture.

A video analyzer introduces the fiducial mark by adding discrete pulses to the video signal. The delay, or interval, between the sync and measuring pulses is controlled by the setting of the potentiometers that move the fiducial marks around the screen. The potentiometer setting, or conversely a voltage that is proportional to the time interval, indicates the exact amount of delay and hence the position of the fiducial mark. The video analyzer is first calibrated with a micrometer placed on the microscope stage (for both the X and Y directions and at several positions on the screen). The setting of the potentiometer can be read off directly, or the voltage output designating the delay can be read with a (digital) voltmeter, or digitized and sent to a computer.

Since the fiducial marks generated by the video analyzer enter the monitor as an integral part of the video signal, we avoid the problem of parallax altogether. We also minimize the influence of picture tube distortion and nonlinearity. Even distortions that appeared in the monitor after the calibration was made need not influence our measurements, although the same would not hold true for the camera or the microscope.

We have used the analog video analyzer extensively for measuring the growth rates of sperm acrosomes only 65 nm in diameter (Figs. 9-11, 9-12). The acrosomal process, which we visualized

*Astronomers routinely couple these approaches with low-dark-current video sensors (SEC tubes, solid-state devices) to detect extremely faint images. The requirements for astronomy and microscopy are similar but not identical, since astronomers tend to be limited by available light while time resolution tends to be more critical in microscopy.

†The experiments with the Newvicon camera (Dage–MTI Model 65), which gave integrated single field images, were performed by Mr. Windham Hannaway at the 1980 Analytical and Quantitative Light Microscopy course, Woods Hole.

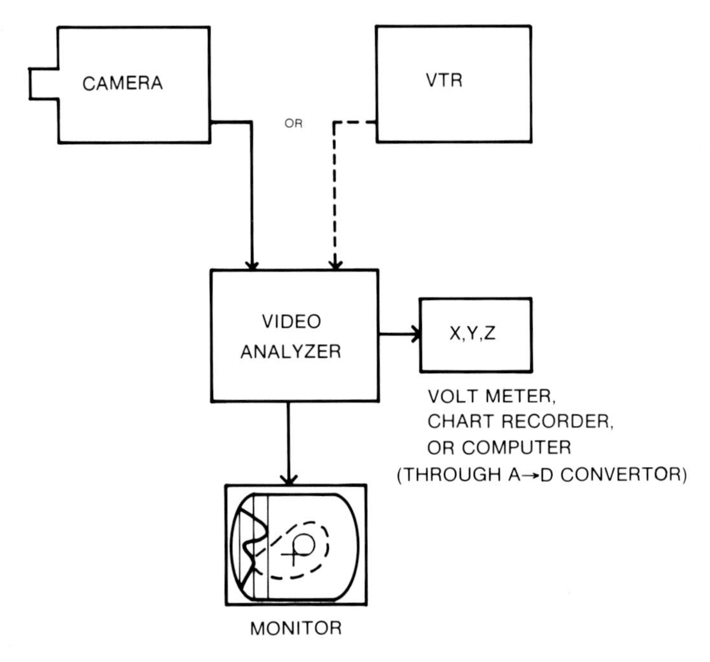

FIGURE 9-9. Hookup for video analyzer.

with Smith DIC optics, grew more than 60 μm in 4 sec. We also measured the movements of echinochrome granules in dividing sea urchin eggs whose cleavage furrows regressed transiently (like a yielding rubber band) by cytochalasin treatment. In both of these examples, we first recorded the rapid movements in the specimen by videotape and then used the video analyzer on the scenes played back field-by-field.

Some video analyzers, known as *video micrometers,* directly indicate the separation between two fiducial lines. Others provide the X and Y positions of the fiducial mark together with the intensity of the signal at that location. The type of video analyzer that provides signal intensities as well as X, Y coordinates can be used to measure intensity distributions, namely as a scanning *photometer*. It is also used to determine geometrical parameters and velocities.

Motion, and other sources of changes in the video signal, can be detected by analog devices by comparing appropriately *delayed signals* with the incoming signal, or by using two "windows" in the video microscope field and observing the *cross correlations* electronically. For example, Biedert *et al.* (1979) used a *video motion analyzer* to measure the velocity of movement of heart cells cultured in monolayer and observed under a microscope at 2000×. Intaglietta *et al.* (1975) used the second method to measure the flow velocity of red cells in capillaries of live animals.

In the analog mode, a video microscope can be used as a simple microphotometer (or in principle as a microspectrophotometer). However, the nonlinear light transfer characteristics (if the gamma is not 1.0), the limited dynamic range (often less than 100 : 1), the sensitometric characteristics, and the inhomogeneity in the sensitivity of the camera tube can limit the precision of *intensity measurements*. Also, the S/N ratio of the camera output and especially of the tape-recorded signal are limited.

Nevertheless, within these constraints, analog video devices can provide simple and convenient means for measuring approximate image intensities, especially for comparing relative intensities within a scene (Fig. 9-13; see also Sections 11.2–11.4). So long as the light level does not change appreciably, we can make intrascene, and to a limited extent interscene, comparisons of intensities even without disabling all of the automatic controls on a video camera.

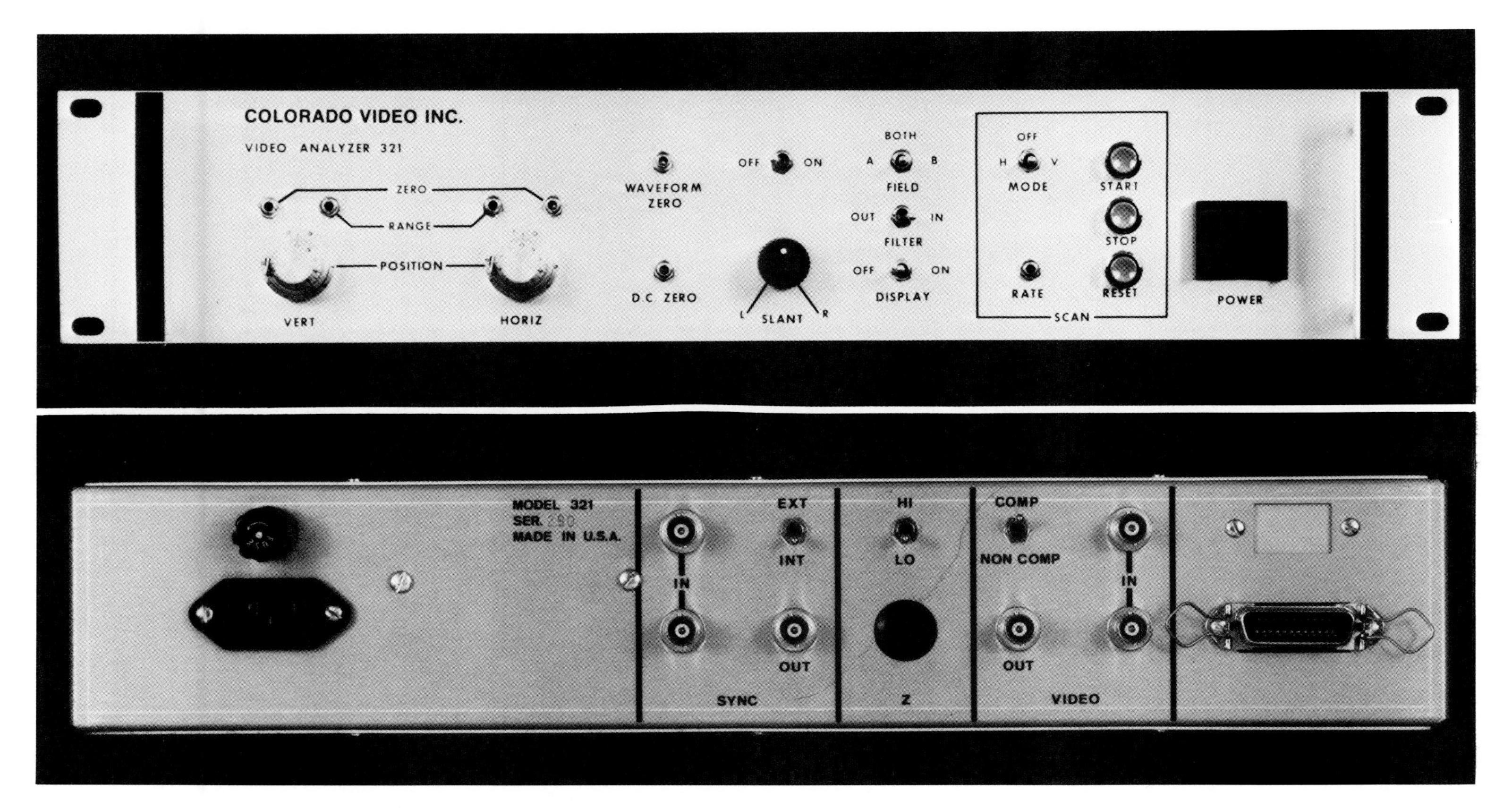

FIGURE 9-10. Video analyzer. Front and rear views of a particularly convenient video analyzer (Colorado Video Model 321). See Section 9.6 for a description of its capabilities. (Courtesy of Colorado Video.)

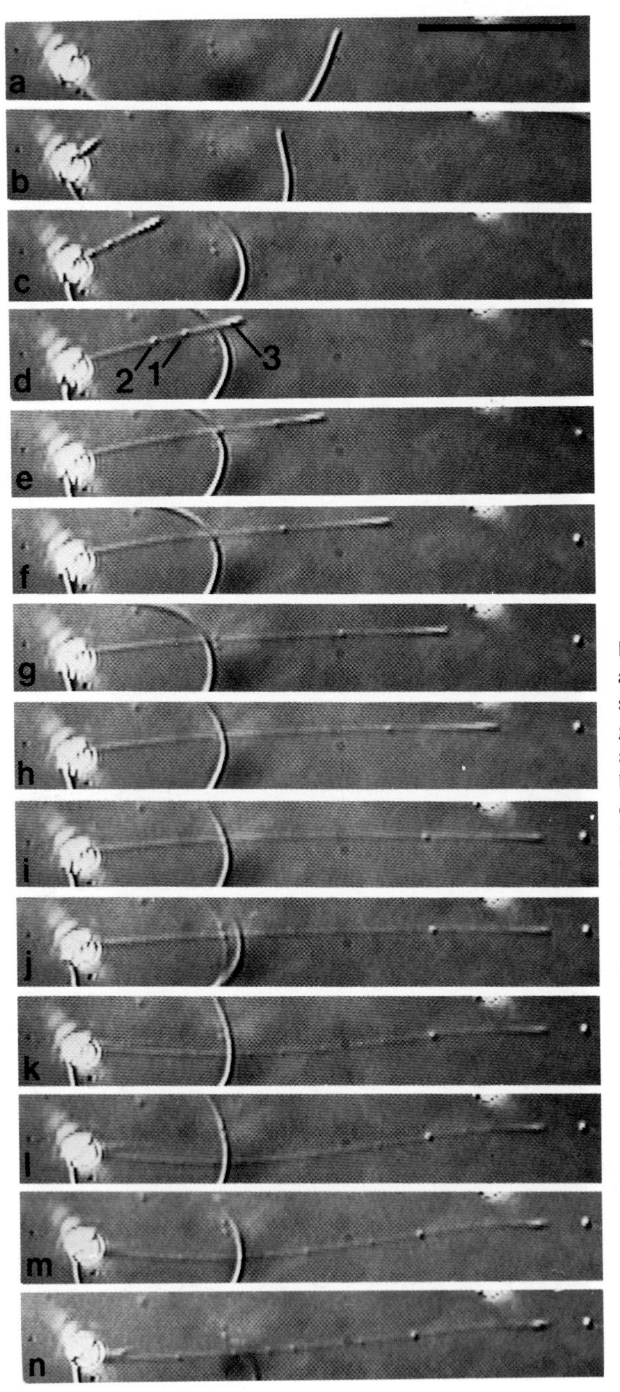

FIGURE 9-11. Elongation of the acrosomal process of a sea cucumber sperm. Stages in the course of elongation of the acrosomal process are shown at intervals of 0.75 sec (top to bottom). In b, the process is just emerging from the bright sperm head. In d, blebs, labeled 1, 2, and 3, can be seen on the process. The prominent curved structure near the center of each picture is the tip of the tail of the same sperm. Rapid changes such as this elongation, can be measured accurately in successive frozen fields (1/60 sec), using the video analyzer shown in Fig. 9-10. The jagged outlines (**jaggies**) of the acrosomal process in frames b and c are artifacts introduced to diagonal outlines in the picture by the sparse scan lines in the frozen video field. Detection of the very thin acrosomal process (65 nm in diameter) was made possible by combining optical contrast enhancement (Smith DIC, in this example) and analog video processing. (From Tilney and Inoué, 1982.)

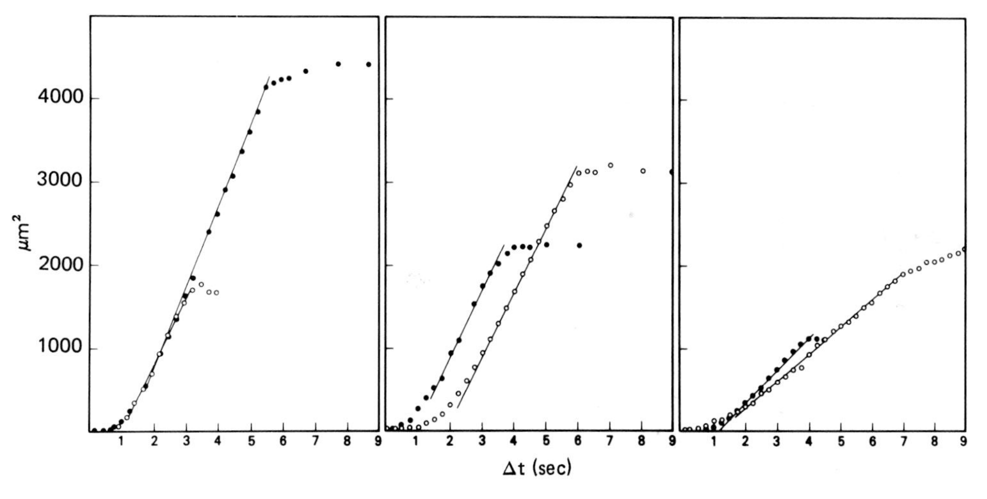

FIGURE 9-12. Time course of acrosomal elongation. The panels are plots of length squared versus time for acrosomal process elongation. The solid circles in the left panel are data points, for the sperm shown in Fig. 9-11, measured with a video analyzer. The other curves are for other sperm. Measurements like these can be made accurately and rapidly, directly from the tape record. These curves measure the very rapid growth kinetics of the actin filaments that extend the acrosomal process. The experiments confirmed the theoretical prediction that length squared increases linearly with time in this diffusion-limited growth process. (From Tilney and Inoué, 1982.)

With the automatic controls disabled, and by reference to an internal optical calibration,* interscene changes in intensities (and hence changes in light absorption, birefringence, fluorescence, etc.) could be measured to a precision of about 3–7%. Naturally, the camera tube cannot be saturated and the light source must be sufficiently stable. Hiramoto *et al.* (1981, 1982) and Salmon *et al.* (1984a) have used an analog video analyzer to measure the distribution of birefringence retardation and its rapid change in the mitotic spindle of sea urchin eggs injected with the antimitotic drug Colcemid (see Fig. 9-14).

By using multiple wavelengths of light, one can overcome a number of limitations inherent in photometry at a single wavelength. Thus, using three closely spaced wavelengths of green light, Pittman and Duling (1975) were able to determine the percentage of oxyhemoglobin in flowing blood in live animals, *independent of vessel diameter,* ranging from 15 to 100 μm. These authors used the analog image analyzer shown in Fig. 9-10 to display the transmission of the vessel together with its video image and measure the percentage of hemoglobin oxidized to a precision of 2–4%.

This excellent paper and a comprehensive review by Spring (1979) are recommended for further information on video quantitation of physiological events. General discussion of microphotometry and morphometry, feature extraction, etc. are beyond the scope of this chapter. The reader is referred to the sources just listed and the two chapters following for further discussion and citations to these subjects.

*For example, see Inoué (1951) and Inoué and Sato (1966) for internally calibrated methods for measuring birefringence with a polarizing microscope. These methods allow precise measurements of birefringence in minute specimen regions, using sensors with limited capacity for absolute intensity measurements (Appendix III). Fluorescence can be calibrated internally by taking advantage of the taper of a glass microneedle filled with fluorescent dye, such as is illustrated in Fig. 9-13. In finite-fringe *interference* microscopy, the fringe displacement, rather than the intensity of the fringes, provides the optical path difference for the monochromatic wavelength used. An analog video analyzer can provide precise measurements of fringe locations, and hence of optical path differences and dry mass of minute specimen regions.

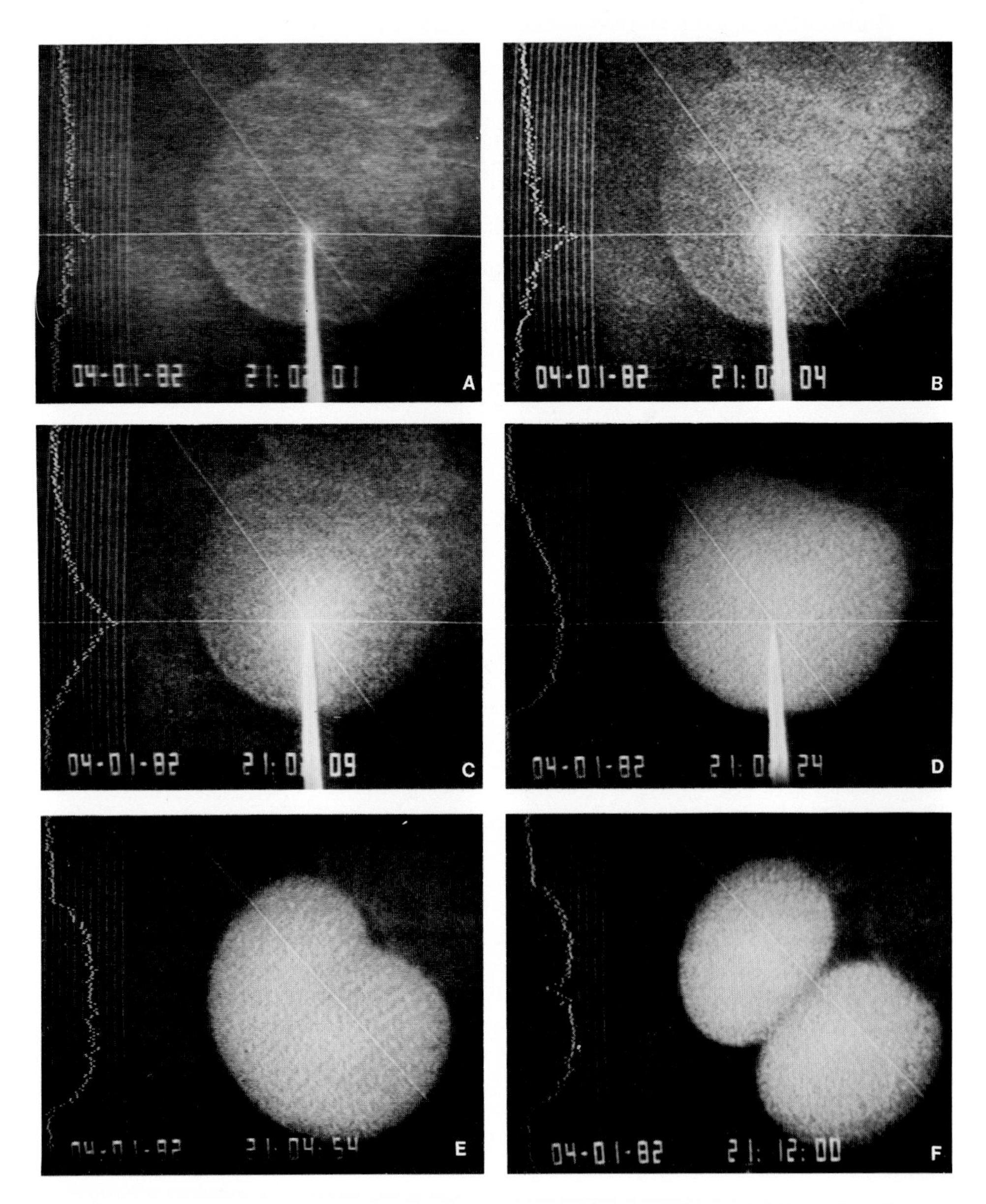

FIGURE 9-13. Diffusion of fluorescent dye microinjected into sea urchin egg. A through F are monitor pictures and intensity scans, along slanted fiducial lines, of a two-cell-stage sea urchin embryo. The process of diffusion of the fluorescent dye, lucifer yellow, is imaged with fluorescent microscopy onto an SIT camera. (A–C) Diffusion of the dye away from the tip of the needle during iontophoretic injection at a constant rate. In D, a few seconds after injection stopped, the dye has diffused uniformly throughout the cell. In E and F, the injected cell is dividing synchronously with its uninjected sister, barely visible at the upper right. Note the change in shapes of the intensity plots in panels D–F (which reflect the shape of the egg cell) compared to panels A–C (which show the steady-state diffusion pattern of the dye in the egg cytoplasm). The sensitivity of the SIT camera allows the use of levels of fluorescent excitation (for dyes such as lucifer yellow) low enough not to damage the cell. The level of fluorescence was too low to be seen through the ocular with a dark-adapted eye. (Inoué, Woodruff, and Lutz, unpublished.)

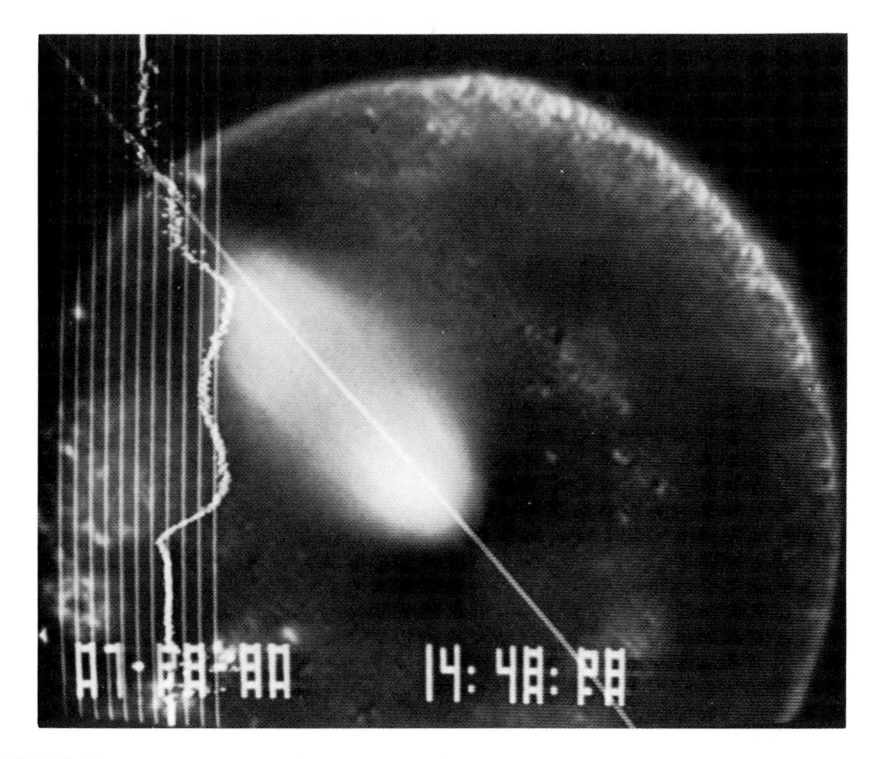

FIGURE 9-14. Real-time display of distribution of birefringence retardation. This monitor picture shows a diagonal line scan displaying the distribution of birefringence retardation (ca. 4 nm) in the meiotic spindle of a living *Chaetopterus* egg. The line scan generated by the analog video analyzer (Colorado Video Model 321) allows one to follow the changing distribution of the weak birefringence in real time. (See also Hiramoto *et al.*, 1981, 1982; Salmon *et al.*, 1984a.) V and H scans can also be output as voltage signals in the slow scan mode for transferring these signals, for example, to a strip chart recorder.

9.6. COMMERCIAL ANALOG PROCESSORS AND ANALYZERS

At every stage of the video train, the equipment processes the signal in one way or another. Much of the processing circuits are factory adjusted, but the camera control box and the monitor, in particular, provide considerable latitude for adjusting the video signal and picture. As discussed above, we can also add a separate processing device. In addition to the applications already discussed, we may need to optimize the picture signal and pedestal levels when we transfer video microscope records from one tape recorder to another (Chapter 12).

Adjustable *analog processors* may reside in video mixers, but are also available commercially as a separate device. In Figs. 9-4 and 9-5 we have already seen how the image contrast can be adjusted with such an analog video processor. Several types of *analog video analyzers,* including video micrometers, motion analyzers, and waveform analyzers, are also available commercially (e.g., from Colorado Video and For-A). The photographs in Figs. 9-4, 9-5, 9-8, 9-13, and 9-14 illustrate the fiducial lines and line scan display (**waveform monitor**) that are simultaneously introduced by an especially convenient commercial video analyzer (Fig. 9-10). This analyzer provides the *X* and *Y* coordinates for the fiducial lines and the intensity (*Z*) of the signal at the *X*–*Y* intercept, each as a 0 to 5.0 DC voltage output. The ''*Y*'' axis can be tilted as in Figs. 9-13 and 9-14. Also, the *X* and the vertical *Y* lines can be made to scan and sweep the picture at adjustable, preset rates. The *Z* output then provides a slow-scan output of the sequential *Y* or *X* lines. These outputs can be recorded on a standard chart recorder (e.g., Intaglietta, 1970), or introduced into a digital computer through an A/D convertor.

These commercial, analog video processors and analyzers can be hooked up to the video circuit very simply (e.g., as in Fig. 9-9); they enhance the image quality and expand the measuring capability of the video microscope quite economically and effectively.

10

Digital Image Processing and Analysis*

ROBERT J. WALTER, JR., and MICHAEL W. BERNS

10.1. INTRODUCTION

New discoveries in the life sciences are often linked to the development of unique optical tools that allow experimental material to be examined in new ways. We as microscopists are constantly searching for new techniques for extracting even more optical information from the material we work with, as the subject of this book aptly demonstrates. It is not surprising then that microscopists have begun to turn to computer technology in order to squeeze more information from their experimental images. Computer processing can be used to obtain numerical information from the microscope image that is more accurate, less time-consuming, and more reproducible than the same operations performed by other methods. Computer processing can be used to enhance the appearance of the microscope image, for example to increase contrast or to reduce noise, in ways that are difficult to duplicate using photographic or video techniques alone. When used to their fullest power, digital image-processing techniques can be used to produce information about the optical image that cannot be obtained in any other way.

The first step in digital image processing is the conversion of the optical image into a form that can be stored in computer memory. This conversion is performed by a light-sensitive system known as an **optical digitizer,** which produces coded numbers that are a measure of light intensity. This system (which in our case incorporates a video camera) samples the optical image over regular intervals, and the electronic output is converted into a string of discrete numerical values† that represent the distribution of intensity in the optical image. This process is called *digitizing,* and the stored numerical representation of the original optical image is called a *digital image.* The digital image is a numerical abstraction that can be manipulated like any other data stored in computer memory.

There are an unlimited number of manipulations that can be performed on a digital image; however, all such manipulations can be placed in one of two general categories: **image analysis** or *image enhancement.* The distinction between these two classes is important, as each type of manipulation may have widely different requirements in terms of the necessary processing equipment and in terms of the expertise required of the user. In addition, many commercially available image-processing systems are designed primarily to perform only one of these two classes of image

*Since Walter and Berns completed this chapter, several commercial systems for digital processing have become available. I have added Appendix IV to reflect these developments. S.I.

†The values are usually digitized in binary values, or expressed as a series of 0's and 1's.

ROBERT J. WALTER, JR., and MICHAEL W. BERNS • Laser Microbeam Program (LAMP), Beckman Laser Institute, University of California, Irvine, California 92717

manipulation; consequently, careful consideration must be given to the requirements of each processing task if optimal use is to made of the available equipment.

The term *image analysis* refers to computer routines that produce descriptive information about the digital image. Typical applications might be the determination of the number of cells in a microscopic field of view or the measurement of the nuclear diameter of a particular cell. This type of numerical analysis is called *feature extraction,* and every quantity that can be used to describe the digital image is called an *image feature*. In the broadest sense, an image feature is any attribute that can be used to describe the digital image, such as the shape of cell boundaries, the distribution of cells in a mixed population, and so on. In addition to feature extraction, routines involving image analysis may also be required to produce "yes or no" type of information about the image being analyzed, such as whether the field contains an interesting cell that should be analyzed further, or whether the field should be rejected and another field of view chosen.

Image-processing routines that perform image analysis can vary greatly in their complexity, and the requirements for equipment differ accordingly. The simplest routines for image analysis are designed to perform tasks that could otherwise be performed by hand, such as counting the number of bacterial colonies on a culture plate, with the advantage of the computer being one of providing greater speed and accuracy. A typical system of this type might consist of only a camera lucida, **digitizing tablet,** and small minicomputer, or alternatively, a video camera, video monitor, and minicomputer. Such a system could be assembled for a few thousand dollars and could be used immediately by untrained personnel. These systems have minimal requirements for computer memory, and because the processing tasks are fairly simple, processing speed is not of critical importance. Commercial systems of this type usually come preprogrammed to perform a few specific tasks, such as measuring areas or preparing histograms; however, they are usually not easy to modify or reprogram for other specific needs.

At the other extreme, routines involving image analysis can become quite complex in terms of both the necessary equipment and the expertise required of the user. An example of a routine involving complex image analysis is one designed to automatically recognize and classify abnormal cells in a histological smear. Such a routine might involve the rapid measurement of many different image features, such as the size and shape of individual cells, their staining densities, number of nearest neighbors, and so on. These data from individual cells might then be compared to data from other cells known to be normal or abnormal by other criteria, resulting in the generation of a probability that each cell being analyzed could be assigned to a normal or an abnormal class. Systems designed for this kind of complex image analysis require at least a high-quality video camera and digitizer, peripheral image storage devices, and a large special-purpose computer processor capable of storing and quickly manipulating large numerical arrays. In addition, many such systems also include special hardware devices that perform image manipulations at a much faster rate than can be performed by programmable general-purpose computers. Complex image-analyzing systems must also be capable of performing complex digital filtering routines on an image (discussed in Section 10.1, 10.12) that are extremely difficult to perform without the aid of a computer. In addition to the high capital cost of such systems, there are additional expenses involved for maintenance, software development, and training of personnel that could easily push the ultimate cost to several hundred thousand dollars.

Many of the routines that perform complex image analysis are part of an area of computer research called *pattern recognition,* which concerns using the computer to mimic, or even improve upon, the cognitive processes a trained observer uses when analyzing complex images. Pattern recognition is one of the more active areas of image-processing research; consequently, most of the state-of-the-art systems that perform this type of analysis are found in large research centers and have been uniquely constructed and programmed for a particular task. The unfortunate consequence of this situation is that an investigator who wishes to pursue a new application of this type is faced with the prospect of building his or her own system from component parts (discussed in Section 10.14), and spending considerable time developing the necessary software. Alternatively, there are some commercially available systems that have been developed for pattern recognition applications; however, most have been designed for the classification of blood smears and are not easy to modify for other purposes.

The second category of routines for manipulating images are those that involve *image enhancement.* Routines for image enhancement differ from those for image analysis in that the object of the processing is not to produce descriptive information about the image, but rather to manipulate the image in such a way that it is more useful to a human observer. Examples of routines that perform image enhancement are: the subtraction of two images captured at different times in order to detect motion (Section 10.8); the substitution of shades of gray in an image with different colors in order to exaggerate contrast; and the use of digital filtering routines to exaggerate cell boundaries (Section 10.11, 10-13). Image enhancement routines can also be used to compensate for degradation of the original image caused by defects or physical limitations in the imaging system, such as aberrations in the optical system or nonuniform response to light of the video system. Since the digital image is an abstraction in the memory of the computer, it can be manipulated in ways that cannot be duplicated by a physical device such as a lens, prism, or filter. Some image enhancement routines involve manipulating the digital image with the mathematical equivalent of a complete optical system that has been carefully designed to correct image degradation. Systems designed for these applications can probably be best described as hybrids between image analysis and image enhancement systems and thus have their own special requirements (Section 10.12).

Many of the processing routines used for image enhancement are also used for image analysis. However, systems designed primarily for image enhancement have requirements in addition to those of analysis systems, in that they must also be flexible and easy to use, i.e., highly **interactive.** An image enhancement system is the type most likely to be used by a researcher who uses a microscope as a daily research tool, and whose processing requirements can change rapidly with time. In this environment, the system must be flexible enough to allow the user to try many different processing routines and observe the results in as short a time as possible. These systems usually have the capability of performing processing in real time, i.e., the digitally processed image is produced as quickly as it can be displayed on the video monitor at video rates of 30 frames/sec. Systems designed primarily for image enhancement, which may not require the speed and extensive memory of a more complex system, are intermediate in price and can be purchased from around $20,000 to more than $100,000, depending on the properties required of the host computer (Section 10.2).

It is apparent, then, that different types of image-processing applications will have different requirements for processing equipment. For the purposes of this discussion, image-processing applications have been grouped into three categories: simple image analysis; complex image analysis involving pattern recognition; and image enhancement. These groupings are entirely arbitrary; in reality, most image-processing systems will be able to perform all three types of processing to a greater or lesser degree. The groupings are useful, however, when considering hardware requirements of different processing applications; and most commercially available image-processing systems fall into one of these three processing categories.

It is our purpose here to present an introduction to the capabilities of image-processing techniques as they apply to video microscopy. We will describe some basic image-processing routines, along with their mathematical descriptions, that are commonly used for image enhancement and analysis. Our discussion is meant to be an introduction only, with reference to source material that provides a more in-depth treatment. We have also compiled a representative listing of the different types of commercially available image-processing equipment so that their features can be compared. It is our hope that this discussion will demonstrate the benefits and power that image-processing techniques bring to microscopy.

10.2. DESIGN OF AN IMAGE-PROCESSING SYSTEM

An image-processing system can be separated into several individual components, each with its own unique function, as demonstrated by the block diagram of an idealized system shown in Fig. 10-1.

The first component utilized in this system is the *optical digitizer,* the light-sensitive device that converts light intensity into a string of digital values. In video, the optical digitizer consists of

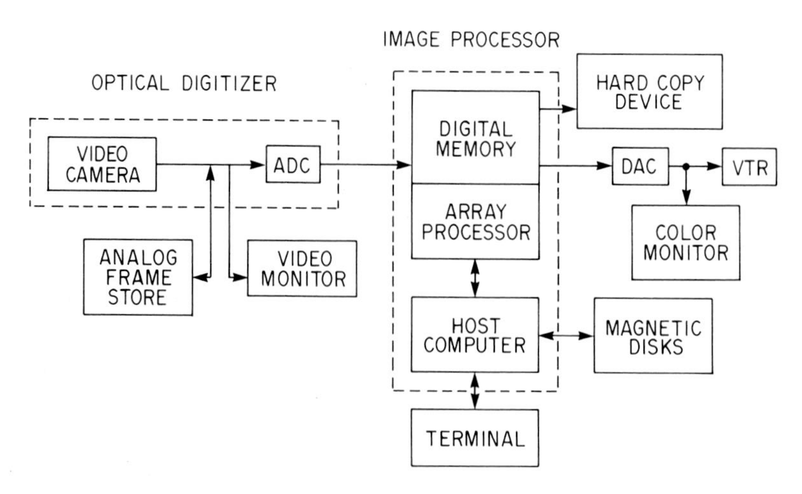

FIGURE 10-1. Block diagram of an idealized image-processing system. ADC: analog-to-digital convertor; DAC: digital-to-analog convertor.

the video camera and a separate electronic digitizer that converts the analog video signal into computer-coded digital information. This electronic digitizer is called an *analog-to-digital converter* (ADC). The ADC may be a separate device or an integral component of the video camera or the image processor itself.

The digitizer samples the video signal at regular intervals along each video scan line, producing an array of digital values that map the optical image into computer memory as shown in Fig. 10-2. Each discrete value in the digital image represents the average light intensity measured over the sampling interval of the digitizer, and is known as a *picture element,* or *pixel.* The pixel is the smallest integral part of the digital image. The properties of the optical digitizer determine how accurately the digital image represents the original optical image; it is a critical component in an image-processing system, and will be discussed in greater detail in Section 10.3.

The next component in the system of Fig. 10-1 is the *image processor,* the device that actually stores and manipulates digital images. Although the makeup of image processors varies from system to system, most contain three distinct components: a *digital image memory,* an **array processor,** and a *host computer,* each of which will be discussed in turn.

A digital image represents a very large amount of computer information that usually must be manipulated all at once. A typical digital image might contain 512 rows of 512 pixels each, where each pixel requires 8 bits (1 byte) or more of computer memory. This corresponds to 262,144 bytes, which is larger than the available memory capacity in most laboratory computers. In addition, since it is often necessary to store several images in memory at one time so that they can be compared, the needed memory can be several times greater. Consequently, most image-processing systems have two distinct digital memory devices: a *digital image memory* used to store one or more digital images, and a *host computer* used to store programs and other data.

When the host computer cannot store an entire digital image in its own memory, it must transfer small blocks of image memory back and forth between the storage device and its own

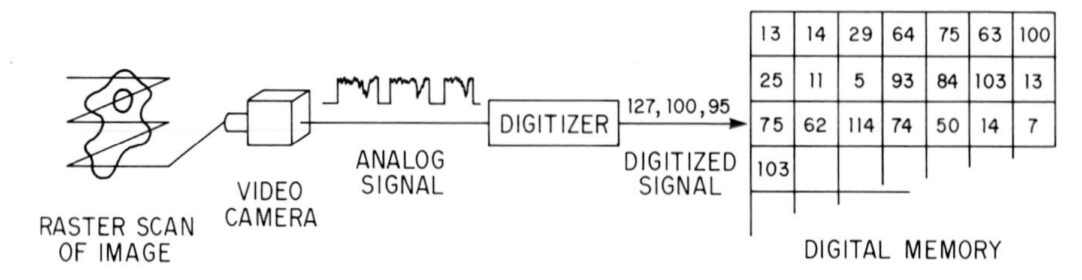

FIGURE 10-2. A video camera used as an optical digitizer.

memory, in order to perform processing routines. This process can take as long as several minutes if the image is large or if the host computer is as (relatively) slow as most microprocessors. The delay caused by the limitations of the host computer can be tedious during complicated processing routines and is intolerable when processing is required in real time.

Much faster processing times can be realized by taking advantage of special-purpose computers known as *array processors,* which can manipulate very large numerical arrays. Array processors can very quickly perform limited types of arithmetic or logical operations on data in their memory, but they do not have the programming versatility or the special functions found in a typical host computer. When using the array processor to manipulate digital images, the host computer only controls the operation of the array processor and does not manipulate image memory directly. Used in this way, a typical array processor can manipulate image memory at very high speeds, often at video input rates of 30 frames/sec. When manipulations beyond the capability of the array processor are required, image memory must be transferred to the host computer and its slower processing speed used.

Usually, the array processor and the digital image memory it manipulates are combined in one device, along with additional memory areas that store **look-up tables** and other data used in manipulating the digital images, as discussed in later sections. Physical devices that are used to store digital images, but which cannot manipulate them, are different from array processors and are known as *digital frame buffers.*

Once a processed image has been produced, it can be routed to several locations for display and storage as shown in Fig. 10-1. The digital image can be reconverted to an analog video format by using a **digital-to-analog convertor** (DAC), and the video image can be displayed on a high-resolution monitor or permanently stored using a video tape recorder (VTR). The video monitor must have display qualities that are equal to or better than the spatial and gray scale resolution of the optical digitizer, if the system is to be put to optimum use.

Images can also be stored in digital form on magnetic disks; however, since a single (512 pixel × 512 line × 8-bit) image contains more than 0.25 megabyte of data, the storage capacity of currently available magnetic disks limits their usefulness for this purpose. Floppy disks can store on the order of 1 megabyte, and hard disks up to several tens of megabytes. Yet the degradation of image quality caused by video tape recording and replay (Chapter 8) can be serious enough for digital processing purposes, that storing images on magnetic disks becomes worthwhile. The cost of magnetic disks compared to videotape must also be considered; currently, a floppy disks costs about $3–$5 and a 10-megabyte hard disk, $150, as compared to about $25 for a 1-hr, ¾-inch videotape cassette, which can store thousands of images.

Photographic copies of digitally processed images can be made directly off of the display monitor using the techniques described in Chapter 12. Alternatively, there are several commercial hard-copy devices available that can make monochrome or color prints of digital images directly from the coded digital information or from the video signal coming from the DAC. These devices include plotters that use multicolor pens to make high-resolution point plots of the digital image; and special small-format high-resolution video monitors, which are photographed. The cost of these devices varies from a low of around $8000 for a point plotter to up to $100,000 for a laser printer.*

In addition to the devices for manipulating and displaying the digital image, Fig. 10-1 shows a video monitor and analog frame storage device between the video camera and the video digitizer. This extra video monitor is convenient for comparing the unprocessed video signal to the same image after computer processing. The analog frame storage device may also be required in situations in which the system is not capable of generating and/or receiving digitized images at video rates. In such situations, the frame storage device can hold a single video frame for the time necessary for digitization to be completed.

10.3. OPTICAL DIGITIZERS

An optical digitizer is any one of a number of devices that convert optical intensity into digital information. An optical digitizer generates a discrete value at each sampling location in the optical

*Recent developments have reduced these prices by at least an order of magnitude (November 1985).

image that is a measure of the average light intensity over that interval. This number is known as the *gray value* or *gray level* of the pixel at that location. Most digitizers are designed to produce binary-coded gray values, that is, the maximum range of gray values in the image is a power of 2. A 6-bit digitizer can produce 64 (= 2^6) different gray values (0 to 63), an 8-bit digitizer can produce 256 gray values (= 2^8), and so on. Several different kinds of optical digitizers have been used in image-processing applications, including rotating drum digitizers used to digitize images captured on film (Marcil, 1975; Sokoloff *et al.*, 1977), flying spot digitizers (Shack *et al.*, 1979; Shoemaker *et al.*, 1982), one-dimensional photodiode arrays (Kohen *et al.*, 1979), as well as video cameras. A detailed discussion of the properties of different optical digitizers can be found in *Digital Image Processing* (Castleman, 1979).

The accuracy of image-processing routines is ultimately determined by the faithfulness with which the optical digitizer can encode the original optical image. When discussing optical digitizers, it is often useful to compare their properties to those that would be expected of an imaginary ideal device. Although all of the many types of light-sensing devices used for digitizing optical images are less than ideal, video cameras probably have some of the least desirable properties. Fortunately, if the shortcomings of video digitizing systems can be characterized, they often can be corrected, with a resulting digital image that closely approximates that that would be expected from an ideal optical digitizing device. This ability to correct for irregularities in the video system, using techniques described in later sections, is just one example of the power of digital-processing techniques.

An ideal optical digitizer measures light intensity at each point of an image along a well-defined sampling grid with rectangular coordinates. The output gray value for each pixel is determined solely by the light intensity at the sampling point and is preferably a linear function of image brightness. If the gray value of an individual pixel in the digital image is known, this *linear transfer function* can be used directly to determine the image brightness that exists at the corresponding point in the original image. When using an ideal optical digitizer, an object found within the optical image will have the same digital form regardless of its position in the original field, except for the translocation. This object will have the same intensity, shape, and size no matter where it is found in the field, and its digital form will not be influenced by other features of the image that may be nearby.

The optical digitizers first used in image processing were scanning devices carefully chosen to have properties that approach those of an ideal device. These devices either physically move the optical sensor over each sampling point in the image or, conversely, illuminate the specimen in such a way that only one point of the image is visible to the sensor at any one time. The scanning

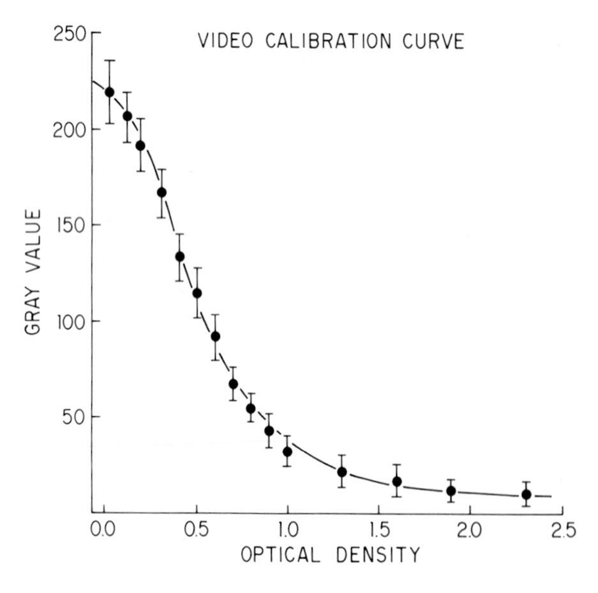

FIGURE 10-3. Response curve of a chalcogenide video camera (Model LST-1 Newvicon, Sierra Scientific) showing the relationship between input light intensity and output gray value for a uniformly illuminated field. Error bars reflect shading distortion produced by the video camera.

mechanisms of these devices are constructed in such a way that the sampling grids are rectangular and well defined. Since the same sensor is used to measure the light intensity at each sampling point, the digitized gray level is a function only of brightness, and is independent of position. Scanning devices typically employ photodiodes or photomultiplier tubes as light sensors that have a well-defined, linear response to light intensity. In some cases, the output voltage of these devices is put through a *logarithmic convertor* so that the output is a linear function of optical density in the original image. The major disadvantage of scanning devices is that they may take several minutes to digitize a single image because of the physical movement that is involved.

In comparison, the properties of video cameras are not so well behaved. For example, the curve in Fig. 10-3 shows that the relationship between input light intensity and output gray value of

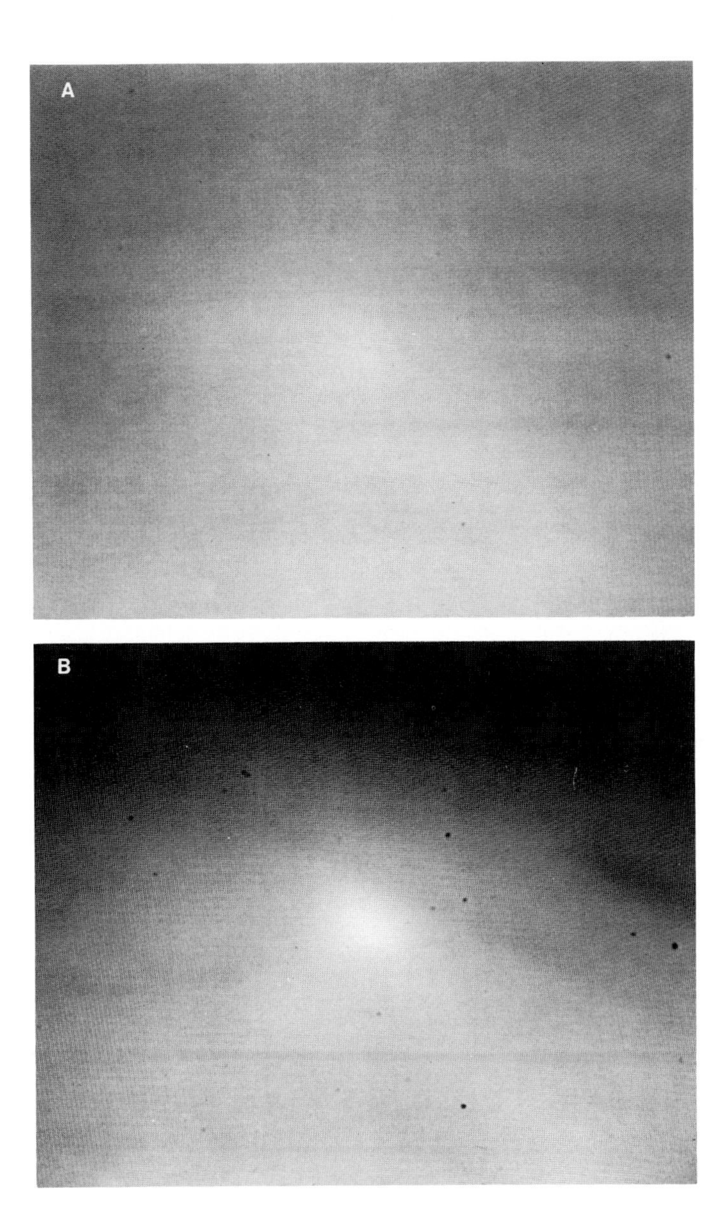

FIGURE 10-4. Uneven image of a uniformly illuminated field caused by shading distortion. (A) Original image. (B) Image after linear contrast enhancement of 3.5 times.

a typical video camera is not a simple function, but instead has a complex shape that is similar to the characteristic curves used to describe photographic film (Shore, 1975). High-contrast images typically have digitized gray values that extend well into one or both of the shoulders of the curve in Fig. 10-3. Digital images obtained with such transfer functions can seriously misrepresent the original optical image because the nonlinear nature of this curve has the effect of suppressing contrast at the highest and lowest gray values (discussed further in Section 10.7). The transfer curve in Fig. 10-3 is not easy to define mathematically, making it difficult to use the digitized gray value to determine brightness in the original image. Consequently, accurate statements concerning relative brightness of different features of the original image are difficult to make.

In contrast to scanning devices that use the same sensor to measure intensity at each sampling point in the optical image, video cameras produce output voltages that are determined by the intensity of illumination falling on different regions of the camera tube target. The relative sensitivity of the target varies from center to edge, principally due to limitations in the scanning properties of the internal electron beam. This variation in sensitivity is called *shading distortion* and is typically 10% or greater, unless corrected (Chapter 7). Figure 10-4 shows a blank video field of uniform illumination that exhibits center to edge shading distortion. Image contrast in Fig. 10-4b has been increased 3.5 times to accentuate this variation. For display purposes, this small amount of shading distortion may not be significant; however, for analytical purposes, it can be a problem because the relative brightness of an object within an image is not independent of its position within the field of view. In addition to this problem, shading distortion often makes the background gray level of the digital image appear uneven, even when the original illumination is uniform; this can disrupt processing routines that require a uniform background level. Different types of video cameras may have other problems, not found in an ideal digitizing device, such as bloom, lag, and geometrical distortion, as discussed in previous chapters.

In spite of these drawbacks, video cameras are often preferred in image-processing applications because of their convenience and speed, particularly for real-time processing. In applications where the quality of the video image is not of critical importance, such as particle counting, distortions in the video image can be ignored and the raw video image can be used unaltered. For other applications, the distortions in the video signal can be reduced to an acceptable level by selecting the type of video camera with properties most appropriate for the task, or by using analog techniques for correcting distortions in the video image as discussed in Chapter 9. Finally, video distortions can be corrected by digital processing techniques, as discussed in later sections.

The usefulness of video in image-processing applications has been spurred by the relatively recent introduction of electronic devices that are capable of digitizing an entire video impage with 8-bit resolution (or more) at video rates. When these devices, known as *video digitizers*, are not used, it is necessary to first capture the video frame on an analog storage device and then to digitize the stored image at rates slower than video rates. Alternatively, other devices can be used that digitize only a single video line or a single pixel from each video frame. A full digitized image is generated by digitizing different video lines or pixel locations from successive video frames until the entire image is captured. Use of these devices requires that the video image remain unchanged for the period of time it takes to produce the entire digital image.

10.4. RESOLUTION OF THE DIGITAL IMAGE

A digital image must be an accurate representation of the original optical image, up to the resolution requirements for each particular application. Resolution of the digital image can be classified into two categories; *gray level resolution*, which describes how accurately the digital image represents differences in intensity within the original image, and *spatial resolution*, which describes how well the digital image represents information about the size and position of features within the original image.

Gray level resolution is determined by the number of different gray values the digitizing device is capable of accurately reproducing, or the memory device is capable of storing, whichever is less. The number of different gray values that are necessary in the digital image is determined by the type of processing that is to be performed. For display applications where the objective is to accentuate subtle differences in image contrast, as many as 256 or more gray values may be required. For an application where the objective is to classify objects based on size alone, information about small differences in contrast may be unimportant, and only a few different gray values may be sufficient to represent the original image. In fact, for many analytical applications, one goal of digital processing is to *reduce* the number of different gray values contained in the image, for the purposes of increasing processing speed and reducing memory storage requirements. If, for a given application, all pertinent information in an 8-bit image can be expressed in four different gray values (a 2-bit image), then the memory storage requirements can be reduced by a factor of four, a considerable savings. Similar considerations apply to the minimum number of pixels needed to reproduce the original optical image in both the horizontal and vertical directions. As will be shown later, much image information can be expressed using only 1-bit gray values. These are called *binary images,* and contain pixels that can only be white or black. Many image-processing applications involve the manipulation of binary images; some image processors have special storage areas and processing routines that are designed specifically for manipulating binary images.

Resolution of gray values is limited by the noise component added to the digital image by the video camera and digitizing electronics. This noise component can be determined by digitizing a video field of uniform illumination and measuring the standard deviation of the digitized gray values about the mean (this assumes that shading distortion has been eliminated). The precision of the digitizing process is usually expressed as the number of bits required to store the standard deviation of the noise component. An 8-bit digitizer with a noise component of 4 gray values has 2 least-significant bits (LSBs) of noise and 6 significant bits of signal. Consequently, an 8-bit digitizer with 2 bits of noise produces no more information than a 6-bit digitizer with a negligible noise component.

Noise can sometimes be reduced by accumulating and averaging several digitized images of the same scene. This manipulation is effective if the digitized image can be modeled as the sum of a true signal and an independent noise function that is normally distributed and additive. Accumulating N frames of such an image will cause the signal component to increase by a factor of N, while the noise component will increase only by a factor of the square root of N (e.g., Castleman, 1979). The signal-to-noise (S/N) ratio in N accumulated images will be $N/\sqrt{N}$; consequently, the S/N ratio will increase approximately by a factor of $\sqrt{N}$ after the accumulation of N images. The S/N ratio can be increased further to a factor of $\sqrt{2N-1}$ if the accumulated image is averaged by dividing it by N (Gould *et al.*, 1981).

The effect of image averaging is shown in the series of photographs in Fig. 10-5. These photographs were made using a Venus LV-2 intensified low-light-level video camera that has a relatively high noise component. Figure 10-5a shows a single digitized image, while Fig. 10-5b through d show the effect of averaging 4, 16, and 64 consecutive video frames, respectively. The averaged image of Fig. 10-5d was produced in approximately 2 sec by using a real-time array-processing system.

When the purpose of digital processing is to create an image for display purposes, the number of different gray values in the digital image should be great enough that the steps between gray values are not visible to the human eye. This number is highly subjective, and is a function of such variables as the complexity of the image, the size of individual pixels, and the closeness of the viewer to the display screen. The human eye can distinguish only about 40–80 different shades of gray within the intensity range of the video monitor (Castleman, 1979); consequently, an image with 6- or 7-bit resolution (64 or 128 gray values) is sufficient for accurate image reproduction. However, 8 bits or more of resolution may be required in the original image if one of the objects of computer processing is to enhance contrast between gray levels in the original digital image, and if gray level steps are to be avoided in the enhanced image. The relationship between the number of gray levels and gray level stepping can be seen in Fig. 10-6, which shows the same image displayed at different resolutions from 6- to 1-bit. Note that gray level stepping becomes obvious first in the background regions where intensity is varying slowly across the image, compared to regions within the cell where intensity varies more rapidly.

The *spatial resolution* of the digital image is determined by the spacing between pixels, known

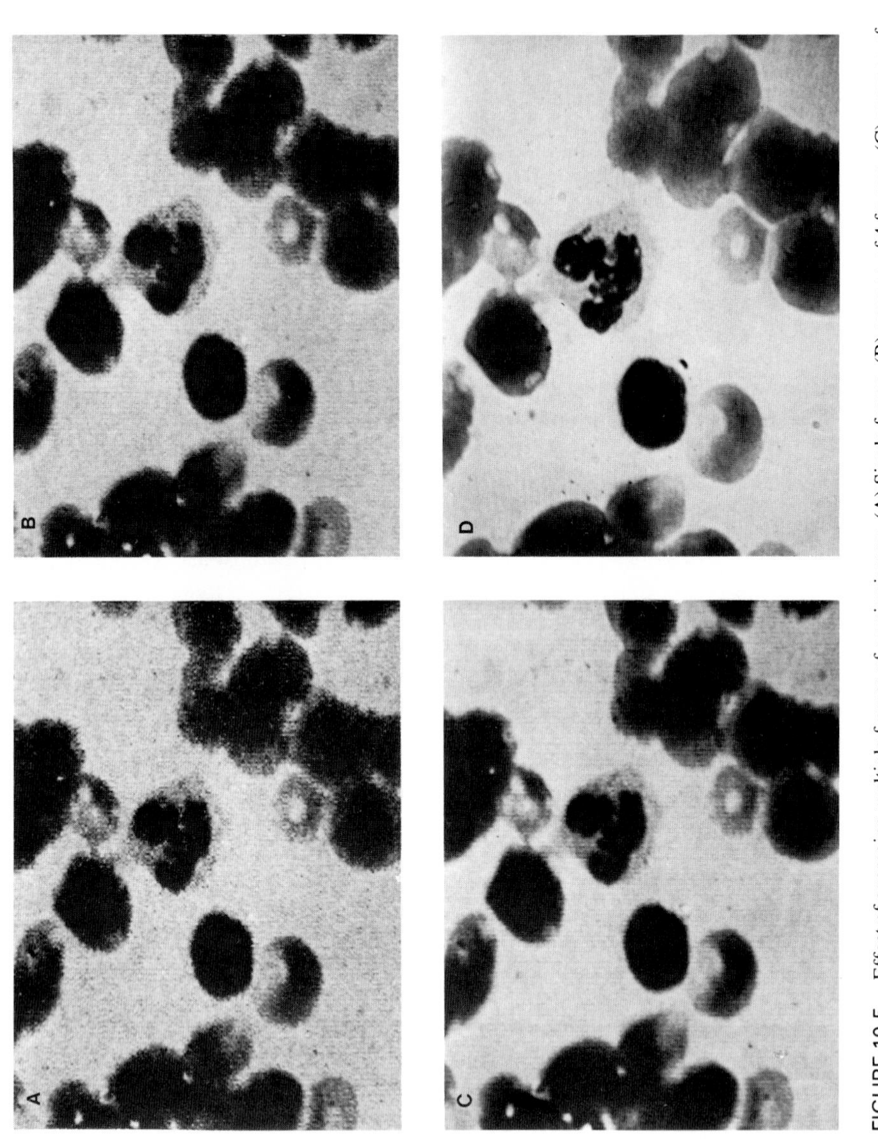

FIGURE 10-5. Effect of averaging multiple frames of a noisy image. (A) Single frame, (B) average of 4 frames, (C) average of 16 frames, (D) average of 64 frames.

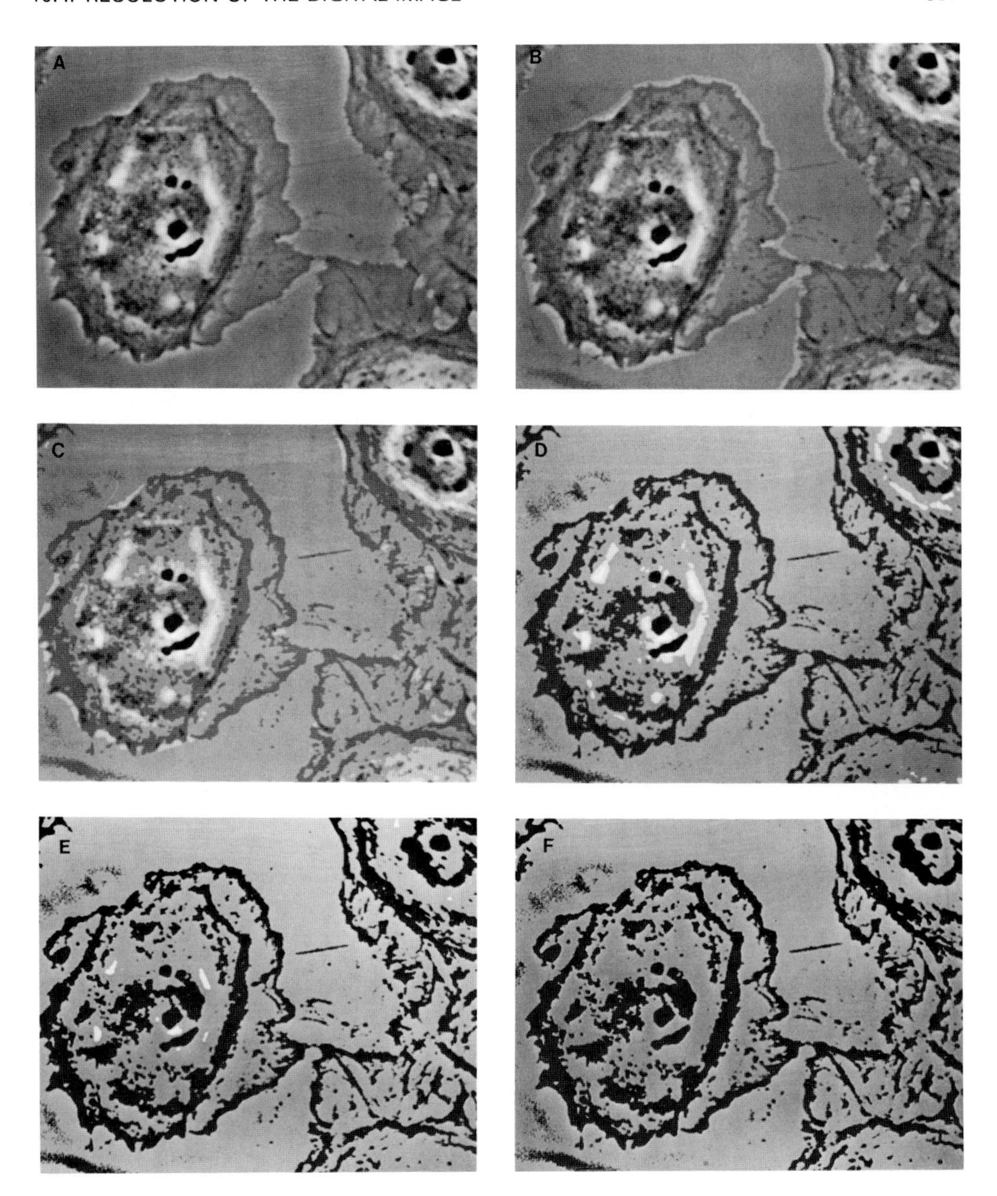

FIGURE 10-6. The effect of gray value resolution on image reproduction. (A) 6-bit image, (B) 5-bit image, (C) 4-bit image, (D) 3-bit image, (E) 2-bit image, (F) 1-bit image.

as the **sampling interval,** and the accuracy of the digitizing device. The gray value of each pixel in the digital image represents the *average* intensity of the optical image measured over the finite sampling interval; consequently, image features that are smaller than the sampling interval will not be represented accurately in the digital image. In order to preserve the spatial resolution of the original image, the digitizing device must use a sampling interval that is, at most, one-half the size of the smallest resolvable feature of the optical image (this is equivalent to sampling at twice the highest spatial frequency and is called the Nyquist criterion for accurately sampling an analog signal). If the smallest resolvable feature in the specimen is 1 μm, then the digitizer must sample at intervals that correspond to 0.5 μm or less in the magnified image; otherwise, the feature will be lost or distorted. When using video, the sampling interval is determined both by the magnification of the optical image onto the face of the video tube and by the specifications of the video digitizer. The vertical sampling interval will be the same as the interval between video scan lines, while the horizontal sampling interval will be determined by the design of the video digitizer and usually cannot be altered. Consequently, the spatial resolution of the digital image must be determined by the magnification at which the microscope image is projected onto the video tube.

Special mention must be made of the relationship between spatial resolution and the averaging technique discussed previously for reducing noise. Under ideal conditions, averaged images must be in perfect register, which means that corresponding locations in the summed images must be exactly superimposed on each other as the images are added. When this is not the case, the averaged image will tend to blur due to misalignment of sequential images. This random misalignment can be observed as a jittering of the video image during normal acquisition and is due to errors in the sync signal produced by the video electronics. We have found that this blurring effect can be quite severe when image averaging is attempted from images digitized from videotape playbacks. Under such circumstances, the resolution of the video image is not determined by the properties of the digitizer, but is limited by the blurring of accumulated images.

As mentioned previously, resolution requirements are likely to be different for every application, and may be reduced to some minimum level to save memory space and increase processing time. For many applications in microscopy, however, the resolution of the digital image must be the best possible and will ultimately be determined by the resolving power of the microscope. In discussing resolution of the digital image we have made the assumption that the microscope and video system will produce an exact reproduction of the original optical image and that the resolution of the digital image will be determined solely by the properties of the digitizing device. This is, of course, not the case, as both the microscope and video system will affect image reproduction, as discussed in Chapters 5 and 7. The type of image modification produced by the optical video system can be described mathematically by two relationships known as the *point-spread function* and the *modulation transfer function,* which are related to each other and are often used in optical design (Appendix II). A detailed description of these functions and their effect on image resolution is beyond the scope of this chapter; however, it should be noted that each of these functions can be estimated using image-processing techniques, and once known, can be used to compensate for image degradation caused by the optical video system (Pratt, 1978; Castleman, 1979). Further detail on these techniques can be found in Section 10.12.

10.5. THE IMAGE HISTOGRAM

One of the most useful functions derived from the digital image is the **gray level histogram.** This is a discrete function that specifies the number of pixels that appear in the image at each gray value. The relationship between an image and its histogram is shown in Fig. 10-7. Figure 10-7A is an image of a PTK_2 tissue culture cell taken with phase-contrast optics, and Fig. 10-7B the gray level histogram derived from this image.

Examination of the gray level histogram can yield important information about the image from which it was derived and show whether the optical digitizer is being used effectively. The histogram of Fig. 10-7B shows that the pixels in the image of Fig. 10-7A are only found over a narrow range of gray values (see **contrast range**), with a fairly even distribution around a peak value of 130. No pixels in the image have gray values near or equal to the digitizing limits of 0 and 255 (for the remainder of this chapter, we will assume that each image has 8-bit resolution with gray values from 0 to 255, where 0 is black and 255 is white). The narrow range of gray values found in this histogram is typical of a low-contrast image. There is a high probability that each pixel in the image will have neighboring pixels with gray values that are equal or nearly equal to its own. Contrast between neighboring pixels, defined as the difference between their respective gray values, will be low under these circumstances. The histogram of Fig. 10-7B also shows that there are many possible gray values that are not represented in the image. These unutilized gray values indicate that the full dynamic range of the digitizer is not being used and that a higher-contrast image might possibly be obtained by making adjustments to the optical system or video camera for a more optimal response. These compensating adjustments will only have a limited effect, however, if the

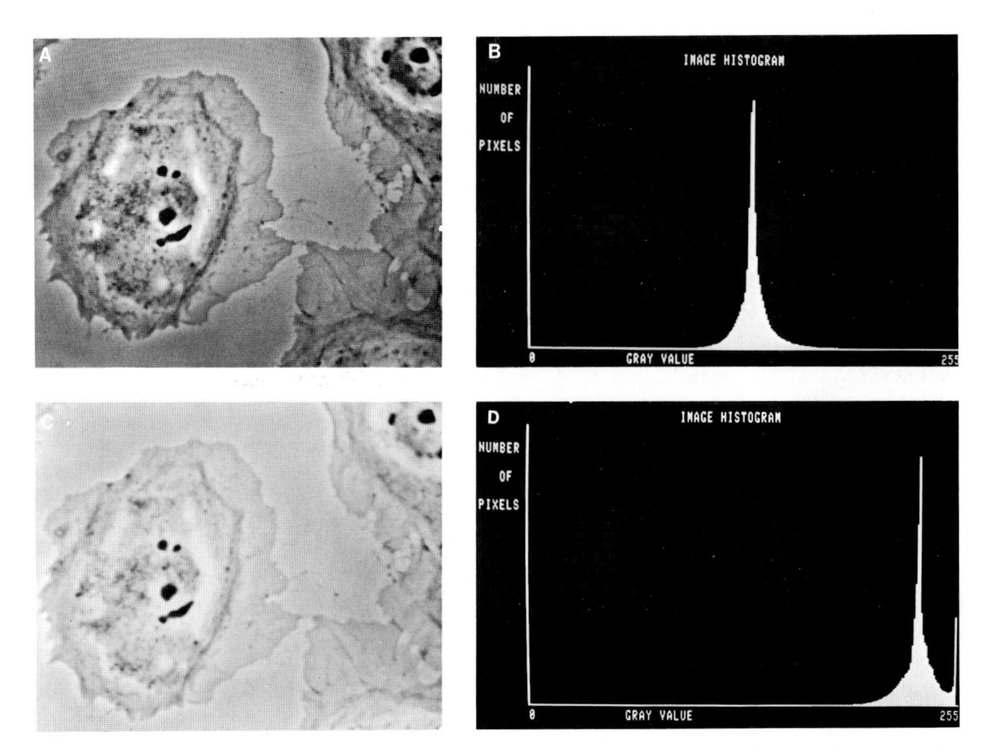

FIGURE 10-7. Examples of image histograms. (A) Image of a PtK-2 tissue culture cell. (B) Gray value histogram of A. (C) Image captured during saturation of the video signal. (D) Histogram of C showing clipping at the highest gray values.

original specimen cannot generate a high-contrast image, and a point will be reached where increasing the sensitivity of the system will be offset by a concomitant increase in image degradation caused by added noise. Experience has shown that the gray level histogram of Fig. 10-7B is typical of those taken from images using phase-contrast, DIC and polarizing optics, even under the best digitizing conditions.

The gray level histogram can also be used to show whether the dynamic range of the digitizing device has been exceeded during the formation of the digital image. Figure 10-7C shows another image of the cell in Fig. 10-7A taken under conditions where the brightest areas in the image have saturated the video signal. Figure 10-7D shows that there is a peak in the histogram of this image at the 255 gray level where the saturated video signal has been given the maximum gray value. This situation is termed *gray level clipping* and usually means that some detail has been lost in the digital image because areas of the original image that might have different intensities have each been assigned the same gray value. Clipping of the histogram may be acceptable in some circumstances if detail is lost only from unimportant parts of the image. Such a situation might occur, for example, if the system has been adjusted to maximize the contrast of a stained cytological smear under brightfield optics, with the clipping occurring in bright background regions where there are no cells. It is good practice in general, however, to avoid the possibility of clipping of the gray level histogram by adjusting the optical and video systems in such a way that no pixels at the upper and lower extremes of the gray level range are produced during the digitization process.

There is a unique gray level histogram for every digital image. For many applications, the histogram contains all of the useful information necessary to describe the digital image. For example, it may be desirable to determine the area occupied by a darkly stained cell in an image with a light background as shown in Fig. 10-8A. The customary approach to this problem might be to outline the boundary of the cell and to count the number of pixels that fall within that boundary, which can be a complicated procedure. A much simpler approach is evident after examining the

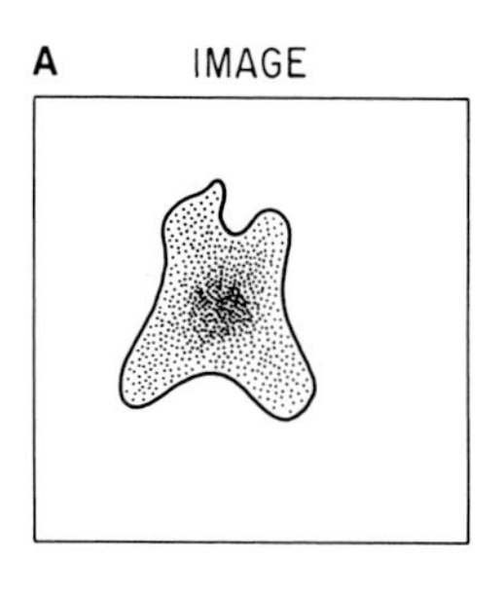

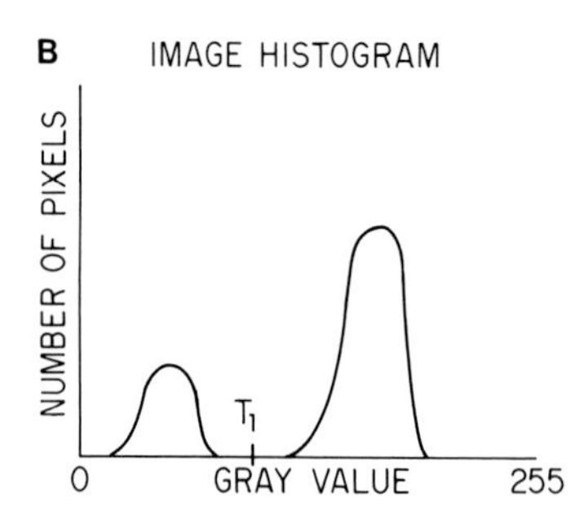

FIGURE 10-8. A bimodal histogram. (A) Idealized image of a darkly stained cell against a light background. (B) Histogram of A. The threshold level T_1 can be used to distinguish gray values of cell regions from background regions.

gray level histogram of this image as shown in Fig. 10-8B. This histogram shows two distinct peaks, the large peak on the right corresponding to the gray values of the background regions and the smaller peak on the left corresponding to the gray values of the darkly stained cell. The gray value of 100 marks a point, T_1, in the histogram above which only background pixels can be found, and below which only pixels from the cell image can be found. Given the reasonable assumption that there is no overlap between the gray values of the cell and of the background, the area of the cell is simply the number of pixels under the smaller peak of the histogram. Numerically, this is shown by the relationship

$$\text{Area} = \sum_{GV=0}^{100} H(GV) \tag{10-1}$$

where GV is the gray value of each pixel and H(GV) is the gray level histogram. Gray values used to separate different regions of a histogram, such as the value of 100 used to distinguish between gray values of cell and background areas in this example, are called *threshold gray values* and are used in many types of image manipulations (Weszka, 1978). A more general formula for finding areas between threshold boundaries of the histogram is

$$\text{Area}(T_1,T_2) = \sum_{GV=T_1}^{T_2} H(GV) \tag{10-2}$$

where T_1 and T_2 are the upper an lower thresholds that define the region of interest in the histogram. It should be noted that

$$\text{Area}(0,255) = \sum_{GV=0}^{255} H(GV) \tag{10-3}$$

is simply the number of pixels in the entire digital image.

Another useful value derived from the image histogram is the *integrated optical density* (IOD), which is defined by the relationship

$$\text{IOD} = \sum_{GV=0}^{255} H(GV) \times GV \tag{10-4}$$

This value is the weighted sum of the histogram, where each term in the histogram has been multiplied by the gray value it represents. The IOD represents the total image "brightness" and is often referred to as the "mass" of the image. When using threshold boundaries, the IOD is expressed

$$\mathrm{IOD}(T_1,T_2) = \sum_{GV=T_1}^{T_2} H(GV) \times GV \tag{10-5}$$

If the dark color in the image of Fig. 10-8A was due to some specific histological stain, then the IOD of this image (taken between the boundaries of 0 and 100 to distinguish cell from background) is a measure of the total amount of stained material. Precise quantitative measurements can be obtained using the IOD if the response of the digitizer has been calibrated, as discussed in the section on densitometry. A measure of the average density of areas within the image can be calculated from the histogram by utilizing both the area function and the IOD in the form

$$\text{average density}(T_1,T_2) = \frac{\text{mass}}{\text{area}} = \frac{\text{IOD}}{\text{area}} = \frac{\sum_{GV=T_1}^{T_2} H(GV) \times GV}{\sum_{GV=T_1}^{T_2} H(GV)} \tag{10-6}$$

The IOD and average density functions are frequently used to assess the quality of staining of histological smears for diagnostic purposes.

Two additional functions that are often used to describe digital images are the *probability density function* and the *area function*. The probability density function is simply the image histogram normalized to unit area,

$$\mathrm{pdf} = H(GV)/A \tag{10-7}$$

where A is the area of the digital image described by equation (10-3). The probability density function describes the probability of finding a particular gray value within an image. The cumulative area function, P(GV), is defined as

$$P(GV) = \sum_{k=0}^{GV} \mathrm{pdf}(k) \tag{10-8}$$

which is the normalized area function of equation (10-2), with the lower threshold set to zero.

When describing calculations based on digital images, it is necessary to use discrete forms of arithmetic as was done in the preceding examples. It is often more convenient, however, to model the digital image as a continuous function and to use mathematical forms that can be treated rigorously to describe manipulations of the image. For example, the continuous form of the cumulative area function is defined as

$$P(GV) = \int_0^{GV} \mathrm{pdf}(x)\, dx \tag{10-9}$$

where $p(x)$ is the continuous form of the probability density function. Continuous mathematics is used extensively in the literature of image processing, particularly in books and journals intended primarily for engineers or mathematicians, where it is often the case that the discrete form of an image manipulation will only be described after the manipulation has been justified or proved using a continuous model.

In the preceding discussion, methods have been described for deriving several kinds of useful information about the digital image from the gray value histogram. It should be noted that for each of these examples, all of the information was derived directly from the histogram without any direct reference to the image it represents. Derivation of the image histogram is a convenient way to reduce the complexity of the digital image when only these types of manipulations are required. When an image can be completely represented by its histogram, storage requirements and processing times can be greatly reduced. The gray value histogram is easy to derive; furthermore, several commercially available array processors can automatically calculate the image histogram in real time during the digitization process. Examination of the gray value histogram can also be a useful way to assess the effect of routines that affect the appearance of the image (such as routines for contrast enhancement) as discussed in the next section.

Some image-processing routines involve the comparison of images from the same specimen taken under different conditions. Such conditions might be the comparison of images of the same specimen taken at different times, images captured before and after treatment with a drug, or images taken while using different colored filters to distinguish the specimen's spectral qualities (such as using red and green filters to distinguish cells in a mixed population that have been specifically labeled with green fluorescein or red rhodamine dyes). The histograms of each of these images are then combined on a multidimensional plot and analyzed to provide new information about the specimen that is not available from any one image or image histogram. This technique is called *multispectral analysis* and is used frequently in image-processing applications (Castillo *et al.*, 1982; Garbay *et al.*, 1981; Preston and Dekker, 1980).

10.6. HISTOGRAM TRANSFORMATIONS AND CONTRAST ENHANCEMENT

Once an image has been converted to digital form, it can be manipulated within the memory of the computer in ways that affect image contrast and overall brightness. Such manipulations affect only the gray values of individual pixels and do not alter the spatial information of the digital image. Given an original digital image, I_1, such manipulations will produce a new image, I_2, as shown in Fig. 10-9. The function f(I) is called an **intensity transformation function (ITF)** and specifies the way the gray value of each pixel in I_1 will be modified to produce a gray value for the corresponding output pixel in the image I_2. This function takes the form

$$I_2(i,j) = f[I_1(i,j)] \tag{10-10}$$

where $I_1(i,j)$ is the input pixel at row i and column j in the original image, $I_2(i,j)$ is the corresponding output pixel in the modified image, and f(I) is the ITF. It should be noted that the output gray value of pixel $I_2(i,j)$ is dependent only on the input gray value of the corresponding pixel in I_1 and is unaffected by the properties of surrounding pixels. This type of manipulation is called a **point operation,** which means that the output gray value of the ITF is solely dependent on input gray value and that the spatial information of the image is irrelevant. Given this property, the specification for individual pixels can be dropped from equation (10-10) and the ITF can be expressed as

$$GV_2 = f(GV_1) \tag{10-11}$$

where GV_2 is the output gray value of image I_2 and GV_1 is the input gray value of image I_1.

A commonly used type of ITF is the linear function of the form

$$GV_2 = mGV_1 + b \tag{10-12}$$

When the slope m of this function is set to unity, a positive value for the intercept b will increase the gray value of each output pixel, which results in a brighter total image. Conversely, under the same conditions, a negative value for b will decrease the gray value of each pixel and produce a darker output image. This effect can visualized by examination of an image histogram before and after transformation by the ITF as shown in Fig. 10-10. When the slope is set to 1, a positive value for b shifts the image histogram to the right, while a negative value for b shifts the histogram to the left.

The effect of varying the slope of the linear ITF can also be visualized by examining the image histogram shown in Fig. 10-11. A slope greater than 1 will tend to increase the range of gray values

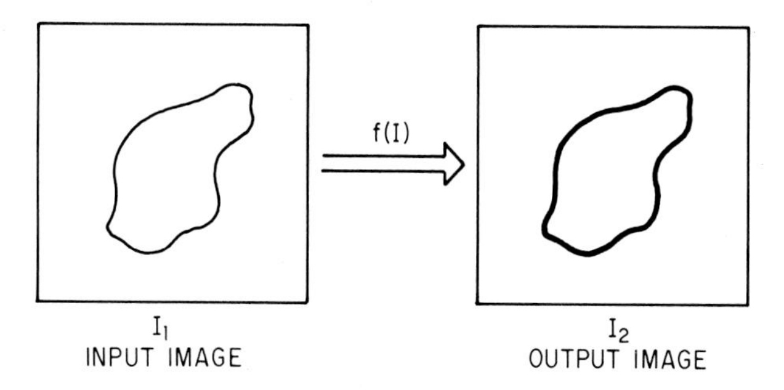

FIGURE 10-9. Histogram transformation. The gray value of each pixel in image I_1 is transformed by the function f(I) to produce image I_2.

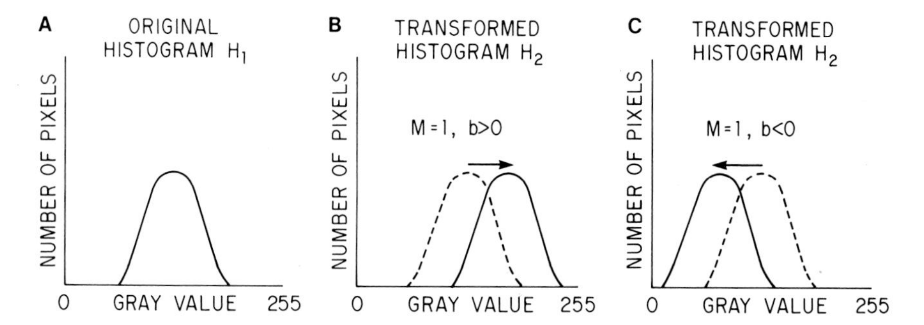

FIGURE 10-10. Changes in the gray level histogram resulting from variations in the intercept of a linear ITF. Positive intercept values shift the histogram to the right, negative intercept values shift the histogram to the left.

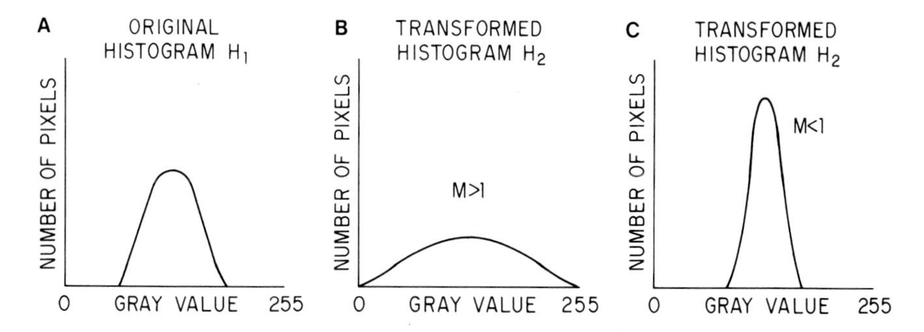

FIGURE 10-11. Changes in the gray value histogram resulting from variations in the slope of a linear ITF. Slope values greater than 1 broaden the histogram, slope values less than 1 narrow the histogram.

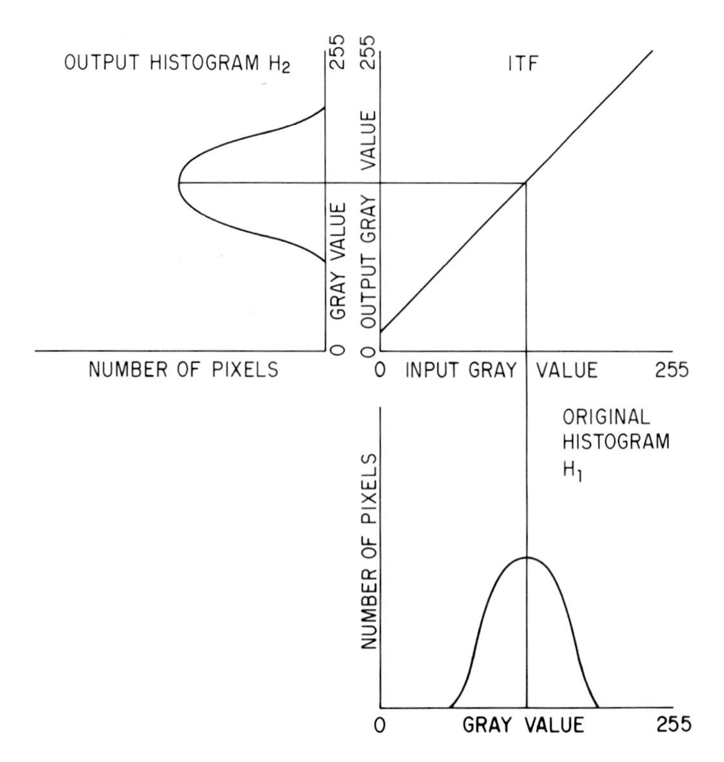

FIGURE 10-12. Relationship between the input histogram, ITF, and output histogram. Each gray value in H_1 is mapped by the ITF to a unique gray value in H_2. (After Castleman, 1979.)

in the output image, as seen by the broadening of the histogram, which tends to produce higher contrast between pixels in the output image. Likewise, a slope of less than 1 will produce a narrowing of the histogram and tend to decrease contrast in the output image.

A linear ITF with a slope of 1 and an intercept of 0 is the identity function that produces an output image identical to the input image. The relationship between the two images I_1, I_2 and the ITF is shown more fully in Fig. 10-12.

The slope of the linear ITF is known as the *contrast enhancement factor,* and describes the amount of contrast enhancement shown in the output image. This factor is analogous to the contrast-transferring characteristics of a video camera or photographic film. The amount of contrast enhancement desired in the output image I_2 can be controlled by fixing the value for m. If the average brightness of the input and output image is to remain constant after the contrast enhancement, then the value of b must be set according to the following equation:

$$b = (\text{mean gray value of } I_1) \times (m - 1) \tag{10-13}$$

In the examples presented thus far in this section, the ITF and image histograms have been represented as continuous functions, meaning that the gray values of the histograms H_1 and H_2 vary continuously between 0 and 255. Under these conditions, a given gray value from the histogram H_1 will map through the ITF to a gray value that exists in the output histogram as shown in Fig. 10-13, as long as the output gray values are between 0 and 255. In this continuous case, transformation of the image through the linear ITF will affect the absolute contrast between pixels but does not change the relative contrast, and the original image can be re-created by mapping the output image I_2 backwards through the ITF. When dealing with digital images, however, the gray values of the input and output images can only exist at discrete integer values; consequently, in the majority of cases a continuous ITF will map the integer values of H_1 to a nonexistent fractional gray value of

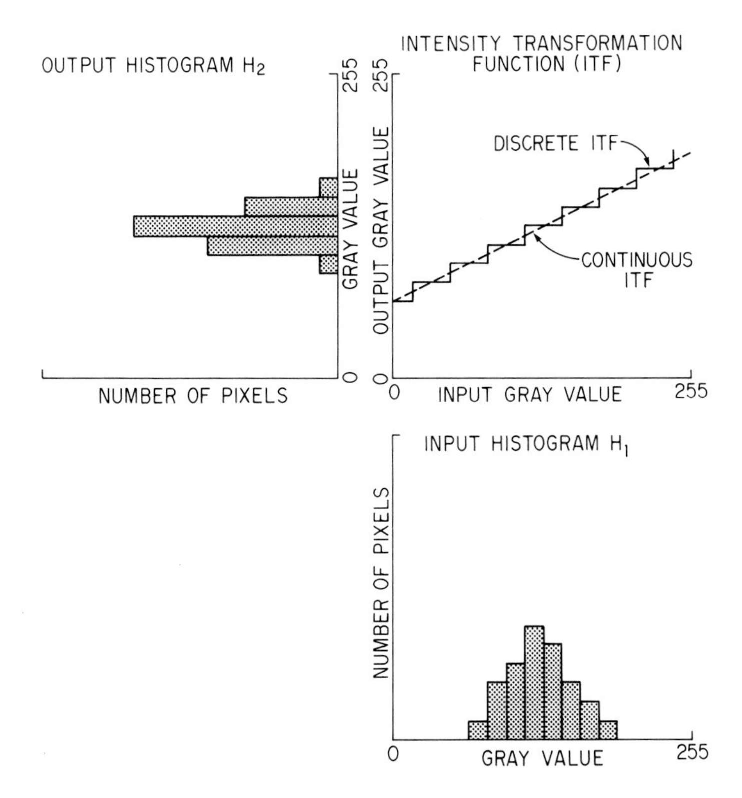

FIGURE 10-13. The discrete ITF. The discrete form of an ITF with slope less than 1 can produce an output histogram with fewer gray values than found in the input histogram. The assignment of two or more gray values to the same value in the output histogram leads to a loss of image resolution.

H_2. The use of discrete gray values requires that the continuous ITF be replaced by a discrete approximation of the ITF which is no longer continuous. The relationship between the continuous ITF and its discrete approximation is shown in Figs. 10-13 and 10-14.

Figure 10-13 shows that when using discrete approximations of an ITF with a contrast enhancement factor of less than 1, two or more gray values of the input histogram H_1 may map to the same gray value in the output histogram H_2. Unlike the continuous case, relative contrast between these pixels has been lost as a result of the transformation, and the original image cannot be re-created by writing backwards through the ITF. This loss of image information due to the compression of gray values may be acceptable or even desirable, if the gray values being compressed do not represent interesting information within the image, as discussed in Sections 10.7 and 10-13.

Figure 10-14 shows that a discrete ITF with a contrast enhancement factor greater than 1 causes a spreading of the gray values in the output histogram in a fashion similar to that for the continuous case. However, unlike the continuous case, there are unrepresented gray values within the range of the output histogram because of the discrete nature of the ITF. The actual number of gray values represented in the input and output histograms of Fig. 10-14 is constant; only the spacing has been changed. It should be noted that under some circumstances, both the spreading and compression effects shown in Figs. 10-13 and 10-14 can occur during the same image transformation (Section 10.7).

While the linear form of the ITF described in this section can be useful for manipulating image brightness and contrast, nearly the same effects can be achieved by adjusting the controls of the video camera directly. Adjustments in the DC OFFSET (BRIGHTNESS) control have the same effect as varying the value of b in equation (10-13), while adjustments of the GAIN (CONTRAST) control have a similar effect as varying the value of m. In addition, contrast enhancement achieved by adjustment

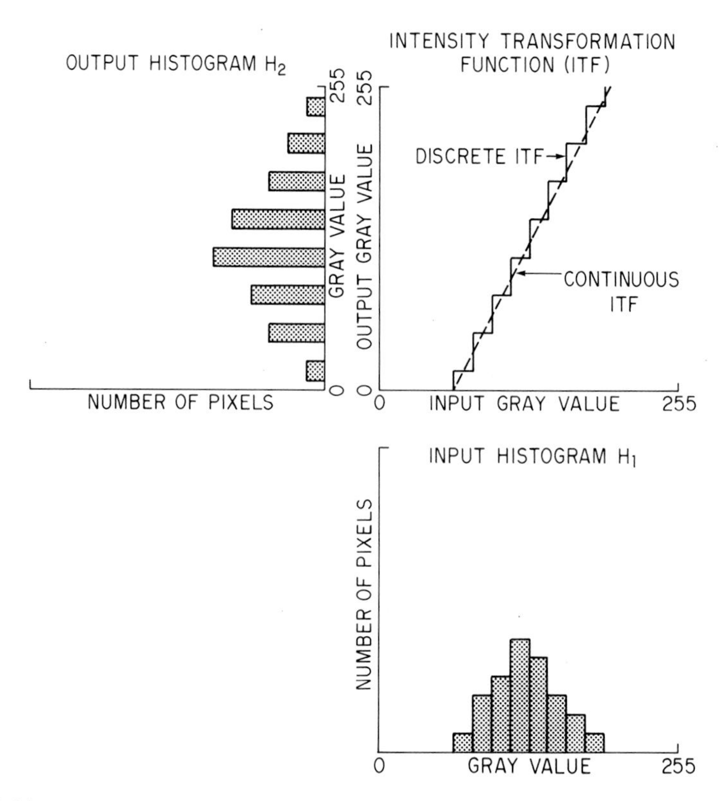

FIGURE 10-14. The discrete ITF. A discrete ITF with a slope greater than 1 causes a spreading of gray values in the output histogram; however, the number of different gray values in the input and output histograms is constant.

of the GAIN control will tend to fill in the intermediate gray values that are missing from the output histogram of Fig. 10-14; consequently, contrast enhancement achieved with the analog video signal may produce more information than the equivalent transformation on the digital image. The amount of enhancement that can be performed on the video signal is of course limited by the physical characteristics of the system, and there will be a point beyond which the video system can enhance the image no further. While digital forms of linear contrast enhancement can often surpass these physical limitations of the video system, the real advantage of digital ITFs lies in the use of nonlinear ITFs as described in the next section.

The conclusion to be reached from these examples is that the use of discrete ITFs does not add any information to the digital image and, in some circumstances, can actually cause information within the image to be lost. In spite of these facts, ITFs are widely used in image-processing applications for several reasons. While contrast-enhancing ITFs do not add any information to the digital form of the image, they can be used to help a human observer distinguish between gray values in the image that might be visually imperceptible in their original form. In addition, well-defined ITFs can be used to match the histograms of two images that have been digitized under different conditions so that the images can be directly compared. Nonlinear forms of the ITF can also be used to calibrate the response of the camera and digitizer system so that gray values from the digital image can be used to directly measure some quality of the original image such as optical density. Techniques using ITFs for this purpose are discussed in further detail in Section 10.10.

Most special-purpose image-processing systems that perform transformations using an ITF take advantage of a specific form of computer memory called an **intensity transformation table** (ITT) or simply a *look-up table* (LUT). The look-up table is used to store the contents of the discrete form of the ITF so that the input gray values GV_1 are matched directly to the output values to which

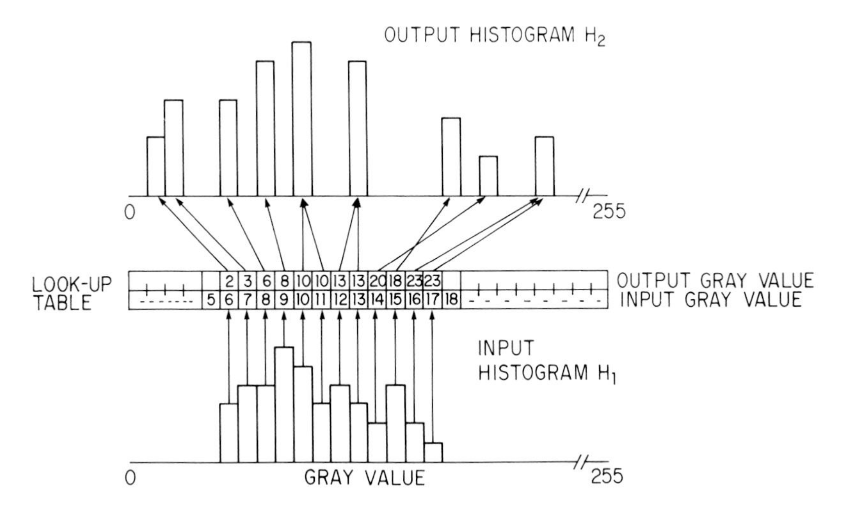

FIGURE 10-15. Implementation of a discrete ITF using a look-up table. (After Pratt, 1978.)

they map as shown in Fig. 10-15. The hardware of the image-processing system is constructed in such a way that the digital image can be passed through the look-up table resulting in the gray value of each pixel in the input image being converted to the output gray value specified by the ITF. Since these transformations are performed by hardware, they can be performed very quickly, usually in real time at video rates when high-speed array processors are used.

Image transformations that involve look-up tables can be implemented in two separate ways as shown in Fig. 10-16. In the first example, the contents of the digital memory are routed through the look-up table in such a way that the modified image is written back into computer memory after the transformation is made; consequently, the original unmodified image is lost. This permanent type of transformation is desirable when the modified digital image is to be used subsequently by the computer for some quantitative purpose such as densitometric measurements, or when the modified image is to be compared with another digital image within computer memory. In the second example of Fig. 10-16, the image transformation is performed just before the digital image is converted to analog form for display on a video monitor. The transformation specified by the ITF stored in the look-up table is displayed visually on the monitor; however, the digital form of the image in computer memory is not altered. This type of arrangement is useful when an enhanced view of an image is needed for display purposes, while the original form of the image may be needed for other purposes. This arrangement is also very helpful when using an interactive image-processing system where several different image transformation functions may be examined on the same image in order to find one that is most meaningful to the observer. Since the digital form of the image is not altered with this arrangement, these different ITFs can be tried without concern for the types of image degradation or modification that the transformation might produce in the original

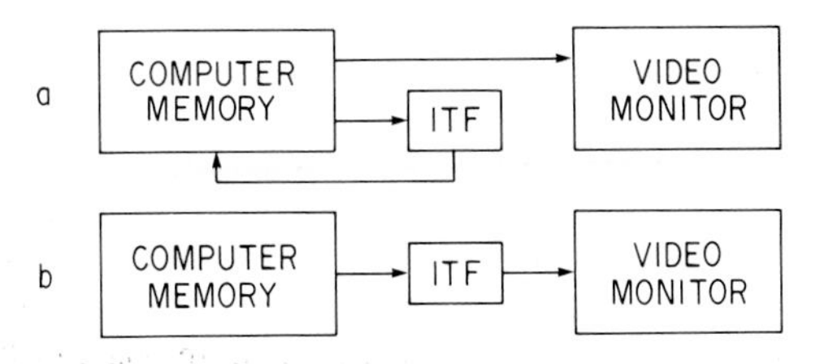

FIGURE 10-16. Two methods for implementing ITFs in an image processing system. In (a), the transformed image is written back to computer memory, destroying the original image. In (b), the transformed image is displayed; however, the original image is unaltered.

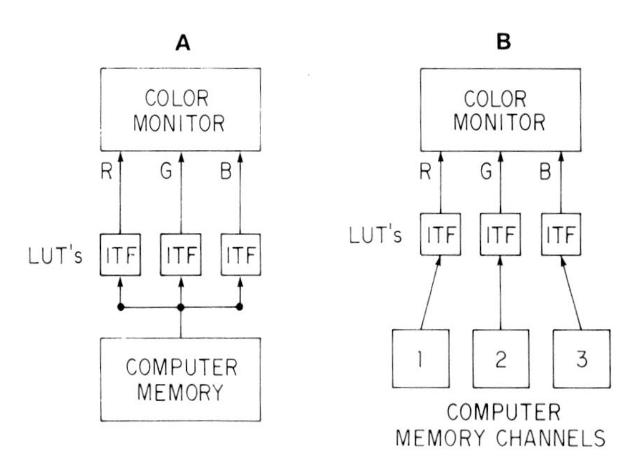

FIGURE 10-17. Two methods for creating pseudocolor displays. In (A), a single image is transformed by three different ITFs. In (B), three separate images are each modified by a single ITF.

image. A well-designed image-processing system should have the capability for implementing both of the arrangements shown in Fig. 10-16.

ITFs loaded into look-up tables are also used to produce color displays from monochrome (black and white) images. These transformations, called *pseudocolor displays,* can be implemented in two separate ways, as shown in Fig. 10-17. In the first case, the digital image is routed simultaneously through three different look-up tables, which, in turn, are directed to the red, green, and blue color guns of the color monitor. The human eye can distinguish many more different colors than it can shades of gray; consequently, pseudocoloring can be used to produce images with more contrast than is possible with monochrome transformations alone. In the second example of Fig. 10-17, each of the look-up tables for the color guns acts upon a separate digital image within the memory of the computer. This arrangement is used frequently in multispectral analysis where each of the digital memories contains an image of the same scene that was captured under different conditions. The ITFs loaded into the look-up tables are then constructed so that color and contrast differences in the displayed image accentuate the differences between each of the stored images. It should be noted that most pseudocolor displays are totally arbitrary functions and do not reveal any information about the true color properties (if there are any) of the original specimen. True color images can only be stored in digital form by capturing multiple frames of the same scene using red, green, and blue color filters. Each of these primary color images is then stored in a separate area of computer memory, as shown in Fig. 10-17B, and the true color image can then be displayed on the color monitor.

Look-up tables can also be used to perform mathematical functions on a digital image, such as multiplication by a constant or the calculation of logarithms. The output gray value of the chosen operation is calculated for each of the possible 256 input gray values and loaded into the appropriate register of the look-up table. The operation can then be performed on the image at video rates simply by writing the image through the look-up table. When a look-up table is used in this fashion, it is sometimes referred to as a *function table.*

10.7. NONLINEAR CONTRAST ENHANCEMENT

One of the powerful features of digital processing that cannot be easily duplicated through analog techniques is the use of *nonlinear* ITFs. The nonlinear nature of the ITF means that contrast can be enhanced over one range of gray values in an image while, at the same time, contrast over another range can be reduced or left unaltered. The ITF can be custom tailored to selectively enhance the contrast of important features in an image while reducing the contrast of other unimportant ones. Nonlinear ITFs can be used to calibrate the light response of the optical system or to correct for other sources of nonlinear response.

The ability of a nonlinear ITF to selectively enhance contrast within an image is shown graphically in Fig. 10-18. As in the earlier example of Fig. 10-8, the input histogram in Fig. 10-18

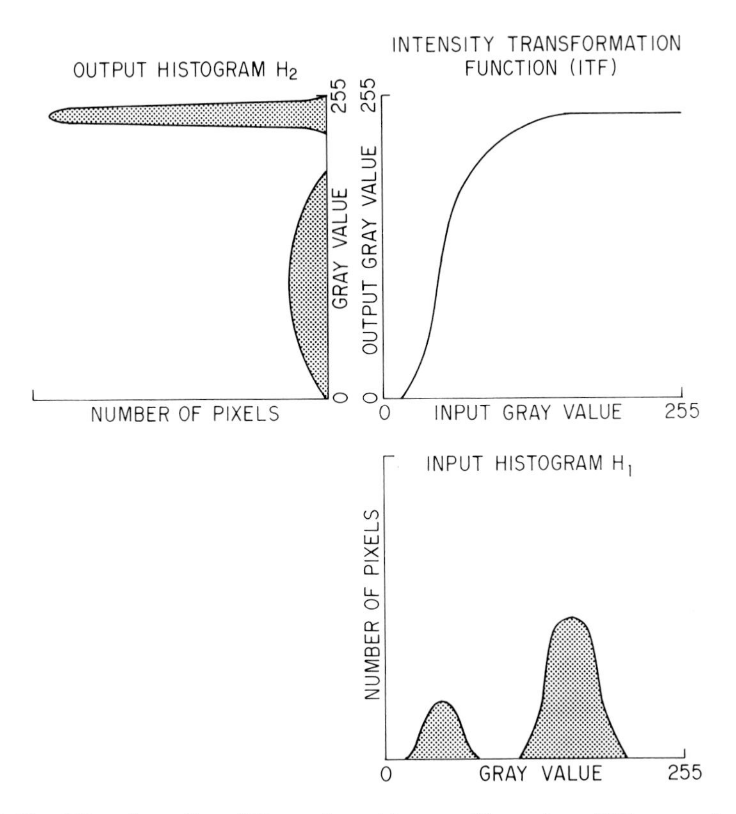

FIGURE 10-18. Effect of a nonlinear ITF on an image histogram. The nonlinear ITF has caused a broadening of the left peak in the output histogram and a simultaneous compression of the right peak.

is assumed to come from a microscope image of darkly stained cells with the small peak on the left corresponding to the cell-containing regions of the image, and the larger peak on the right corresponding to non-cell-containing regions of background. The nonlinear nature of the ITF in Fig. 10-18 has resulted in a broadening of the smaller peak in the output histogram and a concurrent narrowing of the larger peak. The broadening of the smaller peak produces greater contrast between these pixels in the image (which represent the information-carrying portions of the image), while the narrowing of the larger peak produces a reduction of contrast in the background areas that are devoid of interesting information. As was the case using linear ITFs, the contrast enhancement factor of a nonlinear ITF is equal to the slope of the function, which is the first derivative of the ITF at the chosen gray value.

It may still be desirable to use a nonlinear ITF even when all regions of an image contain information of equal importance. The reason for using a nonlinear ITF in this situation can be visualized by examining the histogram in Fig. 10-19. The two darkened zones show areas of the histogram that contain equal numbers of pixels. In the center of the histogram, where there is the greatest number of pixels per gray value, only a few different gray values are used to produce an area with the required number of pixels. At the edge of the histogram, there are fewer pixels at any given gray value; consequently, a larger range of gray values is needed to produce a slice of equal area. By analogy with equation (10-7), this variation in the width of equal-area slices within the image histogram means that any pixel within the image has a higher probability of having neighbors with gray values near the center of the histogram than having neighbors with gray values near the edges. Consequently, the contrast between neighboring pixels in the image will be lower for gray values near the central peak of the histogram than for pixels near the edge of the histogram (Sklansky, 1978b). If all of the pixels in the image represented by this histogram contain information of equal importance, this contrast bias results in pixels with gray values near the edge of the

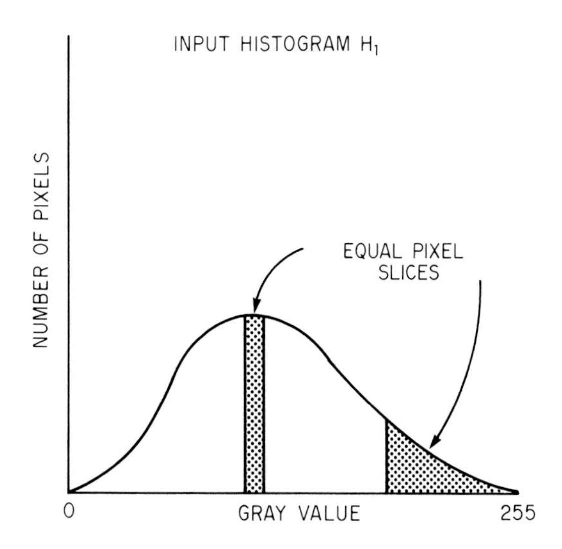

FIGURE 10-19. Histogram showing regions with an equal number of pixels. (After Sklansky, 1978b.)

histogram being perceived more prominently than those near the peak, which is not desirable. This contrast bias can be avoided if the image histogram has a flat distribution with an equal number of pixels at each gray level. If the image does not have this flat distribution originally, it can be transformed through a nonlinear ITF to produce a flat histogram as shown in Fig. 10-20.

In order to transform an input image into an output image with a predetermined histogram, it is

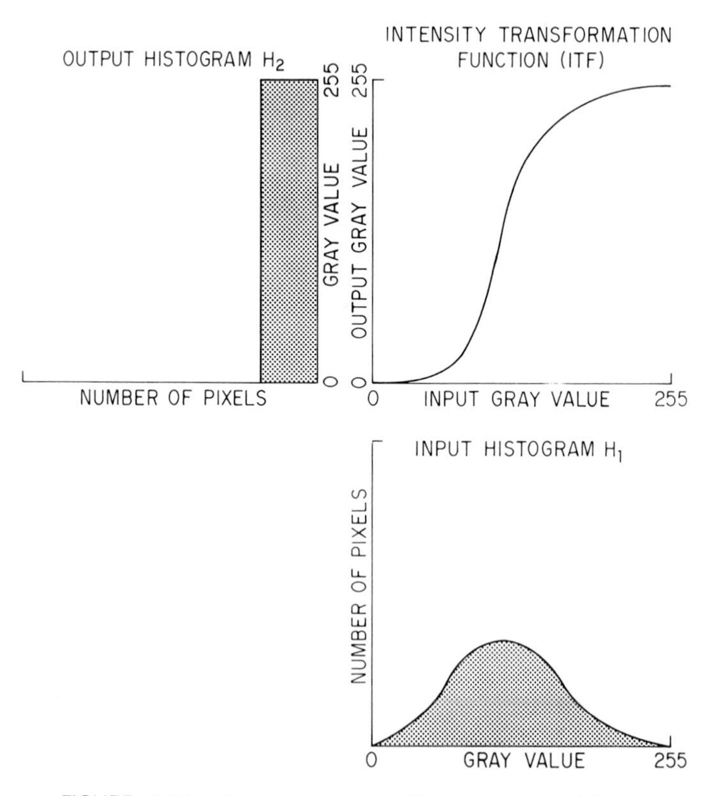

FIGURE 10-20. Use of a nonlinear ITF to produce a flat histogram.

necessary to find a relationship that defines the ITF as a function of input and output histograms. It can be shown that whenever the ITF is a monotonically increasing function, there exist two gray values, GV_1 and GV_2, such that

$$P_2(GV_2) = P_1(GV_1) \tag{10-14}$$

where P(GV) is the cumulative area function described in Section 10.5. This relationship states simply that GV_1 and GV_2 delineate equal areas under their respective cumulative histograms, which follows from the one-to-one mapping of gray values specified by equation (10-11). In the discrete case, equation (10-14) can be written as

$$\sum_{k=0}^{GV_2} H_2(k) = \sum_{j=0}^{GV_1} H_1(j) \tag{10-15}$$

where H_1 and H_2 are the image histograms, and it is assumed that the input and output images have the same area. This relationship can be used directly to derive the discrete form of the ITF by first

```
C
C       THIS SUBROUTINE CREATES AN INTENSITY TRANSFORMATION FUNCTION (ITF)
C       THAT MATCHES THE INPUT HISTOGRAM OF ANY IMAGE TO A SECOND HISTOGRAM
C       OF PREDETERMINED SHAPE.  THE ARRAY <H1> CONTAINS THE INPUT HISTOGRAM,
C       THE ARRAY <H2> CONTAINS THE OUTPUT HISTOGRAM.  THE INTEGER ARRAY
C       <LUT> CONTAINS THE INTENSITY TRANSFORMATION FUNCTION THAT CONVERTS
C       <H1> INTO <H2>. THE INDEX OF <LUT> IS ONE PLUS THE INPUT GREY VALUE
C       FROM <H1>, THE CONTENTS OF <LUT> ARE THE OUTPUT GREY VALUES FOR <H2>.
C       THE ARRAYS <H1>, <H2>, AND <LUT> MUST BE SPECIFIED IN A COMMON
C       STATEMENT OF THE CALLING PROGRAM. NOTE THAT <H1> AND <H2>
C       MUST CONTAIN THE SAME NUMBER OF PIXELS.
C
        DIMENSION H1(256),H2(256),LUT(256)
        COMMON H1,H2,LUT
C
        DO 10 J=1,256
10      LUT(J)=0                                  !INITIALIZE <LUT>
C
        I1=1                                      !GREY VALUE POINTER FOR <H1>
        I2=0                                      !GREY VALUE POINTER FOR <H2>
        CUMM1=0                                   !CUMMULATIVE AREA OF <H1>
        CUMM2=0                                   !CUMMULATIVE AREA OF <H2>
C
20      I2=I2+1
        IF(I2.GT.256) I2=256
        CUMM2=CUMM2+H2(I2)
30      DEL=CUMM2-CUMM1
        IF(DEL.LT.H1(I1)*0.5) GOTO 20
        LUT(I1)=I2-1                              !GREY VALUES ARE ZERO TO 255
        CUMM1=CUMM1+H1(I1)
        I1=I1+1
        IF(I1.GT.256) RETURN
        IF(CUMM1+0.1.GT.CUMM2) GOTO 20            !0.1 CORRECTS FOR ROUNDING ERRORS
        GOTO 30
C
        END
```

FIGURE 10-21. The MATCH program.

performing the summation on the right-hand side of this equation for each value of GV_1, and then finding the value of GV_2 that produces a summation on the left-hand side with the closest value. The FORTRAN language subroutine shown in Fig. 10-21 uses this approach to calculate a discrete ITF for any pair of input and output histograms.

Nonlinear ITFs are typically designed to effect image contrast in a specific way, based on some model of visual perception or on theoretical considerations concerning the information content of the image (e.g., see Pratt, 1978, pp. 317–318). While these transformations may have a solid mathematical basis, the success of any enhancement routine is highly subjective and will ultimately depend on the preference of the observer and the type of image being modified. Figures 10-22 and 10-23 give parallel examples of two images and their histograms that have been modified with different ITFs designed to affect the images in a particular fashion. Figure 10-22 is an image of a tissue culture cell taken with phase-contrast optics while Fig. 10-23 is a stained peripheral blood smear taken with brightfield optics. Figures 10-22C and 10-23C show transformations performed

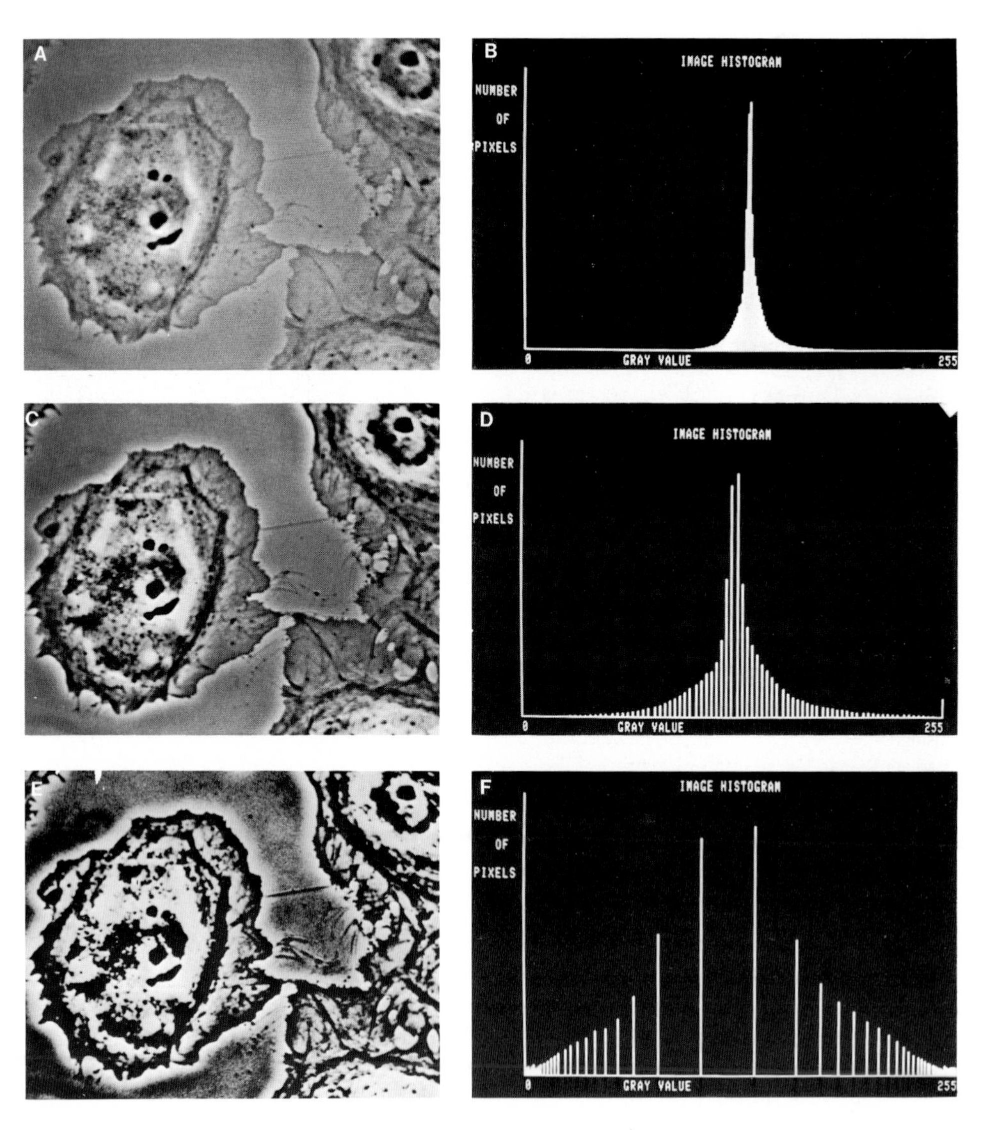

FIGURE 10-22. Effect of contrast enhancement on image–histogram pairs. (A, B) Original image and histogram. (C, D) Linear contrast enhancement. (E, F) Histogram flattening.

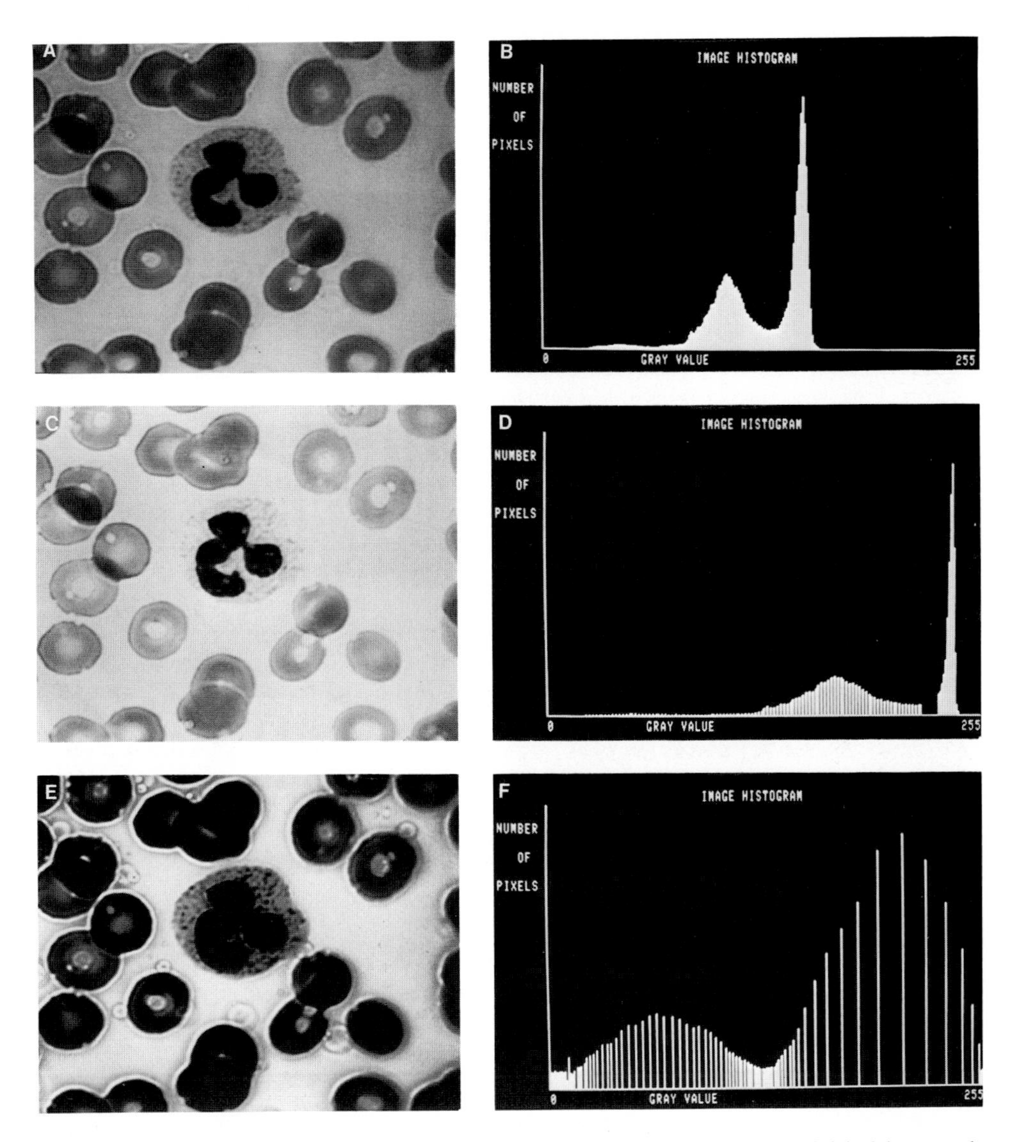

FIGURE 10-23. Effect of contrast enhancement on image–histogram pairs. (A, B) Original image and histogram. (C, D) Linear contrast enhancement. (E, F) Histogram flattening. (Stained specimen courtesy of Barbara Walter, Department of Pediatrics, California College of Medicine, University of California, Irvine.)

using a linear ITF. In Fig. 10-22C, this linear ITF has been designed to spread the image histogram evenly over the full range of gray values. In Fig. 10-23C, the linear ITF has been designed to spread only those gray values found in the smaller peaks of the image histogram. Figures 10-22E and 10-23E show transformations designed to produce images with flat histograms. The ITFs used to make these transformations were generated using the algorithm of Fig. 10-21.

Special consideration must be give to microscope images with histograms similar to that of Fig. 10-22B. This image was made using phase-contrast optics, and unlike the examples of Figs. 10-8 and 10-23, the background regions of this image do not produce a separate peak in the histogram that can be segregated from the gray values of the specimen. Background regions in a phase-contrast image have a refractive image equal to that of the bulk media and produce a middle gray value (phase neutral). Boundaries of the specimen that have a refractive index higher or lower than that of the bulk media have gray values that are lower (phase dark) or higher (phase light) than the

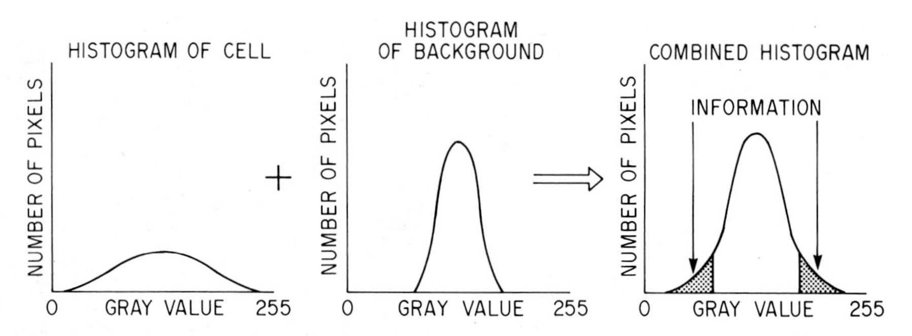

FIGURE 10-24. Typical histogram of phase-contrast, polarization, and DIC images. Overlap between the histogram of the image background and the histogram of the specimen results in the most useful image information being found in the tails of the combined histogram.

background (Chapter 5) and are distributed on both sides of the central peak. Furthermore, only a small part of the specimen is actually visible as phase light or dark in these images, with much of the area within the specimen boundaries being indistinguishable from background. These conditions produce an image histogram where most of the useful image information is found only in a relatively small number of pixels with gray values on the tails of the histogram as shown in Fig. 10-24. Similar histograms are obtained for images captured with polarization and DIC optics. Nonlinear ITFs can cause special problems when used to manipulate images with this type of histogram. For example, an ITF designed to flatten the histogram will tend to increase the contrast in background regions of such images because they contain the most pixels in the image, and can actually cause a simultaneous reduction in the contrast of information-containing regions. This effect can be seen in Fig. 10-22F.

Optimum contrast can be obtained in information-containing regions of such images by selecting an ITF that compresses the contrast in center regions of the histogram, while at the same time increasing contrast in the tail regions. We have found that ITFs of the form

$$GV_2 = \frac{128}{(b-c)^a} * [\ (b-c)^a - (b-GV_1)^a + (GV_1+c)^a] \qquad (10\text{-}16)$$

which have a sigmoid shape, are effective for enhancing images of this type. In this equation, b is the highest gray value of the input histogram and c is the lowest value. a is an arbitrary constant that is set empirically for maximum effect. For values of a approaching 1, this function approximates a linear ITF. As a increases, the function becomes more sigmoid, and greater compression occurs at the middle gray values. Figure 10-25 shows several manipulations of the same image using different values of a in this equation.

As was the case in the previous section, nonlinear ITFs can only be utilized as discrete functions. Discrete ITFs can produce effects that are different from that of their continuous equivalents as seen by the histogram-flattening routines of Figs. 10-22F and 10-23F. Due to the discrete nature of the digital image, this transformation has not produced a histogram with a flat appearance but instead has caused an uneven expansion of the original histogram with the gray values of highest frequency being spaced the farthest apart. This arrangement approximates a flat histogram if the number of pixels per gray value is measured over a broad-enough range. A true flat histogram can be produced from a digital image only if the output image has fewer gray values than the input image (lower dynamic range) or if a randomizing function is used to arbitrarily reassign gray values from the input histogram to the nearest unrepresented gray values in the expanded histogram.

Discrete ITFs can also be used to advantage in that they can be given an entirely arbitrary form that is unrelated to any continuous model. Figure 10-26 shows the image of Fig. 10-22A after transformation with a noncontinuous ITF that reassigns background regions of the histogram to a new gray value of 0, and the tail regions to a new gray value of 255. This transformation has the

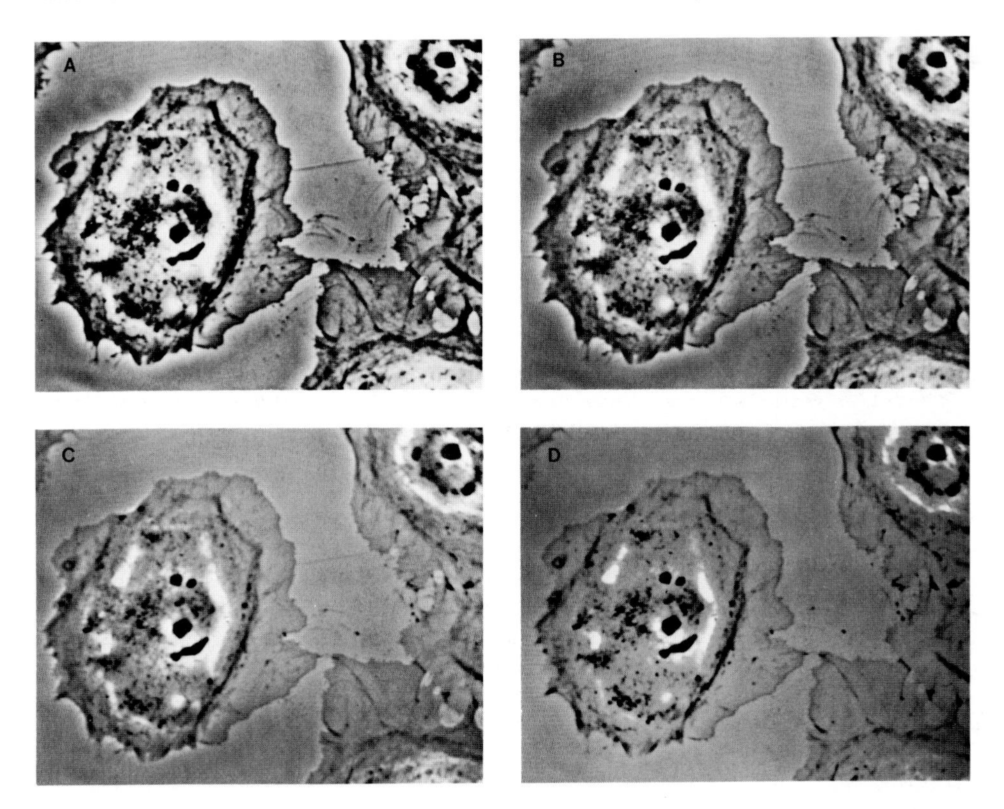

FIGURE 10-25. Contrast enhancement using sigmoid-shaped ITFs generated using different values for a in equation (10-16). (A) $a = 2.0$, (B) $a = 4.0$, (C) $a = 6.0$, (D) $a = 8.0$.

effect of maximizing the contrast between the background of the image and cell boundaries. This transformation can be useful for delineating edges in the image for use with routines that analyze shape or involve pattern recognition. Note that the noncontinuous ITF of the transformation has produced an image with only two gray values; consequently, the image has become a binary image (Section 10.4) and can be manipulated by routines designed specifically for this type of image.

In each of the examples presented in this and the previous section, the input histogram of the original image has not utilized the full dynamic range of the display device; consequently, there have been unrepresented gray values in the input image that could be filled to increase contrast. When an image utilizes the full dynamic range of the display device, there are no unrepresented gray

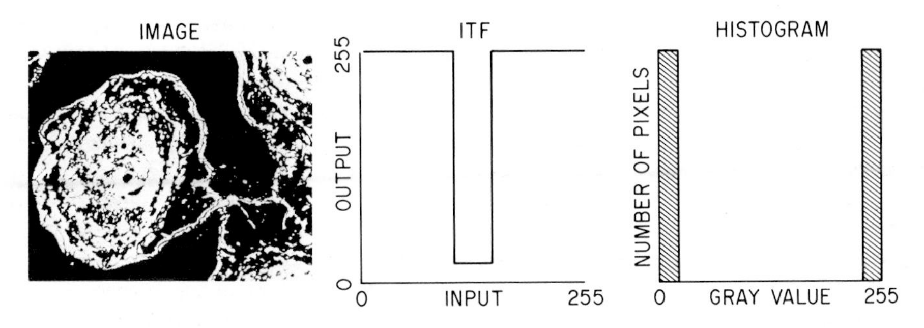

FIGURE 10-26. The binary image. The discontinuous ITF produces an output image with only two gray values. The image histogram shows two peaks, one at 0 gray value and one at 255.

values and contrast cannot be enhanced over one range of gray values in the image without reducing contrast an equal amount over some other range. This limitation can be overcome by using pseudocolor transformation techniques that reassign gray values to new color values. The human eye can distinguish many more colors than shades of gray; consequently, a color transformation can increase contrast over any and all ranges of gray values in the original image without affecting contrast over another. Pseudocolor transformations are a particular type of nonlinear ITF and are used routinely in image processing (Goochee *et al.*, 1980; Berns *et al.*, 1981; Gallistel *et al.*, 1982). Examples of pseudocolor transformations are shown in Fig. 10-55 (see color reproduction following page xxvi).

10.8. MANIPULATIONS USING MULTIPLE IMAGES

Many useful image-processing routines involve algebraic manipulation of two images using the relationship

$$C_{(i,j)} = A_{(i,j)} * B_{(i,j)} \tag{10-17}$$

where $A_{(i,j)}$ and $B_{(i,j)}$ are corresponding pixels in the two images and $*$ is an algebraic operator such as addition, subtraction, multiplication, or division. These manipulations also usually include addition or multiplication of a constant in order to avoid the creation of negative or fractional values.

The most frequently used algebraic manipulation is probably one involving the subtraction of two similar images. Subtraction of two images of the same specimen captured at different times can be used to detect motion of the specimen over that time period. Figure 10-27 shows the result of subtracting two images of a tissue culture cell taken 60 sec apart. Regions of the image that are constant over the time period have canceled in the subtracted image, while only areas that have moved, such as the ruffled membranes at the cell periphery, are visible.

Image subtraction can also be used to remove uneven background shading from an image. This can be accomplished by first capturing an image of a blank field (such as a region of a microscope slide devoid of cells) and then subtracting this image from subsequent frames of the actual specimen. Shading correction is often required before performing routines for pattern recognition of **edge detection** as demonstrated by the example of Fig. 10-28, which shows a hypothetical specimen with a uniform intensity profile that has been superimposed on a background of varying intensity. Before this varying background is added to the image, the specimen can be distinguished from background by the threshold value T_1. After the varying background has been added to the image, the specimen no longer has a uniform profile and there is no intensity level that can be used to distinguish object from background as demonstrated by the thresholds T_1 and T_2. Image subtraction causes a reversal of the sequence in Fig. 10-28 and produces the desired uniform background.

The example of Fig. 10-28 assumes that the specimen has a uniform intensity that can be additively superimposed onto the background. In reality, the intensity of a specimen under a microscope is not additive but is proportional to the background intensity at each point. For example, under brightfield optics the intensity of the specimen is the product of its transmittance times the intensity of the background. Similar considerations apply to images collected with fluorescence, phase contrast, DIC, and other types of optical systems. The effect of this proportionality can be seen in Fig. 10-29. The specimen there has a profile of uniform transmittance; however, when illuminated by a varying background, the right side of the specimen has a slightly higher intensity that is proportional to the background level. If background subtraction techniques are applied to this image, the background will assume a uniform intensity; however, the right side of the specimen will still have a higher apparent transmittance than the left because its initial intensity was proportional to the original level of illumination and was not simply additive.

The most effective way to correct for shading distortion in this case is to calculate the ratio of the specimen and background by dividing one image into the other. The division of two images to correct for background illumination is necessary when accurate quantitative information is needed about the intensity of the specimen, and is discussed further in Section 10.10.

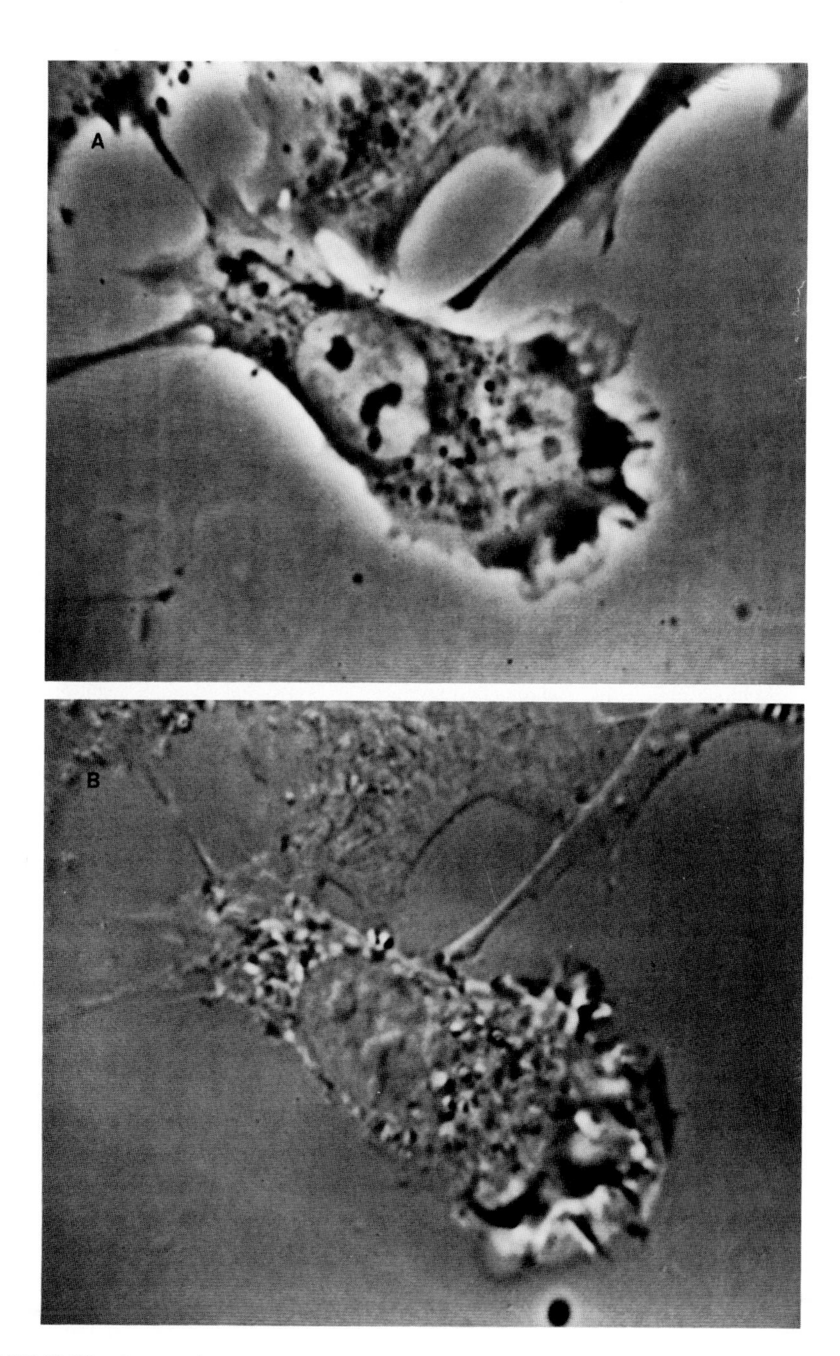

FIGURE 10-27. Image subtraction to detect motion. (A) Original image captured at time zero. (B) Difference image created by subtracting image (A) from a second image of the same field captured 60 sec later. (Specimen courtesy of Dean Sadamune, Department of Developmental and Cell Biology, University of California, Irvine.)

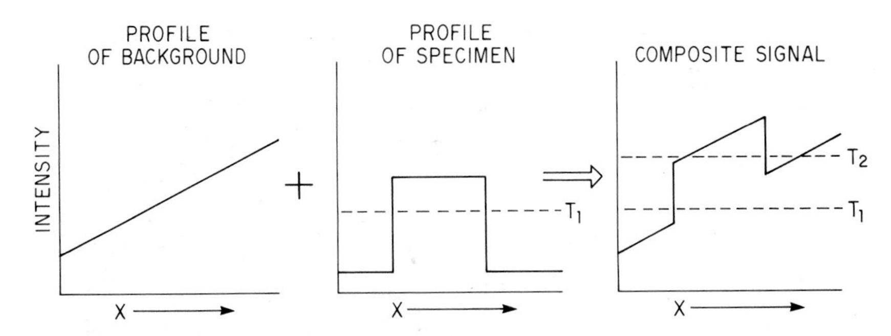

FIGURE 10-28. Effect of nonuniform background on a specimen intensity profile. Before adding the background, the specimen can be distinguished by the threshold level T_1. After the addition of uneven background, there is no effective threshold level for identifying the specimen.

Another useful image manipulation is one that involves subtraction of identical images where one image has been slightly offset in the x or y direction from the other. This manipulation has the form

$$C_{(i,j)} = A_{(i,j)} - B_{(i-x,j-y)} \tag{10-18}$$

where x and y are the amounts of offset. For small offset values, this manipulation approximates the partial spatial derivative of the image in the direction of the offset and is useful for detecting areas of the specimen where the intensity is changing rapidly, such as might occur at an object boundary. Figure 10-30 shows the effect of this offset subtraction on the image of Fig. 10-22. Note that when using phase-contrast optics, this manipulation produces an image that contains much of the same information as would be found in a DIC image of the same specimen (Walter and Berns, 1981).

10.9. GEOMETRICAL MANIPULATIONS

The size and shape of an object in a video image is often an imperfect replica of the original specimen, due to distortion in the optical system, scanning irregularities in the video camera, or mismatch between the aspect ratios of the video camera and display device. This degradation is called *geometrical distortion* and means that image magnification varies with position in the image.

Geometrical distortion is usually not obvious in a well-designed video system and can be ignored for most purposes. This distortion can be a serious problem, however, when making quantitative measurements of shape and size from the video image where it must be established that the measurements are independent of the position of the specimen in the field of view. When geometrical distortion in an image is unacceptable, its effects can be reduced in the image through the use of any of several techniques known as *geometrical decalibration*. The first step in the decalibration procedure is to determine the extent of the distortion by imaging a test pattern of

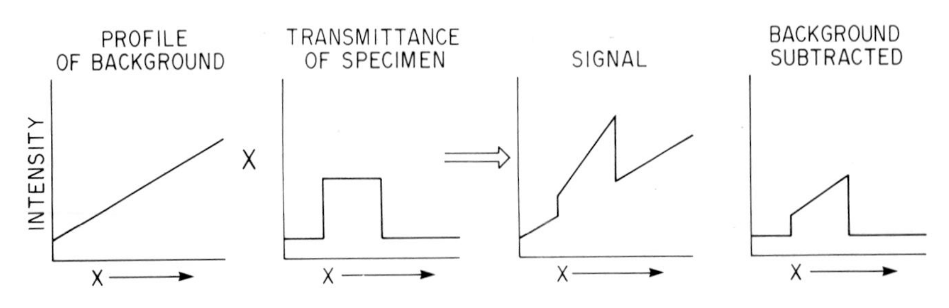

FIGURE 10-29. Effect of background subtraction on a specimen of uniform transmittance.

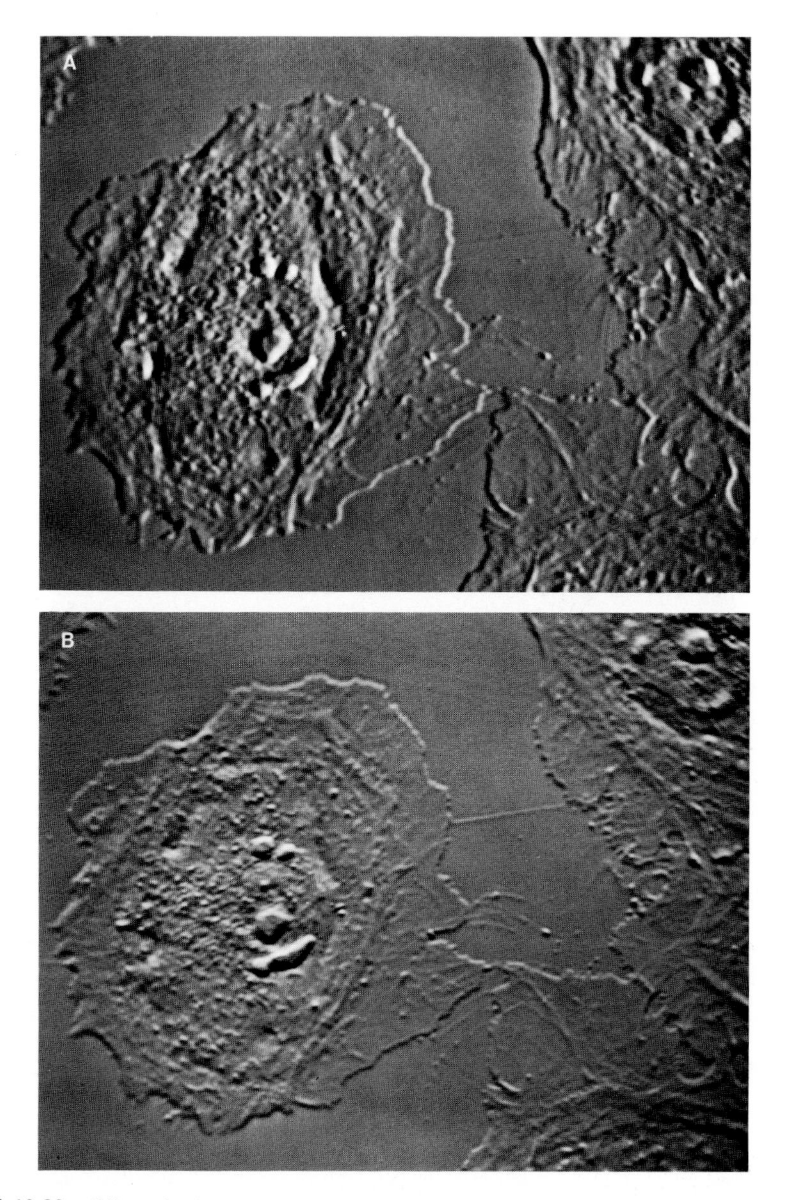

FIGURE 10-30. Effect of subtracting two slightly offset images. (A) Two-pixel offset in the horizontal direction. (B) Two-pixel offset in the vertical direction.

known dimensions, such as a rectangular array of spots. The position of each spot in the test pattern, known as a *control point,* is measured in the distorted image in relation to its expected position in an undistorted image. This mapping of control points between an ideal and a distorted image produces a *mapping function* that describes the distorting properties of the imaging system. Alternately, the mapping function can be determined without actual measurements by creating a mathematical model of the system that is a function of the distorting properties of individual components such as lenses, video cameras, etc. The relationship between the mapping function and the distorted image is shown in Fig. 10-31.

The most efficient method for generating a decalibrated image is to choose a pixel location in the corrected image and then find the corresponding pixel in the distorted image as defined by the

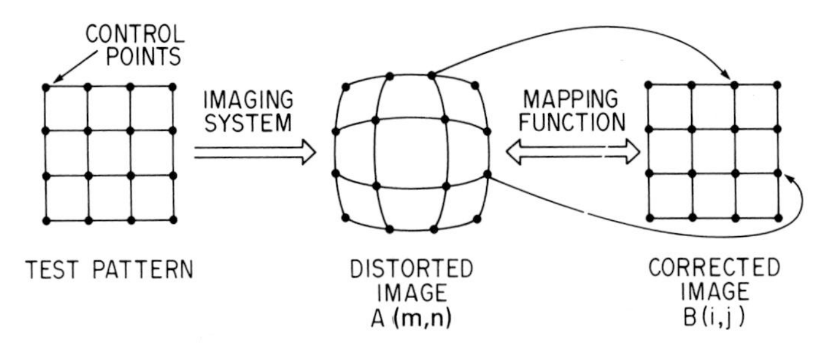

FIGURE 10-31. Use of a mapping function to correct geometrical distortion.

mapping function. The gray value of the chosen pixel in the corrected image is then replaced by the gray value of the corresponding pixel in the distorted image. This relationship takes the form

$$B_{(i,j)} = A_{(m,n)} \tag{10-19}$$

where $B_{(i,j)}$ is the pixel in the corrected image at row i, column j, and $A_{(m,n)}$ is the corresponding pixel in the distorted image. The indexes in this equation are related by the relationship

$$(m,n) = \mathrm{MF}(i,j) \tag{10-20}$$

where MF is the mapping function. It should be noted that the mapping function defines a one-to-one relationship between pixels in the corrected and distorted images, and thus can be used to find pixel locations in either image.

For most mapping functions, the pixel locations specified in the distorted image by equation (10-20) will only be integer values at the control points, and will be fractional at all other locations. This means that equation (10-19) must be restated as

$$B_{(i,j)} = A_{(x,y)} \tag{10-21}$$

where the indexes x and y have been substituted for m and n to show that the corresponding pixel locations in the distorted image are nonintegers. This situation is shown graphically in Fig. 10-32.

When the mapping function specifies a pixel location that is between actual pixels in the distorted image, an estimating procedure must be used to find a gray value to be assigned to the corrected image. The simplest procedure is to select the gray value of the pixel nearest to the fractional location specified by equation (10-21). This procedure is relatively fast; however, it may produce an undesirable ''staircase'' effect in image boundaries if the mapping function involves image rotation. Another approach is to use an interpolative procedure that calculates a gray value for the output pixel based on the gray values of neighboring pixels in the distorted image. One

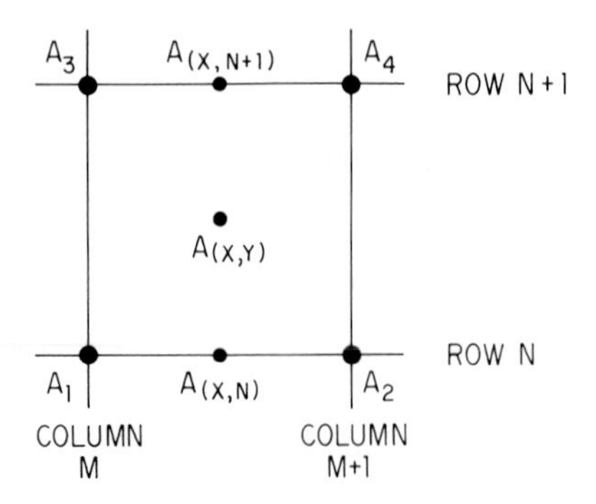

FIGURE 10-32. Coordinate locations used during bilinear interpolation.

procedure that uses the gray values of the four nearest neighboring pixels is called a *bilinear interpolation*, performed as follows. An estimation is first made in the horizontal direction for the rows above and below the mapped pixel using a linear interpolation. These interpolations are defined as

$$A_{(x,n)} = (A_2 - A_1)(x-m) + A_1 \quad (10\text{-}22)$$

and

$$A_{(x,n+1)} = (A_4 - A_3)(x-m) + A_3 \quad (10\text{-}23)$$

as shown in Fig. 10-32.

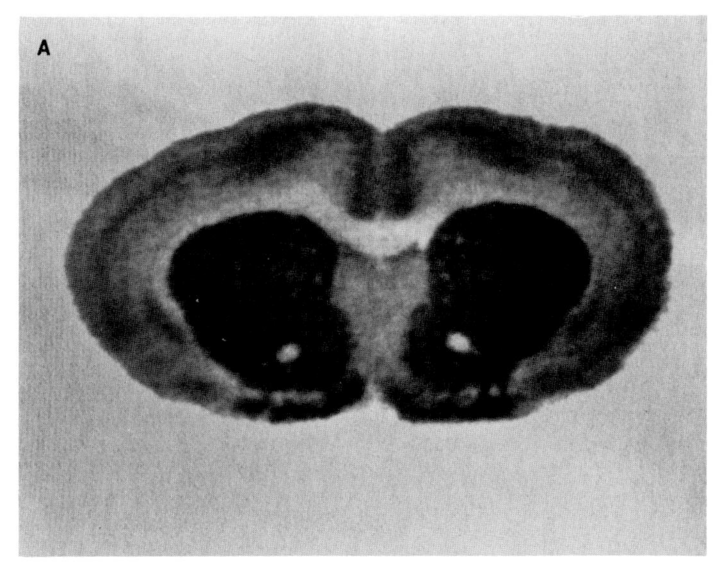

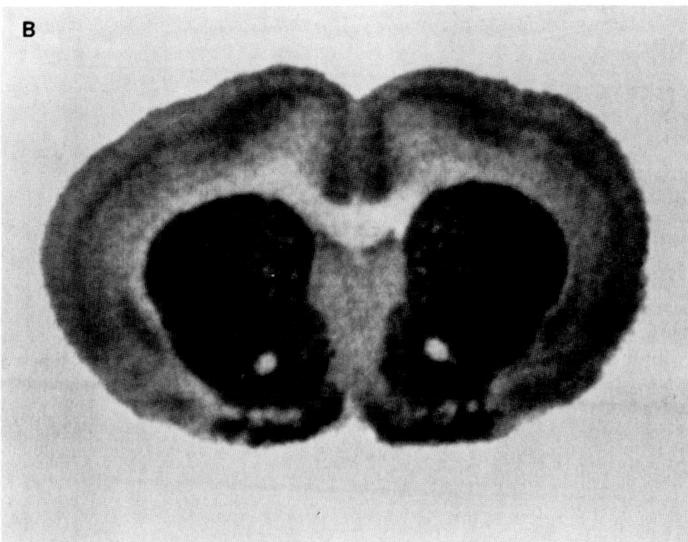

FIGURE 10-33. Geometrical decalibration used to correct uneven image magnification in the vertical and horizontal directions. (A) Original image. (B) Corrected image after magnification 1.19 times in the vertical direction. (Specimen and digital image courtesy of Anthony Altar, Department of Psychobiology, University of California, Irvine.)

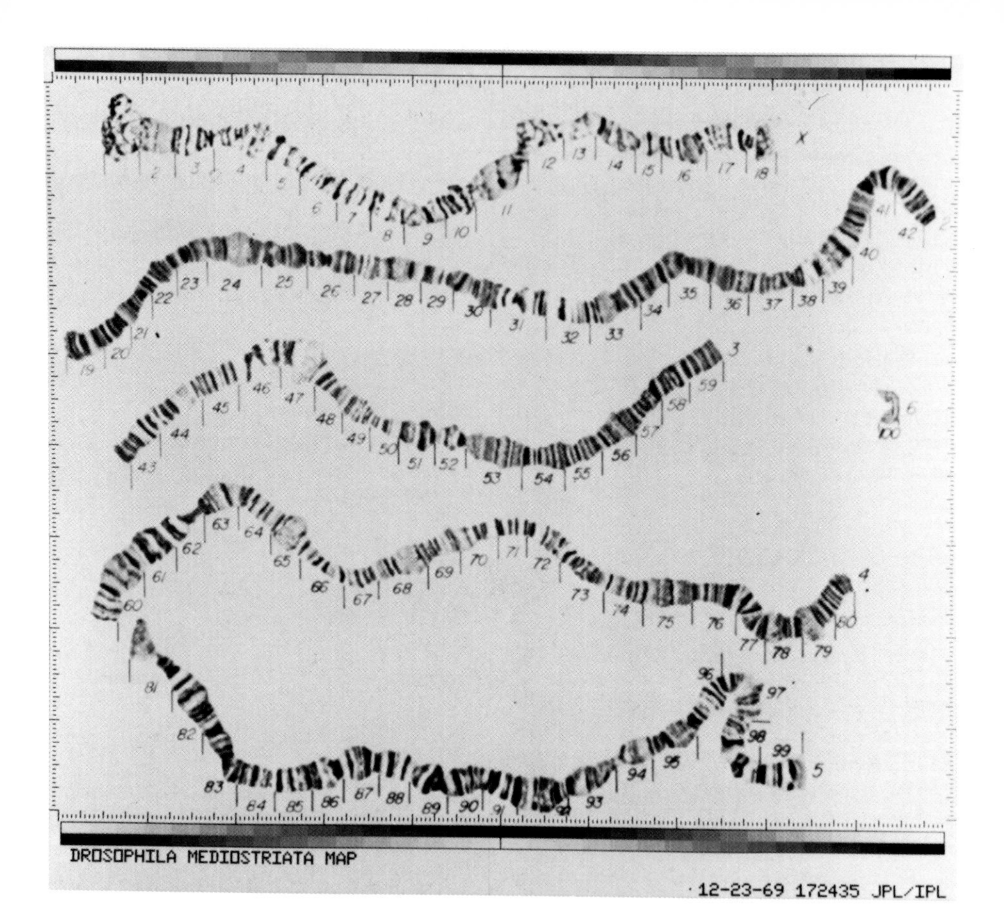

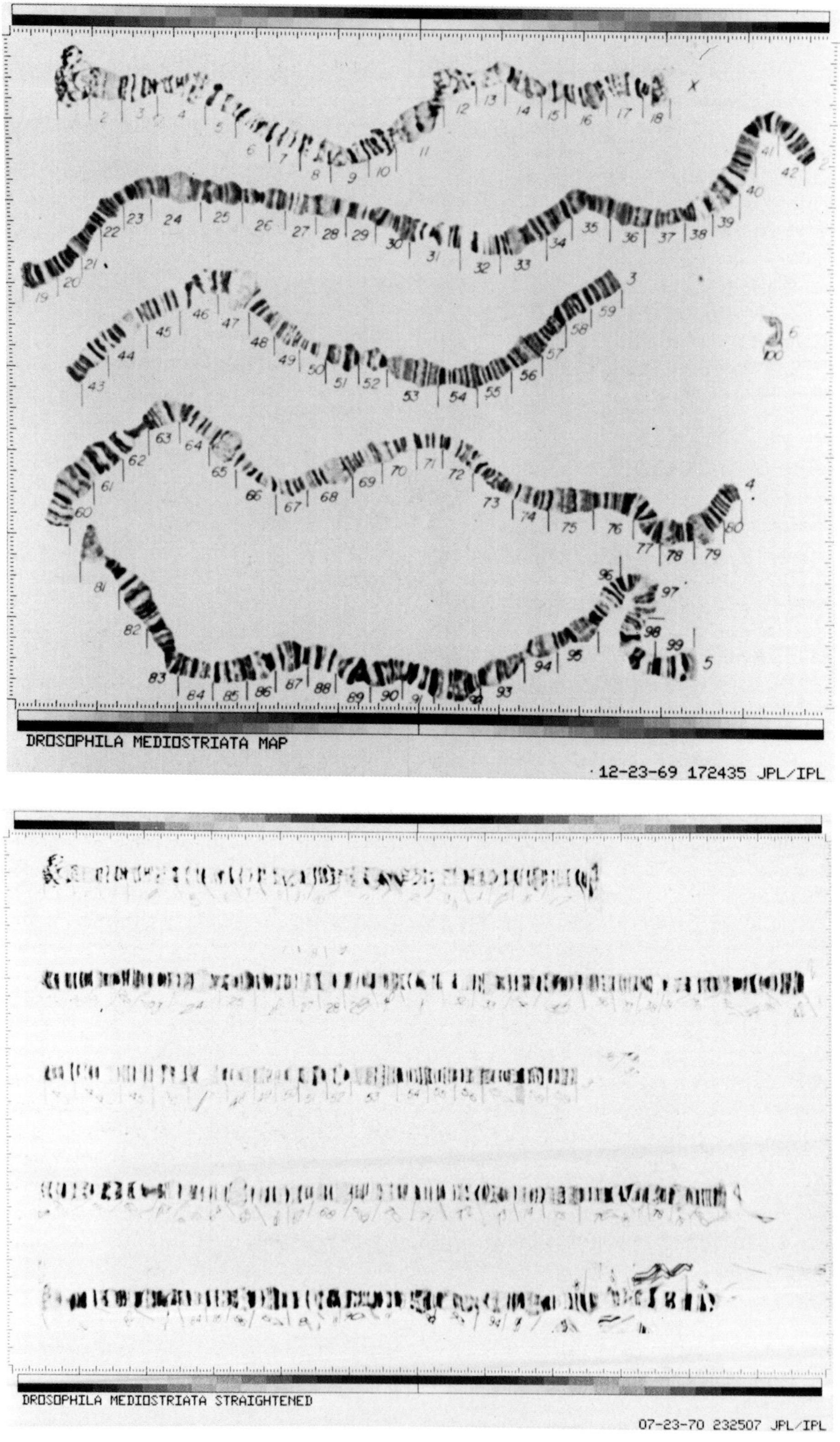

DROSOPHILA MEDIOSTRIATA MAP
12-23-69 172435 JPL/IPL
DROSOPHILA MEDIOSTRIATA STRAIGHTENED
07-23-70 232507 JPL/IPL

FIGURE 10-34. Use of a mapping function to prepare straightened karyograms of *Drosophila* salivary gland chromosomes. (Upper) Original image. (Lower) Straightened image. (Reprinted with permission from Kenneth Castleman, Jet Propulsion Laboratory, Pasadena, California.)

The gray value of the output pixel is then found as

$$A_{(x,y)} = [A_{(x,n+1)} - A_{(x,n)}](y-n) + A_{(x,n)} \quad (10\text{-}24)$$

Analysis will show that the same result will be obtained if the interpolation is performed in the vertical direction first, followed by a horizontal interpolation of the calculated values. Figure 10-33 shows the results of an interpolative procedure used to correct uneven vertical magnification in a digitized video image.

Geometrical manipulations can also be used for other purposes besides correcting for geometrical distortion. Mapping functions can be derived for displacing and rotating objects within the digital image so that related scenes can be displayed in a similar format. Figure 10-34 shows a graphic example of the use of a mapping function for generating straightened images of *Drosophila* polytene chromosomes.

10.10. DENSITOMETRY USING THE DIGITAL IMAGE

Digital processing is ideally suited for applications involving the measurement of intensity information from the original specimen. The tremendous speed of the video camera and digitizer produces an almost instantaneous measurement of image brightness at each sampling point in the image. The only requirement remaining is that the data must be converted from units of gray value into some more meaningful measure such as optical density or phase retardation, so that the values can be determined directly from the digital image.

The first step in using the video system as a densitometer is to determine a calibration function that describes the relationship between the gray values of the digital image and the optical density of the specimen. This function can be determined by placing a series of graded neutral-density filters in the object plane of the microscope and averaging the digitized gray value of the entire image. Figure 10-3 shows a typical calibration curve measured with a chalcogenide (Newvicon) video camera and external shading corrector. This calibration is not a linear function, but has a complex shape that can sometimes be approximated by a logarithmic function. The error bars at each data point in Fig. 10-3 show that the response of the video system is not constant across the video field. This is mostly due to variation in the illumination across the specimen and from the shading distortion introduced by the video camera that cannot be removed from the optical video system. This variation in intensity across the image is not visible to the human eye, but represents a variation of about 8 gray values around the mean at the highest intensities.

If this amount of shading distortion can be accepted, the calibration curve of Fig. 10-3 can be used directly to measure the optical density of individual pixels in any digital image captured by the system under the same conditions. This curve *cannot,* however, be used to determine the average optical density of a *region* of the image by finding the average gray value of the region and then using the calibration curve to find the average optical density. To demonstrate this, consider three pixels that represent optical densities of 0.1, 0.2, and 0.3 in the specimen, and have gray values of 200, 150, and 70, respectively. The average optical density of these pixels is 0.2; however, their average gray value is only 140 instead of 150. This discrepancy arises because the calibration curve is not a linear function. The nonlinear nature of the calibration curve will affect individual gray levels disproportionately according to their position on the calibration curve, and errors will result. Similar problems arise if the raw image is used for image manipulations involving subtraction or division of two images.

The most straightforward way to avoid problems caused by a nonlinear response curve is to create an ITF (Section 10.6) that converts the raw image into a second image with gray values that are a linear function of optical density. This transforming function that can "linearize" raw images is created by using the following procedure.

The gray values in the original raw image can be expressed as a function of optical density according to the relationship

$$GV_1 = f(OD) \quad (10\text{-}25)$$

where f(OD) is the calibration function. This equation can also be used to express the optical density as a function of raw gray value by calculating its inverse

$$OD = f^{-1}(GV_1) \tag{10-26}$$

We need to produce an output image where gray value is a linear function of optical density with the form

$$GV_2 = m \times OD + b \tag{10-27}$$

so by direct substitution of equation (10-26) into (10-27) we get the desired relationship,

$$GV_2 = m\,[f^{-1}(GV_1)] + b \tag{10-28}$$

This equation shows that the inverse of the nonlinear function can be used to correct the nonlinearity.

The ITF defined by equation (10-28) can be easily found by direct substitution using the data from the calibration curve. The slope and intercept of this function can be arbitrarily set so that the brightest and darkest areas of the original image fill the full dynamic range of the digital memory. Once the linearizing function has been determined, it can be loaded into a look-up table (Section 10.6) and used to modify the stored form of the digital image. Once this linearizing function has been calculated and loaded into a look-up table, it can also be used to linearize any subsequent image that is digitized under the same conditions; consequently, the linearization of additional images can be accomplished with little additional effort and can be done even in real time. Figure 10-35 shows the effect of this linearizing procedure on an autoradiographic image of a tritium-labeled rat forebrain section before and after the transformation. Densitometric measurements are made from the linearized image by finding the average gray value of desired areas in the digital image and substituting the average gray value into equation (10-27) (Altar *et al.*, 1984).

When shading distortion in the image is unacceptable, the distortion must first be removed from the image before densitometric measurements can be made. An effective method for removing shading distortion from the video image is to divide each raw image by a second image of a blank uniformly illuminated field (Schultz *et al.*, 1974). When performing densitometry, this background image will always be brighter than the specimen; consequently, the ratio of the images must also be multiplied by a contant value of 255 to avoid the production of fractional values and to preserve the full dynamic range of the digital memory. When shading distortion is severe, this procedure may be more effective if both of the images are first converted to linear functions of neutral density before the ratio procedure is performed. Figure 10-36 shows the effect of the ratio routine on the shading distortion of a uniformly illuminated field of view.

The relationship between the ratio procedure and the linearization procedure is shown in Fig. 10-37. The open circles in this figure show a calibration curve that has been prepared from ratio-corrected images. The variation in gray value has been reduced about 50% in comparison with uncorrected values. The closed circles show a calibration curve prepared from images that were linearized before they were ratio-corrected. The variation in gray value of this curve has been reduced to a level of less than 1 gray value at most intensity levels except the lowest, approximately a two- to threefold improvement over corresponding values that have not been linearized. Note that the variation in gray value becomes largest at the highest neutral densities in the linearized curve. This is due to the shallow slope of the original calibration curve at this level, which means that the camera is relatively insensitive to changes in optical density at this range.

The linearization and image-ratio procedures are designed to correct for the shading distortion and nonlinearities in the response of the video camera. If the ultimate goal of the densitometric measurement is not to measure optical density of the specimen but, instead, some other quantity related to image brightness, such as stain density or the number of autoradiographic grains, then a second calibration procedure must be followed to convert optical density in the specimen to the desired units. This calibration curve must be determined from images of known standards that have been linearized and corrected for shading distortion as described. The calibration curve can then be

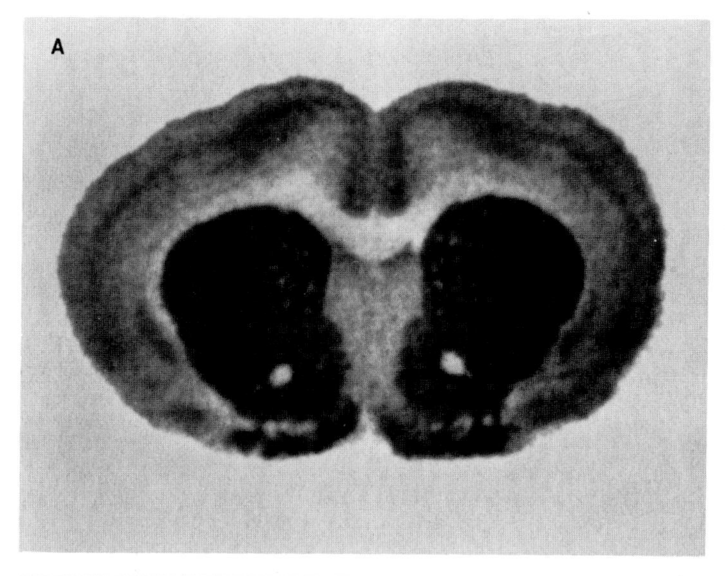

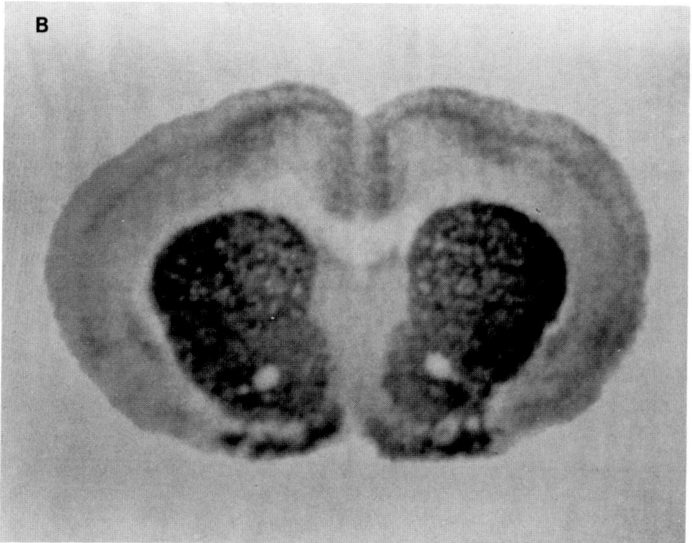

FIGURE 10-35. Transformation of image gray value to a linear function of optical density. (A) Original image. (B) Linearized image produced using equation (10-28). (Specimen and digital image courtesy of Anthony Altar, Department of Psychobiology, University of California, Irvine.)

used to make a second ITF that can be used to convert the image to the desired units directly. This procedure can be avoided if it can be established beforehand that the quantity being measured varies linearly with optical density.

10.11. DIGITAL FILTERING AND SPATIAL CONVOLUTIONS

Each of the digital manipulations presented thus far in this chapter has been an example of a *point operation;* i.e., the gray value of each output pixel of an operation is determined only by the gray value of a corresponding input pixel, and the properties of surrounding pixels are unimportant. Point operations do not alter the spatial information in an image, and can only affect other properties

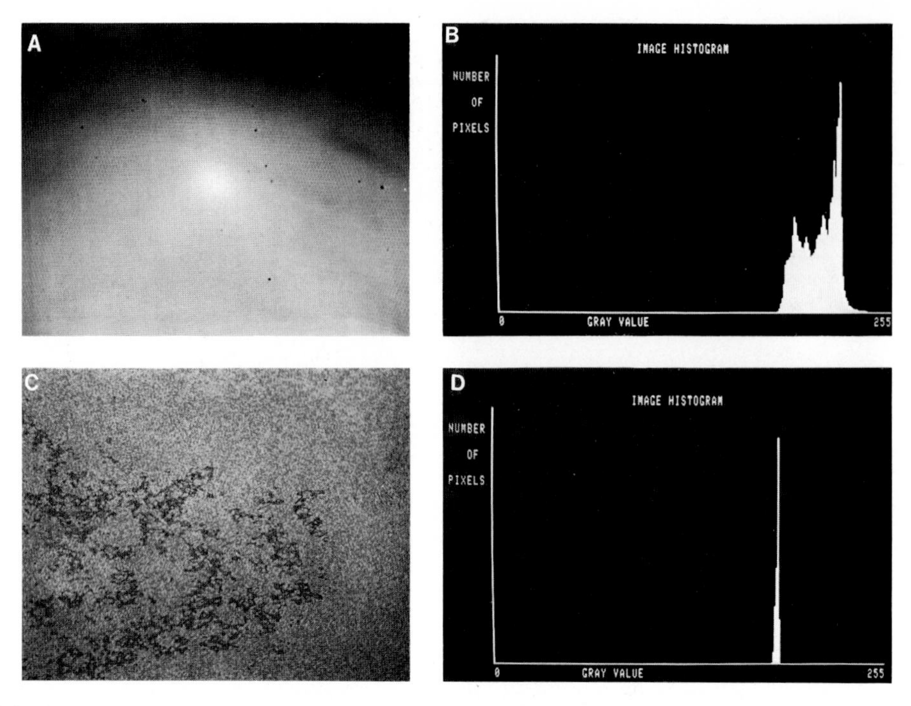

FIGURE 10-36. Effect of the ratio routine on shading distortion. (A) Original image of a uniformly illuminated field of view. Image contrast has been linearly enhanced 3.5 times. (B) Gray value histogram of (A) before enhancement. Standard deviation of image gray value is 12.3 units. (C) Image of (A) after dividing by a second image of brighter illumination. Image has been linearly enhanced 25 times. (D) Gray value histogram of (C) before enhancement. Standard deviation of image gray value is 1.0.

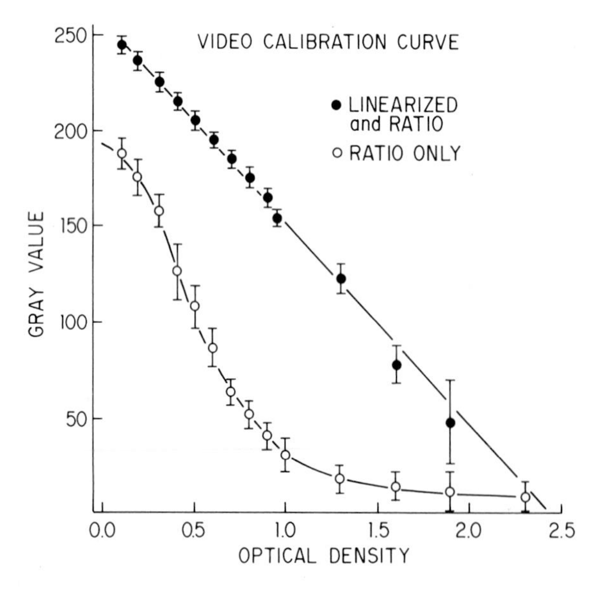

FIGURE 10-37. Effect of the linearization and ratio procedures on variations in the camera response curve due to shading distortion. Open circles: response curve prepared from ratio-corrected standards. The standard deviation of each measurement is between 2 and 6 gray value units. Closed circles: Response curve after linearization followed by ratio-correction. The standard deviation of each measurement is less than 1 gray value unit except at the lowest values. Error bars have been equally exaggerated in both curves to provide greater clarity.

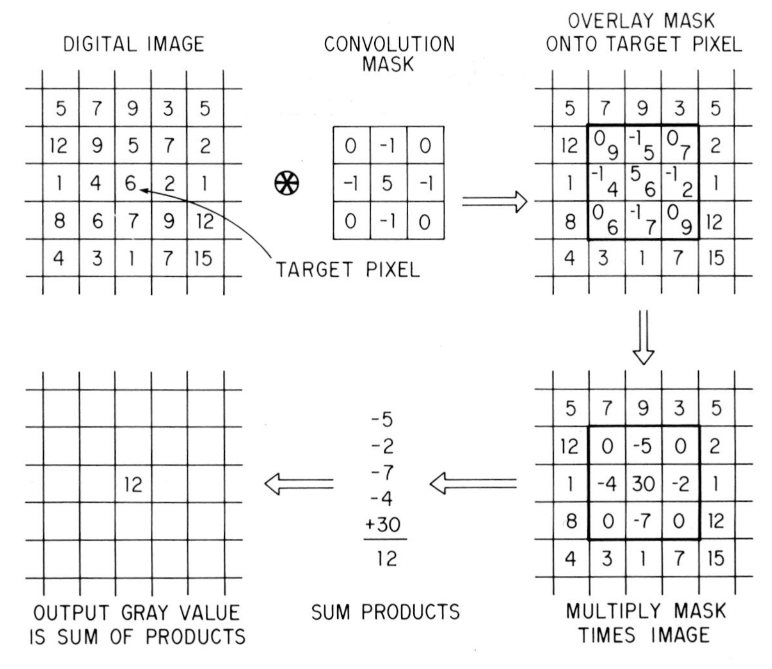

FIGURE 10-38. The convolution operation.

such as brightness and contrast. The term *digital filtering* is used to describe other types of manipulations that are used to affect the spatial information of a digital image, and thus are to be contrasted to point operations. Digital filtering routines are used for such tasks as the preferential enhancement of object boundaries in an image or the suppression of added noise. Digital filtering routines are also used extensively in applications involving computerized tomography and three-dimensional image reconstruction.

An example of a digital filtering routine is *spatial convolution.* A **convolution** operation on a digital image involves a point-for-point multiplication of selected pixel values of the input image times the corresponding pixels in a second image known as a **convolution mask** or a **kernel.** Convolution masks typically contain an odd number of rows and columns and are usually square, with a 3×3 mask being a common form. The convolution operation is performed individually on each pixel of the original input image and involves three sequential operations as shown in Fig. 10-38.

In the first step, the convolution mask is overlayed on the original image in such a way that the center pixel of the mask is matched with the single pixel location to be convolved in the input image, called the target pixel. Second, each pixel value in the original image is multiplied by the corresponding value in the overlying mask. Finally, the gray value of the target pixel is replaced by the sum of all the products determined in the second step. To perform a convolution on an entire image, this operation must be repeated for each pixel in the original image. It can be seen from these steps that a convolution mask with a value of 1 at its center pixel and values of 0 elsewhere, is the identity function, and will produce an output image that is identical to the input image.

The convolution operation is described in its discrete form by the relationship

$$B_{(i,j)} = \sum_{m} \sum_{n} I_{(i-m,j-n)} M_{(m,n)} \tag{10-29}$$

where $B_{(i,j)}$ is the pixel at row i and column j in the output image; $I_{(i,j)}$ is the corresponding pixel in the input image; and $M_{(m,n)}$ is the pixel at row m and column n in the convolution mask. This operation is often represented by the shorthand form

$$B = I \circledast M \tag{10-30}$$

where $\circledast$ represents the convolution operation.

Careful examination of equation (10-29) will reveal that the terms $(i-m)$ and $(j-n)$ actually cause a rotation of the convolution mask 180° around its central pixel before it is overlayed on the target pixel; thus, the mask is inverted before it is used. Many convolution masks are symmetrical about their axis, however, and this rotation has no effect. Excellent discussions on the mathematics of both the continuous and discrete forms of the convolution operation can be found in the books by Castleman (1979), Pratt (1978), and Brigham (1974).

An example of a convolution mask used to reduce noise in an image, called a *smoothing filter,* is shown in Fig. 10-39. When the mask of Fig. 10-39 is convolved with an image, the gray value of each pixel is replaced by the average intensity of its eight nearest neighbors and itself. If the gray value of any of these nine pixels is unusually high or low due to the effect of added noise, this averaging will tend to reduce the effect of the noise by distributing it among all of the neighboring pixels.

Convolution with a smoothing mask can be interpreted as an operation that suppresses the contribution of high spatial frequencies in an image. The term *spatial frequency* is analogous to frequency with respect to time (temporal frequency), and describes how rapidly a signal changes with respect to position in the image. A function (such as a cosine wave) has a low spatial frequency if it shows only a few cycles across the width of an image, while a function has a high spatial frequency if it shows many cycles across the image. The highest spatial frequency that can be displayed in a digital image has a period with a length equal to the width of two pixels.

Random noise in an image usually has a high spatial frequency, and can be effectively removed by a smoothing convolution. Other desirable image features, such as object boundaries, may also have high spatial frequencies and may also be suppressed by a smoothing filter. Consequently, the smoothing mask has the sometimes undesirable effect of blurring an input image, and the larger the mask the more severe this blurring will be. For many applications, the size and form of the smoothing mask must be carefully chosen in order to minimize this trade-off between noise reduction and image degradation. The effect of smoothing filters of several different sizes is shown in Fig. 10-40.

The nine terms of the smoothing mask in Fig. 10-39a add to a value of 1. This mask has been given this property so that the convolution operation will produce an output image with an average brightness that is equal to that of the input image (in some cases this may only be approximate). The sum of terms in most convolution masks will add to a value of between 0 and 1 so as not to create an output image with gray values that exceed the dynamic range of the memory device.

It should be noted that the two smoothing masks in Fig. 10-39 have the same effect and thus are equivalent. In Fig. 10-39a, the image is convolved with a mask of fractional values equal to one-ninth. In the case of Fig. 10-39b, the image is convolved with a mask with values equal to 1 and is

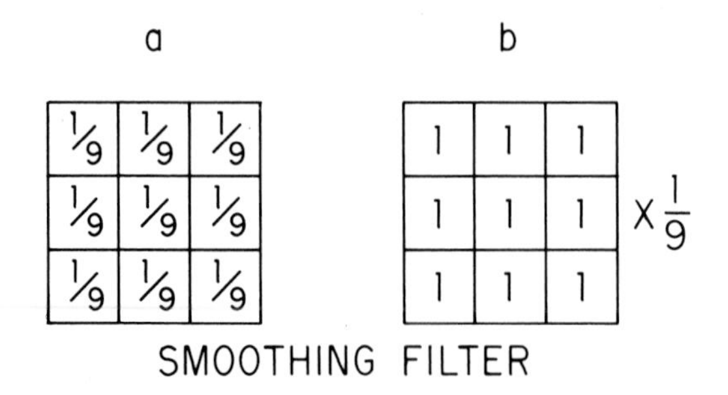

FIGURE 10-39. Equivalent 3 × 3 smoothing masks.

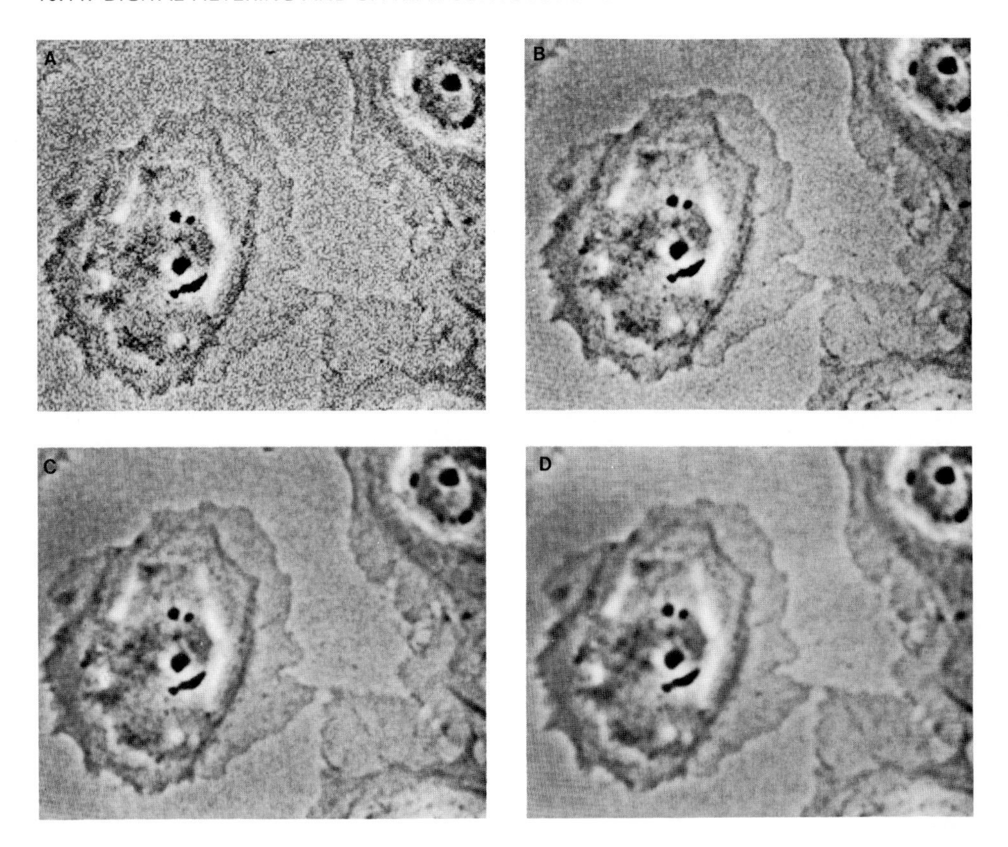

FIGURE 10-40. Effect of smoothing masks of different sizes. (A) Original image with added random noise; (B) 3 × 3 smoothing; (C) 5 × 5 smoothing; (D) 7 × 7 smoothing.

then divided by a constant value of 9 to avoid affecting the brightness of the output image. While the operation of Fig. 10-39a is more direct, fractional values are often difficult to use in an array processor that deals exclusively with integer values. Consequently, it is more common to use convolution masks that have integer values for each term, followed by a division of the output image by a constant equal to the sum of all the terms in the mask, in order to normalize image brightness.

A second type of convolution mask is designed to enhance the contrast of edges and boundaries. Examples of such masks, known as *sharpening filters* and **Laplacian filters,** are shown in Figs. 10-41 and 10-42. The effect of each of these masks on an image can be seen in the examples of Fig. 10-43. In contrast to the smoothing filters, sharpening and Laplacian masks are examples of

1

0	-1	0
-1	5	-1
0	-1	0

2

-1	-1	-1
-1	9	-1
-1	-1	-1

3

1	-2	1
-2	5	-2
1	-2	1

SHARPENING FILTERS

FIGURE 10-41. Sharpening filters.

4

0	-1	0
-1	4	-1
0	-1	0

5

-1	-1	-1
-1	8	-1
-1	-1	-1

6

1	-2	1
-2	4	-2
1	-2	1

LAPLACIAN FILTERS

FIGURE 10-42. Laplacian filters.

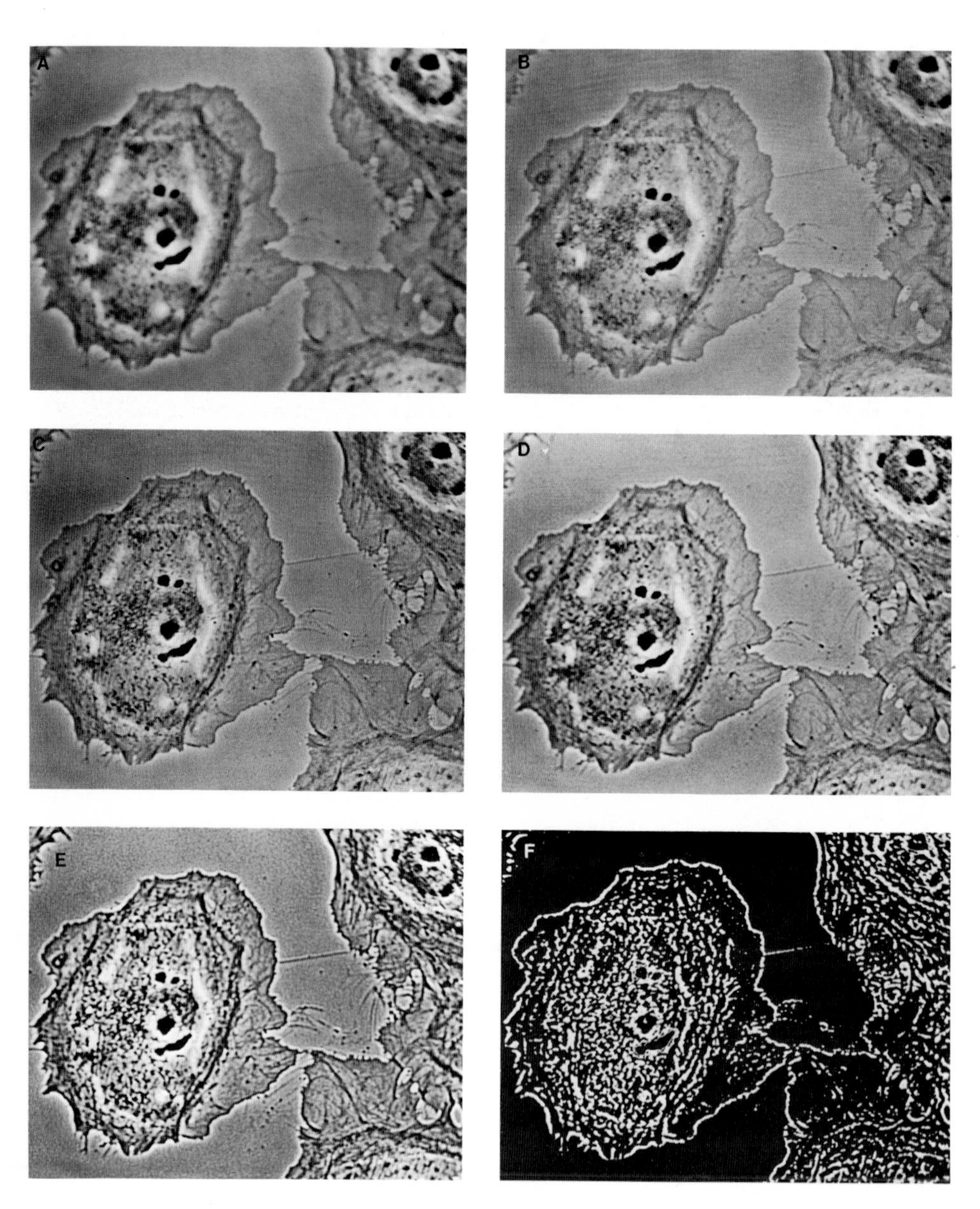

FIGURE 10-43. Effect of image convolution with sharpening and Laplacian masks. (A) Original image; (B) mask 1; (C) mask 2; (D) mask 1, scaled to 6 × 6; (E) mask 2 scaled to 6 × 6; (F) mask 5, scaled to 6 × 6 and thresholded to detect cell boundaries.

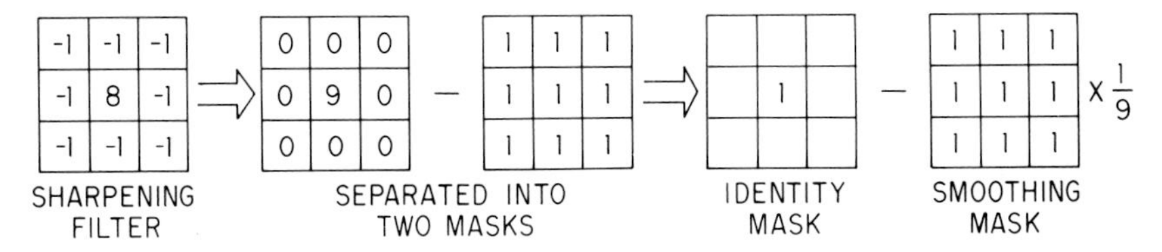

FIGURE 10-44. Separation of a sharpening mask into two masks.

filters designed to enhance the higher spatial frequencies in an image, while suppressing lower frequencies. In addition to enhancing object boundaries, such filters also have the effect of removing slowly varying background shading in an image, and thus can be used to correct for shading distortion in an image without the need for image subtraction involving a second image as described in Section 10.8. However, these filters may also have the undesirable effect of enhancing noise.

Sharpening and Laplacian filters can be adjusted in size in order to have maximum effect on different spatial frequencies. The 3 × 3 masks shown in Figs. 10-41 and 10-42 have their greatest effect on image features that vary greatly over the spacing of a single pixel. In order to enhance spatial frequencies that are varying more slowly in an image, these filters can be scaled to match the desired frequency (Sklansky, 1978b). To double the size of a 3 × 3 mask, each coefficient is replaced by a 2 × 2 matrix having the same value (creating a 6 × 6 mask) and so on. The effect of mask scaling in shown in Fig. 10-43D,E.

Many of the convolution masks commonly used in image processing have been determined empirically for maximum effect and do not have a strong basis in theory, although some attempts have been made to do so (Brooks, 1978). One explanation for the effect of sharpening and Laplacian filters can be given in relation to Fig. 10-44. It is seen that such filters can be separated into two different masks, one of which is subtracted from the other. The first of the two masks shown in Fig. 10-44 is the identity function described earlier, which produces an output image equal to the input image after convolution. The second mask in Fig. 10-44 is the smoothing filter discussed above. As a result, the sharpening and Laplacian filters can be thought of as having the effect of *subtracting a blurred copy of an image from the original image,* as shown in Fig. 10-45. This effect can actually be demonstrated by subtracting a slightly out-of-focus microscope image from an in-focus image, using the techniques of Section 10.8, which produces an image similar to that achieved with a sharpening filter.

The Laplacian filters have the effect of calculating the second-order gradient of intensity in an

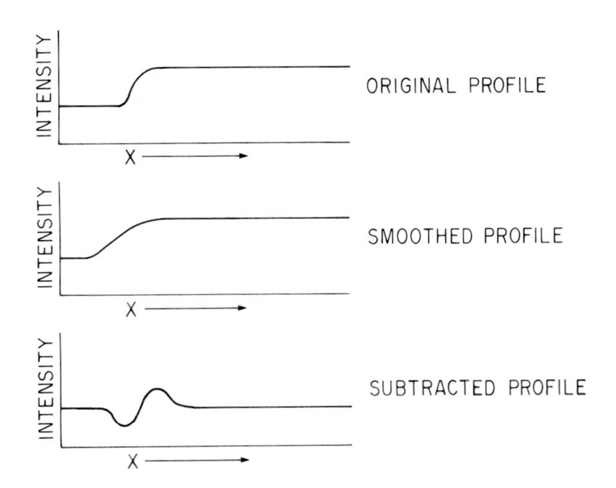

FIGURE 10-45. Effect of subtracting a smoothed profile from itself.

image, which is equivalent to the second derivative of the image function. The sharpening filters have the effect of adding the Laplacian to the original image so that, where no gradient is detected, the intensity of the output image is equal to that of the input image. The Laplacian filters produce an output image that is the second-order derivative only; consequently, the output image has zero value (black) where no gradient is detected. Note that the blanking effect of the Laplacian masks can be predicted by observing that the sum of the terms in these masks is 0.

There are many other convolution masks and filters that are used to enhance high-frequency information in an image. Some of these have come into common use and have been named after their inventors, such as the Sobel, Roberts, and Kirsch operators. Some of these filters are nonsymmetric and will produce different results depending on their orientation. Others involve nonlinear operations between pixels such as the calculation of logarithms and square roots. An excellent comparison and review of several of these filters can be found in Pratt (1978, pp. 479–499).

Convolution operations are computationally expensive in the sense that many operations are required to find the gray value of a single pixel. When using a 3 × 3 mask, as many as nine multiplication and eight additional operations can be required to find the intensity of a single target pixel, which can mean that several million operations are necessary to convolve even a relatively small image. It is for operations such as these that dedicated array processors become most useful for image-processing purposes. By taking advantage of the parallel-processing capability and other special features of an array processor, the time required for the convolution operation can be reduced, possibly to a fraction of a second when using a small mask. Some dedicated array processors exist that can convolve an image with a smoothing mask of variable size, or are limited to a 3 × 3 or 5 × 5 mask, but can perform these convolutions in real time.

10.12. DIGITAL FILTERING INVOLVING FOURIER TRANSFORMS

The single most powerful techniques of digital processing are those that involve Fourier analysis. The mathematical basis for these operations is fairly complex and cannot be adequately discussed within the scope of this chapter. Although Fourier techniques are somewhat complicated, they are also quite elegant, and offer a useful perspective for approaching and understanding the fundamentals of image formation. Convolution and Fourier techniques apply directly to the imaging properties of all types of microscopes, and no microscopist should feel competent at the craft without at least a working knowledge of these techniques.

The basis for Fourier techniques is the *Fourier transform,* which is an operation based on a theorem that states that *any harmonic function can be represented by a series of sine and cosine functions, which differ only in frequency, amplitude, and phase.* The Fourier transform is a function

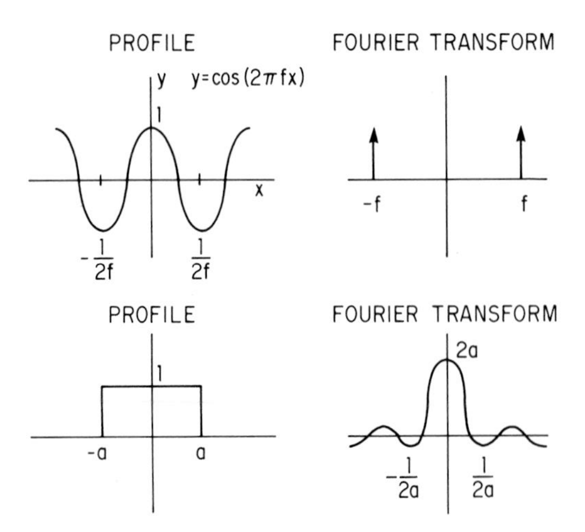

FIGURE 10-46. Two functions and their Fourier transforms. The horizontal axis of the Fourier transforms show the distribution of spatial frequencies in the original functions.

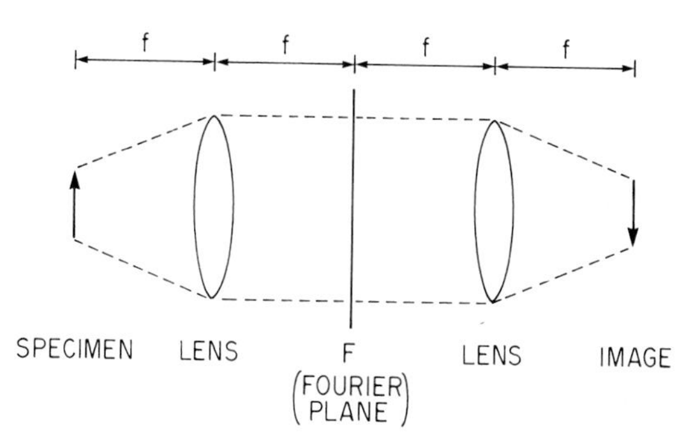

FIGURE 10-47. An optical system for imaging the Fourier transform of an object.

that shows the frequency and amplitude relationship of the harmonic components of the original function from which it was derived. Examples of two functions and their Fourier transforms are given in Fig. 10-46. The main point to be noted there is that the Fourier transform converts a function that varies in space to another function that varies with frequency. It should also be noted that the highest frequencies of the original function are found the farthest away from the origin in the Fourier transform. Excellent discussions on the properties of the Fourier transform can be found in Brigham (1974) and Liu and Liu (1975).

The application of the Fourier transform to optics can be demonstrated by the imaging system shown in Fig. 10-47. The specimen is brightly illuminated from behind, and the light reaching the first lens is an intensity profile that corresponds to the light-transmitting properties of the specimen. The first lens in this system is situated exactly at its focal length from the specimen plane; consequently, light rays leaving the back of the lens travel parallel to each other. The second lens focuses these parallel rays and produces the inverted image seen at the image plane.

If the pattern of light leaving the back of the first lens is examined at the position marked F in Fig. 10-47, which is one focal length from the lens, it will be discovered that the image at this point is the Fourier transform of the intensity profile of the specimen [a full derivation of this relationship can be found in Hecht and Zajac (1974) or Castleman (1979)]. The position where the Fourier transform is found behind the lens is called the *Fourier plane* or *transform plane,* and will always show the Fourier transform of the specimen regardless of the distance of the specimen from the lens (Hecht and Zajac, 1974). The second lens in this system also produces a Fourier transform; however, the image that this lens "sees" is the Fourier transform produced by the first lens. This second Fourier operation forms the inverted image of the specimen found in the image plane. Thus, two Fourier transforms performed sequentially on the same function will reproduce the original function, which has been inverted but is otherwise unaltered.

Every converging lens produces a Fourier transform. For most lens systems, the Fourier plane will be formed only at one position in space and may actually be inside the lens under some circumstances. Within the microscope, the Fourier plane is the same as the back focal plane of the objective; consequently, the Fourier transform of a specimen can actually be visualized by using a phase telescope and looking down the microscope barrel.

It is important to note that the Fourier transform of an image contains all of the information found in the original image. If the information represented in the Fourier transform can be preserved during its transmission through the microscope, the image formed at the image plane will be an exact replication of the original specimen. No optical system is perfect, of course, and some modification of the image will always occur. Such modifications produced by the optical system can be modeled by a series of Fourier operations as shown in Fig. 10-48. This type of modeling is called linear system analysis, and is used extensively in optical applications.

The imaging system in Fig. 10-48 is assumed to produce an ideal Fourier transform of the specimen at the back focal plane of the first lens. This ideal transform is then modified by a nonuniform filter that represents the nonideal properties of the real optical system. This nonuniform

SPECIMEN → IMAGING LENS → IDEAL FOURIER TRANSFORM → NONLINEAR FILTER → ALTERED FOURIER TRANSFORM → IMAGING LENS → IMAGE

FIGURE 10-48. System model for an optical imaging system.

filter can be thought of as a partially transmitting piece of glass or plastic placed in the back focal plane of the microscope. This filter produces an altered Fourier transform by a point-for-point multiplication of the intensity of the ideal transform by the transmittance of the corresponding point in the nonlinear filter. The altered Fourier transform is then transformed a second time by the second lens and an altered image is formed in the image plane.

The utility of this linear system model can be demonstrated by reconsidering the resolution limitations of the microscope as discussed in Chapter 5. In relation to the linear system model of Fig. 10-48, the primary lens of the microscope can be considered a perfect device that images an ideal Fourier transform of the intensity function of the specimen, as shown in Fig. 10-49. This ideal Fourier transform extends to infinity, with the highest spatial frequencies (representing the smallest, least resolvable features of the specimen) being found the farthest from the optical axis. The nonlinear filter in this system is the *aperture function* of the microscope, which completely blocks all light from the ideal Fourier transform that is beyond the diameter of the aperture. Since the highest frequencies in the Fourier transform are found farthest from the optical axis, the aperture function has the effect of removing these high frequencies from the transform. When the image of the specimen is re-formed by the second Fourier operation, the smallest features in the image will be missing, and resolution will be reduced. This analysis predicts that the wider the aperture function of the microscope, the higher the frequencies allowed to pass from the ideal Fourier transform, and the better the resolution of the microscope will be. This is the same conclusion reached by other means in Chapter 5.

The use of partially transmitting filters in the Fourier plane for the purpose of affecting spatial information in the image plane is called *spatial filtering*. This type of manipulation can be performed with real filters on a real imaging system, and can be a fascinating exercise for examining the properties of the microscope or other optical device. (See Appendix I.) Several excellent photographs showing the effects of real spatial filters can be found in Hecht and Zajac (1974) and Walker (1982).

Fourier techniques are used in digital image processing according to the linear system model shown in Fig. 10-50. This system is similar to that of Fig. 10-48 with the exception that a second filter has been added to the Fourier plane. This second filter is an artificial one, created by digital techniques, and is designed to exactly compensate for the degrading effects caused by the first filter (which represents the natural defects of the optical system). This model is implemented as follows. First, the Fourier transform is computed for the raw digital image. This Fourier transform represents the ideal image, which has been degraded by the optical system. Second, each pixel of the Fourier

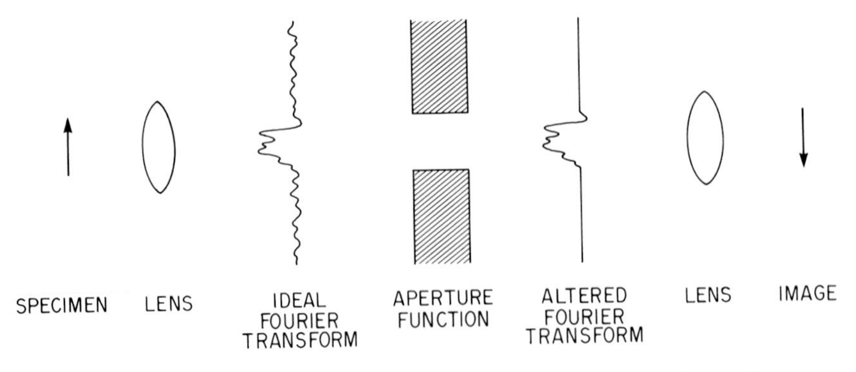

FIGURE 10-49. Model of a miscroscope imaging system utilizing Fourier transforms.

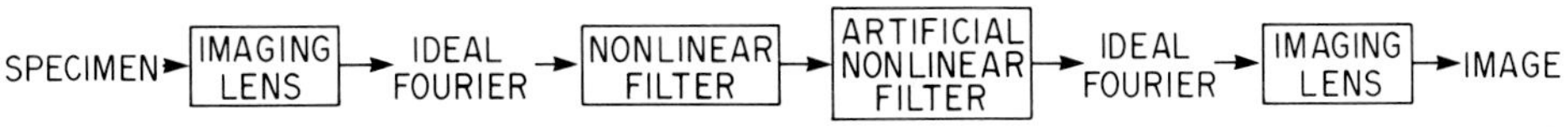

SPECIMEN → IMAGING LENS → IDEAL FOURIER → NONLINEAR FILTER → ARTIFICIAL NONLINEAR FILTER → IDEAL FOURIER → IMAGING LENS → IMAGE

FIGURE 10-50. Use of artificial Fourier filters to correct for degradation in an optical system.

transform is multiplied by its corresponding value in the artificial filter, which has been carefully constructed to complement the nonideal properties of the optical system. Finally, a second Fourier transform is calculated to re-form the original image that is free of the effects of deggradation (in reality, this last operation is a separate operation called the inverse Fourier transform, which has the same effect except that it does not invert the image).

Spatial filtering involving Fourier techniques can be a very useful way for manipulating images. For example, high- or low-spatial-frequency information can be deleted from an image by designing a Fourier filter that is nontransmitting at the location that these frequencies appear in the Fourier transform. This technique is especially useful for removing harmonic noise from an image such as the ''herringbone'' pattern sometimes seen in video images. Since this added noise is harmonic, it will be found in localized discrete parts of the Fourier transform. When these local peaks are removed from the transform, the re-formed image is essentially unaltered except that the herringbone pattern is gone. An especially good photographic example of this type of filtering can be found in Castleman (1979, pp. 391–392).

In order for Fourier filtering to be most effective, a precise description of the degrading properties of the optical system must be known. This description can be obtained theoretically by making mathematical models of the optical system, or can be measured directly by imaging a known specimen and examining the difference between the known and observed properties. As an example, consider a Fourier filtering routine designed to compensate for a nonideal modulation transfer function (MTF) of a real optical video system (Chapter 5, Appendix II). This compensation is carried out as follows. First, the MTF of the optical video system is measured by using a sine wave test pattern and examining the pattern in the back focal plane of the microscope (this pattern is the MTF). Second, the measured MTF is compared to its ideal equivalent in order to calculate the form of a Fourier filter that models the measured degradation. Third, an artificial filter is created that exactly compensates for the first calculated filter. Finally, this artificial filter is used on all subsequent images captured with the optical system according to the Fourier filtering routine of Fig. 10-50.

Fourier techniques are also used for applications involving three-dimensional reconstruction. For example, the upper panels of Fig. 10-51 show a series of photographs of fluorescently labeled chromatin taken at several focal planes within the intact nucleus of a cell within a *Drosophila* salivary gland. When the Fourier transform of these images are calculated, it can be shown that information that comes from in-focus regions of each image changes most rapidly between adjacent photographs, while the Fourier components that are from out-of-focus features are nearly the same in each image (Agard and Sedat, 1983; Gruenbaum *et al.*, 1984). This quality allows the construction of Fourier filters that selectively filter out-of-focus information from each image while retaining the in-focus information, as shown in the lower panels of Fig. 10-51. The relationship between the Fourier transform and the three-dimensional properties of a microscopic specimen can also be used to advantage in the construction of automatic focusing devices for microscopes. In such devices, the lengthy computing time that would normally be required for the calculation of the Fourier transforms can be avoided through the use of charge-coupled devices that can generate a Fourier transform directly from the optical image (Kowel *et al.*, 1976).

Another useful feature of the Fourier transform stems from its relationship to the convolution operation discussed in the last section. It will be remembered that the convolution operation involved several multiplication and addition operations, according to the contents of a convolution mask, to find the intensity of each target pixel. This can be contrasted to Fourier filtering where each value in the Fourier filter is simply multiplied by its corresponding pixel in the Fourier transform of the image. The two operations are related because *the convolution operation is identical to the Fourier filtering operation when the Fourier filter is the Fourier transform of the convolution mask.*

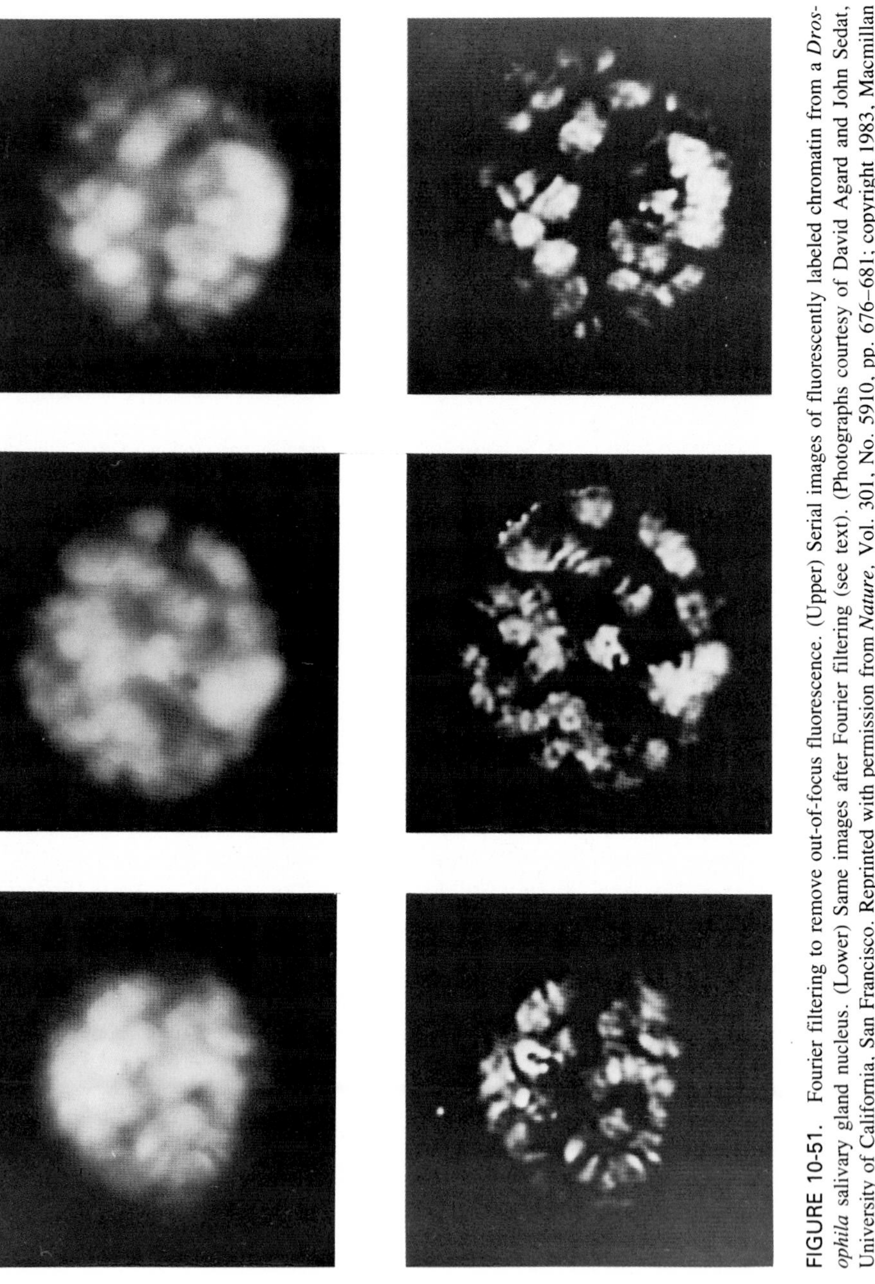

FIGURE 10-51. Fourier filtering to remove out-of-focus fluorescence. (Upper) Serial images of fluorescently labeled chromatin from a *Drosophila* salivary gland nucleus. (Lower) Same images after Fourier filtering (see text). (Photographs courtesy of David Agard and John Sedat, University of California, San Francisco. Reprinted with permission from *Nature*, Vol. 301, No. 5910, pp. 676–681; copyright 1983, Macmillan Journals Limited.)

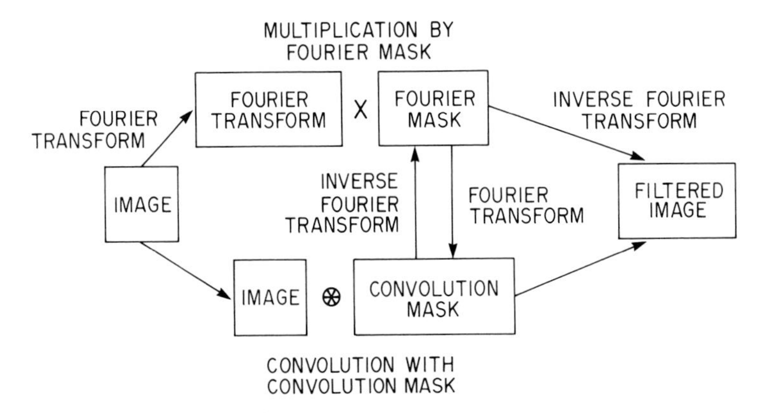

FIGURE 10-52. Equivalent Fourier filtering and convolution operations.

This means that either of these two techniques can be used to obtain identical results from an image, depending only on whether one wants to work in "image space" or "Fourier space." The equivalency of these operations is shown in Fig. 10-52.

The decision to use Fourier filtering or the convolution operation depends on the application being considered. The Fourier transform is an involved operation that takes considerable computer time and memory, and is much slower than a convolution operation using a small mask. The Fourier filtering technique is generally faster, however, than the equivalent convolution operation when the convolution mask is large and approaches the size of the original image. Switching between equivalent Fourier and convolution operations may also be used to reduce the complexity of their respective masks. For example, the Fourier transforms of the convolution masks presented in the last section are large and complex; consequently, it is easier to implement these masks by convolution. Conversely, a simple Fourier filter, like one to remove harmonic noise, would produce a large and complex convolution mask that would be difficult to use.

Fourier techniques also present intriguing possibilities for improving the resolving power of the optical microscope. There are theoretical models that suggest that since the Fourier transform of an image is a harmonic function, the harmonics measured within the aperture of the microscope can be used to predict the shape of the Fourier transform outside of the aperture boundary. This extrapolation, in effect, enlarges the size of the aperture and thus increases the resolving power of the system. This "infinite aperture" technique has been called *superresolution*, and on a theoretical basis it has been predicted that such techniques can be used to increase the resolving power of an optical system to one-fifth the Rayleigh criterion (Harris, 1964; Pask, 1976). Unfortunately, no practical demonstration of this improvement has been presented to date.

10.13. FEATURE EXTRACTION

The ultimate goal of any image-processing routine is to derive useful information from the original image. When the purpose of the image processing is to derive numerical information from the image, the routine can be termed *image analysis*, and the quantities being measured are called image features. For a microscopist, some of the most useful image features are those that describe the size and shape of individual objects within the image, for example, single cells on a culture dish. Examples of interesting features derived from this type of image are cell area and perimeter, circularity, number of touching neighbors, and so on. A representative list of commonly derived image features used to describe individual objects in shown in Fig. 10-53.

Image-processing routines designed for image analysis usually involve three sequential operations, *preprocessing*, **image segmentation,** and *feature extraction* (Sklansky, 1978b). The first operation, preprocessing, describes individual manipulations that are used to convert the original

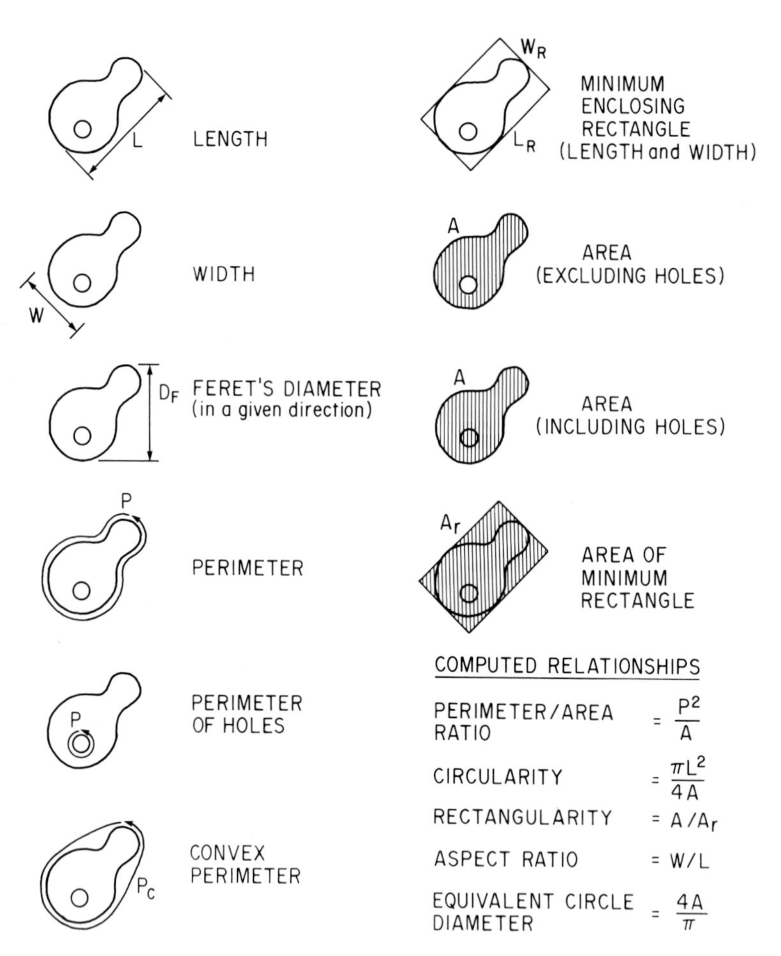

FIGURE 10-53. Examples of commonly calculated image features.

raw image into a form that is more suitable for detailed processing. Examples of preprocessing routines are manipulations to correct shading and geometrical distortion, intensity transformations to increase contrast or correct nonlinear transfer functions, histogram equalizations to put unlike images in a similar format, filtering operations to enhance edges and boundaries, or image subtraction routines to distinguish differences between two similar images. Most of the manipulations described in this chapter are examples of routines that can be used for image preprocessing.

Once the image has been converted to a suitable form, the second step in image analysis is image segmentation. Segmentation describes operations that identify regions of an image that have the same properties while at the same time separating those regions from dissimilar areas. Segmentation usually involves the identification of boundaries in the image that separate one image feature from another. Examples of segmentation boundaries are outlines that separate the image of a cell from cell-free regions, an outline that separates cell nucleus from cytoplasm, or an outline that separates an area of malignant cells from normal cells in a histological section.

Segmentation outlines are created by first identifying a region of common characteristics and then finding a set of connected pixels that completely surround the region, called an *edge map*. An edge map has a width of one pixel, and each pixel in the edge map is called an *edge element*. Several different approaches can be used to find an edge map. If the region boundary can be distinguished from surrounding regions on the basis of gray value alone, a simple approach is to convert the image into a binary image using a gray level thresholding transformation (Sections 10.5, 10.6), followed by a boundary-following routine that marks the areas of transition between black and white pixels. An example of an edge map produced by a threshold-boundary-following procedure that separates the image of a cell from the surrounding background is shown in Fig. 10-54. Examples of other routines used to find edge maps are thresholding based on multidimensional histograms (Preston and Dekker, 1980), thresholding and filtering routines based on local properties of the image (Lipkin *et al.*, 1979; Davis and Rosenfeld, 1975; Shaw, 1979; Schwartz and Soha, 1977; Weszka, 1978; Weszka and Rosenfeld, 1979), region growing (Kirsch, 1971; Zucker, 1976), and boundary-following convolution operations (Diller and Knox, 1983). In many cases, it is impossible to create an entire connected edge map using thresholding or edge-enhancing procedures due to the fact that the region boundary of interest does not have constant properties for its entire perimeter. This is particularly true of the boundaries of cells that are reasonably well attached to a substrate and are being viewed with phase-contrast, DIC, or polarization optics. When thresholding procedures are used on these images, the best boundaries that can be distinguished are characterized by short regions of detected boundaries separated by gaps. These unconnected regions of putative boundary are called *edge segments* and some method must be found to connect them before a true edge map can be generated (Lemkin *et al.*, 1979).

One method for connecting disjointed edge segments is the expansion–contraction routine shown in Fig. 10-54. In this routine, disconnected edge segments are made to **dilate** by arbitrarily surrounding each white pixel in the binary image with eight neighboring pixels that are also white. If the gaps between edge segments are small, a few cycles of boundary growing will cause all of the edge segments to eventually contact another segment nearby. Once a connected boundary has been formed, the boundary is then contracted to a one-pixel-wide edge map using an **erosion** routine that deletes white pixels from the edges of the expanded boundary, but will not cause a break between connected pixels. While these techniques are fairly easy to implement, expansion–contraction routines may not be effective if edge segments are separated by large gaps or if there are false edge segments in the image that do not represent true region boundaries.

Under these circumstances, more sophisticated procedures must be followed in order to accurately segment an image. Examples of such procedures are the Hough transform (Sklansky, 1978a), which calculates the most probable location of the missing edge segments based on the shapes of known segments, Fourier descriptors of shape (Persoon and Fu, 1977; Zahn and Roskies, 1972), or local convolution operations involving variable masks (Diller and Knox, 1983).

Another interesting approach to image segmentation (and image processing in general) are the heuristic processing routines. These routines can be contrasted to the linear operations described thus far where every region of the image is processed in the same fashion. When using heuristic processing, separate regions of an image may be processed differently according to some prior decision about the information content of that part of the image. For example, a heuristic routine might perform an edge-enhancing convolution operation in a region where an edge is suspected of occurring, while performing a smoothing convolution where edges are assumed not to exist in order to reduce noise. The heuristic processing sequence must be determined empirically for each type of image, but can be very effective when the properties of the image to be analyzed can be reasonably well predicted (Martelli, 1976; Ballard and Sklansky, 1976; Ashkar and Modestino, 1978).

Many of the features used to describe images are derived from object boundaries, and consequently concern the edge maps derived during image segmentation. One goal of image analysis is to find a method for describing the edge map so that it can be transferred to the host computer and used efficiently to derive statistics such as the ones in Fig. 10-53. The most obvious way of describing an edge map is to compile a sequential list of each of the X and Y coordinates of the edge elements. A more efficient method comes from the realization that each edge element in an edge map can only have X and Y coordinates that are no greater than one more or one less than the coordinate of the element proceding it. Efficient listing can be achieved by storing the coordinates of the starting point of the edge map, and then listing each subsequent coordinate in reference to its relative displacement from the element preceding it. This type of listing is called *chain coding*, and results in a very efficient way for describing edge maps. Many efficient computer algorithms exist for deriving image statistics based on chain-coded edge maps (Freeman, 1979; Cederberg, 1979; Young *et al.*, 1981).

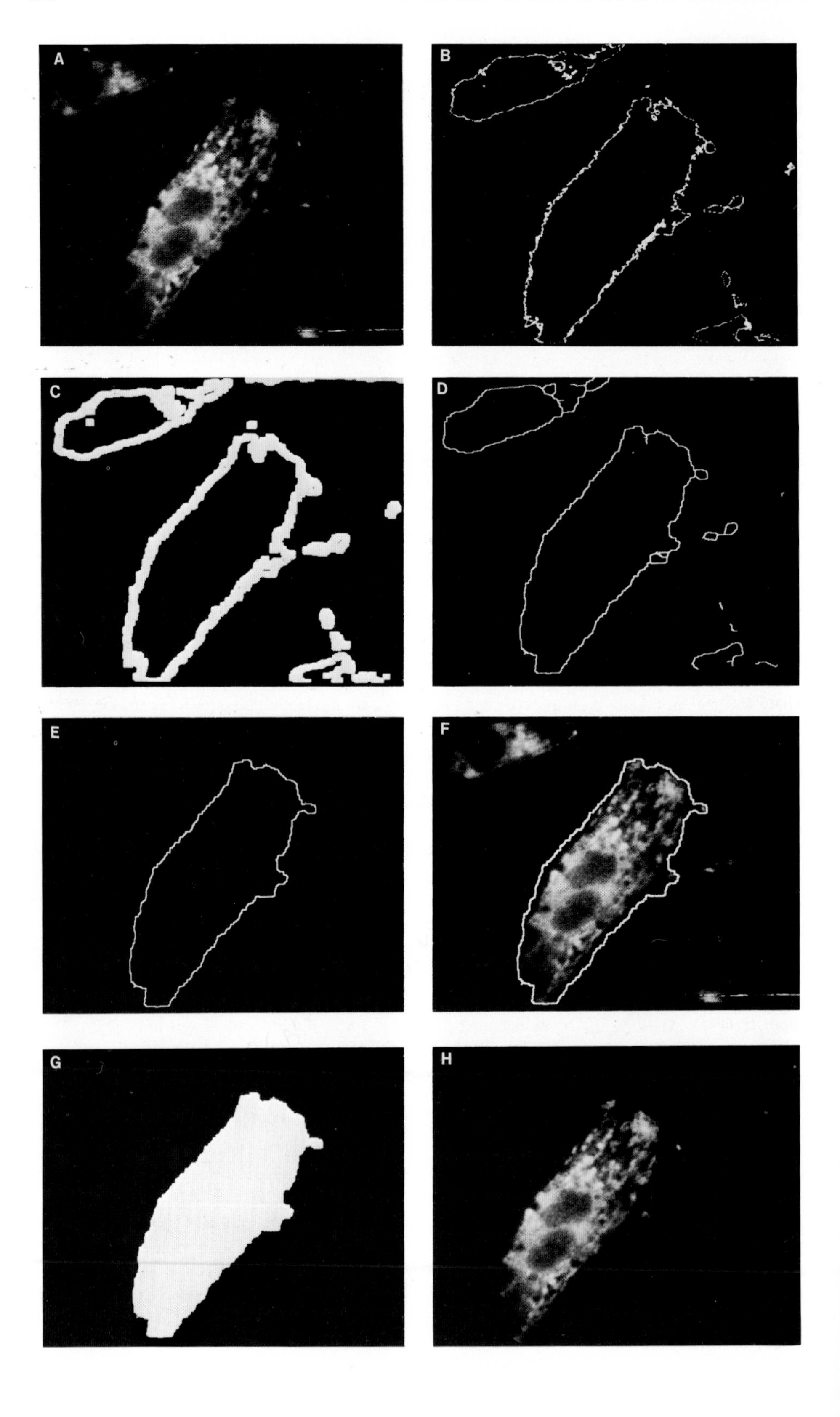
A
B
C
D
E
F
G
H

Edge maps that list every pixel in the segmentation boundary may be overly detailed for many processing applications. This added complexity can mean that too much memory space is being used to store the map, and that unnecessarily long processing times will be involved. Some methods for describing edge maps try to minimize this complexity by reducing the edge map to a set of simple lines and curves that can be used to describe the location of many elements. Examples of such coding routines are polygonal approximations where the edge map consists of the vertices of a polygon that completely encloses the segmented region (Sklansky and Gonzalez, 1979; Pavlidis and Horowitz, 1974), the use of eroded stick descriptors (Blum, 1964), the use of Fourier descriptors of shape (Persoon and Fu, 1977), and others (Paton, 1975; Harlow and Connors, 1979).

The third operation in image analysis is *feature extraction,* which is the actual calculation of numerical information that can be used to describe an image. Most image features are derived from the edge map as described above, or are determined from the properties of the pixels that are enclosed by an edge map (see reviews by Pavlidis, 1978, 1979). These are specially defined regions of the image, which means that in many cases the actual calculations must be performed in the memory of the host computer. This requirement arises because most image processors either do not have the computational power to perform the necessary calculations, or are difficult to use to analyze an irregularly shaped image region.

In spite of these limitations, image processors can sometimes be used to find image features by using what appear to be overly involved and indirect procedures. While these routines may seem more complicated than necessary, the fact that they can be performed by an image processor instead of a host computer usually means that they can be accomplished with greatly increased speed. The two following examples demonstrate how an image processor can be used to perform routines of feature extraction and image analysis with minimum involvement of a host computer.

The first example concerns the analysis of fluorescently labeled cells *in vitro,* such as the ones shown in Fig. 10-54a. These cells are cultured neonatal rat myocardial cells that have been labeled with the mitochondrion-specific vital dye, rhodamine 6G (Johnson *et al.*, 1981). In a previous study, we had shown that the fluorescent intensity of these cells varied rapidly with time and was characterized by a flashing phenomenon where the fluorescence of single mitochondria would rise to a high level and then return to baseline within a period of a few seconds (Siemens *et al.*, 1982). We were able to quantify this flashing effect using a spot photometer; however, it was not possible to measure the fluorescence properties of more than a single area using this equipment. We wished to know if the flashing of a single mitochondrion was correlated to the flashing of other mitochondria, and whether the fluorescence of the entire cell, including mitochondria-free regions, was related to the properties of any of the mitochondria.

An additional study was undertaken to address these questions using a real-time image-processing system (Berns, Siemens, and Walter, 1984). The first step in this study was to image the fluorescence of an entire labeled cell using a Venus TV-2M low-light-level camera so that fluorescent properties of several cell regions could be analyzed at the same time. The video signal was videotaped, and the taped images were digitized and displayed on the image-processing system using a pseudocolor intensity transfer function as shown in Fig. 10-55. This real-time pseudocolor transformation allowed the operator to view subtle changes in fluorescence intensity with greater clarity while the image was being displayed on the monitor. When an interesting event was observed, such as a fluorescent flash, the image was frozen and stored in its digitized form in the memory of the image processor.

FIGURE 10-54. Detection and isolation of fluorescent cell images. (a) Original image. (b). Binary image of (a) after thresholding to detect cell boundaries. (c) Joining of disconnected edge elements after three rounds of dilation. (d) One-pixel-wide boundary resulting from image erosion. (e) Edge map of selected cell. (f) Edge map overlayed on the original image. (g) Binary mask created from the edge map of (e). (h) Deletion of noncell regions of the image after a logical AND operation between the mask of (g) and the original image. (Digital image courtesy of Ann Siemens, Department of Developmental and Cell Biology, University of California, Irvine.)

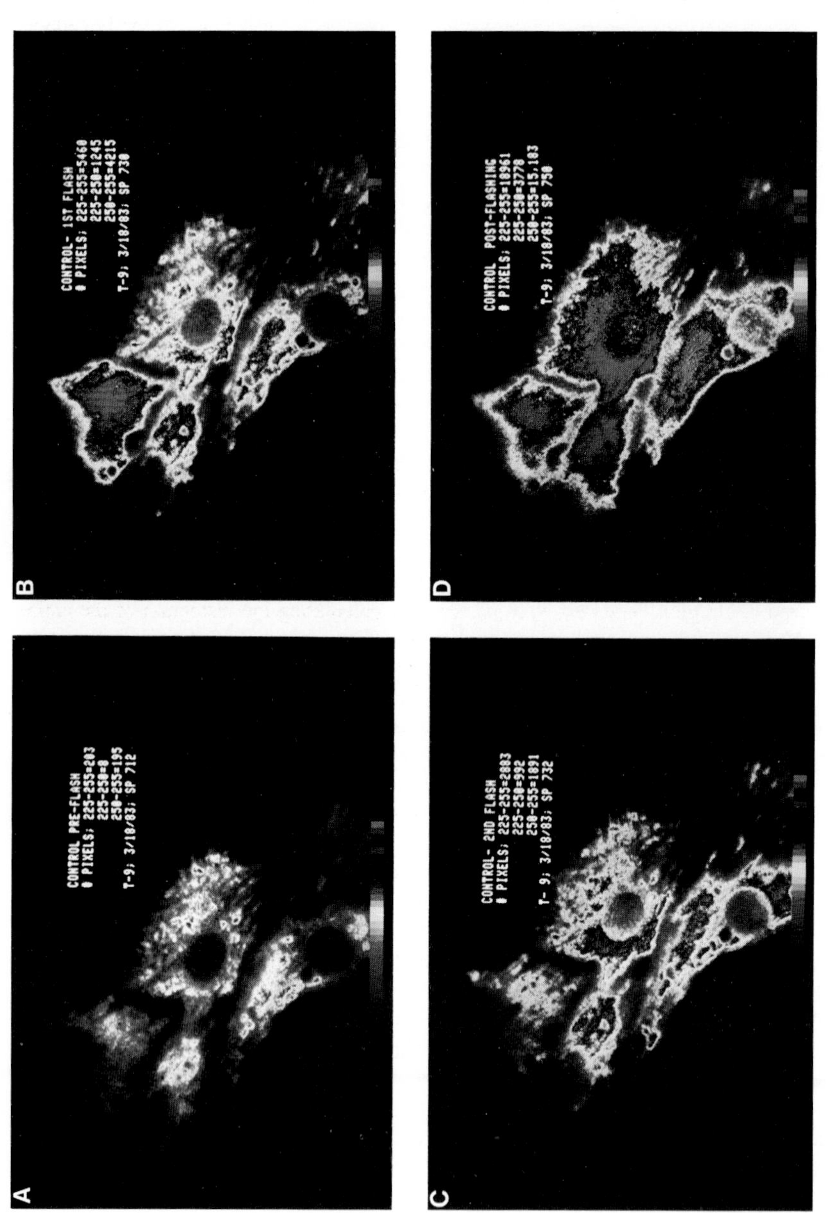

FIGURE 10-55. Use of pseudocolor transformations to enhance contrast in fluorescent cell images. Monochrome images similar to that of Fig. 10-54a were subjected to pseudocolor enhancement to accentuate changes in cellular fluorescence with time. Panels (A)–(D) show changes in fluorescent patterns occurring over a period of 45 sec. For color reproduction of Fig. 10-55 see color plate following page xxvi. (Photographs courtesy of Ann Siemens, Department of Developmental and Cell Biology, University of California, Irvine.)

The next step in the processing was to identify individual cell regions so that their fluorescence intensity could be compared to other images taken at different times. This is a segmentation and feature extraction procedure that was performed as follows. First, an edge map was generated that surrounded the region to be measured. This was done by either using a joystick-controlled cursor that allowed the operator to interactively outline the region, or was done using a thresholding and boundary-following procedure as shown in Fig. 10-54b–e. The thresholding procedure could be performed in real time using the image processor and a joystick controller, while the boundary-following routine was performed in the memory of the host computer and required from 5 to 30 sec depending on the size of the outline.

Once the edge map had been derived, the next step in the procedure was to measure the average gray values of the pixels within the edge map so that they could be used to calculate fluorescence intensity. This calculation could be performed directly in the memory of the host computer; however, the calculation would require several minutes because of the large amount of data and the delays caused during data transfer. A faster procedure for finding the average gray value of outlined regions was developed using the image processor as follows. First, a mask was created from the edge map that was white (gray value of 255) within the map and black (gray value of 0) elsewhere, as shown in Fig. 10-54g. This mask could be created by the image processor in about 20 sec using a "dilation" procedure similar to that shown in Fig. 10-54c, except that the dilation occurs only within the boundary of the edge map. This mask was stored as a separate image in one of the multiple memories of the image processor, and was used to delete all regions of the original fluorescent image that were outside of the edge map, as shown in Fig. 54h. This masking was accomplished in real time using a logical AND operation between the mask image and the digitized image of the fluorescent cell. Once the mask was stored in computer memory, it could be applied repeatedly to other images taken of the same cell, ensuring that the same area was measured in each image.

The final step in this analysis was to use the image processor to measure the average gray value within the masked cell region. This was accomplished by calculating the histogram of the entire image using a histogram feature of the image processor. This histogram calculation could not be used directly on the original image because the image processor always calculates the histogram of the entire image, and thus would have counted regions outside of the segmentation boundary. This inability to manipulate only specific regions of the digital image is typical of most image processors.

Once the image histogram was calculated, it was transferred to the memory of the host computer for final analysis (this is only 256 data points and thus can be performed very quickly). This histogram has a special property in that all of the pixels outside of the outlined region are found in the 0 gray-value channel of this histogram. This is true because the digitizer controls had previously been set so that no pixel in the original video image had a gray value of 0, which is the routine practice to avoid clipping. Since the masking procedure set the gray value of all the pixels outside of the segmented region to 0, all of these pixels represented in the 0 gray-value channel of the image histogram were from outside of the segmented area. This means that the average gray value of the segmented area could be determined directly using equation (10-6), with thresholds set at 1 and 255. The average gray value was then converted to a measure of absolute intensity using a predetermined calibration curve, obtained according to the procedures outlined in Section 10.10 (Berns, Siemens, and Walter, 1984). This process was repeated for different cell areas and for images taken at different times so that changing patterns of fluorescence could be evaluated. By using the image processor, the time required to find the average gray value of any outlined region could be reduced to approximately 30 sec. Other examples of computer analysis of fluorescent images can be found in Benson *et al.* (1980), Rich and Wampler (1981), and D'Ormer *et al.* (1981).

The procedure above can be summarized as follows: (1) a binary mask was created that marked the outlined region; (2) a gray value histogram of the original image was calculated within the masked region; and (3) the average gray value within the masked region was calculated from the histogram. The method described is somewhat complicated because we had to store the binary mask in a separate memory channel normally used to store 8-bit images, and had to perform the logical AND operation to isolate the masked region for the histogram calculation. This procedure can be simplified by taking advantage of a special feature found in some image processors called an *image overlay*. An image overlay is a separate area of image processor memory used to store 1-bit binary images such as the binary mask described in this procedure. Image processors that have special image overlays usually also have the ability to perform manipulations on a separate image based on the contents of the binary overlay. For example, the binary mask could be loaded in an image overlay and the image processor could be used to calculate the histogram of the original fluorescent image wherever the overlay was set to 1. This avoids using a true image memory channel for storing the mask and also eliminates the need for performing the logical AND operation. Image overlays can also be used with ITFs, where the transformation is only performed where the overlay is set to 1. Some image processors have several overlays that can each be used with the same image to perform several different operations in separate regions of the image.

The second example of feature extraction involves the real-time tracking of moving cells under the microscope. This is a common application for studying motility of living cells or simply for keeping a particular cell within the field of view during long-term photomicrography or videotaping. The tracking system described here has been used to measure the motility of normal and transformed cells in culture (Berns and Berns, 1982), and to track normal and laser-irradiated *Taricha granulosa* eosinophil cells (Koonce *et al.*, 1984).

The experimental procedure for tracking cells *in vitro* is to: (1) place a cell of interest in the center of the field of view, (2) wait a short period of time for the cell to move, (3) detect the new position of the cell, and (4) move the target cell back to the center of the field of view using the **stepping motor** on a computer-controlled motorized stage. The movement detected in step 3 is recorded by the computer and used as a measure of the motility of the cell. The image processor is used in this procedure to identify the new location of the target cell so it can be moved back to the field center and its translocation recorded. This identification has to be performed fairly quickly, between 15 and 60 sec, in order to avoid having the cell move too far and possibly being lost.

Special problems arose in these studies in relation to defining what constitutes cell movement. The cell types mentioned above undergo either a fibroblastic or amoeboid type of motility, i.e., they may show a marked change in shape over time. This shape change involves the extension and contraction of cell processes, which can be confused with cell movement when in fact no net movement of the cell has actually occurred. In order to minimize this problem, it was decided that the most accurate method for specifying the location of a cell was to measure its "center of mass," or centroid, and to use displacement of this centroid as an indication of net movement. Simpler tracking procedures involve following the movement of a leading edge (Bartels *et al.*, 1981). An example of a cell used in the tracking procedure is shown in Fig. 10-56.

In order to calculate the center of mass of the target cell, the following procedure was used. First, the image was convolved using a large sharpening mask (Section 10.11) to enhance cell boundaries. Second, a thresholding procedure was used to make a binary image where the cell boundaries were white and all other areas were 0. The convolution operation of the first step also helped to reduce shading distortion in the image and thus made this thresholding step more efficient. It was also found that the intensity of the tungsten illuminator varied noticeably with time, which meant that threshold levels could vary between images. The convolution operation reduced the effect of this variation considerably, making threshold levels more reproducible. An example of a threshold-generated binary image is shown in Fig. 10-56b.

The third step in the procedure was to calculate the centroid based on the boundaries detected in the thresholding procedure. One method for calculating this center is to use a boundary-following routine as was done in the last example, and calculate the cell center from the chain-coded edge map using the host computer. This procedure would take approximately 20 sec; however, it requires an unbroken boundary around the cell as a result of the thresholding procedure, which cannot always be guaranteed. In order for the boundary-following routine to be completely successful, a boundary-closing routine would have to be added to the procedure, such as the expansion–contraction method described previously, which would increase the processing time even more.

In order to bypass this problem, another, much faster, procedure was used to find the centroid that required only the image processor. Figure 10-56b shows the boundary-thresholded transform of Fig. 10-56a with a large box drawn around the cell image. This box is generated by the image processor and its size and location are controlled by an interactive joystick. A special feature of this box (called a cursor box) is that the image processor can count the number of white pixels

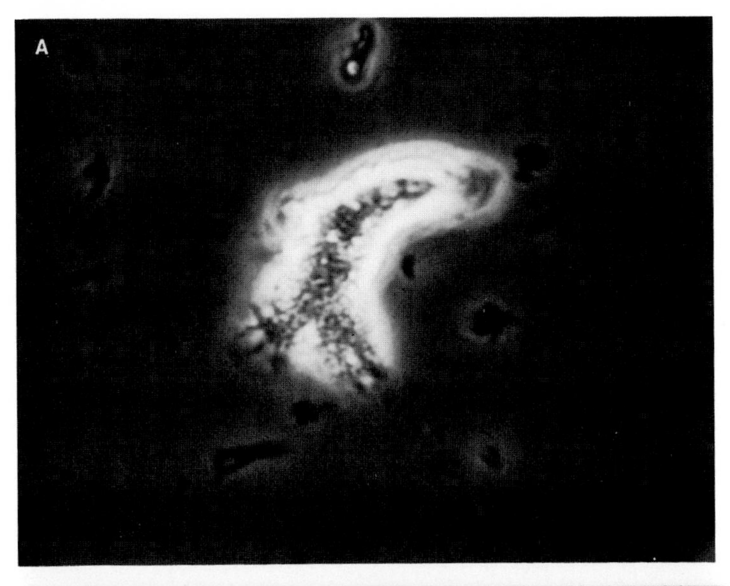

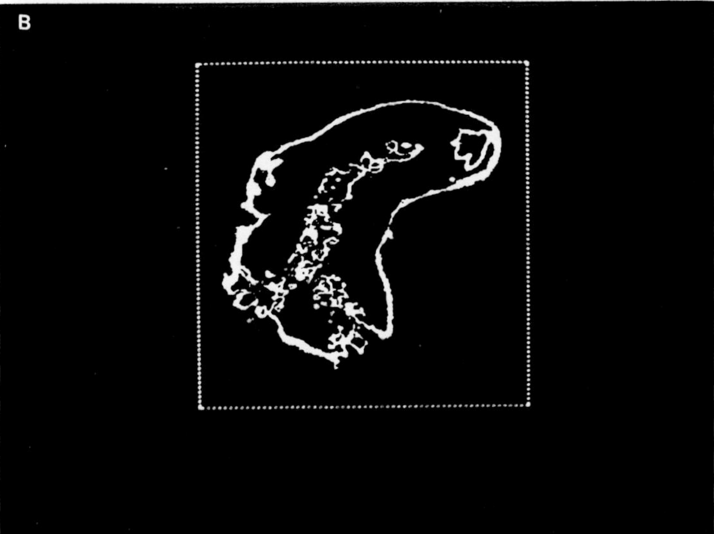

FIGURE 10-56. Transformations used in computer tracking of cell motility. (A) Original phase contrast image of a *Taricha granulosa* eosinophil. (B) Same image after convolution and thresholding to detect cell boundaries. The image processor can count the number of white pixels within the cursor box in one video frame time. (Specimen courtesy of Michael Koonce and Jasmine Chow, Department of Developmental and Cell Biology, University of California, Irvine.)

within its boundaries in one video frame time. At the beginning of every tracking session, the cursor box was placed around the target cell, and was made large enough that the cell would not migrate out of the box during the tracking procedure. After the threshold image had been generated, the number of white pixels were counted within the cursor box. This number is the total number of edge elements detected within the cell. The cursor box was then divided in half in the horizontal direction and the number of white pixels within the box was counted again. If this number was larger than one-half of the number of pixels contained in the original cursor box, then the cell center was contained in the new box. If the number of pixels in the new box was less than one-half the original box, then the cell center was contained in the part of the original box excluded from the second count. The size of the cursor box was then changed by a factor of one-fourth (larger or smaller depending on the outcome of the previous step) and the counting procedure was repeated.

This procedure is known as a *binary search,* and is a very efficient way for finding the cell centroid. The size of the cursor box is changed by 1 over a power of 2 with each iteration until exactly one-half of the pixels are contained in the cursor box or until the increments in box size become less than the width of one pixel. By performing a binary search in the horizontal direction, the X coordinate of the cell center is found as the position of the horizontal line that exactly divides the number of edge elements in two. An analogous search in the vertical direction yields the Y coordinate of the cell center. It can be shown that in the worst case this procedure will find the cell center after $\log_2(i) \times \log_2(j)$ iterations, where i and j are the height and width of the original cursor box. For a 512×512 cursor box, the worst case would only require 18 iterations to find the cell center. Since the image processor performs these operations at video rates of 30 per sec, this calculation can always be made in less than 1 sec.

Once the displacement of the cell had been calculated using this procedure, it was stored in the memory of the host computer and the cell was moved back to the center of the field of view using a motorized stage. The convolution, thresholding, and binary search procedure were repeated automatically every 30 sec using values stored in the memory of the host computer, and thus could be run for long periods without human supervision. At the conclusion of each tracking session, the total cell movement detected could be graphed on the display monitor (Berns and Berns, 1982).

The two routines just presented are simplified examples of a type of computer image processing known as *pattern recognition,* which is a field of computer programming where the computer is asked to make higher-level decisions about the information content of an image. Examples of pattern recognition routines are the classifying of cells as malignant or benign (Pycock and Christopher, 1980), or the preparation of chromosome karyotypes from metaphase squashes (Castleman and Wall, 1973). Research in this type of processing usually involves the search for a set of individual image features that can be used to unambiguously classify images, or objects within images, as being only one type of image out of several different possibilities. One tedious but straightforward method for finding such a set is to make use of a special group of images known as a *learning set.* A learning set is a group of images that have been classified *a priori* as belonging to a particular image class, such as a collection of histological smears that have been classified as malignant by a trained pathologist. A large number of image features are measured from the learning set, as many as possible of the type that might have even a slight usefulness for classifying the image. Measurements are made of the same features from other learning sets that are known to belong to another class of image.

Once this large group of image features has been measured, the task is reduced to finding the smallest group of image features that can be used unambiguously to classify images with the least effort. An example of this type of analysis is the now-classic work of Prewitt and Mendelsohn (1965) who showed that stained human lymphocytes could be classified unambiguously by using three different image features: optical density, cytoplasmic area, and nuclear area. A more recent example of this approach is the work of Olson *et al.* (1980) who measured 37 different image features from learning sets of three different types of tissue culture cells and showed that each cell type could be distinguished from the others reasonably well by comparing a set of only eight discriminating features. Other examples of pattern recognition routines involving microscope images can be found in Young *et al.* (1974), Bowie and Young (1977a,b), Vossepoel *et al.* (1979), Bertram and Rogers (1981), and Marino *et al.* (1981). Examples of pattern recognition involving the Fourier transforms of cellular images can be found in Kopp *et al.* (1974) and Miles and Jaggard (1981).

10.14. COMMERCIAL IMAGE-PROCESSING EQUIPMENT

In this section we present a comparison of several commercially available image-processing systems (Table 10-1). We have not included systems used primarily for clinical diagnosis, as the features of these systems have been compared elsewhere (Preston, 1978, 1979). It should be noted that the specifications in Table 10-1 may no longer be accurate due to the rapid development of these devices since this chapter was written (see Postscript and Appendix IV).

Explanations for each of the descriptive headings follow.

Number of images. This column lists the number of images each system is capable of storing in a standard 512 × 512 × 8-bit format (some systems may have a slightly different standard format, e.g., 640 × 480 × 8-bit). Where a range of values is shown, the lower value is the number of image memories available in a minimum system, with the higher number being the maximum number of images that can be stored after the addition of optional memory.

Image format. Some systems have the capability of displaying and manipulating images in more than one format. For example, eight 512 × 512 × 8-bit image memories might be combined to form a single 1024 × 1024 × 16-bit image that can be treated as a single unit. This column shows the number of different image formats that are available with each system, or when the number of different formats is large, the term "variable" is used.

Host computer. Some systems are supplied with a host computer system while others are designed to interface with a particular type of host computer supplied by the user. The type of host computer will influence the speed and versatility of the image processor, as well as the type of operating environment the user will have to use.

Some image-processing systems, such as the Vickers Magiscan and the Zeiss IBAS-II, are designed with a "menu-oriented" operating system, which means that the unit has been pre-programmed to respond to simple commands or keystrokes entered by the operator. Menu-oriented systems are the easiest to use because they do not require any specialized knowledge of programming languages or computer operations. This type of system may not be as versatile as a fully programmable system, when the development of customized processing routines is required.

Digitizer. This column describes whether or not the system is available with a video digitizer, and if so, the number of bits of resolution.

Look-up tables. This column shows the number of look-up tables each system has available for manipulating image gray values. Systems that have other than 8-bit image resolution or use the look-up tables as function tables may have uneven resolution within look-up tables, e.g., 10-bits in and 12-bits out, meaning an image with 10-bit resolution (1024 gray levels) can be written through the look-up table to generate an output image with 12-bit resolution (a range of 4096 gray values). Note that in some systems some look-up tables can only be used for display purposes, e.g., pseudocoloring, and cannot be used actually to manipulate the stored form of the digital image.

ALUs. In order to perform numerical manipulations at video rates, an image processor must have a device known as an **arithmetic–logic unit** (ALU). Without this device, manipulations such as overlay masking or the addition and subtraction of images would have to be performed by a host computer at rates much slower than video rates. An ALU has two inputs, which can usually be any combination of image memories, live digitized video, or stored constants. The ALU can be programmed to perform an operation on these two inputs (add, subtract, logical AND, etc.) the result of which becomes the ALU's output. The operations are done in pipeline fashion, one pixel at a time, between corresponding pixels in the two input images.

The ALUs column of Table 10-1 describes the type of manipulations each system is capable of performing at video rates using its ALUs. When more than a single ALU is available, note is made as to whether the ALUs can be used in series (the output of one ALU can be used as one of the inputs of a second ALU), in parallel (each ALU can perform independent operations on different images), and whether conditional operations are allowed (the operation of an ALU can be made to occur only when some other event has also occurred). The availability of multiple, interactive ALUs is a very powerful feature of an image-processing system, and greatly extends the speed and complexity of the type of operations it can perform. Figure 10-57 shows an example of the use of two interactive ALUs to perform an image subtraction operation only where a third image, used as an image overlay, has been set to a given value. In Fig. 10-57, the first ALU has been set to test whether each pixel of image number 1 (which is the overlay mask) is equal to a constant. The inputs to ALU 2 are image number 2 and image number 3, which are to be subtracted if the overlay mask is equal to the constant value. ALU 2 has been programmed to perform the subtraction operation only when the

TABLE 10-1. Selected Image Processors and Systems[a]

Manufacturer/ model	No. of images	Image format	Host computer	Real-time digitizer	Real-time ALUs[b]	Real-time functions
Bausch & Lomb Omnicon 3500	1–3	1066 × 480 × 2 512 × 512 × 16 variable	Eclipse S-120	6-bit 8-bit	1 ALU 1 Multiplication	Multiplication, count, add, subtraction, logic
Arlunya TF4000/CTM	1	512 × 512 × 8 512 × 256 × 8	Supplied by user	8-bit	None	Subtraction, averaging
Quantex DS-50	1	512 × 512 × 12	Supplied by user	8-bit	1 ALU	Add, subtraction, averaging
Quantex DS-20F	1	512 × 512 × 8	Supplied by user	8-bit	Subtraction	None
Quantex DF-80	0	512 × 512 × 8	Supplied by user	8-bit	None	Convolve (smooth / sharp only) 3 × 3 to 15 × 15
Quantex QX-2000	1–4	512 × 512 × 16 256 × 256 × 8 variable	Motorolla 68000	8-bit 9-bit	2 par./ser. cond.	Convolve, add / subtraction, multiplication, division, logic, LUTs
International Imaging Systems 66A	1	512 × 512 × 12	Optional[c]	None	1 ALU	Add, subtraction, logic, LUTs
International Imaging Systems 64	4–64	512 × 512 × 16 512 × 512 × 8	Optional[c]	8-bit 10-bit	2 ALUs par./ser.	Add, subtraction, logic, LUTs, histograms
International Imaging Systems 75	4–16	2048 × 2048 × 16 512 × 512 × 8 variable	Optional[c]	8-bit 10-bit	1 ALU	Add, subtraction, logic, LUTs, histograms
Grinnell GMR 260	1–6	1024 × 1024 × 8	Supplied by user	No	None	Overlays, LUTs
Grinnell GMR 275	1–5	512 × 512 × 8 480 × 512 × 8	Digital LSI-11/23 optional	6-bit 8-bit	1	Addition, subtraction, logic, LUTs, multiplication, histograms
Grinnell 2800-30	2–40	512 × 512 × 8 640 × 480 × 8	Digital LSI-11/23 optional	6-bit 8-bit	1–4	Addition, subtraction, multiplication, shift, logic, LUTs
Imaging Technology IP-512	1–40	512 × 512 × 8 1024 × 1024 × 16 variable	Supplied by user	4-bit 6-bit 8-bit	1 ALU 1 Multiplication	Addition, subtraction, multiplication, histograms, logic, LUTs
Recognition Concepts Trapix-5500	2–16	1024 × 1024 × 4 512 × 512 × 16 256 × 256 × 64 variable	Digital LSI-11/23 optional	8-bit 10-bit	1 cond.	Addition, subtraction, logic, LUTs multiplication, histograms
DeAnza Systems IP8500	1–20	512 × 512 × 8 2048 × 2048 × 8 variable	Digital PDP-11 optional	6-bit 8-bit (1–3 ea.)	4 ALUs, 4 multiplication, par./ser., cond.	Addition, subtraction, multiplication, counting, logic, LUTs, histograms
EOIS 1100	1	240 × 256 × 2 480 × 512 × 1	Intel 8085 / 8088	8-bit	None	None
EOIS 1800	1	240 × 246 × 3 480 × 512 × 8	Intel 8085 / 8088	8-bit	None	None

Look-up tables	Scroll/ pan	Zoom	Overlays	Interactive devices	Built-in functions	Packaged software	General description
Shading correction	Yes	2×	Variable	Joystick, light pen	Keyboard functions, function menu cursor/ROIs, image statistics	Convolutions general-purpose package/custom software available	Stand-alone image analyzer / real-time image processor
None	No	None	None	Trackball	Image averaging, peak, detection, variable window, asynchronous analog x, y, z inputs	None	Digital frame store / image averager scan convertor
None	No	None	None	None	Image averaging	General-purpose package	General-purpose image processor
None	No	None	None	None	Image subtraction	None	Digital frame store
None	No	None	None	None	Sharpening and smoothing convolutions	None	Real-time image convolver
3 ea. 12 × 16 9 ea. 8 × 8	Yes	2× 4×	8 × 1-bit alphanum.	Trackball, keyboard	Menu touchscreen, 5 × 5 convolver, keyboard functions, cursor, ROIs, histogram	General-purpose package/image analysis	General-purpose image processor
1 ea. 8 × 12	Yes	2× 4× 8×	4 × 1-bit alphanum.	Joystick, trackball, tablet	Splitscreen, grid overlay	Custom software available	General-purpose image processor / graphics display
4-8 × 8 2-10 × 16 1-12 × 16	Yes	2× 4× 8×	4 × 1-bit alphanum.	Joystick, trackball, tablet	ROIs, grid overlay	Utilities package	OEM-oriented image-processing system
96-8 × 12	Yes	2× 4× 8×	7 × 1-bit alphanum.	Joystick, trackball, tablet	ROIs, grid overlay	Extensive package of general-purpose software	General-purpose image processor
12 × 12, or 12 × 24, or 3 ea. 12 × 8	Yes	1×–16×	2 × 1-bit alphanum.	Tablet, joystick, trackball	Graphics, cursors, text, ROIs	Data base, general purpose	Frame buffer
3 ea. 10 × 8	Yes	2× 4× 8×	4 × 1-bit alphanum.	Joystick, trackball	Graphics, text, cursors, ROIs, splitscreen, histogram	General-purpose software packages	General-purpose image processor
4 ea. 12 × 8	Yes	1×–16×	Uses image memory	Tablet, joystick, trackball	Graphics, text, cursors, ROIs	Data base, general purpose	General-purpose image processor
16 ea. 8 × 8	Yes	2×	1 × 1-bit alphanum.	None	None	Image averaging, general utilities	General-purpose image processor, OEM oriented
3 ea. 8 × 8, or 1 ea. 10 × 12	Yes	1×–8×	alphanum.	Joystick, trackball	Cursors, counter, histograms, ROIs, time-base corrector	System utilities image statistics	General-purpose image processor
16 ea. 12 × 16	Yes	1×–8×	4 × 1-bit alphanum.	Joystick, trackball, tablet	Histograms, counter, cursors, ROIs, splitscreen	Extensive package of general-purpose software	Large image processing system
No	No	None	1 × 1-bit	Light pen	Pseudocolor optional	Custom available	Low-priced frame buffer, computer
No	No	None	1 × 1-bit	Light pen	Pseudocolor optional	System utilities, custom available	Low-priced frame buffer, computer

(continued)

TABLE 10-1 (*Continued*)

Manufacturer/ model	No. of images	Image format	Host computer	Real-time digitizer	Real-time ALUs[a]	Real-time functions
Zeiss IBAS-II	4–63	512 × 512 × 8 256 × 256 × 8	Zilog Z80A	6-bit 8-bit	1 ALU cond.	Addition, subtraction, multiplication, logic, histograms
MCI/LINK Intellect 100	1	512 × 512 × 8	Motorolla 68000	8-bit	1	Addition, subtraction, logic, LUTs
MCI/LINK Intellect 200	4–12	512 × 512 × 8 1024 × 1024 × 8 variable	Digital LSI-11/23	8-bit	1–3	LUTs
Vicom Systems Vicom	1–16	1024 × 1024 × 8 512 × 512 × 16 variable	Motorolla 68000	8-bit	1	Addition, subtraction, logic, LUTs 3 × 3 convolver
Spatial Data Eyecom II	1	640 × 480 × 8	LSI-II/73[d]	8-bit	None	LUTs
Spatial Data Eyecom III	4–32	640 × 480 × 8 640 × 480 × 16	Digital LSI-11/23	8-bit 10-bit	4 par. / ser.	Addition, subtraction, logic, LUTs
Interpretation Systems, Inc.	1–4	variable	Digital LSI-11/23	8-bit	None	None
Vickers Magiscan	6–24	1024 × 1024 × 8 512 × 512 × 8 variable	Custom	6-bit	None	None

[a]See also Postscript and Appendix IV: Addendum on Digital Image Processors.
[b]Abbreviations: par., parallel operation; ser., serial operation; cond., conditional operation; alphanum, alphanumeric.
[c]Supports Digital PDP-11/23 and VAX-11 series, Masscomp H68000, and Hewlett–Packard HP3000.
[d]These entries updated November 1985.

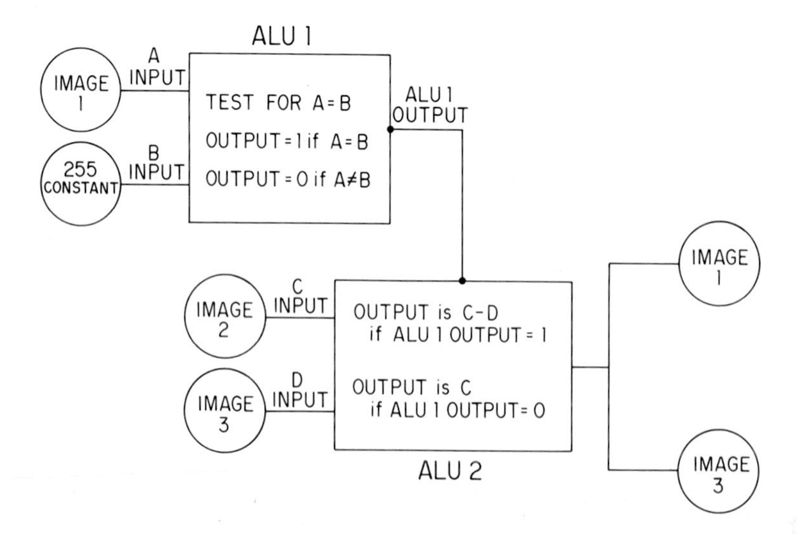

FIGURE 10-57. Example of the use of multiple interactive ALUs to perform conditional image manipulations in real time.

Look-up tables	Scroll/ pan	Zoom	Overlays	Interactive devices	Built-in functions	Packaged software	General description
4 ea. 8 × 8	Yes	2×	4 × 1-bit alphanum.	Tablet	Function menu, image statistics, digital filters	Extensive package, image statistics, chromosome karyotyping	Menu-oriented image analyzer
3 ea. 8 × 8	Scroll	2× 4×	1 × 1-bit alphanum.	Keyboard functions	Keyboard functions	General purpose, image statistics	Menu-oriented image analyzer
3–9 ea. 8 × 8	Yes	1×–16×	1 × 1-bit alphanum.	Trackball, joystick, keyboard	Fourier transforms, 2-D autocorrelator, keyboard functions	General purpose, crystallography	Menu-oriented, image processor, image analyzer
1 ea. 10 × 12 3 ea. 10 × 8	Yes	2× 4×	4 × 1-bit alphanum.	Trackball, joystick, tablet, keyboard	3 × 3 convolver, keyboard functions	General-purpose software package	Menu-oriented general-purpose image processor
3 ea. 8 × 8	Scroll	1×–128×[d]	alphanum.	Tablet, keyboard, joystick	Function keys, cursors, video camera	System utilities, general purpose image statistics	Menu-oriented, image processor frame buffer
16 ea. 10 × 12	Yes	1×–8×	8 × 1-bit alphanum.	Tablet, joystick, trackball	Function keys, cursors, histograms	General-purpose software package	Menu-oriented general-purpose image processor
8 ea.	Yes	1×–16×	9 × 1-bit alphanum.	Trackball	Function menu, function keys	Extensive package of general-purpose software	Stand-alone turnkey image processor
None	Yes	None	Variable	Light pen, keyboard	Digital filtering, keyboard functions	Image analysis, chromosome karyotypes	Menu-oriented image analyzer

output of ALU 1 shows that the test output is true, otherwise the output of ALU 2 is the unaltered form of image 3. Used in this fashion, the masking and subtraction operation can be performed in one video frame time, or 1/30 sec.

Systems that use ALUs usually also have other devices that can manipulate images at video rates, such as *shift registers* (sometimes called scale registers) that can multiply or divide an image by powers of 2, *look-up tables* as described previously, *bit plane masks* that can be used to protect individual bit values of an image from being altered, **multiplexers** that allow the output of the ALU operations to be routed to several different memory locations, and *accumulators* that can be used to count events based on the conditional output of an ALU, such as during the calculation of an image histogram. The combination of the ALUs and these ancillary devices is sometimes given a single name, such as a **pipeline processor** or a *digital video processor*.

Overlays. This column shows the number of overlays each system has available (Section 10.13). Some systems also have alphanumeric overlays that can be useful for placing text on the displayed image for identification purposes.

Special features. In addition to (or in place of) the ALUs, some systems also include special devices that can used to manipulate digital images. These include hardware devices for performing convolutions, Fourier transforms, and calculating histograms at video rates.

Zoom, scroll, and pan. Some systems have the capability of enlarging the image display (zoom) by integer magnification factors. When an image has been zoomed, only a portion of it is usually visible on the display screen; consequently, a scroll (**pan**) function is usually also used that allows any given portion of an image to be centered on the display screen. The scroll function can sometimes also be used during operation of the ALUs to perform manipulations on two pixels that are normally offset from one another. This is a useful feature for performing convolutions.

When a system has the capability of storing multiple images, different image memories are sometimes used to store adjacent portions of a larger image. In order to visualize adjacent regions of different image memories, some systems come with a *roam* feature that allows the scroll operation to treat several images as if they were parts of a larger image.

Interactive devices. Some systems come with interactive **pointing** devices such as joysticks, light pens, trackballs, and/or graphic tablets that can be used with the system. These devices can be used to control the scroll, pan, and roam functions or can be used to control computer-generated cursors and regions of interest (ROIs) that are used for outlining, identifying objects or positioning cursor-generated boxes as described in the previous section.

ACKNOWLEDGMENTS. The authors gratefully acknowledge the helpful suggestions provided by Barbara N. Walter, William Wright, Steven Vaughn, C. Anthony Altar, and Gregory Berns during the preparation of the manuscript, and Beverly Hyndman for excellent secretarial assistance. We are especially indebted to Professor Jack Sklansky, Department of Electrical Engineering, University of California, Irvine, for providing lecture material on digital image processing and for critical reading of the manuscript. Computer processing was performed at the Laser Microbeam Program (LAMP) facility of the National Institutes of Health Biotechnology Resource Program located at the University of California, Irvine.

11

Applications of Video to Light Microscopy

11.1. INTRODUCTION

In this chapter we will survey the applications of video to various modes of light microscopy including low-light-level applications, contrast enhancement, quantitation, scanning, and three-dimensional reconstruction. Rather than limit the discussion to applications that have been published, I shall also include some ongoing developments and potential applications of video that could further improve the image or otherwise expand the utility of the light microscope. Some recent applications of video to biological light microscopy are reviewed comprehensively by Spring (1979, 1983) (see also footnote ‡ on page 397 and the Postscript).

Table 11-1 lists the primary modes of contrast generation for light microscopy. I have indicated the approximate levels of luminances commonly encountered together with typical video cameras used at those light levels. References describing the various modes of contrast generation are listed in Section 5.6. The general basis for selection of cameras is discussed in Section 7.3. More specific examples of video microscopy and choices of equipment are discussed in Chapters 9, 10, and in the following sections.

11.2. LOW-LIGHT-LEVEL MICROSCOPY

As shown in Table 11-1, applications involving *moderately low* levels of luminance are found in most modes of light microscopy. At moderately low light levels, image detail can be seen better through the ocular with the room light dimmed, although the room need not be darkened completely. For some modes of microscopy, the luminance is *very low,* and the image is clearly visible through the ocular only with the room darkened completely. When the luminance is *extremely low,* we can barely detect, or not see the image through the ocular at all, even with the eye fully dark adapted. At the very low and extremely low levels of luminances discussed in this section, intensifier video cameras can substantially improve the visibility of the image and reduce the exposure needed to record a picture photographically.

Luminescence

A number of substances can be made to *luminesce,* or emit "cool light" in the visible wavelength range. Some chemical reactions produce chemiluminescence, which, in the case of bioluminescence, usually involves an oxidative process mediated by the enzyme luciferase. Fireflies and the "phosphorescence" of the summer seas are familiar examples. Breaking chemical bonds mechanically (crushing sugar candy, stripping adhesive tape) can give rise to triboluminescence,

TABLE 11-1. Microscope Contrast Modes; Luminance Levels; Typical Video Cameras

	Luminance (approx. cd/m²)					
	Extremely low ($<10^{-4}$)	Very low ($10^{-4} \sim 10^{-2}$)	Moderately low ($10^{-2} \sim 10$)	Average ($10 \sim 10^{2}$)	Bright ($10^{2} \sim 10^{4}$)	Extremely bright
	Typical camera[a]					
Microscopy contrast mode[b]	I^3V,[c] ISIT,[c] I^2N,[c] microchannel plate[c]	I^2V, SIT, image isocon	Newvicon, Chalnicon, CCDs[d]	Plumbicon, Saticon	Sulfide vidicon	
Luminescence						
Chemi-, electro-, sono-, thermo-, tribo-, radioluminescence						
phosphorescence						
Fluorescence						
Natural						
Applied						
Indicator						
Antibodies						
Darkfield						
UV–IR						
Polarization						
Polarized scattering						
Reflection depolarization						
Dichroism						
Birefringence						
Intrinsic						
Induced						
Form						

Phase contrast	————
Dark	
Light	
Polanret	
Interference	————
Interferometry	
Reflected light	
DIC	————
Smith	
Nomarski	
Single sideband	
Hoffman modulation contrast	
SSEE	
Anaxial	
Brightfield	————
Absorption	
Refraction	
Scattering	
Reflection	

[a]See Table 7-4.
[b]See also Section 5.6.
[c]May require extended (digital or photographic) frame summation.
[d]For static images permitting long integration time, chilled CCDs can be used down to extremely low light levels.

and concentrated sound waves can induce sonoluminescence.* Some materials thermoluminesce when they are mildly heated and the stored bond energy is released. Light-emitting diodes efficiently convert low-voltage electricity into light by electroluminescence. Luminescence also accompanies radioactive decay and electrostatic discharge, e.g., when we rapidly unwind a roll of photographic film stored in a dry environment. A previously illuminated object may emit light for hours or days by phosphorescence (Harvey, 1957, 1965; Glass, 1961).

Since the mid 1960s, Reynolds and co-workers have explored the use of image intensifiers to capture weakly luminescent events observed through the microscope. With the intensifier, they were able to photograph the dim flashes of light given off by a variety of bioluminescent organisms and cells. In the early 1970s, Reynolds coupled the output of image intensifiers to video cameras via low-**f number** lenses, and recorded the camera output on a VTR. Magnified images of the low-level luminescence could thus be observed, recorded, and analyzed in detail (reviews: Reynolds, 1972; Reynolds and Taylor, 1980).

An interesting example of chemiluminescence takes place in the photoprotein aequorin, extracted from the luminescent jellyfish *Aequoria aequoria*. Aequorin gives off a blue glow (peak at 480 nm) when exposed to micromolar concentrations of calcium ions. Aequorin can be microinjected and the luminescence used to pinpoint the presence and distribution of free Ca^{2+} in living cells. Using this approach, Rose and Loewenstein (1975, 1976) showed that the millimolar quantity of calcium present in the whole living cell is mostly sequestered, and present as free Ca^{2+} ions permeating the cytoplasm only at concentrations well below 1 μM. The concentration of free Ca^{2+} rose as high as, or above, 1 μM, causing aequorin to glow *locally* only when Ca^{2+} was microinjected or otherwise released into the cytoplasm. In the absence of metabolic inhibitors, the free Ca^{2+} was sequestered immediately and the aequorin luminescence disappeared rapidly. The whole process was followed under a microscope coupled to an intensifier video camera.

In a striking experiment, Gilkey *et al.* (1978) fertilized medaka eggs (transparent fish eggs a little over 1 mm in diameter) that had been microinjected with aequorin. As the sperm entered, the egg immediately started to glow at the point of sperm entry. The aequorin glow, signifying the release of Ca^{2+}, then spread, and slowly traversed the surface of the egg, as an expanding ring of light (Fig. 11-1). Once past the equator, the ring of light started to shrink again and finally disappeared at the opposite pole of the egg some 2.5 min after the sperm had entered the egg. This dramatic scene demonstrates a propagated surface wave of Ca^{2+} release following fertilization (see also Eisen *et al.*, 1984). The scene was captured through a low-power microscope objective coupled to a four-stage image intensifier (EMI 9694) or Sniperscope (Machlett ML-8685) followed by a Plumbicon video camera and recorded on a ½-inch VTR.

Recently, we (Johnson *et al.*, 1985) studied the bioluminescent flash given off from the small marine dinoflagellate *Gonyaulux polyedra*. With a 40/0.75 objective lens and a three-stage intensifier coupled with fiber optics to a vidicon camera (I^3V, Zeiss–Venus TV-3), we were able to capture the brief luminescent event with high spatial and temporal resolution. As shown in Fig. 11-2, the spatial distribution of the luminescent flash (B) is nearly identical to the distribution of the fluorescence that could be excited in discrete organelles in the cell before the flash was triggered (A). From a ¾-inch videotape record, we were able to track the time course of light emission from each of the organelles together with the rate of decay of their luminescence and fluorescence.†

These low-light-level flashes lasted for only a few to several hundred milliseconds. Therefore, to record the flash and its kinetics, we had to choose a camera that gave adequate sensitivity and low enough noise without frame summation. The SIT (Dage–MTI 65 series) camera simply did not provide enough sensitivity. ISIT (intensifier SIT, Dage–MTI) and I^2N (two-stage intensifier Newvicon, Cohu) cameras showed too many dancing bright spots of light (noise) in the picture. The random dancing bright spots disappeared after one or a few seconds (32 to 128 frames) of

*Walton and Reynolds (1985) have recently observed the spatial distribution of sonoluminescence in water using a four-stage intensifier coupled to a SIT camera.

†For an earlier, detailed study on the luminescent flash from a larger dinoflagellate (*Noctiluca miliaris*) imaged with a multistage intensifier on photographic film, see Eckert and Reynolds (1967).

summation, but by that time the brief luminescent flash was long gone. Of the cameras we tested, only the I^3V gave enough sensitivity and low enough noise without frame summation.

This example illustrates the choice that one may have to make in selecting a high-sensitivity video camera for a particular type of low-light-level application. One is confronted with a choice, or compromise, between spatial and temporal resolution on the one hand, and sensitivity and noise characteristics on the other. While the extremely sensitive I^3V with its fairly homogeneously distributed noise pattern captured the luminescent flashes of *Gonyaulux*, a somewhat less-sensitive (and less expensive) camera may be preferred if the low-light-level image does not require the extra threshold sensitivity of the I^3V, and if the image remains unchanged for several seconds. In those cases, an ISIT, I^2N, or SIT camera, with the frames summed for a few seconds, could provide a considerably better-resolved image than the I^3V. In general, the picture tends to be noisier, and the resolution lower, as the threshold sensitivity of the camera rises. Another important consideration is that the gamma of the ISIT, SIT, and I^2N is approximately unity, making them directly suitable for photometry. In contrast, the sulfide vidicons used in the I^2V and I^3V have a gamma of 0.65. While providing a greater dynamic range, the intensifier Vidicon cameras cannot be used directly for photometry unless corrections are made for their nonlinear light response characteristics. Nevertheless, at extremely low light levels, or for dynamic scenes precluding extended signal integration, one would need to use an I^3V (or four-stage intensifier) with their higher threshold sensitivities.*

Fluorescence

Today, there is a major surge of interest in *fluorescence microscopy*. Among biologists, the interest stems from: the chemical specificity of the signal; the extreme sensitivity possible with fluorescence techniques; the excellent S/N ratio achievable with modern (epi)fluorescence equipment; and the striking high-resolution, high-contrast image attainable. Indeed, one only needs a few thousand fluorochrome molecules to produce a detectable signal; the fluorochromes can be conjugated to (monoclonal) antibodies that seek out the target molecules with a high degree of specificity; and one can measure the microenvironment of the fluorochrome by the change in fluorescence wavelength, the efficiency of fluorescence, or fluorescence depolarization. Furthermore, a fluorescent object is essentially self-luminous, so that its image resolution is governed by twice the NA of the microscope objective lens.†

Despite these major advantages, some shortcomings limited the effectiveness of fluorescence microscopy until quite recently. Although there were several pioneering studies in the 1960s and early 1970s, as we saw in Chapter 1, long photographic or video exposures were needed to capture the dim fluorescent images. The dose of short-wavelength excitation needed to record the fluorescence tended to bleach the fluorochrome and kill or harm a live specimen. It is no wonder, then, that low-noise, low-light-level video cameras were a welcome addition.‡

In 1978, Willingham and Pastan reported the use of an RCA SIT camera to successfully visualize, and follow for over 24 hr, the pinocytotic uptake and transport of rhodamine-labeled proteins into tissue-cultured cells. In this paper, they predicted that video intensification microscopy would find widespread application in the study of living cell behavior. Their prediction is more than borne out by the rapidly increasing number of published reports using video intensification for fluorescence microscopy (e.g., Beckerle, 1984; Bestor and Schatten, 1981; Cheng *et al.*, 1980;

*In comparing the sensitivities of video devices to low light levels, be sure to consider the wavelength characteristics of the source and the pickup device (Section 7.2).

†With fluorescence microscopy, the minimum distance resolvable (d) is given by $d = 1.22\lambda_0/2NA_{obj}$, instead of equation (5-9), since the light emitted by the fluorescent object is basically incoherent. Also, with epifluorescence the brightness of the image rises with the fourth power of the NA of the objective, since the light-gathering power of the condenser and objective lens each rises with the square of the lens NAs, and the objective lens serves both functions. Naturally, lens transmittance also plays a critical role.

‡See Palevitz *et al.* (1981), Pastan and Willingham (1985), Wampler (1985), and Willingham and Pastan (1983), who have reviewed low-light-level video systems and application to fluorescence microscopy.

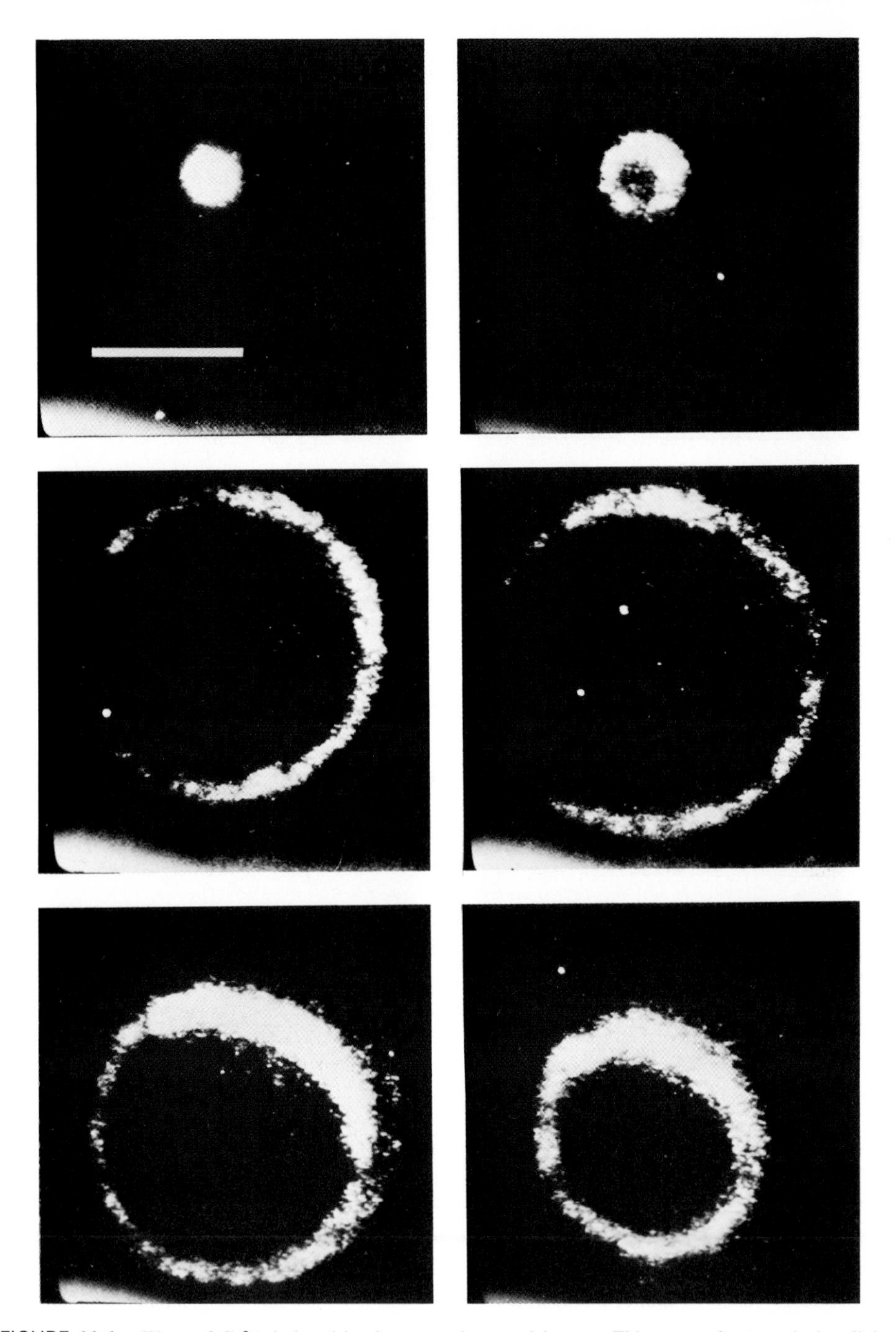

FIGURE 11-1. Wave of Ca^{2+}-induced luminescence in a medaka egg. This wave of extremely low-light-level luminescence was captured with a vidicon camera optically coupled to a four-stage image intensifier (EMI four-stage No. 9694). The luminescence is given off by the photoprotein aequorin, which was microinjected into the medaka egg before it was fertilized. Aequorin luminesces when exposed to micromolar concentrations of Ca^{2+} (Shimomura and Johnson, 1976; Ashley, 1970). The pattern of luminescence illustrated shows the propagated wave of Ca^{2+} release that takes place in the cortex of the egg cytoplasm starting with the point of sperm entry upon fertilization. Successive photographs were taken off the monitor from a ½-inch videotape record, every 10 sec. Scale = 500 μm. (Courtesy of L. Jaffe, from Gilkey *et al.*, 1978.)

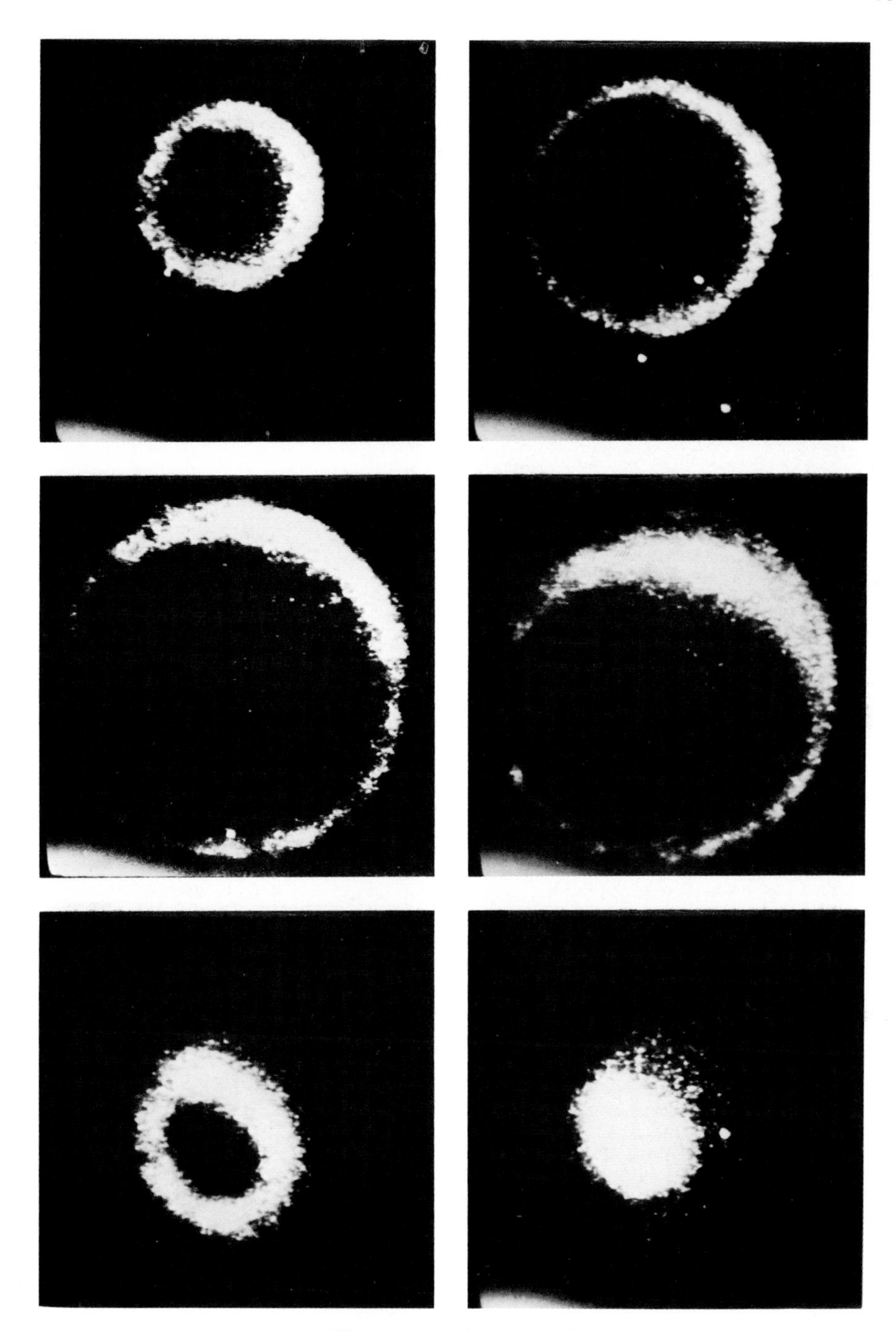

FIGURE 11-1 (*Continued*)

Einck and Bustin, 1984; Ellis *et al.*, 1988; Forman and Turriff, 1981; Gundersen *et al.*, 1982; Kater and Hadley, 1982; Kires and Birchmeier, 1980; Palevitz *et al.*, 1981; Sanger *et al.*, 1984; Toda *et al.*, 1981; Wang, 1984; Willingham and Pastan, 1982).

As reviewed by Taylor and co-workers, various fluorochromes and conjugated fluorochromes have been introduced into living cells and used as probes (Tsien, 1983) for: localizing constituent molecules, for measuring local pH and Ca^{2+} levels, redox potentials, fluidity of membranes,

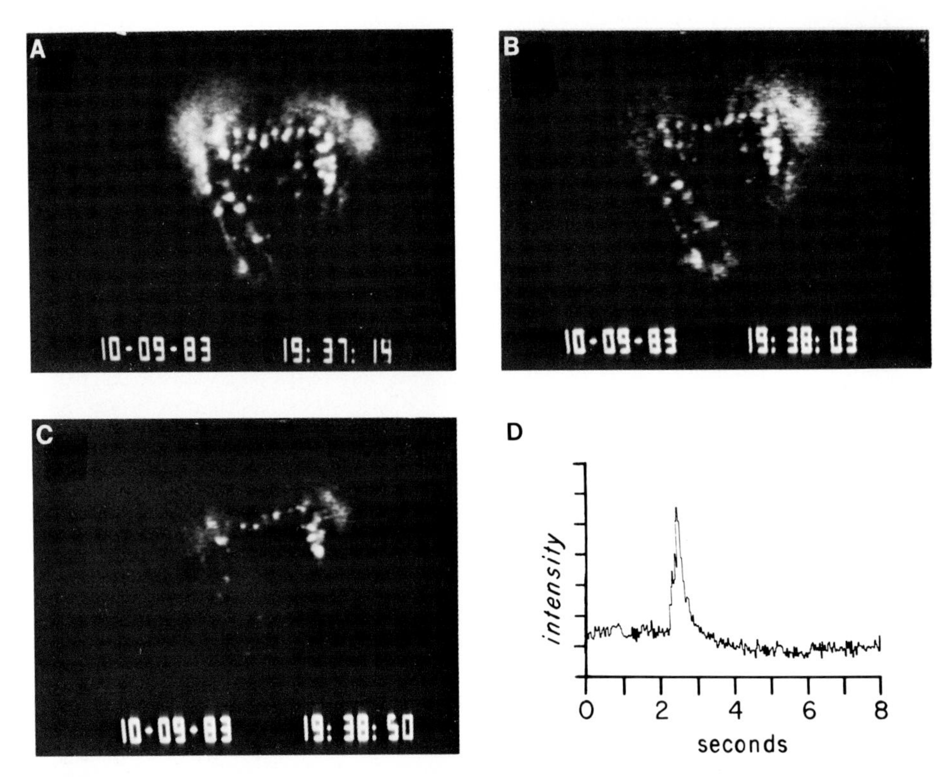

FIGURE 11-2. Bioluminescent flash and autofluorescence in a marine dinoflagellate. (A) Autofluorescence of cytoplasmic particles (450- to 560-nm emission upon excitation with light of 395–420 nm) before bioluminescence was induced. (B) Frozen video field photographed at the peak of the bioluminescent flash, which was induced by excess Ca^{2+} (no fluorescence excitation). (C) Autofluorescence 45 sec after the bioluminescent flash ended; the monitor was photographed with an exposure much longer than in A, in order to capture the weak residual fluorescence. (D) Time course of bioluminescent flash from one of the cytoplasmic particles shown in B. The fluorescence and rapid flash were captured with a three-stage intensifier Vidicon camera (Zeiss–Venus TV-3). The luminescence intensity was measured by placing the coordinate marker of a video analyzer (Colorado Video Model 321) on the image of a single particle, and recording the voltage signal of that point, output from the analyzer onto a strip chart recorder. (From Johnson *et al.*, 1985.)

turnover of molecules, etc., in minute parts of living cells using video-intensified fluorescence microscopy (e.g., Koppel *et al.*, 1982; Poenie *et al.*, 1985; Tanasugarn *et al.*, 1984; Taylor and Wang, 1980; Tycko *et al.*, 1983; Wang *et al.*, 1982; Wang, 1984). *Individual* threads of DNA and F-actin molecules, only a few nanometers thick, have also been visualized under the fluorescence video microscope by coupling these molecules to ligands containing multiple fluorochromes (e.g., Matsumoto *et al.*, 1981; Yanagida *et al.*, 1984).

In fluorescence microscopy, much can be accomplished with the sensitivity and image resolution provided by the 1-inch SIT camera (e.g., Cohu, Dage–MTI, RCA) or the I^2V (e.g., Zeiss–Venus TV-2) without further enhancement. Depending on the emission spectra of the fluorochrome, these cameras can substantially improve on the sensitivity and resolution of the human eye and, especially, a photographic emulsion. In particular, the extended red and infrared sensitivities of modern intensifier cameras open up exciting new possibilities for fluorescence microscopy with very low levels of excitation energy. Camera tubes with higher sensitivity (ISIT, I^2N, I^3V, and the new microchannel-plate intensifier cameras) permit the use of even less excitation and detection of lower concentrations of fluorochrome. When frame-summed or -averaged with a digital image processor, some low-light-level video cameras can provide images with good spatial resolution (Figs. 1-4, 11-3).

In addition to improving the sensitivity and reducing specimen damage, low-light-level video cameras coupled to analog or digital analyzers offer much hope for quantitation of fluorescence and its dynamic changes as indicated in the reviews cited above (see also Barrows *et al.*, 1984; Benson *et al.*, 1980, 1985; De Weer and Salzberg, 1986; McGregor *et al.*, 1984; McNeil *et al.*, 1984; Poenie *et al.*, 1985; Salmon *et al.*, 1984b).

Darkfield

Darkfield is another mode of low-light-level microscopy that is becoming increasingly popular. By precluding the high-NA illuminating beam from entering the objective lens, and producing contrast primarily by light scattering, one derives contrast from specimens whose sizes are far below the limit of resolution of the light microscope. Thus, the Brownian motion of nanometer-sized particles and the motion of bacterial flagella have long been visualized with darkfield microscopy.

As biological macromolecules have become more highly purified, and interest in their assembly properties and polymeric conformation has increased, darkfield microscopy has come to be used more frequently. In 1971, Summers and Gibbons demonstrated the relative sliding among doublet microtubules induced by ATP in trypsin-treated flagellar axonemes. In this landmark demonstration of flagellar motility, these authors recorded the darkfield image of the sliding microtubule doublets (made up of two 24-nm-diameter, tubular polymers of the protein tubulin), illuminated with a high-intensity arc source, on high-speed photographic film.

With the introduction of the SIT camera, several workers adapted it to darkfield microscopy. The process of polymerization of bacterial flagellin was studied (Ishihara and Hotani, 1980), and a conformational change that *propagates* along the length of the helical polymer has recently been observed (Hotani, 1982; Kamiya *et al.*, 1982). Likewise, the flexural rigidity of single microtubules and actin filaments has been visualized, and quantitated, with a SIT camera coupled to a darkfield microscope (Nagashima and Asakura, 1980; Yamazaki *et al.*, 1982; Yanagida *et al.*, 1984).

As in fluorescence microscopy, intensifier video cameras substantially improve the speed by which one can capture the dim images in darkfield. Also, as in fluorescence, it is possible to raise the contrast and sensitivity in darkfield microscopy by suppressing the low-level background video signal and raising the picture contrast electronically.

UV and IR

The absorption and transmission characteristics of *ultraviolet* (UV) and *infrared* (IR) waves have been used in microscopy for many years. Achromatic mirror objectives and **catadioptric** lenses were designed for microscopy in the UV and IR (e.g., see Walker, 1956). Quartz (Leitz, Olympus) and quartz–fluorite (Ultrafluar by Zeiss) objectives were developed for parfocal observation and microspectrophotometry in the UV and visible wavelengths.

Since the late 1930s, Caspersson and co-workers have exploited the high UV absorbance of nucleic acids (relative to proteins) at 260-nm wavelength, pioneered the development of UV microspectrophotometry, and provided a cytological foundation for what has now come to be called molecular genetics (Caspersson, 1950).

For studies of polymers, IR frequencies have been used to measure bond stretching frequencies (Blout *et al.*, 1950), and entomologists and arachnologists have used IR microscopy to see through the cuticular walls that mask the internal structures of beetles and mites in visible light. The earlier images in UV and IR microscopy were generally visualized by photography, although image converter tubes were also used occasionally.

With the extended wavelength characteristics of some photocathodes and video targets, UV and IR images can now be captured directly with video cameras.* Low levels of UV and IR that previously required long photographic exposure can now be captured in real time, providing

*UV vidicons with CdSe target and fused silica faceplate can respond down to ca. 200 nm (e.g., Hamamatsu N983), and PbO–PbS target IR vidicons up to 2.0 μm (e.g., Hamamatsu N214). Cameras using Pyroelectronic vidicons are said to produce IR images up to 32-μm wavelengths (e.g., Videotherm from ISI Group).

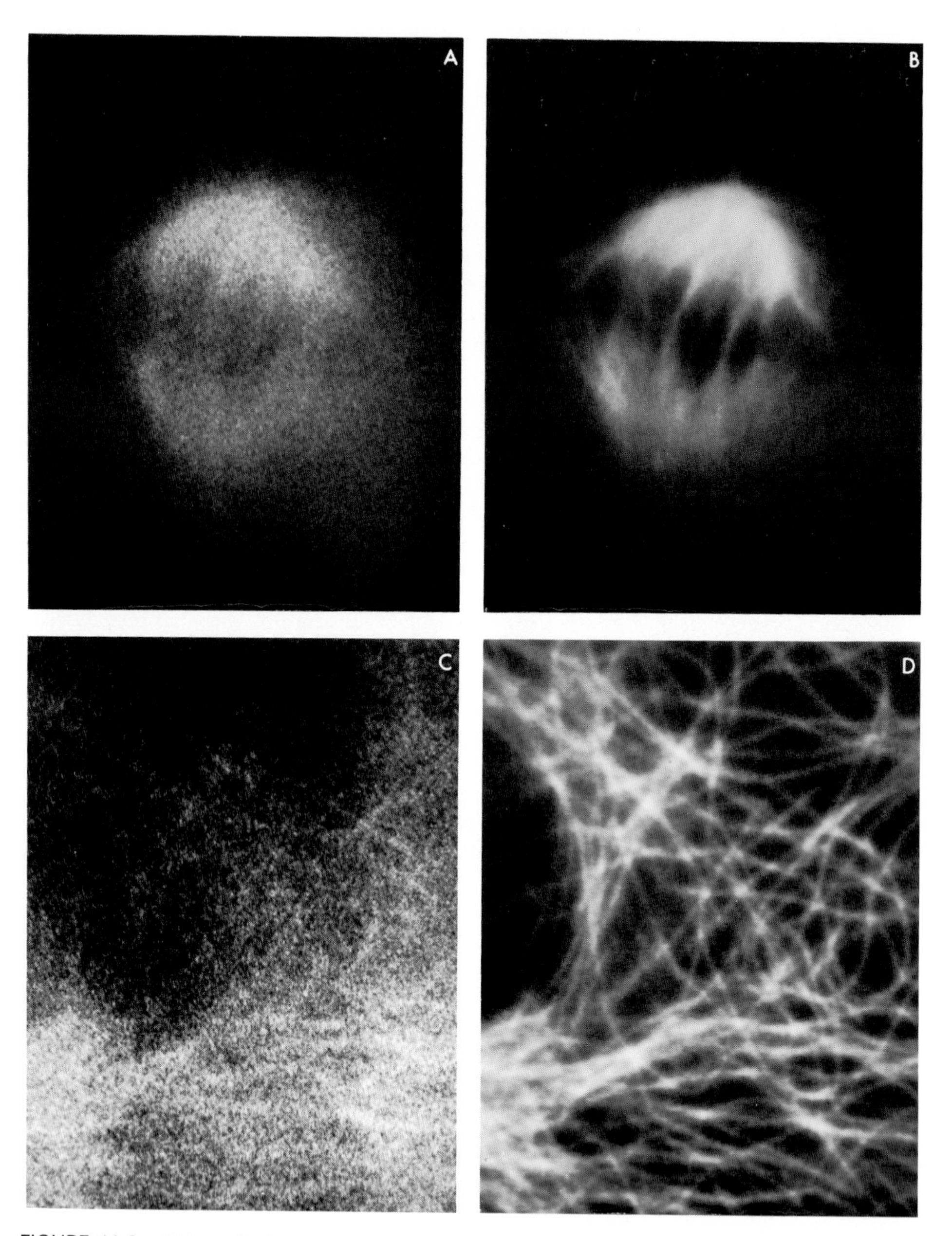

FIGURE 11-3. Noise reduction by digital frame-summing of very-low-light-level fluorescent images. (A) Antitubulin (rhodamine) fluorescence in a dividing PTK_2 cell. The frozen frame of very-low-light-level fluorescence (excitation intensity reduced by an ND-2 filter) was captured with a SIT camera. (B) The S/N ratio was greatly improved by digital summing over 256 frames. The microtubular bundles in the spindle are now evident. (C) Single frozen frame of an interphase cell. (D) After 256-frame digital averaging. The individual microtubules are now clearly resolved.

Fluorescence microscopy with Zeiss ICM microscope using a Plan Apo 63/1.4 objective. The image was captured with a Dage–MTI Model 66 SIT camera, and averaged with an Image-I real-time digital processer. (S. Inoué and K. Fujiwara, unpublished data.)

pictures that can be seen directly on a video monitor, and signals that can be analyzed quantitatively. Also, UV flying-spot microscopy, which can be achieved with video techniques, provides special advantages for observing live cells (see Section 11.5).

Today, video is coupled with IR microscopy to see through silicon semiconductor wafers. As the size of the **integrated circuits** diminishes and their complexities increase, it has become increasingly important to be able to examine the silicon chips thoroughly but rapidly.

The extended IR sensitivity of modern video cameras is not only an advantage for inspection of silicon wafers, covert surveillance, and thermal imaging applications, but also for physiological studies of vision. Excised retinal rods and cones bleach rapidly in the visible wavelength range, even with rather dim illumination, but not in the near IR. To study the small retinal outer segments by microspectrophotometry, one needs to prepare and accurately position the structures without exposure to visible light. MacNichol and others routinely use visible-light cutoff filters and IR-sensitive video cameras to visualize the retinal outer segments under their microscopes (e.g., Dage–MTI 57 M Ultricon camera with an 800-nm-cutoff Kodak Wratten 87 C infrared filter; F. Harosi, personal communication).

Using video cameras with extended wavelength responses, we can now probe, in real time, the microscopic world imaged with IR or UV (and even X-rays), which otherwise provoke little luminance sensation in our eyes. Taking advantage of the high sensitivity of the video cameras, capabilities of image processing, and if necessary under intermittent illumination, we can make our observations with reduced risk of photon or thermal damage to the specimen. The spectrum with which we can explore the dynamic properties and chemical events in living cells and other microscopic samples can be significantly expanded by video.

11.3. CONTRAST ENHANCEMENT

In the previous section, we emphasized the use of video for raising the luminance of low-light-level microscope images. In these applications, video was also used to integrate noise, enhance contrast, and otherwise process or analyze the image.

In the modes of microscopy to be discussed below, we still use video to raise the luminance level of the image in some applications, but the main role of video shifts more toward contrast enhancement, image processing, and analysis.

Polarization Microscopy

Among the modes of contrast generation listed in Table 11-1, we perhaps encounter the greatest range of image luminance in *high-extinction polarization microscopy*. At the low end, weakly birefringent (biological) specimens observed between crossed polars may raise the image brightness by only 10^{-4} to $10^{-5} \times I_p$, the intensity of light that would traverse the microscope if the polars were oriented parallel to each other.*

*The brightness I of a birefringent specimen whose retardance is $\Delta°$ and oriented at 45° to the axes of the crossed polars is

$$I = I_p \sin^2(\Delta/2 + 1/\mathrm{EF})$$

where the extinction factor (EF) is given by

$$\mathrm{EF} = I_p/I_s$$

with I_s being the brightness of the field in the absence of a specimen with the polars crossed (Swann and Mitchison, 1950).

With polarization microscopy limited to low NAs, or using rectified lenses at high NAs, we can obtain EFs in the range of 5×10^3 to 2×10^5 (Inoué and Hyde, 1957). I_s is therefore of the order of 10^{-4} to $10^{-5} \times I_p$ for high-extinction polarization microscopy.

On the high end, a specimen, or contaminant (such as a speck of lint), with a quarter-wave ($\Delta = 90°$) to a half-wave retardance, would shine against the dark background between the crossed polars with a brightness of (½ to 1) × I_p. Thus, either intentionally or accidentally, one encounters luminance ranges that span 4 to 5 decades or more in high-extinction polarized-light microscopy.*

Polarized light is not just a special type of filtered light, but one that expresses a fundamental property of electromagnetic waves. The interaction of polarized light and crystalline matter not only provides surprising contrast and colors, but when viewed with an understanding of its physical optical basis, signals the electronic configuration and fine structure of matter at nanometer and smaller levels, far below the wavelength of light. In other words, polarized-light techniques provide a scalpel far more finely honed for dissecting structure than ordinary diffraction methods that rely mainly on the azimuth-averaged wave nature of light.

Applied to microscopy, we use polarized light to detect and measure anisotropic optical properties of minute regions in the specimen. In crystallography and petrography, the angles of the crystalline faces, together with the refractive indices and absorbance values (measured with the electric vector of the light waves oriented along the major crystalline axes), permit identification of the crystals (Wahlstrom, 1960; Wright, 1911). The difference of refractive indices for waves vibrating along different crystallographic axes, or *birefringence,* and the difference of absorption curves for the different axes, or *dichroism* (pleochroism), reflect the anisotropy of the constituent molecules and their alignment (Appendix III; Chamot and Mason, 1958; Hartshorne and Stuart, 1960).

Applied to biology, polarized-light microscopy not only allows one to distinguish crystals and fibers formed within, or by, living cells (Bennett, 1950; Okazaki and Inoué, 1976; Schmidt, 1924), but also reveals minute regions of cytoplasm or nucleoplasm where small numbers of molecules are regularly aligned (Schmidt, 1937; Inoué and Sato, 1966, 1967). For example, the form birefringence of mitotic spindle fibers (which emerge each time a cell reproduces) signals the dynamic, reversible assembly and alignment of the microtubules that organize the living cytoplasm and distribute the chromosomes for the crucial task of mitosis and cell division (Inoué, 1964, 1981a; Sato *et al.,* 1975).†

Over the years, we and others have managed to improve the contrast, achievable resolution, and image brightness of weakly birefringent objects viewed with polarized-light microscopy. With a light source of high luminous density, **rectified optics,** and stress-free lenses arranged for Koehler illumination, and using quality polars and low-retardation compensators, we can use the highest-NA achromatic objective lenses with matched condenser NA to detect retardations down to $10^{-4}\,\lambda$ and less, with an image resolution of slightly better than 0.2 μm and an image free from *anomalous diffraction* (Inoué, 1961; Inoué and Hyde, 1957; Inoué and Kubota, 1958; Inoué and Sato, 1966; Kubota and Inoué, 1959; Appendix III).

Today, the sensitivity for detecting oriented molecules and *ordered fine structure* with polarized-light microscopy is very high; for the diameter of the Airy disk, one-fifth of an oriented monolayer of B-form DNA molecules (i.e., a layer 4–5 Å thick) provides image contrast that can be detected with the human eye. Furthermore, such sensitive detection and measurements of retardation in polarization microscopy do not require absolute intensity measurements; only the specimen orientation and compensator setting that signals the null points of compensation need be detected by observing contrast changes (e.g., see Inoué and Sato, 1966; Sato *et al.,* 1975).

With video, one can further improve the sensitivity for detecting small retardations with polarized-light microscopy, while concurrently improving the image quality, achievable resolution, and speed of recording.‡

*Note also that the luminance of the field of view varies with magnification and NA. Contaminants can also blast the image tube in darkfield and fluorescence video microscopy. High-sensitivity video cameras should be equipped with automatic shutoff circuits to protect against such mishaps.

†Appendix III provides an expanded coverage of the basis of polarized-light microscopy and its application to analysis of biological fine structure.

‡Some have argued that video enhancement dispenses with the need for optical correction devices such as the polarization rectifier. Such an argument is patently wrong, as evidenced by the fact that, in the absence of a rectifier, the microscope has to be refocused each time the compensator axes are reversed. This demonstrates that the diffraction pattern is, in fact, different for the two compensation orientations. The waves traversing

I, and Allen *et al.*, have shown that the approximately linear light transfer characteristics of some video cameras can be used to advantage to gain a high-contrast picture of weakly retarding objects. The background field brightness introduced by the compensator is set to a level substantially greater than that optimum for visual observation or photography.* To the eye or photographic emulsion, which responds to the log (or ratio) of intensities (Fig. 11-4B), the image through the ocular is then washed out. But the absolute value of the video signal introduced by the specimen has actually become greater (Fig. 11-4A). Therefore, by adjusting the pedestal level to reduce the luminance of the background, the video picture can be presented with much-impaired contrast (Allen *et al.*, 1981a,b; Inoué, 1981b).

While Allen's and my views differ as to the exact amount of compensation that gives rise to optimum enhancement of the polarized-light image, there is no question that polarized-light and related forms of microscopy benefit greatly by video enhancement. Plan Apo objectives with nearly matched NA condensers can now be used to provide well-resolved images under conditions that, without video enhancement, give but a ghost of an image even after rectification (Fig. 1-5). The birefringence of beating cilia can be captured in a 16.7-msec single video field (Fig. 1-6) rather than the many seconds that would have been needed for photography. Without video, we have been unable to photograph the weakly birefringent, rapidly beating cilia in polarized light. Even the rotation of bacterial flagella has been visualized taking advantage of the contrast reversal exhibited by differently oriented portions of the helical flagella viewed between crossed polars in the presence of a suitable compensator (Fig. 11-5).

One can use moderately sensitive Newvicon, Chalnicon, or CCD cameras (with automatic or manual pedestal and gain controls) at elevated bias retardations to enhance the polarized-light image in the manner just discussed. Alternatively, one can use a more sensitive camera, such as an I^2V (e.g., Zeiss–Venus TV-2) with bias retardation levels of a few nanometers, levels which are optimum for visual observation or photographic recording.† Used on a high-extinction (rectified) system, the latter combination gives an image with reasonably high contrast and moderately good resolution (Fig. 11-6).

With either type of camera, contrast can be boosted, shading and background noise reduced, and the image filtered, convoluted, and analyzed, or displayed in pseudocolor, as needed by analog or digital means (Chapters 9, 10; see also Allen and Allen, 1983; Ellis *et al.*, 1988). Figure 11-7 shows examples of a digitally processed image of a critical test sample observed by polarized-light microscopy.

While we can thus enhance the moderately low- to very-low-light-level images of weakly retarding objects with appropriate monochrome video cameras, we can also use color cameras to capture the *interference colors* produced by dichroic, or optically active, materials, or by birefringent specimens whose retardations exceed a large fraction of a wavelength. The vivid orthoscopic and conoscopic microscope images displayed on a color monitor are at the very least pedagogically effective.‡ They also provide convenient means for inspecting the crystallographic details of natural and synthetic crystals and fibers.

adjacent quadrants of the objective aperture are shifted in phase by 180°, a consequence of the optical rotations that give rise to anomalous diffraction (Inoué and Kubota, 1958; Kubota and Inoué, 1959). Thus, video enhancement as such does not correct for the anomalous diffraction introduced by nonrectified objectives of higher NAs (Appendix III).

*For *visual* observation and photography of weakly birefringent objects with a high-extinction polarizing microscope, a compensator retardation of 2–5 nm provides optimum contrast. With *video*, Allen *et al.* (1981a) report (using a Zeiss Axiomat microscope, coupled to a Hamamatsu C-1000 Chalnicon camera) that a bias retardation of from a $\lambda/9$ to a $\lambda/4$ range is optimal. I find (with Dage–MTI 65 through 68 series Newvicon cameras, on a high-extinction polarizing microscope equipped with calcite polars) that a bias retardation of up to about 10 nm ($\lambda/50$) provides maximum contrast. See p. 355 of Inoué (1981b) for further discussions. See also the following footnote regarding the I^2V camera.

†Note that the I^2V with the vidicon tube responds logarithmically with a gamma of 0.65 rather than linearly as do the other cameras mentioned here. The higher sensitivity of the I^2V and its logarithmic response probably explain why *its* optimum bias retardation is more similar to our eye's.

‡The Leitz Prado Pol kit is very effective for demonstrating these phenomena, live to an audience.

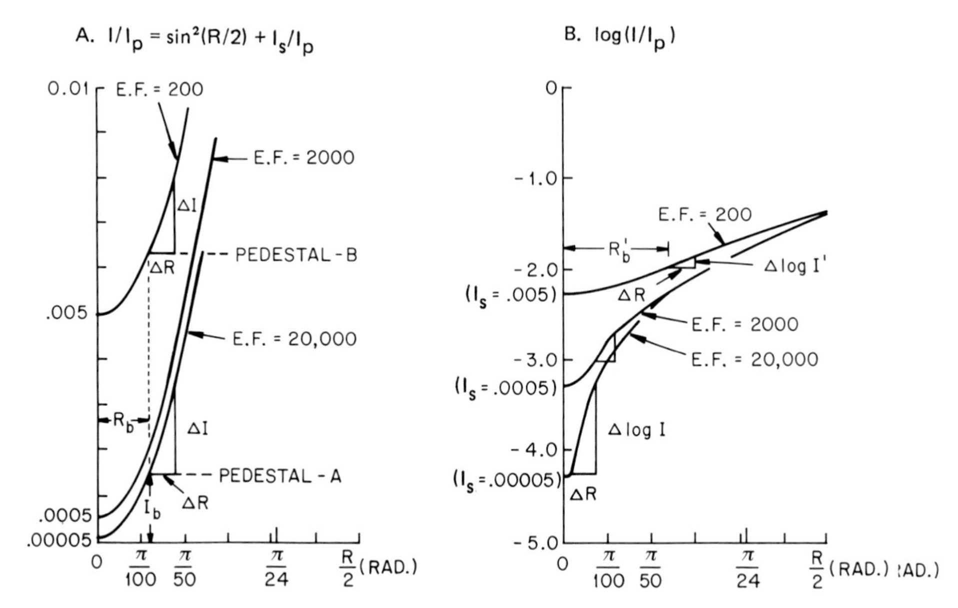

FIGURE 11-4. Response curves of linear and logarithmic detectors to birefringence retardation. (A) Response of idealized photodetector. I is the brightness of retarders placed between crossed polars. For different extinction factors (EF $= I_p/I_s$, where I_p is the field brightness with the polars oriented parallel to each other, and I_s is the brightness with zero retardation and the polars crossed; Appendix III), the curves become displaced vertically without change in shape. The increment of brightness, ΔI, introduced by an increment of retardation, ΔR (in the presence of a given bias retardation, R_b), is not affected by the extinction factor. For a very small ΔR, ΔI converges to 0 at $R_b = 0$. ΔI increases with R_b up to a maximum at $R_b = \lambda/4 = \pi/2$. Thus, with an *idealized linear photodetector* (one that responds linearly without limit to the light intensity), ΔR would give rise to the same incremental signal independent of EF at a given R_b, and the signal (but not necessarily the S/N ratio) would be maximal at $R_b = \lambda/4$. The background brightness I_b could be subtracted electronically by selecting appropriate pedestal levels. *Actual* vidicon cameras have a limited dynamic range and their responses follow curves lying partway between those in panels A and B.

(B) Log plots of brightness of retarders placed between crossed polars. With the intensity plotted on a log scale, the curves for different EFs take on different shapes and slopes. *Contrast detectors* such as the eye and photographic emulsions respond more or less linearly to the incremental brightness (ΔI) divided by the background brightness (I_b see panel A), or to the *log* of brightness. Thus, $\Delta\log I/\Delta R$ determines the sensitivity for such detectors (Inoué and Dan, 1951). For contrast detectors, $\Delta\log I/\Delta R$ (where $\Delta R \rightarrow 0$) is again 0 at $R_b = 0$. The contrast reaches a maximum at $I_b = 2I$. For systems with different EFs, the maximum $\Delta\log I/\Delta R$ as well as the optimum R_b vary considerably. (From Inoué, 1981b; see also Appendix III.)

With the very high sensitivity, and image-processing and wavelength-converting capabilities of video, coupled with a good-quality polarizing microscope, many polarization-dependent phenomena in wavelength ranges within and beyond the visible can now be explored with good spatial and temporal resolution. Much challenge and rich opportunities lie ahead for the life and materials scientists who venture into these relatively uncharted waters.

Phase Contrast

In Section 5.6 we examined the principles of *phase-contrast microscopy*. We saw how Zernike's exposition of the phase contrast principle laid an essential foundation for the modern understanding of microscope image formation, thus breathing new life into the earlier pioneering investigations of Abbe.

Before phase contrast, pure phase objects were visible in ordinary light microscopy *only when the specimen was somewhat out of focus*. With the advent of phase contrast, followed by interference, differential interference contrast, and the various modes of single-sideband microscopy, it became possible to form an intensity-modulated, high-contrast image of the phase specimen *in focus*. In these modes of microscopy, the light waves diffracted by the phase object destructively or constructively interfere with the phase-shifted, (**attenuated**) illuminating wave and form the visible image (Section 5.6).

Since we can optically generate excellent contrast images with phase microscopy, video becomes more useful for *extracting features* and *analyzing* the image than for enhancing contrast in this mode of microscopy. Nevertheless, we can use video to enhance the phase-contrast microscope image even further and improve the detection of objects that are extremely small or thin, or whose refractive index is very close to the ambient, i.e., objects that produce but a minute phase retardation (e.g., Koonce and Schliwa, 1985; Walter and Berns, 1981). Video enhancement should also be useful for *measuring* small amounts of phase retardation or light absorption with Osterberg's Polanret variable-phase contrast system (Osterberg, 1955).

The high-contrast image produced in weakly retarding or absorbing objects by phase microscopy provides interesting opportunities for digitized image analysis, manipulation, contouring, sorting, feature extraction, etc. As described in Chapter 10, Berns and Berns (1982) have devised a computer-aided video system that recognizes the cell's nucleolus and automatically follows the reproduction and migration of tissue-cultured cells for many days. The computer commands the

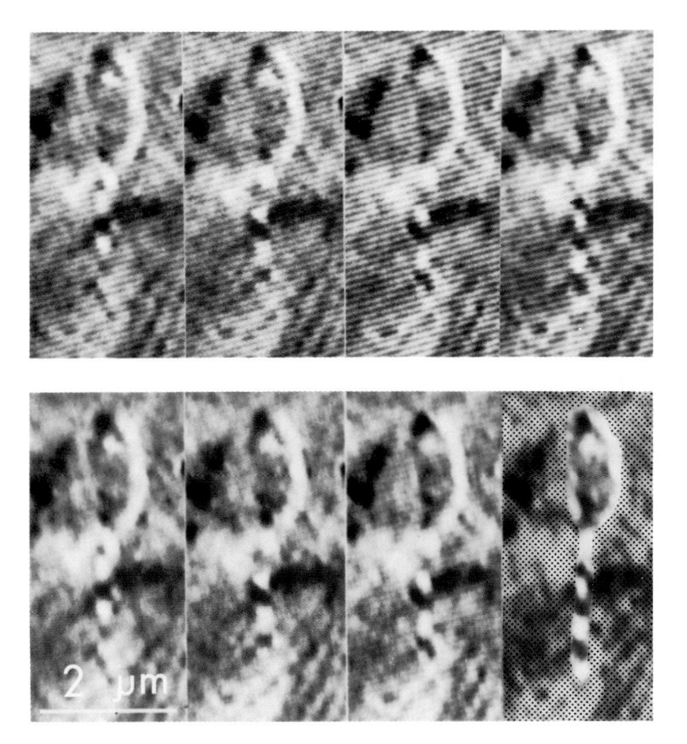

FIGURE 11-5. Birefringence of rotating flagella of a bacterium. (In the last frame, a stippled mask outlines the bacterium and its flagellar bundle, which appears like a barber pole in these figures.) The bacterium was observed between crossed polars in the presence of about $\lambda/20$ compensation through the author's high-extinction, rectified polarizing microscope. The amplitude of the flagellar wave is less than 100 nm. Despite the amplitude, which is smaller than the resolution limit of the microscope, those regions of the flagella tilted to the right and the left show in black and white, in reverse compensation. The contrast reflects their tilt relative to the axes of the compensator (see Appendix III). Time interval between the first two frames, 17 msec; frames 2 and 3, 17 msec; frames 3 and 4, 33 msec. The frozen-field pictures in the top row were taken off the monitor, played back from a ¾-inch VCR (Sony TVO-9000). Those in the bottom row were obtained by photographing the top row of pictures through a Ronchi grating to eliminate the intrusive scan lines (see Section 12.1). The video camera (Dage–MTI series 65) was equipped with a 1-inch Newvicon tube. × 9500. (From Inoué, 1981b.)

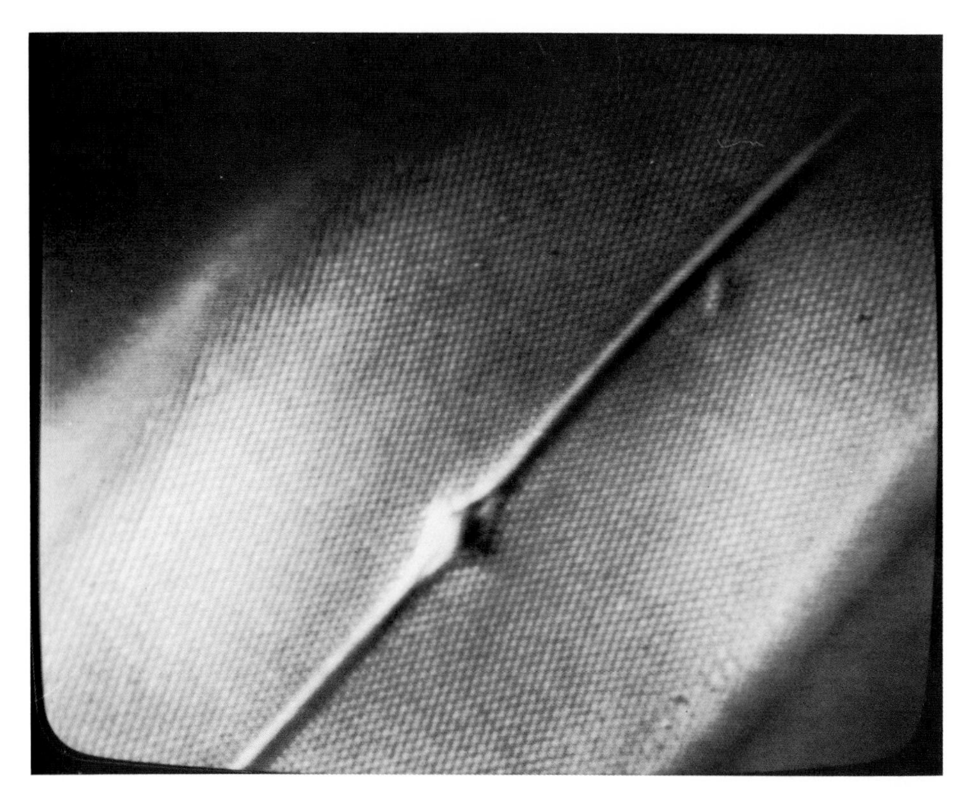

FIGURE 11-6. Diatom in high-extinction DIC taken with a two-stage intensifier camera. This low-light-level image of a diatom (*Pleurosigma angulatum*), produced with a 100/1.35 Plan Apo objective coupled to a 1.35-NA recitified condenser (Nikon) at full aperture, was captured with a two-stage intensified vidicon camera (Zeiss–Venus TV-2). The image, which was not digitally processed, is quite good despite the two stages of intensification. The high-extinction image was produced at a setting near the extinction point of the rectified DIC system.

stage of the servo-driven microscope to track the centroid and the replication and movement of the nucleolus, which acts as a phase-dense marker for each cell.

The phase-contrast microscope produces excellent image contrast at moderately high condenser NA, and, therefore, with good resolution. Moreover, contrast in phase optics is highest at optical boundaries, and decreases away from the boundary. Thus, a phase-contrast microscope acts as a high-spatial-frequency pass filter that helps to accent fine specimen details and isolate very fine structures. But with the phase-contrast system, a distracting halo appears (in reverse contrast) around the phase object. The phase halo can be so pronounced as to conceal the interior details of small, refractile objects such as the head of a spermatozoan.

Interference

Unlike phase contrast, a prominent halo is absent in the image produced by an *interference microscope*. In interference microscopy, the gray scale in monochromatic light, or interference color in white light, directly reflects the actual phase retardation (optical path difference) produced by the specimen. A uniformly thick, homogeneous region of the specimen thus appears with uniform contrast or color, whereas with phase contrast, the edges would appear graded.

While phase contrast reveals fine specimen detail, interference microscopy is better suited for *measurement* of optical path difference and determination of dry mass (e.g., see Barer, 1956). With interference microscopy in the "uniform field" mode, the interference colors signal the exact optical path for each region of the specimen. In the "nonuniform field" mode, the periodic dark fringes provide full wavelength calibrations that can be used for measuring optical path differences.

Most interference microscopes work better when the specimen is illuminated with a nearly collimated beam, i.e., when the condenser NA is very low. As the condenser NA is raised, the image contrast drops. Therefore, the practical image resolution is somewhat limited in the interference microscope. Also, without substantial optical paths, the gray level or color contrast can be quite low.

Video processing can improve the resolution and sensitivity of an interference microscope just as in a polarizing microscope. With the contrast-enhancing and image-thresholding capabilities of video, one should be able to use the condenser at higher NAs to achieve better resolution. Likewise, we should be able to substantially increase the sensitivity for measuring small retardations. The interference fringes can be sharpened by video-image processing, and their locations pinpointed

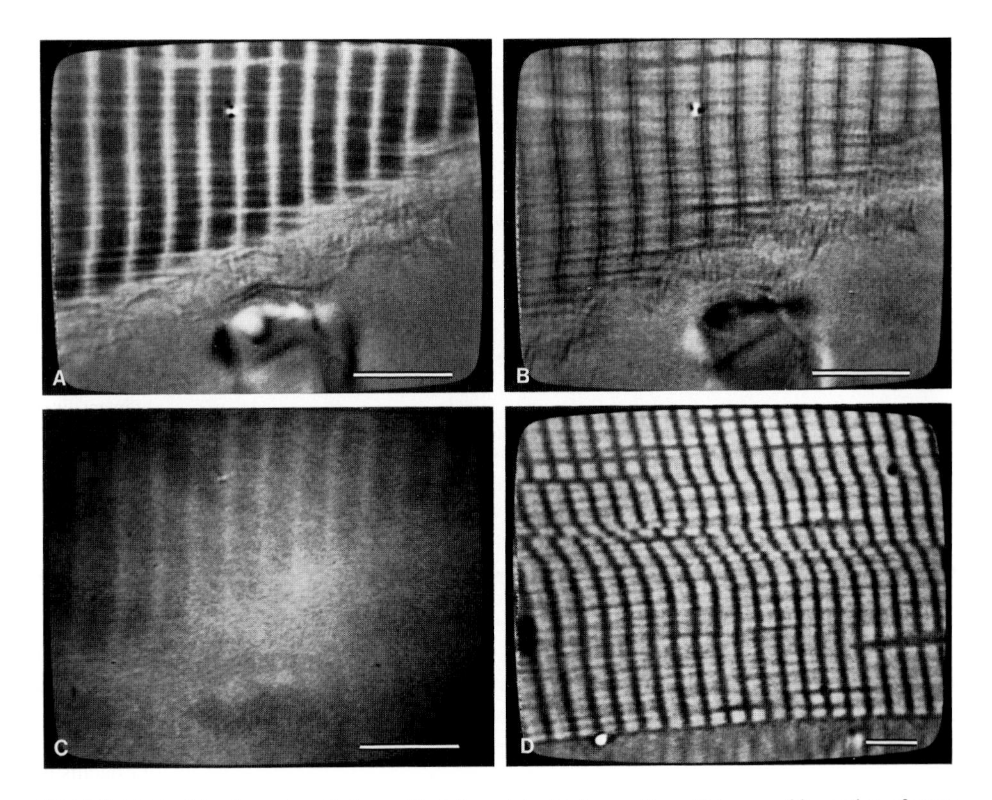

FIGURE 11-7. Muscle section (360 nm thick) in matched-index medium. This very thin section of cross-striated muscle remains embedded in epoxy resin, and is immersed in a medium of matched refractive index. Here, the image is observed, between crossed polars in the presence of about 5-nm compensation through a 100/1.35 Plan Apo objective (Nikon) used with a 1.35-NA rectified condenser at full aperture, with a SIT camera (Dage–MTI Model 66). Without image processing (C), there is very little contrast through the Plan Apo objective lens (which contains crystalline heterogeneity). (A) Subtractive- and (B) additive-compensation. After 128 frame-summation, background subtraction, and contrast enhancement by digital processing (Image-I system; cf. Fig. 6-24), the A- and I-bands and Z-line (with progressively lower birefringence) are clearly visible. With a 40/0.95 Plan Apo objective, the nonbirefringent Z-line is nearly undetectable (D). This very-low-contrast specimen with fine periodicities provides a critical test for the performance of various microscope optics and video enhancing systems. Without video the specimen can just barely be seen through the ocular when strain-free rectified objectives are used critically. Scale = 5 μm.

with a video analyzer (Section 11.4). These improvements could benefit interference microscopy in both the transmitted- and reflected-light modes.

Reflected-light interference microscopy is used in materials research for inspecting surface details of opaque or thick objects. Cline *et al.* (1982) describe a method for reconstructing surface contours from interference patterns obtained with a reflection interference microscope with the aid of video. It is also used in biology, in conjunction with the phenomenon of frustrated total internal reflection, to study how migrating living cells adhere to their substrates (e.g., Gingell and Todd, 1979; Izzard and Lochner, 1976; Sullivan and Mandell, 1983).

Differential Interference, Single-Sideband, Anaxial Illumination

In *differential interference contrast* (DIC), image contrast reflects the gradient of optical paths along a specific shear direction, namely that of the Wollaston beam-splitting prisms. Thus, in both Smith- and Nomarski-type DIC, the image has a shadow-cast appearance resembling the contrast of scenes familiar in daily life.*

Not only does the image in DIC give a ''natural'' appearance, but the optical section that makes up the focused image is remarkably shallow. Furthermore, unlike phase contrast, the DIC image is free of halos and disturbances from out-of-focus objects. Thus, DIC provides an image with a pleasing contrast, neatly sliced and isolated out of a complex three-dimensional phase object (e.g., Schatten *et al.*, 1982). This property of DIC is used to obtain optical sections of cell outlines in complex tissues (e.g., DiBona, 1978; Persson and Spring, 1982; Spring, 1979).

Conventional DIC optics give good contrast with the condenser iris diaphragm partly stopped down, or at moderately high resolution. By using a rectified condenser coupled with Plan Apo objectives, one can obtain DIC images with excellent contrast and outstanding resolution at full condenser NA (Fig. 8-19).

Video enhances the contrast and provides clear DIC images of extremely fine objects. In Fig. 9-11, we have seen serial video micrographs of the rapidly growing, 65-nm-thick, acrosomal process of *Thyone* sperm enhanced with analog camera circuits. Allen, co-workers, and others have used digital video enhancement and background mottle subtraction to study streaming organelles in neurons, amoeba axopodia, and plant cytoplasm (Fig. 1-7; Allen and Allen, 1983; Allen *et al.*, 1982; Brady *et al.*, 1982; Forman *et al.*, 1983). They and others have also visualized particle transport and motility of individual microtubules *in vivo* and *in vitro* (Fig. 11-8; Hayden and Allen, 1984; Schnapp *et al.*, 1985; see also Koonce and Schliwa, 1985).

In these examples, the improved contrast, detection of finer specimen detail, and faster speed of recording were achieved with Newvicon (Dage-MTI) or Chalnicon (Hamamatsu) video cameras. Earlier, Dvorak *et al.* (1975) used an image orthicon on a DIC microscope to obtain low-light-level sequential images of malaria parasites invading a human red blood cell (see cover pictur on *Science*, February 28, 1975, by Dvorak *et al.*).† The image orthicon, now superseded by the image isocon (an isocon with a built-in image intensifier section), provides a very high sensitivity, high resolution, a wide dynamic range, and excellent S/N ratio (Table 7-2). Were it not for the camera's high cost and sensitivity to abuse, the image isocon might find considerably wider use in microscopy.

*The resemblance of the DIC image to the familiar scenes encountered in daily life can help our interpretation of the three-dimensional aspects of the specimen. On the other hand, the familiar appearance can also deceive us, e.g., into seeing a lenticular region with a lower refractive index as if it were a convex surface (see discussions in Inoué and Tilney, 1982).

†Until Dvorak *et al.* devised and applied low-light-level DIC video microscopy, malarial merozoites had never been observed in the process of invading a red cell; the parasites are highly light sensitive. In addition to using an image orthicon, and later an image isocon, these authors developed a sophisticated video-processing and recording system at the National Institutes of Health.

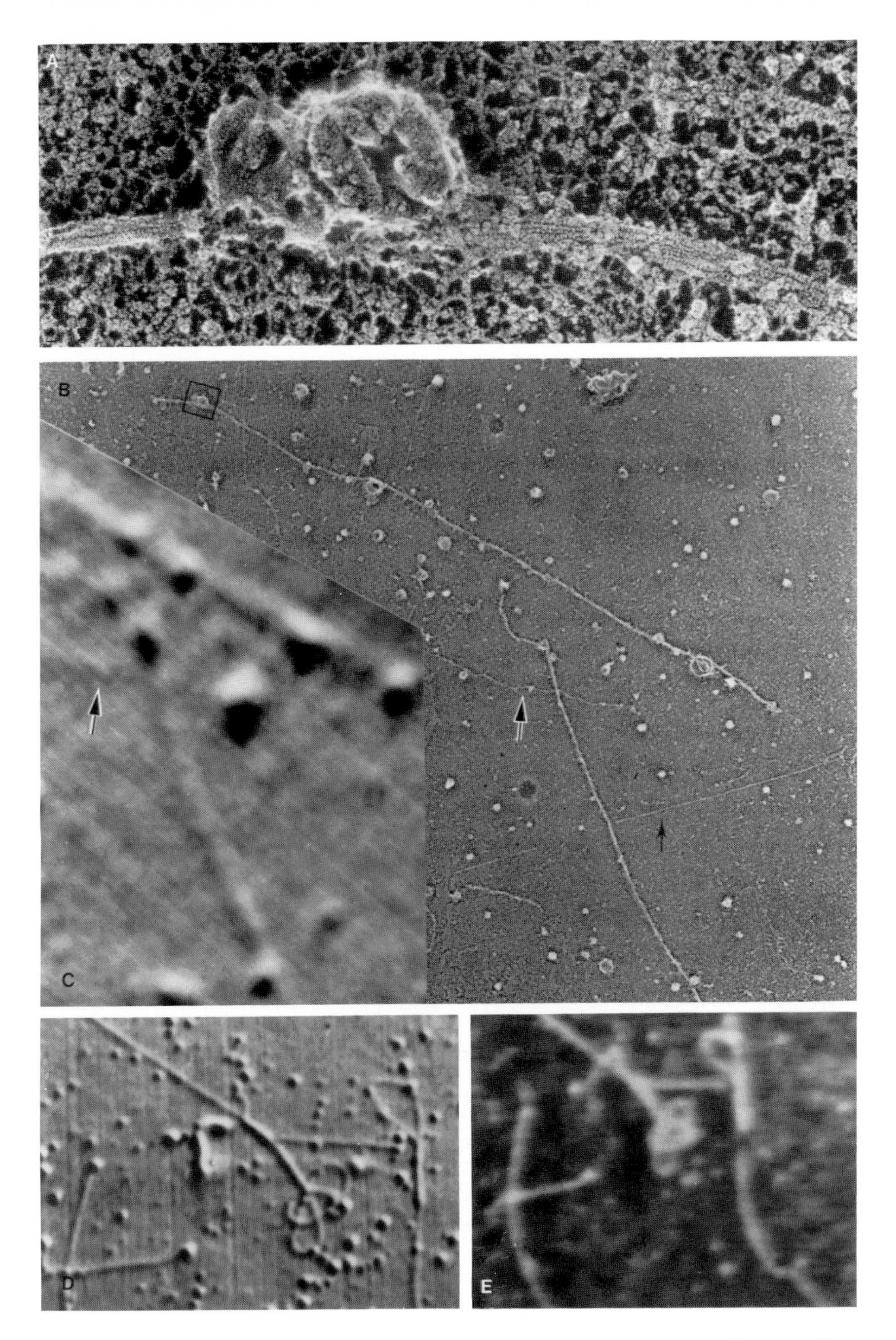

FIGURE 11-8. Single microtubules visualized with video-enhanced microscopy. (A) High-magnification electron micrograph of area indicated by the box at the upper left of B, demonstrating that the transport filaments are microtubules; × 190,000. (B) Low-magnification electron micrograph of the transport filaments (the two thicker filaments without the arrows). × 12,000. (C) Video-enhanced, light micrograph (DIC) of the same region at the same magnification as B. In live images, particles can be seen moving along the filaments. The diameter of the filaments and particles is inflated by the diffraction of light; × 12,000. (D) Video-enhanced DIC image; × 3900. (E) Video-enhanced fluorescence microscope picture of the same field shown in D. The filaments have been stained with rhodamine-conjugated fluorescent antibody against α-tubulin. × 2800. (Courtesy of R. Vale; after Schnapp *et al.*, 1985; see also Fig. 1-7.)

In addition to the Smith and Nomarski DIC microscopes employing Wollaston and Nomarski prisms between two polarizing elements, the modes of microscopy that use *single-sideband modulation* also produce "shadowed" images. These include oblique (or anaxial) illumination, Hoffman modulation contrast, and Ellis's single-sideband edge-enhancement system (SSEE). Since these latter do not rely on polarized light for contrast generation, as do the DIC systems, they can be used with or without the addition of polarization-induced contrast.

B. Kachar, combining oblique illumination at nearly full condenser NA with video enhancement (by regulating pedestal and gain control), has recently obtained striking, thin optical sections of membrane vesicles (Kachar, 1985; Kachar *et al.*, 1984). As shown in Fig. 11-9, the video-enhanced optical sections display fine cellular details and nuclei with good clarity and resolution. Pictures taken at several focal levels show how little the structures present above and below intrude in the optical section in focus.

Video enhancement should raise the sensitivity of SSEE even higher. Even without video enhancement, SSEE has been the only mode of microscopy, besides rectified polarization optics, that clearly displays the boundaries of the tightly packed, anisotropic DNA microdomains in living sperm (see Ellis, 1978; Inoué and Sato, 1966; Appendix III).

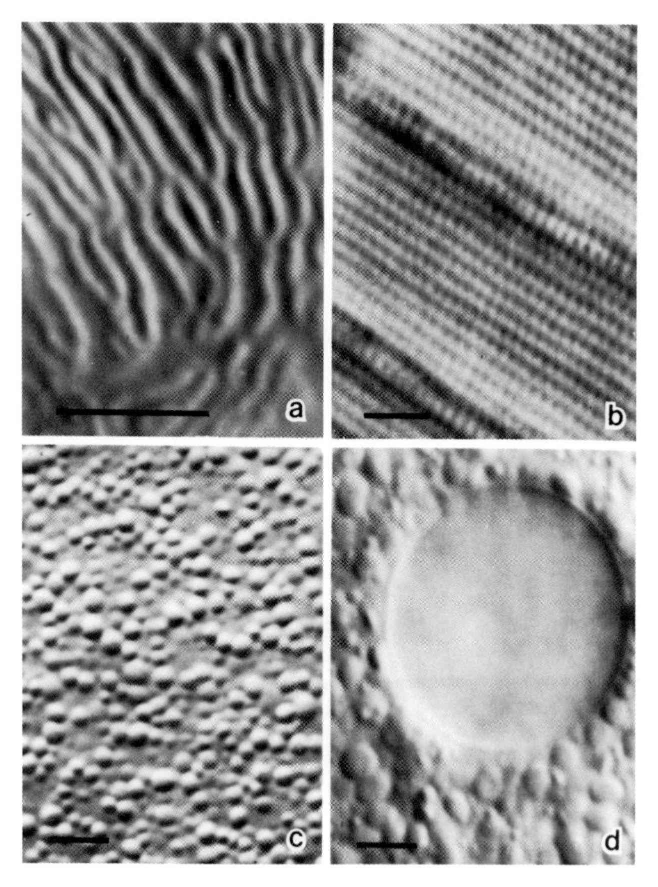

FIGURE 11-9. Video-enhanced anaxial illumination at full condenser NA. (a) Ridges on buccal epithelial cell, Plan Apo 100/1.3. (b) Lattice (0.41-μm period) on the frustule of a diatom, *Surirella gema;* blue monochromatic illumination, 40/1.10 objective. (c) Cortical layer of the sea urchin egg shown in d. (d) Nucleus in 100-μm-thick, living sea urchin egg. The resolution and optical sectioning capabilities of this mode of video microscopy are impressively illustrated. Scale = 5 μm. (Courtesy of B. Kachar; from Kachar, 1985.)

Brightfield

A brightfield image *under a dissecting microscope* appears three-dimensional, natural-looking, and easily interpreted. A relatively large, opaque specimen, illuminated from the same side as the objective lens, absorbs, reflects, and scatters light, and forms an image with color, highlight, shadow, sheen, and texture similar to those observed in the macroscopic world. However, as we raise the magnification to delve deeper into the microscopic world, the character of the image changes dramatically. A white sheet of paper transforms into matted threads of transparent fibers, bright red blood turns into a suspension of pale amber disks, and a shiny metal surface becomes a mosaic of slates covered with scratches and pits. The familiar scenes change as the magnification rises, in part because the wavelength of our probe is no longer much smaller than the dimensions of the specimen being probed.

With transillumination, *brightfield microscopy* brings us farther into an unfamiliar world. The tip of a microneedle drawn out of a black glass rod is as clear as a needle made of clear glass; images of fine phase objects show only by their schlieren pattern, and disappear when the microscope is brought exactly into focus. At exact focus, few other than carbon particles and fine layers of metal appear dark, and only dyes with exceptionally high extinction appear to stain our otherwise invisible specimen.*

Despite the strangeness of the transilluminated microscopic world, or perhaps partly blind to this strangeness, much of light microscopy relies on brightfield transillumination.†

As this discussion suggests, the brightfield microscope is suited primarily for contrast production by absorption or reflection of light. Much of early biological and medical microscopy depended on development of effective stains, and the time-consuming methods of fixation, embedding, sectioning, staining, and mounting. The pioneers painstakingly reconstructed the third dimension from serial sections, seriated time sequences from many samples, and deduced tissue and organellar composition from reactions to chemicals and stains. Thus, with laborious effort, they portrayed life's microscopic architecture that explains how we sense, move, grow, reproduce, and become diseased.

Metallographic, and some of petrographic, microscopy is based on the image differentially *reflected* from polished and etched specimens. While polarized light is used to reduce glare, it is generally the texture, reflectance, and color that clue the microscopist to the nature of the different sample regions. Thus, contrast generation in reflected-light microscopy does not appreciably differ from that used in brightfield methods.‡

In brightfield microscopy, differences in absorption and reflection can profitably be enhanced by video. Rather than with standard color cameras, which do not appreciably enhance color contrast, one can enhance subtle differences in color or absorption with video in monochrome. Coupled with appropriate filters to elevate the contrast of a weakly colored or stained specimen, analog and digital enhancement can do wonders to bring out image details that otherwise would not be visible.

Vital indicator dyes, too pale at nontoxic concentrations to be useful in the past, can now be used to measure concentrations of various ions and chemicals. Natural dyes and pigments can be detected and measured in concentrations or quantities that earlier were far too small to be noticed through the microscope. Subtle reflectance differences, and illusive textures and color differences,

*In the submicrometer world that makes up our microscopic specimen, only metals that contain freely mobile electrons, or materials such as dyes, graphite and iodine crystals with high concentrations of conjugated bonds, absorb enough energy from the illuminating beam to produce perceptible contrast.

†It would not be fair to say that the earlier microscopists were unaware of the strangeness, or unaware of the danger of naively interpreting the (transilluminated) microscopic image. Indeed, many were well aware that diffraction could give rise to phantom membranes and filaments. They even reported that the contrast of the denser A-band in cross-striated muscle appears lighter or darker than its surroundings depending on whether the microscope is focused above or below the unstained muscle fibrils.

‡IR and UV microscopy, as commonly practiced, are also forms of brightfield microscopy. However, because our eyes do not normally see in those wavelengths, I classified them earlier together with other low-luminance microscopic methods.

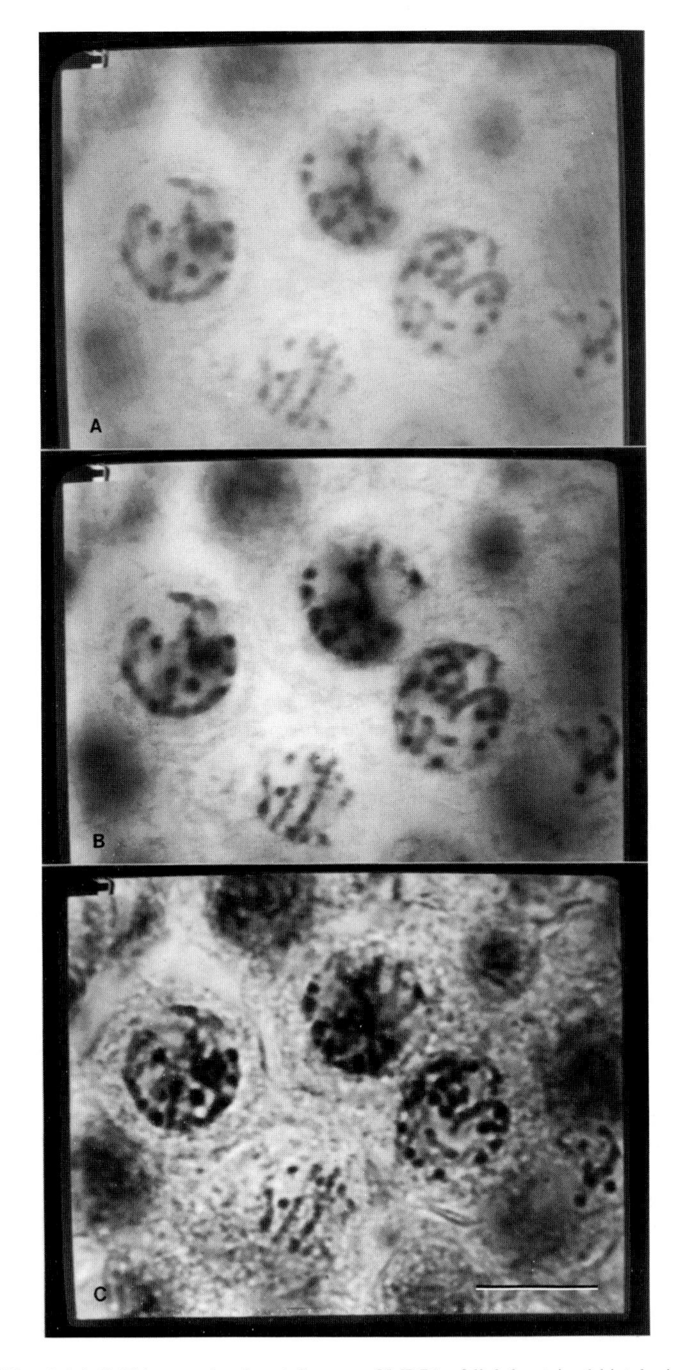

FIGURE 11-10. Brightfield images (retrieved from an OMDR) of lightly stained histological slide. Histological section of (25-year-old) human testis, lightly stained for chromatin (Heidenhein's hematoxylin), showing the prophase of meiosis I. The objective was a CF 100/1.35 Plan Apo lens (Nikon). A and B were taken with an oil-contacted condenser at its full NA of 1.35; × 2500. (A) Unenhanced image. The full NA of the condenser provides high resolution and thin optical sectioning, but low visual contrast. (B) Contrast enhanced by digital processing (Image-I system). (C) Same as A, but condenser iris closed to NA 0.45. The contrast of the microscope image rises with reduced condenser NA, but the accompanying loss of resolution and diffraction from out-of-focus structures leads to a confused image. Scale = 10 μm.

The high quality of these micrographs is due in part to the fact that they were made using a new light-pipe scrambler (Fig. III-21). The scrambler produces uniform full-aperture illumination of the condenser, with little loss of the field luminance.

can also be accentuated by video enhancement and add to the many morphometric capabilities already available in video microscopy.

We should especially note, as in other forms of microscopy, that video enhancement allows the use of brightfield microscopy with high condenser NAs. Since increased condenser NA raises the resolving power and reduces depth of field, both lateral and axial resolution can be dramatically improved by the use of video. Figure 11-10 illustrates the improvement achieved in brightfield microscopy by analog and digital video enhancement. The improvements in contrast and resolution are in addition to the gain in sensitivity, which can be especially great at the extreme ends of the spectrum, where our eye fails more rapidly the farther we depart from the green peak of our sensation.

The sensitivity of video cameras to the red end of the spectrum results in an interesting phenomenon when we apply video to brightfield microscopy. As an incandescent illuminator is dimmed by reducing the electric current, the color temperature of the illuminator drops and the field acquires an orange tint. As the source is dimmed further, the image turns deep orange and eventually becomes invisible to the eye; but a Newvicon camera, for example, can still yield a surprisingly respectable image. The difference in spectral sensitivities of the eye (Chapter 4) and the camera tube (Chapter 7) makes the camera tube appear very sensitive when the color temperature of the illuminator drops.

11.4. IMAGE QUANTITATION BY VIDEO

Photometry, Densitometry

Light microscopes have long been used as integral parts of microphotometers and microdensitometers. These instruments measure the light emitted from, or transmitted through, specified small areas of the specimen seen through a microscope. In some cases, the intensity or density scanned along a particular axis or area of the specimen is recorded (e.g., Joyce-Loebl Scanning Recording Microdensitometer; see Fig. III-24). Often, a densitometer is combined with a spectroscope or monochromator at the illuminating, imaging, or both ends of the microscope. Several manufacturers, including Leitz, Olympus, Vickers, and Zeiss, have produced scanning microspectrophotometers (and fluorometers) that automatically scan the spectrum from the UV to the near-IR.* Hárosi and MacNichol have constructed an automatic recording, polarizing microspectrophotometer that scans the entire visible spectrum forward and backward in less than 2 sec. With this highly sensitive instrument, they have been able to measure the absorption curve and dichroism on isolated retinal rods and cones in 10-μm-diameter areas without appreciably bleaching or damaging the light-sensitive retinal elements (e.g., Hárosi, 1982).

In virtually all of these devices, low-noise *photomultiplier* tubes have been used as the primary sensor. Photomultipliers, especially when chilled, can be used to count down to a few photons and to measure low levels of light with very short time constants. Photomultipliers provide an internal gain of 10^5 to 10^7, with a response that can span as much as ten orders of magnitude.†

The noise in some intensifier video cameras is also limited by photon statistics (Sections 7.2, 7.3; Gruner *et al.*, 1978; Reynolds, 1966; Rose, 1973). Compared to photomultipliers, the dynamic range of the video imaging devices (except for some chilled CCDs) is quite limited, but the distribution of light intensities for a two-dimensional array of some 10^5 to 10^6 points is provided in a fraction of a second. Therefore, the spatial and spatiotemporal distribution of light can be measured conveniently and with considerable sensitivity with an appropriate video camera. Video is particularly advantageous where the overall speed, and convenience for measuring many image points simultaneously, outweigh the need for the high precision and short response time that the photomultiplier can provide for measuring single image areas at a time.

*See, e.g., Piller (1977) for a comprehensive discussion on photometry of microscopic areas.

†According to Levi (1980), a photomultiplier tube can be operated with the response current deviating a maximum of 3% from linearity over six orders of magnitude of the light flux entering the tube. See Also Moore *et al.* (1983) for useful suggestions on the use of photomultipliers.

Where a dynamic range and speed of response greater than those achieved by video are needed, but with many more image points than can be measured conveniently with photomultipliers, investigators are using *photodiode arrays* coupled to microscope and the other imaging systems. For example, Roos and Brady (1982) used a 1728-element linear-array CCD to image muscle fibers, and Huxley *et al.* (1981) described an ingenious device for registering the diffraction pattern and measuring the spatial periodicity and minute movements in muscle. Grinvald *et al.* (1981), Salzberg *et al.* (1977), and others have measured the action potentials generated sequentially by many cells in a ganglion by the change in fluorescence of "reporter dyes" taken up by the neuronal cell membranes.

Examples of video photometry for measuring luminescence, fluorescence, birefringence, etc. were mentioned in the previous section. Among these, fluorescence cytochemistry is a rapidly growing area with wide, exciting potentials. By digitally processing and analyzing the fluorescent video microscope image, interesting opportunities are promised for measuring chemical changes in minute parts of functioning living cells. Especially noteworthy is a recent report by Benson *et al.* (1985) who have extrapolated the digitized video microscope image backwards to "recover" the fluorescent image of a cell before any *bleaching* had taken place. They observed that individual fluorescent particles in single living cells were bleached by the light used to excite fluorescence, exponentially with time but at quite different rates. To correctly assess the relative fluorescence of the particles, they extrapolated the exponential decay curves backwards to zero time for each image point, which allowed them to synthesize an image showing the distribution of fluorescence as it would have appeared in the absence of photobleaching.

Fluorescence recovery after photobleaching (FRAP) is another new technique that is evoking much interest and activity. In the FRAP technique, one determines the anchorage and mobility of molecules by measuring the time course of fluorescence recovery in a region of a living cell bleached with a laser microbeam. The fluorescence recovery signals the diffusion of the labeled molecules back into the region zapped with the microbeam (e.g., see Jacobson *et al.*, 1983; Leslie *et al.*, 1984; Salmon *et al.*, 1984b; Saxton *et al.*, 1984). Several laboratories are now exploring the use of video photometry in FRAP experiments (e.g., Watt Webb, personal communication, Applied Physics, Cornell University; C. Glabe, Worcester Foundation; see also McGregor *et al.*, 1984).

Morphometry, Sorting, Tracking

Over the past decade, much expert effort has gone into the development of digital image-analyzing video systems that are applied to light microscopy. These devices have been used to identify, measure, and classify cells, organelles, metallurgical specimens, IC chips, and other microscopic specimens or features. Morphometry, karyotyping, automatic cell sorting, industrial inspection, as well as counting, sorting, and tracking, are among some of their applications (see Mayall and Gledhill, 1979). The reader is referred to Chapter 10, where the principles of digital image processing and analysis, as well as applications in biology, are extensively discussed (see also Bartels *et al.*, 1984; Bradbury, 1983; Glaser *et al.*, 1983; Sauer, 1983; Wenzelides *et al.*, 1982).

11.5. SPECIAL TYPES AND APPLICATIONS OF LIGHT MICROSCOPY

Scanning Light Microscopes

When Nipkow invented television in 1883, he scanned an image of the scene in a raster with a spinning disk perforated with a series of small holes (Chapter 1). The 50-cm-diameter *Nipkow disk* was perforated by thirty 0.8-mm-diameter holes spaced 12° apart and arranged in an Archimedes spiral, ranging in radii from 216 to 241 mm. Behind the disk spinning at 25 Hz (half of the power-line frequency used in Germany), he placed a photoelectric cell, which gave rise to photocurrents corresponding to the image scanned 25 times a second in a raster slightly over 1 inch wide. A neon bulb, blinking brighter and dimmer following the photocurrent, gave rise to the video picture when viewed through the scanning holes in the same Nipkow disk.

A Nipkow disk, placed at the plane of the field diaphragm of a light microscope adjusted for Koehler illumination, yields a flying spot of light that scans the microscope field in a raster. A photocell or photomultiplier, responding to the rapidly changing amounts of light that come through the objective lens, would then provide a picture signal of the specimen. The signal sent to a monitor with synchronized sweep would produce a video image of the microscope scene. Petráň used a second Nipkow disk at the image plane to reduce the flare originating from light scattering in the microscope optics [(Petráň *et al.*, 1968); this is a form of confocal microscopy (see below)].

This principle of *field-scanning microscopy* was applied in modified form for UV microscopy by Freed and Engle (1962) and Montgomery *et al.* (1956). Rather than placing a mechanical Nipkow disk in front of a UV light source, these investigators used a UV-emitting cathode-ray tube (CRT) driven by an electronic raster generator. The CRT provided a point source of UV that scanned out a raster. The UV-transmitting condenser lens focused the raster onto the specimen plane. A UV-sensitive photocell placed behind the objective lens captured the picture signal. The signal sent to the monitor, scanned in synchrony with the UV-emitting CRT, displayed the UV-absorbance image of the specimen. In addition, the UV emitted at any desired region of the source CRT could be made substantially brighter locally, or flashed. The brighter area of the CRT served as a source for *microbeaming* the specimen with a small defined spot of UV focused onto the specimen by the condenser.

As predicted by Young and Roberts (1951), a tissue-cultured cell observed with the UV flying-spot microscope remained viable and continued to divide more or less normally for a long time. Time-lapse movies of dividing cells, displaying the distribution of the UV-absorbing chromosomes, could be filmed off the monitor for many hours. In contrast, photographing only a few UV-absorption pictures directly through a regular UV microscope would very quickly kill the same cell. This is a striking example of the greater sensitivity achieved by electronic detectors as compared to photographic emulsion.*

Some electronic imaging devices use a single or an array of light detectors and slow-scan video to produce a linear sequence of electronic signals for facsimile transmission.† In its modern form, the detector can be a linear array of photodiodes, on which the focused image of the specimen is swept (at right angles to the length of the array).‡ By using a linear array of, e.g., 1024, 1720, etc., CCD photosensor elements in place of the video horizontal scan line, one can obtain slow-scan video signals with very high resolution, without taxing the electronic circuits. Such a camera should be able to capture a large area of the microscope image at high resolution and low geometrical distortion if the specimen is stationary (over the several seconds required to scan a frame) and the image is sufficiently bright.

More recently, field-scanning microscopes are being developed or used for other purposes. E. Hansen and R. D. Allen of Dartmouth College (personal communication) are using laser field-scanning microscopy to determine the advantages of selectively modulating the phase of the beam that passes different parts of the specimen. Cox and Sheppard (1983) and Wilson and Sheppard

*Scanning may provide another benefit. The damage by UV to the living cell may be less because the cell constituents appear to "recover" during the interval in which they are not irradiated by the scanning UV.

UV and laser microbeam irradiation have been used in cell biology for many years (e.g., see Berns, 1974). Recently, video has been used in conjunction with microbeam irradiation, in part to avoid exposing the observer's eye to the intense irradiation, but more importantly to track the event taking place in the irradiated cell. Berns and Berns (1982) used digitally processed video images to track the migration and division of living cells after they had irradiated local regions of the cell (Section 10.13). For some experiments, they sensitized the target with light-absorbing reagents that selectively bound to the target. Leslie and Pickett-Heaps (1983) used a Venus DV-2 intensifier Vidicon camera to follow the effects of UV microbeam irradiation on the birefringence and mophology of mitotic spindles in dividing diatoms. Salmon *et al.* (1984a) and Hamaguchi and Hiramoto (personal communication) also followed the effect of UV microbeams on the birefringence of mitotic spindles in dividing sea urchin eggs with video.

†Slow-scan video uses a lower number of frames per second and a lower scanning speed to transmit image information without taxing the frequency-handling capability of, e.g., a regular telephone line.

‡Such arrangements are used, e.g., in the cameras manufactured by Data Copy Corp. and EG & G Reticon (Table 7-6).

(1984) scan the specimen, rather than the illuminating beam, through the axis of a *confocal microscope* system.

Conceptually, the confocal system is made up of two identical microscopes with their objective lenses pointing at each other and focusing on the same point in the specimen. One of the microscopes, used on the illuminator side, focuses the image of a small pinhole into the middle of the field. The specimen and/or the lenses are moved along precise paths that scan the specimen through the focused image of the pinhole.* A second pinhole is located at the conjugate of the first.

With confocal optics, Wilson and Sheppard (1984) have shown that the limit of resolution of the (fluorescence) microscope can be raised by a factor of two above the classically accepted limit.† In addition, the two pinholes, placed at the entrance and exit fields of the confocal microscope, accentuate the contrast of the in-focus image while blocking out the contrast-reducing rays from the out-of-focus regions (Egger and Petráň, 1967). Therefore, if the specimen is translated along the optical axis of the microscope, the image in a (reflecting) confocal microscope does not become blurred as in a regular microscope. Instead, the image of the vertically translated specimen is sharp throughout, showing good contrast and resolution wherever the specimen comes into focus. This feature of the confocal microscope is particularly useful for high-resolution stereo microscopy (see Wilson and Sheppard, 1984, Chapters 5, 6).

Since the specimen, rather than the point of light, can be scanned in the confocal system, one can obtain an image with a very large field of view (even though the scanning time is naturally considerably slower than for regular video). The potential of the new resolution limit, optical sectioning capability, extended field of view, and other capabilities offered by the confocal microscope system are indeed intriguing.

A somewhat different scanning system is used by Koester in his wide-field *specular microscope* (Koester, 1980; Koester *et al.*, 1980). This ophthalmoscope is an epi-illuminating microscope made to observe the endothelial cell and other layers of the cornea in patients at intermediate magnification in real time. Koester's microscope is equipped with an entrance slit (S1) at the field plane, a polygonal oscillating mirror (M1) near the aperture (A) plane, and an exit slit (S2) placed symmetrically with the entrance slit (Fig. 11-11). A stationary mirror (M2), just behind the exit slit, reflects the rays back onto M1. As M1 oscillates, adjacent slit-shaped regions of the cornea (C) are illuminated in succession. The image for each slit-shaped region is reflected by M1 through the exit slit, which occludes the out-of-focus rays. In the final image plane (F), the sequential slit-shaped regions appear side-by-side, and together form an image of an optically-sectioned plane. The scanned image clearly shows the corneal epithelial cells and the defects in the stroma, relatively unperturbed by light scattering from the out-of-focus regions. This is important since the cornea of patients being examined for ocular problems often exhibit a considerable amount of light scattering. The Koester microscope is illuminated with a high-intensity xenon arc lamp, and the image is displayed through a video camera and monitor. By scanning with a slit of light rather than a point of light, this system provides considerably greater illumination and circumvents the low luminance problem inherent in the Petráň-type system.

Tomography, Three-Dimensional Reconstruction

A number of workers have explored the use of video for *tomography* with the light microscope. By using high-NA objective and condenser lenses of the light microscope, one not only obtains good resolution but also an optical section with a shallow depth of field. Agard and Fay and their respective co-workers have reconstructed the three-dimensional distribution of fluorescent chromosomes (Fig. 10-51) and muscle fiber components from optical sections, taking into account the contribution of light from out-of-focus images (Agard, 1984; Agard and Sedat, 1983; Fay *et al.*,

*The scanner in a laser video (or audio) disk player is essentially a confocal reflecting microscope. An acoustical microscope is also designed as a confocal microscope (e.g., see Foster, 1984; Quate, 1980; Thaer *et al.*, 1982).

†The resolution of a diffraction-limited optical system can be raised beyond the classical Abbe limit if one is prepared to narrow down the field of view (McCutcheon, 1967). In the confocal microscope the field of view is a diffraction pattern of the pinhole source formed by the "condensing" lens.

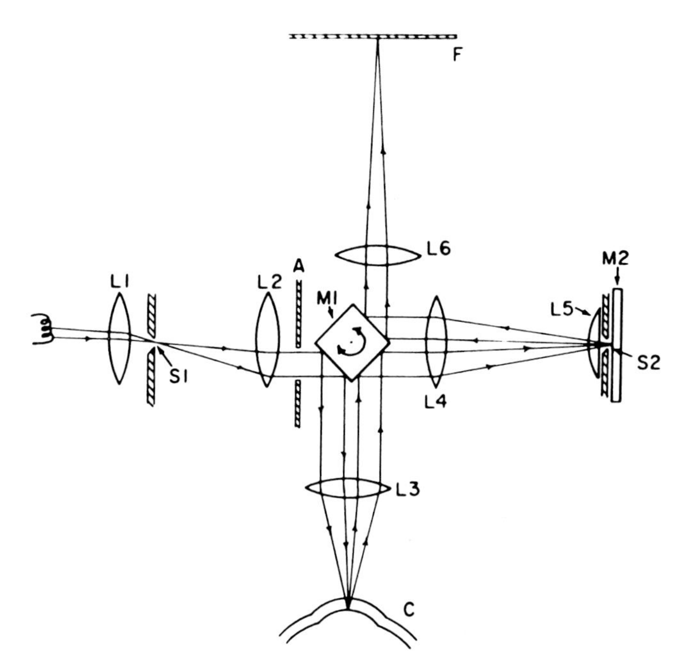

FIGURE 11-11. Schematic diagram of slit-scanning mirror microscope. With this ingenious form of confocal microscopy, coupled to a video system, Koester is able to display a clear optical section of the corneal endothelium, even when pathological conditions cause light scattering in the cornea. L1 to L6 = lenses. See text. (Courtesy of C. J. Koester; from Koester, 1980.)

1985; Mathog *et al.,* 1984). Spring (1983) and others have also used video-enhanced microscopy combined with optical sectioning to trace the outline of cells, vessels, etc., and to measure their area and volume changes under various physiological conditions (Section 11.3).

Tomography using the shallow depth of field in the light microscope differs from tomography with electron microscopy, X-ray, and other forms of high-energy beams. The probing beam in the latter have low angular dispersion and a very large depth of field. Nevertheless, a number of the techniques and analytical methods developed for electron microscopic and X-ray tomography can also be helpful in tomographic and three-dimensional image reconstruction with the light microscope (e.g., Agar, 1983; Agard and Stroud, 1982; Andrews and Ottensmeyer, 1982; Baker *et al.*, 1983; Boyes *et al.,* 1982; Crewe and Ohtsuki, 1982; Gordon and Palatini, 1982; Lamy *et al.,* 1982; Panitz and Ghiglia, 1982; Reedy *et al.*, 1983; Reid, 1982; Smith *et al.*, 1984; Williams, 1977).

In the light microscope, *holography* can be used to increase the depth of field, or the information along the axis of observation. As pointed out by Ellis (1966), one can observe, in one hologram made through the microscope, the images of the condenser, the specimen, the body tube, the ocular, and even the barrel of the camera lens. Not only can one focus through the specimen after taking the hologram, but one can also ''reach in'' and modify the state of the light waves at the back aperture of the objective lens during reconstruction. Thus, one can transform the image, originally taken in brightfield, to darkfield, bright- and dark-phase contrast, interference contrast, etc. during holographic reconstruction, long after the original hologram was made through the microscope. The coherence length of the light waves emitted by the laser used to produce the hologram, and thus the depth of field of the hologram, is in fact so great that blemishes on any optical element throughout the optical train can enter and confuse the reconstructed image. As far as I am aware, video image processing has not been applied to holographic microscopy.*

In contrast to holography, *aperture scanning* introduces incoherence to the microscope field

*See von Bally and Greguss (1982) for generally related articles, although not dealing directly with microscopy.

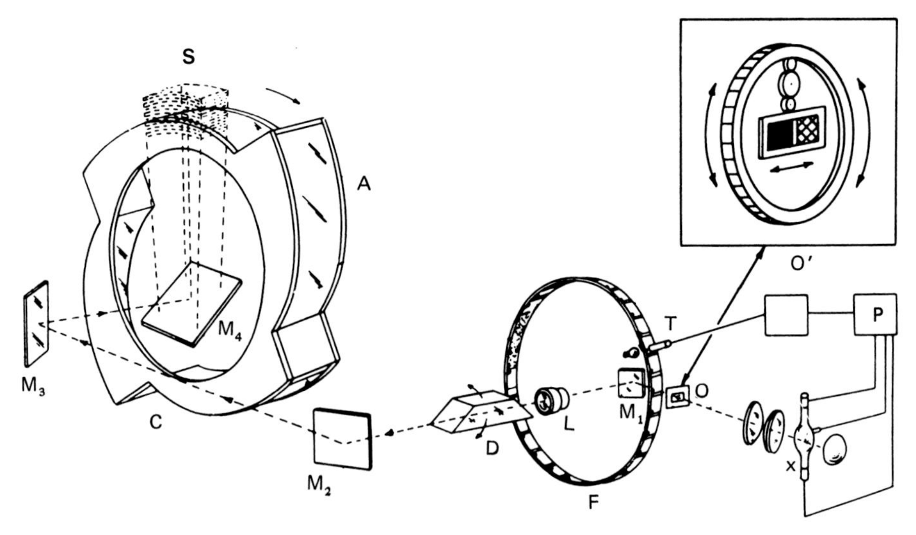

FIGURE 11-12. Schematic diagram of "Synthalyzer" for true three-dimensional display of images. In this interesting device, a three-dimensional display is achieved by synchronizing the position of a moving projection screen (A) with that of a series of images taken at sequential focal levels and recorded on a film strip (F). X, xenon flash tube and condensing optics; T, exciter lamp and photosensor for triggering the flash tube; P, power supply; L, lens; M_1–M_4, first-surface plane mirrors; C, stepped cylinder having translucent Archimedes spiral segments (A); S, three-dimensional synthesis; O, O′, optical dissector; D, dove prism for rotating the slit-shaped image generated by the optical dissector. See text. (From de Montebello, 1969.)

illuminated with a laser. G. W. Ellis of the University of Pennsylvania is exploring the advantages of selectively modifying the aperture function using aperture-scanning microscopy (personal communication). Aperture scanning, combined with digital image processing, promises interesting opportunities for video microscopy. One should be able to interactively modify the MTF curve, by manipulating the optical contrast and frequency response characteristics almost at will. The combination of aperture scanning and digital image processing should also be applicable to three-dimensional image reconstruction and stereoscopic displays.*

Many approaches have been used over the years to *reconstruct* the *three-dimensional* structure of microscopic objects. In both electron and light microscopy, three-dimensional models have been constructed by stacking an array of manually traced serial sections drawn on glass or plastic plates. Alternatively, cardboard or wax plates were cut out, stacked, and cemented together to form three-dimensional models. Also, serial focal pictures were filmed and projected as a motion picture focusing through the specimen. These and many other approaches to three-dimensional reconstruction and imaging, including stereoscopy, are discussed in an excellent book by Gaunt and Gaunt [1978; see Section 12.7 (following chapter), for discussions on stereoscopic video microscopy].

In using serial sections to build three-dimensional models, exact registration between successive sections can become a real problem. Several approaches for overcoming this problem have been devised, including the use of registration markers, and the use of a "blinker," or "flicker scope" (Levinthal and Ware, 1972; Levinthal *et al.*, 1974; Wrigley *et al.*, 1982). Others use epi-illuminating microscopes to observe the stable surfaces of a (frozen) block as the sections are successively removed (Macfarlane *et al.*, 1983).

de Montebello (1969) proposed an interesting method for three-dimensional optical reconstruction and dissection (Fig. 11-12). He made a projection screen around the surface of a drum (C)

*Split-aperture and related forms of stereoscopic microscopy usable at high NAs with video are described in Section 12.7. For a theoretical treatment of three-dimensional image processing and stereoscopic image display, see Chapter 17 in Castleman (1979).

whose radius varies as an Archimedes spiral (A). As the drum spins, the plane of the focusing screen (S) moves closer and farther away from the observer. An image projected on the drum would thus oscillate back and forth. In his "Synthalyzer," photographs of serial sections were recorded on an endless loop of 16-mm motion picture film. Each frame was serially scanned by a vibrating slit [the optical dissector (O,O′)]) and projected onto the oscillating screen with whose movement the dissector was synchronized. Thus, one could achieve a three-dimensional display or could "dissect away" certain parts of the structure, optically.

Others have displayed frozen video images directly in "objective three-dimensional space" (as distinguished from the viewing of stereo pair images, which gives rise to a subjective sensation of three-dimensional space). One interesting approach is the *varifocal mirror* or Spacegraph system, which forms a three-dimensional virtual image with good contrast (Fuchs *et al.*, 1982a,b; Muirhead, 1961; Sher and Barry, 1985).

As shown in Figs. 11-13 and 11-14, a vibrating mirror consists of a film of tautly stretched aluminized mylar (Fuchs's varifocal mirror) or an aluminized glass plate (Sher's Spacegraph System), placed in front of a loudspeaker. When the mirror is flat, we see, behind the mirror, a virtual life-size image of a cathode-ray tube that is rigidly secured in front of the mirror. When an electric current is passed through the loudspeaker coil, the speaker drives the mirror and changes its curvature. The mirror, which is now slightly concave or convex, displaces the image of the CRT back and forth along the Z axis away from its stationary image (Fig. 11-14). For each Z value, the CRT (with a low-persistence phosphor) plots out the intensity j for each of the X, Y positions (i.e.,

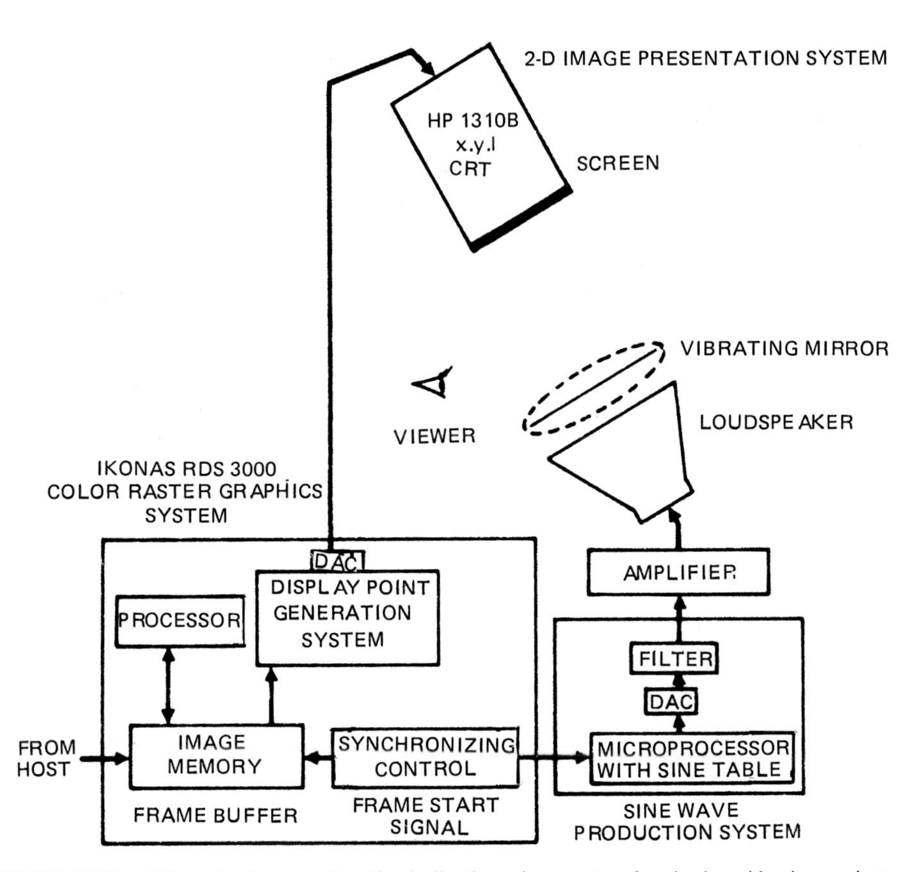

FIGURE 11-13. Schematic diagram of varifocal vibrating mirror system for viewing video images in true three dimensions. This system, developed at the University of North Carolina, takes advantage of the NTSC encoding system for efficient handling of signals for the X, Y, Z spatial coordiates. See text. (From Fuchs *et al.*, 1982b.)

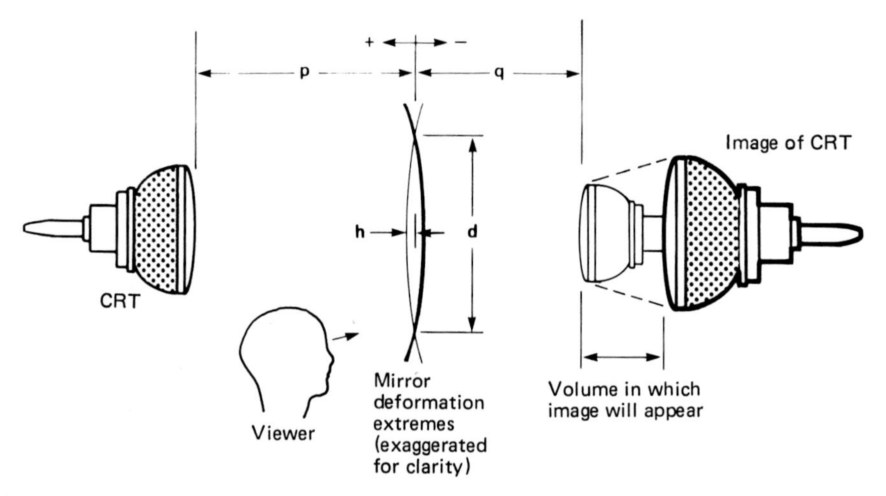

FIGURE 11-14. The vibrating mirror SpaceGraph system for viewing video images in true three dimensions. This system, developed at BBN Laboratories (marketed for a time as the Genisco System), uses glass as the mirror substrate, instead of a stretched, mylar film, as in the varifocal system. The geometrical relationships shown here apply as well to Fig. 11-13. (From Sher and Barry, 1985.)

the two-dimensional cross section of the three-dimensional picture cut through at height Z) commanded by a point-position-display controller. As Z is varied, a new set of X, Y, j points are plotted out on the CRT. For three-dimensional display, a 30-Hz sine wave drives the loudspeaker. Thus, a series of X, Y, j images on the CRT is projected into the Z-axis space by the vibrating mirror, completing its cycle once every 1/30 sec. Since all of the points are presented in rapid succession, the eye perceives the points coherently in the X, Y, Z, three-dimensional image space.

The vibrating mirror display can be examined from different angles by moving the head or by walking around the display. One can also rotate the display itself around the X, Y, Z, or a fourth axis in real time by manipulating a four-axis joystick. Also, one can display an image with a portion of the structure sliced off, or alternatively, as a section sliced through the object.

So far, the varifocal mirror, or SpaceGraph, system works best when displaying bright points, lines, and surfaces, such as in displaying molecular models and dendritic structures. It is also being used to examine the three-dimensional architecture of helical nucleosomes by displaying a series of Fourier-processed, tilted images of a polytene chromosome obtained through a high-power scanning–transmission electron microscope (Olins *et al.*, 1983). Hopefully, we will soon see the varifocal mirror or an alternative system applied to three-dimensional reconstruction of light microscope images, preferably presenting the picture in near real time.*

*In 1981, one varifocal system was marketed at over $100,000. In 1983, a varifocal display for under $10,000 is said to be operable with any raster graphics system with 16 or more bits per pixel (Fuchs *et al.*, 1982b).

There are some inherent problems when one tries to display transilluminated *halftone* images with varifocal-type systems. The number of volume-image elements (voxels) that can be displayed is limited by the memory size of the frame buffers and the high frequency required for the video signal. Furthermore, the image itself is not too easy to interpret (Sher and Barry, 1985). However, fluorescence, darkfield, and certain polarized-light images of fibrils and membranes should display well. Perhaps one could also generate more interpretable image planes by making them interact with each other through logical operations, or one could primarily use epi-illuminated scenes. The new systolic array processors (Davis and Thomas, 1984; Hannaway *et al.*, 1984) may provide a part of the answer for handling such three-dimensional logic operations.

12

Presentation of Video Data

This chapter will examine how to prepare video data for publication and for presentation to an audience, including stereoscopic displays.

12.1. PUBLICATION OF VIDEO MICROGRAPHS

Introduction

We can copy, edit, and narrate videotape of dynamic scenes captured through the microscope and produce videotape for distribution. Similarly, we can copy the edited material on 16-mm motion picture or video disks for distribution.* We can also photograph still pictures off a video monitor playing back a videotape or disk and publish them in lieu of regular micrographs. Often, *still photographs* are the only means to document and communicate the findings made via video microscopy, a point well to keep in mind when we make video recordings of scenes through the microscope.

For example, picture defects such as focused particles of dirt, hot spots, blooming, streaking, ringing, noise, distortion, and shading all get carried over to the still photograph once they are recorded on the original videotape. Also, we must carefully frame the scene from the beginning, since there is much less leeway for cropping or enlarging a part of the scene in video than, say, in 35-mm photography. In other words, the specimen must be centered and *fill* the scene so that we can take advantage of the full resolution that is provided by video, which is especially limited with most VCRs or even with an OMDR (Section 8.9).

Also, we should record the images of micrometer scales for each magnification on the tape, at least once a day. In addition, with the large number of frames that may have to be identified on the tape, and the opportunity for using the tape record to measure time of occurrence, velocity, rates of changes, etc., one should routinely include time–date records right on the scene. Other relevant records, such as temperature, experimental conditions, time of specimen preparation, lenses used, etc., could be read onto the audio channel (as well as written in a notebook), and in some cases, montaged with the scene.† At the time of the original observation, we can record lower-magnification views to provide overviews, and higher magnifications to provide close-ups without great difficulty, but we cannot easily extract such views later from tape recordings made at another magnification.

*Alan R. Liss (150 Fifth Avenue, New York, N.Y. 10011) publishes an annual video disk supplement to the journal *Cell Motility*. The first number appeared in December 1984.

†I cannot overemphasize the need to keep good records on the tape, as well as in a notebook, showing the date and time for each entry. If for no other reason, the notebook saves a great deal of time in searching for particular scenes. Also, it is essential that every cassette or reel, and its container, be labeled *before* use, at least with the date of recording.

We can photograph the video microscope scenes at three stages: live from the monitor before recording, with or without signal processing; from a scene played back from a VTR or OMDR and from a processed image of the signal played back from an OMDR or VTR.†

As stressed in Chapter 7 and elsewhere, the video picture played back from a standard ¾- or ½-inch VCR provides, at best, only half the resolution of a 1-inch camera of moderately high resolution. At the center of the picture in a good monochrome system, the horizontal resolution of the camera and monitor commonly reaches 600–800 TV lines (300–400 line pairs) per picture height. Even in color, a good-quality camera and RGB monitor provide resolutions of 500–600 TV lines. Three-quarter-inch recorders generally reduce the resolution to 320–340 TV lines in monochrome, and 240–260 TV lines in color. High-resolution monochrome U-format recorders and optical memory disk recorders can provide resolutions approaching or exceeding 450 TV lines (Sections 8.4, 8.9) Clearly, for video microscopy these are preferred over the standard decks.

At very low light levels, the limiting resolution and MTF of the camera tube can become so low, or the picture noise so high, that the picture may not be appreciably degraded by recording on a ¾-inch or even a ½-inch VCR. On the other hand, after digital averaging by image-processing computers, even low-light-level scenes can contain sufficient detail to tax the resolution of a standard ¾-inch VCR.

Whether the signal comes directly from the camera or is processed, one may wish to avoid the loss of resolution that results from VCR recording. If the scene is stationary, we may be able to photograph a monitor that is directly connected to the camera or processor before its output enters the recorder. But for dynamic scenes, we may only be able to photograph a scene that was first recorded.*

For rapid events, we can even copy segments of tape onto a motion analyzer, then select individual fields to photograph, with or without digital enhancement (Figs. 1-6, 9-11). For example, the velocity of beating cilia and flagella is not always uniform, so that certain fields show considerably less blur than others and provide more suitable still records. Also, we can play back scenes recorded on some time-lapse VCRs at a rate considerably slower than the recording speed, and photograph the desired scene (field) by pausing the recorder. Depending on the VCR, the noise bar may be eliminated automatically in the paused picture, or we may have to ease it out (or onto a less significant part) of the picture, manually (Section 8.5).

By using *image-processing computers* with frame buffers, real-time A/D and D/A convertors, and logic units, we can freeze single fields *or* frames. We can also sum specific numbers of frames or freeze a picture displaying the weighted running average (Figs. 1-7, 1-8; Chapter 10). Within certain limits, the scene can first be recorded, then averaged, frozen, or digitally enhanced during playback of the OMDR or VCR.

Regardless of the stage of video display, whether before recording, recorded, or recorded and processed, the procedure for photographing the monitor is essentially the same.

Photographing the Monitor

In photographing the monitor, we should be aware of the following. With video and motion picture viewed at frame rates above the threshold for flicker fusion, our eye and brain fill in parts of

†Instead of photographing a monitor, it is also possible in principle to print out a slow-scan video directly onto photographic emulsion or other media. Today, one finds an increasing variety of equipment for directly printing out slow-scan video, and for converting standard-speed video to slow scan. But here we will mainly discuss how to photograph a standard monitor to produce still photographs appropriate for high-quality halftone reproduction.

*One alternative for time-lapse recording would be to insert a high-resolution, standard-speed VTR (¾- or 1-inch) or OMDR ahead of a ¾- or ½-inch VCR. The high-resolution VTR or disk recorder would be used to record limited sequences to provide high-resolution playback and analysis, while the smaller VCR could be used to capture time-lapsed and extended records at considerably lower cost. Both 1-inch VTRs and laser disk recorders are, however, very expensive. A laser disk recorder (that can hold 16,000 monochrome video frames with a resolution of 450 lines PPH) is now available for about $24,000 (Section 8.9). As shown earlier, monitor pictures photographed off an OMDR can be very respectable (e.g., Figs. 8-19, 8-20).

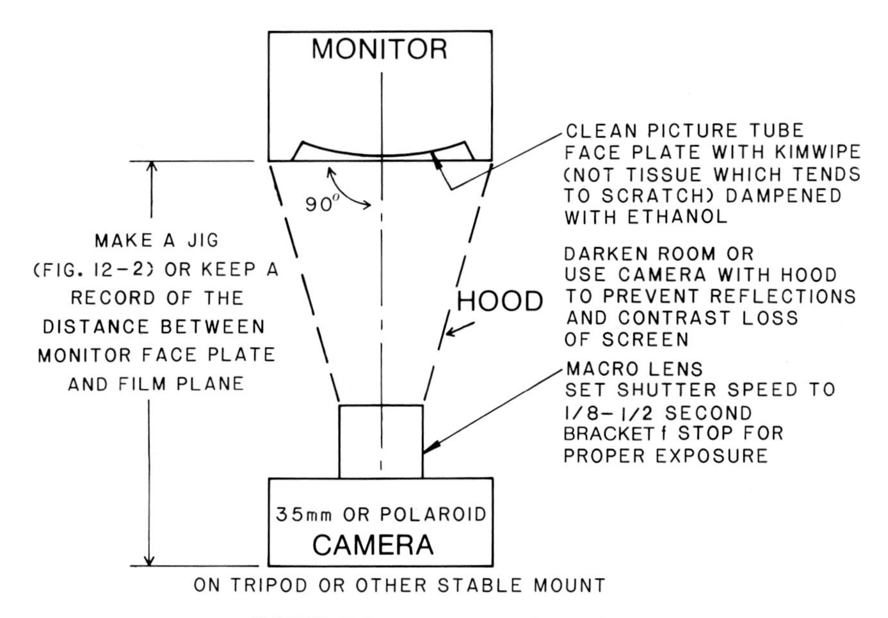

FIGURE 12-1. Photographing the monitor.

the picture that are missing for short times or distances. By filling in, or integrating the scenes, we average out the noise and smooth out the incrementally displaced images. We perceive the discontinuous images as though they were four-dimensional, contiguous, moving objects. Only when we freeze a frame or field do we realize the limited quality of the individual frames and fields, which, when integrated together over time, produce quite respectable pictures with few glaring defects.

When possible, we should photograph the monitor with the tape running at normal speed, rather than as a frozen frame or field. By photographing the picture from the running tape, random noise is reduced by *photographic integration.* Also, many tape decks provide better-quality reproduction with the tape running at a standard, constant speed than when the tape is stopped. In helical scan recorders, the tension on the tape, stability of tape location, scan angle, head travel, and aerodynamics of the thin air film between the tape and the spinning head drum, are all equilibrated *dynamically* for the tape traveling at a given constant speed. We simply cannot expect these parameters to remain unchanged when we stop the tape transport and freeze the frame or field. So long as the scene is changing slowly enough so that it is stationary for at least ⅛ sec, it is clearly better to photograph the monitor with the tape running at standard speed.

To photograph a scene that is changing, or a specimen that is moving, too rapidly, we would have to place the recorder in the freeze-frame or freeze-field mode. With most VTRs, the field can be frozen simply by depressing the PAUSE button. The frozen field is usually displayed with half the number of scan lines of the frame, although in some cases the field may be interlaced with itself to fill in the gap left by the undisplayed field. With *freeze frame,* one has available the full interlaced set of scan lines, so that the vertical resolution is maintained and defects in the pixels, or picture noise, are less conspicuous. However, if the specimen is moving too rapidly, the odd and even fields would have registered two different images, so that moving objects jitter up and down in the frozen frame. With the *video field* frozen, time resolution is roughly doubled compared so that moving objects jitter up and down and the frozen frame would appear to flicker. Vertical resolution and S/N ratio are, however, inevitably sacrificed.

We can photograph a monitor quite simply with the arrangement shown in Fig. 12-1.* First, center a suitable photographic camera mounted on a sturdy tripod or other support in front of the monitor. The camera can be any standard- or instant-type format. The camera should be equipped

*See also "Photographing Television Images" (Eastman Kodak Co.) and Ortner (1984).

with a good-quality lens that can focus the image of the monitor faceplate to nearly fill the camera format. Also, the camera should be equipped with a manual shutter-speed control and an exposure meter or auto-iris control.*

Load the camera with moderately fast, fine-grain film (e.g., Kodak Plux-X, Agfa Super-Pan), and set the shutter speed to ⅛ (or ¼) sec. With too short an exposure, the upper or lower part of the photograph, exposed to fewer cycles of the video scan than the rest of the monitor, appears noticeably darker. With too long an exposure, dynamic parts of the scene become blurred.

Once you find the monitor-to-camera distance that fills 80–90% of the camera format with the image of the monitor, measure the distance and note it on a label attached to the camera. Since it is necessary to keep the camera stably centered, and the face of the monitor maintained at right angles to the axis of the camera, it is worth constructing a simple (plywood) *copy stand* of the type shown in Fig. 12-2.

If it is possible to choose, one would use a monitor with a flat, rather than curved, faceplate, and with good linearity and low distortion. Special flat-faced, low-distortion monitors are available for photographing the display, but most monitors can provide reasonably undistorted pictures that are adequate for general display. Naturally, the aspect ratio, H- and V-sync hold controls, as well as monitor orientation and camera centration must be adjusted carefully.

The monitor can be photographed in the standard overscanned mode to eliminate possible glitches at the edges of the picture. Or, if one needs to recover the full frame (where the edge or a scene may be clipped), one can photograph the monitor in the underscan mode. With underscan, distortions of the picture arising from unevenness of beam sweep can become prominent (Fig. 7-36), but distortion and parallax due to curvature of the monitor screen can be reduced.

Recall that the vertical resolution of the displayed picture is dependent on the accuracy of the interlace scan lines (Fig. 6-16). If the scan lines become paired, the vertical resolution drops to one-half the maximum. It is therefore *very important to adjust the* V-HOLD *control critically*.

As we adjust the V HOLD control on the monitor, exact 2 : 1 interlace is achieved just before the picture becomes unstable. Trim the V HOLD control delicately while inspecting the interlace of the scan lines with a magnifier (see footnote* on p. 256).

With the room darkened, adjust the CONTRAST and BRIGHTNESS controls on the monitor so that (1) the contrast, as well as detail in the highlights and shadows of the picture, appear optimally, and (2) there is enough light to expose the camera at ⅛ (or ¼) sec near its optimum f stop setting. Contrast adjusted very slightly lower than for optimal viewing of the monitor by eye seems to work out quite well for photography. One does, however, have to keep in mind that the gray scale, or tonal range, that can be reproduced in the halftone press is limited, compared to photographic prints; thus we must generally photograph pictures for publication with a somewhat compressed tonal range. Naturally, we must also be careful not to overdrive the monitor since the highlights can bloom and the picture can become distorted.

In copying from a monitor, be sure to darken the room or use a hood, for two reasons: (1) scattering of room light in the phosphor reduces contrast, and (2) reflections off the monitor faceplate can appear in the photographs. If the lights are not turned off or masked, reflections of the lamps or bright objects are likely to appear on the picture, but only to be discovered after the photograph has been processed.

A video microscope picture photographed off the monitor is treated like any other micrograph, except that it cannot be enlarged very far. Figure 11-5 shows successful examples of rather great enlargements of single field pictures. However, scan lines usually intrude and resolution is lost if we attempt very great enlargements or extreme cropping.

In preparing original video micrographs for publication, we should generally not reproduce the full standard monochrome monitor picture at larger than about 4×5 inch^2. With halftone reproduction using very fine screens, one can obtain excellent reproduction at just under 7×9 cm^2 for a full-monitor scene displaying a single *field* played back from ¾-inch videotape (e.g., see Figs. 1-6,

*I generally use a Nikon FM 35-mm camera with a 55-mm-focal-length Macro lens. In the viewfinder of this camera, all three LEDs (instead of the usual single one) light up when the exposure controls are set optimally for the monitor picture.

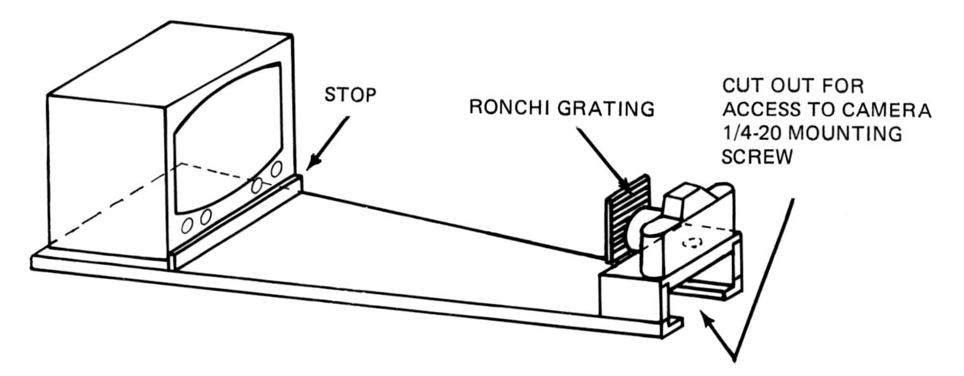

FIGURE 12-2. Copy stand for photographing the monitor. A simple jig of this type can be constructed from ⅜-inch plywood. (See Fig. 12-1.)

9-11). For a single *frame,* one can use twice that size (Figs. 8-19, 8-20). The *RCA Electro-Optics Handbook* (1974), Section 13, provides detailed data relevant to photographing the display of a cathode-ray tube, including the video monitor. Color monitor pictures (Fig. 1.8 E, F) can be conveniently copied with one of the portable hood–camera assemblies manufactured, for example, by Kodak and Polaroid.

Scan Line Removal

Video scan lines limit the vertical resolution, intrude into the picture, and hide image details. The discrete scan lines also make a diagonal line jagged, or stepped, especially if a line is tilted several degrees to the horizontal (e.g., see Fig. 9-11A,B).

Ideally, one would eliminate scan lines by making them so fine that they are no longer visible. If, at the same time, we could proportionately reduce the size of the scanning apertures in the camera and picture tubes, the picture resolution would be much improved, every point of the image would be scanned uniformly, diagonal lines would not be stepped, and no distracting scan lines would be present. We could then enlarge the video photographs considerably, and extract a large amount of detail from a single frame. The 2000- to 2500-scan-line picture used in the scanning electron microscope (SEM) attests to the quality of the picture that one can attain with tightly spaced scan lines.*

To achieve the excellent resolution that is potentially attainable with a scan rate as high as 2000–2500 scan lines (with a proportionate increase in horizontal resolution), we would need to increase the bandwidth by 15–25 times, or reduce the scanning speed to 1/15th to 1/25th of the standard video rate. At present, such extended bandwidth or reduced scanning rates would not be very feasible or practical for video microscopy. How, then, can we minimize the intrusion of the scan lines in conventional video display?

For relatively static scenes, we could in principle dither the scan lines up and down synchronously in the camera and monitor. With the rasters appropriately dithered, the vertical resolution would not be limited by the discrete scan lines, a diagnoal line would not be stepped, and the distracting scan lines would be absent. Actually, I do not know whether such dithering techniques have been applied to video microscopy.

Gordon W. Ellis, in our laboratory, has developed a very effective way to *filter out video scan lines* selectively. As described in Appendix I, Ellis first produces a diffraction pattern of the video picture with an optical system that is essentially a scaled-up light microscope. A 35-mm transparency of the video picture is placed in the specimen plane. At the plane corresponding to the back

*In addition to the large number of scan lines, the SEM also uses a very-slow-speed scan for photographing, and a photomultiplier as the detector, which provides a greatly expanded dynamic range compared to standard video camera tubes.

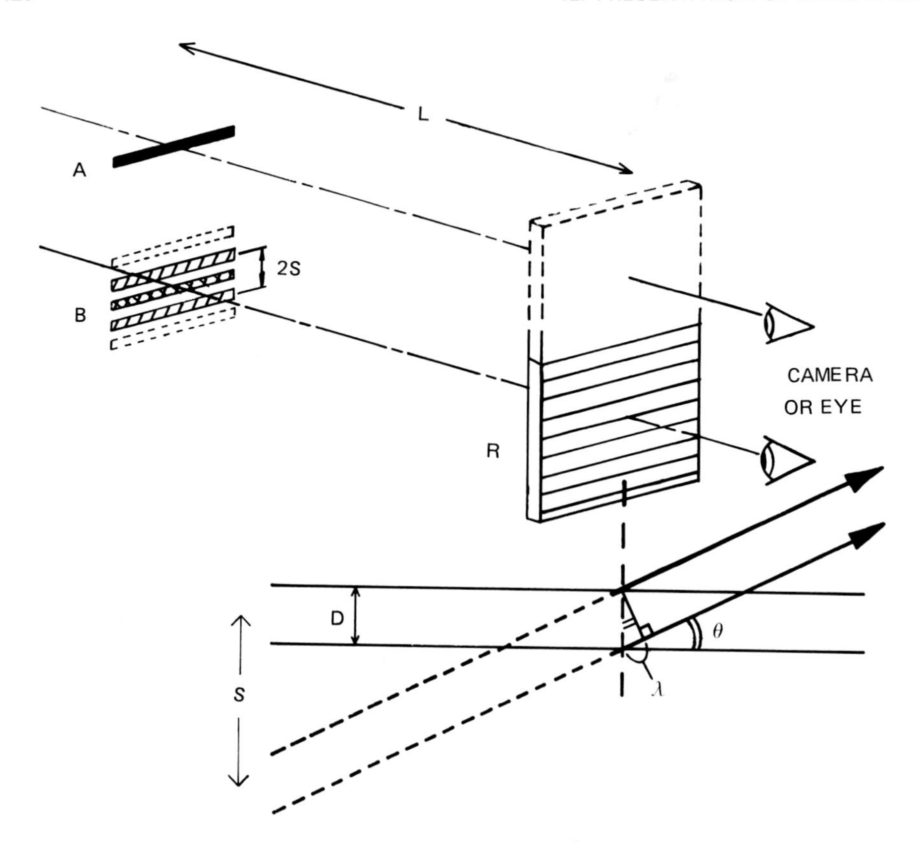

FIGURE 12-3. Function of Ronchi grating. A Ronchi grating may be used to remove video scan lines from a photograph of, or when photographing, the monitor. It is especially useful for photographing single fields (in which scan lines are prominent). See text.

aperture of the objective lens, he selectively masks out the diffraction spots that are produced by the scan-line periods. In other words, he selectively filters out the spatial frequency corresponding to the period of the scan lines from the image-forming waves. By this application of *spatial filtration,* only the periodic scan lines are removed, and Ellis achieves the remarkable improvement in picture quality that is shown in Appendix I.

It is surprising that so much detail can emerge after spatial filtration, and that the discontinuities of the oblique lines can disappear altogether. But the photographs taken before and after spatial filtration clearly show these results. When the scan lines are discrete, they inevitably distract from the fine detail that is hidden in the picture.

We can also reduce the prominence of the video scan lines by another diffraction method, by using a Ronchi grating* with 50–100 rulings/inch (Inoué, 1981b). The method is especially effective for photographs of single fields where the scan-line interval is quite coarse.†

The Ronchi grating conceals the video scan lines in the following manner (Fig. 12-3). A grating conceals the video scan lines in the following manner (Fig. 12-3). A grating with a periodic

*The Ronchi grating is made up of an alternating set of absorbing and transmitting lines of equal width, ruled on the surface of a transparent substrate. The grating is commonly used for testing the optical quality of lenses and mirrors. Earlier, I used the Ronchi grating for yet another purpose; enhancing and determining the period of structures whose periodicity was barely perceptible in micrographs (e.g., Erickson, 1973; Gilev, 1979). Ronchi gratings are available from Rolyn Optics Co. and Edmund Scientific Co.

†For *properly interlaced full frame* pictures, the scan lines should be barely perceptible (e.g., Figs. 8-19, 8-20) and one should not have to use a Ronchi grating.

spacing of D diffracts the light in directions perpendicular to its rulings by: $\sin \theta = \lambda/D$, where θ is the angle of deviation of the diffracted beam and λ is the wavelength of light. If the grating were purely sinusoidal, only the first-order diffraction pattern would appear. With a Ronchi ruling, which is made up of discrete absorbing and clear spaces of equal widths, we also see a weak second-order pattern in white light.

Through a Ronchi grating (R) whose rulings are oriented horizontally, the diffraction pattern B replaces a single scan line A, with parasitic scan lines appearing above and below the scan lines (Fig. 12-3). The parasitic scan lines (with the second-order line showing in much reduced contrast) are spaced at a distance S from the original scan line. S is given by: $S = L \sin \theta$, where L is the (radial) distance between the grating, R, and A. If S is set at one-half of the scan-line interval, the first-order diffraction patterns of the scan lines would overlap and fill the darker space between the scan lines.

In practice, we adjust the distance between the picture and the Ronchi grating until the number of scan lines exactly doubles, then offset it slightly. At that point, the scan lines become inconspicuous (Fig. 11-5). The Ronchi grating can be placed directly in front of the lens of the camera photographing the monitor (Fig. 12-2); in front of the enlarging lens during photographic enlargement; or on the lens of a camera copying the print.

While this method is simple to apply, the Ronchi grating does blur the image vertically by a small amount, whereas Ellis's spatial-filtration method truly eliminates the scan lines, without affecting coarser *or* finer picture details. However, in order to optically filter pictures even as small as a 35-mm negative, one must use special large-diameter lenses, precision support for a rather large projection distance, and an intense laser source that provides reasonable exposure times.

12.2. OPTICAL COPYING WITH VIDEO

We can, as described above, copy titles, still photographs or motion pictures, or scenes on a video monitor *optically* onto video tape or disk. We can also use a video camera to **optically copy** and *insert* notations, arrows, event markers, oscilloscope displays, the dial of a micrometer or thermometer, etc. onto our video micrograph. The inserted image can be superimposed or montaged with the live scene from the camera, or the signal played back from a VTR. A *special effects generator* is used to mix these signals (see Section 12.4).

For optically copying video, one could use the instrumentation camera normally used on the microscope or a simple surveillance camera. For optically recording an event marker, counter, meter, etc., a compact solid-state camera (Fig. 7-8A,B) could provide extra flexibility.

Depending on the image detail required, the standard lens on the surveillance camera may be adequate, or one may need to use a higher-quality lens. Sixteen-millimeter cine or video camera lenses can be used conveniently, since they can screw directly into the C-mount thread of the camera (Section 3.3). Many video camera lenses have zoom capability, which is helpful for filling the frame with the essential part of the scene or for providing proper perspectives.

Some 35-mm camera and enlarger lenses, which may already be on hand, can provide the very sharp images that are desired for copying with video cameras. I use a Nikon 55-mm Macro lens or a Vivitar 70–150-mm zoom lens on our 1-inch Newvicon video camera (interchangeably with the Nikon FM 35-mm camera that I use for photographing still pictures off the monitor). One can quickly adapt a 35-mm camera lens to the video camera by using a camera lens to C-mount adapter, which is available commercially (Fig. 12-4). With these adapters, one can mount a 35-mm camera lens on a video camera at the proper focus and with the needed antireflection baffles already in place. EL Nikkor enlarger lenses, which are corrected for moderately close projection distance and flat field, provide exceptionally sharp images for copying.

We can copy titles, graphs, and still photographs with a video camera in the following manner. A professional copy stand with illuminators is handy but not essential. Instead, the video camera can be supported horizontally with a sturdy tripod, or on a solid box or book on a desktop or laboratory bench. Attach the item to be copied with Scotch Cover-up Tape or drafting tape (which are designed to peel off cleanly) onto the wall. Align the camera axis perpendicular to the subject. For illumina-

FIGURE 12-4. C-mount adapter. The photograph shows an example of a commercially available adapter for mounting a 35-mm camera lens (in this case, a Nikon F-type bayonet mount) to the video camera. (Courtesy of Nikon Inc., Instrument Division, Garden City, N.Y.)

tion, use one or two desk or flood lamps, a quartz–halogen microscope lamp without the projection lens, or a slide projector. Position the illuminators to provide uniform illumination without glare or *reflections* of the source (Fig. 12-5).

Photographic prints, or even black paper that is not particularly lustrous, can curl or otherwise become oriented at grazing angles and produce unexpected reflections of the source. Glossy paper, and glass or plastic covers, can reflect room scenes including ceiling and window light, shiny parts of the camera, the faceplate of the monitor, and even highlights off the operator. As in photographic slide making, these reflections introduce unexpected, obtrusive ghosts. While one should be able to notice such distracting images on the video monitor, it is safer to dim the room light or arrange to copy in a location that is least likely to pick up unwanted glare.

We can also copy *transparencies* placed on a light box. With video, one can obtain a positive copy directly from a *negative* if a contrast inversion circuit is available on the video camera, monitor, or mixer. We can also copy images *projected* with a lantern slide-, overhead-, or movie-projector (the original tape for Fig. 8-7, which does not show the noise bars at standard playback speed, was copied off a projeted, time-lapse movie screen).

Unlike copying from video to movie film (V → F), or optically copying a video monitor with a video camera, copying from movie film to video (F → V) is surprisingly simple and effective. The movie can be projected from behind on a rear projection screen (via a first-surface mirror to correct the left–right inversion), or simply projected from the front. For front projection, the axes of the projector and video camera should be kept close together to prevent picture distortion. Select the surface for projection and the projected picture size, so that the detail and brightness do not fall off at the periphery of the picture. Using a video camera equipped with a 1-inch Newvicon tube, I find that a video scene that occupies approximately three-fourths of the full frame of a 16-mm movie (taken on Plus-X film) gives a good match between film granularity and video resolution.

In F → V transfer, we do not generally encounter the black stripe that intrudes when one tries to transfer video to movie film without proper synchronization. The target in the vidicon camera tube stores the charge and memorizes the image for the duration of a video frame. Therefore, in F → V transfer, one does not have to be concerned with synchrony between the movie and video.

Since we can readily adjust the black level and image contrast in video, we can use F → V transfer to bring out image features that were not at all obvious on the original film. I found, in a time-lapse movie of dividing cells that I had recorded in polarized-light microscopy and had reviewed scores of times, surprising new information that had gone unnoticed until I transferred the film and reviewed it as video.

Sometimes it is necessary to *optically copy the video monitor with a video camera*. One may wish to enlarge a portion of the picture, but lack the equipment for electronic zooming. One may wish to record an RGB color picture onto tape, but lack the encoder that changes the RGB signal to an NTSC signal. (One can record NTSC but not RGB onto a regular VCR.) Or, the sync signal on a time-lapsed tape may be adequate for triggering a monitor with a fast AHFC without flagging, but inadequate for a second recorder.

We can use the same camera and lenses used for copying titles, etc. to optically copy a video monitor to video. We can copy the full picture displayed on the monitor (in the underscan mode), or only parts of the picture, so that they may be magnified or zoomed up. If we wish to copy two parts of the picture, for example, a magnified portion of the screen and the time–date record montaged into a single picture, we can do so with a special effects generator.

In optically copying video, we should *synchronize* the camera with the monitor being copied. If the monitor and camera are not exactly synchronized, the V blank of the two would be at a random phase relative to one another, so that a dark horizontal stripe could intrude and wander up or down the picture. Many VTRs cannot be synchronized externally in playback, so that if we wish to optically copy a scene from a VTR, it has to be the source of the sync signal. Figure 12-6 shows an arrangement for providing this sync.

Synchronization eliminates the dark stripe. However, one may still encounter moiré fringes that arise between the scan lines of the monitor and the camera. This can be especially troublesome when copying from color to monochrome, since the RS-170A color signal is repeated at a slightly different frequency than the 60 Hz used for the monochrome RS-170 signal (Section 6.6). We can reduce or alter the moiré pattern by adjusting the camera angle and position, the V HOLD control on the monitor, the ratio of the original picture size to the copy, and by slightly defocusing the camera.

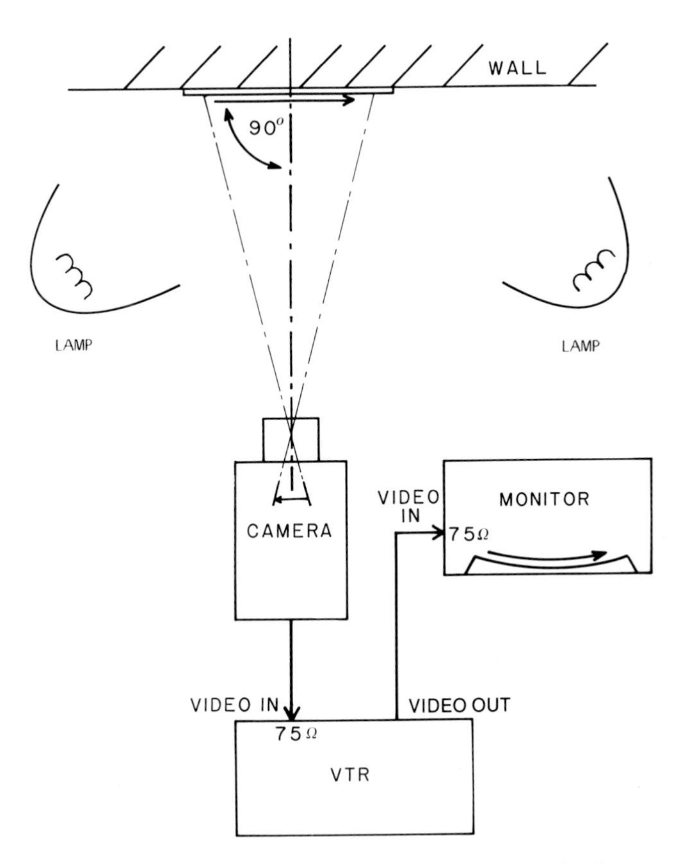

FIGURE 12-5. Video copying of titles, charts, and photographs. See text.

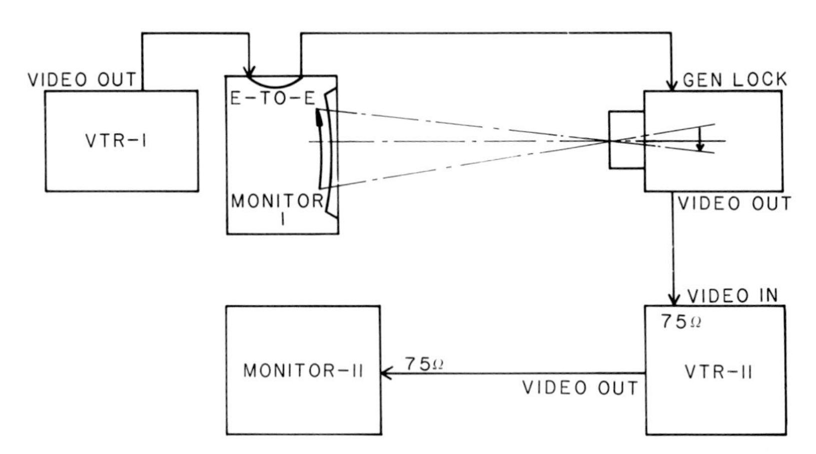

FIGURE 12-6. Optically copying the output of a VTR. See text.

When optically copying a video picture off a monitor, dim the room lights to prevent reflection off the monitor faceplate. Also, be careful not to raise the monitor CONTRAST too high, since the luminosity range of the monitor can easily exceed the intrascene dynamic range of the camera. One can use a standardized monitor, an oscilloscope, or waveform monitor to monitor the strength of the signal leaving the camera.

12.3. COPYING AND EDITING ONTO VIDEOTAPE

Occasionally, we may demonstrate the live scene in a video microscope directly to an audience. More commonly, we would first record the scene and sound on a VTR during an experiment or a period of concentrated observation. Later, we might edit, narrate, and copy the video data onto another tape for presentation in a compact package that displays particularly interesting scenes or conveys a message.

In preparing the video data for presentation, we can slow down or speed up events, add titles and explanations, montage scenes for comparison, and adjust the gray levels and image contrast to optimize the picture for presentation or to emphasize important features.

We will now examine how to copy, prepare, and edit videotape.

The Equipment

We can copy and edit video and audio information from one videotape to another with quite simple equipment. All that is required is the VTR on which the original recording was made, a second VTR with the desired format for the copy (and a pause function as described below), and a monitor (Fig. 12-7). Professional studios use expensive *editing decks* with time-base correctors and capabilities for *insert editing*, but the simple system described below should be adequate for many applications in video microscopy.*

The first VTR should preferably be the one used to record the original videotape, but can be another VTR (even one made by a different manufacturer), so long as the playback *speed* and recording *format* are identical to the original VTR.

Within limits, the format of the second VTR need not be identical to the first. For example, one can generally copy or edit from ¾ onto ½ inch, from color to monochrome, from time-lapse to standard speed, and even, in some cases, from noninterlaced 256/60 to interlaced 525/60 as in transferring microcomputer output to VTR (see Fig. 12-10). However, we can generally not copy

*Several new ½-inch VCRs now provide editing capability as described below.

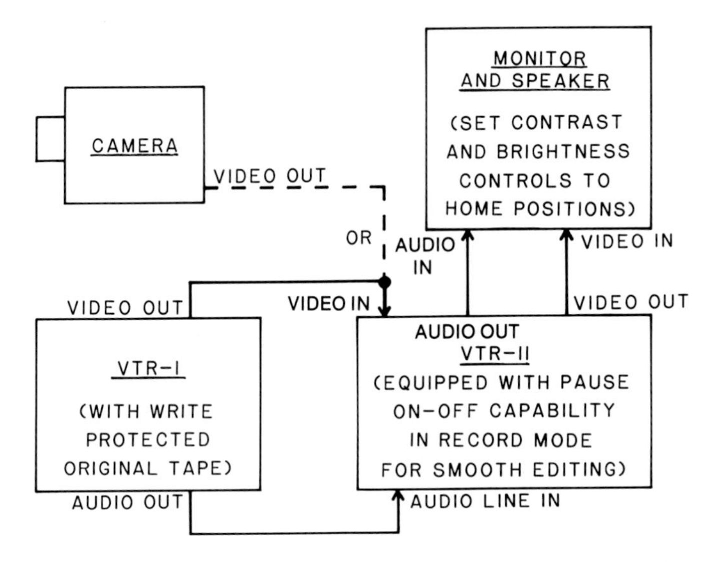

FIGURE 12-7. Minimum arrangement for copying/editing from tape to tape. See text.

the European formatted 625/50 SECAM/PAL tape to 525/60 NTSC, and vice versa, without special equipment.*

If we wish to not only copy but also *edit* the tape, we need to use either an editing deck or a second VTR equipped with a PAUSE button that works in the record mode (e.g., the Sony VO-4800, 5800, etc.). I will mainly discuss the use of the latter type of ¾-inch VCRs.

The arrangement shown in Fig. 12-8 is convenient for editing scenes whose brightness or background levels must be balanced. One can put together this arrangement from standard video components described in earlier chapters. We use the video processor (Colorado Video Model 604) to adjust the picture and sync signals to their proper voltages and the pedestal to appropriate levels. We can make these adjustments either by viewing a monitor whose CONTRAST and BRIGHTNESS controls are placed in their home positions (Section 3.2), or by displaying the signal on a cathode-ray oscilloscope or waveform monitor (Section 6.3). If a time-base corrector is available, placing it between VTR I and VTR II can help improve the stability of the edited picture.

Selecting and Organizing the Scenes

Before discussing the mechanics of editing onto tape, we will briefly consider how to select and prepare the video material to be edited. I will assume that most of the video microscope scenes will be selected from a store of data already on hand.

First, think through the *message* that is to be conveyed and select the scenes accordingly. For most presentations, a series of brief active scenes that clearly show the features of interest are more effective than long, drawn-out scenes. Twenty-five to forty-five seconds per scene is usually adequate. Repeat a scene if necessary. Avoid being cute, and concentrate on the story.

To organize the scenes into a streamlined sequence, prepare several Xerox copies of pages blocked out with 6 to 12 blank TV screens, arranged as in a comic strip (Fig. 12-9). Try filling in the blocks with simple sketches of each scene. Likewise, fill appropriate blocks (or file cards) with titles, including authorship, institution, date, conclusions, and acknowledgments. For each block or

*Common examples of *incompatibility* between tape and VTRs are: tape recorded on a time-lapse VTR versus standard VTR; 525/60 versus 625/50 format; VHS versus Beta. In some cases, the cassettes will not fit into an incompatible VCR. Others, for example the ¾- and ½-inch time-lapse VCRs, use standard-sized cassettes but run them at nonstandard speeds. Tape recorded on such equipment fits physically, but cannot be played back on standard-speed machines.

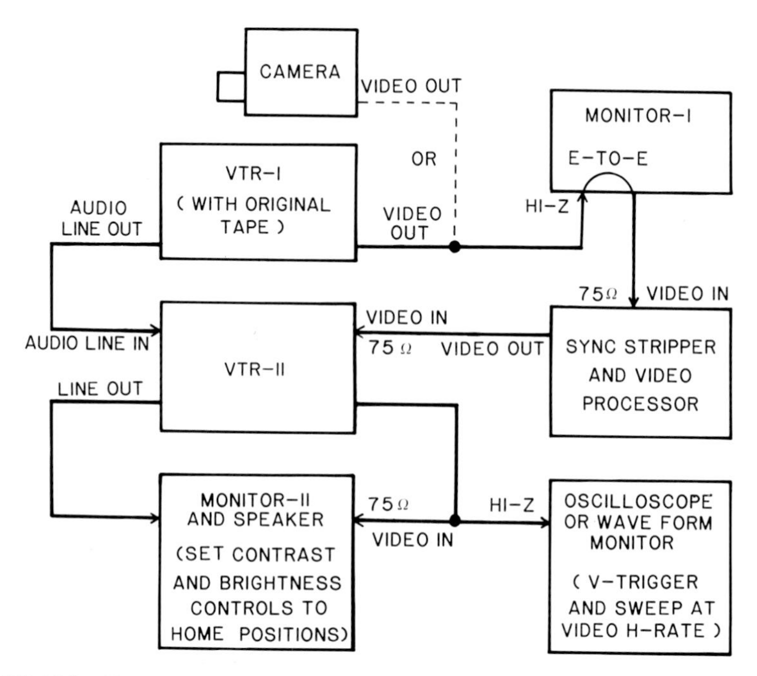

FIGURE 12-8. Tape-to-tape copying/editing with provision for regulating signal levels. See text.

FIGURE 12-9. Blank screens for scene planning. See text.

file card, add the cassette and counter numbers for the beginning and end of each scene. Where applicable, also record the minute and second mark and sound cues.

Make a table of the titles and scenes, serially numbered, and the number of seconds each is to run. A title should run for about twice the time required to read it out loud quickly. If audio is used to narrate the scenes, make up a *script* before finalizing the sequence; the length of a scene may have to be adjusted to accommodate the narrative. Narrations may have to accompany previous scenes in order to prepare the viewer for a sudden event. Add up the times for all the titles and scenes, and delete or adjust their lengths to fit the total time allotted. Update the cues for the beginning and end of each scene.

Organizing the material in this fashion helps to cut down the time spent on the VTRs, reduces errors, and improves the coherence and impact of the edited tape. Additional tips and suggestions can be found in video magazines and home video handbooks.

Title Making

We can use a video camera to copy titles optically or we can generate titles *electronically* with a video title maker or a simple computer.

We can make titles to be copied optically using transfer letters, magnetic boards and letters, or even by handwriting. If a close-up lens is available for the video camera, titles can also be prepared on an electric typewriter. A black carbon paper, inserted behind the typing paper with the carbon facing forward, provides a dark imprint and a title with good contrast. As in lantern slides, each line should contain no more than about 7 words or 40 characters, in order to keep the title clearly legible.

We can record electronically produced titles directly onto videotape. *Video title markers* are available for this purpose, but some *home computers* serve quite well. Figure 12-10 shows a title made with an Atari 800 computer hooked up in place of the camera in Fig. 12-8. A simple computer program can automatically center the lines and title.

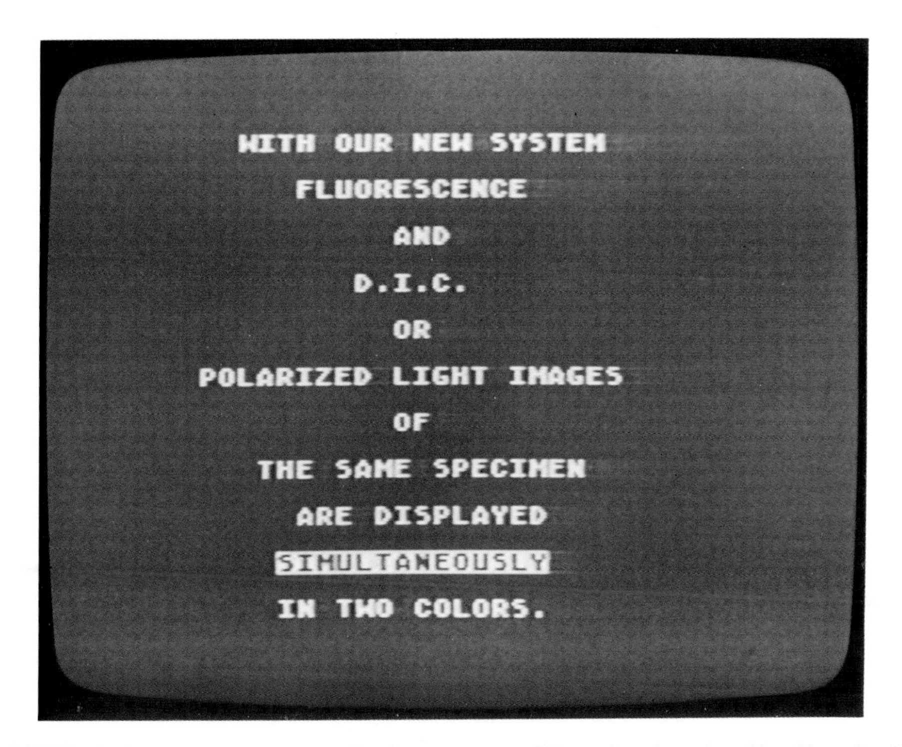

FIGURE 12-10. Video title produced with a home computer. This monitor picture is a video title made with an Atari 800 computer. (Not all home computers provide signals that are compatible with video tape recording.) This method is very much less expensive than using commercial video title makers.

One can generate and store the titles as a group on a VTR (VTR I in Figs. 12-7, 12-8) before the actual task of editing. Or, for simple sequences, one can copy prepared titles optically with a video camera, or generate them electronically, according to prepared wording, as the editing progresses.

The Mechanics of Editing

We will now discuss the mechanics of copying and editing video from tape to tape using the arrangements shown in Fig. 12-7 or 12-8.

For the arrangement shown in Fig. 12-7, connect the VIDEO OUT terminal of VTR I to the VIDEO IN terminal of VTR II with a 75-ohm shielded cable. For the arrangement in Fig. 12-8, make similar connections with terminations as indicated in the figure. With either arrangement, set the VIDEO–TV select switch on VTR II to VIDEO. Set the CONTRAST and BRIGHTNESS controls on monitor II to their home positions, and turn the DC reset switch to ON. Make sure that monitor II is properly terminated at 75 ohms.

To transfer *audio,* connect the AUDIO **LINE OUT** jack of VTR I to the AUDIO **LINE IN** jack, rather than the MIKE IN jack, of VTR II; otherwise, the signal from VTR I will be too strong and distort the recorded sound. If a MIKE–LINE select switch is present on VTR II, set it to LINE. Turn on the AGC (automatic gain control) for the audio channel on VTR II [unless some unusual aspects of the incoming audio have to be corrected by manual control of the audio level, guided by the audio level meter (**VU meter**)].

In VTRs with *audio dub* capability in one of the two audio channels (usually channel 1), connect the audio cable from VTR I to the channel that cannot be used for dubbing (i.e., to channel 2); in this way, the other channel is reserved for adding comments and sound later. In the audio dub mode, the VTR can be used to narrate or add comments to the tape without erasing the video scene or recording over the audio signal on the alternate audio channel. The two audio channels can be played back together or separately.

Before copying or editing video, prepare the tape *cassettes* as follows. First, *write protect* the original tape from accidental erasure. For ¾-inch cassettes, remove the *red button* in the bottom of the cassette (Fig. 8-17). With ½-inch cassettes, break off the write-protect *plastic tab* from the cassette. Make sure the tape is not slack in the cassette. Then place the original tape in VTR I. Now place the tape cassette, onto which the copy is to be made, into VTR II. But before writing onto the tape or disabling the write-protect feature on the cassette, play back the tape to make sure that you are not recording over an important record.

If the tape going into VTR II had been used and write protected earlier, the tape cannot be recorded over without disabling the protection. For a used ¾-inch cassette that no longer needs to be protected, reinsert the red button. For ½-inch cassettes, disable the protection by covering up the hole behind the missing tab with Scotch tape, or by filling the hole with a snugly fitting cube carved out of a rubber eraser. Affix a label to the cassette *before* inserting it into the VCR, to prevent later confusion.

Before recording onto used or new tape, it is good practice to run the tape once completely through VTR II in the *record play* mode. Replace VTR I with a standard video camera with its power on, but with the lens capped to provide blank dark frames and sync pulses. This operation erases the old data and lays down clean blank frames and sync pulses, so that noise does not appear on frames that are not recorded on. The tape is also burnished and the frequency of dropouts is reduced. After running the tape all the way forward in VTR II, rewind it without interruption to the beginning. The tape, whose edge plays a critical role in aligning the tracks, is thus rewound evenly (Chapter 8).

Once the cassettes are prepared, the connections established as in Fig. 12-7 or 12-8, and the equipment powered up, the picture and sound on the tape played on VTR I should appear on monitor II when the RECORD button is depressed on VTR II. So long as you depress *only* the RECORD and not the PLAY button on VTR II, it does not record, so we can *preview* the tape that is in VTR I. We can also press the PAUSE, RECORD, and PLAY buttons on VTR II *simultaneously* to preview the scene played back on VTR I.

In previewing, locate the proper scene and make sure that the video and audio circuits are working properly. With the setup shown in Fig. 12-8, adjust the pedestal and signal levels to their optimum levels by controlling the video processor (leave monitor II controls in their home position) while checking the signal voltage on the oscilloscope and/or examining the picture on monitor II.

Now review the material previously recorded on VTR II, including blank fields used as head and tail leaders, by depressing only the PLAY button on VTR II.

VTR II *records* the material played off of VTR I when we simultaneously depress the PLAY *and* RECORD buttons, so long as VTR II is not in pause. With the type of recorders we are discussing, depressing the PAUSE button once freezes the recording without unlatching the PLAY and RECORD buttons. When we depress the PAUSE button again, the recording continues exactly where it left off before.

If, instead of pause, we stop VTR II by depressing the STOP button, we activate a series of events that interfere with smooth editing onto the tape. When we depress the STOP button, all other buttons are inactivated and pop up. With STOP, the tape is wound away from the recording heads and drum, back into the cassette in some VCRs; in others the tape is not wound back into the cassette, but the drum and heads stop turning. In contrast, when we place VTR II into the pause mode, the tape transport mechanism is inactivated but the tape stays in place around the video drum and heads, which continue to rotate. When we depress the PAUSE button again, the tape transport mechanism starts up directly (or with a brief delay), and the tape moves forward synchronized with the rotating heads. Thus, by using the PAUSE button in the record play mode, we can record titles and scenes consecutively on VTR II with *smooth transitions* free from glitches.*

In many VCRs, a protective circuit switches the recorder out of *pause into stop* after the pause operation has lasted for several minutes. This safety feature protects the tape from excess wear and stretch, but the tape is no longer positioned at the desired transition point for editing. Thus, only if the tape can be edited in quick succession, can we pause VTR II at the end of the scene we are recording, and then resume the recording by pressing PAUSE again as just described.

If the editing cannot progress fast enough for the pause period allowed by VTR II, we should use the procedure shown in Fig. 12-11. Record a scene for several seconds past the *cue* (1E) at which one desires to end that scene. Then rewind the tape on VTR II briefly, stop, and depress the RECORD button. Next, preview VTR I. Readjust the video processor if necessary, locate the desired starting point of the next scene (cue 2S and 2E), and memorize the exact cues for the scene. The cues can be the minutes and seconds mark on the time–date record or, more conveniently, particular comments on the audio channel. After clearly establishing the cues, wind VTR I back to a point 10–20 sec before cue 2S, and stop. Now play the tape on VTR II forward to cue 1E, and pause. Depress the RECORD and PLAY buttons together (*without* releasing the PAUSE button) to prepare VTR II for recording. Play back VTR I and, exactly at cue 2S, depress the *pause* button on VTR II to continue the recording. Continue the recording to a point several seconds past cue 2E, and repeat the above procedure for the next scene or title. By using this procedure, one can change the tape in VTR I, or search for the starting point for the next scene, without rushing. By using these editing methods, a new scene (or title, etc.) can be dubbed to the end of the previous scene from VTR I or from a video camera smoothly, without introducing noise or unsightly glitches. This method of adding new scenes only to the tail of the previous scene is known as *assemble editing*.

With the simple arrangements described, we *cannot insert* a scene or title (between scenes that have already been recorded) without introducing noisy frames. In conventional recorders, the erase head precedes the recording head by some distance (Chapter 8), so that an insert replaces the beginning of the succeeding scene with several erased, noisy frames. For insertion without such major defects, one needs to use proper editing decks, which are generally expensive.† However, as

*With color recording, some moiré fringes may appear for a few frames following the splicing point with this type of editing. With the Sony VO-4800 or VO-5800 used as VTR II, we find that the edited transitions can be nearly perfect for monochrome scenes.

†Several new four-head ½-inch VCRs in the $1000 price range now permit insertion video editing and sound-on-sound dubbing. Some also allow still-frame (or field), noiseless frame-by-frame advance, slow motion, and rapid picture search (tabulated in *Video,* Nov. 1983, pp. 60–65). These consumer VCRs may serve as convenient and economical devices for editing, screening, and presenting tape, so long as the edited product does not have to be copied again. The quality of tape copied from ½-inch VCRs, in contrast to ¾-inch VCRs, tends to be quite poor.

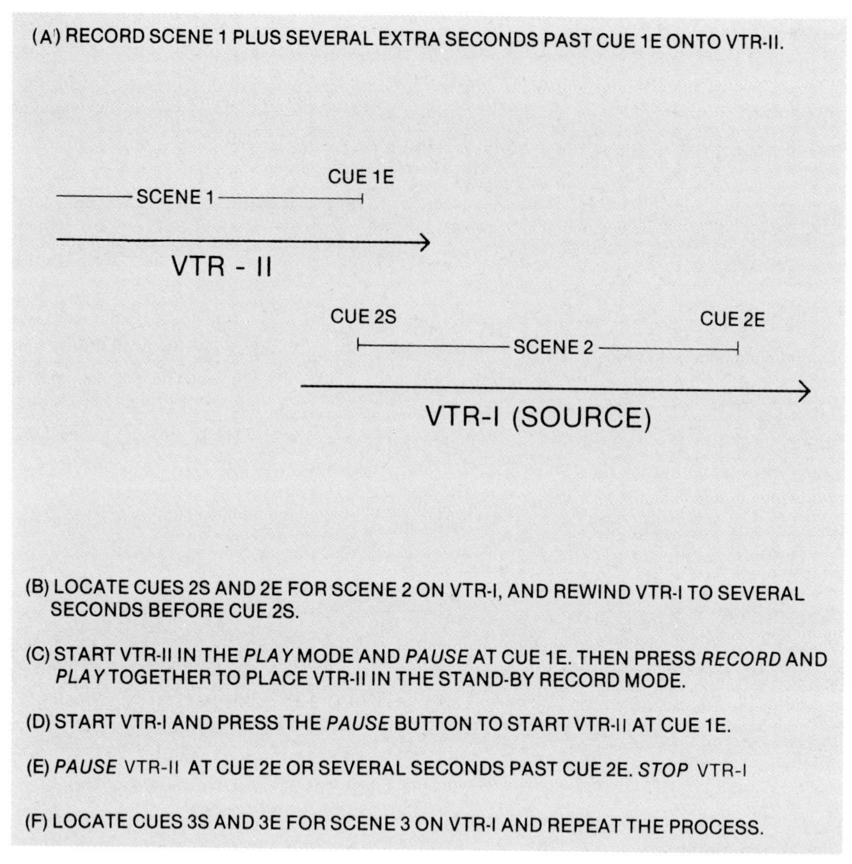

FIGURE 12-11. Assemble editing onto a VCR equipped with pause function in the record mode.

long as one plans the editing ahead of time so that the scenes can be assemble edited tail-to-head in sequence, the simple system described above should work adequately. While we cannot usefully insert video with the simple system, sound can be dubbed in the audio dub channel after the tape has been edited.

12.4. SPECIAL EFFECTS

While copying one videotape to another, we can introduce a wide range of effects. For example, in Figs. 12-7 and 12-8, we can pause VTR I to record a frozen frame (or field) onto VTR II. If VTR II has time-lapse capability, we can use it to speed up motion recorded on the original tape, or to achieve greater time lapse than could be achieved by single time-lapse recording. We can connect a motion analyzer (e.g., Sony SVM 1010; Section 8.4) in place of, or directly after, VTR I in Fig. 12-7 or 12-8. Then, we can play back the sequence stored in the motion analyzer field-by-field, or backwards and forwards in slow motion, and record the slowed-down motion onto VTR II. Likewise, we can place a video analyzer after VTR I and add fiducial marks or intensity traces to the signal played back from the recorder (Figs. 9-13, 9-14).

In addition, by using a *special effects generator,* or a *video mixer,* one can mix or superimpose video scenes in a variety of ways. The special effects generator (Fig. 12-12) synchronizes, appropriately keys, and adds together the video signals from two or more sources to be mixed. The sources can be two cameras, a camera and a VTR, a camera and the output from a computer, etc.

Figure 12-13 illustrates how we can combine the outputs of a VTR or computer and a camera by using a special effects generator. When a VTR or computer provides one of the signals, that source must generally also provide the sync. Connect the composite video from the sync source to the AUX IN terminal of the mixer. If desired, a monitor (A) wired E-to-E could be placed between the source and the AUX IN terminal. The slave camera is synchronized through its genlock, or H and V drive inputs, by the sync signals extracted from the sync source by the video mixer. Feed the camera output to one of the VIDEO IN terminals of the mixer, again if desired E-to-E through another monitor (B).

Connect VTR II, onto which you wish to record the mixed signal, to one of the VIDEO OUT channels or the PREVIEW terminal of the mixer. With the appropriate switches on the mixer activated, we should see the combined picture on monitor C (and its signal or waveform on the optional cathode-ray oscilloscope) when the RECORD button on VTR II is depressed for preview.

With this arrangement, one can mix or montage the signal from VTR I (or an alternate source of video signal and sync) and the camera in various ways, depending on the settings of the video mixer. We can *superimpose* the two signals with the brightness of each picture adjusted to the desired level, or as a *fade* or **dissolve** with one picture fading away and/or the other gradually appearing. The picture *polarity switch* allows one to reverse the contrast and make a negative of one of the scenes.

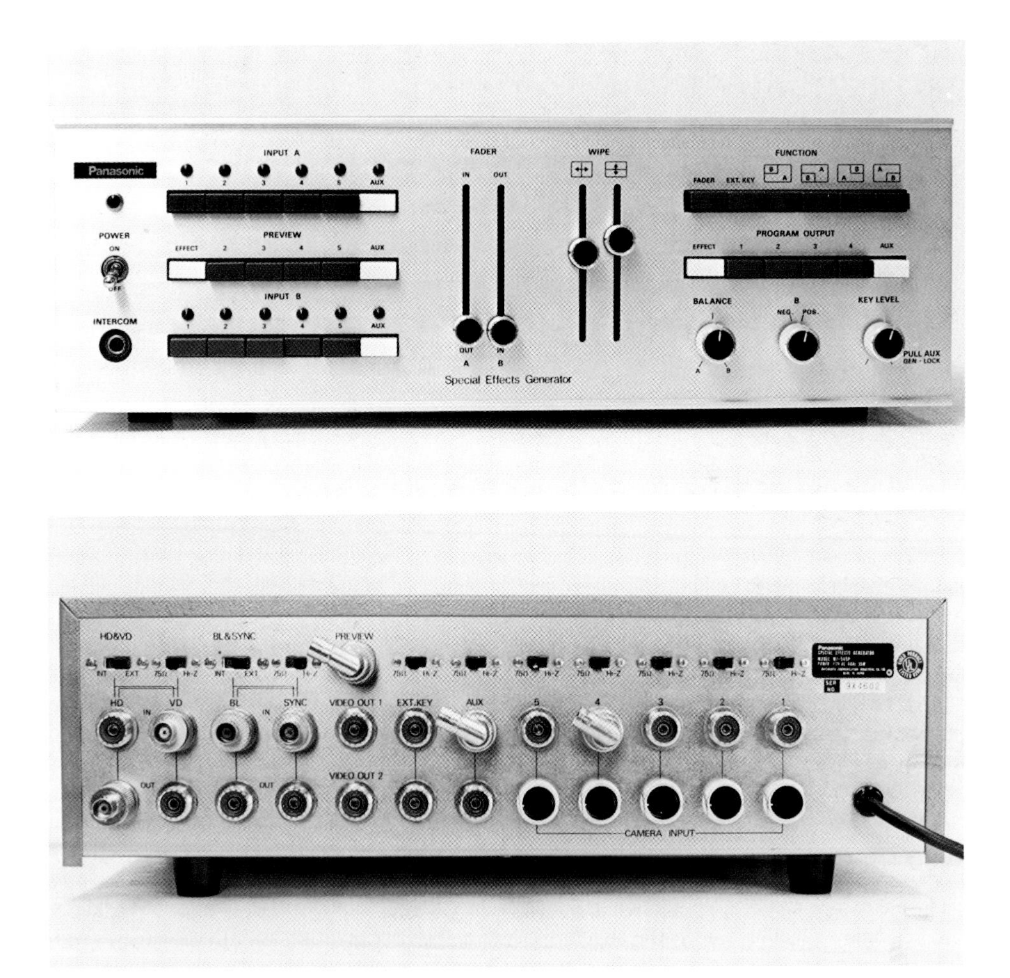

FIGURE 12-12. Special effects generator: front and back views. Special effects generators, or video mixers, can handle input from several cameras and a tape recorder, OMDR, or digital image processor to provide fades, keying, and wipes. A monochrome model (Panasonic WJ-545P) is shown here.

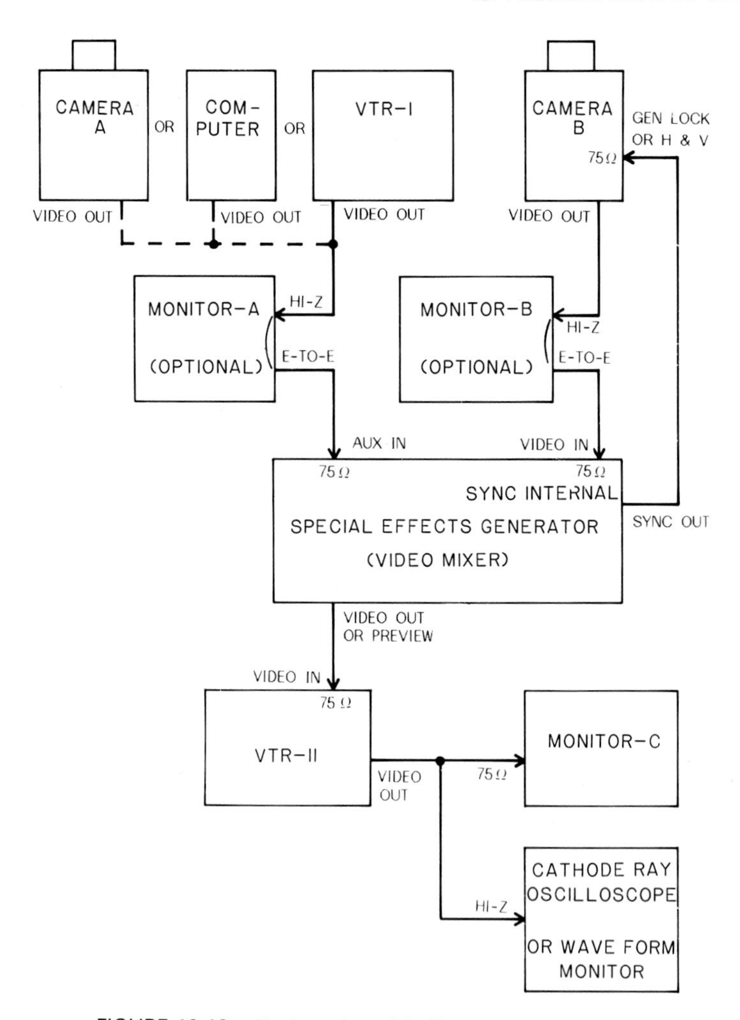

FIGURE 12-13. Hookup of special effects generator. See text.

We can *montage* the two pictures so that a part of the scene is occupied by one of the pictures and the remainder by the other, again with their brightnesses adjusted as desired, with or without contrast reversal of one of the two scenes (Fig. 12-14). We can *wipe* the boundaries of the pictures so that one picture is wiped away while the other takes over.

Also, by using a scene with a dark background, one can use the shape of the brighter subject as a *key*. The key blanks out the part of the picture that it covers. One can use a separate camera to provide the key, or one of the two inputs can be connected E-to-E to the KEY INPUT terminal and serve as the key. The keyed shape can then be used to form the boundary of a montage that includes one of the two pictures.

One can think of many applications of video mixing using a special effects generator, either in the process of tape editing or while gathering the data. A white arrow, circles, captions, etc., on a black background (or vice versa) can be captured with a key camera. The symbol superimposed on a video microscope scene calls attention to points of interst, allows one to place a fiducial marker or scale, or follow specific items in the picture. We can make various quantitative measurements with a digitizer and computer and display the raw scene together with the graph of the measurement. We can use event markers to signal an electrical stimulation, the beginning of an injection or perfusion, or other events. We could include faces of an oscilloscope, or electrical gauges and **digital voltmeters** or **multimeters,** to show action potentials and rapid changes of physiological or environmental parameters. The image of the micrometer scale on the microscope fine focus can give the plane of focus, the reading of a photometer the intensity of the illumination, and so on. We could also convert some of this information to frequency-modulated sound to be stored on one of the two audio tracks.

Thus, as we transfer video recordings from one VTR to another or as we record the original scene, we can combine different signals, montage spatially separated scenes, as well as freeze, stretch, or compress time.

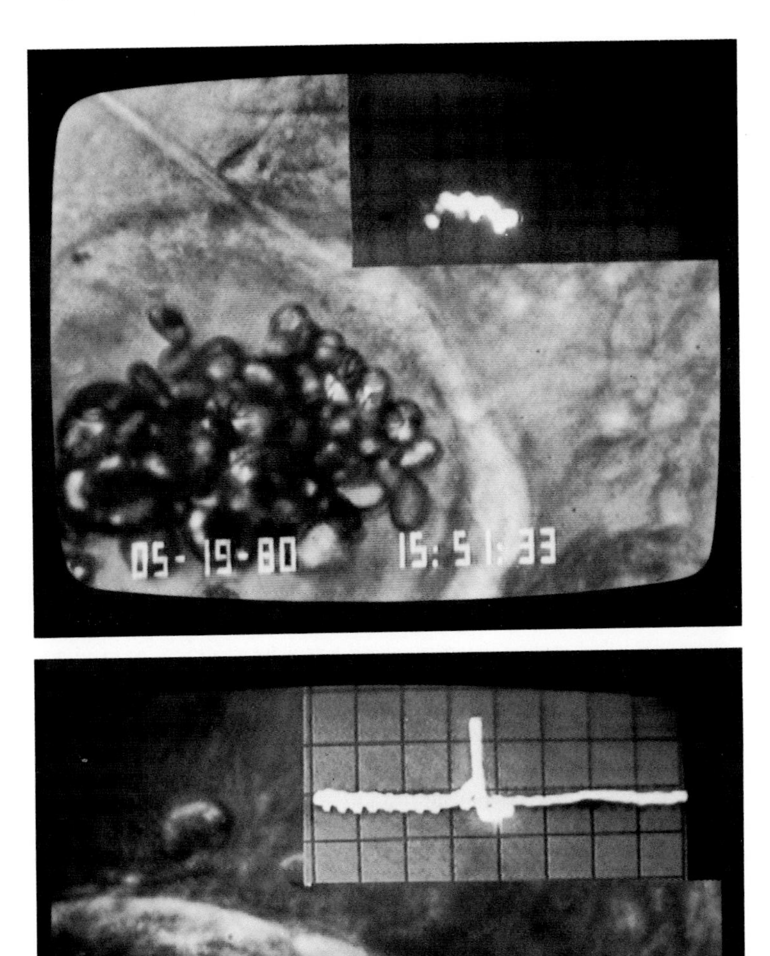

FIGURE 12-14. Montaging with a special effects generator. Top and bottom: two monitor views of a statocyst from the marine snail *Hermissenda*. The microelectrode (just out of focus at the upper left) is impaled in one of the statocyst's epithelial cells. The microelectrode records electrical activity produced when cilia are bent by the statoconia (dark pebblelike objects). The insets are concurrent oscilloscope traces showing the electrical activity, montaged and displayed on the monitor with a special effects generator. (From E. Stommel, R. Stephens, and S. Inoué, unpublished.)

12.5. VIDEO PROJECTION

To critically observe a video microscope scene displayed on a 9-inch monitor, one tends to choose a viewing distance of about 2.5–4 feet. I am assuming that the signal came to the monitor directly from a 1-inch monochrome camera (with a resolution ≥ 700 TV lines PPH), that the resolution of the microscope image was not appreciably below that of the camera tube, that the monitor was properly adjusted to provide 2 : 1 interlace, and that the viewer had normal vision.

In general, one examines a video microscope scene on a monitor from a viewing distance that is not too close to make the scan lines conspicuous, nor too far to comfortably view the finest picture detail. For standard video with 480 active scan lines and a 4 : 3 (H:V) aspect ratio, properly interlaced scan lines are spaced 0.3 mm apart on a monitor with a 9-inch picture diagonal (see Fig. 7-32). On the same monochrome monitor, 600 TV lines of horizontal resolution corresponds to a horizontal period of 0.5 mm/line pair. For well-lit, high-contrast targets, we can just resolve 0.5 mm/line pair at a distance of 1 m. The viewing distance of 2.5–4 feet thus satisfies the conditions for critical, yet comfortable viewing of monochrome camera signals on a 9-inch monitor. We will call the distance fitting these criteria the *optimum viewing distance*.

If a monochrome video signal does not come directly from a camera but rather is being played back through a standard ¾-inch VCR, a viewing distance of 4–6 feet might be more appropriate. A similar distance is appropriate for viewing a scene captured through an NTSC color camera and displayed directly on a 9-inch color monitor. With ¾-inch VCR recordings, the horizontal resolution drops to 320–340 TV lines in monochrome and 240–260 TV lines in color. Standard NTSC color cameras tend to provide a horizontal resolution of 320–340 TV lines, so that in these cases the optimum viewing distance of monitors with the same size screen is roughly doubled, compared to viewing a monochrome monitor displaying a signal directly from a monochrome camera.

As the size of the monitor increases, the optimum viewing distance increases more or less proportionately. For a monitor with a 9-inch picture diagonal, there is room for critical observation by one to a few individuals, while with a 15- to 19-inch monitor a dozen or so individuals could examine the picture critically.

With the development of more efficient monitors, excellent-quality color monitors with screen sizes of over 30 inches diagonal have become available, providing an optimum viewing distance of some 10–20 feet for color. Nevertheless, for monochrome presentation to a group of over a dozen or so individuals, or color to an audience of over three or four dozen, one has to rely on multiple monitors or use a video projector.

With *multiple monitors*, one can obtain pictures with reasonable luminosity, but they have several drawbacks for displaying video microscope scenes. The quality of the pictures in a multi-monitor installation tends to be inadequate for critically viewing monochrome microscope scenes. It is also difficult to optimize the contrast and brightness of the monitors to the scene. Furthermore, one finds few installations with an easily adjustable video pointer or some other internal marker that can be moved about on all of the monitors simultaneously. Such pointers are needed for emphasizing important features in a dynamic video scene.

Rather than using multiple monitors, we can present dynamic video microscope scenes to a large audience more effectively with a *video projector*. With a video projector, the audience focuses on a single, large, projected picture. By using the single projector, we can adjust the contrast and brightness to match the needs of the scene and easily point to important features on the screen.

For projecting video onto an auditorium-sized screen, two types of projectors are currently available. The **Eidophor,** or light valve projector, projects a schlieren image of an oil film whose surface is textured according to the video picture. In the projector,* the oil film is continuously refreshed onto a slowly revolving, flat glass plate in an evacuated chamber equipped with a pair of glass windows (Fig. 12-15A). An electron beam modulated by the video signal scans the surface of the oil film in a video raster. As the beam scans the oil film, each small area of its surface is charged proportionately with the current of the impinging electron beam. The charge reduces the surface

*These light valve projectors are manufactured, e.g., by General Electric Co., Projection Display Products Operation.

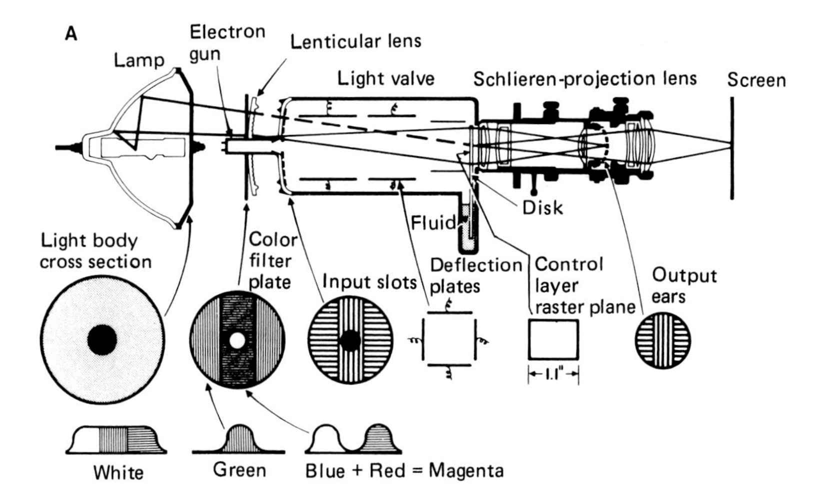

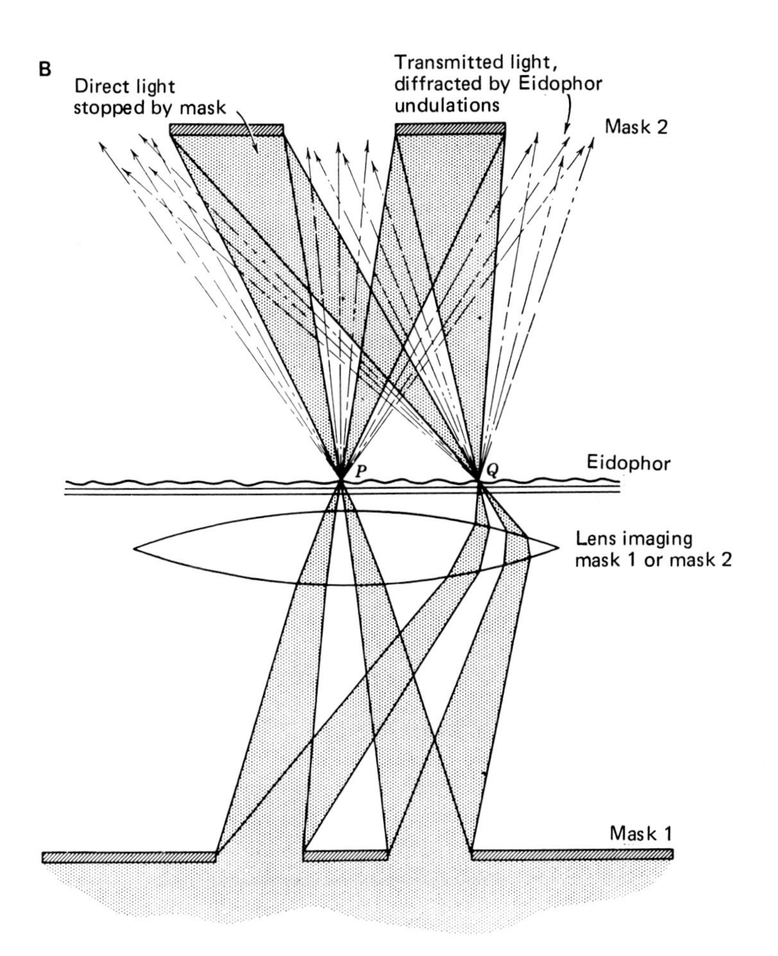

FIGURE 12-15. Schematic of light valve video projector. In a light valve, or Eidophor, projector (A), the electron beam embosses the image on an oil film instead of drawing on a fluorescent screen. A separate, high-intensity light source projects a schlieren image of the oil film; thus, the brightness of the projected image is independent of the image-forming process. See text. (A; Courtesy of General Electric Co.; B: from Zworykin and Morton, 1954, p. 285.)

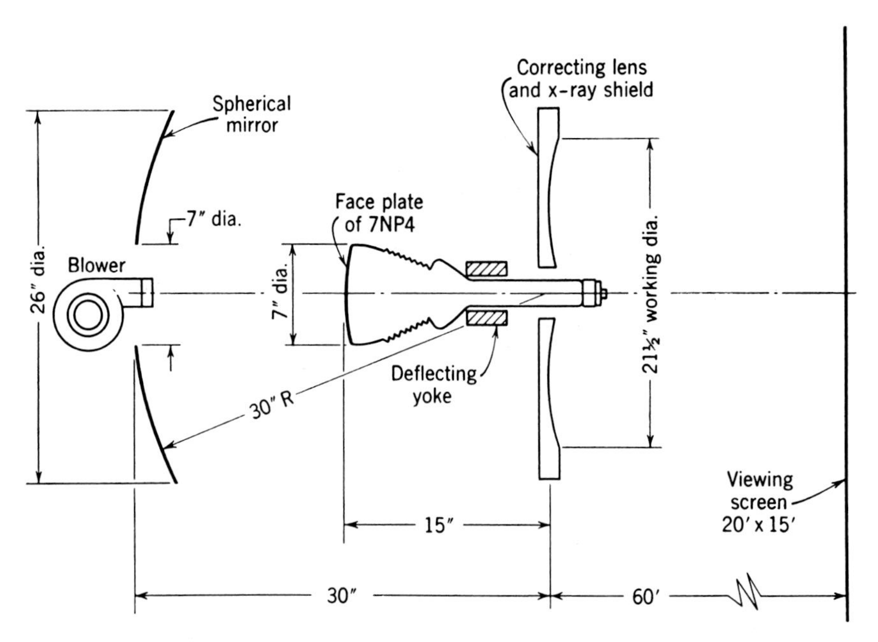

FIGURE 12-16. Schematic of a catadioptric video projector. In a catadioptric projector, a high-intensity picture tube produces a very bright picture. The picture is projected on a distant screen by a spherical mirror whose aberrations are corrected by a Schmidt correction plate. (From Zworykin and Morton, 1954, p. 438.)

tension of the film, which forms a miniature, lenticular bulge. The textured oil film lies at the focal plane of a schlieren optical system illuminated by a high-intensity xenon lamp. The light beam, deflected by the textured oil film, projects a schlieren image of the video according to the charge laid on the film by the scanning electron beam (Fig. 12-15B).

For color video, the General Electric system modulates the R, G, and B color signals with three high-frequency waves. The electron beam carrying the modulated signals produces three sets of grating-shaped ripples on the surface of the oil film. The gratings deflect the white illuminating beam from the xenon lamp into three different directions. The deflected light beam for each of the three primary-color signals then separately passes through its schlieren knife edge and a color filter (Fig. 12-15A). The three beams focused together produce the colored picture on the screen.

Being illuminated by a high-intensity xenon lamp, the light valve projector can yield a bright image on a large projection screen, up to 25 feet wide. The image is of remarkably high quality, but the equipment is expensive (about $100,000) and requires a trained engineer to operate.

Instead of using a separate illuminator as in the light valve projector, some picture tubes (specially designed for projection video) can be made so bright that an image on the faceplate can be projected directly by a lens or mirror optical system. Most projection video systems sold for use in the home or for small gatherings are of this type. Many use front projection, but some use rear projection screens to improve the picture brightness.

These projectors may be of the three-tube type, with each tube providing the picture for one of the three primary colors, or the single-tube type used in monochrome and some color projectors. With the three-tube color projectors, registration of the three projected images is often inadequate and the screen contrast too low for viewing video microscope scenes.

On the other hand, we find that a monochrome video projector (Fig. 12-16), using a single projection picture tube, spherical mirror, and Schmidt correction plate, can deliver a remarkably clear and bright image, projected foward up to about a 12-foot picture diagonal. Over the past several years, we have regularly used such a projector* to display standard-speed and time-lapse

*These catadioptric video projectors are produced by Image Magnification Systems and priced under $10,000.

monochrome video microscope images to an audience of a few dozen to a few hundred individuals. While the equipment is not easily transportable, and performs best after about a half-hour of warm-up, it is simple to operate, provides **keystone** correction and a good range of controls for contrast, brightness, and image sharpness, appropriate for displaying video microscope scenes in monochrome to a large audience.

12.6. VIDEO-TO-FILM TRANSFER

As described earlier, motion picture film to video (F → V) transfer is surprisingly simple. On the other hand, it is not a simple matter to obtain good-quality transfer of video microscope records to motion picture film (V → F).

One could simply point a (16 mm) movie camera at a monitor and copy the picture in much the same way that one makes still photographic records of the monitor. However, unless we use an animation camera to copy off sequential single frames of video, a horizontal dark stripe tends to appear and wander up and down the movie.

In a monitor showing a running video scene being copied onto film, the persistence of the phosphor has to be short enough to prevent the appearance of afterimages and comet tails. With short persistence of the phosphor, the exposure of each movie frame must be closely matched to the video frame. The blind period of the motion picture camera during film transport must also be short, and in reasonable phase relative to the vertical blanking period of the video. Without proper synchrony between the video and motion picture camera, parts of the video picture are, in fact, not photographed. The projected movie then shows the conspicuous dark stripe that wanders up and down the scene.

In order to gain proper sync between video and film, a special arrangement, known as a *kinescope,* is used. In brief, a kinescope operates as shown in Fig. 12-17. The output of the VTR is displayed on a low-distortion monitor with a flat faceplate. The rapid film transport of a special movie camera (the kinescope) is synchronized to the vertical sync pulse from every other field of the video. Thus, the kinescope optically copies each frame displayed on the monitor faceplate in synchrony onto the movie film.

In commercial V → F transfers, special cameras are used to eliminate every fifth frame of the video, so that the video running at 30 fps (frames per second) is transferred to film at the standard movie rate of 24 fps. In this way, the natural speed of the events on videotape is maintained in the

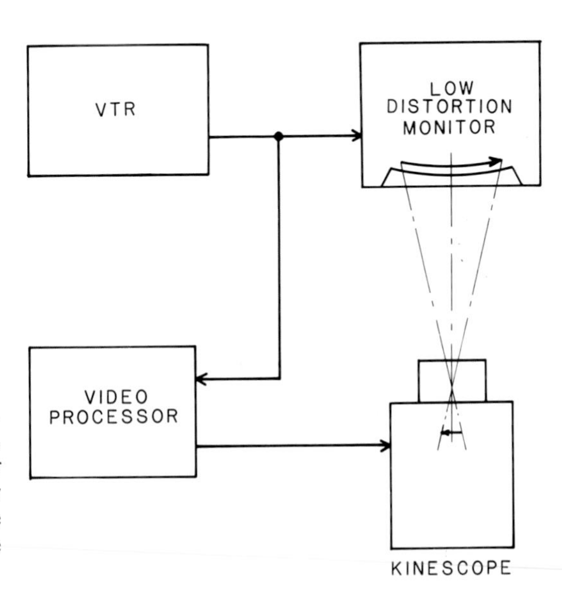

FIGURE 12-17. Schematic of kinescoping. In transferring video to motion picture film, synchrony is a major problem, since their frame rates are different. It may be solved by using a special video processor to synchronize the motion picture camera (kinescope) to the monitor display. See text.

film copy. Some high-quality transfers are made with *laser scanners* that draw out the picture onto film in synchrony with the video raster. With laser transfer, the color picture quality is said to be superb.

For V → F transfer of video microscope scenes, it is sometimes more desirable to retain all 30 frames of the video, rather than to adjust the speed for a 24-fps projection. Also, proper adjustment of contrast and black level is essential to a microscopist.

Many video labs around the country provide quick (but fairly expensive) services for V → F transfer, but virtually all of the facilities today are set up for color work and not for monochrome. These facilities use color monitors, which generally do not provide monochrome copies with adequate contrast or resolution for video microscopy. Even if a monochrome monitor were used, the kinescope operator is generally unfamiliar with the features of the microscope scene that need to be brought out, so that important features tend to become lost in the highlight or shadow, or the contrast is altogether too low. Some video labs use color film for copying monochrome scenes, an unsatisfactory choice for copying monochrome video microscope scenes.

At this point, probably the best way to obtain satisfactory monochrome V → F transfers is for the microscopist to control the monitor at the kinescope facility in a university or industrial audiovisual department.

Fortunately, when a motion picture made by V → F transfer is projected, video scan lines tend not to appear obtrusively.

12.7. STEREO VIDEO MICROSCOPY

We discussed some aspects of three-dimensional image reconstruction in Section 11.5. We also touched on the basic principles of stereoscopy in Section 4.8 [see also the classic text by Judge, 1926, and interesting accounts of stereo- and 3D-(video)projection by Lipton, 1982]. Before discussing how to use video to obtain stereoscopic images through the microscope, we will first examine how we can use the light microscope to obtain stereoscopic images in general.

Stereoscopy with the Microscope

For relatively low magnifications, we form stereoscopic images with a *binocular stereo*, or *dissecting microscope*. With these instruments, a pair of objective lenses view the specimen from two different angles, each providing an erect image into its corresponding ocular (Fig. 12-18). The axes of the left and right objectives and oculars converge by about ±7°. The oculars (Oc_L and Oc_R) maintain their convergence as we adjust their separation to the interpupillary distance of the individual observer. To change the magnification on these microscopes, the objective lenses or oculars are switched.

On modern dissecting microscopes, one may find a single, large-aperture objective lens between the specimen and Ob_L and Ob_R or in place of these lenses (Fig. 12-18). In these microscopes, the axes of the oculars may not converge appreciably. A zoom lens commonly allows continuous adjustment of magnification.

We can illuminate the specimen in a dissecting microscope from above by epi-illumination, or from below by transillumination. With epi-illumination, the highlights and shadows, as well as the features that are hidden from view, are similar to those that we experience daily in the three-dimensional world. With transillumination, many of these cues are lost, although oblique illumination provides highlights and shadows that can add to a sense of three-dimensional relief (though not necessarily correctly). With both epi- and transillumination, binocular parallax provides the dominant visual cue for the three-dimensional shape and arrangement of structures under the dissecting microscope.

With a regular *compound microscope*, we can produce stereoscopic pairs through single high-NA objective lenses in several different ways.* First, we can tilt the specimen to produce *stereo*

*See Gaunt and Gaunt (1978) for an excellent account of the history and principles of three-dimensional reconstruction and stereoscopy applied to various types of microscopes. They illustrate the excellent stereo

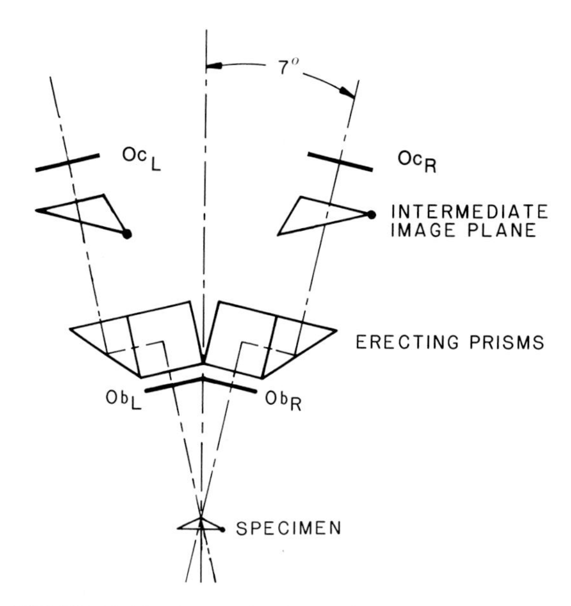

FIGURE 12-18. Schematic of dissecting microscope. See text.

pairs (Fig. 12-19A). Specimens in a polarizing microscope are sometimes sandwiched at the center of two glass hemispheres of a *universal stage* (Fig. 12-20). By using the goniometer of the universal stage, together with the matching UM, UMK, or UD series long-working-distance objective lenses, one gains a wide choice of tilt axes and viewing angles. We can tilt the specimen on the universal stage to practically any desired angle and produce stereo-pair photographs, although the NA of the long-working-distance objective and condenser lenses combined with the glass hemispheres is usually limited to about 0.65.

As an alternative, we can translate the specimen laterally as shown in Fig. 12-19B. We can use this method with higher-NA objectives, but the stereo pairs are not as effective as those produced by tilting the specimen.

Another method of producing stereo pairs through single, high-NA objective lenses is by tilting the observing or illuminating beam rather than the specimen. To tilt the observing beam, we place a *partial mask* C or D at the back aperture of the objective lens as shown in Fig. 12-21A. Mask C occludes the left half of the aperture and limits the clear aperture to the right half (as seen looking down the microscope). Assume that point E in the specimen lies closer to the objective lens than points F and G, which are in focus. Without the mask, the rays from E focus at a point E′ above the intermediate-image plane. At the intermediate image plane, they produce an out-of-focus disk, or a circle of confusion. When we insert mask C, the circle of confusion in the image plane shrinks down to E″ and shifts toward G′. Thus, the axis of the image-forming beam effectively tilts clockwise. Mask D shifts E″ toward F′, and the beam effectively tilts counterclockwise.

Osborn *et al.* (1978) have used this method* to produce stereo-pair photographs of microtubules decorated with fluorescent antibodies. In their micrographs, which show outstanding resolution, the long microtubules, convoluted in three dimensions, shine brilliantly against a dark back-

pairs that can be produced by electron beams and X-rays whose short wavelengths and low cone angles provide great depths of field. With the longer wavelengths used in the light microscope, the cone angle has to be increased in order to achieve high resolution. Thus, the depth of field becomes very shallow, and effective stereo pairs become more difficult to produce.

*Strictly speaking, Osborn *et al.* used a combination of the methods shown in Fig. 12-21A and B since they employed epifluorescence.

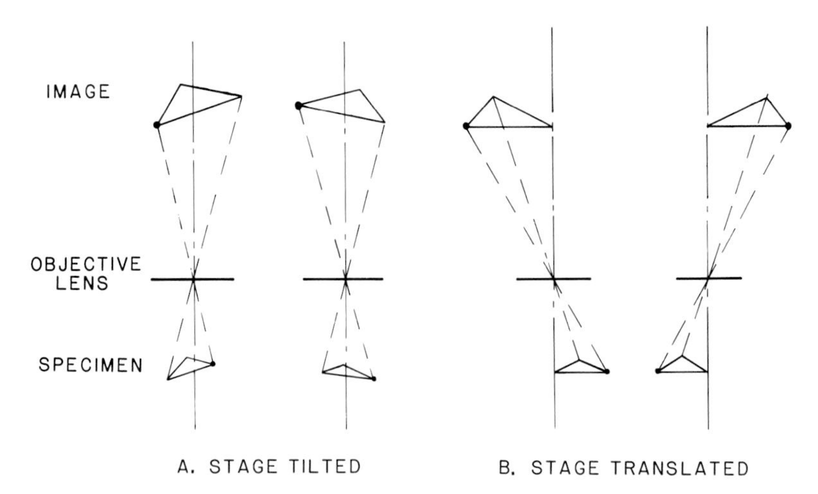

FIGURE 12-19. Two methods for producing stereo pairs with a conventional, compound microscope. See text.

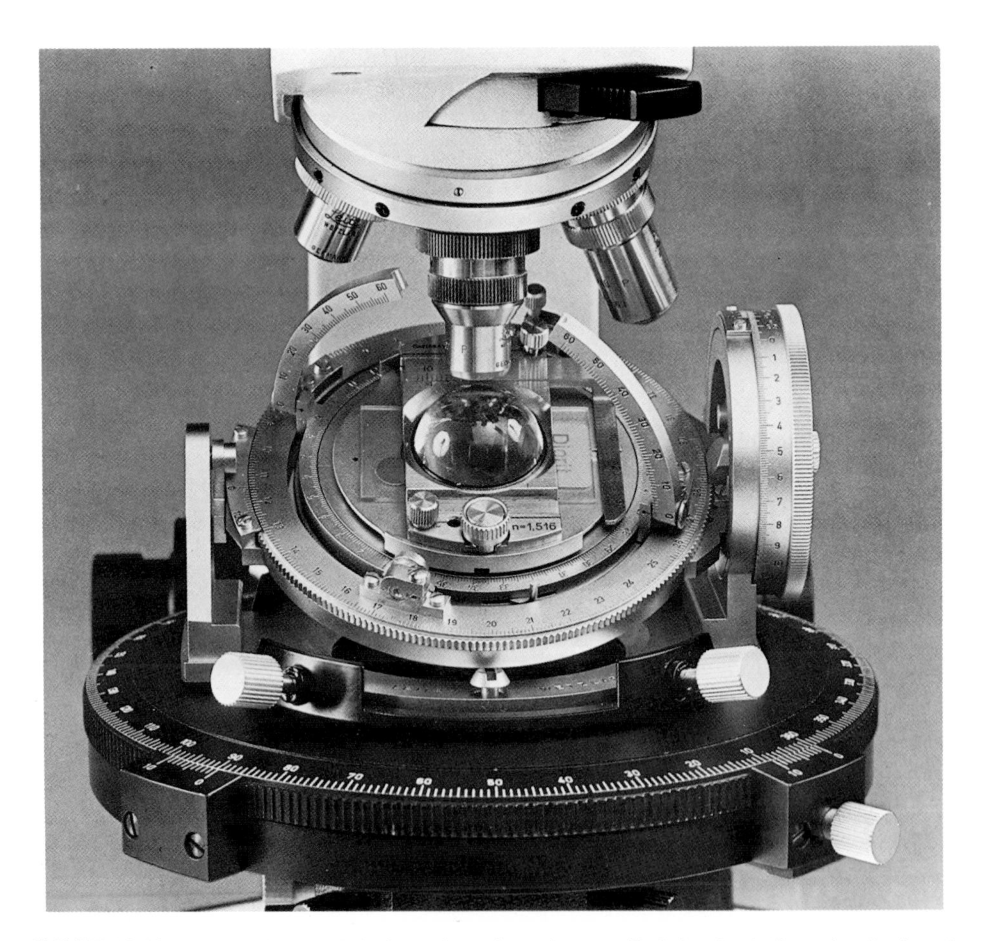

FIGURE 12-20. Universal stage. This four-axis goniometric stage (E. Leitz, Inc.) allows the viewing of specimens from many known tilt angles. Such stages have been used for many years to orient crystallographic specimens so that they could be viewed (conoscopically) along selected optical axes. (Courtesy of E. Leitz Inc.)

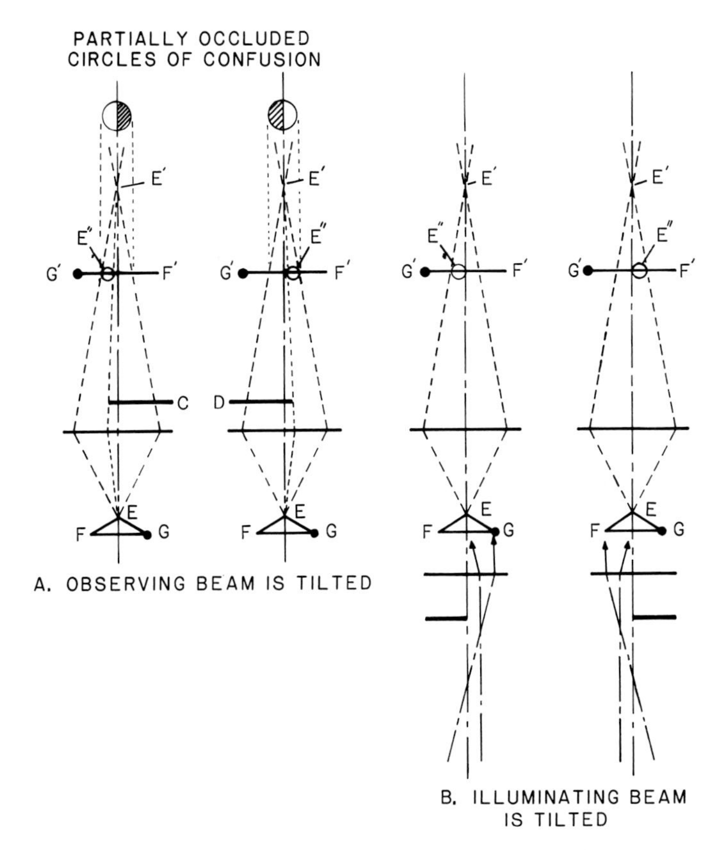

FIGURE 12-21. Principles of producing stereo pairs by partial masking of objective (A) or condenser (B) aperture. See text.

ground. These authors took the fluorescent micrographs in succession with a 1.4-NA objective lens screwed into a slotted collar that accepted the partial masks.

Instead of placing the masks at the objective back aperture, we can tilt the observing beams with masks placed at the eyepoint (with the left and right masks reversed). The eyepoint is conjugate with the back aperture of the objective lens and is physically more accessible (Section 5.3; Fig. 5-14).

Alternatively, we can place the masks at another conjugate plane, the plane of the condenser iris. Here, the masks provide oblique illumination that effectively tilts the major axes of the illuminating beams (Fig. 12-21B). Whether placed above the objective lens or below the condenser, these partially occluding masks shrink the circle of confusion for the out-of-focus images and shift them laterally to provide the apparent parallax needed for stereoscopy.*

For specimens that are not stationary, the stereo pair must be produced either simultaneously or in very rapid succession. We can form *simultaneous stereo pairs,* or present the tilted beams to the left and right eyes simultaneously, through a single objective lens and binocular tubes. For simultaneous viewing, masks can be attached to partially occlude the eyepoints.† Alternatively, we can

*Strictly speaking, geometrical optics does not properly portray image formation with high-NA lens systems; each "point" in the image consists of a series of diffraction fringes that repeat both vertically and laterally (Section 5.5). However, the predictions of geometrical optics work reasonably well for stereoscopy, especially for three-dimensional specimens with high contrast such as Golgi-stained neurons, fluorescent antibody-stained filaments and microtubules.

†As recognized for many years, the edges of the observer's pupils can also serve this function.

place a pair of "masks," or *half-shade filters,* made out of polarizing filters whose transmission axes lie at right angles to each other, at the back aperture of the objective lens (Fig. 12-22). Another set of polarizing filters placed in front of the eyes extinguishes light from each half of the filter. Thus, each half of the half-shade filter acts as though it was an opaque mask for the left or right eye (Inoué, 1944). In place of polarizing half-shade filters, a pair of complementary color filters, such as red and green (or blue) filters, can be used to mask out the appropriate parts of the aperture and provide the tilted beam for the left and right eyes simultaneously.

Rather than view stereo pairs through the microscope or as photographic prints, one can project slides of stereo-pair micrographs that the audience views through polarizing or red–green (blue) glasses complementing the projector(s). With polarized stereo, one can display color, but beaded projection screens depolarize light and are unusable. Some white painted surfaces and lenticular aluminized projection screens do not appreciably depolarize light and are used instead.

The use of holography and varifocal mirrors for three-dimensional display of microscope images was discussed in Chapter 11.

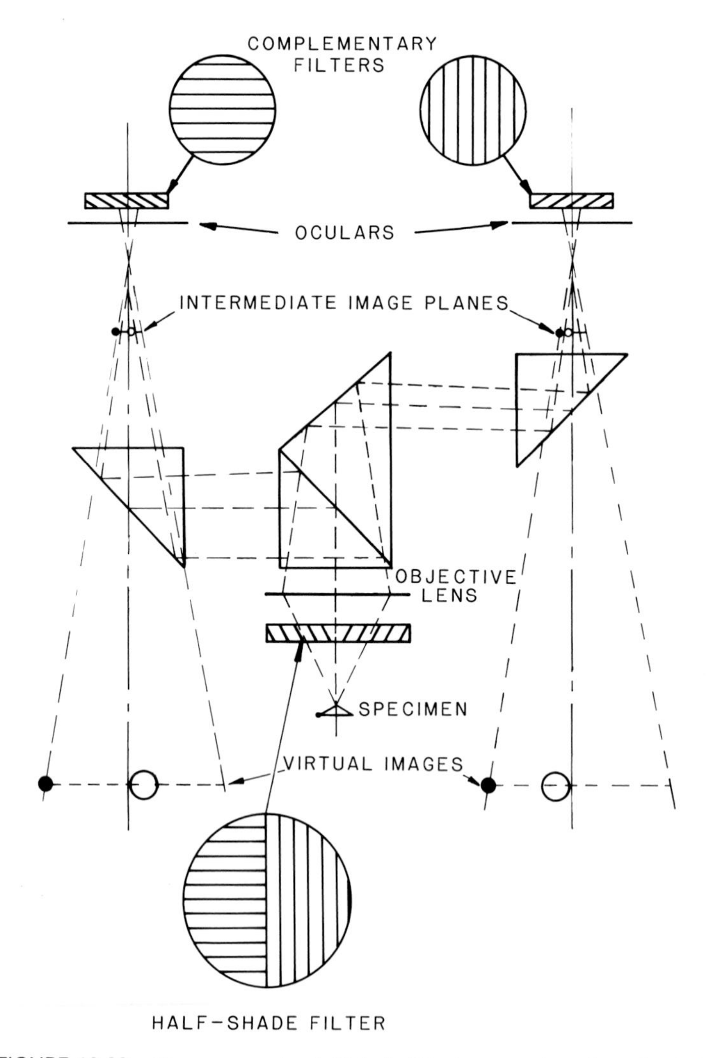

FIGURE 12-22. Stereoscopy using half-shade filter in an aperture plane. See text.

Stereo Video Display

We can adapt several of the methods discussed above to produce stereoscopic microscope images with video. For example, we could use a video color camera on a monocular compoound microscope with a single objective lens to capture a red–green **anaglyph,** or stereo pair. We place a red and green half-shade filter of the type shown in Fig. 12-22 at the objective back aperture or one of its conjugate planes to generate the red and green tilted beams. The superimposed red–green images formed by the two tilted beams are captured by the color camera. Its output can be recorded or displayed directly on a color monitor or projector. The audience views the stereo image through a pair of red–green glasses.*

In this configuration, we find that the stereo image appears best when the red and green half-shade filter is placed at the plane of the condenser iris rather than right above the objective or at the eyepoint. With this method, we have observed striking stereoscopic images of sea urchin sperm swimming freely in large helical paths, a pattern quite different from what we usually see when the sperm are swimming near the surface of the slide or coverslip. We presented the tape-recorded stereo scenes at the August 1981 General Scientific Meetings of the Marine Biological Laboratory, Woods Hole (Inoué *et al.*, 1981).

On a dissecting microscope, one could mount a pair of genlocked, matched monochrome video cameras above the left and right oculars. The synchronized outputs of the two cameras could then be fed to the red and green channels of an RGB monitor or projector and viewed through complementary colored glasses to provide intermediate-magnification stereo images in real time (Fig. 12-23).

With a binocular compound microscope with a single objective lens, one can also use a pair of matched monochrome cameras on the two eyetubes. One places a half-shade filter at the condenser aperture plane or its conjugate and complementary polarizing or red–green filters above the oculars (Fig. 12-22). The cameras are wired as in Fig. 12-23 to display a red–green stereo image on the screen of an RGB monitor or projector.

Instead of using a pair of matched cameras, we can use a single camera on a monocular compound microscope to produce monochrome or color stereo pairs. We generate the stereo pairs on alternate fields of the video by the masks shown in Figs. 12-24 or 12-25. The opaque masks occlude the observing or illuminating beams in rapid succession at the eyepoint or another conjugate plane (see Fig. 12-21). The opaque masks can be two sides of a slit-shaped opening that vibrate sideways and alternately occlude the beam (Fig. 12-24). Or the masks can be parts of a revolving sector wheel that alternately occlude the aperture as shown in Fig. 12-25. We spin the sector wheel with a servo motor, which we drive with a special *video processor* in phase with the video sync pulses from the camera. Alternate fields of the camera then see the image formed by beams tilted to the left and the right (Fig. 12-26; Inoué *et al.*, 1981).

One can view the alternate video fields on a single monitor through a pair of goggles that contain mechanical or (**PLZT**)† electro-optical shutters. These shutters alternately transmit light to the left and right eyes in synchrony with the video fields (Figs. 12-27, 12-28).‡

Instead of using goggles to view the alternate fields of the single monitor, we can display the left and right images on two identical monitors (but with the H sweep reverse in one) (Fig. 12-29). The monitors receive either the active first- or second-field signal from the video processor. The two pictures are orthogonally polarized relative to each other and then combined optically by a beam-

*Note that contrary to common belief, color vision is not involved with complementary color stereoscopy. The complementary colored filters cut out the beams that form the unwanted images before they enter the eye. Therefore, color blindness should not affect this mode of stereoscopy, although focal differences in the red and green images may. Regardless, 5–10% of the population is apparently unable to use binocular parallax for stereoscopy.

†Cutcheon *et al.* (1973). PLZT electro-optic materials and devices, 1- and 4-inch circular wafers, are available from Motorola, Inc., Communications Systems Division.

‡To drive the PLZT goggles in synchrony with the video played back from a VCR, we amplified the output of the video processor that drives the sector wheel (Fig. 12-26).

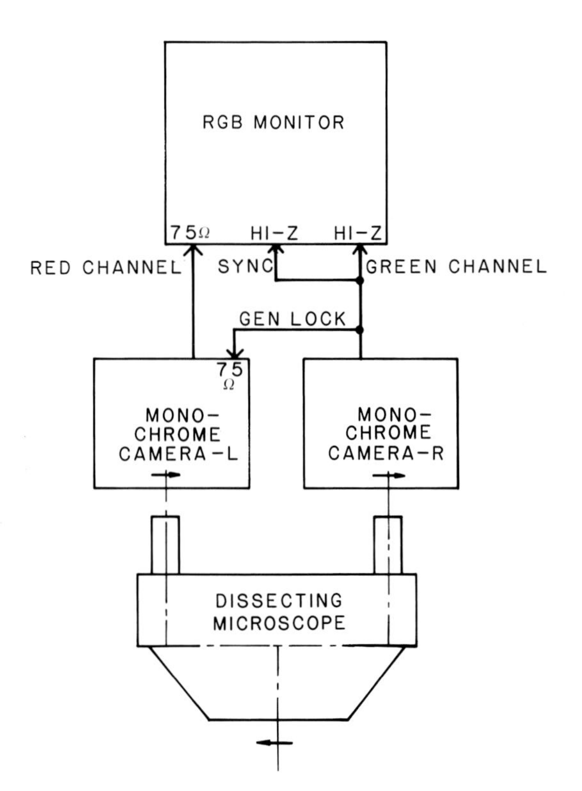

FIGURE 12-23. Use of video to produce red–green anaglyphs for stereo viewing through a dissecting microscope.

splitting mirror. Complementary polarizing glasses worn by the audience provide the pictures to the left and right eyes, respectively.

With the alternate-field video display, one can in principle obtain stereoscopic images in color as well as in monochrome. The alternate-field stereo can be recorded on a standard VTR. We recorded and demonstrated the feasibility of the monochrome system with the sector wheel placed at the eyepoint. Diatom shells, imaged in polarized light through a 40/0.95 Plan Apo objective lens on a Newvicon camera, showed quite strikingly in three dimensions. The stereo image should improve further with a better match between the two monitors and by the use of a camera tube with shorter persistence and target storage. One annoying feature of the alternate-field stereo display is the flicker, which becomes pronounced at higher luminosities of the monitor, since each eye sees an image interrupted 30 times/sec.

FIGURE 12-24. Vibrating aperture mask for producing alternate-field stereo pairs. See text.

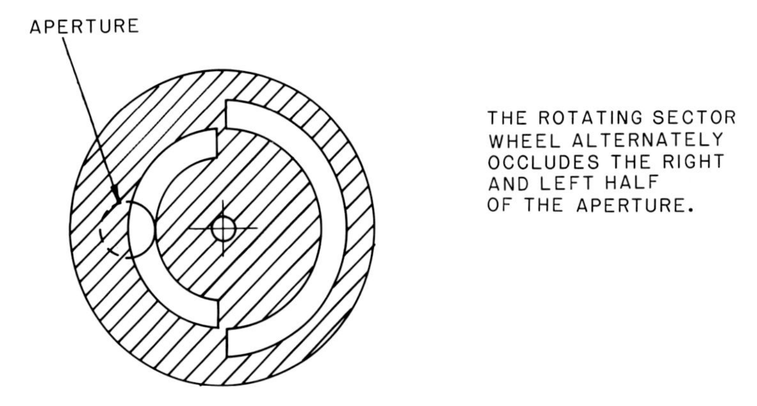

FIGURE 12-25. Revolving sector-wheel mask for producing alternate-field stereo pairs. See text.

To overcome the 30-Hz flicker problem, a system developed by Stereographics Corp. uses PLZT goggles triggered at 60 Hz with the video left–right images presented at 120 Hz. Their control box takes the 60-field inputs from the left and right image camera and outputs the left–right alternating 120-Hz signal to which the PLZT goggles are synchronized (Lipton, 1984).

Instead of alternate-field display, Butterfield (1978) records two micrographs side-by-side on a

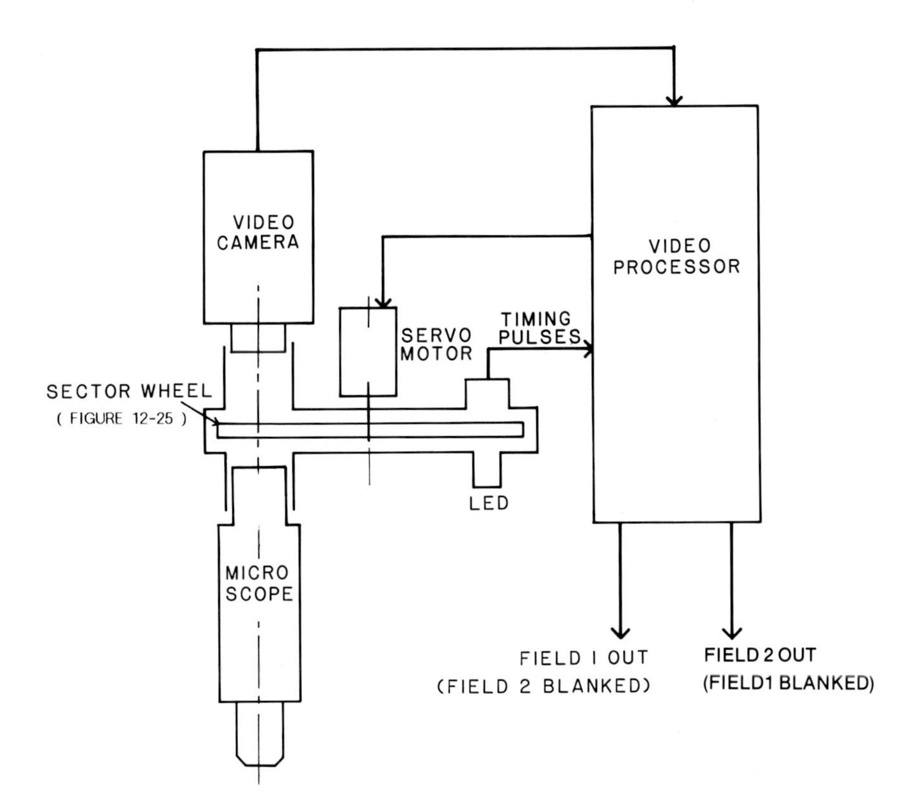

FIGURE 12-26. Scheme for stereoscopy by presenting alternate-aperture masked scenes to view alternate video fields. In this example, the aperture masking takes place at the eyepoint, but other aperture planes can be used. See text.

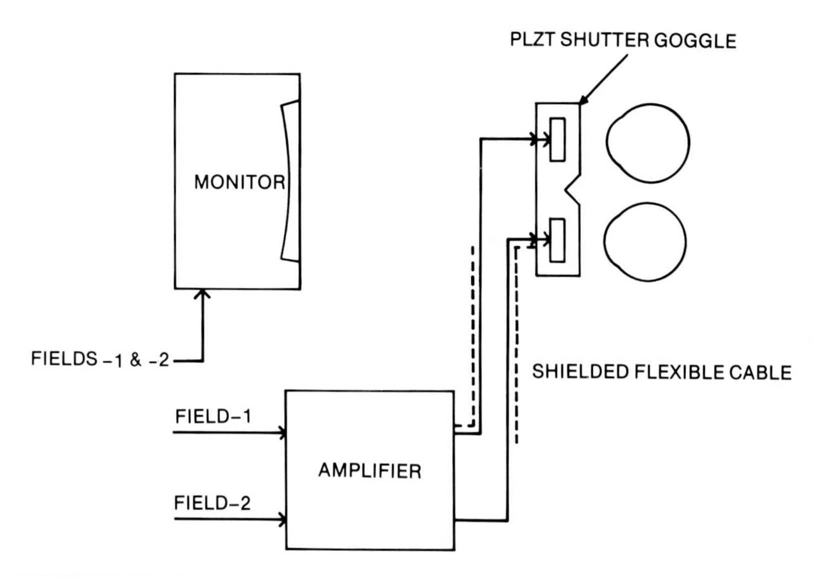

FIGURE 12-27. Schematic for alternate-field stereo using PLZT shutter goggles. See text.

single TV frame. His three-dimensional video microscope magnifies the specimen up to 120× on a monitor displaying a 125-mm-square picture, or projects the picture onto a 24 × 24-inch non-depolarizing screen. Butterfield says: "These side-by-side pictures are optically superimposed on the special screen by the stereo projector and each picture is polarized with quality polarizers. A convergence control provides proper stereo positioning on the screen" (from literature on Stereo-Color TV Projection System from 3D Technology Corp.)

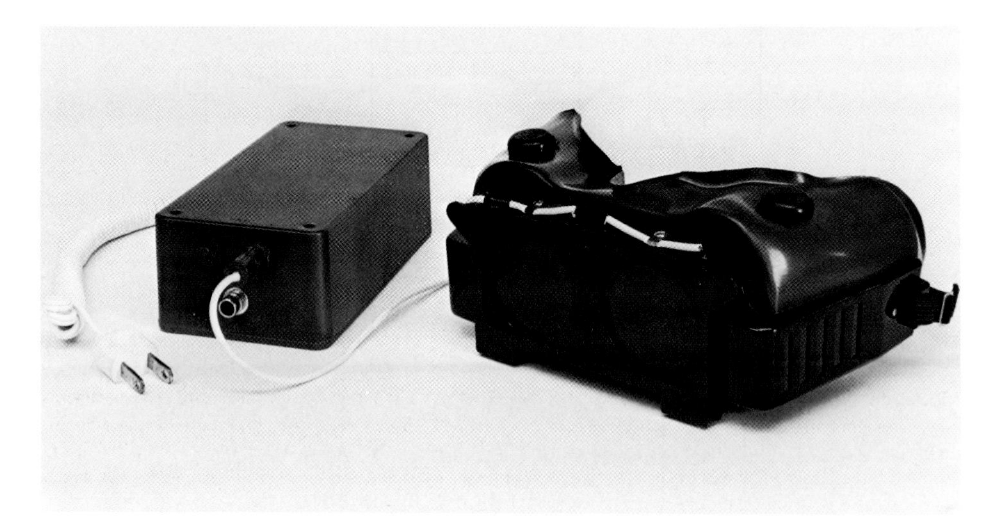

FIGURE 12-28. Photograph of PLZT shutter goggles. These goggles contain laminates of an electro-optic ferroelectric ceramic, PLZT (lanthanum-modified lead zirconate titanate), coated with transparent metallic conducting films, and sandwiched between crossed polarizers. When a high voltage (about 1 kV) is applied to the PLZT material, it is strongly birefringent, allowing light to pass through the crossed polarizers. The box at the left supplies the high voltage, alternately to the left and right wafers, in synchrony with the video signals.

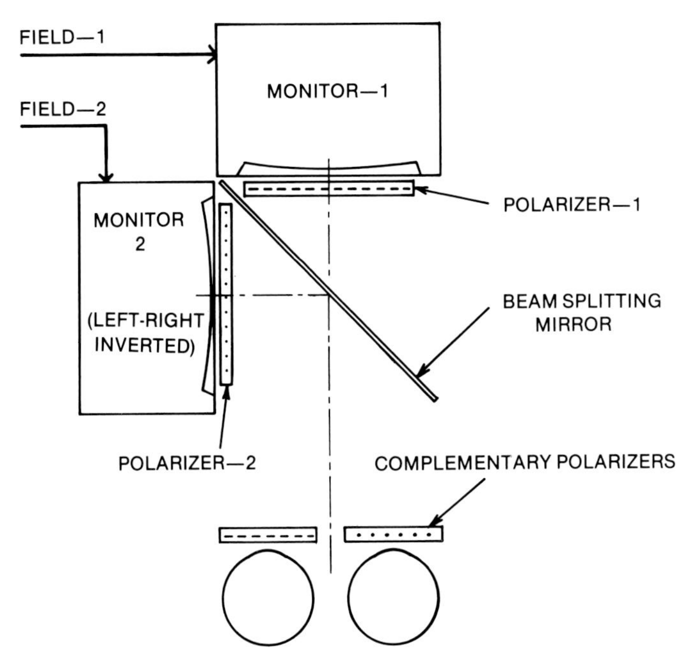

FIGURE 12-29. Double-monitor scheme for producing stereo images with video. See text.

By producing stereo pairs side-by-side on single video frames, rather than on alternate fields, one escapes the flicker problem associated with the latter method. Also, the electronic system is simpler and color can be used more readily. In principle, the horizontal resolution could drop to one-half since each half of the video frame is used for the whole scene, but this seems to be compensated in the Butterfield system by raising the video bandwidth to double the resolution.*

*The stereoscopic video display system developed recently by Stereographics Corporation uses special cameras and a monitor multiplexer and allows *recording on standard VTRs* and *flicker-free playback* in color or monochrome on a 120 Hz-field rate monitor. The image can also be projected and viewed stereoscopically through Polaroid viewing glasses. In these standard-video-compatible systems, vertical resolution is halved but horizontal resolution is fully maintained.

We have just developed a digital processing algorithm for converting through-focal optical sections of very high resolution video microscope images into stereoscopic pairs in a fraction of a second (Inoué *et al.*, 1985). The stereo pairs are synthesized in the Image-I processor from optical sections stored on an optical memory disk recorder. The display can be made dynamically, i.e., as stereoscopic pairs that focus through different focal levels of the specimen. We plan to extract stereo pairs serially along the X, Y, and time axes; thereby securing a four-dimensional display as well. (November 1985.)

Postscript

When I started writing this book a little over three years ago, a bright future was clearly on the horizon for video microscopy. What I did not anticipate then was the continuing, incredible growth of video and image-processing technology, and the number of new products that would be introduced in so short a time. Integrated circuits became ever more powerful as they also shrank in size. New generations of stand-alone processors now handle megabit computations in a flash, free from large host computers.*

Solid-state image pickup devices and lasers, Gen II intensifiers, liquid crystal displays, and even vidicons and SIT tubes continue to be improved at impressive rates. Just as quickly as the new components have appeared, they have been integrated into video cameras and recorders, as well as image processing, analyzing, storage, display, and hard-copying devices. Many new models continue to appear at dizzying rates.† The new generations of equipment tend to provide improved sensitivity, better and less noisy pictures, a greater array of functions, and are generally more compact and easier to use. Often, they are also significantly less expensive to purchase, operate, and maintain.

On the one hand, this is an opportune and exciting time to explore new applications of video microscopy. On the other hand, what I have written in these pages will have to be updated rather frequently by new information. Table 2 lists some magazines that report on developments in related technology and describe new products. (Free) trade magazines, such as those listed in Table 3, trade shows and seminars sponsored by the manufacturers, distributors, or publishers (note especially the annual Electronic Imaging Exposition and Conference sponsored by Morgan-Grampian), and Product Announcements or Descriptions distributed by the manufacturers can also be informative.

Applications of video to microscopy were described throughout the text, particularly in Chapters 9–11. Articles involving video microscopy now appear regularly in journals such as: *Analytical and Quantitative Cytology, The Biophysical Journal, Cell, Cell Motility, IEEE Transactions, Journal of Cell Biology, Journal of Histochemistry and Cytochemistry, Journal of Microscopy, Journal of Molecular Biology, Microscopica Acta, Nature, Proceedings of the National Academy of Sciences, Science, Ultramicroscopy,* etc.‡ The book *Optical Methods in Cell Physiology* (P. De Weer and B. Salzberg, eds., Wiley, New York), recording the September 1984 Symposium of the Society of General Physiologists in Woods Hole, is a convenient source summarizing some recent

*Many (new) companies now market *digital image-processing* programs and hardware (Appendix I). Table 1 complements the list of manufacturers of image processors and systems given in Table 10-1. See also Appendix IV. A comprehensive introductory text on digital image processing has appeared recently in paperback (Baxes, 1984).

†TV receivers too are "going digital" (Mitchell, 1983; Weber, 1984). Once the IC chips developed for this market are perfected, image fidelity of CCTV should improve further, while the cost of frame buffers, A/D and D/A convertors, and digital processors could drop dramatically. In the meantime, solid-state pickup devices with larger arrays of pixels have become more commonly available (Lewis, 1985), making the resolution of the chips more competitive with the vidicons.

‡See also the publication list compiled in *Microscopica Acta* **87:**183–226 (1983).

TABLE 1. Digital Image Processors—Supplementary List of Suppliers

Cambridge Instruments, Ltd. Rustat Road, Cambridge CB1 3QH, United Kingdom. Produces Quantimet image-analyzing system.

Comtal/3M Image Processing Systems 505 West Woodbury Road, Altadena, CA 91001. Produces Vision Ten/24 high-resolution image-processing system.

Datacube 4 Dearborn Road, Peabody, MA 01960. Produces image-processing system and supplies single-board frame buffers, processors, etc.

Eigen Video P.O. Box 848, Nevada City, CA 95959. Produces image processors and video disk recorders.

E. Leitz, Inc. Rockleigh, NJ 07647. Produces DADS and TAS image-analyzing systems.

Hughes Aircraft Company Image and Display Products, 6155 El Camino Real, Carlsbad, CA 92008. Produces or markets image-processing devices.

Interactive Video Systems, Inc. 45 Winthrop Street, Concord, MA 01742. Produces Image-I, an interactive, real-time, video image-processing system incorporating programs developed for microscopy by Ted Inoué of Universal Imaging Corporation.

Matrox Electronic systems, Ltd. 5800 Andover Avenue, T.M.R., Montreal, Quebec H4T 1H4, Canada. Supplies single-board real-time image-processing system.

MCI/LINK Link Systems (USA), Inc. 3290 West Bayshore Road, Palo Alto, CA 94303. Markets Crystal, Intellect 100, and other digital image processors.

Megavision, Inc. P.O. Box 60158, Santa Barbara, CA 93160. Produces 1024 XM high-resolution image-processing system.

Mercury Computer Systems 600 Suffolk Street, Lowell, MA 01854. Produces ZIP image-processing library and hardware.

Microtex Digital Imaging 80 Trowbridge Street, Cambridge, MA 02138. Produces digital image-processing system, single-board subsystems, and microchannel intensifier systems.

Motion Analysis Corporation 1211 North Dutton Avenue, Suite E, Santa Rosa, CA 95401. Produces Expert Vision system for motion analysis.

Optronics International. Inc. 7 Stuart Road, Chelmsford, MA 01824. Produces high-resolution image-scanning and analysis system.

PEL Picture Element Limited, 635 Waverley Street, Palo Alto, CA 94301. Produces digital image-processing and storage devices.

Perceptive Systems, Inc. 3655 Figueroa Street, Glendale, CA 91206. Produces image-processing systems with software for microscope image analysis and karyotyping, developed by K. R. Castleman.

Photonics Microscopy, Inc. 2625 Butterfield Road, Suite 204-S, Oak Brook, IL 60521. Markets the Hamamatsu Photonic microscopy system developed in consultation with R. D. Allen.

Pie Data Medical Riviera Beach, FL 33404. Markets VIP and VIS X-ray video image-processing systems.

Wyndham Hannaway and Associates 3038 Ninth Street, Boulder CO 80302. Produces innovative image-processing software and custom-fabricated control equipment for video microscopy.

TABLE 2. Some Magazines That Report on Developments Related to Digital Image Processing

Computer Graphics World 1714 Stockton Street, San Francisco, CA 94133.

Computers and Electronics (Ziff–Davis Publishing Company, 3460 Wilshire Boulevard, Los Angeles, CA 90010.

EITV (Education and Industrial TV) C. S. Tepfer Publishing Company, Inc., 51 Sugar Hollow Road, Danbury, CT 06810.

High Technology High Technology Publishing Corporation, P.O. Box 2810, Boulder, CO 80322.

Photomethods (Including "Imaging Technology in Research and Development") P.O. Box 5860, Cherry Hill, NJ 08034.

TABLE 3. Selected Trade Magazines Relevant to Digital Image Processing

Broadcast Engineering P.O. Box 12902, Overland Park, KS 66212.
Digital Design (since the May 1985 issue, incorporates *Electronic Imaging*) Morgan–Grampian Publishing Company, 1050 Commonwealth Avenue, Boston MA 02215.
Laser Focus/Electro-Optics Penwell Publishing Company, 119 Russell Street, P.O. Box 1111, Littleton, MA 01460.
Lasers and applications 23717 Hawthorne Boulevard, Suite 306, Torrance, CA 90505. (See especially annual "Designers' Handbook and Product Directory")
Photonics Spectra P.O. Box 1146, Pittsfield, MA 01202.
Video systems Intertec Publishing Corporation, 9221 Quivira Road, P.O. Box 12901, Overland Park, KS 66212.

applications of video and related approaches to light microscopy. Also an extensive volume, *Applications of Fluorescence in the Biomedical Sciences* [D. L. Taylor, A. S. Waggoner, R. F. Murphy, F. Lanni, R. R. Birge (eds.), Liss, New York (1986)], based on an international conference held at the Universtiy of Pittsburgh in April 1985 covers Fluorescence Probes and Spectroscopy, Imaging and Quantitative Fluorescence Microscopy, and Flow Cytometry.

An excellent two-day course on digital image processing is offered two to three times a year by Ken Castleman and Don Winkler. The lectures are based on their extensive experience with image processing at NASA, and provide considerable emphasis on applications in microscopy. (Perceptive Systems, Inc., 5231 Whittier Oaks, Friendswood, Tex. 77546.) Week-long short courses on advanced microscopy, emphasizing hands-on experience with the latest microscope and video equipment, are offered by myself, R. D. Allen, and others at the Marine Biological Laboratory. (Admissions Officer, MBL, Woods Hole, Mass. 02543.)

Supplemented by information from these sources, I hope the chapters collected here will imbue the reader with enthusiasm, and provide the background needed to adopt the powerful techniques of video microscopy. Perhaps some of our readers will even find new ways to harness the evolving powers of video to reach further into the dynamic, submicrometer world.

Appendixes

I

Spatial Filtration for Video Line Removal

GORDON W. ELLIS

The growing popularity in microscopy of video recording and image-processing techniques presents users with a problem that is inherent in the method—horizontal scan lines. These lines on the monitor screen can be an obtrusive distraction in photographs of the final video image.

Care and understanding in making the original photograph can minimize the contrast of these lines. Two simple, but essential, rules for photography of video images are: (1) Use exposures that are multiples of the video frame time (1/30 sec in the USA). An exposure time less than this value will not record a completely interlaced image.* (2) Adjust the V HOLD control on the monitor so that the two fields that make up the frame are evenly interlaced. Alternate scan lines should be centered with respect to their neighbors (a magnifier is helpful here).†

Following these rules will often result in pictures in which the scan lines are acceptably unobtrusive without recourse to further processing. If the subject matter is such that the remaining line contrast is disturbing, Inoué (1981b) has described a simple technique that can often yield satisfactory results using a Ronchi grating. However, on occasion, when important image details are near the dimensions of the scan lines, the slight loss in vertical resolution resulting from this diffraction-smoothing method may make it worth the effort to remove the lines by *spatial filtration*.

The technique of spatial filtration, pioneered by Maréchal, is described in many current optics texts. A good practical discussion of these techniques is found in Shulman (1970). To produce the pictures for the present article, I used a somewhat simpler apparatus than the optical correlator illustrated in Shulman (see also Walker, 1982).

In the present instance, the apparatus consists of: a monochromatic light source; two surplus aerial camera lenses (Kodak Aero-Ektar 12-inch f/2.5); an enlarger's negative carrier; and a 35-mm single-lens reflex camera with an extension bellows and a long-focal-length lens (in this case an ancient 7-inch Kodak Rapid-Rectilinear objective borrowed from a Kodak Model 3A Folding Pocket Camera, ca. 1910). The camera must have a plain ground-glass focusing screen with a clear central disk bearing a cross for parallax checking. I use a Nikon F with a type "C" screen.

The light source for the optical processor is a vibrating light-fiber conveying light from an argon ion laser (Ellis, 1979). The source does not need to be a laser, but the requirements are easily met by the laser source described and it was available. The requirements are: high intensity, monochromaticity, and small size (but not too small). A 100-watt HBO 100 superpressure mercury arc has about the right source size. If fitted with a heat filter and a 546-nm interference filter and

*See also page 426 regarding minimum suggested exposure time.

†See footnote * on page 256.

GORDON W. ELLIS • Department of Biology, University of Pennsylvania, Philadelphia, Pennsylvania 19104

housed in a suitable box (for safety as well as stray light control) without a lens, it would provide a more economical source for a dedicated optical processor.

The two Aero-Ektars are mounted on a common axis with the light source and the camera. The light source is at the entrance pupil (film plane) of the first Aero-Ektar, which is face to face with its mate and about 13 cm away. The negative carrier is mounted between them about 10 cm from the second lens. Ideally, these lenses would be mounted with their front focal planes coinciding with each other and the film carrier, but this is not a necessary condition for this purpose. The closer spacing was dictated by space limitations. The spatial filter is mounted on axis at the back focal plane of the second Aero-Ektar and is carried on the barrel of the 35-mm camera's lens. The camera is focused on the object transparency in the negative carrier.

The action of the spatial filter is selectively to prevent the light diffracted by the scan lines in the transparency from reaching the camera film plane where they would produce line images by interference with the undiffracted zeroth order. The simplest spatial filter that would eliminate the lines (as in the classic Abbe–Porter experiments) is an iris diaphragm closed down to the point that the first orders of the line diffraction pattern are just blocked. At first glance, this might seem the optimum filter for this purpose; however, video camera modulation transfer functions are not necessarily the same for vertical and horizontal detail. In fact, today's instrumentation cameras typically offer higher resolution in the horizontal direction (parallel to the scan lines) than can be achieved vertically within the constraints of the standard (USA) 525-line format. Consequently, in front of the camera lens, I use a glass plate bearing opaque spots about 50% larger than the light source image and spaced to occlude all the diffracted orders from the scan lines while leaving the zeroth order and most of the aperture unobstructed. Then, to reduce stray light, I close the lens iris down to the level of the second orders of the diffraction pattern.

Two accessories simplify adjusting the processor. They are an auxiliary focusing magnifier (actually a telescope), used to aid parallax focusing, and an aperture-viewing magnifier. The latter is

FIGURE I-1. Original.

used to examine the exit pupil of the camera's viewfinder, which provides a clear and magnified image of the filter and the diffracted light that it must be positioned to block.

Photographically, there are several options in going from the original lined photograph to the filtered negative for the final print. One could, in principle, treat the optical processor as a special giant enlarger and project unlined prints from lined negatives. Unfortunately, this would move the printing easel well into the next room, in addition to requiring impractically long exposures. Otherwise, the shortest path in steps, though not necessarily in time, is to photograph the spatially filtered image on SO-185 Rapid Processing Copy film (EKC). Because this a self-reversing film, the developed SO-185 image is a filtered negative that can be used in an ordinary enlarger. I have found that this works well using the 488-nm line of the argon ion laser, but is about 10,000 times slower using the 546-nm light from a 100-W mercury arc, and will not work at all with a helium–neon laser. An equally direct alternative I have not yet tried is to use SO-185 to make the original photograph of the video monitor. The resulting positive transparency could then be filtered in the optical processor and photographed on Panatomic-X, Plus-X, or other panchromatic film to produce the filtered negative. Using a CRT with a highly actinic phosphor might make this a practical approach. Conventional monitors would probably require excessively long exposures.

For the photographs shown, I have followed a more complex path, but one that allows use of panchromatic film and short exposures. The original negative was exposed on Pan-X at 1/30 sec. This negative was copied onto Kodak Technical pan and developed in 1 : 3 Microdol-X to produce a long-scale positive transparency for use in the processor. The filtered negative was then made on Pan-X. Here the exposure was 1/125 sec using 0.5 W from the 514-nm argon laser. Figure I-1 is a print from the original negative. The photograph does not show optimal adjustment of the interlace, but instead represents a less favorable case to demonstrate the power of the filtering technique. The result, the spatially filtered print in Fig. I-2, is entirely free of the video scan lines. Note the subjective impression that the delined print looks sharper than the original.

FIGURE I-2. Spatially filtered reproduction.

The subject of the video micrograph is a diatom frustule that is one of 50 on an arranged slide from Turtox (No. B1.434). It is viewed in polarized light through a 43/0.65-NA Leitz stress-free achromat and displayed on a 9-inch Panasonic monochrome monitor by a Dage–MTI Model 65 camera equipped with a Newvicon camera tube. Magnification of the image on the monitor was 1980× and covers 93 μm horizontally, edge to edge.

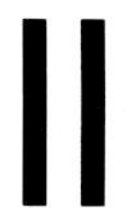

Modulation Transfer Function Analysis in Video Microscopy

ERIC W. HANSEN

II.1. INTRODUCTION

The modulation transfer function (MTF) is a powerful tool for the analysis of imaging systems. It describes in quantitative terms the relationship of object and image, and provides a convenient framework for analyzing the performance of complex imaging systems, which may combine optical and electronic components. This appendix first presents a brief tutorial introduction to the MTF, and then illustrates its use with an analysis of a typical video microscope system.

II.2. MODULATION TRANSFER

To begin, let us consider the imaging system shown schematically in Fig. II-1. An object with a sinusoidal spatial variation of intensity, $I_o(x_o)$ along the x_o axis, is imaged with unity magnification, producing a sinusoidal image intensity $I_i(x_i)$. (The subscripts o and i denote object and image, respectively.) The peak-to-peak distance along the sinusoid* is its spatial period; the reciprocal of the period is the spatial frequency, f_x, expressed in units of cycles per unit length (e.g., cycles/mm). In general, we may observe three things about the image. First, the spatial frequency of the image is the same as that of the object. Second, the contrast, or modulation, defined as

$$\text{modulation} = (I_{max} - I_{min})/(I_{max} + I_{min}) \qquad \text{(II-1)}$$

is typically less in the image than in the object. Third, the object is centered about $x_o = 0$, but the image is displaced by an amount Δx_i.

If we image several such test objects with different spatial frequencies, we will find that the image modulation and position shift will vary as a function of spatial frequency. The *modulation transfer function*, $M(f)$, is defined

$$M(f) = \text{image modulation/object modulation} \qquad \text{(II-2)}$$

The MTF expresses the way the imaging process alters the contrast of a sinusoidal object, as a function of spatial frequency. The position shift Δx_i is equivalent to a *phase* shift $\Delta\phi_i$ of the

*The sinusoid is a sine or cosine function.

ERIC W. HANSEN • Thayer School of Engineering, Dartmouth College, Hanover, New Hampshire 03755

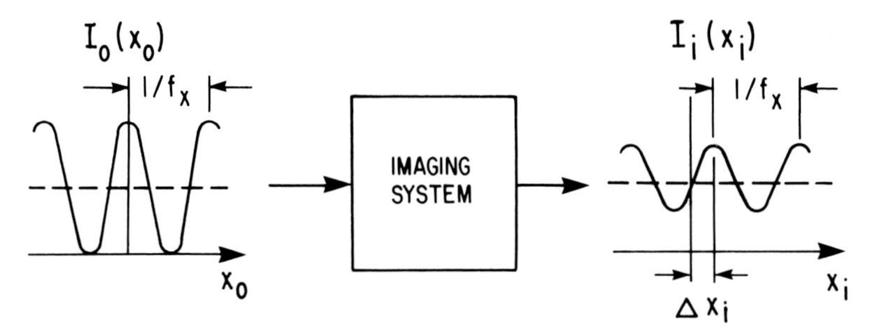

FIGURE II-1. Imaging a sinusoidal test object.

sinusoid; we relate the two by noting that one period of the sinusoid (a distance $X_i = 1/f_x$) corresponds to 2π radians of phase. Hence, we have the proportion

$$\Delta\phi_i/2\pi = \Delta x_i/(1/f_x) = f_x\Delta x_i$$

from which it follows that $\Delta\phi_i = 2\pi f_x \Delta x_i$. We may define a *phase transfer function* [or system phase response, $\phi(f_x)$, as a function of frequency] by

$$\phi(f_x) = \Delta\phi_i = 2\pi f_x \Delta x_i(f_x) \tag{II-3}$$

In mathematical terms, the intensity of the sinusoidal test object as a function of position is

$$I_o(x_o) = a + b\cos(2\pi f_x x_o)$$

and that of its image is

$$I_i(x_i) = a + M(f_x)\, b\cos[2\pi f_x x_i + \phi(f_x)] \tag{II-4}$$

For notational simplicity, we shall consider one-dimensional systems whenever possible; it should be remembered, however, that in an actual imaging system the MTF and phase response depend on spatial frequencies in both the horizontal and vertical directions, f_x and f_y. This is especially true in video, where the raster scanning process causes the horizontal and vertical MTFs to differ.

An alternative definition of modulation transfer, used in the television industry to evaluate video components, is based on the response to square-wave (bar) patterns, rather than sinusoids. Using the Fourier series, whereby a square wave is expressed as a sum of sinusoids, the square-wave response $C(f)$ (also called the contrast transfer function, CTF) is approximately related to the MTF $M(f)$ by the equation (Schade, 1975, p. 6)

$$M(f) = (\pi/4)\,[C(f) + C(3f)/3 - C(5f)/5 + C(7f)/7 + C(11f)/11 - C(13f)/13] \tag{II-5}$$

It can be shown that, except at very low spatial frequencies, the square-wave response and the MTF are virtually identical, and we shall use them interchangeably.

The MTF and phase response are sometimes combined into a single mathematical quantity, called the *optical transfer function* (OTF), denoted $H(f)$:

$$H(f) = M(f)\exp[j\phi(f)]$$

where j is the (imaginary) square root of -1, and exp[.] is the exponential function. The MTF and phase response are the modulus (or magnitude) and phase, respectively, of this complex-valued function.

We proceed from the simple sinusoidal object to real objects via Fourier theory, which says that an arbitrary object may be expressed as a combination of sinusoids of various frequencies, amplitudes, and phases. The imaging system is understood as weighting and shifting the sinusoidal components of the object according to their spatial frequencies. A good example is the ''softening'' of sharp edges by a low-NA system. The sharpness of the edge is due to the presence of high-spatial-frequency components. These are attenuated relative to the lower-frequency components by the MTF, which decreases in value to zero as spatial frequency is increased. The loss of sharpness (reduced resolution) reflects the general ''low pass'' nature of the MTF.

The phase response of an *ideal* imaging system is *linearly* proportional to frequency. Thus:

$$\phi(f_x) = \phi_o f_x \tag{II-6}$$

The relationship between phase and position (II-3) shows that a linear phase results in a position shift that is *independent* of frequency:

$$\Delta x_i = \Delta\phi_i/2\pi f_x = \phi_o f_x/2\pi f_x = \phi_o/2\pi$$

All the sinusoidal components of the image are displaced the same amount, hence the image merely undergoes a position shift, without degradation of image quality. If the phase deviates from this ideal linear characteristic, then some sinusoidal components will be shifted more than others, and a degradation of the image will result. Interestingly, *phase is often more important than magnitude in preserving image fidelity* (Shack, 1974; Oppenheim and Lim, 1981). In an aberration-free optical system with a symmetric aperture (e.g., round and centered on the optical axis), the phase transfer function is identically zero for all spatial frequencies in all directions, and only the MTF is important. However, the electronics in a video system possess less ideal phase characteristics, and can noticeably degrade image quality.

The MTF model of imaging is particularly valuable for analyzing multicomponent systems, e.g., a microscope followed by a video camera and a monitor. The system MTF is simply the *product* of the MTFs of the individual components:

$$M_{sys}(f) = M_{mon}(f) \times M_{cam}(f) \times M_{mic}(f) \tag{II-7}$$

The system phase response is the *sum* of the individual phases:

$$\phi_{sys}(f) = \phi_{mon}(f) + \phi_{cam}(f) + \phi_{mic}(f) \tag{II-8}$$

It is important to note, however, that (square wave) contrast transfer functions do not multiply in this way, except for those spatial frequencies where the MTF and CTF are essentially identical.

The optical transfer function is essentially determined by two things: the entrance and exit pupils of the optics (e.g., the condenser and objective NAs in a microscope), and aberrations in the optical elements, which are considered to introduce phase errors in the optical fields. Even in the absence of aberrations, diffraction at the pupils will limit the spatial frequency response, and hence the spatial resolution of the instrument. [Figures 5–24, 5–25; see also pp. 318–324 of Smith (1966) for graphs of MTFs under various conditions.] Since the effects of diffraction establish the ultimate limits to resolution, a perfectly aberration-free instrument is termed *diffraction limited*. The MTF of a diffraction-limited microscope with circular pupil, under conditions of incoherent illumination ($NA_{cond} \simeq NA_{obj}$), is shown in Fig. II-2. Image contrast is highest for low spatial frequencies, and decreases to zero as spatial frequency increases. The frequency at which contrast finally goes to zero is called *cutoff*, and is related to system parameters by the formula

$$f_c = 2NA_{obj}/\lambda$$

where λ is the wavelength of light. This expresses, in spatial frequency terms, the well-known fact that resolution increases with larger NAs and shorter wavelengths.

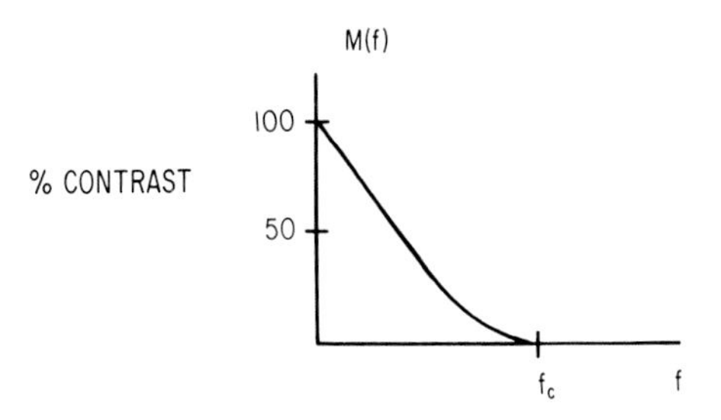

FIGURE II-2. Modulation transfer function of a diffraction-limited microscope with circular pupil, incoherent illumination.

II.3. THE MTF AND THE DIFFRACTION PATTERN

The MTF is closely related to a second quantity, the *point-spread function* (PSF). The PSF is simply the image of a point object (Fig. II-3). Due to aberrations and diffraction at apertures, a point object is never perfectly imaged, in the geometric optics sense, but is spread into a finite-sized distribution of intensity. The image of an arbitrary object is a superposition of PSFs, and appears blurred in proportion to the spatial extent of the PSF. In microscopy, the PSF is more commonly referred to as the ''diffraction pattern'' (Section 5.5); for a diffraction-limited instrument, the PSF is the well-known Airy pattern.

To completely understand the connection between the MTF and PSF requires Fourier theory. [The mathematically inclined reader is referred to the excellent books by Bracewell (1978), Goodman (1968), and Castleman (1979) for details.] However, an intuitive appreciation of this significant relationship is easily obtained. In Abbe's theory of the microscope, the back aperture determines the highest spatial frequency passed by the system; as we saw in the previous section, this is the cutoff frequency of the MTF, which is proportional to NA. The other well-known measure of resolution—Rayleigh's two-point criterion—is based on the narrowness of the PSF. Two adjacent points in an object are spread by the optics into two PSFs. The narrower the PSF, the closer the points can be to one another and still be resolved by the unaided eye. Taken together, the Abbe and Rayleigh statements imply that having a *wide* MTF (high-spatial-frequency ''bandwidth'') is the same as having a *narrow* PSF (fine spatial resolution). In the language of Fourier theory, the MTF (more correctly, the OTF) and the PSF are each other's *Fourier transform.* A fundamental property

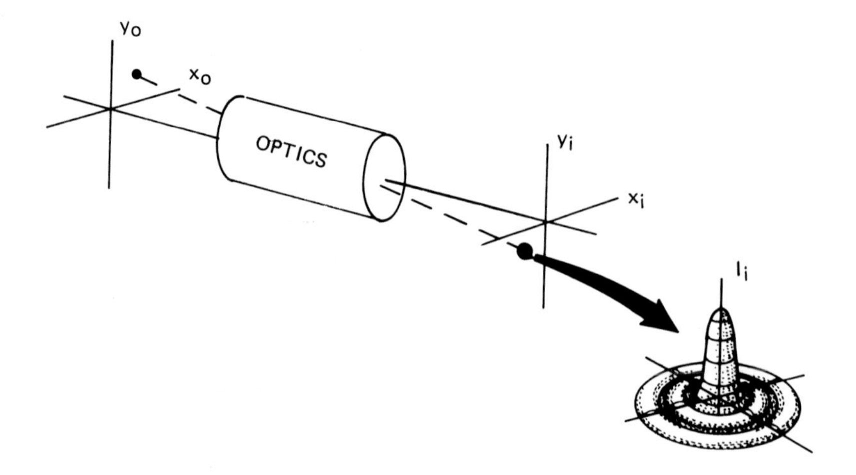

FIGURE II-3. The point-spread function is the image of a point object.

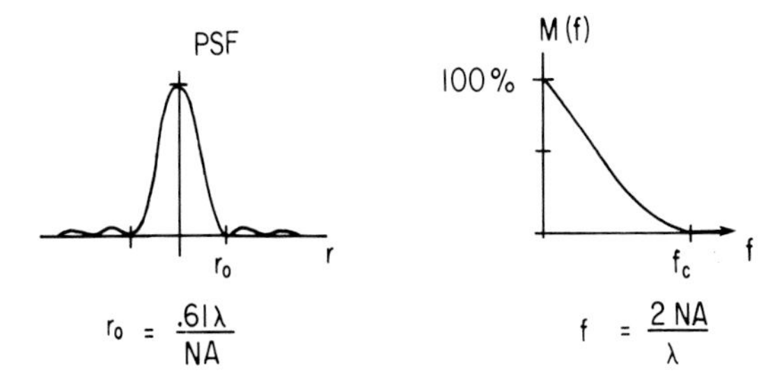

FIGURE II-4. Relationship between diffraction-limited MTF and point-spread function.

of Fourier transforms is that the width of a function is inversely proportional to the width of its transform (Section 10.12).

This relationship is illustrated, for a diffraction-limited incoherent system with circular aperture, in Fig. II-4. The cutoff frequency of the MTF is

$$f_c = 2\mathrm{NA}/\lambda$$

and the radius of the first dark ring of the PSF (Airy disk) is

$$r_o = 0.61\lambda/\mathrm{NA}$$

We observe that f_c and r_o are indeed inversely proportional.

An alternative definition of resolution is provided by the **Sparrow criterion.** Unlike the Rayleigh criterion, two points separated by the Sparrow distance are barely resolved, if at all; however, any separation beyond the Sparrow distance produces a minute dip in intensity between the two points, which may be visualized using contrast enhancement techniques. With video contrast enhancement, resolution beyond the Rayleigh limit and close to the Sparrow limit has been achieved in practice. The Sparrow distance for a diffraction-limited system is $0.5\lambda/\mathrm{NA}$, precisely the inverse of the MTF cutoff frequency.

This same reciprocal relationship between spatial resolution and spatial frequency response carries over to analog electronic systems, where the PSF is replaced by the time response to a very short electrical impulse, and the OTF is replaced by the system's response, in magnitude and phase, to a sinusoidal electrical signal. The shapes of the impulse response and transfer function are quite different from PSFs and OTFs; in particular, the impulse response lacks the symmetry of an optical PSF, so there will be nonnegligible phase effects, which are seldom linear. However, the underlying concepts are the same, and this permits combined optical–electronic systems to be analyzed within a common framework.

II.4. SAMPLING AND DIGITIZED VIDEO

In video image processing, sampling comes into play in two ways. First, the television raster scan samples the image vertically, converting the two-dimensional scene into discrete rows—the raster lines. Second, when a video image is converted to digital form for computer processing, each row of the video image—each horizontal scan—is converted to a set of discrete sample values, called picture elements, or *pixels*. The sampling process does not alter the MTF of the imaging system *per se*. However, the question of how many samples are necessary to accurately represent an object imposes a limit on the spatial frequency content of the image, and so it is appropriate to discuss it here.

Figure II-5 shows several sampling situations. The objects consist of alternating light and dark bars of various spatial frequencies. The samples are regularly spaced in position, and are indicated

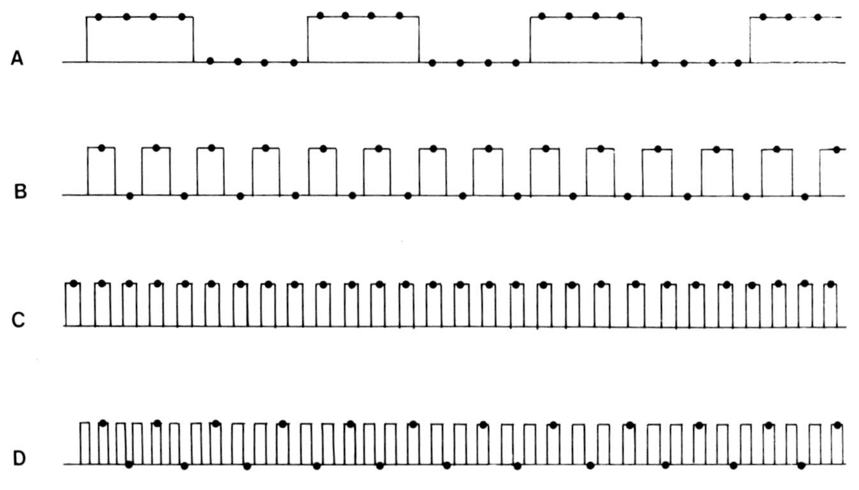

FIGURE II-5. Sampling a bar pattern.

by the heavy dots. In Fig. II-5a there are four samples per bar; the original pattern is easily reconstructed from the samples. In Fig. II-5b the spatial frequency of the pattern has increased, so there is only one sample per bar, but the original pattern will still be reconstructable from the samples. In Fig. II-5c the spatial frequency has increased further, to where the dark bars fall between the samples, and all the samples are light. Based on the samples, one would be led to conclude that the object was uniform in intensity rather than a fine bar pattern. Evidently the spatial frequency information necessary to reconstruct the pattern has been lost in the sampling process. Finally, Fig. II-5d shows the case of one sample every third bar. The samples alternate light and dark, just as in Fig. II-5b, and would be recontructed as a bar pattern of *lower* spatial frequency. This is the phenomenon of *aliasing,* so called because a higher spatial frequency masquerades as a lower one. It can be a serious source of error in digitized images.

It is evident from this example that sampling establishes an upper limit to the spatial frequency content of an image. From Fig. II-5, it may be concluded that this limit is one sample per bar, or two per cycle. This same limit may be established by more rigorous mathematical arguments based on Fourier theory and sinusoidal rather than bar patterns. The fundamental conclusion, however, is the same: to recover an image from its samples, the sample density must be at least two per cycle of the highest spatial frequency present in the image. Turning this around, if a digital image memory consists of 512 × 512 pixels, the image must be limited in both directions to 256 cycles (line pairs), or 512 lines of resolution. Likewise, the raster scan of a video camera imposes a vertical spatial frequency limit around 240 cycles, or 480 lines. For a variety of practical reasons, however, it is advisable to stay below the upper limit and "oversample" by a factor of two or so (see Figures 6-15, 6-16).

II.5. SYSTEMS ANALYSIS OF A VIDEO MICROSCOPE

To illustrate the power of MTF methods in systems analysis, we consider a practical example—a microscope with video camera and digital frame memory.

We assume the microscope is diffraction limited, and incoherent ($NA_{cond} \simeq NA_{obj}$). The MTF, shown in Fig. II-2, has the mathematical description:

$$M_{mic}(f) = \begin{cases} \frac{2}{\pi}\left[\cos^{-1}(f/f_c) - (f/f_c)\sqrt{1-(f/f_c)^2}\,\right] & \text{for } f \leq f_c \\ 0 & \text{for } f > f_c \end{cases} \qquad \text{(II-9)}$$

The cutoff frequency, f_c, is given by

$$f_c = 2NA_{obj}/\lambda$$

For a 40×/0.95-NA objective in green light (546 nm), the spatial frequency cutoff is 3480 cycles/mm; the Rayleigh resolution distance is 0.35 μm.

Resolution for the video camera is considered separately for the vertical and horizontal directions. The limit to vertical resolution is the number of lines in the raster scan, which we denote N_V. For conventional television, N_V is around 480. Horizontal resolution is limited by the dimensions and profile of the scanning electron beam. It is expressed as a number (here denoted N_H) of TV lines per picture height (TVL PPH), which is interpreted as the number of resolvable lines in a horizontal distance equal to the height of the picture. Since a conventional television image has a 4 : 3 aspect ratio (width : height), the *actual* number of resolvable lines in the horizontal direction is 4/3 N_H. (Recall that in video terminology, two "lines" comprise one "cycle" or "line pair" in optical terminology.)

MTF curves for vidicon tubes are approximately Gaussian. They are measured with square-wave test objects, with the response to a fully modulated 15-line object taken to be maximum contrast (unity MTF). Two critical frequencies typically appear in camera specs—the limiting resolution, and a midrange value, which may be either the point of 50% contrast or the percent contrast at 400 lines. The Fourier transform of a Gaussian function is itself a Gaussian, so the (horizontal) MTF may be modeled by a Gaussian function:

$$M_{cam}(f) = \exp(-0.7f^2/f_o^2) \quad \text{(II-10)}$$

where f_o = 50% contrast frequency (TVL PPH). Note that when $f = f_o$, we have $M_{cam}(f_o) = \exp(-0.7) = 0.496$, approximately 50%. This model is a very good fit to the published MTF curves for most common vidicon-type tubes. The MTF of a type S4076 1-inch Newvicon tube (Panasonic) is shown in Fig. II-6. Typical limiting resolution is 800 lines, with 50% contrast at 370 lines. There are about 480 active lines vertically, and 4/3 × 800 = 1067 (barely) resolvable lines across the full width.

It should be kept in mind, though, that the camera tube is only one component of the camera system. The camera MTF is significantly affected by the operating voltages supplied to the tube, the characteristics of the magnets that help scan and focus the electron beam, and the frequency response of the electronic circuits that amplify the video signal from the tube. In some high-performance cameras, the resolution may be much better than that predicted by the tube specifications. In addition, tube-to-tube variability may cause two "identical" cameras to differ in performance. Nevertheless, the data provide helpful indicators of average behavior, and are useful for the illustrative example that follows.

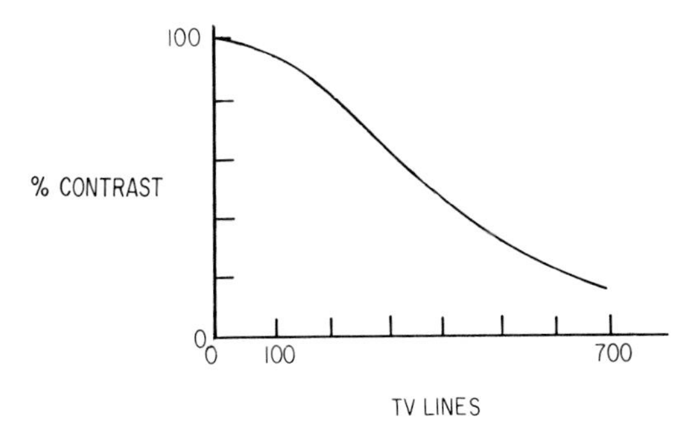

FIGURE II-6. MTF of Panasonic S4076 1-inch Newvicon. (See also Fig. 7-9C.) Information courtesy of Matsushita Electronics Corp., Takatsuki, Japan.

The simple model shown in Fig. II-7 is used to evaluate the combined MTF of the microscope–camera system. The specimen is imaged via the microscope optics (magnification M_{obj}) to the primary image plane, and perfectly transmitted to the vidicon camera face, with magnification M_{rel}, by the relay optics, which may include an ocular, a low-magnification relay such as a Zeiss Optovar, and/or a zoom lens.

The microscope MTF is described in terms of cycles per millimeter of resolution at the specimen plane, while the camera MTF is expressed in terms of lines per picture height at the camera face. The two planes are related by the net magnification, $M_{obj}M_{rel}$. A spatial frequency f_{sp} at the specimen plane appears as a lower frequency $f_{sp}/M_{obj}M_{rel}$ at the camera target.

The active scanned area of the vidicon target has a 4 : 3 aspect ratio with a diagonal measurement D, therefore its dimensions are $0.8D \times 0.6D$. In a "1-inch" tube, $D = 0.625$ inch = 15.875 mm. The camera field has N_V lines of resolution vertically and $\frac{4}{3} N_H$ lines horizontally, thus there are $N_V/0.6D$ lines/mm vertically and $\frac{4}{3} N_H/0.8D = N_H/0.6D$ lines/mm horizontally. (The fundamental relationship is the same for horizontal and vertical.) Dividing by 2 to convert from lines to cycles (line pairs) gives the key expression relating spatial frequency at the specimen plane, f_{sp}, to lines of resolution, N (in either direction), at the camera face:

$$f_{sp} = M_{obj}M_{rel}N/1.2D \tag{II.11}$$

Substituting the Newvicon resolution limits, $N_H = 800$ and $N_V = 480$, into this expression, and assuming a 40×, 0.95-NA objective with unity magnification relay, we find that the maximum (specimen) spatial frequencies that will be passed by the tube are $f_H = 1680$ cycles/mm and $f_V = 1009$ cycles/mm. This is considerably less than the cutoff of 3480 cycles/mm computed earlier for the microscope, and demonstrates that significant resolution is lost in the video camera.

The solution to this problem is to magnify the image ($M_{rel} > 1$), reducing the effective field size and increasing the "density" of TV lines in the field. Twofold magnification with increase f_H to 3360, which nearly matches the microscope's cutoff. The field diameter is reduced twofold. However, this will still not recover all the microscope's resolution, since the *system* MTF, which is the product of the microscope and camera MTFs (equations II-9 and II-10), is depressed below the diffraction-limited curve by the camera response. Figure II-8 shows the system MTF plotted for various magnification factors. The graphs show that M_{rel} of between 4 and 8 is required for the microscope–camera system to approach the diffraction-limited performance of the microscope alone.

A similar analysis may be applied to digitization of the video signal. A typical frame memory is 512×512 pixels, and if aliasing is to be avoided, spatial frequencies in the image cannot exceed the equivalent of 512 lines. This will always be met in the vertical direction, since $N_V = 480$, but it may

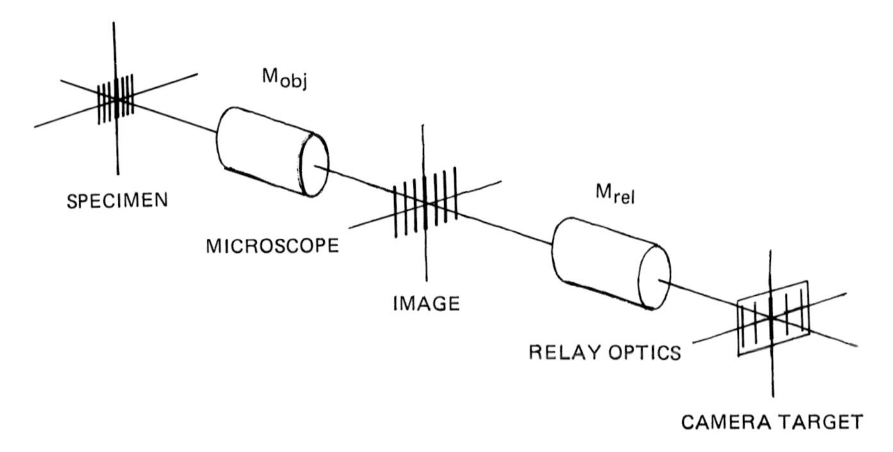

FIGURE II-7. Model for analysis of microscope–camera system.

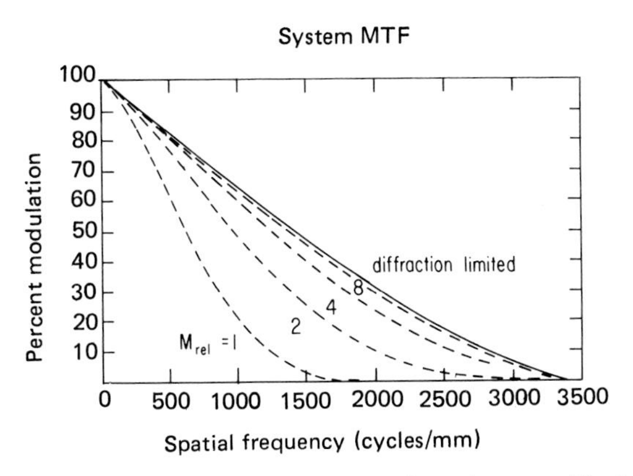

FIGURE II-8. MTF curves of a microscope–camera system, for various magnifications in the relay optics.

constrain the horizontal direction. To determine the horizontal limits, we use equation (II-11); letting $f_{sp} = 3480$, the spatial cutoff of the microscope optics, and $N = 512$, we obtain a value $M_{rel} \approx 5$. Since $M_{rel} \geq 8$ is required to achieve full resolution, we are not constrained by the digitizer in this example.

Likewise, we may include the effects of a video tape recorder on the system MTF. The (horizontal) resolution limit of a monochrome unit is typically 320 lines PPH. (Color recorders are limited to 240 lines, and many have this limit even when operating monochrome, if the color circuits are not bypassed.) Again using equation (II-11), at the specimen plane, the 320-line cutoff corresponds to

$$f_H = \frac{40 \times 320 \times M_{rel}}{1.2 \times 15.875} = 672 M_{rel} \text{ cycles/mm}$$

A 5× magnification will match the cutoff of the VTR to the optics, but more will probably be required to completely overcome the recorder's limitations depending on its MTF curve.

A further consideration is image brightness, which is proportional to the area of the field. If the field diameter is reduced by M_{rel} to improve resolution, the image brightness will go down by a factor of M^2_{rel}. In some situations this may be unacceptable, and the microscopist will be forced to trade resolution for brightness.

This example is illustrative of the insight into system behavior that can be gained by MTF analysis. The method, expressions, and plots are sufficiently general to allow other microscope–camera combinations to be analyzed. In any actual experiment, the "best" system configuration will depend on the specimen and the goals of the experiment, as well as the instrumental characteristics. The results of these analyses, e.g., the relay magnification, should therefore be regarded as nominal design points about which the microscopist may search for the best overall imaging performance.

An Introduction to Biological Polarization Microscopy*

III.1. INTRODUCTION

Electron micrographs show fine structural detail and organization in thin sections of fixed cells. Time-lapsed motion pictures show the continually changing shape and distribution of organelles in living cells. How can we combine these observations and follow the changing fine structure in a single living cell as it undergoes physiological activities or experimental alterations?

The relevant dimensions are too small for regular light microscopy since many of the changes of interest take place in the 0.1 to 100 nm range, i.e., the size range of atoms to macromolecular aggregates. From wave optics, the resolution of a light microscope is limited by its NA and the wavelength of light according to equation (5-9, p. 115).† On the other hand, the electron microscope with its very high resolution, practically 1–2 nm for many biological specimens, does not usually permit nondestructive study of cells.

One effective way of closing this gap is to take advantage of anisotropic optical properties, which, albeit at a rather low level, are often exhibited by cellular fine structures. Anisotropic optical properties, such as birefringence and dichroism, measure molecular and fine structural anisotropy as we shall soon see. These properties can be studied *nondestructively* in individual living cells with a sensitive polarizing microscope, so that the time course of fine structural changes can be followed in small parts of a single cell.

From equation (5-9), the smallest distance resolvable with any microscope with a 1.40-NA objective (with a nearly matched NA condenser at $\lambda = 550$ nm) is $\simeq 0.2$ μm, or 200 nm. With a sensitive polarizing microscope, we can determine molecular alignment or changes in fine structure taking place within each resolvable unit area that is smaller than 0.2×2 μm, or less than 1 μm^2.

*Revision of a manuscript originally prepared for a NATO Advanced Summer Institute held at Stresa, Italy, in 1969, but which was never published. Lectures related to this were delivered to the Physiology course and the short course on Analytical and Quantitative Light Microscopy in Biology, Medicine, and the Materials Sciences at the Marine Biological Laboratory, Woods Hole, Massachusetts, and the University of Pennsylvania. The active discussions and willing cooperation of my students and colleagues guided me in developing these lectures and demonstrations. Several of the illustrations in this appendix were initially prepared by Richard Markley, Dr. and Mrs. Hidemi Sato, and Hiroshi Takenaka. They have been refined and finished by Bob and Linda Golder and Ed Horn of the MBL. Their efforts, and support by NIH Grant 5R01-GM-31617 and NSF Grant PCM 8216301, are gratefully acknowledged.

†This does not mean that we cannot form an image of individual organelles, filaments, or particles that are smaller than half the wavelength of light. If they provide enough contrast, and are separated from other comparable structures by a distance greater than the resolution limit, they can be visualized (by their image expanded to the width of their diffraction pattern; Figs. 1-7, 11-3, 11-5, 11-8).

III.2. ANISOTROPY

Anisotropy relates to those properties of matter that have different values when measurements are made in different directions within the same material. It is the opposite of isotropy, where the property is the same regardless of the direction of measurement. A common example of anisotropy is found in a piece of wood. Wood is much stronger parallel to the grain than across. As the humidity changes, wood shrinks and swells across the grain much more than it does along the grain. Also, wood splits more easily along the grain. In these examples, we might have anticipated the anisotropic properties from the grain structure of wood, but there are other anisotropic properties that are not quite so obvious.

Related to the fact that it is more difficult to stretch wood along the grain than across the grain—that is, the coefficient of elasticity is greater parallel to the grain than across—sound waves travel faster parallel to the grain than across the grain. Tyndall pointed out many years ago that, depending on the type of wood, the velocity of sound may be three times as great when the sound is traveling parallel to the grain than across the grain.

We can demonstrate this phenomenon by taking two blocks of wood cut from a single plank (Fig. III-1). The two blocks are made to the same dimensions but one (P) is cut parallel to the grain and the other (S) across. When we tap on the end of a block, a standing wave is formed. The compressive sound wave travels back and forth quickly, and since the ends are open, a longitudinal standing wave is formed with its loops at the ends of the block and a node in the middle. This standing wave takes on a particular pitch or tone. The tone tells us how quickly the wave is traveling in the block. Parallel to the grain we get one tone. Across the grain we get another. In fact, the velocity of sound across the grain is much lower, so that the tone is quite a bit lower (by an octave in one example; see legend to Fig. III-1). This is a dramatic demonstration of acoustic anisotropy, or anisotropic propagation of sound waves. The demonstration tells us how structural anisotropy can relate to wave propagation.

In this case, the grain structure of the wood was obviously visible, and it is not difficult to grasp the reason for the anisotropy; but anisotropy can exist even if we cannot see heterogeneous structures. The anisotropic properties can lie in the material that appears homogeneous at the microscopic level but which is anisotropic at the atomic or molecular levels.

III.3. SOME BASIC FEATURES OF LIGHT WAVES

Before discussing anisotropy further, we will review some basic physical optics needed in the subsequent discussions.

Light is a form of electromagnetic wave. Assume a capacitor that is charged and let it discharge as a spark through two electrodes as shown in Fig. III-2. In the spark, a current flows

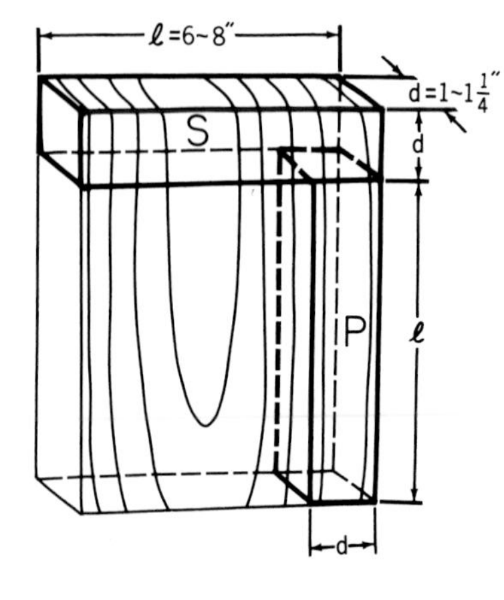

FIGURE III-1. Wooden blocks for demonstrating acoustic anisotropy. Blocks P and S are finished to the same dimensions, but P is cut parallel to the grain while S is cut across grain. The frequency of the standing wave, generated by gently holding onto the middle of the block and tapping (with a steel ball with a flexible handle) from an end, measures the acoustic velocity for the compressive wave that travels through the block. In one example (8-inch-long blocks made of hard oak), P resonated at 4500 Hz and S at 2500 Hz. In this case, the compressive sound wave traveled nearly two times faster parallel to the grain than across.

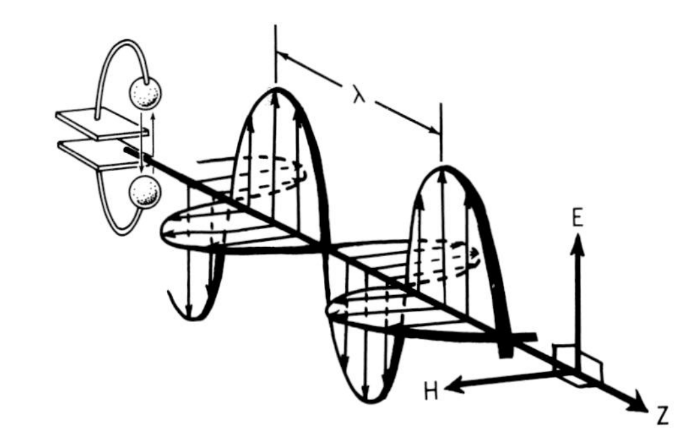

FIGURE III-2. Propagation of electromagnetic wave. The wave is generated by the discharging spark (or oscillating molecular dipole) to the left. The spark current oscillates at a frequency ν characteristic of the circuit. The resultant electromagnetic disturbance is propagated with the electric (E) and magnetic (H) vectors vibrating perpendicular to each other and to the direction of propagation Z. ν is determined by the oscillator, and the wavelength (λ) is given by $\lambda = v/\nu$, where the velocity (v) is given by equation (III-1) or (III-2).*

down for a short time, slows down, but because of the inductance of the circuit, flows back upwards, recharging the capacitor again. The current thus oscillates back and forth at a frequency (ν) that is characteristic of the circuit. The electric current oscillating up and down in the spark gap creates a magnetic field that oscillates in a horizontal plane. A changing magnetic field in turn induces an electric field so that we end up with a series of electrical and magnetic oscillations that propagate as an electromagnetic wave.

As seen in Fig. III-2, the electric field in an electromagnetic wave vibrates with its vectorial force growing stronger and then weaker, pointing in one direction, and then in the other direction, alternating in a sinusoidal fashion. The magnetic field oscillates perpendicular to the electric field, at the same frequency. The electric and the magnetic vectors, which express the amplitude and vibration directions of the two waves, are oriented perpendicular to each other and to the direction of propagation.

From the relationships defining the interaction of the electric and magnetic fields, one can deduce the velocity of the resulting electromagnetic wave. The equations of Maxwell show that the velocity (v) is exactly c (the velocity of light in vacuum = 3×10^{10} cm/sec) divided by the square root of the dielectric constant (ξ) of the medium times the magnetic permeability (μ) of the medium. Thus,

$$v = c/\sqrt{\xi \cdot \mu} \qquad \text{(III-1)}$$

For most materials that occur in living cells, namely for nonconducting material, $\mu = 1$, so that

$$v = c/\sqrt{\xi} \qquad \text{(III-2)}$$

Empirically, we also know that the velocity of light is inversely proportional to the refractive index (n) of the material through which it propagates, i.e.,

$$v = c/n \qquad \text{(III-3)}$$

Equations (III-2) and (III-3) tell us that the refractive index is equal to the square root of the dielectric constant of that material (if the measurements are both made at the same frequency, ν). Thus,

$$n_\nu = \sqrt{\xi_\nu} \qquad \text{(III-4)}$$

This tells us that *optical measurements are, in fact, measurements of electrical properties of the material.* The dielectric properties in turn directly reflect the arrangement of atoms and molecules in a substance.

The direction of interaction between an electromagnetic field and a substance can be consid-

*In the equations, $\xi = \mu = 1$ for a vacuum, and very close to 1 for air.

ered to lie in the direction of the electric vector. That is so whether we consider the electric or the magnetic vectors, since what matters is the effect of the electric or magnetic fields on the electrons in the material medium. (The magnetic field affects those electrons which move in a plane perpendicular to the magnetic field.) From here on, then, we will represent the vibrations of an electromagnetic wave by indicating the direction of the electric field alone. (The magnetic field is, of course, still present perpendicular to the electric field, but we will not show it in our vector diagrams, in order to avoid confusion.)

III.4. OPTICAL ANISOTROPY

The dielectric constant is anisotropic in many substances. Take a cube of such a substance and place it on a Cartesian coordinate, with the sides of the cube parallel to the *X, Y, Z* axes. The dielectric constant for an electric field measured along the *X* axis—that is, ξ measured by placing the plates of a capacitor on the faces of the cube parallel to the *Y–Z* plane (the plane that includes the *Y* and *Z* axes, or perpendicular to the *X* axis)—would have one value, ξ_x. Along the *Y* axis, it would have another value, ξ_y, and along the *Z* axis, a third value, ξ_z.

We now take the square roots of these three dielectric constants and make the lengths of the *X, Y,* and *Z* axes proportional to these values. Then we draw an ellipsoid whose radii coincide with the *X, Y,* and *Z* axes as just defined (Fig. III-3). What we obtain is known as the Fresnel ellipsoid. This ellipsoid describes the dielectric properties measured in all directions in the material. Each radius provides the $\sqrt{\xi}$ value for an electromagnetic wave *whose electric vector lies in the direction of that radius.*

Since $\sqrt{\xi} = n$ from equation (III-4), the Fresnel ellipsoid is, in fact, a refractive index ellipsoid. $\sqrt{\xi_x}$ is the refractive index for waves whose electric fields are *vibrating* in the *X*-axis direction, $\sqrt{\xi_y}$ for waves with electric fields *vibrating* along the *Y* axis, and $\sqrt{\xi_z}$ for waves with electric fields *vibrating* along the *Z*-axis direction.

The value of the refractive index given by the radius of the Fresnel ellipsoid is valid for all waves whose electric vector vibrates in the direction of that radius regardless of the wave's direction of propagation. For example, waves with their electric vector in the *X*-axis direction may be propagating along the *Y*- or *Z*-axis direction. Therefore, given the direction of the electric vector

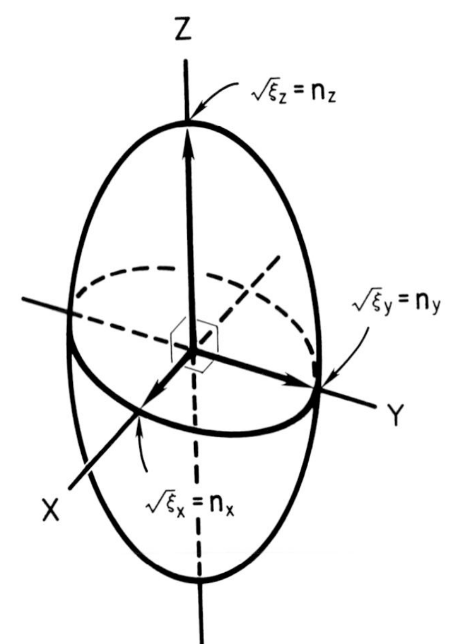

FIGURE III-3. The refractive index, or the Fresnel, ellipsoid. The radius of the Fresnel *ellipsoid* gives the refractive index (n), or the square root of the dielectric constant ($\sqrt{\xi}$), for waves whose electric displacement vectors lie in the direction of the radius of the ellipsoid within an anisotropic medium.

A cross section through the center of the ellipsoid gives the refractive index *ellipse* for waves traveling normal to that section. The major and minor axes of the ellipse denote the refractive indices encountered by the slow and fast waves, which vibrate with their electric displacement vectors along those two axes.

(or more properly, the **electric displacement vector** in the material medium), the refractive index suffered by a wave is defined regardless of its direction of propagation.

III.5. BIREFRINGENCE

How do we use this ellipsoid? If light is traveling along the X-axis direction, for example, then the refractive indices for the waves vibrating in the material are given by an *index ellipse,* which is *the cross section of the three-dimensional ellipsoid cut perpendicular to the direction of propagation* (the X-axis direction in this example). For light traveling in any direction, we obtain an index ellipse by making a major cross section of the ellipsoid perpendicular to that direction of travel. The index ellipse gives us the refractive indices and shows the vibration direction (of the electric vector) of the light waves, because *the waves vibrate with the electric vectors along the major and minor axes of the index ellipse. The refractive indices that the two waves suffer are given by the major and minor radii of the ellipse.* In other words, *for each direction of travel in an anisotropic medium, there are two plane-polarized light waves that vibrate perpendicular to each other and travel at different velocities.* This is, in fact, the phenomenon of birefringence.

To restate what we have just seen, as light travels through an anisotropic material, the wave gets split into two vibrations. The two vibrations (remember, we are just considering the electric vector directions and not the magnetic vectors) are mutually perpendicular to each other and perpendicular to the direction that the wave travels. The wave whose electric vector vibrates along the major axis of the index ellipse is called the slow wave, because the refractive index for this wave is greater than the refractive index for the other wave.* The wave vibrating perpendicular to the slow wave is the fast wave.

Each index ellipse then provides the slow and fast vibration axes for waves traveling perpendicular to the plane of the ellipse. The index ellipse (the cross section of the index ellipsoid) is commonly used to designate the birefringence of a material observed with light propagating perpendicular to that plane (e.g., Figs. III-14, III-17).

We shall now examine another manifestation of birefringence, taking advantage of another demonstration. We place a piece of cardboard with a small hole in front of a light source and on top of the hole, a crystal of calcite. Calcite is a form of calcium carbonate, whose rhombohedral shape is given in Fig. III-4. When we place the calcite crystal on top of the hole, we find that the image seen through the crystal is doubled. As we turn the crystal on the cardboard, we find that one of the images is stationary, while the other one precesses around the first.

This observation is explained by the birefringence of calcite. The birefringence is so strong that we not only get two waves, but even the directions of travel of the two waves become separated (Fig. III-5). One of the waves is traveling straight through—its image remains stationary when the crystal is turned. That is called the ordinary ray, or the o-ray, because it behaves (is refracted) in an ordinary fashion. The other wave, the precessing one, refracts in an extraordinary fashion, and is called the extraordinary ray, or the e-ray.

Not only is the light split into two, but as we argued before, each must be vibrating in a unique direction. The two must also be vibrating with their electric vectors perpendicular to each other. It turns out that the e-ray always vibrates in the plane that joins it and the o-ray. The o-ray always vibrates at right angles to this, as we will now verify.

We orient the crystal so that the e-ray image (which is the one that precesses) appears on top (Fig. III-4). The direction of vibration of the e-ray then should be in the direction joining it and the o-ray image, namely up and down. The o-ray should have its electric vector horizontally. If we use a polarizer that transmits waves with horizontally oscillating electric fields, the top image should disappear; if we use one that transmits electric fields with vertical vibration, the bottom image should disappear.

One handy, inexpensive standard polarizer, or polar, is a pair of Polaroid sunglasses. It shows no apparent anisotropy if we examine an ordinary source of light. No difference is apparent as we

*Equation (III-3) shows the relation of the refractive index to the (phase) velocity of travel.

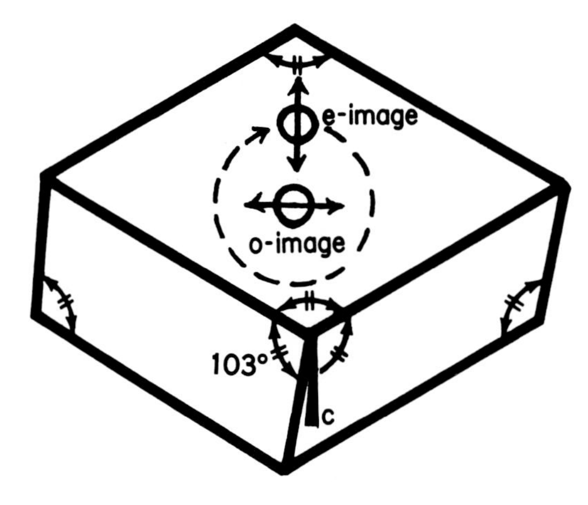

FIGURE III-4. Double image of a single light spot seen through a calcite crystal. As the crystal is turned, the e-ray image precesses around the o-ray image. The e-wave vibrates in the plane that includes the c-axis (the principal section). The o-wave vibrates perpendicular to it. The c-, or optic-axis, of the crystal is indicated by c. In calcite, it is the axis of threefold symmetry. It makes an equal angle with all three of the crystal faces that join at the two corners, where all edges lie at 103° angles with each other. The c-axis lies in the direction of the semi-ionic bond that links the planar CO_3 groups and Ca atoms in the calcite ($CaCO_3$) lattice. (Evans, 1976; Inoué and Okazaki, 1978.)

turn the sunglasses in different directions. But if we look at the surface of water, or at the glare on the road or painted surfaces, the reflected light is cut out as the Polaroid glasses are intended to do. Light that is reflected from the surface of water or any nonconducting material is plane polarized, and especially strongly so at a particular angle of incidence—the Brewster angle. It is polarized so that the electric vector is vibrating parallel to the surface from which it is reflected, and not vibrating perpendicular to the surface. This behavior can be explained from a series of equations of Fresnel, but an easier way is to use the stick model proposed by Robert W. Wood. Consider a stick of wood in place of the electric vector. If it hits the water surface at an angle, the stick goes into the water and is not reflected. If the stick comes in parallel to the surface, it can bounce back. Since in nature we are dealing with horizontal surfaces, it will be a horizontal vibration that is reflected. We want to cut the glare, so Polaroid sunglasses are made to remove the horizontal vibrations and transmit the

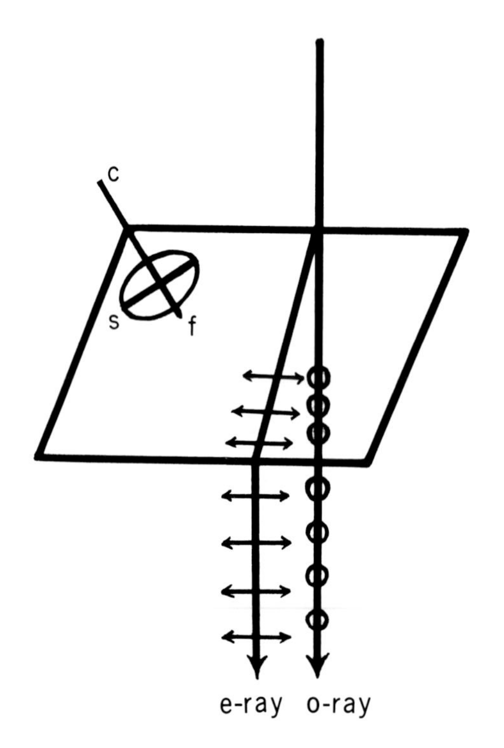

FIGURE III-5. Path of light rays through a calcite crystal. The crystal is shown through a principal section—that is, through a plane including the optic axis, c. The o-ray is refracted as though it were traveling through an isotropic medium; hence, it proceeds without deviation after normal incidence. The e-ray direction is deviated (in the principal section) even for normal incidence, hence the name extraordinary ray. The c-axis is the axis of symmetry of the index ellipsoid. Calcite is a negatively birefringent crystal so that the o-ray is the slow wave. The o-wave always vibrates in a plane perpendicular to the principal section while the e-wave vibrates in the principal section.

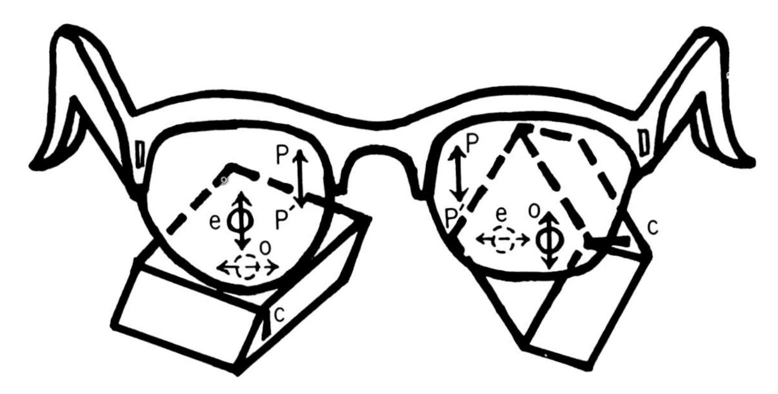

FIGURE III-6. Determining the electric vector directions for the e- and o-ray in calcite with Polaroid sunglasses. Polaroid sunglasses are made to absorb and remove the horizontal electric vectors and transmit the vertical vibrations, PP′. On the left side, the e-ray image from the calcite crystal is transmitted but the o-ray image is absorbed. On the right side, the calcite crystal is turned 90° and the e-ray image is absorbed. At an intermediate angle, both images are partially transmitted following the cosine squared law.

electric vector that is vibrating vertically. In these glasses, then, we have a standard for transmission of the electric vector (Fig. III-6).

Returning to the calcite crystal (Fig. III-4), the top image has its e-vector vertically, so that when we look through our sunglasses, the bottom image should disappear (Fig. III-6, left). We find that is exactly the case. If we turn the sunglasses or the crystal slowly, two images appear for a while, until at 90° the ''top'' image becomes extinguished (Fig. III-6, right). Thus, with birefringence a light beam is split into two waves, each of the waves vibrates with its electric vector perpendicular to the other, and they travel at different velocities. (Close inspection will reveal that the e-ray image appears farther through the crystal than the o-ray image, indicating that the o-ray image suffered greater refraction, or that $n_e < n_o$ in calcite.)

Calcite crystals can be used as very effective polarizers; in fact, Nicol, Glan–Thompson, Ahrens, and other prisms are made of calcite to isolate and transmit one of the polarized waves.*

For UV microbeam work, it is much less expensive and more effective to use a simple cleaved piece of calcite crystal that has been optically polished on two opposite faces. Placed in front of the microbeam source, the crystal splits the beam into two and gives two equally intense images of the microbeam source (Fig. III-7). One can choose an image with known vibration (or use both) because the e-ray, whose electric wave vibration joins the two images, can be identified by its precession when the crystal is turned around the axis of the microscope. The o-ray vibration is perpendicular to it. One now has two polarized microbeam sources with equal intensities, which can be used for analyzing e.g., the arrangement of DNA bases as discussed later.

Down to which wavelength will calcite transmit? Generally, about 240 nm, but this depends on the calcite. One can get calcite that transmits to somewhat shorter wavelengths, depending on the source and purity of the crystal.

In material such as calcite, the two beams were visibly split into two, but for most material, especially the kind of biological material making up cytoplasmic or nuclear structures in a living cell, the birefringence is so weak that the two beams are not visibly split. Even when they are not visibly split, the extraordinary and the ordinary waves are present, with their individual vibration planes and separate velocities. Before we look into the properties of these two waves and how they can be used to detect birefringence in living cells, let us briefly consider the molecular and atomic meaning of *dichroism,* another form of optical anisotropy.

*It is unfortunate that the wave transmitted by these otherwise superior polars is the e-wave. The velocity of the e-wave, or refractive index of the e-ray, varies with direction of propagation, so that these calcite prisms introduce *astigmatism* unless all the beams travel parallel to each other through the crystal.

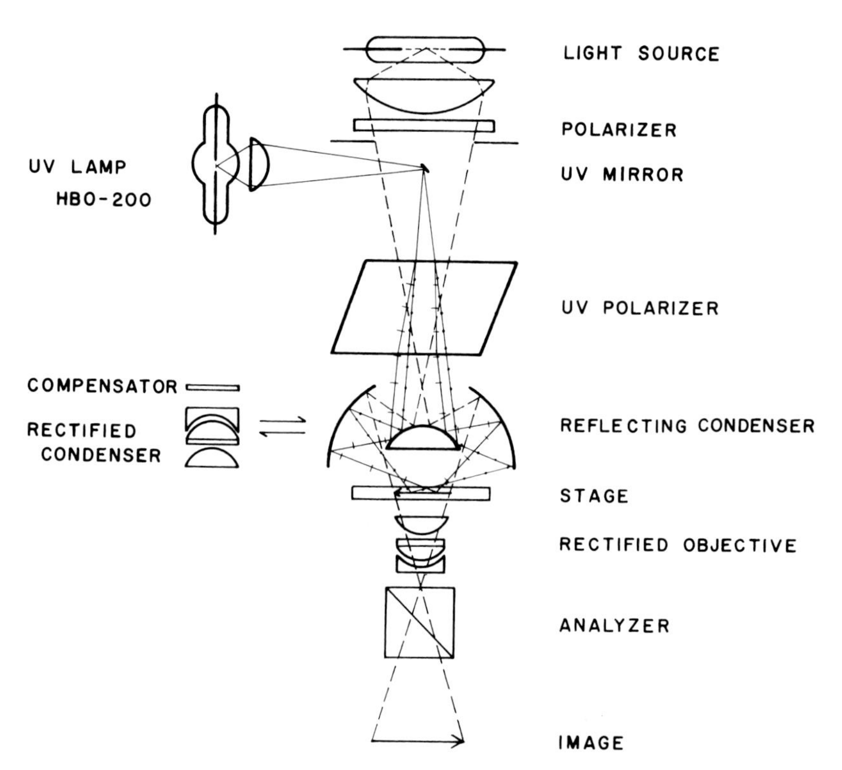

FIGURE III-7. Schematic diagram of polarized UV microbeam apparatus. Compare with schematics of inverted polarizing microscope (Fig. III-21) and photograph of rectified instrument (Fig. III-22). Each ray from the UV source (a small, first-surface mirror mounted on a dovetail slide beneath the polarizer for visible light illumination), as well as from the visible light source diaphragm (which lies between the polarizer and the UV mirror), is split into e- and o-rays by the UV polarizer. The UV polarizer is a cleaved crystal of calcite optically polished on the top and bottom faces. The reflecting condenser focuses the double image of the UV mirror and the visible light source diaphragm onto the specimen plane, each image with its electric vector as defined in Figs. III-4 to III-6. The image of the UV mirror with the desired polarization is superimposed on the specimen to be microbeamed. (From Inoué and Sato, 1966b.)

III.6. DICHROISM

Again, we start with a demonstration. A small source of light, say a flashlight bulb, is placed behind a crystal of tourmaline. Tourmaline is a prismatic crystal with a threefold rotational symmetry, and the type we use for this demonstration is a clear green crystal. Tourmaline crystals are dichroic and transmit in the green (or blue) as long as the axis of symmetry is oriented perpendicular to the light beam (Fig. III-8). But if we turn the crystal and look down the axis of symmetry, no light, or very little light with a brownish tint, is transmitted.

We can understand the wave-optical basis of these observations by returning to an analysis similar to the one used for birefringence. Dichroic crystals are also birefringent, so that light is again split into two waves in the crystal. As shown in Fig. III-9, for each direction of travel, the light wave is split into two waves, whose electric vectors are oriented perpendicular to each other and to the direction of travel. For light traveling along the Z axis, we have one wave (Y) with its electric vector vibrating in the Y–Z plane and another (X) with its electric vector vibrating in the X–Z plane. Both of these must be absorbed because light does not come through the Z-axis direction. This means that the electronic disturbances caused by both the y electric vector and the x electric vector result in

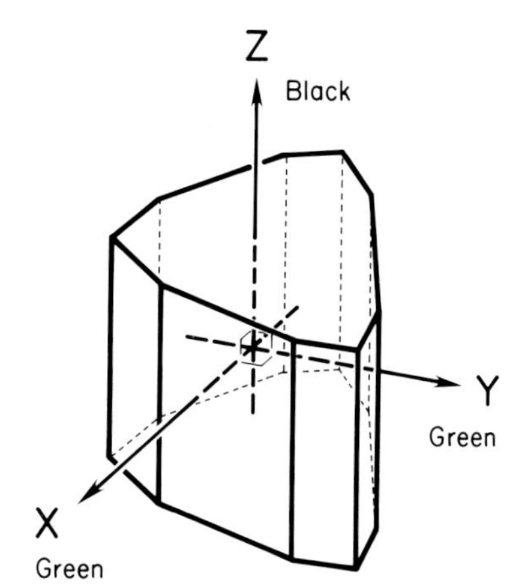

FIGURE III-8. Transmission of light through a dichroic crystal—in this illustration, a clear green crystal of tourmaline. Viewed along the *X* and *Y* axes, the crystal is clear green. Viewed along the *Z* axis, no light is transmitted and the crystal is black or dark brown, even when the crystal is considerably shorter along the *Z* axis than along the *X* and *Y* axes.

absorption of the energy. The chromophores in the crystal are oriented such that they absorb light whose electric fields are vibrating in the *X–Y* plane.

If we examine the light traveling along the *X* direction, we again know that light must vibrate in two directions perpendicular to that—the *Z* direction and the *Y* direction. We already know that the *Y* vibration is absorbed, and therefore the green light we see must have its electric vector in the *Z* direction. Likewise for the *Y* propagation, the *x* electric vectors are absorbed and only the *z* vectors must make up the transmitted green light. We thus deduce that the light coming through the tourmaline crystal is plane polarized with its electric vector parallel to the crystalline (*Z*) axis of symmetry. This is readily confirmed with our standard Polaroid sunglasses.*

This is the basis of dichroism. As we saw, both dichroism and birefringence can be manifested as macroscopically recognizable optical properties, which vary with the direction of propagation of light, but they are at the same time a reflection of fine-structural, molecular, or atomic anisotropies. For dichroism it is the orientation of chromophores that determines anisotropic absorptions. For birefringence it is the anisotropy of dielectric polarizabilities. The anisotropic polarizability can originate in the fine structure or within the molecules.

III.7. DETERMINATION OF CRYSTALLINE AXES

We are now almost ready to apply these principles to the polarizing microscope to study biological material. Let us place two pieces of plastic sheet Polaroid in front of a diffuse light source and orient them at right angles to each other so that the light is extinguished. In between the "crossed polars" we insert our samples, just as we do with a polarizing microscope. When, for example, a piece of cellophane is placed between crossed polars, we observe light through that area (Fig. III-10). The phenomenon is due to birefringence, as explained below, even though we do not see a double image.

As in Fig. III-11, let the light (with an amplitude of OP) from the polarizer (PP′) vibrate vertically. Upon entering the specimen. OP is vectorially split into two vibrations, OS and OF. These vibration directions are uniquely defined by the crystalline structure of the specimen since

*Polaroid sunglasses and filters are themselves made from a dichroic material. One common form uses a stretched film of polyvinyl alcohol (PVA) impregnated with polyiodide crystals. In the stretched PVA film, the micelles and molecular backbones are oriented parallel to the direction of stretch (Fig. III-17). The needle-shaped, dichroic polyiodide crystals are deposited in the interstices, parallel to the PVA molecules. All of the polyiodide crystals, therefore, become oriented parallel to each other, and one obtains a large sheet of dichroic material (Land, 1951; Land and West, 1946).

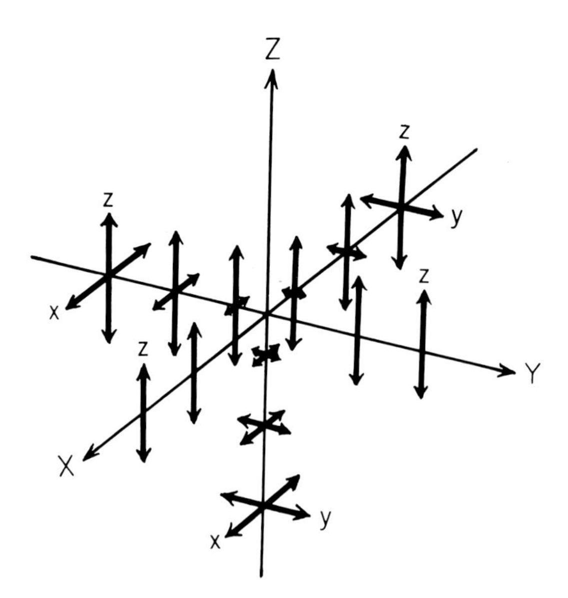

FIGURE III-9. Explanation of dichroic absorption in tourmaline. The x and y electric vectors are absorbed regardless of the direction of propagation. Only the z vectors are not absorbed, so that the transmitted light is plane polarized parallel to the Z axis.

they lie in the directions of the major and minor axes of the index ellipse. In the crystal (cellophane), these two waves travel at different velocities so that when they come out of the crystal, they (O′S′ and O′F′) are out of phase (Δ) relative to each other.* The combination of these two waves, which remain out of phase and continue to oscillate in planes perpendicular to each other in air, yields an elliptically polarized wave (O″P″–O″A″).

Light that is elliptically polarized no longer vibrates only along the PP′ axis, as it has a component that exists along the AA′ axis. That component (O″A″) of the elliptically polarized light can pass through the second polar (the analyzer) and give rise to the light that we observe.

For the birefringent specimen between crossed polars, the elliptically polarized wave is produced by the splitting of the original plane-polarized wave into two vectors. Therefore, we should not get elliptical polarization if, for some reason, one of the two vectors were missing. This happens when the slow or fast specimen axes become oriented parallel to the polarizer axis. At that orientation of the crystal, OP′ cannot be split into two vectors, so OP′ emerges from the crystal unaltered. This wave is then absorbed completely by the analyzer and no light comes through. Therefore, by rotating the specimen between crossed polars and observing the orientation where the specimen turns dark (Fig. III-10), we can determine its crystalline axes. *When the specimen turns dark between crossed polars, the orientations of the major and minor axes of the index ellipse are parallel to the axes of the polars.* (Note that these axes may or may not coincide with the geometrical axes or cleavage planes of the specimen.)

We have now established the directions of the two orthogonal axes in the crystal, but we still do not know which is the fast axis and which is the slow axis. In order to determine this, we

*Δ, expressed in fraction of wavelength (of the monochromatic light wave in air; λ_0), is proportional to the thickness of the crystal (d) and to its "coefficient of birefringence" ($n_e - n_o$). Thus,

$$\Delta = (n_e - n_o)d/\lambda_0 \qquad \text{(III-5)}$$

where n_e is the refractive index for the extraordinary ray, and n_o that of the ordinary ray. The e-ray may either be the slow or fast wave depending on the crystal type. If the e-ray is the slow wave (as it is in quartz and cellophane), the crystal is said to be positively birefringent; if the e-ray is the fast wave (as it is in calcite), the crystal is said to be negatively birefringent. The Fresnel ellipsoid for a positively birefringent material is prolate, while the one for a negatively birefringent material is oblate.

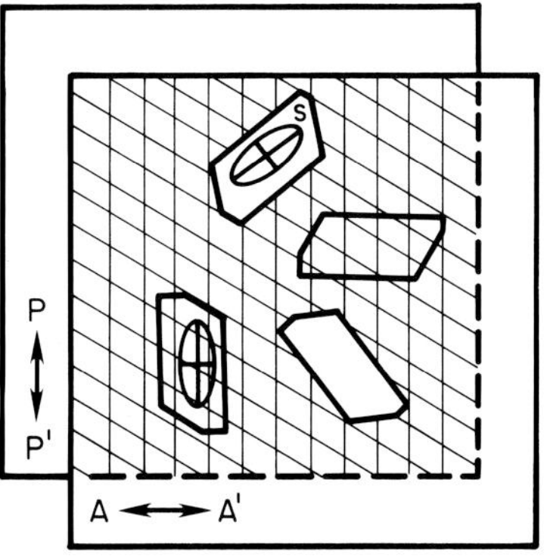

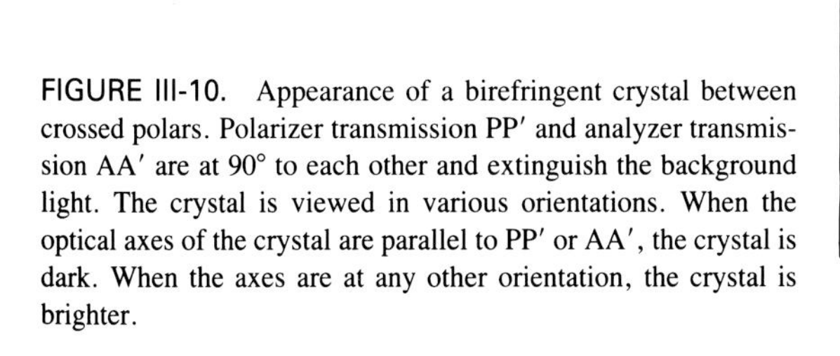

FIGURE III-10. Appearance of a birefringent crystal between crossed polars. Polarizer transmission PP′ and analyzer transmission AA′ are at 90° to each other and extinguish the background light. The crystal is viewed in various orientations. When the optical axes of the crystal are parallel to PP′ or AA′, the crystal is dark. When the axes are at any other orientation, the crystal is brighter.

superimpose a crystalline material whose slow and fast axis directions are already known. Such a crystal is known as a compensator.

Let us assume in our next demonstration that a particular piece of plastic—the compensator—has its slow and fast axes oriented as shown in Fig. III-12. We superimpose this on an unknown sample; both are placed between a pair of crossed polars. If the slow axis of the unknown sample and that of the compensator lie perpendicular to each other, then the two effects cancel each other out. Where the compensator and the sample are superimposed on each other, we observe that the light is extinguished; the specimen appears dark (Fig. III-12, the crystal in the lower right position). What has happened is that the elliptically polarized light produced by the specimen was restored by the compensator to the original plane-polarized light (Fig. III-13). Therefore, no vectorial component is present in the analyzer transmission direction, so that all the light is absorbed by the analyzer.

On the other hand, if the specimen and the compensator slow axes were parallel to each other, the two retardations would add up together, and we would get more light through the analyzer than before (Fig. III-12, the crystal oriented as in the top). By finding the orientation of the compensator

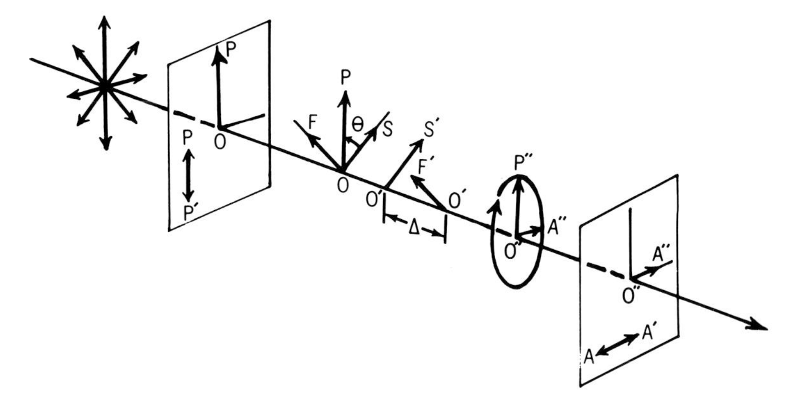

FIGURE III-11. Explanation of the phenomenon in Fig. III-10. So long as the angle θ between the slow axis (OS) of the crystal and OP (the electric vector from the polarizer) is not 0 or 90°, OP is split into vectors OS and OF by the birefringent crystal. Passing through the crystal, they acquire a phase difference, Δ, which remains constant once in air (an isotropic medium) up to the analyzer. Viewed from the analyzer direction, the O′S′ and O′F′ vibrations combine to describe an elliptically polarized wave, O″P″–O″A″. The O″P″ component is absorbed by the analyzer, but the O″A″ component is transmitted, and hence the crystal appears bright on a dark background. Where θ is 0 or 90°, the OF or OS vector is missing, so that O″P″ = OP and O″A″ → 0. Hence, no light passes the analyzer.

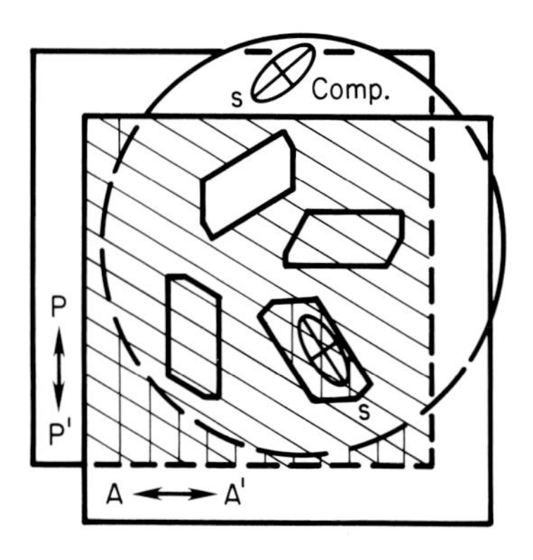

FIGURE III-12. Birefringent crystal between crossed polars, in the presence of a compensator. The compensator (large circle) introduces some light between crossed polars similar to the crystal in Figs. III-10 and III-11. On this gray background, the crystal appears darker (lower right orientation) when its *s* axis is in the opposite quadrant from the *s* axis of the compensator (subtractive orientation; see Fig. III-13). It is completely extinguished when the conditions in equation (III-6) or (III-7) are satisfied. When the crystal and compensator *s* axes are in the same quadrant (top orientation), the crystal appears considerably brighter than the gray background. The contrast of the crystal against the gray background disappears when the axes of the crystal coincide with PP′ and AA′.

that extinguishes or increases light from a birefringent specimen, we can then *determine the slow and fast axis directions* of the specimen.*

At this point I would like to point out that birefringence is not to be confused with *optical rotation*. Optical rotation is a phenomenon that we would encounter, for exmaple, if we observed a solution of sugar instead of our crystalline specimen (or if we observed quartz along its optical axis, the axis of symmetry). A solution of sugar has no directionality, but it can twist polarized light (in a left-handed or right-handed direction depending on its chemical nature). The result is as though the polarizer PP′ were turned; so the light can be *extinguished by turning the analyzer* AA′ by that same amount. With birefringence, you can generally not extinguish the light by turning the analyzer, but with optical rotation you can (for monochromatic light). Optical rotation reflects the three-dimensional asymmetry of the individual molecules, which themselves can be randomly oriented (as in the solution of sugar). The dispersion (variation with different wavelengths of light) is usually very substantial for optical rotation but rather small (parctically negligible) for birefringence.

III.8. MOLECULAR STRUCTURE AND BIREFRINGENCE

The next demonstration will let us correlate molecular and micellar structure with optical and mechanical anisotropy. We take a sheet of polyvinyl alcohol (PVA) produced by casting a hot water sol of the polymer onto a clean piece of glass. The basic structure of PVA is shown in Fig. III-14. In a cast (dried gel) sheet of PVA viewed normal to the surface, the polymers and their micelles would

*Quantitatively, we obtain extinction when

$$\sin\frac{\Delta}{2}\,\sin 2\theta = -\sin\frac{R_c}{2}\,\sin 2\theta_c \qquad \text{(III-6)}$$

where Δ and R_c are the retardances of the specimen and the compensator, respectively, expressed in degrees of arc ($1\lambda = 360°$), and θ and θ_c are respectively the angle between the slow axes of the specimen and of the compensator relative to OP, the polarizer transmission direction (Fig. III-13). We use this formula extensively for the photographic photometric analysis of retardance and azimuth angles of DNA microcrystalline domains in a cave cricket sperm (Figs. III-24, III-25; Inoué and Sato, 1966b).

Equation (III-6) can be simplified as follows. The specimen is generally oriented so that $\theta = 45 \pm 5°$. When, in addition, Δ and R_c are both not greater than 20°, or less than approximately $\lambda/20$, then to a very close approximation, equation (III-6) can be written:

$$\Delta = R_c \sin 2\theta_c \qquad \text{(III-7)}$$

This is the principle by which specimen retardance is measured with the *Brace–Koehler compensator* (see also Salmon and Ellis, 1976).

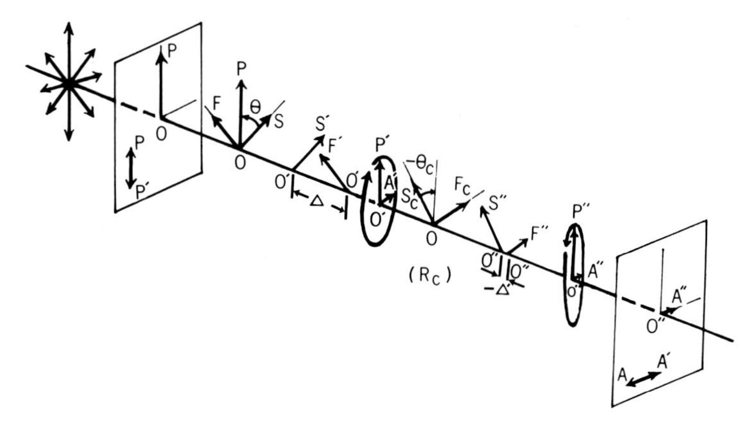

FIGURE III-13. Compensation. The left half of the figure is identical to most of Fig. III-11. However, after the light becomes elliptically polarized O′P′–O′A′) by the specimen, a compensator with a retardance, R_c, is added to the light path, with OS_c, the compensator slow axis, lying in the subtractive orientation, i.e., in the quadrant opposite to OS, the specimen slow axis. R_c is adjusted so as to restore the phase difference, Δ, between O′S′ and O′F′ that was introduced by the specimen. Then, the phase difference, $-\Delta'$, between O″S″ and O″F″ in the waves emerging from the compensator becomes zero and O″S″ + O″F″ = O″P″. The conditions in equation (III-6) or (III-7) are now satisfied—that is, we have achieved compensation. At compensation, O″A″ → 0 and O″P″ = OP, and hence the specimen is extinguished (Fig. III-12, lower right orientation). If $-\Delta'$ is not 0, O″A″ is also not 0, and light whose amplitude is O″A″ passes the analyzer. The sine squared relation in equation (III-9) explains why the crystal is much brighter than the background when OS and OS_c are parallel. [Substitute $(\Delta + R_c)$ for Δ in equation (III-9).]

be expected to be arranged helter-skelter as shown in Fig. III-15. Such a structure would be optically isotropic, as we can see by the lack of birefringence—no light passes the cast PVA sheet placed between crossed polars at any orientation. Such a structure should also be mechanically isotropic and difficult to tear in any direction, since covalent molecular backbones traverse in every direction. Indeed, cast sheets of PVA are very tough.

The micelles and molecular chains of the PVA sheet can be aligned by stretching a sheet that has been softened by mild heating with an electric iron. Figure III-16 shows the film before (A) and after (B) stretching. The arrangement of the micelles and molecular backbones after stretch is shown in Fig. III-17. The stretched film is highly birefringent and produces several wavelengths of

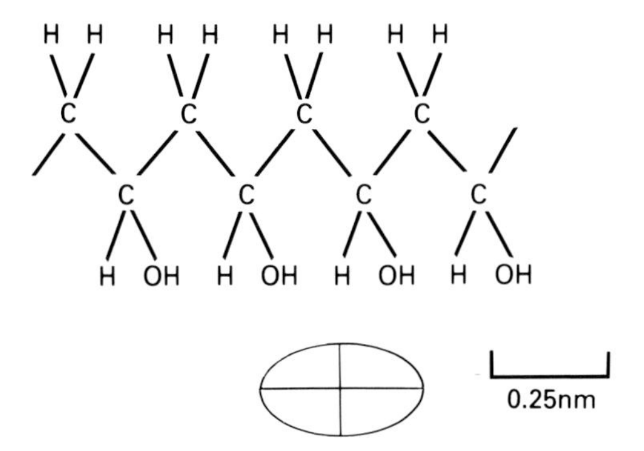

FIGURE III-14. Chemical structure of PVA. PVA is a long, linear, polymeric chain of $-(CH_2-CH{\cdot}OH)-$. In the absence of external constraints, rotation around the C–C bonds allows the chain to fold into random-shaped strands, except where adjacent chains lie in close proximity parallel to each other and form a minute crystalline domain or a ''micelle'' (Fig. III-15).

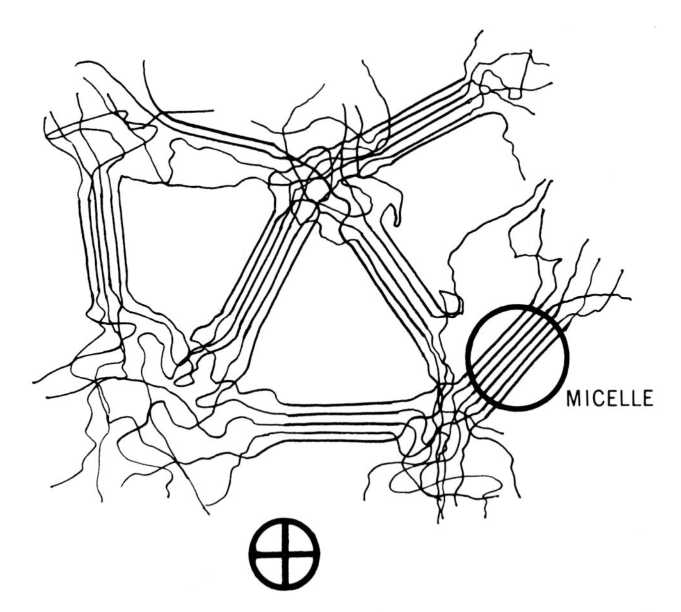

FIGURE III-15. "Fringed micelle" structure of a cast gel of PVA viewed from its face. The circle to the right indicates a micelle where several PVA molecules run parallel and form a minute crystalline region. The micelles, and the polymer chains of PVA in between, are randomly oriented. The structure is isotropic as indicated by the index ellipse at the bottom.

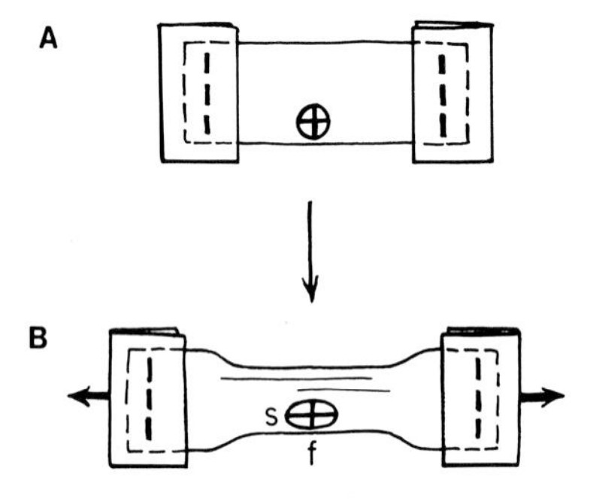

FIGURE III-16. Stretching PVA film. (A) A thin rectangular sheet of isotropic PVA is supported on both ends, taped and stapled onto pieces of cardboard. (B) The sheet is mildly heated, for example with a tacking iron in the middle, and then stretched. The stretched portion becomes optically anisotropic (birefringent) with the slow axis in the stretch direction. It is also mechanically anisotropic and tends to split parallel to the stretch direction.

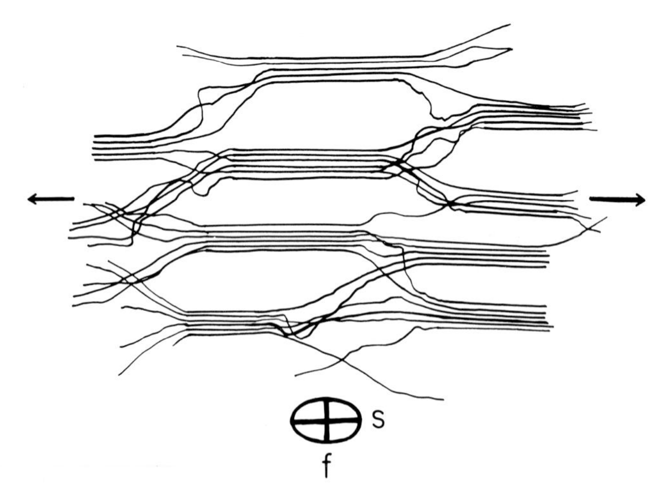

FIGURE III-17. Fringed micelles in stretched PVA. As manifested in the optical and mechanical anisotropy (Fig. III-16), the fringed micelles and the PVA molecules in general have become aligned with their long axes parallel to the stretch direction.

retardation as seen from its color* between crossed polars. In spite of the strong birefringence, light is extinguished when the film is oriented with its stretch direction parallel to the axis of the crossed polars.

Compensation reveals that it is the slow vibration axis that lies parallel to the stretch direction. This agrees with the molecular polarizability expected in PVA. In a polymer, with a covalently linked backbone and with small side groups† as in PVA, the polarizability (hence, the dielectric constant) is considerably greater parallel to the covalent chain backbone than across it.

A striking mechanical anisotropy of the stretched film of PVA also reflects the micellar and molecular arrangements. The film is even tougher than before when one attempts to tear across the stretch direction. However, in response to a rap on the tautly held film, it readily splits into strips parallel to the stretch direction as shown in Fig. III-16B.

The mechanical and optical anisotropies and the directions of their axes reflect the arrangements and anisotropic properties of the underlying molecules. In some biological samples, the molecules may take on a more complex arrangement than the homogeneous distribution seen in this PVA model. Nevertheless, the dielectric and optical properties specific to the particular molecular species provide important clues regarding the biological fine structure.

For example, in a protein molecule with extended chains in the β-form, or with a collagen helix, the slow axis lies parallel to the long axis of the polypeptide chain; the molecular polarizability is considerably greater in that direction than across the chain.

In B-form DNA, the slow axis is perpendicular to the backbone of the Watson–Crick helix. The conjugated purine and pyrimidine bases exhibit a greater UV absorbance and electrical polarizability in their plane than perpendicular to the planes. Because the base planes in B-form DNA are oriented at right angles to the molecular backbone, they show a characteristic UV negative dichroism and strong negative birefringence.

In lipids, the slow birefringence axis of the molecule lies parallel to its backbone. But since they tend to make layered sheets with the molecular axes lying perpendicular to the plane of the sheets, the layered structure also introduces *form birefringence*.‡ In form birefringence of platelets, the slow axis is parallel to the plane of the plates. The axis of symmetry is perpendicular to the plates, and hence platelet-form birefringence always has a negative sign (greater refractive index perpendicular to the axis of symmetry, or $n_e < n_o$). Lipid bilayers (or multilayers in the Schwann sheath of myelinated nerve) show a combination of intrinsic positive and form negative birefringence. Depending on the refractive index of the imbibing medium, the negative-form birefringence of the plates may become so strong as to overcome the intrinsic positive birefringence of the lipid molecules.

*See equation (III-5). In white light that is composed of various wavelengths (λ_0's), the thickness (d) of the crystal that gives rise to a full-wavelength retardation [$(n_e - n_o)d = N\lambda_0$, where N = 1, 2, 3 . . .] varies with λ_0. Where the retardation equals N λ_0, the light exiting the crystal returns to a plane-polarized wave identical to the one before it entered the crystal. Here it is extinguished by the analyzer and that wavelength is missing from the light transmitted through the analyzer. Thus, we get white minus that wavelength, or a series of "interference colors," from strongly birefringent specimens observed between crossed polars in white light.

†As the side groups become larger, or if there are side groups with greater polarizabilities—often associated with light-absorbing *conjugated bonds*—the birefringence of the polymer becomes weaker, and can even reverse in sign. For example, the degree of nitration of nitrocellulose is monitored by observing its birefringence, which changes from positive to zero to negative with increased nitration. The birefringence of DNA is strongly negative as discussed later.

‡Form birefringence, also known as textural birefringence, arises when platelets or rodlets of submicroscopic dimensions are stacked. The platelets or rodlets must be regularly aligned, with spacings that are considerably smaller than the wavelength of light. As shown by Wiener (1912) and by Bragg and Pippard (1953), such bodies generally exhibit anisotropy of electrical polarizability and therefore of refractive index. The anisotropy is stronger, the greater the difference of refractive index between the medium lying between the rodlets or platelets and the rodlets or platelets themselves. The anisotropy disappears or becomes least strong when the refractive index of the medium matches that of the rodlet or platelet. Form birefringence is due to the shape and orientation of molecules or molecular aggregates, and is independent of the birefringence that is intrinsic to the constituent molecules themselves. In contrast to form birefringence the value of *intrinsic birefringence* usually does not change with the refractive index of the imbibing medium.

In contrast to platelet-form birefringence, form birefringence of rodlets is positive. Positive-form birefringence is observed in microtubules, which are thin, elongated structures approximately 24 nm in diameter, made up of rows of globular protein molecules approximately 5 nm in diameter, or in actin, a twisted double cable of globular protein molecules, which themselves possess low molecular asymmetry.

III.9. THE POLARIZING MICROSCOPE

All the principles discussed so far provide the basis for analyzing molecular arrangements and cellular fine structure with the polarizing microscope. A polarizing microscope is nothing more than an ordinary microscope equipped with a polarizer underneath the condenser, an analyzer above the objective, and somewhere in the path, in a convenient place between the polarizer and analyzer, a compensator (Fig. III-18). The compensator can come before or after the specimen. The specimen is supported on a rotatable stage. In principle, that is all there is to a polarizing microscope.

However, if one decides to study cytoplasmic or nuclear structure with a polarizing microscope and tries with one borrowed from a friend in crystallography or mineralogy, chances are that it will be a quite disappointing experience. Other than a few crystalline inclusions, little of interest is seen in the living cell. One reason is that the amount of light that strays through an oridinary polarizing microscope between crossed polars often cannot be reduced enough to see the weak birefringence exhibited by the filaments and membranes of interest to us.

The degree to which we can darken the field is expressed quantitatively as the *extinction factor* (EF), which is defined as

$$EF = I_p/I_s \tag{III-8}$$

where I_p is the intensity of light that comes through a polarizing device when the polarizer and analyzer transmission directions are parallel, and I_s the minimum intensity that can be obtained when the polarizer and analyzer are crossed. For an ordinary polarizing microscope, the EF is often of the order of 10^3 or even lower with high-NA lenses. *What we need for cell study is an EF of at least* 10^4.

In order to achieve this high an EF, we must attain a number of conditions concurrently. The polarizer and analyzer themselves obviously must be of good quality. They can be of selected sheet Polaroids, as far as the extinction goes (but their transmittance is low, and the rippled surface can deteriorate the image). We cannot use optical elements between the polars that are strained. As seen by stressing a piece of glass or plastic between crossed polars, the strain introduces birefringence. This must not take place because that birefringence can be much higher than the specimen birefringence. So we must use lenses, slides, and coverslips that are free from strain. A little speck of airborne lint (e.g., from our clothing or lens tissue) can be highly birefringent, so we must keep our system meticulously clean. Also, since almost all the light is extinguished by the analyzer, the light source must be very bright (yet harmless to our specimen), and we should work in a darkened area to improve the sensitivity of our eye. Alignment of the optical components is critical, and Koehler illumination must be used to gain minimum stray light coupled with maximum brightness of the field.

The reason for all of this care is that the light that comes from the specimen is only a minute fraction of the original light. If Δ is the retardance of the specimen and θ its azimuth orientation, both expressed in degrees, the luminance (I) due to specimen birefringence (Δ) is given by

$$I = I_p \sin^2 \frac{\Delta}{2} \tag{III-9}$$

where I_p is the luminance of the field with the polarizer and analyzer transmission parallel. I turns out to be of the order of 10^{-4} to $10^{-6} \times I_p$ for the range of specimen retardations of interest to us. So we are trying to see very dim light from the specimen—as though we were trying to see starlight

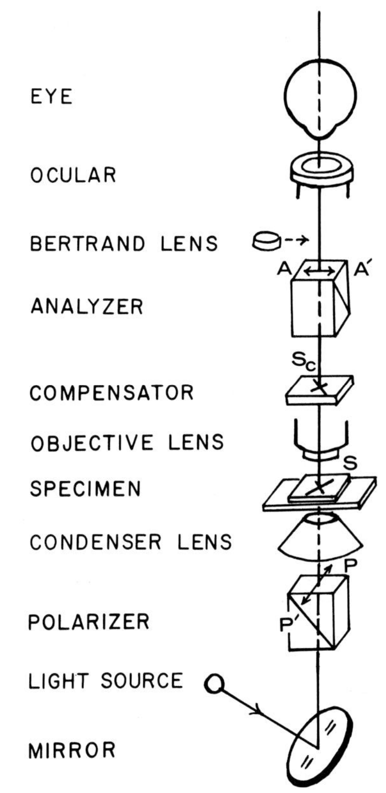

FIGURE III-18. Arrangement of conventional polarizing microscope. The polarizer was commonly oriented with its transmission axis PP′ north–south to the observer. AA′ was oriented east–west.* The Bertrand lens converts the ocular into a telescope and allows one to view an image of the objective lens back aperture. This provides a ''conoscopic'' image of the specimen, showing the interference pattern of convergent polarized light passing through the specimen at different angles. The Fraunhofer diffraction pattern produced by the specimen is also seen there. With the Bertrand lens out of the light path, the polarizing microscope gives a regular ''orthoscopic'' image.

during the daytime. The light is there, but there is so much more light around our specimen that we cannot see it. We have to somehow make the background dark enough so we can see the object. We must increase the EF and improve the specimen contrast.

The situation is improved by good use of a compensator. This point can be demonstrated by placing, between crossed polars, a weakly birefringent specimen that we can just barely see. The birefringent region is barely brighter than the background (Fig. III-10). When we introduce the compensator, the background is now gray and the specimen is darker or much brighter than the background (Fig. III-12). In the presence of the compensator, the image brightness and the contrast can be vastly improved as discussed in Chapter 11 (Fig. 11-4B), and images of weakly birefringent objects can now be displayed clearly (Figs. III-19, III-20, III-25).

III.10. RECTIFICATION

There is yet one more improvement required for detecting weakly birefringent specimens at high resolution in polarization microscopy. Following equation (5-9), we require high objective *and* condenser NA to obtain high resolving power. But *each time the NA is increased by 0.2, stray light increases tenfold even when we use strain-free lenses.* With objective and condenser NAs at 1.25, the EF may drop to nearly 10^2. This lowered extinction (increased stray light) at high NA is due to the rotation of the plane of polarization at *every* oblique-incidence interface between the polarizer and the analyzer. It is an inherent physical optical phenomenon that had been considered uncorrectable until we developed the polarization rectifier (Figs. III-21; Inoué and Hyde, 1957; see also Huxley, 1960). With the rectifier the extinction is very good up to high NAs, and we are finally able

*More recently, the orientations of PP′ and AA′ have been reversed in most commercial instruments.

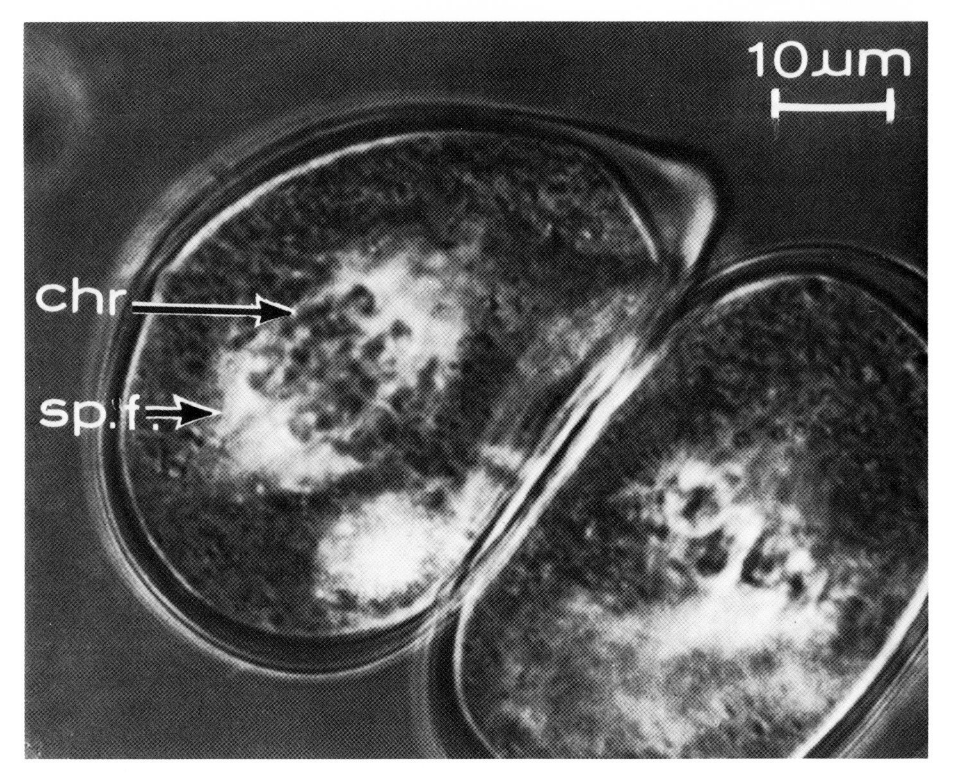

FIGURE III-19. Living pollen mother cell of Easter lily (*Lilium longiflorum*). Spindle fibers (sp.f.) show stronger birefringence adjacent to kinetochores (spindle fiber attachment point) of helical chromosomes (chr) which show little birefringence. The maximum retardance of the spindle is less than $\lambda/100$. (From Inoué, 1953.)

FIGURE III-20. Birefringent spindle fibers in living oocyte of marine worm (*Chaetopterus pergamentaceus*). Chromosomal spindle fibers, white; astral rays, oriented at right angles to spindle axis, dark. The maximum retardance of these spindle fibers is about 3.5 nm. (From Inoué, 1953.)

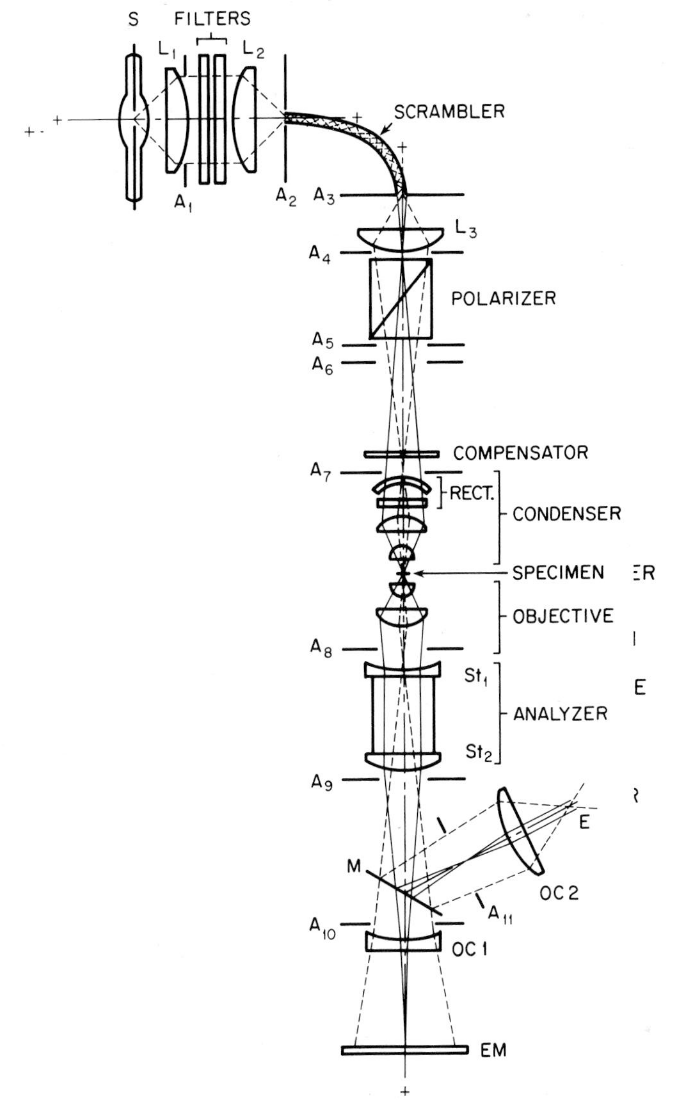

FIGURE III-21. Schematic optical path of transilluminating, universal polarizing microscope designed to provide maximum sensitivity and superior image quality. Inverted system with light source (S) on top and detectors (EM and E) at the bottom. Light from a high-pressure arc lamp is filtered (to remove infrared and provide monochromatic illumination) and focused by L_1 and L_2 onto a fiber-optic light scrambler at A_2. The fiber scrambles the image of concentrated mercury arc (footnote p. 127) and provides a uniform circular patch of light that acts as the effective light source at A_3. This source, projected by the zoom lens (L_3), is made to just fill the condenser aperture diaphragm (A_7). Illuminance of the field can be regulated without affecting the cone angle of illumination (or disturbing the color temperature in white light) by adjusting the iris (A_1). The polarizing Glan–Thompson prism is placed behind stop A_5 away from the condenser to prevent light scattered by the polarizer from entering the condenser. Half-shade and other special plates are placed at level A_6, and compensators above the condenser. Depolarization by rotation of polarized light in the condenser and objective lenses, slide, and coverslips are corrected by the rectifier (RECT.). The image of the field diaphragm (A_4 or A_6) is focused on the specimen plane by the condenser, whose NA can be made equal to that of the objective. Stigmatizing lenses (St_1, St_2), which minimize the astigmatism that is otherwise introduced by the calcite analyzer, are low-reflection coated on their exterior face and cemented directly onto the analyzing Glan–Thompson prism to protect the delicate surfaces of the calcite prism. Stops (A_1–A_{11}), placed at critical points, minimize scattered light from entering the image-forming system. The final image is directed by OC_1 onto a photographic, video, or other sensor (e.g., photomultiplier; EM), or to the eye (E) via mirror (M) and ocular (OC_2). Components A_3 through EM are aligned on a single optical axis to minimize degradation of the image.

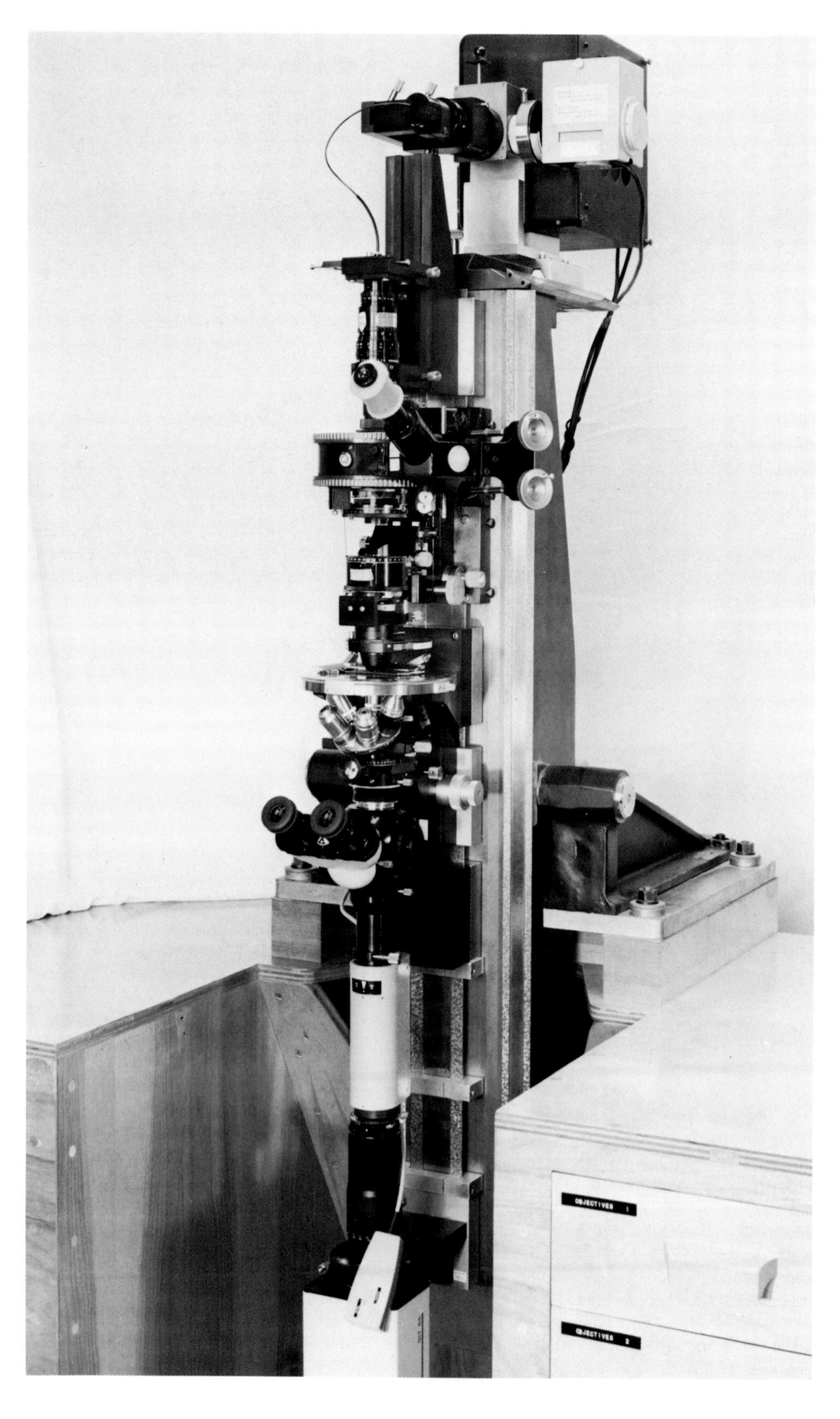
OBJECTIVES 1
OBJECTIVES 2

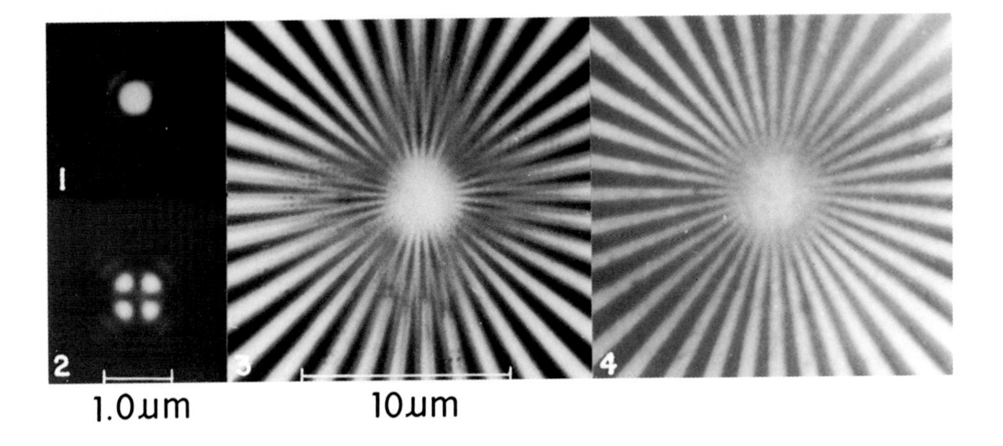

FIGURE III-23. The effect of rectification on image reliability in the polarizing microscope. (1) The Airy disk diffraction pattern of a pinhole viewed through crossed polars with a rectified polarizing microscope of $NA_{obj} = NA_{cond} = 1.25$. The pattern is identical with that obtained with a microscope in nonpolarized light. (2) The same as 1, but with nonrectified lenses between crossed polars. The diffraction pattern is anomalous; each bright point in the specimen is represented in the image by a bright four-leaf clover pattern surrounding a dark cross. (3) A Siemens Test Star pattern viewed between crossed polars in the absence of rectification. Contrast is reversed and spurious toward the center of the image. (4) The same test pattern viewed with the same optics after rectification. The aperture function of the lens is now uniform and the diffraction anomaly has disappeared. (From Inoué and Kubota, 1958.)

←

FIGURE III-22. Rectified, universal polarizing microscope designed by author, G. W. Ellis, and Ed Horn (see also footnote on p. 145). Optical layout as shown schematically in Fig. III-21. The components, in eight units, are mounted on slides that ride on a 1/10,000th-inch-precision, 3-inch-wide dovetail, (4V + 1H) feet long. The well-aged cast iron (Mehanite) dovetail bench is mounted on a horizontal axis and can be used vertically, horizontally, or at angles in between. Two smaller dovetails, built into the same casting. run precisely parallel to the central dovetail and provide added flexibility (micromanipulators, UV microbeam source, etc. are mounted on these dovetails). Supported on the sturdily built wooden bench, by the horizontal axis near the center of gravity of the massive optical bench, and designed for semikinematic support of the components wherever practical, the microscope is quite immune to vibration.

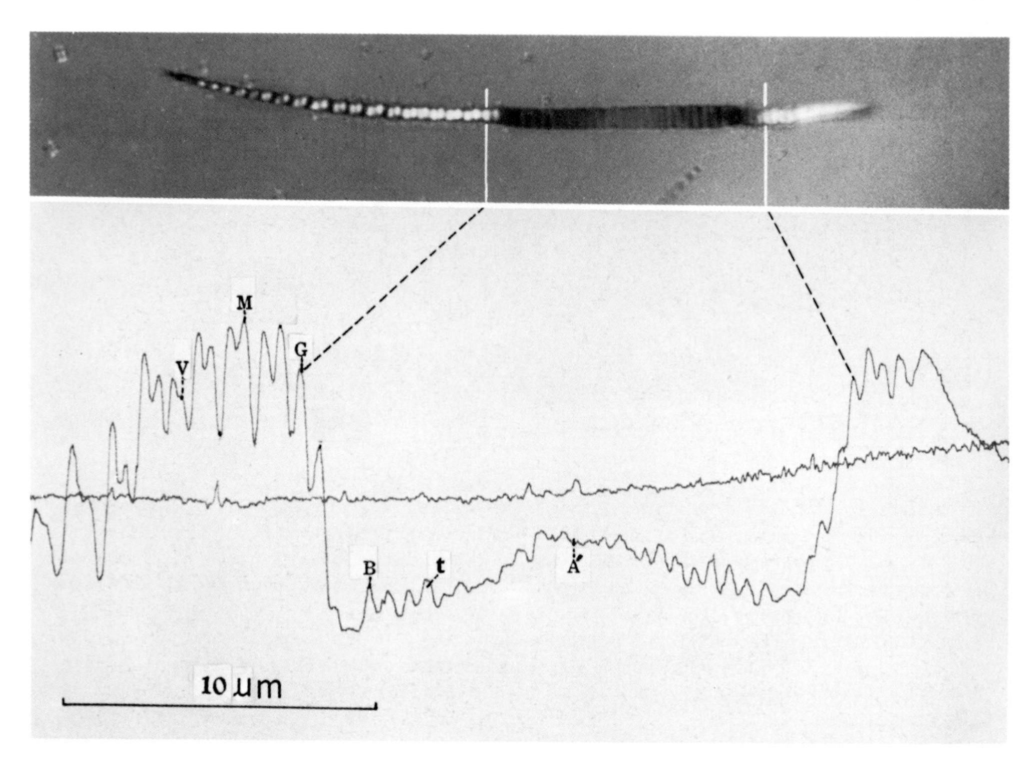

FIGURE III-24. Photograph of living cave cricket sperm in rectified polarizing microscope, with densitometer trace. The broad dark band in the middle of the sperm (where the densitometer trace dips below background) is where the birefringence of the DNA bases were selectively perturbed by irradiation with polarized UV microbeam. The compensator is in the subtractive orientation so that the sperm head is darker than the background except where the specimen retardation is greater than the compensator (especially at the small dumbbell-shaped white patches, which give rise to the M shapes on the densitometer trace). In these regions, the birefringence and optical axes of the "microcrystalline domains" of the sperm DNA are such as to overcome the subtractive effect of the compensator. (From Inoué and Sato, 1966b.)

to use the polarizing microscope to study weakly retarding specimens at the theoretical limit of microscopic resolution. Diffraction image errors, which can be present without rectification, are also corrected with the rectifier (Fig. III-23).*

*As shown in Fig. III-23, rectification corrects for the diffraction error that is introduced by conventional lenses when we observe low-retardation objects between crossed polars. However, the sensitivity for detecting weak retardations at high resolution is so improved by rectification that another hitherto unnoticed optical phenomenon becomes apparent. At the edges of any specimen, including isotropic materials, light is diffracted as though each edge were covered with a double layer of extremely thin, birefringent material; the slow axis on the high index side lying parallel, and on the low index side lying perpendicular, to the edge. We have named this phenomenon *edge birefringence*. It is found at all sharp boundaries (edges) whether the two sides of the boundary are solid, liquid, or gas, so long as there exists a refractive index difference on the two sides of the boundary (e.g., see Fig. 1-7D,F). It is clearly not based on the presence of a membrane at the optical interface, but is a basic diffraction phenomenon taking place at every edge. The electric vectors parallel and perpendicular to the edges must contribute to diffraction in a slightly asymmetric way on both sides of the edge and give rise to edge birefringence. Edge birefringence disappears when the refractive indices on both sides of the boundary are matched. It reverses in sign when the relative magnitude of refractive indices on the two sides of the boundary are reversed.

This behavior of edge birefringence is distinct from form birefringence. Form birefringence becomes zero (reaches a minimum value for rodlets, or maximum absolute value for platelets) when the refractive index of the immersion medium matches that of the rodlets or platelets, but then rises again (parabolically) when the refractive index of the immersion medium exceeds the match point (Ambronn and Frey, 1926; Sato *et al.*, 1975).

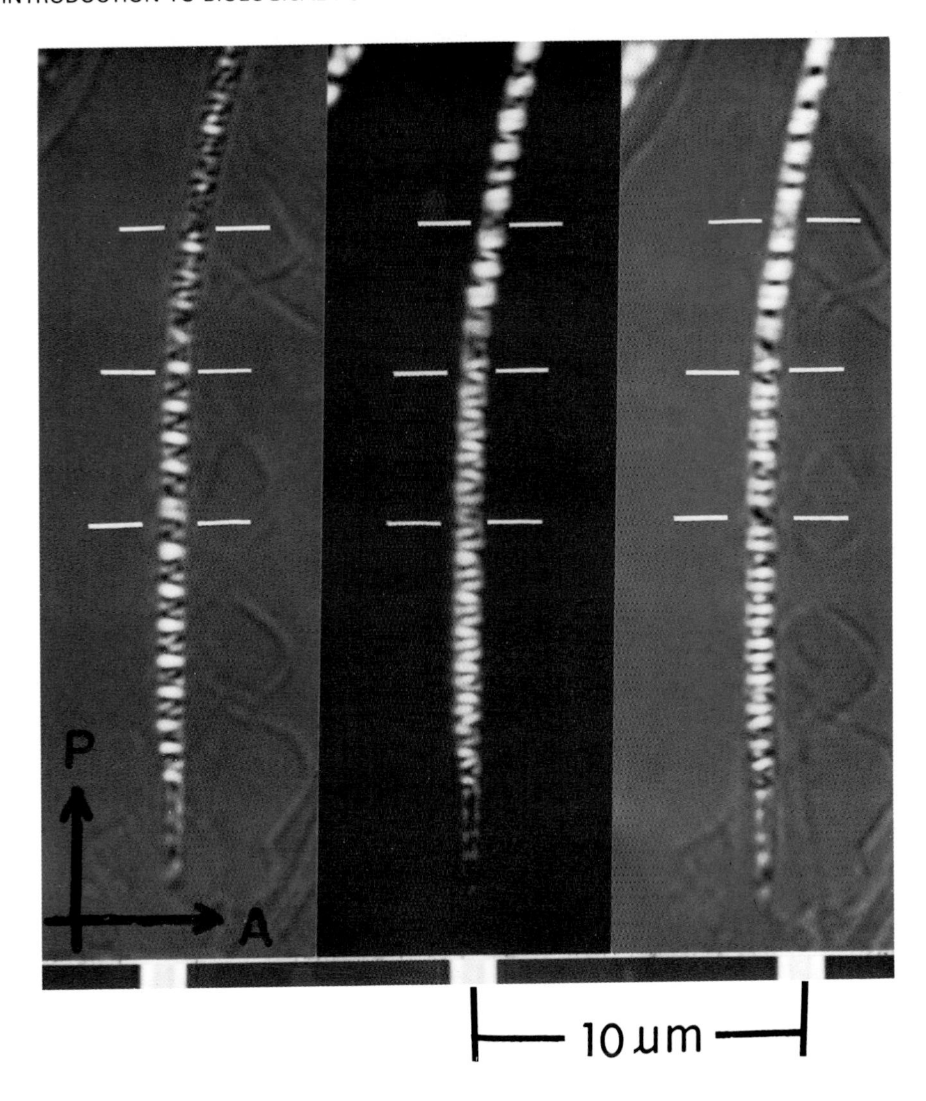

FIGURE III-25. Sperm head of a cave cricket (*Ceunthophilus nigricans*) observed with the rectified polarizing microscope at three different settings of a mica compensator. The detailed distribution of intrinsic birefringence in these chromosomes is shown with great clarity in the specimen immersed in dimethylsulfoxide ($N_D{}^{20}$ = 1.475). Horizontal white bars: positions of helix "breaks" that correspond to the ends of chromosomes. (From Inoué and Sato, 1966b.) Prior to these studies, few chromosomes had been seen in mature sperm of any species.

The improvements achieved with the rectified polarizing microscope have made possible a determination of the complex alignment of DNA molecules in each unit diffraction area of a live insect sperm (Figs. III-24, III-25). The birefringence and azimuth orientation of each DNA micro-domain were measured (both to a precision of 0.1°) from microdensitometer traces (of the type shown in Fig. III-24) by correlating the traces taken at several compensator settings and stage orientations. The changes in birefringence and azimuth angles, following polarized UV microbeam irradiation (which selectively abolishes the birefringence contribution by those bases that absorbed the UV), reveal the helical packing arrangement of the DNA molecules within the microdomains (Inoué and Sato, 1966b).

The rectified polarizing microscope offers many unique opportunities for studying molecular organization, and its changes, in a single living cell undergoing physiological activities and developmental changes.

ANNOTATED BIBLIOGRAPHY*

*Italicized reference numbers indicate material that may be of special interest to the reader.

BIBLIOGRAPHY

1. Allen, R. D. (1972) Pattern of birefringence in the giant amoeba, *Chaos carolinensis. Exp. Cell Res.* **72:**34–45.
2. Allen, R. D., and N. Kamiya, eds. (1964) *Primitive Motile Systems in Cell Biology.* Academic Press, New York.
3. Allen, R. D., and L. I. Rebhun (1962) Photoelectric measurement of small fluctuating retardations in weakly birefringent, light-scattering biological objects. I. The revolving tilted compensator method. *Exp. Cell Res.* **29:**583–592.
4. Allen, R. D., J. Brault, and R. D. Moore (1963) A new method of polarizing microscopic analysis. I. Scanning with a birefringence detector system. *J. Cell Biol.* **18:**223–235.
5. Allen, R. D., J. W. Brault, and R. Zeh (1966) Image contrast and phase modulation light methods in polarization and interference microscopy. *Adv. Opt. Electron Microsc.* **1:**77–114.
6. Ambronn, H., and A. Frey, (1926) *Das Polarisationsmikroskop, seine Anwedung in der Kolloidforschung in der Farbarei.* Akademische Verlag, Leipzig.
7. Ambrose, J., A. Elliot, and R. B. Temple (1949) New evidence on the structure of some proteins from measurements with polarized infra-red radiation. *Nature* **163:**859–862.
8. Anderson, S. (1949) Orientation of methylene blue molecules absorbed on solids. *J. Opt. Soc. Am.* **39:**49–56.
9. Andrews, C. L. (1960) *Optics of the Electromagnetic Spectrum.* Prentice–Hall, Englewood Cliffs, N.J.
10. Aronson, J. F. (1961) Sarcomere size in developing muscles of a tarsonemid mite. *J. Biophys. Biochem. Cytol.* **11:**147–156.
11. Aronson, J. F. (1963) Observations on the variations in size of the A regions of arthropod muscle. *J. Cell Biol.* **19:**359–367.
12. Aronson, J. F. (1967) Polarized light observations on striated muscle contraction in a mite. *J. Cell Biol.* **32:**169–179.
13. Aronson, J., and S. Inoué (1970) Reversal by light of the action of N-methyl N-desacetyl colchicine on mitosis. *J. Cell Biol.* **45:**470–477.
14. Asakura, S., M. Kasai, and F. Oosawa (1960) The effect of temperature on the equilibrium state of actin solutions. *J. Poly. Sci.* **44:**35–49.
15. Bajer, A., and J. Molé-Bajer (1972) Spindle dynamics and chromosome movements. *Intntl. Rev. Cytol.* suppl. **3**:1–271.
16. Barer, R., and V. E. Cosslett, eds. (1966) *Advances in Optical and Electron Microscopy,* Volume 1. Academic Press, New York, pp. 77–114.
17. Bear, R. S., and F. O. Schmitt (1936) Optical properties of the axon sheaths of crustacean nerves. *J. Cell Physiol.* **9:**275–288.
18. Bear, R. S., and F. O. Schmitt (1936). The measurement of small retardation with the polarizing microscope. *J. Opt. Soc. Am.* **26:**363–364.
19. Bear, R. S., F. O. Schmitt, and J. Z. Young (1937) The sheath components of the giant nerve fibres of the squid. *Proc. R. Soc. London* **123:**496.

19a. Begg, D. A., and G. W. Ellis (1979) Micromanipulation studies of chromosome movement. I. Chromosome–spindle attachment and the mechanical properties of chromosomal spindle fibers. *J. Cell Biol.* **82:**528–541.

19b. Begg, D. A., and G. W. Ellis (1979) Micromanipulation studies of chromosome movement. II. Birefringent chromosomal fibers and the mechanical attachment of chromosomes to the spindle. *J. Cell Biol.* **82:**542–554.

20. Bendet, I. J., and J. Bearden, Jr. (1972) Birefringence of spermatozoa. II. Form birefringence of bull sperm. *J. Cell Biol.* **55:**501–510.

21. Bennett, H. S. (1950) The microscopical investigation of biological materials with polarized light. In: *Handbook of Microscopical Technique* (C. E. McClung, ed.). Harper & Row (Hoeber), New York, pp. 591–677.

22. Bernal, J. D., and I. Fankuchen (1941) X-ray and crystallographic studies of plant virus preparations. I. Introduction and preparation of specimens. II. Modes of aggregation of the virus particles. *J. Gen. Physiol.* **25:**111–146 and 4 plates.

23. Born, M., and E. Wolf (1980) *Principles of Optics,* 6th ed. Pergamon Press, Elmsford, N.Y.

24. Bozler, E. (1937) The birefringence of muscle and its variation during contraction. *J. Cell. Comp. Physiol.* **10:**165–182.

25. Bragg, W. L., and A. B. Pippard (1953) The form birefringence of macromolecules. *Acta Crystallogr. Sect. B* **6:**865–867.

26. Brewster, D. (1916) On new properties of heat, as exhibited in its propagation along plates of glass. *Philos. Trans.* **106:**46–114 and Plates II–V.

27. Brewster, D. (1818) On the laws of polarization and double refraction in regularly crystallized bodies. *Philos. Trans.* **108:**199–273 and Plates XV–XVI.

28. Bryan, J., and H. Sato (1970) The isolation of the meiosis I spindle from the mature oocyte of *Pisaster ochraceus. Exp. Cell Res.* **59:**371–378.

29. Campbell, R. D., and S. Inoué (1965) Reorganization of spindle components following UV microirradiation. *Biol. Bull.* **129:**401.

29a. Cande, W. Z., and K. L. McDonald (1985) *In vitro* reactivation of anaphase spindle elongation using isolated diatom spindles. *Nature* **316**:168–170.

30. Carolan, R. M., H. Sato, and S. Inoué (1965) A thermodynamic analysis of the effect of D_2O and H_2O on the mitotic spindle. *Biol. Bull.* **129:**402; **131:**385.

31. Cassim, J. Y., and E. W. Taylor (1965) Intrinsic birefringence of poly-γ-benzyl-L-glutamate, a helical polypeptide, and the theory of birefringence. *Biophys. J.* **5:**531–551.

32. Cassim, J. Y., P. S. Tobias, and E. W. Taylor (1968) Birefringence of muscle proteins and the problem of structural birefringence. *Biochim. Biophys. Acta* **168:**463–471.

33. Chamot, E. M., and C. W. Mason (1958) *Handbook of Chemical Microscopy,* 2nd ed. Wiley, New York.

34. Chinn, P., and F. O. Schmitt (1937) On the birefringence of nerve sheaths as studied in cross section. *J. Cell. Comp. Physiol.* **9:**289–296.

35. Cohen, L. B., B. Hille, and R. D. Keynes (1969) Light scattering and birefringence changes during activity in the electric organ of *Electrophorus electricus. J. Physiol. (London)* **203:**489–509.

36. Cohen, L. B., and R. D. Keynes (1969) Optical changes in the voltage-clamped squid axon. *J. Physiol. (London)* **204:**100–101.

37. Cohen, L. B., R. D. Keynes, and B. Hille (1968) Light scattering and birefringence changes during nerve activity. *Nature* **218:**438–441.

38. Colby, R. H. (1971) Intrinsic birefringence of glycerinated myofibrils. *J. Cell Biol.* **51:**763–771.

39. Configurations and Interactions of Macro-Molecules and Liquid Crystals (1958) *Discuss. Faraday Soc.* **25:**1–235.

40. Dan, K., and K. Okazaki (1951) Change in the birefringence of the cortical layer of sea-urchin eggs induced by stretching. *J. Cell. Comp. Physiol.* **38:**427–435.

41. deSénarmont, H. (1840) Sur les modifications que la réflexion spéculaire à la surface des corps métalliques imprime à un rayon de lumière polarisée. *Ann. Chim. Phys.* **73**(Ser. 2):337–362.

42. Dietz, R. (1963) Polarisationsmikroskopische Befunde zur chromosomeninduzierten Spindelbildung bei der Tipulide. *Zool. Anz. Suppl.* **26:**131–138.

43. Ditchburn, R. W. (1963) *Light.* Blackie, London.

44. Dreizel, P. B., and A. Pfleidener (1959) Histochemische und Polarisationsoptische untersuchungen am Amyloid. *Arch. Pathol. Anat. Pathol. Physiol.* **332:**552.

45. Dvorak, J. A., T. R. Clem, and W. F. Stotler (1972) The design and construction of a computer-compatible system to measure and record optical retardation with a polarizing or interference microscope. *J. Microsc.* **96:**109.

45a. Dustin, P. (1978) *Microtubules.* Springer-Verlag, New York.
46. Eberstein, A., and A. Rosenfalck (1963) Birefringence of isolated muscle fibres in twitch and tetanus. *Acta Physiol. scand.* **57**:144–166.
47. Eisenberg, D., and W. Kauzmann (1969) *The Structure and Properties of Water.* Oxford University Press (Clarendon), London.
48. Ellis, G. W., and D. A. Begg (1981) Chromosome micromanipulation studies. In: *Mitosis/Cytokinesis.* (A. M. Zimmerman and A. Forer eds.). Academic Press, New York, pp. 155–179.
49. Engelmann, T. W. (1873) Mikroskopische Untersuchungen über die quergestreifte Muskelsubstanz. I, II. *Pfluegers Arch.* **7**:33–71, 155–188.
50. Engelmann, T. W. (1875) Contractilitat and Doppelbrechung. *Pfluegers Arch.* **11**:432–464.
51. Engelmann, T. W. (1906) Zur Theorie der Contractilitat. *Sitzungsber. K. Preuss. Akad. Wiss.* **39**:694–724.
52. Evans, R. C. (1976) *An Introduction to Crystal Chemistry.* Cambridge University Press, London.
53. Fauré-Fremiet, E. (1970) Microtubules et Mecanmismes Morphopoietiques. *Ann. Biol.* **9**:1–61.
54. Fischer, E. (1936) The submicroscopic structure of muscle and its changes during contraction and stretch. *Cold Spring Harbor Symp. Quant. Biol.* **4**:214–221.
55. Fischer, E. (1938) The birefringence of smooth muscle (*Phascolosoma* and *Thyone*) as related to muscle length, tension and tone. *J. Cell. Comp. Physiol.* **12**:85–101.
56. Forer, A. (1965) .Local reduction in spindle fiber birefringence in living *Nephrotoma suturalis* (Loew) spermatocytes induced by ultra-violet microbeam irradiation. *J. Cell Biol.* **25**(Mitosis Suppl.):95–117.
57. Forer, A. (1966) Characterization of the mitotic traction system, and evidence that birefringent spindle fibers neither produce nor transmit force for chromosome movement. *Chromosoma* **19**:44–98.
58. Forer, A. (1966) A simple conversion of reflecting lenses into phase-contrast condensers for ultraviolet light irradiations. *Exp. Cell Res.* **43**:688–691.
59. Forer, A. (1969) Chromosome movements during cell division. In: *Handbook of Molecular Cytology* (A. Lima-de Faria, ed.). North-Holland, Amsterdam, pp. 553–601.
60. Forer, A. (1976) Actin filaments and birefringent spindle fibers during chromosome movement. In: *Cell Motility,* Volume 3 (R. D. Goldman, T. D. Pollard, and J. L. Rosenbaum, eds.). Cold Spring Harbor Laboratory, Cold Spring Harbor, N.Y., pp. 1273–1293.
61. Forer, A., and R. D. Goldman (1972) The concentrations of dry matter in mitotic apparatuses in vivo and after isolation from sea-urchin zygotes. *J. Cell Sci.* **10**:387–418.
62. Forer, A., and A. M. Zimmerman (1976) Spindle birefringence of isolated mitotic apparatus analysed by pressure treatment. *J. Cell Sci.* **20**:309–327.
63. Forer, A., and A. M. Zimmerman (1976) Spindle birefringence of isolated mitotic apparatus analyzed by treatments with cold, pressure, and diluted isolation medium. *J. Cell Sci.* **20**:329–339.
64. Forer, A., V. I. Kalnins, and A. M. Zimmerman (1976) Spindle birefringence and isolated mitotic apparatus: Further evidence for two birefringent spindle components. *J. Cell Sci.* **22**:115–131.
65. Francis, D. W., and R. D. Allen (1971) Induced birefringence as evidence of endoplasmic viscoelasticity in *Chaos carolinensis. J. Mechanochem. Cell Motil.* **1**:1–6.
66. Françon, M. (1961) *Progress in Microscopy.* Row, Peterson, Evanston, Ill.
67. Frey-Wyssling, A. (1943) Doppelbrechung und Dichroismus als Mass der Nukleinsaure-Orientierung in Chromosomen. *Chromosoma* **2**:473–481.
68. Frey-Wyssling, A. (1953) *Submicroscopic Morphology of Protoplasm.* Elsevier, Amsterdam.
69. Frey-Wyssling, A. (1964) Quantitative Bestimmung des Formdichroismus. *Z. Wiss. Mikrosk.* **66**:45–53.
70. Gahm, J. (1964) Quantitative polarisationsoptische Messungen mit Kompensatoren. *Zeiss Mitt. Fortschr. Tech. Opt.* **3**:153–192.
71. Geil, R. H. (1963) *Polymer Single Crystals.* Interscience, New York.
72. Gibbons, I. R. (1967) The organization of cilia and flagella. In: *Molecular Organization and Biological Function* (J. Allen, ed.). Harper & Row, New York, pp. 211–237.
72a. Gibbons, I. R. (1975) The molecular basis of flagellar motility in sea urchin spermatozoa. In: *Molecules and Cell Movement* (S. Inoué and R. E. Stephens, eds.). Raven Press, New York, pp. 207–232.
73. Goldman, R. D., and L. I. Rebhun (1969) The structure and some properties of the isolated mitotic apparatus. *J. Cell Sci.* **4**:179–209.
74. Gray, G. W. (1962) *Molecular Structure and the Properties of Liquid Crystals.* Academic Press, New York.
75. Haase, M. (1961) Optische Eigenschaften neuerer Polarisationsfilter. *Zeiss Mitt. Fortschr. tech. Opt.* **2**:173–181.
76. Hallimond, A. F. (1953) *Manual of the Polarizing Microscope.* Cooke, Troughton & Simms, York.
77. Hárosi, F. (1975) Linear dichroism of rods and cones. In: *Vision in Fishes* (M. A. Ali, ed.). Plenum Press, New York, pp. 55–65.

77a. Hárosi, F. (1981) Microspectrophotometry and optical phenomena: Birefringence, dichroism, and anomalous dispersion. In: *Springer Series in Optical Sciences,* Volume 23 (J. M. Enoch and F. L. Tobey, Jr., eds.). Springer-Verlag, Berlin, pp. 337–399.
78. Hart, R., R. Zeh, and R. D. Allen (1977) Phase-randomized laser illumination for microscopy. *J. Cell Sci.* **23:**335–343.
79. Hartshorne, N. H., and A. Stuart (1960) *Crystals and the Polarising Microscope: A Handbook for Chemists and Others,* 3rd ed. Arnold, London.
80. Hartshorne, N. H., and A. Stuart (1964) *Practical Optical Crystallography.* American Elsevier, New York.
81. Higashi, S., M. Kasai, F. Oosawa, and A. Wada (1963) Ultraviolet dichroism of F-actin oriented by flow. *J. Mol. Biol.* **7:**421–430.
82. Hiramoto, Y., Y. Hamaguchi, Y. Shoji, and S. Shimoda (1981) Quantitative studies on the polarization optical properties of living cells. I. Microphotometric birefringence detection system. *J. Cell Biol.* **89:**115–120.
83. Hiramoto, Y., Y. Hamaguchi, Y. Shoji, T. E. Schroeder, S. Shimoda, and S. Nakamura (1981) Quantitative studies on the polarization optical properties of living cells. II. The role of microtubules in birefringence of the spindle of the sea urchin egg. *J. Cell Biol.* **89:**121–130.
83a. Hiramoto, Y., and Y. Shoji (1982) Location of the motive force for chromosome movement in sand-dollar eggs. In: *Biological Functions of Microtubules and Related Structures* (H. Sakai, H. Mohri, and G. G. Borisy, eds.). Academic Press, New York, pp. 247–249.
84. Hsien Y. H., M. Richartz, and K. L. Yung (1947) A generalized intensity formula for a system of retardation plates. *J. Opt. Soc. Am.* **37:**99–106.
85. Hughes, A. F., and M. M. Swann (1948) Anaphase movements in the living cell: A study with phase contrast and polarized light on chick tissue culture. *J. Exp. Biol.* **25:**45–70.
86. Huxley, A. F. (1960) British patent specification 856,621. Improvements in or relating to polarizing microscopes (applied July 20, 1956).
87. Inoué, S. (1951) A method for measuring small retardations of structures in living cells. *Exp. Cell Res.* **2:**513–517.
88. Inoué, S. (1952) The effect of colchicine on the microscopic and sub-microscopic structure of the mitotic spindle. *Exp. Cell Res. Suppl.* **2:**305–318.
89. Inoué, S. (1952) Effect of temperature on the birefringence of the mitotic spindle. *Biol. Bull.* **103:**316.
90. Inoué, S. (1952) Studies on depolarization of light at microscope lens surfaces. I. The origin of stray light by rotation at the lens surfaces. *Exp. Cell Res.* **3:**199–208.
91. Inoué, S. (1953) Polarization optical studies of the mitotic spindle. I. The demonstration of spindle fibers in living cells. *Chromosoma* **5:**487–500.
92. Inoué, S. (1960) Birefringence in Dividing Cells. Time-lapse motion picture.
93. Inoué, S. (1961) Polarizing microscope: Design for maximum sensitivity. In: *The Encyclopedia of Microscopy* (G. L. Clarke, ed.). Reinhold, New York, pp. 480–485.
94. Inoué, S. (1964) Organization and function of the mitotic spindle. In: *Primitive Motile Systems in Cell Biology* (R. D. Allen and N. Kamiya, eds.). Academic Press, New York, pp. 549–598.
95. Inoué, S. (1969) The physics of structural organization in living cells. In: *Biology and the Physical Sciences* (S. Devons, ed.). Columbia University Press, New York, pp. 139–171.
96. Inoué, S. (1976) Chromosome movement by reversible assembly of microtubules. In: *Cell Motility,* Volume 3 (R. D. Goldman, T. D. Pollard, and J. L. Rosenbaum, eds.). Cold Spring Harbor Laboratory, Cold Spring Harbor, N.Y., pp. 1317–1328.
97. Inoué, S. (1981) Cell division and the mitotic spindle. *J. Cell Biol.* **91**(Part 2)**:**131s–147s.
97a. Inoué, S. (1981) Video image processing greatly enhances contrast, quality, and speed in polarization-based microscopy. *J. Cell Biol.* **89:**346–356.
98. Inoué, S. (1982) The role of self-assembly in the generation of biologic form. In: *Developmental Order: Its Origin and Regulation* (P. B. Green, ed.). Liss, New York, pp. 35–76.
99. Inoué, S., and A. Bajer (1961) Birefringence in endosperm mitosis. *Chromosoma* **12:**48–63.
100. Inoué, S., and K. Dan (1951) Birefringence of the dividing cell. *J. Morphol.* **89:**423–456.
101. Inoué, S., and W. L. Hyde (1957) Studies on depolarization of light at microscope lens surfaces. II. The simultaneous realization of high resolution and high sensitivity with the polarizing microscope. *J. Biophys. Biochem. Cytol.* **3:**831–838.
102. Inoué, S., and D. P. Kiehart (1978) In vivo analysis of mitotic spindle dynamics. In: *Cell Reproduction: In Honor of Daniel Mazia* (E. Dirksen, D. Prescott, and C. F. Fox, eds.). Academic Press, New York, pp. 433–444.
103. Inoué, S., and C. Koester (1959) Optimum half shade angle in polarizing instruments. *J. Opt. Soc. Am.* **49:**556–559.

104. Inoué, S., and H. Kubota (1958) Diffraction anomaly in polarizing microscopes. *Nature* **182:**1725–1726.
105. Inoué, S., and K. Okazaki (1978) Biocrystals. *Sci. Am.* **236**(4)**:**82–92.
106. Inoué, S., and H. Ritter, Jr. (1975) Dynamics of mitotic spindle organization and function. In: Molecules and Cell Movement (S. Inoué and R. E. Stephens, eds.). Raven Press, New York, pp. 3–30.
107. Inoué, S., and H. Sato (1962) Arrangement of DNA in living sperm: A biophysical analysis. *Science* **136:**1122–1124.
108. Inoué, S., and H. Sato (1966) Arrangement of DNA molecules in the sperm nucleus: An optical approach to the analysis of biological fine structure. In: *Biophysical Science Series 3, Progress in Genetics III* (Japanese Biophysical Society ed.). Yoshioka, Kyoto, pp. 151–220.
109. Inoué, S., and H. Sato (1966) Deoxyribonucleic acid arrangements in living sperm. In: *Molecular Architecture in Cell Physiology* (T. Hayashi and A. G. Szent-Gyorgyi, eds.). Prentice–Hall, Englewood Cliffs, N.J., pp. 209–248.
110. Inoué, S., and H. Sato (1967) Cell motility by labile association of molecules: The nature of mitotic spindle fibers and their role in chromosome movement. *J. Gen. Physiol.* **50:**259–292.
111. Inoué, S., and L. G. Tilney (1982) The acrosomal reaction of *Thyone* sperm. I. Changes in the sperm head visualized by high resolution video microscopy. *J. Cell Biol.* **93:**812–819.
112. Inoué, S., G. G. Borisy, and D. P. Kiehart (1974) Growth and lability of *Chaetopterus* oocyte mitotic spindles isolated in the presence of porcine brain tubulin. *J. Cell Biol.* **62:**175–184.
113. Inoué, S., G. W. Ellis, E. D. Salmon, and J. W. Fuseler (1970) Rapid measurement of spindle birefringence during controlled temperature shifts. *J. Cell Biol.* **47:**95a.
114. Inoué, S., J. Fuseler, E. D. Salmon, and G. W. Ellis (1975) Functional organization of mitotic microtubules—Physical chemistry of the in vivo equilibrium system. *Biophys. J.* **15:**725–744.
115. Inoué, S., D. P. Kiehart, I. Mabuchi, and G. W. Ellis (1979) Molecular mechanism of mitotic chromosome movement. In: *Motility in Cell Function, First John M. Marshall Symposium in Cell Biology* (F. A. Pépe, J. W. Sanger, and V. T. Nachmias, eds.). Academic Press, New York, pp. 301–311.
116. Inoué, S., H. Ritter and D. Kubai (1978) Mitosis in *Barbulanympha.* II. Dynamics of a two-stage anaphase, nuclear morphogenesis, and cytokinesis. *J. Cell Biol.* **77:**655–684.
116a. Itoh, T. J., and H. Sato (1984) The effects of deuterium oxide (^2H_2O) on the polymerization of tubulin in vitro. *Biochim. Biophys. Acta* **800**:21–27.
116b. Izutsu, K. (1961) Effects of ultraviolet microbeam irradiation upon division in grasshopper spermatocytes. I. Results of irradiation during prophase and prometaphase I. *Mie Med. J.* **11**(2)**:**199–212.
116c. Izutsu, K. (1961) Effects of ultraviolet microbeam irradiation upon division in grasshopper spermatocytes. II. Results of irradiation during metaphase and anaphase I. *Mie Med. J.* **11**(2)**:**213–232.
117. Jagger, W. S., and P. A. Liebman (1970) Birefringent transients in photoreceptor outer segments. *Biophys. Soc. Abstr. 14th Annu. Meet.* p. 59A.
118. Jelley, E. E. (1945) Microscopy. In: *Physical Methods of Organic Chemistry* (A. Weissberg, ed.). Interscience, New York, pp. 451–524.
119. Jenkins, F. A., and H. White (1957) *Fundamentals of Optics,* 3rd ed. McGraw–Hill, New York.
120. Jerrard, H. G. (1948) Optical compensators for measurement of elliptical polarization. *J. Opt. Soc. Am.* **38:**35–59.
121. Jerrard, H. G. (1954) Transmission of light through birefringent and optically active media: The Poincaré sphere. *J. Opt. Soc. Am.* **44:**634–640.
122. Johansen, A. (1918) *Manual of Petrographic Methods,* 2nd ed. McGraw–Hill, New York.
123. Kautz, J., Q. B. De Marsh, and W. Thornburg (1957) A polarizing and electron microscope study of plasma cells. *Exp. Cell Res.* **13:**596–599.
123a. Kiehart, D. P. (1981) Studies on the in vivo sensitivity of spindle microtubules to calcium ions and evidence for a vesicular calcium-sequestering system. *J. Cell Biol.* **88:**604–617.
124. Kiehart, D. P., I. Mabuchi, and S. Inoué (1982) Evidence that myosin does not contribute or force production in chomosome movement. *J. Cell Biol.* **94:**165–178.
125. Kobayasi, S. (1964) Effect of electric field on F-actin orientated by flow. *Biochim. Biophys. Acta* **88:**541–552.
126. Kobayasi, S., H. Asai, and F. Oosawa (1964) Electric birefringence of actin. *Biochim. Biophys. Acta* **88:**528–540.
127. Köhler, A. (1921) Ein Glimmerplattchen Grau. I. Ordnung zur Untersuchung sehr schwach doppelbrechender Praparate. *Z. Wiss. Mikrosk.* **38:**29–42.
128. Kubo, R. (1947) Statistical theory of rubber-like substances. *J. Colloid Sci.* **2:**527–535.
129. Kubota, H., and S. Inoué (1959) Diffraction images in the polarizing microscope. *J. Opt. Soc. Am.* **49:**191–198.

130. Kunitz, M. (1930) Elasticity, double refraction and swelling of isoelectric gelatin. *J. Gen. Physiol.* **13:**565–606.
131. LaFountain, J. R. (1972) Changes in the patterns of birefringence and filament deployment in the meiotic spindle of *Neophrotoma suturalis* during the first meiotic division. *Protoplasma* **75:**1–17.
132. LaFountain, J. R. (1974) Birefringence and fine structure of spindles in spermatocytes of *Nephrotoma suturalis* at metaphase of first meiotic division. *J. Ultrastruct. Res.* **46:**268–278.
133. LaFountain, J. R. (1976) Analysis of birefringence and ultrastructure of spindles in primary spermatocytes of *Nephrotoma suturalis* during anaphase. *J. Ultrastruct. Res.* **54:**333–346.
134. LaFountain, J. R., and L. A. Davidson (1979) An analysis of spindle ultrastructure during prometaphase and metaphase of micronuclear division of *Tetrahymena. Chromosoma* **75:**293–308.
135. Langford, G. M., and S. Inoué (1979) Motility of the microtubular axostyle in *Pyrsonympha. J. Cell Biol.* **80:**521–538.
136. Lambrecht, K. See catalog "Polarizing Optics," 4204 N. Lincoln Ave., Chicago, Ill. 60618.
137. Land, E. H. (1951) Some aspects of the development of sheet polarizers. *J. Opt. Soc. Am.* **41:**957–963.
138. Land, E. H., and C. D. West (1946) Dichroism and dichroic polarizers. In: *Colloid Chemistry,* Vol. 6 (J. Alexander, ed.). Reinhold, New York, pp. 160–190.
138a. Leslie, R. J., and J. D. Pickett-Heaps (1983) Ultraviolet microbeam irradiations of mitotic diatoms: Investigation of spindle elongation. *J. Cell Biol.* **96:**548–561.
138b. Leslie, R. J., and J. D. Pickett-Heaps (1984) Spindle microtubule dynamics following ultraviolet microbeam irradiations of mitotic diatom. *Cell* **36:**717–727.
139. Luthy, H. (1946) Dichroismus der einzelnen Nervenfaser. *Helv. Physiol. Pharmacol. Acta* **4:**50.
140 MacInnes, J. W., and R. B. Uretz (1967) Thermal depolarization of fluorescence from polytene chromosomes stained with acridine orange. *J. Cell Biol.* **33:**597–604.
141. Maestre, M. F., and R. Kilkson (1965) Intrinsic birefringence of multiple coiled DNA, theory and applications. *Biophys. J.* **5:**275–287.
142. Makas, A. S. (1962) Film polarizer for visible and ultraviolet radiation. *J. Opt. Soc. Am.* **52:**43–44.
143. Malawista, S. E., K. G. Bensch, and H. Sato (1968) Vinblastine and griseofulvin reversibly disrupt the living mitotic spindle. *Science* **160:**770–772.
144. Mandarino, J. A. (1959) Absorption and pleochroism: Two much-neglected optical properties of crystals. *Am. Mineral.* **4:**65–76.
145. Marek, L. F. (1978) Control of spindle form and function in grasshopper spermatocytes. *Chromosoma* **68:**367–398.
146. Mark, H. F. (1957) Giant molecules. *Sci. Am.* **197**(Sept.)**:**80–89.
147. Martin, L. C. (1966) *The Theory of the Microscope.* Blackie, London.
148. Mazia, D. (1961) Mitosis and the physiology of cell division. In: *The Cell* (J. Brachet and A. Mirsky, eds.). Academic Press, New York, pp. 77–394.
148a. Mazia, D., and K. Dan (1952) The isolation and biochemical characterization of the mitotic apparatus of dividing cells. *Proc. Natl. Acad. Sci. USA* **38:**826–838.
149. McIntosh, J. R., W. Z. Cande, and J. A. Snyder (1975) Structure and physiology of the mammalian mitotic spindle. In: *Molecules and Cell Movement* (S. Inoué and R. E. Stephens, Raven Press, New York, pp. 31–76.
150. McIntosh, J. R., P. K. Hepler, and D. G. van Wie (1969) Model for mitosis. *Nature* **224:**659–663.
151. Menke, W. (1943) Dichroismus and Doppelbrechung der Plastiden. *Biol. Zentralbl.* **63:**326–349.
152. Mercer, E. H. (1952) The biosynthesis of fibers. *Sci. Mon.* **75:**280–287.
152a. Michael, K. (1981) *Die Grundzüge der Theorie des Mikroskops in elementarer Darstellung.* Wissenschaftliche Verlag, Stuttgart.
153. Mitchison, J. M. (1952) Cell membranes and cell division: The structure of the cell membrane. *Symp. Soc. Exp. Biol.* **6:**105–127.
154. Mitchison, J. M. (1953) A polarized light analysis of the human red cell ghost. *J. Exp. Biol.* **30:**397–432.
155. Mitchison, J. M., and M. M. Swann (1952) Optical changes in the membranes of the sea-urchin egg at fertilization, mitosis and cleavage. *J. Exp. Biol.* **29:**357–362.
156. Monné, L. (1939) Polarizationoptische Untersuchungen über den Golgi-Apparat und die Mitochondrien männlichen Geschlechtszellen eineger Pulmonaten Arten. *Protoplasma* **32:**184.
157. Monné, L. (1944) Cytoplasmic structure and cleavage pattern of the sea urchin egg. *Ark. Zool.* **35A:**1–27.
158. Monné, L. (1945) Investigations into the structure of the cytoplasm. *Ark. Zool.* **36A:**1–29.
159. Nakajima, H. (1964) The mechanochemical system behind streaming in *Physarum.* In: *Primitive Motile Systems in Cell Biology* (R. D. Allen and N. Kamiya, eds.). Academic Press, New York, pp. 111–123.
160. Nicklas, R. B. (1970) Mitosis. In: *Advances in Cell Biology, Volume II* (D. M. Prescott, L. Goldstein,

and E. H. McConkey, eds.). Appleton–Century–Crofts, New York, pp. 225–297.

160a. Nicklas, R. B. (1983) Measurements of the force produced by the mitotic spindle in anaphase. *J. Cell Biol.* **97:**542–548.

161. Noll, D., and H. H. Weber (1935) Polarisationsoptik und molekularer Feinbau der Z-Abschnitte des Froschmuskels. *Pfluegers Arch.* **235:**234–246.

162. Okazaki, K., and S. Inoué (1976) Crystal property of the larval sea urchin spicule. *Dev. Growth Differ.* **18:**413–434.

163. Okazaki, K., K. McDonald, and S. Inoué (1980) Sea urchin larval spicule observed with the scanning electron microscope. In: *The Mechanisms of Biomineralization in Animals and Plants* (M. Omori and N. Watanabe, eds.). Tokai University Press, Tokyo, pp. 159–168.

164. Oncley, J. L. (1959) Biophysical sciences—A study program. *Rev. Mod. Phys.* **31:**1–568.

165. Oosawa, F., and S. Asakura (1975) *Thermodynamics of the Polymerization of Protein.* Academic Press, New York.

166. Oster, G. (1955) Birefringence and dichroism. In: *Physical Techniques in Biological Research,* Volume I (G. Oster and A. Pollister, eds.) Academic Press, New York, pp. 439–460.

167. Pattri, H. O. E. (1932) Über die Doppelbrechung der Spermien. *Z. Zellforsch. mikrosk. Anat.* **16:**723–744.

168. Perutz, M. F., M. Jobe, and R. Barer (1950) Observations of proteins in polarized ultra-violet light. *Discuss. Faraday Soc.* **9:**423–427.

169. Perutz, M. F., and J. M. Mitchison (1950) State of haemoglobin in sickle cell anaemia. *Nature* **166:**677–682.

170. Pfeiffer, H. H. (1938) Double refraction measurements and structural changes in mitotic spindles disturbed by centrifugal force. *Biodynamica* **2:**1–8.

171. Pfeiffer, H. H. (1941) Experimentalle Beitrage zur submikroskopischen Feinbaukunde (Leptonik) undifferenzierten Cytoplasmas. *Ber. Dtsch. Bot. Ges.* **59:**288.

172. Pfeiffer, H. H. (1941) Mikrurgisch-polarisationsoptische Beitrage zur submikroskopischen Morphologie larvaler Speicheldrusenchromosomen von *Chironomus. Chromosoma* **2:**77–85.

173. Pfeiffer, H. H. (1942) Polarisationsmikroskopische Dehnungs und Kontraktionsversuche an Chromatinabschnitten von Chara-Spermatozoiden. *Chromosoma* **2:**334–344.

174. Pfeiffer, H. H. (1951) Polarisationsoptische Untersuchungen am spindellapparat mitotischen Zellen. *Cytologia* **16:**194–200.

175. Pfeiffer, H. H. (1949) *Das Polarisationsmikroskop als Messinstrument in Biologie und Medizin.* Vieweg, Braunschwieg.

176. Picken, L. (1960) *The Organization of Cells and Other Organisms.* Oxford University Press (Clarendon), London.

176a. Pickett-Heaps, J. D., and D. H. Tippit (1978) The diatom spindle in perspective. *Cell* **14:**455–467.

177. Platt, J. R., ed. (1964) *Systematics of the Electronic Spectra of Conjugated Molecules.* Wiley, New York.

178. Ramachandran, G. N., and S. Ramaseshan (1961) Crystal optics. In: *Handbuch der Physik,* Volume XXV/I (S. Flugge, ed.). Springer-Verlag, Berlin, pp. 1–217.

179. Rebhun, L. I., and G. Sander (1967) Ultrastructure and birefringence of the isolated mitotic apparatus of marine eggs. *J. Cell Biol.* **34:**859–883.

180. Rebhun, L. I., and N. Sawada (1969) Augmentation and dispersion of the in vivo mitotic apparatus of living marine eggs. *Protoplasma* **68:**1–22.

180a. Rebhun, L. I., J. Rosenbaum, P. Lefebvre, and G. Smith (1974) Reversible restoration of the birefringence of cold-treated, isolated mitotic apparatus of surf clam eggs with chick brain tubulin. *Nature* **249:**113–115.

181. Rees, A. L. G. (1951) Directed aggregation in colloidal systems and the formation of protein fibers. *J. Phys. Chem.* **55:**1340–1344.

182. Richartz, M., and Y. H. Hsu (1948) Analysis of elliptical polarization. *J. Opt. Soc. Am.* **39:**136–157.

183. Rinne, F. W. B., and M. Berek (1953) *Anleitung zu optischen Untersuchungen mit dem Polarizationmikroskop.* Stuttgart, Schweizerbart.

184. Ritter, H., S. Inoué, and D. Kubai (1978) Mitosis in Barbulanympha. I. Spindle structure, formation, and kinetochore engagement. *J. Cell Biol.* **77:**638–654.

184a. Roberts, K., and J. S. Hyams, eds. (1979) *Microtubules.* Academic Press, New York.

185. Ruch, F. (1956) Birefringence and dichroism of cells and tissues. In: *Physical Techniques in Biological Research,* Volume III (G. Oster and A. Pollister, eds.). Academic Press, New York, pp. 149–176.

186. Rünnstrom, J. (1928) Die Veränderungen der Plasmakolloide bei der Entwicklungs-Erregung des Seeigeleies. *Protoplasma* **4:**338–514.

187. Salmon, E. D. (1975a) Pressure-induced depolymerization of spindle microtubules. I. Spindle birefringence and length changes. *J. Cell Biol.* **65:**603–614.
188. Salmon, E. D. (197b) Pressure-induced depolymerization of brain microtubules in vitro. *Science* **189:**884–886.
189. Salmon, E. D. (1975c) Spindle microtubules: Thermodynamics of in vivo assembly and role in chromosome movement. *Ann. N.Y. Acad. Sci.* **253:**383–406.
190. Salmon, E. D. (1984) Tubulin dynamics in microtubules of the mitotic spindle. In: *Molecular Biology of the Cytoskeleton* (G. Borisy, D. Cleveland, and D. Murphy, eds.). Cold Spring Harbor Laboratory, Cold Spring Harbor, N.Y., pp. 99–109.
191. Salmon, E. D., and D. A. Begg (1980) Functional implications of cold-stable microtubules in kinetochore fibers of insect spermatocytes during anaphase. *J. Cell Biol.* **85:**853–865.
192. Salmon, E. D., and G. W. Ellis (1975) A new miniature hydrostatic pressure chamber for microscopy: Strain-free optical glass windows facilitate phase contrast and polarized light microscopy of living cells. Optional fixture permits simultaneous control of pressure and temperature. *J. Cell Biol.* **65:**587–602.
193. Salmon, E. D., and G. W. Ellis (1976) Compensator transducer increases ease, accuracy and rapidity of measuring changes in specimen birefringence with polarization microscopy. *J. Microsc.* **106:**63–69.
194. Salmon, E. D., M. McKeel, and T. Hays (1984) Rapid rate of tubulin dissociation from microtubules in the mitotic spindle *in vivo* measured by blocking polymerization with colchicine. *J. Cell Biol.* **99:**1066–1075.
195. Sato, H. (1969) Analysis of form birefringence in the mitotic spindle. *Am. Zool.* **9:**592.
196. Sato, H. (1975) The mitotic spindle. In: *Aging Gametes* (R. Blandau, ed.). Karger Basel, pp. 19–49.
197. Sato, H., and J. Bryan (1968) Kinetic analysis of association–dissociation reaction in the mitotic spindle. *J. Cell Biol.* **39:**118a.
198. Sato, H., and S. Inoué (1964) Condensation of the sperm nucleus and alignment of DNA molecules during spermiogenesis in *Loligo pealei. Biol. Bull.* **127:**357.
199. Sato, H., G. W. Ellis, and S. Inoué (1975) Microtubular origin of mitotic spindle form birefringence. *J. Cell Biol.* **67:**501–517.
199a. Sato, H., T. Kato, T. C. Takahashi, and T. Ito (1982) Analysis of D_2O effect on *in vivo* and *in vitro* tubulin polymerization and depolymerization. In: Biological Functions of Microtubules and Related Structures (H. Sakai, H. Mohri, and G. G. Borisy, eds.). Academic Press, New York. pp. 211–226.
200. Sawada, N., and L. I. Rebhun (1969) The effect of dinitrophenol and other phosphorylation uncouplers on the birefringence of the mitotic apparatus of marine eggs. *Exp. Cell Res.* **55:**33–38.
201. Schmidt, W. J. (1924) *Die Bausteine des Tierkorpers in polarisiertem Lichte.* Cohen, Bonn.
202. Schmidt, W. J. (1937) *Die Doppelbrechung von Karyoplasma, Zytoplasma und Metaplasma. Protoplasma-Monographien* Volume 11. Borntraeger, Berlin.
203. Schmidt, W. J. (1939) Doppelbrechung der Kernspindel und Zugfasertheorie der Chromosomenbewegung. *Chromosoma* **1:**253–264.
204. Schmidt, W. J. (1941) Einiges uber Optische Anisotropie und Feinbau von Chromatin and Chromosomen. *Chromosoma* **2:**86–100.
205. Schmidt, W. J. (1944) Die Doppelbrechung des Protoplasmas und ihrer Bedeutung fur die Erforschung seines submikroskopischen Baues. *Ergeb. Physiol.* **44:**27.
206. Schmidt, W. J. (1951) Polarisationsoptische analyse der Verknüpfung von Protein-und Lipoidmolekeln, erlautert am Aussenglied der Sehzellen der Wirbeltiere. *Publ. Staz. Zool. Napoli* **23:**158–183.
207. Schmidt, W. J. (1952) Polarisationmikroskopische Beobachtungen an Zellmembrane, Kern, Cytoplasma Tentakel der *Noctiluca miliaris* Suriray. *Zool. Anz.* **148:**115–131.
208. Schmidt, W. J. (1957) Instrumente und Methoden zur mikroskopischen Untersuchung optisch anisotroper Materielen mit Ausschluss der Kristalle. In: *Handbuch der Mikroskopie in der Technik,* Volume I (H. Freund, ed.). Umschau Verlag, Frankfurt am Main, pp. 147–314.
209. Schmidt, W. J. (1964) *Wissenschaftliche Veröffentlichungen.* Justus Leibig-Universität, Giessen/Lahn.
210. Schmitt, F. O. (1955) Cell constitution. In: *Analysis of Development* (B. Willier, P. Weiss, and V. Hamburger, eds.). Saunders, Philadelphia, pp. 39–69.
211. Schmitt, F. O., and N. Geschwind (1957) *The axon surface. Prog. Biophys.* **8**:165–215.
212. Schrader, F. (1953) *Mitosis—The Movements of Chromosomes in Cell Division.* Columbia University Press, New York.
213. Sebera, D. (1964) *Electronic Structure and Chemical Bonding.* Gina (Blaisdell), Boston.
214. Seeds, W. E. (1953) Polarized ultraviolet microspectrography and molecular structure. *Prog. Biophys.* **3:**27–46.
215. Seeds, W. E., and M. H. F. Wilkins (1959) Ultra-violet microspectrographic studies of nucleoproteins and crystals of biological interest. *Biochim. Biophys. Acta* **11:**417–427.
216. Shillaber, C. P. (1944) *Photomicrography in Theory and Practice.* Wiley, New York.

216a. Shurcliff, W. A. (1962) *Polarized Light: Production and Use*. Harvard University Press, Cambridge, Mass.
217. Skinner, C. A. (1925) A universal polarimeter. *J. Opt. Soc. Am.* **10:**491–520.
217a. Sluder, G. (1976) Experimental manipulation of the amount of tubulin available for assembly into the spindle of dividing sea urchin eggs. *J. Cell Biol.* **70:**75–85.
217b. Sluder, G. (1979) Role of spindle microtubules in the control of cell cycle timing. *J. Cell Biol.* **80:**674–691.
218. Smith, L. W. (1960) Diffraction images of disk-shaped particles computed for full Köhler illumination. *J. Opt. Soc. Am.* **50:**369–374.
218a. Soranno, T., and E. Bell (1982) Cytostructural dynamics of spreading and translocating cells. *J. Cell Biol.* **95:**127–136.
219. Stephens, R. E. (1965) Analysis of muscle contraction by ultraviolet microbeam disruption of sarcomere structure. *J. Cell Biol.* **25:**129–139.
220. Stephens, R. E. (1969) Factors influencing the polymerization of outer fiber microtubule protein. *Q. Rev. Biophys.* **1:**377–390.
221. Stephens, R. E. (1972) Studies on the development of the sea urchin *Strongylocentrotus droebachiensis*. II. Regulation of mitotic spindle equilibrium by environmental temperature. *Biol. Bull.* **142:**145–159.
222. Stephens, R. E. (1973) A thermodynamic analysis of mitotic spindle equilibrium at active metaphase. *J. Cell Biol.* **57:**133–147.
223. Strong, J. (1958) *Concepts of Classical Optics*. Freeman, San Francisco.
224. Sutherland, G. B. B. M., and M. Tsuboi (1957) The infra-red spectrum and molecular configuration of sodium deoxyribonucleate. *Proc. R. Soc. London Ser. A.* **239:**446–463.
225. Swann, M. M. (1951) Protoplasmic structure and mitosis. I. The birefringence of the metaphase spindle and asters of the living sea-urchin egg. *J. Exp. Biol.* **28:**417–433.
226. Swann, M. M. (1951) Protoplasmic structure and mitosis. II. The nature and causes of birefringence changes in the sea-urchin egg at anaphase. *J. Exp. Biol.* **28:**434–444.
227. Swann, M. M., and J. M. Mitchison (1950) Refinements in polarized light microscopy. *J. Exp. Biol.* **27:**226–237.
228. Swann, M. M., and J. M. Mitchison (1958) The mechanism of cleavage in animal cells. *Biol. Rev.* **33:**103–135.
229. Takahashi, W. N., and T. E. Rawlins (1933) Rod-shaped particles in tobacco mosaic virus demonstrated by stream double refraction. *Science* **77:**26–27.
229a. Takashima, S. (1968) Optical anisotropy of synthetic polynucleotides. I. Flow birefringence and π-electron polarizability of bases. *Biopolymers* **6:**1437–1452.
230. Tasaki, I., A. Watanabe, R. Sandlin, and L. Carnay (1968) Changes in fluorescence, turbidity and birefringence associated with nerve excitation. *Proc. Natl. Acad. Sci. USA* **61:**883–888.
231. Taylor, D. L. (1975) Birefringence changes in vertebrate striated muscle. *J. Supramol. Struct.* **3**:181–191.
232. Taylor, D. L. (1976) Quantitative studies on the polarization optical properties of striated muscle. I. Birefringence changes of rabbit psoas muscle in the transition from rigor to relaxed state. *J. Cell Biol.* **68:**497–511.
233. Taylor, D. L. (1976) Motile models of amoeboid movement. In: *Cell Motility*, Volume 3. (R. Goldman, T. Pollard, and J. Rosenbaum, eds.). Cold Spring Harbor Laboratory, Cold Spring Harbor, N.Y., pp. 797–821.
234. Taylor, D. L. (1976) The contractile basis of amoeboid movement. IV. The viscoelasticity and contractility of amoeba cytoplasm in vivo. *Exp. Cell Res.* **105:**413–426.
235. Taylor, D. L., and R. M. Zeh (1976) Methods for the measurement of polarization optical properties. I. Birefringence. *J. Microsc.* **108:**251–259.
236. Taylor, D. L., R. D. Allen, and E. P. Benditt (1975) Determination of the polarization optical properties of the amyloid congo red complex by phase modulation microspectrophotometry. *J. Histochem.* **22:**1105–1112.
237. Taylor, D. L., J. S. Condeelis, P. L. Moore, and R. D. Allen (1973) The contractile basis of amoeboid movement. I. The chemical control of motility in isolated cytoplasm. *J. Cell Biol.* **59:**378–394.
238. Taylor, D. L., J. A. Rhodes, and S. A. Hammond (1976) The contractile basis of amoeboid movement. II. Structure and contractility of motile extracts and plasmalemma-ectoplasm ghosts. *J. Cell Biol.* **70:**123–143.
239. Taylor, E. W. (1965) The mechanism of colchicine inhibition of mitosis. *J. Cell Biol.* **25:**145–160.
240. Taylor, E. W., and W. Cramer (1963) Birefringence of protein solutions and biological systems. I. *Biophys. J.* **3:**127–141.
241. Thornburg, W. (1957) The form birefringence of lamellar systems containing three or more components. *J. Biophys. Biochem. Cytol.* **3:**413–419.

242. Thornburg, W., and E. DeRobertis (1956) Polarization and electron microscope study of frog nerve axoplasm. *J. Biophys. Biochem. Cytol.* **2**:475–482.
243. Tilney, L. G., and K. R. Porter (1965) Studies on microtubules in Heliozoa. I. *Protoplasma* **60**:317–344.
243a. Timasheff, S. N., and G. D. Fassman, eds. (1971) *Subunits in Biological Systems.* Marcel Decker, New York.
244. Tippit, D. H., D. Schultz, and J. D. Pickett-Heaps (1978) Analysis of the distribution of spindle microtubules in the diatom *Fragilaria. J. Cell Biol.* **76**:737–763.
245. Tobolsky, A. V. (1967) *Properties and Structures of Polymers.* Wiley, New York.
246. Tuckerman, L. B. (1909) Doubly refracting plates and elliptic analysers. *Univ. Nebraska Stud.* pp. 173–219.
247. Uretz, R. B., and R. P. Perry (1957) Improved ultraviolet microbeam apparatus. *Rev. Sci. Instrum.* **28**:861–866.
247a. Vidal, B. D. C., M. L. S. Mello and E. R. Pimentel (1982) Polarization microscopy and microspectrophotometry of Sirius Red, Picrosirius and Chlorantine Fast Red aggregates and of their complexes with collagen. *Histochem. J.* **14**:857–878.
248. Von Ebner, U. (1882) *Untersuchungen über die Ursache der Anisotropie organischer Substanzen.* Leipzig.
249. Von Hippel, A. R., ed. (1965) *The Molecular Designing of Materials and Devices.* MIT Press, Cambridge, Mass.
250. Von Muralt, A. L., and J. T. Edsall (1930) Flow birefringence of myosin. *Trans. Faraday Soc.* **26**:837–852.
251. Wahlstrom, E. E. (1960) *Optical Crystallography,* 3rd ed. Wiley, New York.
252. Weber, H. H. (1934) Der Feinbau und die mechanischen Eigenschaften der Myosin Fadens. *Pfluegers Arch.* **235**:205–233.
253. Weidner, R. T., and R. L. Sells (1965) *Elementary Modern Physics,* Volume 3. Allyn, Bacon & Cottrell, Boston.
254. Weisenberg, R. C., and A. C. Rosenfeld (1975) In vitro polymerization of microtubules into asters and spindles in homogenates of surf clam eggs. *J. Cell Biol.* **64**:146–158.
255. Weis-Fogh, T. (1961) Molecular interpretation of the elasticity of resilin, a rubber-like protein. *J. Mol. Biol.* **3**:648–667.
256. West, C. D., and R. C. Jones (1951) On the properties of polarization elements as used in optical instruments. I. Fundamental considerations. *J. Opt. Soc. Am.* **41**:976–986.
257. West, S. S. (1970) Optical rotation and the light microscope. In: *Introduction to Quantitative Cytochemistry,* Volume 2 (G. L. Wied and F. G. Bahr, eds.). Academic Press, New York, p. 451.
258. Wiener, O. (1912) Die Theorie Des Mischkörpers für das Feld der stationaren Strömung. I. Die Mittelwertzatze für Kraft, Polarisation und Energie. *Abh. Math. Phys. Kl. Koenigl. Ges. Wiss.* **32**:507–604.
259. Wiener, O. (1926) Formdoppelbrechung bei Absorption. I. Ubersicht. *Kolloidchem. Beih.* **23**:189.
260. Wilkins, M. H. F., and B. Battaglia (1953) Note on the preparation of specimens of oriented sperm heads for X-ray diffraction and infra-red absorption studies and on some pseudo-molecular behaviour of sperm. *Biochim. Biophys. Acta* **11**:412–415.
261. Wilson, E. B. (1925) *The Cell in Development and Heredity,* 3rd ed. Macmillan Co., New York.
261a. Wolman, M. (1970) On the use of polarized light in pathology. In: *Pathology Annual 1970* (S. C. Sommers, ed.). Appleton-Century-Crofts, New York, pp. 381–416.
262. Wolstenholme, G. E. W., and M. O'Connor, eds. (1966) *Principles of Biomolecular Organization.* Churchill, London.
263. Wood, E. A. (1964) *Crystals and Light: An Introduction to Optical Crystallography.* Van Nostrand, Princeton, N.J.
264. Wood, R. E. (1934) *Physical Optics,* 3rd ed. Macmillan Co., New York.
265. Wright, F. E. (1911) *The Methods of Petrographic-Microscopic Research: Their Relative Accuracy and Range of Application.* Carnegie Institution of Washington, Washington, D.C.
266. Zernike, F. (1958) The wave theory of microscopic image formation. In: *Concepts of Classical Optics* (J. Strong, ed.). Freeman, San Francisco, pp. 525–536.
266a. Zirkle, R. E., R. B. Uretz, and R. H. Haynes (1960) Disappearance of spindles and phargmoplasts after microbeam irradiation of cytoplasm. *Ann. N.Y. Acad. Sci.* **90**:435–439.
267. Zirwer, D., E. Buder, W. Schalike, and R. Wetzel (1970) Lineardichroitische Untersuchungen zur DNS-Organisation in Spermienkopfen von *Locusta migratoria* L. *J. Cell Biol.* **45**:431–434.
268. Zocher, H., and F. Jacoby (1972) Flow dichroism. *Kolloid Beih.* **24**:365.

IV

Addendum on Digital Image Processors

In the two years since Walter and Berns completed Chapter 10, several companies have introduced new digital image processors and programs that are applicable to microscopy. Many of the new processors are less expensive, easier to use, and often perform quite complex functions faster than the more elaborate systems. In addition, they can perform many functions in real or near real time on-line, tend to be self-contained, and do not require hookup to, or backup from, centralized computers.

In this appendix, I will survey the types of systems available that complement those in Table 10-1, and briefly discuss how one might get started with digital image processing and analysis for microscopy.

In addition to the type of equipment developed and used at major installations at the National Institutes of Health, University of Chicago, the Lawrence Livermore Laboratory, etc. for large-scale image analysis and cell sorting (Chapter 10, Section 11.4), three types of commercially available digital image-processing and analysis systems are directly applicable to microscopy.

A. Turnkey, Limited-Function Processors

The first group of systems, costing in the $10,000 to $20,000 range, are generally self-contained turnkey units with limited, specific functions. Most of the equipment in this group is made to freeze and store a video frame, either unprocessed, or with digital summing or running-averaging to reduce random noise. Incoming images often can be subtracted from the stored image to produce difference images or to perform background subtraction. They are well suited for eliminating random as well as optical noise and the influence of shading. Some degree of control of gray scale and contrast manipulation is usually available, but most of the processing is limited to point operations. The processing is usually not programmable or expandable by the addition of an external computer. Not only are these processors considerably less expensive than the other types, but they are quite compact and very simple to use. Where the need for image processing is limited to noise averaging, background subtraction, and limited contrast enhancement, such a processor can be very effective. Examples of processors in this group are: Arlunya TF4000 series Temporal Filter TV Store System (The Dindima Group Pty., Ltd.); Hughes Model 794 Television Image Processor (Hughes Aircraft Co., now distributed by Dage/MTI); Crystal digital image-processing system (MCI/Link); VIP-300 (Pie Data Medical).

B. Dedicated Processors with Extended Functions

This group includes many of the new stand-alone processors that take advantage of the large-scale computation that can be carried out rapidly on the specialized programmable hardware that is now available. The built-in, or attached, computer usually needs only to instruct the dedicated hardware to perform the processing tasks, but generally are not themselves used to carry out the large-scale number-crunching operations required for image processing. In addition to the functions mentioned for the first group of processors, point operations allowing arbitrary modification of gray scale and contrast, or display in pseudocolor, are carried out in real time by look-up tables. Logical operations (addition, subtraction, ORing, exclusive ORing, etc.) can be carried out by the ALU in a single frame time between images stored in the (three or so) frame buffers. Convolutions involving comparison of brightness of adjacent pixels, and multiplications by weighting factors that filter, sharpen, edge-enhance, edge-detect, etc., are performed in fractions of a second. Histograms of pixel brightness distribution can be displayed, or used to stretch or equalize gray scale and enhance contrast of regions of interest. These images can be made to zoom, shrink, roam, etc., and a variety of quantitative tasks can be performed. Many of these functions can be performed speedily and, depending on the system, interactively.

These image processors and analyzers are considerably more flexible than those in Group A, but their cost tends to run in the $45,000 to $65,000 range. Most of the units in this group are turnkey or menu-oriented systems and generally do not require knowledge of computer programming or operation. Naturally, some learning is required to master the power of these highly flexible systems.

Processors and analyzers in this group include: Hamamatsu Photonic System; Image-I from Interactive Video Systems; PSICOM 327 from Perceptive Systems; QX-9000 system from Quantex; and Systems by G. W. Hannaway and Associates.

The Hamamatsu Photonic System was specifically designed for microscopy with input from R. D. Allen of Dartmouth College. Several pushbutton functions and adjustments are built into the control unit.

The program used in the Image-I system was developed for the author's laboratory, primarily for real-time and near-real-time image processing and quantitation in microscopy. It features ease of use, versatility, speed, and power. No programming is required since single key strokes trigger most functions, but a C-library package is also available for those wishing to assemble additional software for custom applications.

The Perceptive Systems PSICOM 327 and related processors and analyzers were developed by K. Castleman. They are capable of performing pattern recognition functions such as chromosome karyotyping, quantitative cell image analysis, sandgrain classification, etc.

The Quantex QX-9000 series processor is controlled by a built-in 68000 central processing unit, is versatile, and uses a good mix of hard-wired and software-controlled functions for rapid image processing. It is a natural extension of the earlier (primarily hard-wired) processors produced by Quantex.

The processor and analyzer produced by G. W. Hannaway and Associates has available several innovative programs, but which require some degree of familiarity with computer programming and operation. The system can provide considerable flexibility, depending on the degree of program availability. Again, some of the programs were designed to meet the needs of microscopists.

C. Major Processing and Analyzing Systems

Produced by DeAnza, Grinnell, Kontron, and other companies that, for many years, have specialized in producing general image-processing and analyzing equipment, these general-purpose, modular image processors tend to use powerful main-frame computers to perform a large volume of high-speed image-processing computations. These are the most flexible and powerful systems, but also require greater familiarity with computer operation or programming to exploit their power. The cost of these systems varies, depending on the configuration, but generally falls in the $60,000 to $150,000 range. The basic capabilities of several of these systems are listed in Chapter 10.

REMARKS

In place of acquiring a complete image processing system, it is also possible to purchase advanced components made for **OEMs** (original equipment manufacturers) that can be assembled into, or coupled with, computers for image processing or analysis. This may appear to be an economical alternative at first, but one needs to keep in mind that the apparent savings in cost may be illusory. One needs to add the costs of (1) purchasing and assembling the full set of hardware and peripherals needed; (2) **software** acquisition or development; (3) the lack of the follow-up service that is available for the integrated system. While it is common to underestimate somewhat the cost of the first and third items, the cost of software development is the one most likely to be grossly underestimated.

In a discussion at the Castleman and Winkler course on digital image processing in 1984, the consensus was that the ratio of cost for hardware acquisition and software development was generally greater than 50 : 50, probably around 40 : 60. In other words, $30,000 worth of hardware would probably require another $45,000 of software development to make the system functional (although finished software that functions with specific hardware configuration may be available for the order of one to a few thousand dollars, depending on the type). With very few exceptions, I would advocate the use of an assembled, complete system, which tends to be more economical in the long run, can be used right away, and whose whole system would receive backup service by experienced personnel.

Another general point to consider: The systems in Group A are generally easier to use than those in Group B, and those in Group B easier than in Group C. The speed for the limited tasks in Group A may be just as fast as those in Group B or C; and those in Group B, using dedicated hardware to carry out some computationally intensive tasks, can be much faster than those achieved by the VAX and other powerful computers used in Group C.

The final point to consider is that the price of image-processing and analyzing systems can be expected to drop significantly in the next few years relative to the power and sophistication of function. Powerful new memory and processing chips are being developed, not only for TV applications, as described elsewhere, but also for image processing generally. Already, one-megabyte memory chips and ultra-large-scale integrated circuits (more powerful than the very-large-scale integrated circuits), and compact, high capacity magnetic and optical disks have been announced early in 1985.*

Digital processing provides opportunities not available by analog processing. Video images are grabbed without deterioration in a single frame time; noise is averaged and background subtracted effectively. Various tasks of selective image filtering and enhancement, as well as quantitation, are achieved efficiently and more effectively. (See Baxes, 1984, for a comprehensive, nonmathematical description, illustrated with many photographs, of the type of image processing that can be carried out with modern digital processors.) However, some tasks of contrast enhancement, image improvement, and simple quantitation can be carried out quite well by analog means (Chapter 9). Depending on the nature of the specific image-processing and analysis task, it is well to consider where in the whole system improvements are needed: the microscope optics; the selection of the specimen; video camera, recorder, monitor; analog enhancing devices; or digital processing and analysis.

*At the same time, personal computers are using ever more powerful CPUs. My expectation is that by 1986 or 1987 several modular plug-in image processing boards and powerful software applicable to image processing and analysis in microscopy will become available for the IBM PC XT/AT and comparable microcomputers.

References*

Acker, D. E. (1983) Understanding digital TBC specifications. *EITV* **April:**33–40. [8.5]

Agar, A. W. (1983) The use of a digitizing frame store in microscopy. *J. Microsc.* **130:**339–344. [11.5]

Agard, D. A. (1984) Optical sectioning microscopy: Cellular architecture in three dimensions. *Annu. Rev. Biophys. Bioeng.* **13:**191–219. [11.5]

Agard, D., and J. W. Sedat (1983) Three dimensional architecture of a polytene nucleus. *Nature* **302:**676–681. [10.12; Fig. 10-51]

Agard, D. A., and R. M. Stroud (1982) Linking regions between helices in bacteriorhodopsin revealed. *Biophys. J.* **37:**589–602. [11.5]

Aldersey-Williams, H. (1984) Liquid crystals in flat panel displays. *Electronic Imaging* **3**(11)**:**54–57 [7.5]

Allen, R. D. (1985) New observations on cell architecture and dynamics by video-enhanced contrast optical microscopy. *Annu. Rev. Biophys. Biophysical Chem.* **14:**265–290. [Fig. 1-7]

Allen, R. D., and N. S. Allen (1983) Video-enhanced microscopy with a computer frame memory. *J. Microsc.* **129:**3–17. [3.10, 11.3]

Allen, R. D., G. B. David, and G. Nomarski (1969) The Zeiss–Nomarski differential interference equipment for transmitted-light microscopy. *Z. wiss. Mikrosk.* **69:**193–221. [5.6]

Allen, R. D., J. L. Travis, N. S. Allen, and H. Yilmaz (1981a) Video-enhanced contrast polarization (AVEC-POL) microscopy: A new method applied to the detection of birefringence in the motile reticulopodial network of *Allogromia laticollaris. Cell Motil.* **1:**275–289. [3.1, 11.3]

Allen, R. D., N. S. Allen, and J. L. Travis (1981b) Video-enhanced contrast, differential interference contrast (AVEC-DIC) microscopy: A new method capable of analyzing microtubule-related motility in the reticulopodial network of *Allogromia laticollaris. Cell Motil.* **1:**291–302. [3.1, 11.3]

Allen, R. D., J. Metuzals, I. Tasaki, S. T. Brady, and S. P. Gilbert (1982) Fast axonal transport in squid giant axon. *Science* **218:**1127–1129. [11.3]

Allen, R. D., D. G. Weiss, J. H. Hayden, D. T. Brown, H. Fujiwake, and M. Simpson (1985) Gliding movement of and bidirectional transport along single native microtubules from squid axoplasm: Evidence for an active role of microtubules in cytoplasmic transport. *J. Cell Biol.* 100: 1736–1752. [11.3]

All-in-One Camcorders (1983) *EITV* **July:**35–45. [8.3]

Altar, C. A., R J. Walter, K. A. Neve, and J. F. Marshall (1984) Computer-assisted video analysis of ^{3}H-spiroperidol binding autoradiographs. *J. Neurosci. Methods* **10:**173–188. [10.10]

Andrews, D. W., and F. P. Ottensmeyer (1982) Electron microscopy of the poly-L-lysine alpha-helix. *Ultramicroscopy* **9:**337–348. [11.5]

Ashkar, G. P., and J. W. Modestino (1978) The contour extraction problem with biomedical applications. *Computer Graphics and Image Processing* **7:**331–355. [10.13]

Ashley, C. C. (1970) An estimate of calcium concentration changes during the contraction of single muscle fibers. *J. Physiol. (London)* **210:**133P–134P. [Fig. 11-1]

Bacus, J. W., and E. E. Gose (1972) Leukocyte pattern recognition. *IEEE Trans. Syst. Man Cybern.* **2:**513–526. [1.2]

Baker, T. S., D. L. D. Caspar, and W. T. Murakami (1983) Polyoma virus 'hexamer' tubes consist of paired pentamers. *Nature* **303:**446–448. [11.5]

*See Appendix III (pp. 500–510) for additional references (Bibliography).

Ballard, D. H., and J. Sklansky (1976) A ladder structured decision tree for recognizing tumors in chest radiographs. *IEEE Trans. Comput.* **25:**503–513. [10.13]

Bally, G. von, and P. Greguss, eds. (1982) *Optics in Biomedical Sciences.* Springer-Verlag, New York. [11.5]

Barer, R. (1956) Phase contrast and interference microscopy in cytology. In: *Physical Techniques in Biological Research,* Volume III (G. Oster and A. W. Pollister, eds.). Academic Press, New York, pp. 29–90.[11.3]

Barlow, H. B., and J. D. Mollon, eds. (1982) *The Senses.* Cambridge University Press, London. [4.1]

Barrows, G. H., J. E. Sisken, J. C. Allegra, and S. D. Grasch (1984) Measurement of fluorescence using digital integration of video images. *J. Histochem. Cytochem.* **32:**741–746. [11.2]

Bartels, P. H., and G. L. Wied (1973) Extraction and evaluation of information from cell images. *Proceedings of the First Annual Life Sciences Symposium: Mammalian Cells, Probes, and Problems,* LASL, University of California, Los Alamos, N. M., pp. 15–28. [1.2]

Bartels, P. H., G. B. Olson, H. G. Bartels, D. E. Brooks, and G. V. Seaman (1981) The automated analytic electrophoresis microscope. *Cell Biophys.* **4:**371–388. [10.13]

Bartels, P. H., J. Layton, and R. L. Shoemaker (1984) Digital Microscopy. In: *Computer-Assisted Image Analysis Cytology* (S. D. Greenberg, ed.). Karger, Basel, pp. 28–61. [11.4]

Baxes, G. A. (1984) *Digital Image Processing: A Practical Primer.* Prentice–Hall, Englewood Cliffs, N.J. [1.3, 6.5, Postscript, Appendix IV]

Beck, C. (1938) *The Microscope: Theory and Practice.* Beck, London. [5.1]

Beckerle, M. C. (1984) Microinjected fluorescent polystyrene beads exhibit saltatory motion in tissue culture cells. *J. Cell Biol.* **98:**2126–2132. [11.2]

Bedford, A. V. (1950) Mixed highs in color television. *Proc. IRE* **38:**1003–1009. [4.9, 7.4]

Belling, J. (1930) *The Use of the Microscope.* McGraw–Hill, New York. [5.1]

Bennet, A. H. B., H. Jupnik, H. Osterberg, and O. W. Richards (1951) *Phase Microscopy.* Wiley, New York. [5.6]

Bennett, H. S. (1950) The microscopical investigation of biological materials with polarized light. In: *Handbook of Microscopical Technique* (C. E. McClung, ed.). Harper & Row (Hoeber), New York, pp. 591–677. [5.6, 11.3]

Benson, D. M., J. A. Knopp, and I. S. Longmuir (1980) Intracellular oxygen measurements of mouse liver cells using quantitative fluorescence video microscopy. *Biochim. Biophys. Acta* **591:**187–197. [10.13, 11.2]

Benson, D. M., J. Bryan, A. L. Plant, A. M. Gotto, Jr., and L. C. Smith (1985) Digital imaging fluorescence microscopy: Spatial heterogeneity of photobleaching rate constants in individual cells. *J. Cell Biol.,* 100:1309–1323. [11.2, 11.4]

Berns, G. S., and M. W. Berns (1982) Computer-based tracking of living cells. *Exp. Cell Res.* **142:**103–109. [10.13, 11.3, 11.5]

Berns, M. W. (1974) *Biological Microirradiation: Classical and Laser Sources.* Prentice–Hall, Englewood Cliffs, N.J. [11.5]

Berns, M. W., J. Aist, J. Edwards, K. Strahs, J. Girton, M. Kitzes, J. Hammer-Wilson, L.-H. Liaw, A. Siemens, M. Koonce, R. Walter, D. van Dyk, J. Coulombe, T. Cahill, and G. S. Berns (1981) Laser microsurgery in cell and developmental biology. *Science* **213:**505–513. [10.7]

Berns, M. W., A. E. Siemens, and R. J. Walter (1984) Mitochondrial fluorescence patterns in Rhodamine 6G-stained myocardial cells in vitro. Analysis by real-time computer video microscopy and laser microspot excitation. *Cell Biophys.* **6:**263–277. [10.13]

Bertram, J. F., and A. W. Rogers (1981) The development of squamous cell metaplasia in human bronchial epithelium by light microscopic morphometry. *J. Microsc.* **123:**61–73 [10.13]

Bestor, T. H., and G. Schatten (1981) Anti-tubulin immunofluorescence microscopy of microtubules present during the pronuclear movements of sea urchin fertilization. *Dev. Biol.* **88:**80–91. [11.2]

Biedert, S., W. H. Barry, and T. W. Smith (1979) Inotropic effects and changes in sodium and calcium contents associated with inhibition of monovalent cation active transport by ouabain in cultured myocardial cells. *J. Gen. Physiol.* **74:**479–494. [9.5]

Blifford, I. H., Jr., and D. A. Gillette (1973) An automated particle analysis system. *Microscope* **21:**121–130. [1.2]

Blout, E. R., G. R. Bird, and D. S. Grey (1950) Infra-red microscopy. *J. Opt. Soc. Am.* **40:**304–313. [11.2]

Blum, H. (1964) Transformation for extracting new descriptions of shape. In: *Symposium on Models for the Perception of Speech and Visual Form* (W. Wathen-Dunn, ed.). MIT Press, Cambridge, Mass., pp. 362–381. [10.13]

Bok, D., and R. W. Young (1979) Phagocytic properties of the retinal pigment epithelium. In: *The Retinal Pigment Epithelium* (K. M. Zinn and M. F. Marmor, eds.). Harvard University Press, Cambridge, Mass., pp. 148–174. [4.2]

Born, M., and E. Wolf (1980) *Principles of Optics,* 6th ed. Pergamon Press, Elmsford, N.Y. [5.5; Fig. 5-21]

Bowie, J. E., and I. T. Young (1977a) An analysis technique for biological shape. II. *Acta Cytol.* **21:**455–464. [10.13]

Bowie, J. E., and I. T. Young (1977b) An analysis technique for biological shape. III. *Acta Cytol.* **21:**739–746. [10.13]

Boyes, E. D., B. J. Muggridge, and M. J. Goringe (1982) On-line image processing in high resolution electron microscopy. *J. Microsc.* **127:**321–335. [11.5]

Bracewell, R. N. (1978) *The Fourier Transform and Its Application,* 2nd ed. McGraw–Hill, New York. [II.3]

Bradbury, S. (1983) Commercial image analyzers and the characterization of microscopical images. *J. Microsc.* **131:**203–210. [11.4]

Brady, S. T., R. J. Lasek, and R. D. Allen (1982) Fast axonal transport in extruded axoplasm from squid giant axon. *Science* **218:**1129–1131. [11.3]

Braithwaite, L. (1985) The official *Video* magazine blank tape tests. *Video* **8**(11):78–112. [8.8]

Brigham, E. O. (1974) *The Fast Fourier Transform.* Prentice–Hall, Englewood Cliffs, N.J. [10.11, 10.12]

Brooks, M. J. (1978) Rationalizing edge detectors. *Computer Graphics and Image Processing* **8:**277–285. [10.11]

Butler, J. W., M. K. Butler, and A. Stroud (1966) Automatic classifications of chromosomes. II. In: *Data Acquisition and Processing in Biology and Medicine* Volume 4 (K. Enslein, ed.). Pergamon Press, Elmsford, N.Y., pp. 47–60. [1.2]

Butterfield, J. F. (1978) Video microscopy. *Microscope* **26:**171–182. [12.7]

Cagnet, M., M. Françon, and J. C. Thrierr (1962) *Atlas of Optical Phenomena.* Springer-Verlag, Berlin. [5.5, 5.9]

Carman, G., P. Lemkin, L. Lipkin, B. Shapiro, M. Schultz, and P. Kaiser (1974) A real time picture processor for use in biologic cell identification. II. Hardware implementation. *J. Histochem. Cytochem.* **22:**732–740. [1.2]

Caspersson, T. (1950) *Cell Growth and Cell Function.* Norton, New York. [11.2]

Caspersson, T. (1973) Fluorometric recognition of chromosomes and chromosome regions. In: *Fluorescence Techniques in Cell Biology* (A. A. Thaer and M. Sernetz, eds.). Springer-Verlag, Berlin, pp. 107–123. [1.2]

Caspersson, T., J. Lindsten, G. Lomakka, H. Wallman, and L. Zech (1970) Rapid identification of human chromosomes by TV techniques. *Exp. Cell Res.* **63:**477–479. [1.2]

Castillo, X., D. Yorkgitis, and K. Preston (1982) A study of multidimensional multicolor images. *IEEE Trans. Biomed. Eng.* **29**(2)**:**111–121. [10.5]

Castleman, K. R. (1979) *Digital Image Processing.* Prentice–Hall, Englewood Cliffs, N.J. [1.3, 5.7, 6.5, 10.3, 10.4, 10.11, 10.12, 11.5, II.3; Figs. 10-12, 10-34]

Castleman, K., and R. Wall (1973) Automatic systems for chromosome identification. In: *Chromosome Identification* (T. Caspersson and L. Zech, eds.). Academic Press, New York. pp. 77–84. [1.2, 10.13].

Cederberg, R. L. T. (1979) Chain link coding and segmentation for raster scan devices. *Computer Graphics and Image Processing* **10**:224–234. [10.13]

Chamot, E. M., and C. W. Mason (1958) *Handbook of Chemical Microscopy,* 2nd ed. Wiley, New York. [5.6, 11.3]

Cheng, S.-Y., F. R. Maxfield, J. Robbins, M. C. Willingham, and I. H. Pastan (1980) Receptor-mediated uptake of 3,3′, 5-triiodo-L-thyronine by cultured fibroblasts. *Proc. Natl. Acad. Sci. USA* **77:**3425–3429. [11.2]

Cline, H. E., A. S. Holik, and W. E. Lorensen (1982) Computer-aided surface reconstruction of interference contours. *Appl. Opt.* **21:**4481–4488. [11.3]

Cornsweet, T. N. (1970) *Visual Perception.* Academic Press, New York. [4.1]

Cox, I. J., and C. J. R. Sheppard (1983) Scanning optical microscope incorporating a digital framestore and microcomputer. *Appl. Opt.* **22:**1474–1478. [11.5]

Crewe, A. V., and M. Ohtsuki (1982) A "stand-alone" image processing system for STEM images. *Ultramicroscopy* **9:**101–108. [11.5]

Currier, R. L. (1983) Interactive video disc learning systems. *High Technology* **3**(11)**:**51–60. [8.9]

Cutcheon, J. T., J. O. Harris, Jr., and G. R. Laguna (1973) Electro-optic devices utilizing quadratic PLZT ceramic elements. Report No. SLA 73-0777, Sandia Laboratories, Albuquerque, N.M. [12.7]

Davidson, M. L. (1968) Perturbation approach to spatial brightness interaction in human vision. *J. Opt. Soc. Am.* **58:**1300–1309. [Fig. 4-13]

Davis, L. S., and A. Rosenfeld (1975) A survey of edge detection techniques. *Computer Graphics and Image Processing* **4:**248–270. [10.13]

Davis, R., and D. Thomas (1984) Systolic array chip matches the pace of high-speed processing. *Electronic Design* Oct. 31, pp. 207–218. [11.5]

DeForest, L. (1907) Space Telegraphy. U.S. Patent No. 879,532. [1.1]

de Montebello, R. L. (1969) The RLM Synthalyzer technique and instrumentation for optical reconstruction and dissection of structures in three dimensions. *Ann. N.Y. Acad. Sci.* **157**:487–496. [11.5; Fig. 11-12]

Dempster, W. T. (1944) Principles of microscope illumination and the problem of glare. *J. Opt. Soc. Am.* **34**:694–710. [3.8, 5.3, 5.4]

De Weer, P., and B. M. Salzberg, eds. (1986) *Optical Methods in Cell Physiology.* Wiley, New York, in press. [11.2, Postscript]

DiBona, D. R. (1978) Direct visualization of epithelial morphology in the living amphibian urinary bladder. *J. Membr. Biol.* **40**:45–70. [11.3]

Diller, K. R., and J. M. Knox (1983) Automated computer analysis of cell size changes during cryomicroscope freezing: A biased trident convolution mask technique. *Cryo-Letters* **4**:77–92. [10.13]

D'Ormer, P., K. Pachmann, R. Penning, G. J'Ager, and H. Rodt (1981) Discrimination in immunofluorescence between labeled and unlabeled cells by quantitative microfluorimetry and computer analysis. *J. Immunol. Methods* **40**(2):155–163 [10.13]

Dvorak, J. A., L. H. Miller, W. C. Whitehouse, and T. Shiroishi (1975) Invasion of erythrocytes by malaria merozoites. *Science* **187**:748–750. [11.3; Fig. 1-4]

Dyson, J. (1950) An interferometer microscope. *Proc. R. Soc. London Ser. A* **204**:170–187. [5.6]

Eckert, R., and G. T. Reynolds (1967) The subcellular origin of bioluminescence in *Noctiluca miliaris. J. Gen. Physiol.* **50**:1429–1458. [1.2, 5.6, 11.2]

Edelhart, M. (1981) Optical disks: The omnibus medium. *High Technology* **1**:42–57. [8.9]

Egger, M. D., and M. Petráň (1967) New reflected-light microscope for viewing unstained brain and ganglion cells. *Science* **157**:305–307. [11.5]

Einck, L., and M. Bustin (1984) Functional histone antibody fragments traverse the nuclear envelope. *J. Cell Biol.* **98**:205–213. [11.2]

Eisen, A., D. P. Kiehart, S. J. Wieland, and G. T. Reynolds (1984) Temporal sequence and spatial distribution of early events of fertilization in single sea urchin eggs. *J. Cell Biol.* **99**:1647–1654. [11.2]

Ekstrom, A. (1910) Swedish Patent No. 32,220. [1.1]

Ellis, G. W. (1966) Holomicrography: Transformation of image during reconstruction *a posteriori. Science* **154**:1195–1196. [5.6, 11.5]

Ellis, G. W. (1978) Advances in visualization of mitosis *in vivo.* In: *Cell Reproduction: In Honor of Daniel Mazia* (E. Dirksen, D. Prescott, and C. F. Fox, eds.). Academic Press, New York, pp. 465–476. [5.6, 11.3]

Ellis, G. W.(1979) A fiber-optic phase-randomizer for microscope illumination by laser. *J. Cell Biol.* **83**:303a. [Appendix I; Table 5-1]

Ellis, G. W. (1985) Microscope illuminator with fiber optic source integrator, *J. Cell Biol.* 101:83a. [5.8]

Ellis, G. W., S. Inoué, and T. D. Inoué (1985) Computer aided light microscopy. In: *Optical Methods in Cell Physiology* (P. De Weer and B. M. Salzberg, eds.). Wiley, New York, in press. [3.10, 11.2, 11.3; Fig. 1-8]

Ennes, H. E. (1979) *Television Broadcasting: Equipment, Systems, and Operating Fundamentals,* 2nd ed. Sams, Indianapolis. [Preface, 6.1, 6.3, 6.4, 6.5, 6.6, 7.4, 7.5; Figs. 6-21, 7-27, 7-28, 8-8, 8-9; Table 6-4]

Erickson, R. O. (1973) Tubular packing of spheres in biological fine structure. *Science* **181**:705–716. [12.1]

Faroudja, Y., and J. Roizen (1981) Color-under recorders: Limitations, manual correction, and automated enhancement. *SMPTE J.* **Dec.**:1152–1157. [8.3]

Fay, F. S., K. E. Fogarty, and J. M. Coggins (1985) Analysis of molecular distribution in single cells using a digital imaging microscope. In: *Optical Methods in Cell Physiology* (P. De Weer and B. M. Salzberg, eds.). Wiley, New York, in press. [11.5]

Fiorentini, A. (1964) Dynamic characteristics of visual processes. *Prog. Opt.* **1**:253–288. [4.4]

Flory, L. E. (1951) The television microscope. *Cold Spring Harbor Symp. Quant. Biol.* **16**:505–509. [1.2]

Flory, L. E. (1959) Scanning microscopy in medicine and biology. *Proc. IRE* **Nov.**:1889–1894. [1.2]

Flory, L. E., and W. S. Pike (1953) Particle counting by television techniques. *RCA Rev.* **14**:546–556. [1.2]

Forman, D. S., and D. E. Turriff (1981) Video intensification microscopy (VIM) as an aid in routine fluorescence microscopy. *Histochemistry* **71**:203–208. [11.2]

Forman, D. S., K. J. Brown, and D. R. Livengood (1983) Fast axonal transport in permeabilized lobster giant axons is inhibited by vanadate. *J. Neurosci.* **3**:1279–1288. [11.3]

Foster, J. (1984) High resolution acoustic microscopy in superfluid helium. *17th International Conference on Low Temperature Physics,* Karlsruhe, West Germany. G. L. Report No. 3760. W. W. Hansen Laboratory of Physics, Stanford University. [11.5]

Françon, M. (1961) *Progress in Microscopy.* Row, Peterson, Evanston, Ill. [5.1, 5.5, 5.6; Figs. 5-18 to 5-20]

Franklin, G., and T. Hatley (1973) Don't eyeball noise. *Electron. Des.* **24**:184–187. [6.5]

Freed, J. J., and J. L. Engle (1962) Development of the vibrating-mirror flying spot microscope for ultraviolet spectrophotometry. *Ann. N.Y. Acad. Sci.* **97:**412–448. [1.2, 11.5]

Freeman, H. (1970) Boundary encoding and processing. In: *Picture Processing and Psychopictorics* (B. Lipkin and A. Rosenfeld, eds.). Academic Press, New York, pp. 241–266. [10.13]

Fuchs, H., S. M. Pizer, E. R. Heinz, S. H. Bloomberg, L.-C. Tsai, and D. C. Strickland (1982a) Design and image editing with a space-filling 3-D display based on a standard raster graphics system. *Proc. Soc. Photo. Opt. Instrum. Eng.* **367:**117–127. [11.5]

Fuchs, H., S. M. Pizer, L.-C. Tsai, S. H. Bloomberg, and E. R. Heinz (1982b) Adding a true 3-D display to a raster graphics system. *IEEE Comput. Graphics Appl.* **2:**73–78. [11.5; Fig. 11-13]

Fuseler, J. W. (1975) Mitosis in *Tilia americana* endosperm. *J. Cell Biol.* **64:**159–171. [5.10]

Gabor, D. (1948) A new microscope principle. *Nature* **161:**777–778. [5.6]

Gabor, D. (1949) Microscopy by reconstructed wave-fronts. *Proc. R. Soc. London Ser. A* **197:**454–486. [5.6]

Gabor, D. (1951) Microscopy by reconstructed wave-fronts. II. *Proc. Phys. Soc. London Ser. B* **64:**449–470. [5.6]

Gage, S. H. (1941) *The Microscope*. Cornell University Press (Comstock), Ithaca, N.Y. [5.1]

Galbraith, W. (1982) The image of a point of light in differential interference contrast microscopy: Computer simulation. *Microsc. Acta* **85:**233–254. [5.5]

Galbraith, W., and G. B. David (1976) An aid to understanding differential interference contrast microscopy: Computer simulation. *J. Microsc.* **108:**147–176. [5.6]

Galbraith, W., and R. J. Sanderson (1980) The intensity distribution about the image of a point. *Microsc. Acta* **83:**395–402. [5.5]

Gallistel, C. R., C. T. Piner, T. O. Allen, N. T. Adler, E. Yaden, and M. Negin (1982) Computer assisted analysis of 2-DG autoradiograms. *Neurosci. Biobehav. Rev.* **6:**409–420. [10.7]

Garbay, C., G. Brugal, and C. Choquet (1981) Application of colored image analysis to bone marrow cell recognition. *Anal. Quant. Cytol.* **3:**272–280. [10.5]

Gaunt, W. A., and P. N. Gaunt (1978) *Three-Dimensional Reconstruction in Biology*. University Park Press, Baltimore. [11.5, 12.7]

Ghatak, A. K., and K. Thyagarajan (1978) *Contemporary Optics*. Plenum Press, New York. [Fig. 5-21]

Gilev, V. P. (1979) A simple method of optical filtration. *Ultramicroscopy* **4:**323–336. [12.1]

Gilkey, J. C., L. F. Jaffe, E. B. Ridgway, and G. T.Reynolds (1978) A free calcium wave traverses the activating egg of medaka, *Oryzias latipes*. *J. Cell Biol.* **76:**448–466. [11.2; Fig. 11-1]

Gingell, D., and I. Todd (1979) Interference reflection microscopy: A quantitative theory for image interpretation and its application to cell–substratum separation measurement. *Biophys. J.* **26:**507–526. [11.3]

Glaser, E. M., M. Tagamets, N. T. McMullen, and H. Van der Loos (1983) The image-combining computer microscope—An interactive instrument for morphometry of the nervous system. *J. Neurosci. Methods* **8:**17–32. [11.4]

Glass, B. (1961) *Light and Life*. Johns Hopkins Press, Baltimore. [11.2]

Gonzalez, R. C., and P. Wintz (1977) *Digital Image Processing*. Addison–Wesley, Reading, Mass. [6.5]

Goochee, C., W. Rasband, and L. Sokoloff (1980) Computer densitometry and color coding of [^{14}C] deoxyglucose autoradiographs. *Ann. Neurol.* **7:**359–370. [10.7]

Goodman, J. W. (1968) *Introduction to Fourier Optics*. McGraw–Hill, New York. [6.5, II.3]

Gordon, G., and D. J. Palatini (1982) Enhanced visualization of enzymatic reaction product by means of image analysis. *Histochem. J.* **14:**842–845. [11.5]

Gould, R. G., M. J. Lipton, P. Mengers, and R. Dahlberg (1981) Investigation of video frame averaging digital subtraction fluoroscopic system. *Proc. SPIE Int. Soc. Opt. Eng.* **314:**184. [10.4]

Graff, G. (1984) Liquid crystals. *High Technology* **4**(5)**:**54–68. [7.5]

Gregory, R. L. (1970) *The Intelligent Eye*. McGraw–Hill, New York. [4.8]

Grinvald, A., L. B. Cohen, S. Lesher, and M. B. Boyle (1981) Simultaneous optical monitoring of activity of many neurons in invertebrate ganglia using a 124-element photodiode array. *J. Neurophysiol.* **45:**829–840. [6.2, 11.4]

Gruenbaum, Y., M. Hochstrasser, D. Mathog, H. Saumweber, D. A. Agard, and J. W. Sedat (1984) Spatial organization of the *Drosophila* nucleus: A three-dimensional cytogenetic study. *J. Cell Sci.* Suppl. 1: 223–234. [10.12]

Gruner, S. M., J. R. Milch, and G. T. Reynolds (1978) Evaluation of area photon detectors by a method based on detective quantum efficiency (DQE). *IEEE Trans. Nucl. Sci.* **25:**562–565. [11.4]

Gruner, S. M., J. R. Milch, and G. T. Reynolds (1982) Slow-scan silicon-intensified target-TV X-ray detector for quantitative recording of weak X-ray images. *Rev. Sci. Instrum.* **53:**1770–1778. [9.4]

Gundersen, G. G., C. A. Gabel, and B. M. Shapiro (1982) An intermediate state of fertilization involved in internalization of sperm components. *Dev. Biol.* **93:**59–72. [11.2]

Hall, J. A. (1980) Arrays and charge coupled devices. In: *Applied Optics and Optical Engineering,* Volume VII (R. R. Shannon and J. C. Wyant, eds.). Academic Press, New York, pp. 349–400. [7.1]

Hannaway, W., G. Shea, and W. R. Bishop (1984) Handling real-time images comes naturally to systolic array chip. *Electronic Design* **Nov. 15:** 289–300. [11.5]

Hansen, G. L. (1969) *Introduction to Solid-State Television Systems: Color and Black and White.* Prentice–Hall, Englewood Cliffs, N.J. [Preface]

Harlow, C. A., and R. W. Connors (1979) Image analysis segmentation methods. In: *Biomedical Pattern Recognition and Image Processing* (K. S. Fu and T. Pavlidis, eds.). Verlag Chemie, Weinheim, pp. 111–130. [10.13]

Hárosi, F. (1982) Recent results from single-cell microspectrophotometry: Cone pigments in frog, fish, and monkey. *Color Res. Appl.* **7:**135–141. [11.4]

Harris, J. L. (1964) Diffraction and resolving power. *J. Opt. Soc. Am.* **54:**931–936. [10.12]

Hartshorne, N. H., and A. Stuart (1960) *Crystals and the Polarising Microscope: A Handbook for Chemists and Others,* 3rd ed. Arnold, London. [5.6, 11.3]

Harvey, E. N. (1957) A history of luminescence from the earliest times until 1900. *Mem. Am. Philos. Soc.* **44:**1–692. [11.2]

Harvey, E. N. (1965) *Living Light.* Princeton University Press, Princeton, N.J. [11.2]

Hayden, J. H., and R. D. Allen (1984) Detection of single microtubules in living cells: Particle transport can occur in both directions along the same microtubule. *J. Cell Biol.* **99:**1785–1793. [11.3]

Hecht, E., and A. Zajac (1974) *Optics.* Addison–Wesley, Reading, Mass. [5.7, 10.12]

Hecht, S. (1984) Outlook brightens for semiconductor vector lasers. *High Technology* **4**(1)**:**43–50. [8.9]

Hecht, S., and J. Mandelbaum (1939) The relation between vitamin A and dark adaptation. *J. Am. Med. Assoc.* **112:**1910–1916. [Fig. 4-4]

Hecht, S., and S. Shlaer (1936) Intermittent stimulation by light. V: The relation between intensity and critical frequency for different parts of the spectrum. *J. Gen. Physiol.* **19:**965–977. [Fig. 4-15]

Hecht, S., and E. L. Smith (1936) Intermittent stimulation by light. VI. Area and the relation between critical frequency and intensity. *J. Gen. Physiol.* **19:**979–989. [Fig. 4-15]

Hecht, S., and C. D. Verrijp (1933) The influence of intensity, color and retinal location on the fusion frequency of intermittent illumination. *Proc. Natl. Acad. Sci. USA* **19:**522–535. [Fig. 4-15]

Helmholtz, H. (1866) *Handbuch der Physiologischen Optik.* Voss, Leipzig. [4.7]

Hiramoto, Y., and Y. Shoji (1982). Location of the motive force for chromosome movement in sand-dollar eggs. In: *Oji International Seminar on Biological Functions of Microtubules and Related Structures.* (H. Sakai, H. Mohri, and G. G. Borisy eds.) Academic Press, Japan, pp. 247–259.

Hiramoto, Y., Y. Hamaguchi, Y. Shoji, T. E. Schroeder, S. Shimoda, and S. Nakamura, (1981) Quantitative studies on the polarization optical properties of living cells. II. The role of microtubules in birefringence of the spindle of the sea urchin egg. *J. Cell Biol.* **89:**121–130.

Hirschberg, I. (1985) Advances in high-resolution imagers. *Electronic Imaging* **4:**56–62. [Table 7-3]

Hirshon, R. (1984a) Flat panel display technologies. *Electronic Imaging* **3**(1)**:**40–48. [7.5]

Hirshon, R. (1984b) Optical storage meets increased data demands. *Electronic Imaging* **3**(3)**:**30–35. [8.9]

Hobbs, M. (1982) *Servicing Home Video Cassette Recorders.* Hayden, Rochelle Park, N.J. [8.2, 8.7]

Hoffman, R., and L. Gross (1975) Modulation contrast microscopy. *Appl. Opt.* **14:**1169–1176. [5.6]

Hotani, H. (1982) Micro-video study of moving bacterial flagellar filaments. III. Cyclic transformation induced by mechanical force. *J. Mol. Biol.* **156:**791–806. [11.2]

Huxley, A. F., V. Lombardi, and L. D. Peachey (1981) A system for recording sarcomere longitudinal displacements in a striated muscle fibre during contraction. *Boll. Soc. Ital. Biol. Sper.* **57:**57–59. [11.4]

Inoué, S. (1944) Stereoscopic Apparatus. Japanese Patent No. 166,528. [12.7]

Inoué, S. (1951) A method for measuring small retardations of structures in living cells. *Exp. Cell Res.* **2:**513–517. [9.5]

Inoué, S. (1961) Polarizing microscope: Design for maximum sensitivity. In: *The Encyclopedia of Microscopy* (G. L. Clarke, ed.). Reinhold, New York, pp. 480–485. [11.3]

Inoué, S. (1964) Organization and function of the mitotic spindle. In: *Primitive Motile Systems in Cell Biology* (R. D. Allen and N. Kamiya, eds.). Academic Press, New York, pp. 549–598. [11.3]

Inoué, S. (1981a) Cell division and the mitotic spindle. *J. Cell Biol.* **91**(Part 2)**:**131s–147s. [11.3]

Inoué, S. (1981b) Video image processing greatly enhances contrast, quality, and speed in polarization-based microscopy. *J. Cell Biol.* **89**:346–356. [3.1, 8.4, 11.3, 12.1, Appendix I; Figs. 1-5, 1-6, 11-4, 11-5]

Inoué, S., and K. Dan (1951) Birefringence of the dividing cell. *J. Morphol.* **89:**423–456. [Fig. 11-4]

Inoué, S., and W. L. Hyde (1957) Studies on depolarization of light at microscope lens surfaces. II. The simultaneous realization of high resolution and high sensitivity with the polarizing microscope. *J. Biophys. Biochem. Cytol.* **3:**831–838. [11.3]

Inoué, S., and H. Kubota (1958) Diffraction anomaly in polarizing microscopes. *Nature* **182:**1725–1726. [5.5, 11.3]

Inoué, S., and D. A. Lutz (1982) Simultaneous video display of fluorescence and polarized light, or DIC, microscope images in real time. *J. Cell Biol.* **95:**461a. [Fig. 5-35]

Inoué, S., and H. Ritter, Jr. (1975) Dynamics of mitotic spindle organization and function. In: *Molecules and Cell Movement* (S. Inoué and R. E. Stephens, eds.). Raven Press, New York, pp. 3–30. [5.6]

Inoué, S., and H. Sato (1966) Deoxyribonucleic acid arrangements in living sperm. In: *Molecular Architecture in Cell Physiology* (T. Hayashi and A. G. Szent-Gyorgyi, eds.). Prentice–Hall, Englewood Cliffs, N.J., pp. 209–248. [5.6, 9.5, 11.3]

Inoué, S., and H. Sato (1967) Cell motility by labile association of molecules: The nature of mitotic spindle fibers and their role in chromosome movement. *J. Gen. Physiol.* **50:**259–292. [11.3]

Inoué, S., and L. G. Tilney (1982) The acrosomal reaction of *Thyone* sperm. I. Changes in the sperm head visualized by high resolution video microscopy. *J. Cell Biol.* **93:**812–819. [5.6, 11.3]

Inoué, S., D. Cohen, and G. W. Ellis (1981) High resolution, stereo video microscope. *Biol. Bull.* **161:**306. [12.7]

Inoué, S., T. D. Inoué, and G. W. Ellis (1985) Rapid stereoscopic display of microtubule distribution by a video-processed optical sectioning system. *J. Cell Biol.* **101:**146a. [12.7]

Intaglietta, M. (1970) Graphic display of television raster lines. *Rev. Sci. Instrum.* **41:**1105–1106. [9.6]

Intaglietta, M., N. R. Silverman, and W. R. Tompkins (1975) Capillary flow velocity measurements *in vivo* and *in situ* by television methods. *Microvasc. Res.* **10:**165–179. [1.2, 9.5]

Ishihara, A., and H. Hotani (1980) Micro-video study of discontinuous growth of bacterial flagellar filaments *in vitro*. *J. Mol. Biol.* **139:**265–276. [11.2]

Izzard, C. S., and L. R. Lochner (1976) Cell-to-substrate contacts in living fibroblasts: An interference reflexion study with an evaluation of the technique. *J. Cell Sci.* **21:**129–159. [5.6, 11.3]

Jacobson, K., E. Elson, D. Koppel, and W. Webb (1983) International workshop on the application of fluorescence photobleaching techniques to problems in cell biology. *Fed. Proc.* **42:**72–79. [11.4]

Jenkins, F. A., and H. E. White (1957) *Fundamentals of Optics,* 3rd ed. McGraw–Hill, New York. [5.8]

Johnson, C. H. S., S. Inoué, A. Flint, and J. W. Hastings (1985) Compartmentation of algal bioluminescence: Autofluorescence of bioluminescent particles in the dinoflagellate *Gonyaulux* as studied with image-intensified video microscopy and flow cytometry. *J. Cell Biol.* **100:**1435–1446. [11.2; Fig. 11-2]

Johnson, L. D., M. L. Walsh, B. J. Bockus, and L. B. Chen (1981) Monitoring of relative mitochondrial membrane potential in living cells by fluorescent microscopy. *J. Cell Biol.* **88:**526–535. [10.13]

Judge, A. W. (1926) *Stereoscopic Photography.* Chapman & Hall, London. [12.7]

Julesz, B. (1971) *Foundations of Cyclopean Perception.* University of Chicago Press, Chicago. [4.8]

Kachar, B. (1985) Asymmetric illumination contrast: A new method of image formation for video light microscopy. *Science* **227:**766–768. [11.3; Fig. 11-9]

Kachar, B., D. F. Evans, and B. W. Ninham (1984) Video enhanced differential interference contrast microscopy: A new tool for the study of association colloids and prebiotic assemblies. *J. Colloid Interface Sci.* **100:**287–301. [11.3]

Kamiya, R., H. Hotani, and S. Asakura (1982) Polymorphic transition in bacterial flagella. In: *Prokaryotic and Eukaryotic Flagella* (W. B. Amos and J. G. Duckett, eds.). Cambridge University Press, London, pp. 53–76. [11.2]

Kater, S. B., and R. D. Hadley (1982) Intracellular staining combined with video fluorescence microscopy for viewing living identified neurons. In: *Cytochemical Methods in Neuroanatomy,* Volume 1 (V. C. Palay and S. L. Palay, eds.). Liss, New York, pp. 441–459. [11.2]

Kazan, B., and M. Knoll (1968) *Electronic Image Storage.* Academic Press, New York. [1.1]

Keller, H. E., and W. C. McCrone (1977) Standardization in the design of microscopes. *Microscope* **25:**169–177. [5.9]

Kelly, D. H. (1961) Visual response to time-dependent stimuli. *J. Opt. Soc. Am.* **51:**422–429. [Fig. 4-16]

Kirsch, R. A. (1971) Computer determination of the constituent structure of biological images. *Comput. Biomed. Res.* **4:**315–328. [10.13]

Koester, C. J. (1961) Interference microscopy: Theory and Techniques. In: *The Encyclopedia of Microscopy* (G. L. Clarke, ed.). Reinhold, New York, pp. 420–434. [5.6]

Koester, C. J. (1980) Scanning mirror microscope with optical sectioning characteristics: Applications in ophthalmology. *Appl. Opt.* **19:**1749–1757. [11.5; Fig. 11-11]

Koester, C. J., C. W. Roberts, A. Donn, and F. B. Hoefle (1980) Wide field specular microscopy: Clinical and research applications. *Ophthalmology* **87:**849–860. [11.5]

Kohen, E., C. Cohen, B. Thorell, and P. Bartick (1979) A topographic analysis of metabolic pathways in single living cells by multisite microfluorometry. *Exp. Cell Res.* **119:**23–30. [10.3]

Köhler, A., (1893) Ein neues Beleuchtungsverfahren für mikrophotographische Zwecke. *Z. wiss. Mikrosk.* **10:**433–440. [5.3]

Koonce, M. P., and M. Schliwa (1985) Bidirectional organelle transport can occur in cell processes that contain single microtubules. *J. Cell Biol.* **100:**322–326. [11.3]

Koonce, M. P., R. J. Cloney, and M. W. Berns (1984) Laser irradiation of centrosomes in newt eosinophils: evidence of centriole role in motility. *J. Cell Biol.* **98:**1999–2010. [10.13]

Kopp, R. E., J. Lisa, J. Mendelsohn, B. Pernick, H. Stone, and R. Wohlers (1974) The use of coherent optical processing techniques for the automatic screening of cervical cytologic samples. *J. Histochem. Cytochem.* **22:**598–604. [10.13]

Koppel, D. E., J. M. Oliver, and R. D. Berlin (1982) Surface functions during mitosis. III. Quantitative analysis of ligand-receptor movement into the cleavage furrow: Diffusion vs. flow. *J. Cell Biol.* **93:**950–960. [11.2]

Kowel, S. T., P. G. Kornreigh, and O. Lewis (1976) Focus detection using direct electronic Fourier-transform detectors. *Appl. Photogr. Eng.* **2:**113. [10.12]

Kries, T. E., and W. Birchmeier (1980) Stress fiber sarcomeres of fibroblasts are contractile. *Cell* **22:**555–561. [11.2]

Kristan, J., and M. Blouke (1982) Charge-coupled devices in astronomy. *Sci. Am.* **247**(4)**:**66–74. [7.1]

Kubota, H., and S. Inoué (1959) Diffraction images in the polarizing microscope. *J. Opt. Soc. Am.* **49:**191–198. [5.5, 11.3]

Lamy, J., P.-Y. Sizaret, J. Frank, A. Verschoor, R. Feldmann, and J. Bonaventura (1982) Architecture of *Limulus polyphemus* hemocyanin. *Biochemistry* **21:**6825–6833. [11.5]

Lancaster, D. (1976) *TV Typewriter Cookbook.* Sams, Indianapolis. [9.2]

Land, E. H. (1977) The retinex theory of color vision. *Sci. Am.* **237**(6)**:**108–128. [4.7]

Land, E. H. (1983) Recent advances in retinex theory and some implications for cortical computations: Color vision and the natural image. *Proc. Natl. Acad. Sci. USA* **80:**5163–5169. [4.7]

Land, E. H., and J. J. McCann (1971) Lightness and retinex theory. *J. Opt. Soc. Am.* **61:**1–11. [4.5, 4.7; Fig. 4-14]

Land, E. H., E. R. Blout, D. S. Grey, M. S. Flower, H. Husek, R. C. Jones, C. H. Matz, and D. P. Merrill (1948) Color translating ultraviolet microscope. *Science* **109:**371–374. [1.2]

Lemkin, P., G. Carman, L. Lipkin, B. Shapiro, M. Schultz, and P. Kaiser (1974) A real time picture processor for use in biologic cell identification. I. Systems design. *J. Histochem. Cytochem.* **22:**725–731. [1.2]

Lemkin, P., B. Shapiro, L. Lipkin, J. Maizel, J. Sklansky, and M. Schultz (1979) Preprocessing of electron micrographs of nucleic acid molecules for automatic analysis by computer. II. Noise removal and gap filling. *Comput. Biomed. Res.* **12:**615–630. [10.13]

Leslie, R. J., and J. D. Pickett-Heaps (1983) Ultraviolet microbeam irradiations of mitotic diatoms: Investigation of spindle elongation. *J. Cell Biol.* **96:**548–561. [11.5]

Leslie, R. J., W. M. Saxton, T. J. Mitchison, B. Neighbors, E. D. Salmon, and J. R. McIntosh (1984) Assembly properties of fluorescein-labeled tubulin *in vitro* before and after fluorescence bleaching. *J. Cell Biol.* **99:**2146–2156. [11.4]

Levi, L. (1980) *Applied Optics: A Guide to Optical System Design,* Volume 2. Wiley, New York. [4.1, 4.3, 11.4; Figs. 4-7, 4-10].

Levinthal, C., and R. Ware (1972) Three dimensional reconstruction from serial sections. *Nature* **236:**207–210. [11.5]

Levinthal, C., E. Macagno, and C. Tountas (1974) Computer-aided reconstruction from serial sections. *Fed. Proc.* **33:**2336–2340. [11.5]

Lewis, R. (1985) Solid-state cameras. *Electronic Imaging* **4**(1)**:**44–52. [This article includes a two-page-long vendor list; 7.1, Postscript; Table 7-3]

Lipkin, B. S., and A. Rosenfeld, eds. (1970) *Picture Processing and Psychopictorics.* Academic Press, New York. [1.2]

Lipkin, L., P. Lemkin, B. Shapiro, and J. Sklansky (1979) Preprocessing of electron micrographs of nucleic acid molecules for automatic analysis by computer. *Comput. Biomed. Res.* **12:**279–289. [10.13]

Lipton, L. (1982) *Foundation of Stereoscopic Cinema.* Van Nostrand Reinhold, New York. [12.7]

Liu, C. L., and J. W. S. Liu (1975) *Linear Systems Analysis.* McGraw–Hill, New York. [10.12]

Loveland, R. P. (1981) *Photomicrography: A Comprehensive Treatise,* 2nd ed. Two volumes. Wiley, New York. [5.1, 5.6, 5.9]

Ludvigh, E., and E. F. McCarthy (1938) Absorption of visible light by the refractive media of the human eye. *Arch. Ophthalmol.* **20:**37–51. [Fig. 4-6]

Macfarlane, T. W. R., S. Petrowski, L. Rigutto, and M. R. Roach (1983) Computer-based video analysis of cerebral arterial geometry using the natural fluorescence of the arterial wall and contrast enhancement techniques. *Blood Vessels* **20:**161–171. [11.5]

Marcil, G. (1975) Microdensitometric data acquisition. *Proc. Soc. Photo-Opt. Instrum. Eng.* **48:**23–33. [10.3]

Marino, T. A., D. Biberstein, P. N. Cook, L. Cook, and S. J. Dwyer, III (1981) A quantitative morphologic analysis of interatrial muscle cells in the ferret heart. *Anat. Rec.* **201:**31–42. [10.13]

Marr, D. (1982) *Vision: A Computational Investigation into the Human Representation and Processing of Visual Information.* Freeman, San Francisco. [4.8]

Marriott, F. H. C. (1962) Colour vision: Other phenomena. In: *The Eye,* Volume 2 (H. Davson, ed.). Academic Press, New York, pp. 273–298. [4.7]

Martelli, A. (1976) An application of heuristic search methods to edge-on contour detection. *Commun. ACM* **19:**73–83. [10.13]

Martin, L. C. (1966) *The Theory of the Microscope.* American Elsevier, New York. [5.1, 5.5]

Mathog, D., M. Hochstrasser, Y. Gruenbaum, H. Saumweber, and J. Sedat (1984) Characteristic folding pattern of polytene chromosomes in *Drosophila* salivary gland nuclei. *Nature* **308:**414–421. [11.5]

Matsumoto, S., K. Morikawa, and M. Yanagida (1981) Light microscopic structure of DNA in solution studied by the 4′, 6-diamidino-2-phenylindole staining method. *J. Mol. Biol.* **152:**501–516. [11.2]

Mayall, B. H., and B. L. Gledhill, eds. (1979) *Sixth Engineering Foundation Conference on Automated Cytology. J. Histochem. Cytochem.* **27:**1–641. [11.4]

McCutchen, C. W. (1967) Superresolution in microscopy and the Abbe resolution limit. *J. Opt. Soc. Am.* **57:**1190–1192. [11.5]

McGregor, G. N., H.-G. Kapitza, and K. A. Jacobson (1984) Laser-based fluorescence microscopy of living cells. *Laser Focus/Electro-Optics* **Nov.:**84–93. [11.2, 11.4]

McNeil, P. L., R. F. Murphy, F. Lanni, and D.L. Taylor (1984) A method for incorporating macromolecules into adherent cells. *J. Cell Biol.* **98:**1556–1564. [11.2]

Mendelsohn, M. L., B. H. Mayall, J. M. S. Prewitt, R. C. Bostrom, and W. G. Holcomb (1968) Digital transformation and computer analysis of microscopic images. *Adv. Opt. Electron Microsc.* **2:**77–150. [1.2]

Michael, C. R. (1969) Retinal processing of visual images. *Sci. Am.* **220**(May)**:**104–114. [4.4]

Michel, K. (1981) *Die Grundzüge der Theorie des Mikroskops in elementarer Darstellung.* Wissenschaftliche Verlagsgeselschaft M. B. H., Stuttgart. [5.5]

Middleton, R. G. (1981) *Effectively Using the Oscilloscope.* Sams, Indianapolis. [7.5]

Miles, C. P., and D. L. Jaggard (1981) The use of optical Fourier transforms to diagnose pleomorphism, size and chromatin clumping in nuclear models. *Anal. Quant. Cytol.* **3:**149–156. [10.13]

Miller, K. (1985) Call to the colors: Progress toward a three-dimensional standard. *Photonics Spectra* **19**(2)**:**75–82. [Fig. 4-19]

Mitchell, P. W. (1983) Digital signal processing comes to TV: The signal's unchanged, but new sets will enhance the picture. *High Technology* **Dec.:**16–20. [Postscript]

Montgomery, P. O., F. Roberts, and W. Bonner (1956) The flying-spot monochromatic ultra-violet television microscope. *Nature* **177:**1172. [1.2, 11.5]

Moore, J. H., C. C. Davis, and M. A. Coplan (1983) *Building Scientific Apparatus: A Practical Guide to Design and Construction.* Addison–Wesley, Reading, Mass., pp. 242–271. [11.4]

Mueller, C. G., and V. V. Lloyd (1948) Stereoscopic acuity for various levels of illumination. *Proc. Natl. Acad. Sci. USA* **34:**223–227. [4-21]

Muirhead, T. (1961) Variable focal length mirror. *Rev. Sci. Instrum.* **32:**210–211. [11.5]

Nagashima, H., and S. Asakura (1980) Dark-field light microscopic study of the flexibility of F-actin complexes. *J. Mol. Biol.* **136:**169–182. [11.2]

Nipkow, P. (1884) German Patent No. 30,105. [1.1]

Nomarski, G. (1955) Microinterféromètre difféfentiel à ondes polarisées. *J. Phys. Radium* **16:**S9–S13. [5.6]

Noronha, S. F. R. (1981) Time base correctors: A must for quality correction. *Photomethods* **May:**56–58. [8.5]

O'Kane, D. J., R. E. Kobres, and B. A. Palevitz (1980) Analog subtraction of video images for determining differences in fluorescence distributions. *J. Cell Biol.* **87:**229a. [9.3]

Okazaki, K., and S. Inoué (1976) Crystal property of the larval sea urchin spicule. *Dev. Growth Differ.* **18:**413–434. [11.3]

Olins, D. E., A. L. Olins, H. A. Levy, R. C. Durfee, S. M. Margle, E. P. Tinnel, and S. D. Dover (1983) Electron microscope tomography: Transcription in three-dimensions. *Science* **220:**498–500. [11.5]

Olson, A. C., N. M. Larson, and C. A. Heckman (1980) Classification of cultured mammalian cells by shape analysis and pattern recognition. *Proc. Natl. Acad. Sci. USA* **77:**1516–1520 [10.13]

Oppenheim, A. V., and J. S. Lim (1981) The importance of phase in signals. *Proc. IEEE* **69:**529–541. [II.2]

Oppenheim, A. V., and R. W. Schafer (1975) *Digital Image Processing.* Prentice–Hall, Englewood Cliffs, N.J. [6.5]

Ortner, E. H. (1984) Through the viewfinder. *Popular Science* **April:**64. [12.1]

Osborn, M., T. Born, H.-J. Koitsch, and K. Weber (1978) Stereo immunofluorescence microscopy. I. Three-dimensional arrangement of microfilaments, microtubules and tonofilaments. *Cell* **14:**447–488. [12.7]

Osterberg, H. (1955) Phase and interference microscopy. In: *Physical Techniques in Biological Research,* Volume I (G. Oster and A. W. Pollister, eds.). Academic Press, New York, pp. 377–437. [5.6, 11.3]

Padawer, J. (1967) The Nomarski intereference-contrast microscope. An experimental basis for image interpretation. *J. Roy. Microscop. Soc.* **88:**305–349. [5.6]

Palevitz, B. A., D. J. O'Kane, R. E. Kobres, and N. V. Raikhel (1981) The vacuole system in stomatal cells of *Allium.* Vacuole movements and changes in morphology in differentiating cells as revealed by epifluorescence, video and electron microscopy. *Protoplasma* **109:**23–55. [11.2]

Panitz, J. A., and D. C. Ghiglia (1982) Point projection imaging of unstained ferritin molecules on tungsten. *J. Microsc.* **127:**259–264. [11.5]

Parpart, A. K. (1951) Televised microscopy in biological research. *Science* **113:**483–484. [1.2]

Pask, C. (1976) Simple optical theory of super-resolution. *J. Opt. Soc. Am.* **66:**68–70. [10.12]

Pastan, I. H., and M. C. Willingham (1985) Endocytosis. In: *Morphological Methods in Endocytosis.* Plenum Press, New York, in press. [11.2]

Paton, K. (1975) Picture description using Legendre polynomials. *Computer Graphics and Image Processing* **4:**40–54. [10.13]

Pavlidis, T. (1978) A review of algorithms for shape analysis. *Computer Graphics and Image Processing* **7:**243–258. [10.13]

Pavlidis, T. (1979) Methodologies for shape analysis. In: *Biomedical Pattern Recognition and Image Processing* (K. S. Fu and T. Pavlidis, eds.). Verlag Chemie, Weinheim, pp. 131–152. [10.13].

Pavlidis, T., and S. L. Horowitz (1974) Segmentation of plane curves. *IEEE Trans. Comput.* **23:**860–870. [10.13]

Persoon, E., and K. S. Fu (1977) Shape discrimination using Fourier descriptors. *IEEE Trans. Syst. Man Cybern.* **7:**170–179. [10.13]

Persson, B.-E., and K. R. Spring (1982) Gallbladder epithelial cell hydraulic water permeability and volume regulation. *J. Gen. Physiol.* **79:**481–505. [11.3]

Petráň, M., M. Hadravský, M. D. Egger, and R. Galambos (1968) Tandem-scanning reflected-light microscope. *J. Opt. Soc. Am.* **58:**661–664. [1.2, 11.5]

Pike, W. S. (1962) Some television image enhancement techniques. *Ann. N.Y. Acad. Sci.* **97:**395–407. [1.2]

Piller, H. (1977) *Microscope Photometry.* Springer-Verlag, Berlin. [5.1, 5.4, 5.5, 5.6, 11.4]

Pirenne, M. H., and F. H. C. Marriott (1962) Visual functions in man In: *The Eye,* Volume 2 (H. Davson, ed.). Academic Press, New York, pp. 3–217. [4.2, 4.3; Fig. 4-11]

Pittman, R. N., and B. R. Duling (1975) Measurement of percent oxyhemoglobin in the microvasculature. *J. Appl. Physiol.* **38:**321–327. [9.5]

Poenie, M., J. Alderton, R. Y. Tsien, and R. A. Steinhardt (1985) Changes of free calcium levels with stages of the cell division cycle. *Nature* **315:**147–149. [11.2]

Polyak, S. (1957) *The Vertebrate Visual System.* University of Chicago Press, Chicago. [Figs. 4-1, 4-2]

Pratt, W. K. (1978) *Digital Image Processing.* Wiley, New York. [4.5, 6.5, 10.4, 10.7, 10.11; Figs. 4-12, 4-13, 10-15]

Preston, K. (1978) Biomedical image processing. In: *International Symposium on Advances in Digital Image Processing: Theory, Application, Implementation* (P. Stucki, ed.) Plenum Press, New York, pp. 125–146. [10.14]

Preston, K. (1979) Computer hardware for biomedical pattern recognition. In: *Biomedical Pattern Recognition and Image Processing* (K. S. Fu and T. Pavlidis, eds.). Verlag Chemie, Weinheim, pp. 213–231. [10.14]

Preston, K., and A. Dekker (1980) Differentiation of cells in abnormal human liver tissue by computer image processing: A preliminary investigation to diagnostic microscopy. *Anal. Quant. Cytol.* **2:**203–220. [10.5, 10.13]

Prewitt, J. M. S., and M. L. Mendelsohn (1966) The analysis of cell images. *Ann. N.Y. Acad. Sci.* **128:**1035–1053. [10.13]

Pycock, D., and J. T. Christopher (1980) Use of the Magiscan image analyses in automated uterine cancer cytology. *Anal. Quant. Cytol.* **2:**195–202. [10.13]

Quate, C. F. (1980) Microwaves, acoustic and scanning microscopy. In: *Scanned Image Microscopy* (E. A. Ash, ed.). Academic Press, New York, pp. 23–55. [11.5]

Ramachandran, G. N., and S. Ramaseshan (1961) Crystal optics. In: *Handbuch der Physik,* Volume XXV/I (S. Flugge, ed.). Springer-Verlag, Berlin, pp. 1–217. [5.6]

Raster Graphics Handbook (1980) Conrac Corp., Covina, Calif. (Forthcoming editions will be published by Van Nostrand, Princeton, N.J.) [7.4, 7.5; Figs. 7-30 to 7-34, 7-41; Tables 6-1, 6-3]

RCA Electro-Optics Handbook (1974) RCA Solid State Division, Electro-Optics and Devices, Lancaster, Pa. [4.1, 5.7, 7.2, 12.1; Figs. 4-1, 4-5, 4-17, 4-18, 7-3 to 7-7, 7-12, 7-16, 7-18; Table 7-2]

Reedy, M. C., M. K. Reedy, and R. S. Goody (1983) Co-ordinated electron microscopy and X-ray studies of glycerinated insect flight muscle. II. Electron microscopy and image reconstruction of muscle fibers fixed in rigor, in ATP and in AMPPNP. *J. Muscle Res. Cell Motil.* **4**:55–81. [11.5]

Reid, M. H. (1982) Quantitative stereology and radiologic image analysis. Part 1: Computerized tomography and ultrasound. *Med. Phys.* **9**:346–360. [11.5]

Renwanz, K. (1983) Panasonic field report: Panasonic RECAM Camera Recording System. *Broadcast Engineering* **March**:276–285. [8.3]

Reynolds, G. T. (1964) Evaluation of an image intensifier system for microscopic observations. *IEEE Trans. Nucl. Sci.* **11**:147–151. [1.2]

Reynolds, G. T. (1966) Sensitivity of an image intensifier film system. *Appl. Opt.* **5**:577–583. [11.4]

Reynolds, G. T. (1972) Image intensification applied to biological problems. *Quart. Rev. Biophys.* **5**:295–347.

Reynolds, G. T. (1972) Image intensification applied to biological problems. *Q. Rev. Biophys.* **5**:295–347. [7.1, 11.2]

Reynolds, G. T., and D. L. Taylor (1980) Image intensification applied to light microscopy. *BioScience* **30**:586–592. [11.2]

Rich, E. S., Jr., and J. E. Wampler (1981) A flexible, computer-controlled video microscope capable of quantitative spatial, temporal and spectral measurements. *Clin. Chem.* (N.Y.) **27**:1558–1568. [10.13]

Roos, K. P., and A. J. Brady (1982) Individual sarcomere length determination from isolated cardiac cells using high-resolution optical microscopy and digital image processing. *Biophys. J.* **40**:233–244. [11.4]

Rose, A. (1947) The sensitivity performance of the human eye on an absolute scale. *J. Opt. Soc. Am.* **38**:196–207. [4.1, 4.5]

Rose, A. (1973) *Vision: Human and Electronic.* Plenum Press, New York. [4.1, 4.9, 11.4]

Rose, B., and W. Loewenstein (1975) Calcium ion distribution in cytoplasm visualized by aequorin: Diffusion in cytosol restricted by energized sequestering. *Science* **190**:1204–1206. [11.2]

Rose, B., and W. Loewenstein (1976) Permeability of a cell junction and the local cytoplasmic free ionized calcium concentration: A study with aequorin. *J. Membr. Biol.* **28**:87–119. [11.2]

Salmon, E. D., M. McKeel, and T. Hays (1984a) Rapid rate of tubulin dissociation from microtubules in the mitotic spindle *in vivo* measured by blocking polymerization with colchicine. *J. Cell Biol.* **99**:1066–1075. [9.5, 11.5]

Salmon, E. D., W. M. Saxton, R. J. Leslie, M. L. Karow, and J. R. McIntosh (1984b) Diffusion coefficient of fluorescein-labelled tubulin in the cytoplasm of embryonic cell of a sea urchin: Video image analysis of fluorescence redistribution after photobleaching. *J. Cell Biol.* **99**:2157–2164. [11.2, 11.4]

Salzberg, B. M., A. Grinvald, L. B. Cohen, H. V. Davila, and W. N. Ross (1977) Optical recording of neuronal activity in an invertebrate central nervous system: Simultaneous monitoring of several neurons. *J. Neurophysiol.* **40**:1281–1291. [6.2, 11.4]

Sanger, J. W., B. Mittal, and J. M. Sanger (1984) Interaction of fluorescently labelled contractile proteins with the cytoskeleton. *J. Cell Biol.* **99**:918–928. [11.2]

Sato, H., G. W. Ellis, and S. Inoué (1975) Microtubular origin of mitotic spindle form birefringence: Demonstration of the applicability of Wiener's equation. *J. Cell Biol.* **67**:501–517. [11.3]

Sauer, B. (1983) Semi-automatic analysis of microscopic images of the human cerebral cortex using the grey level index. *J. Microsc.* **129**:75–87. [11.4]

Saxton, W. M., D. L. Stemple, R. J. Leslie, E. D. Salmon, M. Zavortink, and J. R. McIntosh (1984) Tubulin dynamics in cultured mammalian cells. *J. Cell Biol.* **99**:2175–2186. [11.4]

Schade, O. H., Sr. (1975) *Image Quality—A Comparison of Photographic and Television Systems.* RCA Laboratories, Princeton, N.J. [II.2]

Schatten, G. (1981) The movements and fusion of the pronuclei at fertilization of the sea urchin *Lytechinus variagatus:* Time-lapse video microscopy. *J. Morphol.* **167**:231–247. [3.1]

Schatten, G., H. Schatten, T. H. Bestor, and R. Balczon (1982) Taxol inhibits the nuclear movements during fertilization and induces asters in unfertilized sea urchin eggs. *J. Cell Biol.* **94**:455–465. [11.3]

Schmidt, W. J. (1924) *Die Bausteine des Tierkörpers in polarisiertem Lichte.* Cohen, Bonn. [5.6, 11.3]

Schmidt, W. J. (1937) *Die Doppelbrechung von Karyoplasma, Zytoplasma und Metaplasma. Protoplasma-Monographien,* Volume 11. Borntraeger, Berlin. [5.6, 11.3]

Schnapp, B. J., R. D. Vale, M. P. Sheetz, and T. S. Reese (1985) Single microtubules from squid axoplasm support bidirectional movement of organelles. *Cell,* **40**:455–462. [11.3; Fig. 11-8]

Schultz, M. L., L. E. Lipkin, M. J. Wade, P. F. Lemkin, and G. Carman (1974) High resolution shading correction. *J. Histochem. Cytochem.* **22**:751–754. [10.10]

Schwartz, A. A., and J. M. Soha (1977) Variable threshold zonal filtering. *Appl. Opt.* **16**:1779–1781. [10.13]

Shack, R. V. (1974) On the significance of the phase transfer function. *Proc. SPIE* **46**:39–43. [II.2]

Shack, R., N. Baker, R. Buchroeder, D. W. Hillman, R. Shoemaker, and P. H. Bartels (1979) Ultrafast laser scanning microscope. *J. Histochem. Cytochem.* **27**:153–159. [10.3]

Shaw, G. B. (1979) Local and regional edge detectors: Some comparisons. *Computer Graphics and Image Processing* **9:**135–149. [10.13]

Sher, L. D., and C. D. Barry (1985) The use of an oscillating mirror for three-dimensional displays. In: *New Methodologies in Studies of Protein Configuration* (T. T. Wu, ed.). Van Nostrand–Reinhold, Princeton, N.J. in press. [11.5; Fig. 11-14]

Shillaber, C. P. (1944) *Photomicrography in Theory and Practice.* Wiley, New York. [3.7, 5.1, 5.4, 5.5, 5.6, 5.9, 11.5]

Shimomura, O., and F. H. Johnson (1976) Calcium-triggered luminescence of the photoprotein aequorin. *Symp. Soc. Exp. Biol.* **30:**41–54. [Fig. 11-1]

Shlaer, S., E. L. Smith, and A. M. Chase (1942) Visual acuity and illumination in different spectral regions. *J. Gen. Physiol.* **25:**553–569. [4.4; Fig. 4-9]

Shoemaker, R. L., P. H. Bartels, D. W. Hillman, J. Jones, D. Kessler, R. V. Shack, and P. Jukobratovich (1982) An ultrafast laser scanner microscope for digital image analysis. *IEEE Trans. Biomed. Eng.* **29:**82–91. [10.3]

Shore, S. H. (1975) An overview of sensitometry and densitometry for the macro and micro conditions. *Proc. Soc. Photo-Opt. Instrum. Eng.* **48:**2–17. [10.3]

Showalter, L. C. (1969) *Closed Circuit TV for Engineers and Technicians.* Sams, Indianapolis. [Preface 7.5, 9.2]

Shulman, A. R. (1970) *Optical Data Processing.* Wiley, New York, pp. 324–357. [Appendix I]

Siemens, A., R. J. Walter, L.-H. Liaw, and M. W. Berns (1982) Laser stimulated fluorescence of submicron regions within single mitochondria of rhodamine treated myocardial cells in culture. *Proc. Natl. Acad. Sci. USA* **79:**466–470. [10.13]

Sklansky, J. (1978a) On the Hough technique for curve detection. *IEEE Trans. Comput.* **27:**923–926. [10.13]

Sklansky, J. (1978b) Image segmentation and feature extraction. *IEEE Trans. Syst. Man Cybern.* **8:**237–247. [10.7, 10.11, 10.13; Fig. 10-19]

Sklansky, J., and V. Gonzalez (1979) Fast polygonal approximation of digitized curves. *Pattern Recognition* **11:**604–609. [10.13]

Smith, F. H. (1956) Microscopic interferometry. In: *Modern Methods of Microscopy* (A. E. J. Vickers, ed.). Butterworths, London, pp. 76–86. [5.6]

Smith, K. T., S. L. Wagner, and M. J. Bottema (1984) Sharpening radiographs. *Proc. Natl. Acad. Sci. USA* **81:**5237–5241. [11.5]

Smith, W. J. (1966) *Modern Optical Engineering.* McGraw–Hill, New York. [II.2]

Sokoloff, L., M. Reivich, C. Kennedy, M. Des Rosiers, C. Patlak, K. Pethigrew, O. Sakurada, and M. Shinohara (1977) The [^{14}C]-deoxyglucose method for the measurement of local cerebral glucose oxidation: Theory, procedure and normal values in the conscious and anesthetized albino rat. *J. Neurochem.* **28:**897–916. [10.3]

Spencer, M. (1982) *Fundamentals of Light Microsopy.* Cambridge University Press, London. [Preface, 5.1]

Spotlight: The Super Slo-Mo System (1984) *Video Systems* **Sept.:**83–85. [8.6]

Spring, K. (1979) Optical techniques for the evaluation of epithelial transport processes. *Am. J. Physiol.* **237**(3)**:**F167–F174. [9.5, 11.1, 11.3]

Spring, K. R. (1983) Application of video to light microscopy. In: *Membrane Biophysics,* Volume II. Liss, New York, pp. 15–20. [11.1]

Stein, P. G., L. E. Lipkin, and H. M. Shapiro (1969) Spectra II: General-purpose microscope input for a computer. *Science* **166:**328–333. [1.2]

Stroke, G. W. (1969) *An Introduction to Coherent Optics and Holography,* 2nd ed. Academic Press, New York. [5.6]

Strong, J. (1958) *Concepts of Classical Optics.* Freeman, San Francisco. [5.6]

Sullivan, J. A., and G. L. Mandell (1983) Motility of human polymorphonuclear neutrophils: Microscopic analysis of substrate adhesion and distribution of F-actin. *Cell Motil.* **3:**31–46. [11.3]

Summers, K., and I. R. Gibbons (1971) Adenosine-triphosphate-induced sliding of tubules in trypsin-treated flagella of sea urchin sperm. *Proc. Natl. Acad. Sci. USA* **68:**3092–3096. [11.2]

Swann, M. M., and J. M. Mitchison (1950) Refinements in polarized light microscopy. *J. Exp. Biol.* **27:**226–237. [11.3]

Swinton, A. A. C. (1908) Distant electric vision. *Nature* **78:**151. [1.1]

Tanasugarn, L., P. McNeil, G. T. Reynolds, and D. L. Taylor (1984) Microspectrofluorometry by digital image processing: Measurement of cytoplasmic pH. *J. Cell Biol.* **98:**717–724. [11.2]

Tannas, L. E. (1981) *Flat-Panel Displays.* Van Nostrand–Reinhold, Princeton, N.J. [Fig. 7-33]

Taylor, D. L., and Y.-L. Wang (1980) Fluorescently labelled molecules as probes of the structure and function of living cells. *Nature* **284:**405–410. [11.2]

Thaer, A. A., and M. Sernetz, eds. (1973) *Fluorescence Techniques in Cell Biology*. Springer-Verlag, Berlin. [5.6]

Thaer, A., M. Hoppe, and W. J. Patzelt (1982) The ELSAM acoustic microscope. *Leitz Mitt. Wiss. Tech.* **8:**61–67. [11.5]

Tilney, L. G., and S. Inoué (1982) Acrosomal reaction of *Thyone* sperm. II. The kinetics and possible mechanism of acrosomal process elongation. *J. Cell Biol.* **93:**820–827. [Figs. 9-11, 9-12]

Toda, T., M. Yamamoto, and M. Yanagida (1981) Sequential alterations in the nuclear chromatin region during mitosis of the fission yeast *Schizosaccharomyces pombe:* Video fluorescence microscopy of synchronously growing wild-type and cold-sensitive *cdc* mutants by using a DNA-binding fluorescent probe. *J. Cell Sci.* **52:**271–287. [11.2]

Tsien, R. Y. (1983) Intracellular measurements of ion activities. *Annu. Rev. Biophys. Bioeng.* **12:**91–116. [11.2]

Tycko, B., C. H. Keith, and F. R. Maxfield (1983) Rapid acidification of endocytic vesicles containing asialoglycoprotein in cells of a human hepatoma line. *J. Cell Biol.* **97:**1762–1776. [11.2]

van der Horst, C. J. C., C. M. de Weert, and M. A. Bouman (1967) Transfer of spatial chromaticity-contrast at threshold in the human eye. *J. Opt. Soc. Am.* **57:**1260–1266. [Fig. 4-13]

van Wezel, R. (1981) Video Handbook (G. J. King, ed.). Butterworths, London. [Preface, 8.8, 9.2; Fig. 9-7]

von Bally, G., and P. Greguss, eds. (1982) *Optics in Biomedical Sciences*. Springer-Verlag, Berlin. [11.5]

Vossepoel, A. M., A. W. Smeulders, and K. van den Brock (1979) DIODA: Delineation and feature extraction of microscopical objects. *Comput. Programs Biomed.* **10:**231–244. [10.13]

Wahlstrom, E. E. (1960) *Optical Crystallography,* 3rd ed. Wiley, New York. [11.3]

Walker, J. (1982) Simple optical experiments in which spatial filtering removes the "noise" from pictures. *Sci. Am.* **247**(5)**:**194–206. [10.12, Appendix I]

Walker, P. M. B. (1956) Ultraviolet absorption techniques. In: *Physical Techniques in Biological Research,* Volume III (G. Oster and A. W. Pollister, eds.). Academic Press, New York, pp. 401–487. [11.2]

Walter, R. J., and M. W. Berns (1981) Computer-enhanced video microscopy: Digitally processed microscope images can be produced in real time. *Proc. Natl. Acad. Sci. USA* **78:**6927–6931. [10.8, 11.3]

Walton, A. J., and G. T. Reynolds (1984) Sonoluminescence. *Adv. Phys.,* **33:**595–660. [11.2]

Wampler, J. E. (1985) Low-light video systems. In: *Bioluminescence and Chemiluminescence: Instrumentation and Applications,* Volume 2 (K. van Dyke, ed.). CRC Press, Boca Raton, Fla., pp. 123–146. [11.2]

Wang, Y.-L. (1984) Reorganization of actin filament bundles in living fibroblasts. *J. Cell Biol.* **99:**1478–1485. [11.2]

Wang, Y.-L., J. M. Heiple, and D. L. Taylor (1982) Fluorescent analog cytochemistry of contractile proteins. *Methods Cell Biol.* **25B:**1–11. [11.2]

Weber, D. M. (1984) Digital circuits point towards better TV sets. *Electronics Week* **13:**49–53. [Postscript]

Weiner, P. K., S. V. Forgue, and P. R. Goodrich (1950) The Vidicon photoconductive camera tube. *Electronics* **23:**70–73. [1.2]

Wenzelides, V. K., I. Hillmann, and K. Voss (1982) Zur Darstellung der Leberzellgrenzen mit Anilinblau und ihre vollautomatische Erfassung mit einem Bildanalysesystem. *Acta Histochem.* **70:**193–199. [11.4]

Weszka, J. S. (1978) A survey of threshold selection techniques. *Computer Graphics and Image Processing* **7:**259–265. [10.5, 10.13]

Weszka, J. S., and A. Rosenfeld (1979) Histogram modification for threshold selection. *IEEE Trans. Syst. Man Cybern.* **9:**38–52. [10.13]

Wheeler, H. A. (1939) Wide band amplifiers for television. *Proc. IRE* **27:**420. [1.1]

Williams, M. A. (1977) Quantitative methods in biology. In: *Practical Methods in Electron Microscopy,* Volume 6 (A. M. Glauert, ed.) Elsevier-North Holland, New York, pp. 1–234. [11.5]

Willingham, M. C., and I. Pastan (1978) The visualization of fluorescent proteins in living cells by video intensification microscopy (VIM). *Cell* **13:**501–507. [11.2]

Willingham, M. C., and I. H. Pastan (1983) Image intensification techniques for detection of proteins in cultured cells by fluorescence microscopy. *Methods Enzymol.* **98:**266–283, 635 (addendum). [11.2]

Wilson, A. (1984) Solid-state camera design and applications. *Electronic Imaging* **3**(4)**:**38–46. [Table 7-3]

Wilson, T., and C. Sheppard (1984) *Theory and Practice of Scanning Optical Microscopy*. Academic Press, New York. [11.5]

Wright, F. E. (1911)*The Methods of Petrographic-Microscopic Research: Their Relative Accuracy and Range of Application.* Carnegie Institution of Washington, publ. no. 158, Washington, D.C. [11.3]

Wright, W. D. (1940) Colour vision and chromaticity scales. *Nature* **146:**155–158. [4.7]

Wrigley, N. G., R. K. Chillingworth, E. Brown, and A. N. Barrett (1982) Multiple image integration: A new method in electron microscopy. *J. Microsc.* **127:**201–208. [11.5]

Yamazaki, S., T. Maeda, and T. Miki-Nomura (1982) Flexural rigidity of singlet microtubules estimated from

statistical analysis of fluctuating images. In: *Biological Functions of Microtubules and Related Structures* (H. Sakai, H. Mohri, and G. G. Borisy, eds.). Academic Press, New York, pp. 41–48. [11.2]
Yanagida, T., M. Nakase, K. Nishiyama, and F. Oosawa (1984) Direct observation of motion of single F-actin filaments in the presence of myosin. *Nature* **307**:58–60. [11.2]
Young, I. T., J. E. Walker, and J. E. Bowie (1974) An analysis technique for biological shape. I. *Inf. Control* **25**:357–370. [10.13]
Young, I. T., R. L. Pevenini, P. W. Verbeek, and P. J. van Otterloo (1981) A new implementation for the binary and Minkowski operators. *Computer Graphics and Image Processing* **17**:189–210. [10.13]
Young, J. Z., and F. Roberts (1951) A flying-spot microscope. *Nature* **167**:231. [1.2, 11.5]
Young, T. (1802) On the theory of light and colours. *Philos. Trans.* **92**:12–48. [4.7]
Zahn, C. T., and R. Z. Roskies (1972) Fourier descriptors for plane closed curves. *IEEE Trans. Comput.* **21**:269–281. [10.13]
Zernike, F. (1955) How I discovered phase-contrast. *Science* **121**:345–349. [5.6]
Zernike, F. (1958) The wave theory of microscopic image formation. In: *Concepts of Classical Optics* (J. Strong, ed.). Freeman, San Francisco, Appendix K, pp. 525–536. [5.6]
Zucker, S. (1976) Region growing: Childhood and adolescence. *Computer Graphics and Image Processing* **5**:382–399. [10.13]
Zworykin, V. K. (1931) Photocell theory and practice. *J. Franklin Inst.* **212**:1–41. [1.1]
Zworykin, V. K. (1933) Description of an experimental television system and the kinescope. *Proc. IRE* **21**:1655–1673. [1.1; Figs. 1-2, 1-3]
Zworykin, V. K. (1934) The iconoscope—A modern version of the electric eye. *Proc. IRE* **22**:16–32. [1.1; Figs. 1-2, 1-3]
Zworykin, V. K., and L. E. Flory (1951) Television as an educational and scientific tool. *Science* **113**:483. [1.2]
Zworykin, V. K., and G. A. Morton (1954) *Television: The Electronics of Image Transmission in Color and Monochrome,* 2nd ed. Wiley, New York. [1.1; Figs. 1-1, 12-15, 12-16]
Zworykin, V. K., E. A. Ramberg, and L. E. Flory (1958) *Television in Science and Industry*. Wiley, New York. [1.1, 1.2]

CITED PAMPHLETS

Applications Guide for VII Test Instruments. Visual Information Institute, Inc., P.O. Box 33, Xenia, Ohio 45385.
CCD: The Solid State Imaging Technology. Fairchild CCD Imaging, 3440 Hillview Avenue, Palo Alto, California 94304.
Display Devices. RCA Solid State Division, Electro-Optics and Devices, New Holland Avenue, Lancaster, Pennsylvania 17604.
Electro-optic devices utilizing quadratic PLZT ceramic elements. Report No. SLA 73-0777. Sandia Laboratories, Albuquerque, New Mexico 87115.
IEEE Standards on Television: Methods of measurement of aspect ratio and geometric distortion. Std 202-1954. Reaffirmed 1972. Institute of Electrical and Electronic Engineers, Inc., 345 East 47th Street, New York, New York 10017.
Imaging Device Performance in the C-1000 Camera. Hamamatsu Systems, Inc., 332 Second Avenue, Waltham, Massachusetts 02154.
Imaging Devices. Publication IMD 101. RCA Solid State Division, Electro-Optics and Devices, New Holland Avenue Lancaster, Pennsylvania 17604.
Kodak Films for Cathode-Ray Tube Recording. Publication P-37. Eastman Kodak Company, Rochester, New York 14650.
Kodak Professional Black-and-White Films (1976) Publication F-5. Eastman Kodal Company, Rochester, New York 14650.
Performance Test for Medical Imaging Monochrome Display Devices. Visual Information Institute, Inc., P.O. Box 33, Xenia, Ohio 45385.
Photographing Television Images. Publication No. AC-10. Eastman Kodak Company, Rochester, New York 14650.
Sony Basic Video Recording Course. Sony Video Communications, 700 West Artesia Boulevard, Compton, California 90220.

Stereo-Color TV Projection System. 3D Technology Corporation, 4382 Lankershim Boulevard, North Hollywood, California 91602.
Technology in Retrospect and Critical Events in Science (1969). National Science Foundation, Washington, D.C. 20550.
Test Procedure for Precision Raster Scan Displays. Visual Information Institute, Inc., P.O. Box 33, Xenia, Ohio 45385.

Index

Suffixes g,f,n, and t denote Glossary, figure, footnote, and table, respectively.